TROPICAL CROPS
DICOTYLEDONS

TROPICAL CROPS

Dicotyledons

J. W. PURSEGLOVE

LONGMAN

LONGMAN GROUP LIMITED
London

Associated companies, branches and representatives throughout the world

First published 1968
Second impression 1969
Third impression, in one volume, 1974
Reprinted 1976
ISBN 0 582 46666 0

Set in 10 on 12 point Times New Roman
Printed in Hong Kong by
Sheck Wah Tong Printing Press

To

Phyl, Jeremy and Erica

PREFACE

In writing this work I have always kept in mind the situation I found myself in over thirty years ago, when I was appointed an agricultural officer in the Uganda Protectorate. I was posted up-country and put in charge of the agriculture of a district of some 6,000 square miles. There was no officer to take over from and my senior officer was nearly 200 miles away; even some of the crops were unfamiliar to me. The only reference works available were in my own limited library and the notes which I had taken at the Imperial College of Tropical Agriculture in Trinidad, during a nine-months' course in tropical agriculture, superimposed on my training as a botanist and agriculturist in England.

Since that time a great deal of research has been done on tropical crops, more particularly on those of economic importance. Much more still remains to be done. Books have been written on some of the individual crops, but the need for an up-to-date text on the botany and agronomy of many of the crops still exists. I have attempted here to provide this for the dicotyledonous crops.

During the past thirty years I have been privileged to work and travel extensively in the tropics of Africa, the Far East and the New World. I have grown, or seen growing at first hand, most of the many crops which are dealt with in these pages. In the last ten years I have given annual courses on the botany of tropical crops to undergraduate and postgraduate students. I have had at my disposal the very large living collection of tropical crops and the excellent library at the University of the West Indies here at St Augustine. The descriptions and drawings of the crops have been based, in almost all cases, on the examination of living material.

This work is now offered for the guidance of 'the man in the field', for students reading for degrees and diplomas in tropical agriculture or botany, and for all those who are interested in, or need to know something of the botany and agronomy of the tropical dicotyledonous crops. The various aspects of the subject matter are dealt with under specified headings, details of which will be found in the Introduction. This should facilitate easy reference.

Many people have encouraged me to persevere with this work. It is impossible to mention them all by name, except for Dr E. E. Cheesman and Professor Fred Hardy, who taught me many years ago, and my undergraduate and graduate students, past and present, who have urged me to get on with it.

I am grateful to the following people for reading and commenting on various sections of the manuscript: Dr Caroline Allen (various sections); Dr M. H. Arnold (cotton); Mrs Glenys Barker (many sections); Dr B. G. D. Bartley (cocoa breeding); Dr J. W. Blencowe (rubber); Mr N. B. Charles (tomato); Dr E. E. Cheesman (cocoa and other sections); Dr E. M. Chenery (tea); Dr A. Carvalho (coffee); Professor S. C. Harland (cotton); Mr Paul Holliday (cocoa, pepper and rubber); Mr I. Hosein (citrus); Dr E. F. Iton (cocoa); Dr C. A. Krug (coffee); Dr G. K. Maliphant (citrus); Mr A. R. Melville (coffee and other sections); Mr D. B. Murray (cocoa); Dr R. Nichols (nutmeg); Dr W. V. Royes (pigeon pea); Dr P. J. Sale (cocoa); Dr R. E. Schultes (rubber); Dr W. B. Storey (papaya); Dr A. S. Thomas (coffee); Dr C. C. Webster (rubber and tung); Mr D. B. Williams (sweet potato).

I am deeply indebted to Dr F. W. Cope, Senior Lecturer in Botany at the University of the West Indies, Trinidad, for reading the manuscript and for his many valuable suggestions in regard to facts, spelling and grammar. My thanks are also due to Mr D. Rhind, Editor of the Tropical Agriculture Series, for his help from the time this work was first contemplated, and to Mr Paul Niekirk and Mr J. S. W. Gibson of Messrs Longmans, Green and Company. I would stress, however, that all these helpful people mentioned by name are in no way responsible for any mistakes which may still occur; these are mine alone.

I wish to record my appreciation of the excellent work done by Miss Patricia Lalla and Miss Marjorie Wong in the preparation of the drawings. My thanks are also due to Mr M. Bhorai Kalloo and Mr R. Ramkissoon for collecting the living material of the crops; to Dr Alma Jordan and her staff of the library of the University of the West Indies at St Augustine for their help in providing references and other literature; and to Mrs Marjorie Achong, Mrs Tara Bacon, Mrs Edna Herriot and Miss Phyllis Granger for typing the manuscript.

Finally, my sincere thanks are due to the authorities of the University of the West Indies for providing me with the facilities to write this work.

December 1966 J.W.P.

University of the West Indies,
St Augustine,
Trinidad.

CONTENTS

ILLUSTRATIONS

INTRODUCTION

In these pages I have attempted to present the basic information on the botany and agronomy of the tropical dicotyledonous crops. For this purpose I define a crop as a plant which is grown on a field scale, the produce of which is either used locally or is partly or wholly exported. Plants grown on a small scale, usually in gardens or back-yards, and whose products may appear in local markets, may be mentioned under 'Useful Products', if the families in which they occur are included in this work. Wild plants yielding useful products are sometimes dealt with in a similar manner. Plants in other families yielding useful products are given in the final chapter.

The term tropical is more difficult to define. Strictly speaking, of course, this applies to crops which are grown between the tropics of Cancer and Capricorn. Some crops, such as cotton and tobacco, originated and are still grown in the tropics, but now have their major centres of production in the subtropics and warm temperate countries. Such crops are given full treatment in the present publication. Conversely, those crops which originated in the temperate regions, but are currently grown in the lowland tropics, such as Brassicas, are mentioned briefly, since more detailed accounts may be found in works on temperate crops. Most temperate crops can be grown at high altitudes in the tropics, but these are omitted unless their main centres of production are in the tropical highlands. Thus the deciduous fruit crops such as apples, pears, peaches and plums are not included. On the other hand, pyrethrum, *Chrysanthemum cinerariaefolium* (q.v.), and black wattle, *Acacia mearnsii* (q.v.), which are temperate in origin, but whose main production is at high altitudes in the tropics, are included, as accounts of them are difficult to find elsewhere.

The number of crops grown in the tropics is far greater than those of temperate countries. Some of the crops, such as cocoa, coffee, rubber and most spices, can only be grown in the tropics, and temperate countries are dependent upon these regions for their production. Many of the tree crops, and others, are of an horticultural nature. The distinction made between agricultural and horticultural crops in some temperate countries is difficult to apply in the tropics. It has been ignored in the present work. The space given to each crop is dependent upon its importance on a world scale and the amount of research which has been devoted to it. Thus some of the minor crops are covered in one or two pages, whereas major crops are given up to thirty to forty pages.

THE ARRANGEMENT

In the text proper the crops are given under their plant families, which are arranged in alphabetical order. Similarly, the genera are arranged alphabetically within the families. This seems to me to be preferable to grouping the crops according to the type of commodity which they produce, *e.g.* fibre crops, oil seeds, etc. Brief synopses of the families and genera are given presenting the main characteristics of these taxa, thus avoiding needless repetition. It also permits students and others to recognize the botanical affinities of the crops. The information given for most of the crops is arranged in a logical sequence under specified headings; this should facilitate easy reference. In considering each crop in this way information on the various aspects of the crops is not omitted. All pertinent phases of the botany of each crop, including the ecology and structure, are given first, as they determine the way in which the crop is cultivated. The text of each of the major crops is arranged under the headings given below.

USEFUL PRODUCTS

The inclusion of this heading after the synopsis of the family or the genus permits the mention of plants which are not strictly crops by my definition, but which yield minor economic products. Cover crops, green manures, fodders and shade trees may be mentioned in a similar way. The principal ornamental plants within the family or genus are sometimes included. In this way the conspectus of the family is broadened and it permits the students and others to recognize plants which are related to the crops. Other useful products are given in the last chapter.

SYSTEMATICS

I have tried to provide the currently correct scientific name according to the *International Rules of Botanical Nomenclature*. The authorities for the Latin binomials of plants, diseases and pests are given the first time the name is included. In botanical nomenclature, when a species is transferred to another genus, the first authority is given in brackets, followed by the name of the author making the change, e.g. *Cajanus cajan* (L.) Millsp. In zoological nomenclature the name of the first authority is given in brackets when the genus is changed and the name of the author making the change is *not* recorded, e.g. *Ascia monuste* (L.).

The taxonomy of many of the tropical crops is in a very confused state. The species exhibit great variability and plasticity, and man, over the ages, has selected forms suitable for his particular use and environment. Many of the crops have an extensive synonymy and a large number of species, subspecies, varieties and forms have been erected from time to time which are of doubtful validity, as they often cross readily resulting in intermediate forms, some of which depend on single gene differences. Differences which would be sufficient to delimit wild species cannot be used for crop plants. Until such time as the nomenclature of the disputed species has been clarified, preferably by the study of large living collections, it seems desirable to take a wider view of a species and to limit the number as much as possible.

The systematics of some of the crops are discussed in order to clarify the position, although I am not necessarily in agreement with the classifications which have been proposed. No attempt has been made to provide a complete list of synonyms. I only quote those which are often incorrectly given in modern literature.

Keys are provided for the identification of the principal genera and species.

COMMON NAMES

The common names of some of the crops are also numerous. The most widely used English name (or names) is given opposite the scientific name and, in some cases, other common English names are listed below the binomial. Much confusion arises when agriculturists and others rely entirely on the common or vernacular names of the crops, particularly when these are of African, Asiatic, Latin American or other local origin. Thus *Manihot esculenta* Crantz (q.v.) is known as cassava, cassada, manioc, mandioc, tapioca, Brazilian arrowroot, yuca (Mexico), mhogo (Swahili), omuwogo (Luganda), and a vast number of other vernacular names. I make a plea, therefore, to those scientists who write on crops to include the scientific name once in their publications, unless the common name is so well known as to render this unnecessary.

CHROMOSOME NUMBERS

The basic chromosome number (x) of each genus, where known, is given in brackets after the generic name. The somatic chromosome number (2n) of the species appears in brackets after the scientific name.

USES

This is usually the first heading given under each crop. It sets the stage, as it were, for the subsequent sections. It provides information on the products produced by the crop and the parts of the plant which are used. The wide range of products from tropical crops is far greater than those supplied by the crops of temperate countries.

ORIGIN AND DISTRIBUTION

This subject is a particular interest of mine and is sometimes dealt with more fully than some of the other sections. An author should be entitled to his foibles! Recent archaeological excavations have added greatly to our knowledge of the origins of cultivated plants. I have brought together my ideas on the subject as a whole in the next chapter. This includes an account of the possible routes by which the crops were spread.

CULTIVARS

Throughout the text I use the term cultivar (written cv.—singular and cvs—plural) to designate what is still known in many circles as an agricultural or horticultural variety. The use of the term is recommended by the *International Code of Nomenclature for Cultivated Plants*. The cultivar name is given in single quotation marks. The term cultivar, which can be used in all languages, prevents confusion with the botanical taxonomic term variety, an intraspecific unit, which represents a morphological variant of the species, often having its own geographical distribution. Cultivar should not be confused with the term cultigen, which is a crop species not known in a wild state.

A cultivar may be defined as an assemblage of cultivated individuals which is designated by any characters, morphological, physiological, chemical or others, significant for the purpose of agriculture, horticulture or forestry, and which, when reproduced sexually or asexually, retains its distinguishing features. A cultivar may be: (1) a clone, which consists of individuals vegetatively propagated from a single plant and which are, therefore, of genetic uniformity, e.g. *Theobroma cacao* 'ICS 95'; (2) a line, which is an assemblage of sexually reproducing individuals of uniform appearance propagated by seeds, the stability of which is maintained by selection to a standard, *e.g. Phaseolus vulgaris* 'Canadian Wonder'; (3) an assemblage of individuals, usually reproducing sexually and showing genetic differences, but having one or more characters by which it can be differentiated from other cvs, e.g. watermelon 'Dixie Queen'; (4) a uniform group which is a first generation hybrid (F_1) reconstituted on each occasion by crossing two or more breeding stocks maintained either by inbreeding or as clones, *e.g.* maize 'U.S. 13', a double-cross involving four inbred lines. Cvs can also be obtained from selected mutations, e.g. orange 'Washington Navel' or from interspecific crosses, e.g. tangelo (*Citrus reticulata* × *C. paradisi*) 'Ugli'.

The correct name of a cv. is the earliest legitimate name available and should, where possible, be validly registered and published, together with the origin and a description. Under the *Plant Varieties and Seeds Act, 1964*, in the United Kingdom, breeders and other interested persons may now patent new cvs, giving them an exclusive right for a period of 15 years to propagate, or to authorize others to propagate the new cvs for the purpose of selling planting material or produce of gazetted species.

Under the individual crops I have given a classification of the cvs, based on morphological, agronomic or other characteristics, wherever this wil serve a useful purpose. I have refrained from providing long lists of cvs, as new ones are constantly being produced. Occasionally, I have included a few of the better known cvs or those which are particularly suited to the tropics.

ECOLOGY

Crops, like other plants, have distinct ecological preferences, but they can usually be grown over a wider range of environments than their wild relatives, as man, throughout the ages, has selected or bred crops for the region in which he lives. This is particularly true of annual crops, many of which can be grown under very diverse conditions from the tropics to the temperate zones. Man can often modify the environment by cultural and other methods to provide better conditions for growth and productivity. When a crop plant exists in a wild state the conditions under which it occurs naturally are given. Mention is made of the most suitable altitude, climate and soil for the production of the individual crops. The photoperiodic requirements are given where these are known, as the length of day may have a profound effect on growth, flowering and the time to maturity. Aspects of the physiology of the crop are mentioned where applicable, but no attempt has been made to deal with this subject in any detail.

STRUCTURE

The morphological descriptions given under this heading have been based, in almost all cases, on an examination of living material of the individual crops. I was very fortunate in having the large living plant collections of the University of the West Indies in Trinidad available for study. Consequently I did not have to rely on herbarium specimens or published descriptions. Thus any mistakes which may occur are my own and are not copied from other sources.

I have used the botanical 'shorthand', which is usual in such descriptions, in order to conserve space. When I began writing I tried to avoid the use of botanical terms wherever possible, but I quickly realized that it was impossible to be precise without their inclusion. A glossary of the terms will be found in Bailey (1949) and many other botanical works. In the major crops the descriptions are given under the following organs: roots, stems, leaves, flowers, fruits and seeds. With the other crops these are usually combined. Little or nothing is given on the anatomy of many of the crops. This is only included for those organs which affect the growing and production of the crop, such as the bark and latex vessels of Para rubber and lint hairs of cotton.

POLLINATION

The mechanism of pollination of all tropical crops is not known with certainty. The available information is given under a separate heading to facilitate easy reference. It includes details of self-compatibility and self- and cross-incompatibility where these are known to occur. Methods of emasculation and hybridization are sometimes given under this heading and sometimes under 'Improvement'.

GERMINATION

Details are given where possible of the method of germination (epigeal or hypogeal), seed dormancy, period of viability, speed and percentage of germination, and methods of storage.

CHEMICAL COMPOSITION

The chemical composition of plants, organs, and products varies with the cv. grown, the environment, the season and the cultural treatment. A typical composition is given, but the figures can only be used as a guide, as considerable variation occurs.

PROPAGATION

Many tropical crops, including some of the tree crops, are still grown almost entirely from seed, which is the cheapest method of propagation. The seedlings of tree crops are usually raised in nurseries or in containers and are transplanted into the field when they have reached a suitable height. Sometimes they are cut back and planted as stumps. Vegetative propagation requires more skill and is more expensive. It may be used for the following purposes: (1) to establish clones from selected high-yielding parent plants, *e.g.* rubber; (2) to propagate plants which do not produce viable or sufficient seed, *e.g.* seedless citrus; (3) to select for sex in dioecious crops, *e.g.* nutmeg, or in female sterile plants, *e.g.* pimento and tung, thus eliminating a large proportion of useless males; (4) to provide resistance or immunity to diseases and pests by a suitable combination of stock and scion, *e.g.* citrus; (5) to obtain earlier fruiting, increased vigour or dwarfing effects, etc.

The methods of vegetative propagation employed vary with the different crops. The reader is referred to the crops quoted below for a description of various methods. These include: (1) cuttings, which may be leafy semi-hardwood or soft-wood cuttings, *e.g.* cocoa (pp. 583–586), single-leaf cuttings, *e.g.* tea (p. 605), or root cuttings, *e.g.* breadfruit (p. 382); (2) layering or marcotting; (3) grafting, including approach grafting or inarching, *e.g.* mango (pp. 29–30); (4) budding, including shield or T budding, *e.g.* citrus (pp. 512–516), and patch budding with or without the modified Forkert method, *e.g.* rubber (pp. 157–159).

HUSBANDRY

The essential basic information has been provided as a guide to growing the various crops. The methods of cultivation vary widely in different parts of the world, as do yields; thus it is impossible to lay down hard and fast rules. Information is given on the following aspects of husbandry: spacing, seed rate, planting, manuring, shade and cover crops where necessary, after-care and maintenance, pruning where done, harvesting, yields, and processing up to the time the product leaves the farm or estate. The response to fertilizers is variable and it is only possible to give some indication of what is used in some areas and what may be expected to give a profitable return. Herbicides are being used increasingly to control weeds, but no attempt has been made to make recommendations in this rapidly changing field. Some information on cover crops is given under Leguminosae (pp. 217–222) and rubber (p. 160), and on shade trees under Leguminosae (pp. 222–223), coffee (pp. 471–472), cocoa (pp. 587–588), and tea (p. 606).

MAJOR DISEASES AND PESTS

Since the number of diseases and pests which attack tropical crops is legion, reference is made only to some of the more destructive enemies of the individual crops. In the case of the major crops the symptoms are briefly described. Little or no attempt has been made to give control measures, as these change rapidly. Breeding for resistance is given under 'Improvement'.

MINERAL DEFICIENCIES

The symptoms are described for coffee, cocoa, citrus and tea.

IMPROVEMENT

An account is given of the major problems confronting the plant breeder, together with his aims, methods and results. Little work has been done on the improvement of many of the local food crops, while the major economic crops have often received considerable attention.

The first step in any breeding programme is usually the collection, study and sorting out of local and introduced cvs. It is essential that selected cvs should be tested over as wide a range of local environments as possible. In self-pollinating crops pure lines may be established from single-plant selections; different selection techniques will usually be required for partially or wholly cross-pollinated crops. The presence or absence of self- and cross-incompatibility must also be borne in mind. Male sterility, genetic or induced, may be utilized in some crops. Single-plant selections in self-pollinating crops can often provide a considerable advance in genetically mixed populations until the yield potential of the material has been reached; further advance is then only possible following gene recombination resulting from controlled hybridization and by an increase in variability, arising from the introduction of genes from other sources. The 'panmixia' method, in which artificial or natural pollinations are made between a large number of cvs grown together and showing a wide range of desirable characters, is a useful method of maintaining a wide range of variability. Hybridization between cvs and related species, followed by back-crossing to the cv. recurrent parent, is often used to transfer resistance to disease and other characters, at the same time maintaining the desirable characteristics of the original cv. Inbreeding to produce reasonably homozygous lines, followed by crossing of selected inbred parents which combine well, is resorted to in some cases. Mutations, natural or induced, may sometimes prove useful.

The improvement and breeding of tree crops present many problems which are not encountered with annual crops. These problems include: (1) the greater importance of the individual plant; (2) the difficulty of deciding whether a better-than-average performance is due to genetic superiority or to a better-than-average environment (such as a better soil, or the plant occupying more than its fair share of space, for which a correction must be made); (3) the longer life history; (4) the longer time for first bearing and for carrying out adequate tests under different environmental conditions; (5) larger space required; (6) relatively expensive propagation and culture; (7) impossibility of changing quickly the cv. or cvs grown; (8) considerations of quality and the difficulty of forecasting market requirements a long time ahead; (9) heterogeneous nature of most tree crops, many of which are cross-pollinated; (10) the greater importance of pests and diseases, particularly of viruses and new strains; (11) the presence of self- and interclonal-incompatibility; (12) development of successful and economic methods of vegetative propagation; (13) presence of nucellar embryos in some species; (14) greater difficulty of manipulation in hybridization, etc. Tree crops have the advantage over most annuals in that they can usually be propagated vegetatively, but it should be remembered that this does nothing to improve the genotype.

PRODUCTION

Statistics of the production of many of the food crops are not readily available. This is usually remedied in the case of export crops. The trends of production are given rather than the actual figures for individual years which would quickly become out-of-date. The main producing, exporting and importing countries are noted.

REFERENCES

The number of references in the text and at the end of each crop is reduced to the minimum in order to conserve space. A few key references only are given which will permit the reader to obtain a lead into the literature of individual crops. A list of general references is given at the end.

ILLUSTRATIONS

A full-page drawing of most of the crops is provided, showing the general habit and the detailed structure of some organs such as flowers, fruits and seeds. Nearly all the drawings have been made from living specimens under my supervision. The drawings should be consulted in connection with the morphological descriptions given under 'Structure'. Some of the finer detail may be obtained by using a lens.

INDICES

A general index is given at the end. The scientific names of dicotyledons mentioned in the text are given in the Appendix, together with the authorities for the generic and specific names.

THE ORIGIN AND SPREAD OF TROPICAL CROPS

ORIGINS OF AGRICULTURE

When man emerged from the hunting and food-gathering stage of his existence to that of a cultivator certain plants were domesticated, depending on the region in which he lived. An intermediate stage as a herdsman with domesticated animals is interpolated in some areas. Man's first crops were probably volunteer plants which grew on or near the household midden with its higher fertility. The next step would be to cultivate the soil and sow the seeds. The sites included new alluvial soils along rivers which would be free from weeds. Primitive man throughout the ages recognized and used a great number of wild plant species for a wide variety of purposes. A number of crops were taken early into cultivation, some of which were later abandoned.

The first recorded cultivation is in northern Iraq and neighbouring regions was dated at least 7000 B.C., the first crops domesticated being wheat and barley. In the fifth millennium B.C. agriculture was extended to the great river basins of Mesopotamia and Egypt. In the New World the beginnings of agriculture date from about 5000–3000 B.C. In Mexico, some of the earliest crops cultivated were the common gourd (*Lagenaria siceraria*), *Setaria macrostachya* HBK., cucurbits, beans, amaranths, cotton and avocados (*Persea americana*), some of which were grown before maize.

Almost all the early archaeological sites yielding crop remains are found in arid or semi-arid regions. In areas of higher rainfall the conditions for the preservation of plant tissues are much less favourable, so that little is known of the early domesticated plants of these regions, or the time when they were taken into cultivation. The first crops to be grown in order of time were cereals, but not necessarily so in all areas at later dates. In the wet tropics root crops may have been among the early domesticates. The great ancient civilizations, however, appear to have been based on cereals, with wheat and barley in western Asia, Egypt and Europe, rice in south-eastern Asia, and maize in the New World.

CHANGES OCCURRING UNDER DOMESTICATION

All domesticated plants must have arisen originally from wild progenitors.

Nevertheless, a surprising number of crop plants are not now known in a truly wild state. Introgression has occurred from other species and some crops have arisen from the hybridization of wild plants or early cultivars, resulting in some instances in the production of polyploids. Modern weedy species, often growing with, and related to the crop plant, may have developed along with the crop and may not be the ancestor of the crop, as was often supposed in earlier times.

From the time a plant is taken into cultivation, man, consciously or unconsciously, selects for characters which are useful to him, bringing about genetic, physiological and structural changes, which often render the plant less likely to survive in a wild state. These changes include: (1) spread to a greater diversity of environments and a wider geographical range, some of them unsuited to the wild plant; (2) the crop may come to have a different ecological preference; (3) simultaneous seed bearing; (4) absence of shattering or scattering of seeds and sometimes a complete loss of dispersal mechanisms; (5) increase in size of fruits and seeds, often reducing disperal efficiency; (6) conversion of perennials into annuals; (7) loss of seed dormancy; (8) loss of photoperiodic controls; (9) absence of normal pollinating organisms; (10) changes in the breeding system, usually from cross- to self-fertilization; (11) loss of defensive adaptations such as hairs, spines or thorns; (12) loss of protective coverings and sturdiness, combined with a reduction in the development of fibrous tissue; (13) improvement in palatability and chemical composition thus rendering them more likely to be eaten by animals; (14) increased susceptibility to diseases and pests; (15) development of seedless parthenocarpic fruits; (16) selection for double flowers, which may involve conversion of stamens into petals; (17) multiplication by vegetative propagation. These points should be considered in relation to the individual crops.

It will be seen that a number of tropical crops have originated outside Vavilov's (1951) famous centres of origin, which may be centres of cultivation, migration, selection or hybridization and not of the origin of some of the crops. It should be realized that the same crop may have been taken into cultivation in more than one area. It does not necessarily follow that the area in which a crop plant evolved is that most ideally suited to it; optimal conditions for a crop may very well occur in another hemisphere. The wild progenitor of the crop may have a limited natural distribution and may be confined to a particular ecological niche, having evolved in competition with many other species. On domestication the crop is spread widely in diverse environments. Although the original introduction may have been due to a very small number of individuals, and often was, cultivation ensures the build up of vast populations of the species growing in close proximity. In a wild state individual specimens may be widely scattered, as is the case in *Hevea brasiliensis*. When grown under cultivation in pure stands the chance of cross-pollination in out-breeding crops is greatly increased. Furthermore, the crop plant is removed from competition with other species and, by weeding, the use of manures, irrigation, absence or control of diseases and pests, and other cultural treatments, is given a more favourable habitat.

On domestication the crop is immediately subjected to artificial selection by man. As has been pointed out by Hutchinson (1965), the nature of the sample of the population from which the original supply of seed or planting material was taken has always left its mark on the race developed in the new areas. The race is then fashioned by the selective forces of the new environment, natural and artificial, and these have had the most striking effects in a surprisingly short time. Genotypes which would fail in a natural population may survive because of man, or be induced by him, and are given a chance to express themselves in an expanding population in a new environment. In naturally cross-pollinated crops, and this includes most of the tropical tree crops, the individual plants are characteristically very heterozygous, thus providing a wide source of variability, although it is possible to maintain the phenotype in some cases by vegetative propagation.

In their spread to a great diversity of environments crop plants show considerable variability and plasticity; and, as has been shown by Stebbins (1950), a variable environment strongly promotes rapid evolution and may in fact be essential for speeding up evolutionary change. Darlington (1963) states 'movement has a snowball effect on variation, the snowball being the variants that are collected by hybridization and selection on the track of the migrants'. The vast multiplication of progeny from a few individuals must have a profound effect on the subsequent development of the crop, particularly when it is far removed from wild members of the species or allied species. In addition to the artificial selection imposed by man, such a population is subject to disruptive selection and genetic drift, as has been shown by Manglesdorf (1965) for maize. In populations which contract and expand greatly in numbers, many genes are lost and others increase in frequency. Recessive genes masked, or having a low frequency at the centre of origin may attain a high frequency in the new areas and at the periphery of expansion, and may therefore manifest themselves more frequently in homozygous condition. This phenomenon is likely to be accelerated in the case of self-pollinating crops; or when a normally cross-pollinated plant becomes self-fertilizing, as in the tomato; or when self-compatible forms are developed from self-incompatible forms, as in cocoa. New cvs can arise at the periphery of migration which may be useful to the breeder and for other purposes.

In a large and rapidly expanding population the number of naturally occurring mutations will be increased, adding to the gene pool, particularly when the crops are exposed to new and variable environments in which man attempts to grow them as he constantly extends the areas of production. Over 30 mutations have been recorded in *Coffea arabica*, mainly in Brazil. If the mutation is of value to him, man will preserve and increase it in cultivation. In the case of parthenocarpic crops such as the banana and seedless citrus, which are clonally propagated, change is restricted to that which arises as mutations, unless man re-introduces the sexual cycle.

Man is destroying tropical vegetation on an ever-increasing scale. The importance of collecting the wild and related species of crop plants before they are irretrievably lost and of maintaining them in central gene banks, together with as complete a range of cvs as possible, cannot be over-stressed.

ROUTES OF SPREAD

There was a considerable movement of crops within Asia from very early times. Crops from India had reached China some 4,000 years ago. Crops originating in Africa had reached India some 3,000 years ago and others were also taken in the opposite direction. The main route out of Africa was across the southern end of the Red Sea along the Sabaean lane. The Sabaeans in southern Arabia dominated the maritime trade between East Africa and India from the 7th to the 1st century B.C. As I have shown under the individual crops, spices played an important part in this trade, and, due to their high cost and demand in Europe, were largely responsible for European exploration and the early conquests by Spain and Portugal, culminating in the end of the Dutch monopoly of nutmegs and cloves in the Moluccas towards the close of the 18th century.

The Ethiopians, who are of Caucasoid origin, probably reached their present region by the 3rd millennium B.C. Passing through or near Iraq-Kurdistan, they seem to have brought with them emmer-type wheat and other crops. Their conquests of Arabia in the 4th and 6th centuries A.D. would also result in the spread of crops.

From the 7th century A.D. there was a wave of migration from Persia and Arabia to the east coast of Africa, with the establishment of fortified towns from Cape Guardafui to Mozambique. There was little penetration into the African interior except along the slave routes (which permitted the spread of plants such as the mango) until the European explorations of the mid-19th century. European contact with West Africa dates from the mid-15th century. People from Malaysia probably went to Madagascar about A.D. 500 taking with them the banana, which then spread to Africa.

The silk route from China overland to India and inner Asia was established in the 2nd and 1st centuries B.C. China had developed a direct trade with Arabia by the 3rd century A.D., but previous to this was trading with Borneo and the Philippines. With the rise of Islam in the 7th century A.D. and the religious conquests, sweeping north to Constantinople and west across northern Africa to the Mediterranean, many Asian crops, including orange and sugar cane, reached southern Europe.

Recent archaeological excavations in the New World have shown that there has been communication between the peoples of Mexico and coastal Peru since about 1000 B.C., which resulted in the movement of crops between the two regions, although it was less than one might expect.

It can thus be seen that there was ample opportunity for the spread of crops within the Old and New Worlds extending over a considerable period of time.

I am convinced, however, that there was *no movement of crops by man* between the Old and the New World in pre-Columbian times. Studies of distribution show that plants must have been spread between the eastern and western hemispheres in geological time long before the advent of man. Various authorities have invoked Wegener's theory of drifting continents, land bridges, or routes via Antarctica and the Bering Strait. I favour the last two routes.

Three crops only are known to have been in both the Old and New Worlds before 1492. These are the common gourd, *Lagenaria siceraria* (q.v.), which is of very ancient cultivation in both hemispheres, the sweet potato, *Ipomoea batatas* (q.v.), which had reached Polynesia and was taken by the Maoris to New Zealand, and the coconut, *Cocos nucifera*, which was found on the western coast of Panama and could have drifted across the Pacific. In addition, *Gossypium herbaceum* (q.v.) must have reached South America to cross with a wild, diploid New World cotton to produce the tetraploid *G. barbadense* (q.v.). As I have shown elsewhere (Purseglove, 1963, 1965), and under the crops given later in this work, it is *not* necessary to invoke the aid of man in the transport of these crops between the two hemispheres. Compared with these few exceptions, Merrill (1954) states that some 1,800 cultivated species were *not* common to both hemispheres before 1492.

After the discovery of the New World by Columbus in 1492 there was a rapid spread of crops between the New and Old Worlds. Columbus on his second voyage in 1493 took sugar cane, citrus, and other crops to the Caribbean. He also carried maize, sweet potatoes and other crops to Spain. Within less than a century New World crops, such as maize, had reached China.

The main routes along which crops were carried in post-Columbian times include the old Portuguese route which existed from 1500–1665 from Portugal to the East *via* Brazil and the Cape of Good Hope. The Portuguese made settlements in Goa, Malacca and the Moluccas, and reached Canton in 1515, later founding Macao in 1557, from where they traded with Formosa and Japan. They also carried New World plants to West and East Africa. The slave traffic between West Africa and Brazil and other New World territories began in the 16th century, and was also responsible for the spread of crops between these regions. The other important trade route was the Spanish galleon line which operated between Acapulco in Mexico and the Philippines, *via* Guam, from 1565–1815. Thus some crops reached the East *via* the Atlantic and others *via* the Pacific; others were taken by both routes. From the end of the 16th century the Dutch, British and French were also responsible for the spread of many crops throughout the tropics.

Purseglove (1957) has shown the prominent role of the early tropical botanic gardens in the distribution and establishment of crops in tropical countries. The first tropical garden was founded by the French at Pamplemousses in Mauritius in 1735. That at St. Vincent in the West Indies was founded by the British in 1764 and it was to this garden that Bligh carried the breadfruit, *Artocarpus altilis* (q.v.), from Tahiti in 1793. The Royal Botanic Gardens at Kew in England played a notable part in the distribution of tropical crops, including rubber, *Hevea brasiliensis* (q.v.), in which the Singapore Botanic Gardens ably assisted. The botanic gardens were the forerunners of tropical departments of agriculture and they, in their turn, have been responsible for the introduction and spread of crops. The modern research units devoted to various crops, usually receiving financial assistance from industrial concerns and producers, also play their part.

In the transport of crops from one region of the world to another, man has often inadvertently introduced camp-following weeds. When this was done without the attendant diseases and pests, the weeds frequently increased at an alarming rate and often nowadays present a serious problem.

CENTRES OF PRODUCTION

A striking feature of the present-day distribution of tropical crops is the fact that the main areas of production of the major economic crops are usually far removed from the regions in which they originated. This is illustrated by the dicotyledonous crops listed below, but it is also true of many monocotyledonous crops.

Crop	Country of Origin	Country of Maximum Production
Black wattle (*Acacia mearnsii*)	Australia	S. and E. Africa
Citrus (*Citrus spp.*)	S.E. Asia	United States
Clove (*Eugenia caryophyllus*)	Moluccas	Zanzibar
Cocoa (*Theobroma cacao*)	S. America	Ghana
Coffee (*Coffea arabica*)	Ethiopia	C. and S. America
Groundnut (*Arachis hypogaea*)	S. America	India, China
Nutmeg (*Myristica fragrans*)	Moluccas	Grenada, W. Indies
Pyrethrum (*Chrysanthemum cinerariaefolium*)	Dalmatia	Kenya
Rubber (*Hevea brasiliensis*)	S. America	Malaysia
Soya bean (*Glycine max*)	N.E. Asia	United States

Other examples will be found under the individual crops.

Many reasons have been adduced for this (Purseglove, 1963), but undoubtedly the most important is the introduction of crops into the new areas without the major diseases and pests which attack them at their centre of origin. Thus it is almost impossible to grow rubber economically in the New World tropics because of South American leaf blight, *Dothidella ulei* (q.v.), which, fortunately, is not known in the Far East. Coffee in the western hemisphere has surprisingly few diseases and pests; the coffee leaf rust, *Hemileia vastatrix* (q.v.), which caused the complete collapse of the arabica coffee crops in Ceylon and Java in the 1870's, has not yet reached Latin America.*

For an introduction to be successful there must be a place in the agricultural system for the new crop, whether it be an economic or a food plant;

* *Hemileia vastatrix* was first reported in Brazil in 1970.

and, in the case of the latter, there must be a place for it in the diet of the people. Many people are very conservative in their choice of foods. The planters in Malaya were reluctant to give the newly introduced rubber crop a trial, but when coffee was attacked by diseases and pests and the price of this crop slumped, they willingly switched to the new crop for which there was a rapidly increasing demand. Rubber now provides over 50 per cent of Malaya's domestic exports. Maize, beans (*Phaseolus vulgaris*), sweet potatoes and cassava, all of New World origin, are now important staple foods over large areas of Africa, but never achieved the same importance in India.

Other factors which have influenced the present distribution of crops include:

1. Selection and breeding of new cvs suited to new environments and utilization, *e.g.* high-yielding RRIM rubber clones in Malaya.
2. The area of origin may not necessarily provide the optimum environmental conditions.
3. Competition from other species and weeds is removed when crops are grown in pure stands, *e.g.* rubber and oil palm in Malaysia.
4. Improved methods of husbandry. Research stations are often set up in the areas of maximum production which will assist in this and in breeding, *e.g.* Rubber Research Institute of Malaya.
5. Availability of land, which may sometimes be the result of acreage reduction in other crops.
6. The system of agriculture practised and whether this depends on small peasant holdings or large-sized plantations.
7. The cost, supply and efficiency of labour.
8. The availability of capital.
9. The existence of markets, with production keeping pace with demand.
10. The nearness of markets, particularly for perishable produce.
11. Availability of planting material, *e.g.* the plentiful supply of rubber seeds at the Singapore Botanic Gardens when planting was begun.
12. Local knowledge, e.g. Ridley had discovered a system of tapping and preparation of rubber, which was of inestimable benefit to the rubber planters.
13. Political stability.
14. War and the cutting off of existing supplies, *e.g.* soya bean for oil production in the United States after 1942.
15. Government subsidies for increased production and new crops, *e.g.* the guaranteed price of castor seed in 1951–1954 in the United States.
16. Sponsorship by industry which may be willing to provide markets for a new crop, including protection and contracts.
17. Advisory services to the farmers on the new crops and on advances in methods of cultivation.
18. Efficient control of weeds, diseases and pests.
19. Plant introduction and evaluation, *e.g.* the U.S.D.A. Plant Introduction Service.

20. Genetic adjustment of the crop and testing under a wide range of environments.
21. Compulsory cultivation enforced by those in authority, *e.g.* the early history of the clove crop in Zanzibar.
22. Ease of mechanization of the crop, including mechanical harvesting.
23. Lucky chance introduction, *e.g.* the high quality of the rubber trees from Wickham's original collection; Deli oil palms in Indonesia.
24. The role of botanic gardens, agricultural departments and other bodies.
25. The role of the individual, who may be a successful planter or a specialist in popularizing the crop, *e.g.* Ridley and rubber.
26. Early success may be important in the establishment of a new crop.

THE FUTURE

The movement of crops, which was begun several thousand years ago, still continues. The efficiency of modern transport and the development of new techniques have added greatly to the ease with which planting material can be taken from one part of the world to another. Search in inaccessible areas still calls for organized expeditions. Many territories, whose economy is based on one or two economic crops, are attempting to broaden their economy by the introduction of new crops. New industrial crops, particularly for the production of chemicals and drugs, are becoming increasingly important. A search is being made for plants which are rich in alkaloids and steroids, the latter for use in the contraceptive pill, which may help to reduce the present widening gap between the world's population and its food resources.

REFERENCES

BURKILL, I. H. (1966). *A Dictionary of the Economic Products of the Malay Peninsula*, 2 vols. 2nd ed. Kuala Lumpur: Ministry of Agriculture and Cooperatives.

BURKILL, I. H. (1953). Habits of Man and the Origins of the Cultivated Plants of the Old World. *Proc. Linn. Soc., London*, **164**, 12–42.

CALLEN, E. O. (1965). Food Habits of some Pre-Columbian Mexican Indians. *Econ. Bot.*, **19**, 335–343.

DARLINGTON, C. D. (1963). *Chromosome Botany and the Origins of Cultivated Plants.* New York: Haffner.

DE CANDOLLE, A. (1886). *Origin of Cultivated Plants.* 2nd. ed. Reprinted New York: Haffner, 1959.

DRESSLER, R. L. (1953). The Pre-Columbian Cultivated Plants of Mexico. *Bot. Museum Lfts, Harvard Univ.*, **16**, 115–163.

GREENWAY, P. J. (1944–1945). Origins of some East African Food Plants. *E. Afr. agric. J.*, **10**, 34–39, 115–119, 177–180, 251–256; **11**, 56–63.

HARLAN, J. R. (1961). Geographic Origin of Plants useful to Agriculture, in *Germ Plasm Resources*, 3–19, Amer. Assoc. Advanc. Sci.: Washington.

HEISER, C. B. (1965). Cultivated Plants and Cultural Diffusion in Nuclear America. *Amer. Anthropologist*, **67**, 930–949.

HELBAEK, H. (1959). Domestication of Food Plants in the Old World. *Science*, **130**, 365–372.

HUTCHINSON, J. B. (1959). *The Application of Genetics to Cotton Improvement.* Cambridge Univ. Press.

HUTCHINSON, SIR JOSEPH (1965). ed. *Essays on Crop Plant Evolution.* Cambridge Univ. Press.

KNOWLES, P. F. (1960). New Crop Establishment. *Econ. Bot.* **14**, 263–275.

MACNEISH, R. S. (1964). Ancient Mesoamerican Civilization. *Science*, **143**, 531–537.

MANGLESDORF, P. C. (1965). The Evolution of Maize, in *Essays on Crop Plant Evolution*, 23–49, ed. Sir Joseph Hutchinson. Cambridge Univ. Press.

MANGLESDORF, P. C., MACNEISH, R. S., and GALINAT, W. C. (1964). Domestication of Corn. *Science*, **143**, 535–545.

MERRILL, E. D. (1954). *The Botany of Cook's Voyages.* Waltham: Chronica Botanica.

MORISON, S. E. (1963). *Journals and other Documents on the Life and Voyages of Christopher Columbus.* New York: Heritage Press.

MORISON, S. E. and OBREGON, M. (1964). *The Caribbean as Columbus saw it.* Boston: Little, Brown and Co.

POLUNIN, N. (1960). *Introduction to Plant Geography.* London: Longmans.

PURSEGLOVE, J. W. (1957). History and Functions of Botanic Gardens with special reference to Singapore. *Trop. Agriculture, Trin.*, **34**, 165–189.

PURSEGLOVE, J. W. (1963). Some Problems of the Origin and Distribution of Tropical Crops. *Genet. Agraria*, **17**, 104–122.

PURSEGLOVE, J. W. (1965). The Spread of Tropical Crops, in *The Genetics of Colonizing Species*, ed. H. G. Baker and G. L. Stebbins, 375–389. New York: Academic Press.

SMITH, C. E. (1965). The Archaeological Record of Cultivated Plants of the New World Origins. *Econ. Bot.*, **19**, 322–334.

STEBBINS, G. L. (1950). *Variation and Evolution in Plants.* New York: Columbia Univ. Press.

TOWLE, M. A. (1961). *The Ethnobotany of Pre-Columbian Peru.* Chicago: Aldine.

VAVILOV, N. I. (1951). *The Origin, Variation, Immunity and Breeding of Cultivated Plants.* Waltham: Chronica Botanica.

WHITAKER, T. W. and CUTLER, H. C. (1965). Cucurbits and Cultures in the Americas. *Econ. Bot.*, **19**, 344–349.

WHYTE, R. O. (1958). *Plant Exploration, Collection and Introduction.* Rome: FAO.

ZUKOVSKIJ, P. M. (1962). *Cultivated Plants and their Wild Relatives.* Farnham Royal: Commonw. Agric. Bureaux.

ANACARDIACEAE

About 60 genera and 400 spp., mainly tropical. Trees and shrubs with resinous sap. Lvs usually alternate, simple or pinnate; stipules absent. Fls small, in panicles, often honey-scented, usually regular, hermaphrodite or unisexual, often with hermaphrodite and male fls in same inflorescence; sepals 3–7; petals 3–7, rarely 0, free; disc present, usually annular; stamens as many or twice as many as petals, filaments free; anthers 2-celled, opening lengthwise; ovary superior, usually 1-celled; styles 1–5; ovules solitary; fruits mostly drupaceous; seeds with little or no endosperm, cotyledons fleshy.

USEFUL PRODUCTS

EDIBLE FRUITS AND NUTS: *Mangifera indica* (q.v.), the mango, is one of the most widely distributed fruits of the tropics. *Anacardium occidentale* (q.v.) yields the cashew nut. *Spondias cytherea* Sonn. (syn. *S. dulcis* Forst.), the Golden or Otaheite apple, a native of Polynesia, is cultivated in many tropical countries, and its yellow fruits, up to 3 in. in diameter, with a large stone, are eaten fresh or cooked and are often seen in tropical markets. *S. mombin* L. (syn. *S. lutea* L.), the yellow mombin or hog plum, and *S. purpurea* L. (syn. *S. mombin* auth.), the red mombin or Spanish plum, natives of tropical America, are occasionally cultivated for their fruits, which are 1–2 in. long. *Spondias* spp. have pinnate lvs. *Pistacia vera* L., a native of western Asia and cultivated in Mediterranean countries from ancient times, supplies the pistachio or green almond. The fruits of some other members of the family are also edible, *e.g. Dracontomelon mangiferum* Bl., Argus pheasant tree, and *Bouea* spp. in Indo-Malaysia. *Buchanania lanzan* Spreng. in India and Burma produces an edible nut.

TANNIN: Quebracho, the heartwood of two large South American trees, *Schinopsis lorentzii* (Griseb.) Engl. and *S. balansae* Engelm., is one of the world's most important sources of tannin. Argentina and Paraguay are the chief centres of the industry and the main exports are to the United States. The dried leaves of the sumacs, *Rhus* spp., are also a source of tannin, with 3 spp. in use in the United States and one in Sicily.

LACQUER AND MASTIC: Lacquer is a natural varnish obtained by tapping the poisonous sap of *Rhus verniciflua* Stokes, a native of China; Burmese lacquer is obtained from *Melanorrhoea usitata* Wall. Mastic is a resin obtained by tapping *Pistacia lentiscus* L. and yields a pale varnish used for coating pictures, etc.; Bombay mastic is obtained from *P. cabulica* Stocks. *Schinus molle* L., the pepper tree, yields American mastic and is also a very attractive tree for planting at the higher elevations in the tropics; the fruits may also be used as a pepper substitute.

VESICANTS: Several members of the family produce an irritant poisonous sap which sets up severe irritation of the skin resulting in painful blisters and open sores which heal slowly. These include the *rengas* trees of Malaya, belonging to the genera *Gluta, e.g. G. renghas* L., *Melanorrhoea* and *Melanochyla*. They contain red sap which blackens on exposure. The irritant seems to be a volatile aromatic substance and even sheltering under a tree during rain can cause irritation. The poison ivies of the United States are *Rhus* spp., *e.g. R. toxicodendron* L.

Anacardium L. (x = ?)

About 8 spp. of trees and shrubs in tropical America, of which *A. occidentale* (q.v.) is widely cultivated throughout the tropics for its nuts.

Anacardium occidentale L. (2n = 42) CASHEW

USES

The seeds are the source of cashew nuts, produced by shelling the roasted fruits; they are used in confectionery and dessert. They yield an edible oil, but due to the high price of the kernels this is not usually extracted. The shells or pericarps yield cashew-shell oil which is a vesicant and is used as a waterproofing agent and as a preservative. Distilled and polymerized, the oil is used in insulating varnishes and in the manufacture of typewriter rolls, oil- and acid-proof cements and tiles, brake-linings, inks, etc. The cashew apple, which is the fleshy swollen pedicel, is juicy and astringent and is edible; the juice may be fermented and made into wine; the pulp may also be made into preserves. The sap from the bark provides an indelible ink.

ORIGIN AND DISTRIBUTION

The cashew is a native of tropical America from Mexico to Peru and Brazil and also of the West Indies. It was one of the first fruit trees from the New World to be widely distributed throughout the tropics by the early Portuguese and Spanish adventurers. It was introduced into India from Brazil by the Portuguese in the 16th century and it probably reached the East African coast and Malaya about the same time. It has become naturalized in many tropical countries, particularly in coastal areas.

ECOLOGY

The cashew is hardy and drought-resistant, but it is damaged by frost.

It thrives under a variety of climatic and soil conditions and can be grown from sea level to 4,000 ft, but is best suited to the lower elevations. It is grown in areas with rainfall ranging from 20–150 in. per annum. It grows best on sandy soils with good drainage, but it is often grown on hillsides too dry and too stony for other crops.

STRUCTURE

Spreading evergreen tree to 12 m in height. Lvs alternate, simple, leathery, obovate, glabrous; lamina 6–20 cm long, 4–15 cm broad, rounded and often notched at apex, tapering at base; veins prominent, lateral veins spreading, 10–20 pairs; petiole 1–2 cm long, swollen at base, flattened on upper surface. Inflorescence a lax terminal many-flowered panicle to 20 cm long, sweet-scented; bracts pubescent; polygamous with male and hermaphrodite fls in same inflorescence; sepals 5, narrow, green, 5 mm long; petals 5, linear, 1 cm long, reflexed in open flower, opening pale greenish-cream with red stripes, later turning red; stamens 10. In male fl. 9 short stamens 4 mm long, some of which may not be functional, and 1 long stamen 12 mm long with red anther projecting above corolla; male fls have been found with 7 stamens 4 mm long and with 3 stamens 6, 8, and 10 mm respectively. Other variations also occur. In hermaphrodite flower 9 short stamens as in male and one stamen 8 mm long projecting just above corolla; ovary 1-locular with single ovule; style simple 12 mm long and exserted from corolla to same length as long stamen in male fl. Fr. kidney-shaped nut (achene), about 3 cm long and 2·5 cm broad, greyish-brown in colour with hard pericarp, somewhat embedded in enlarged pedicel, receptacle and disc (cashew apple), which is shiny, red or yellow in colour, pear-shaped, thin-skinned, soft and juicy, with characteristic penetrating odour, 10–20 cm long and 4–8 cm broad; seed kidney-shaped with reddish brown testa, 2 large white cotyledons and small embryo.

POLLINATION

The number of hermaphrodite flowers per inflorescence is about 60 and the ratio of male to hermaphrodite flowers is about 6 : 1. Northwood (1966) has shown that most of the flowers open between 6 a.m. and 6 p.m. with a peak opening period between 11 a.m. and 12.30 p.m., that the stigmas are receptive as soon as the flowers open, and that anthesis takes place 1–5 hours later. The flowers are visited by flies, ants and other insects and these are believed to transfer the sticky pollen to the stigma. Bagged inflorescences do not produce fruits, but flowers can be successfully self-pollinated by hand. Not all the flowers pollinated produce mature fruits as there is an

---

Fig. 1. **Anacardium occidentale:** CASHEW. **A**, flowering branch (× ⅓); **B**, male flower in longitudinal section (× 3); **C**, hermaphrodite flower in longitudinal section (× 3); **D**, developing fruit (× 2); **E**, fruit and apple (× ⅓).

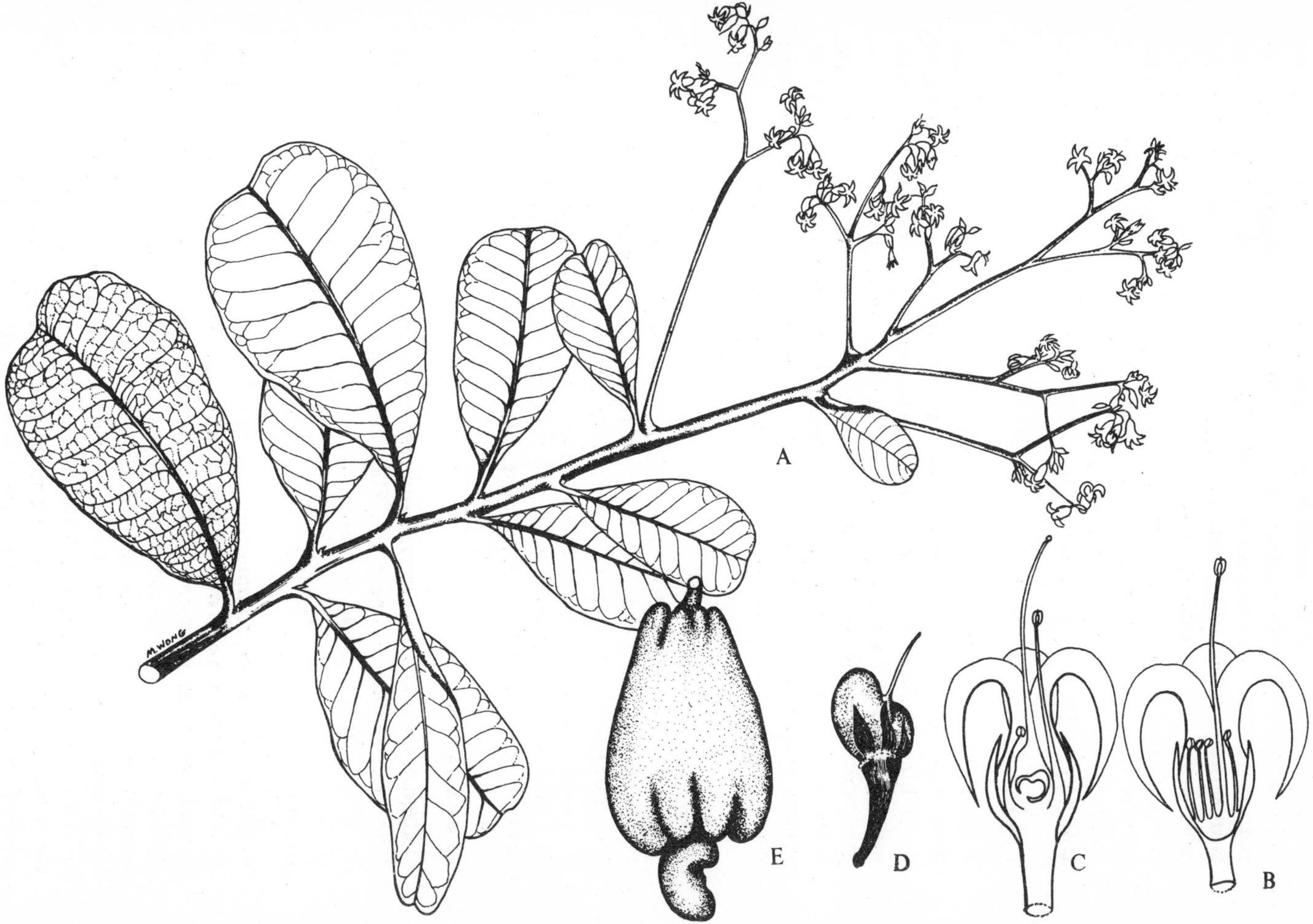
A
E
D
C
B
M.WONG

early fruit fall due to physiological causes. About 10 per cent of the hermaphrodite flowers produce mature fruits, with an average of 5–6 fruits per inflorescence. In southern Tanzania flower production, pollination and fruit setting are efficient and do not normally limit yields, whereas in India pollination was found not to be very efficient and increased fruit-setting was obtained by artificial pollination. The significance of the three lengths of stamens is not yet clearly understood. Northwood (1966) describes a third type of flower in some trees of Jamaican origin in which the reproductive organs are reduced and infertile. Rain at flowering reduces fruit set.

GERMINATION

Germination of the nuts is often poor and rather slow. In the field the first seedlings may appear two weeks after planting, but many nuts take much longer to germinate. Nuts with a high specific gravity germinate more quickly and have a higher total viability than those with a low specific gravity and they also produce more vigorous seedlings. Very large nuts have a low density and give poor germination. Only nuts which sink in water should be planted, but even better results can be obtained by planting those which sink in a solution of 1½ lb of sugar in 1 gallon of water.

CHEMICAL COMPOSITION

The nuts contain: water 5 per cent; protein 20 per cent; fat 45 per cent; carbohydrate 26 per cent; fibre 1·5 per cent; mineral matter 2·5 per cent. The shell (pericarp) contains approximately 50 per cent of cashew-shell oil, a vesicant, composed of 90 per cent anacardic acid and 10 per cent cardol. The cashew apple contains: water 88 per cent; protein 0·2 per cent; fat 0·1 per cent; carbohydrate 11·6 per cent; rich in vitamin C.

PROPAGATION

Cashews are usually grown from seeds planted *in situ*, with 3 seeds per hole, later thinned to one plant. They can also be propagated vegetatively by air-layering, grafting or inarching.

HUSBANDRY

The usual spacing recommended is 30 × 30 ft, thinning, if necessary, to 60 × 60 ft after 10 years or so. Once established, little management is needed beyond weeding. Inter-cropping may be done during the first two years. The trees are shaped by removing the lower branches and the water shoots which come up from the base during the first 3 years. Thereafter little or no pruning is necessary. Economic bearing commences in the 3rd year; the trees are in full production by the 10th year and they continue bearing for a further 20 years. Flowering takes place during the dry season. The shell of the nut reaches its full size 3–4 weeks after flowering, after which the pedicel swells to produce the cashew apple. The fruits mature in 2–3 months and are harvested fully ripe. In Tanzania the nuts are allowed to fall from the tree and are collected from the ground. In dry weather they are left on the ground until the apple dries, but should be collected daily in wet weather. The nuts are removed from the apples and are dried in the sun or on a barbecue, which takes 1–3 days. The yield of nuts per tree varies enormously ranging from 1–100 lb with an average yield per annum per acre of mature trees of 850–1,000 lb.

The nuts from Tanzania are exported to India for shelling, where it is done by hand. They are usually roasted in the shell after which they are cracked. Oil-bath processes give a quick and uniform roasting and a higher recovery of cashew-shell oil. After cracking, the testas are removed and the nuts are vacuum-packed for export. Attempts at mechanical shelling have not been very successful.

MAJOR DISEASES AND PESTS

Few serious diseases or pests have been reported. A die-back occurs in some areas and *Helopeltis* spp. are serious pests in Tanzania.

IMPROVEMENTS

Northwood (1966) has shown that there is a very considerable variation in yield and nut size in Tanzania which should provide ample scope for the selection of superior trees for use in a breeding programme. The yields of the best trees were more than twice that of the mean, but the percentage of such trees was very low and it is also necessary to consider quality. A tree which produces a large number of nuts often has small nuts which are not suitable for the cashew trade.

PRODUCTION

The main areas of production are southern India, Mozambique and Tanzania, the last two countries sending their produce to India for shelling. The stands of wild trees in Brazil have barely been touched, although this country does export some nuts. The first small exports from Tanzania were made in 1938, but were discontinued during the war. In 1960 Tanzania exported 36,718 tons, valued at over £2 million. The international trade in cashew nuts is a recent development. Up to 1925 exports from India did not exceed 50 tons per annum; by 1941 exports had reached nearly 20,000 tons; in 1956 over 31,000 tons of kernels were exported, of which the United States took about 70 per cent.

REFERENCES

MUTTER, N. E. S. and BIGGER, M. (1961). Cashew. *Tanganyika Min. Agric. Bull.*, **11**.

NORTHWOOD, P. J. (1966). Some Observations on Flowering and Fruit-setting in Cashew, *Anacardium occidentale* L. *Trop. Agriculture, Trin.*, **43**, 35–42.

Mangifera L. (x = 10)

About 62 spp. of tall evergreen trees in south-east Asia and Malaysia to New Guinea, with the greatest number of spp. in the Malay Peninsula. 15 spp. yield edible fruits, but only one sp., *M. indica* (q.v.), is widely planted throughout the tropics. The following spp. are among those sometimes cultivated in orchards in the East, but give inferior fruits to *M. indica*: *M. caesia* Jack, *M. foetida* Lour., *M. lagenifera* Griff., *M. odorata* Griff., *M. zeylanica* Hook. f. The sap of some spp., particularly the white latex of immature fruits, has irritant properties.

Mangifera indiça L. (2n = 40) MANGO

USES

'Mango is decidedly the most popular fruit among millions of people in the Orient, particularly in India, where it is considered to be the choicest of all indigenous fruits. It occupies relatively the same position in the tropics as is enjoyed by the apple in temperate America and in Europe' (Singh, 1960). It has been variously described as 'king of all fruits' and 'a ball of tow soaked in turpentine and molasses, and you have to eat it in the bath tub'. The latter may well apply to stringy fruits from unselected seedling trees, which differ greatly from named selected clones in palatability. Ripe fruits are eaten raw as a dessert fruit and are used in the manufacture of juice, squash, jams, jellies and preserves. They can be canned. Unripe fruits are used in pickles, chutneys, and culinary preparations; they are also sliced, sun-dried, and seasoned with turmeric to produce *amchur*, which may be ground into a powder and used in soup, chutneys, etc. In India the seeds are used for human food in times of scarcity and a flour is also made from them. Leaves are fed to cattle in times of shortage, but prolonged feeding may result in death. Urine of cattle fed on mango leaves is used as a yellow dye. The timber is used in many ways and is valued for boats and dug-outs. The mango holds an honoured place in Hindu mythology, religion, ritual, culture, ceremonies and customs. Festoons of mango leaves are used as decoration at almost all Hindu ceremonials and festivals. A mango grove was presented to Lord Buddha.

ORIGIN AND DISTRIBUTION

Mukherjee (1953a) suggests that *M. indica* had an allopolyploid origin through interspecific hybridization, accompanied by chromosome doubling and was further differentiated by gene mutations and hybridization. It probably originated in the Indo-Burma region and grows wild in the forests of India, especially in hilly areas in the north-east. It then spread under cultivation over the Indian sub-continent at an early age, where it has been grown for 4,000 years or more. It has since been taken to all parts of the tropical world and has become naturalized in many areas.

It was probably taken to Malaya and neighbouring East Asian countries by Indians in the 5th or 4th century B.C. and to the East African coast by Persians about 10th century A.D. The Portuguese took mangoes to West Africa and the New World and it had reached Brazil by the beginning of the 18th century. The first introduction into the West Indies was to Barbados about 1742 from Rio de Janeiro. Mangoes were obtained from a French ship sailing between Mauritius and Haiti captured by the British in 1782 off Jamaica and were planted in that island, which later received its first grafted mangoes from India *via* Kew in 1869. An unsuccessful introduction was made into Florida in 1833, but successfully so in 1861, and inarched trees were obtained from Calcutta in 1888.

CULTIVARS

Indian cvs are mainly monoembryonic; in the Philippines and Hawaii they are mainly polyembryonic. Seedling trees from zygotic embryos do not breed true and their fruits are often fibrous and turpentine-flavoured, while seedlings from apomictic embryos will be the same genetically as the parent. Named clones are usually propagated vegetatively. Much confusion still exists in the naming and sorting out of cvs, of which the greatest number occur in India, Indo-China and the Philippines. Cvs are often chosen for colour and flavour rather than yield. Each local area has tended to select cvs suited to its local environment.

In India 'Alphonso' and 'Mulgoa' are well-known cvs; a seedling of the latter in Florida gave rise to 'Haden'. 'Cambodiana' was an apomictic seedling selection in Indo-China. 'Peach' is grown in Queensland and Natal.

In the West Indies two of the most important cvs are: 'Julie', which is slow-growing but bears dependably. It seems to have been brought from Rêunion to Martinique. Fr. green, sometimes tinged red, flat-sided; flesh yellow to orange, sweet, with few fibres; good texture; stones small and thin, adhering to flesh.

'Bombay', known as 'Peter' in Trinidad and elsewhere as 'Apple', 'Mango Blanca'. An Indian cv. It is very vigorous and productive. Fr. large, rounded, green or flushed rose, ripening yellow; flesh yellow, very free of fibres, good flavour; seed very free from flesh.

ECOLOGY

Mangoes are grown at elevations from sea level to 4,000 ft in the tropics, but they do best below 2,000 ft and in climates with strongly marked seasons and dry weather for flowering and fruiting. In the ever-wet tropics heavy rain during this period causes a marked reduction in pollination, fruit set and maturing fruits. They are damaged by frost and young trees are usually killed. The optimum growth temperature is 75–80°F. They are grown in areas with an annual rainfall of 10–100 in., provided there is an adequate dry season. They will thrive on a wide variety of soils, provided they are not too waterlogged, too alkaline or too rocky, but even shallow, impervious soils produce mangoes. A pH of 5·5–7·5 is preferred. Very fertile soils with adequate supplies of water throughout the year may result in luxuriant vegetative growth and poor cropping. Some of the best mango groves in India are on the Indo-Gangetic plain with deep alluvial soils and in areas of lateritic soil.

Mangoes have a decided tendency to biennial bearing and may only produce one good crop every 3–4 years. This is influenced by climate and cv.; a high C/N ratio is required for flower initiation, as is the abundant production of new growth during an 'off' year. Flowering may sometimes be induced by smudge fires.

STRUCTURE

An erect, branched, evergreen tree, 10–40 m high, living for 100 years or more, with dense dome-shaped canopy.

ROOT: Long tap-root up to 6 m in depth and dense surface mass of feeding roots.

STEM: Pronounced trunk with greyish-brown, fissured bark; branchlets rather stout.

LEAVES: Spirally arranged, glabrous, exstipulate, produced in flushes; young lvs usually reddish in colour, later turning dark shiny green and remaining on tree for a year or more. Petiole 1–10 cm long, somewhat flattened on upper surface, and with marked pulvinus at base. Lamina 8–40 × 2–10 cm, narrowly elliptic or lanceolate, somewhat leathery, apex acuminate, base tapering, margin usually undulate; midrib prominent and up to 30 pairs of lateral veins; stomata on both surfaces, but with greater number on lower surface.

INFLORESCENCE: Widely branched terminal panicle, 10–60 cm in length, with 1,000–6,000 fls borne on new growth. Branches often tinged red and usually pubescent. Polygamous with male and hermaphrodite fls in same infl., with 1–36 per cent of latter, although higher figures recorded in a few cvs.

FLOWERS: 5–8 mm in diameter in cymes on branchlets; subsessile, sweet-scented. Sepals usually 5 (rarely 4–7), free, concave, yellowish-green, hirsute. Petals usually 5 (rarely 4–7), twice as long as calyx, cream with 3–5 darker yellow ridges on inner surface, petals later becoming pinkish. Annular, fleshy, 5-lobed disc between corolla and androecium. Stamens

---

Fig. 2. **Mangifera indica**: MANGO. **A**, inflorescence (× ½); **B**, male flower with one fertile stamen (× 6); **C**, male flower with two fertile stamens in longitudinal section (× 6); **D**, hermaphrodite flower (× 6); **E**, hermaphrodite flower in longitudinal section (× 6); **F**, shoot with fruit (× ⅓); **G**, seed enclosed in endocarp (× ½); **H**, fruit in longitudinal section (× ½).

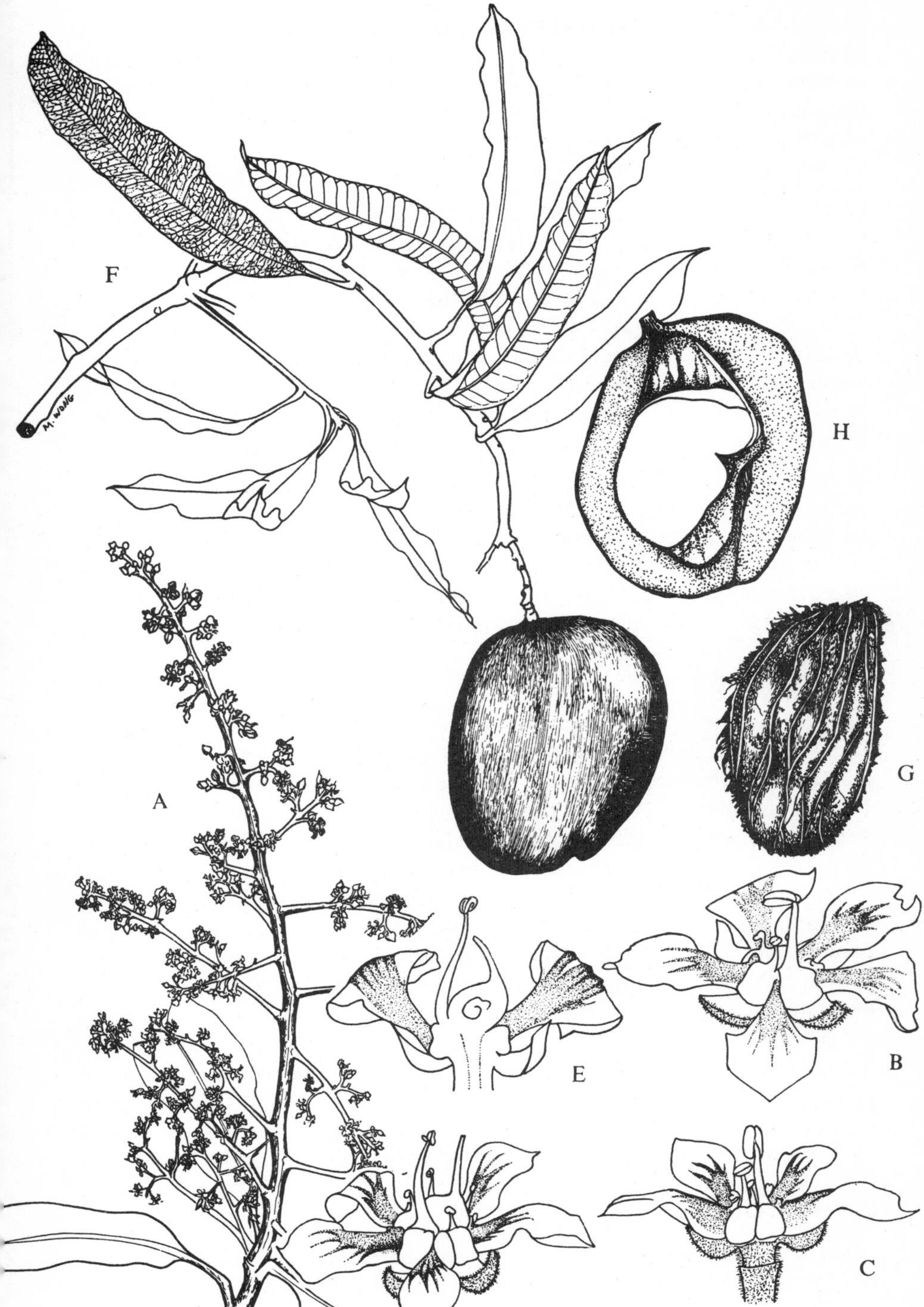
F
M. WONG
H
G
A
E
B
D
C

usually 5 (rarely 3–7) inserted on outer margin of disc, of which 1 or occasionally 2 are fertile, about 2 mm long with pink anthers turning purple on anthesis, and remainder are staminodes. Pistil abortive in male fls. In hermaphrodite fls stamens as above, and with sessile, oblique, one-celled ovary set on disc with lateral style and small simple stigma, approximately same length as fertile stamen; single anatropous ovule.

FRUIT: Fleshy drupe, variable in size from 2·5–30 cm long, in shape from rounded to ovoid-oblong and sometimes laterally compressed, and in colour with varying mixtures of green, yellow and red. Pistillate area near proximal end may develop into a conical projection or beak, above which there is a pronounced sinus. Basal end may be depressed, elevated or intermediate. Exocarp fairly thick and gland-dotted; edible mesocarp of varying thickness, texture and flavour from soft, free of fibres, sweet and juicy to fibrous and turpentine-flavoured; endocarp thick, woody and fibrous, and may be free from mesocarp or with fibres extending into it.

SEED: Inside stony endocarp, variable in size; testa and tegument represented as two papery layers; exalbuminous; two fleshy cotyledons. Some seeds monoembryonic with zygotic embryo only (occasionally apomictic); others polyembryonic with 2–12 embryos in which apomictic embryos are produced from the epidermal cells of the nucellus and in which the zygotic embryo may or may not be suppressed.

POLLINATION AND FRUIT SET

In the northern hemisphere floral buds are usually initiated in October and November, flowering taking place over a period of 2–3 weeks in January to March. In the southern hemisphere flowering is usually in June–August. Flowers usually open early in the morning with maximum anthesis between 8–12 a.m.; stigmas appear to be immediately receptive. Nectar is secreted by the disc; the flowers are visited by insects, mainly flies, and pollination is mainly entomophilous. Pollination is essential for fruit setting and the development of seed, even when all the embryos are apomictic. Bagged panicles do not set fruit. It has been shown that 65–85 per cent of the hermaphrodite flowers remain unpollinated and that only 0·1–0·25 per cent of them reach the harvesting stage, fruit drop occurring at all stages. Trees are generally self-compatible* and self-fertilization by pollen from the same flower is possible. Rain and high humidity at blossoming reduces pollination and fruit setting. The time of development after fertilization to maturity of fruit is 2–5 months, depending on cv. and temperature. Some cvs in addition to the main fruiting season set a few fruits throughout the year.

GERMINATION

Seeds are usually planted with the endocarp attached and will remain viable for about a month after extracting from fully ripe fruits. They are best stored in charcoal. Seeds are sometimes shelled (endocarp removed)

* A number of cvs have now been found to be self-incompatible in India.

and this hastens germination. Germination of seed stones takes approximately 20 days and seedlings reach a height of about 6 in. at 6–8 weeks, and 9–12 in. at 10–12 weeks. If planted in a nursery, seedlings are usually transferred to pots when the first reddish leaves turn dark green and with the cotyledons still attached to the root collar.

Zygotic embryos in monoembryonic seeds do not breed true. Apomictic embryos in polyembryonic seeds will have the same genetical constitution as the parent; but zygotic seedlings, if present, cannot be distinguished from the others. Occasionally seedlings with multiple shoots from buds in the axils of the cotyledons are produced, but these will have only one root.

CHEMICAL COMPOSITION

Edible portion of ripe fruit is 60–75 per cent by weight. A typical composition of the mesocarp is: water 84 per cent; sugar 15 per cent (varies from 10–20 per cent); protein 0·5 per cent. Unripe fruit are rich in starch, which is hydrolysed to sugars during ripening. The fruit is an important source of vitamin A, fair in vitamin B and with varying quantities of vitamin C. The seeds contain: carbohydrate 70 per cent; fat 10 per cent and protein 6 per cent.

PROPAGATION

Mangoes can be propagated by seed and most of the world's mango trees are of seedling origin. Except in the case of apomictic seedlings from polyembryonic seeds, plants must be propagated vegetatively to produce clonal material true to type. Seeds may be planted in nurseries, sometimes subsequently planted into pots, or planted directly into pots of bamboo or other material and then planted on the permanent site when 6–12 months old. They may also be planted direct at stake. Stocks are also raised in this way and the seed for them should be selected from strong-rooted and vigorous cvs. Little is known of the effect of stock on scion or of stock selection.

The commonest method of vegetative propagation in India is approach-grafting, also known as inarching, where it has been used since ancient times; it is also used in Trinidad. It is slow, laborious, cumbersome and expensive. In most other areas, e.g. Hawaii, Philippines and Florida, budding is practised and it is a cheaper and quicker method.

In the approach-grafting method stocks 6–18 months old are placed in pots on scaffolding close to the scion tree and approach-grafted to shoots of approximately pencil thickness, the union being covered with waterproof tape. The scion is half cut away at 8–9 weeks, completely severed at 10–11 weeks, and the original top of the stock is then removed. After hardening off for a month, they are then planted in the field, some 12–24 months after sowing the seed. The successful take is 90–95 per cent.

A modified system (Bharath, 1958) is now available in which 6-weeks'-old seedlings, with their seeds and roots covered with wet moss and wrapped in polythene, are tied to scion shoots and approach-grafted, union being effected in 4 weeks. They are then severed from the parent tree, potted up and, after 1 month's hardening off, are ready for planting, some 3 months after sowing the seed. In Trinidad this method has given 90–100 per cent successful takes and is more rapid and cheaper than using older seedling stocks and is more economic in scion material.

In budding methods, budwood is obtained from the ends of young branches with fully-grown leaves which are defoliated 2–3 weeks before taking, during which time the petiole stubs absciss. Plump buds are then budded on to stocks which are coming into a flush, as the bark is more easily separated at this time. The seedling stocks are usually raised in nurseries, the seeds being planted in rows 1 ft apart.

Shield budding, using the T or inverted T method on to 1-year-old seedlings at about 8 in. above ground level, is the usual method practised in Florida and Hawaii. If the bud is green after 3–4 weeks, indicating that union has taken place, the stock is ringed or cut back to force the bud into growth, and is finally removed a few weeks later when the bud has produced 3–5 branches, 8–10 in. in length. The budded seedlings are usually transplanted with a ball of earth attached. The percentage success is usually lower than with approach grafting.

In Trinidad (Hosein, 1958) seedlings, 3–4 months old, have been successfully budded by the T method, 3–4 in. above ground level, and covered with waterproof tape or polythene, including the bud. After 21 days the wrapping material is removed from the bud only and the stock cut back $\frac{1}{3}$ of its length. It is cut back further after 2 months, and finally removed entirely together with the wrapping material at 3 months. Lifting is done in two stages, taking 1 month, the budded seedlings are then transplanted in a ball of earth, 7–8 months after sowing the seeds.

Flap- and patch-budding have also been used in some countries, as has the Forkert method in which no wood is left on the bud. Vegetative propagation by cuttings, layering and marcotting has not been very successful in mangoes.

Old, off-type and unproductive trees may be renovated by top-working, using either approach-grafting, side- or cleft-grafting.

HUSBANDRY

The usual spacing for mangoes is 30–35 ft apart. Planting holes 2–3 ft wide and deep should be prepared 6 months before planting, farmyard manure, bone meal and wood ash being incorporated with the soil. It is usual to supply irrigation where necessary and manure during the first 4–5 years of life, when the tree should be encouraged to make vigorous growth; thereafter these are often discontinued. Any subsequent manuring should maintain a suitable balance between K and N. Intercropping in the early years with vegetables, legumes and pineapples is often practised. After-cultivation consists mainly in keeping down weed growth and weeding round the trees. Pruning is not usually done, except for the removal of weak growth and dead or diseased wood and parasitic Loranthaceae; it may assist in shaping the tree early in life. Grafted or budded mangoes usually produce a few fruits (15–20) in the 4th or 5th year, but flower buds are usually removed before this. Seedling mangoes take longer to come into bearing. Yields then gradually increase. Singh (1960) states that by the 10th year trees bear 400–600 fruits in the 'on' year, with further increase up to the 20th year and with yields declining after the 40th year. Yields will, of course, depend on the cv. and the environmental conditions. For local consumption fruits are picked almost fully ripe and before abscission; for export they should be picked when they have reached their full size but before softening. Cold storage at temperatures of 45–50°F, depending on cv., will keep fruits for 2–4 weeks.

MAJOR DISEASES

Anthracnose, *Colletotrichum gloeosporioides* Penz., is the most serious and widespread disease of mangoes, particularly in moist climates. It causes leaf-spot, wither-tip of young twigs, blossom-blight and fruit-rot, the last appearing as black spots, and is particularly harmful in export fruit. Control is possible by copper fungicides, application of which should be begun before the flowers open. Powdery mildew, *Oidium mangiferae* Berthet, causes losses to flowers and young fruits in India.

MAJOR PESTS

Mango-hopper or jassid, *Idiocerus* spp., is the most serious pest of mango blossoms in India. Some control is possible with DDT. Fruit-flies, including the Mediterranean fruit-fly [*Ceratitis capitata* (Wied.)], Mexican fruit-fly [*Anastrepha ludens* (Loew.)], South American fruit-fly [*A. fraterculus* (Wied.)], West Indian fruit-fly (*A. monbinpraeoptans* Sein.) and *Dacus* spp., attack mangoes in various parts of the world by laying eggs in mature fruits on which the larvae feed rendering them useless for human consumption. Some control is possible by the destruction of fallen infected fruits. Strict quarantine measures are adopted in many countries against the entry of fruit-flies, notably in the United States. Mango weevils, *Cryptorrhynchus* spp., damage the seeds. Mangoes are also attacked by thrips, scale insects and mealy bugs.

IMPROVEMENT

Little has yet been done to improve the crop except for the selection of clones from existing seedling populations, in which colour and flavour have received the most attention, the clones then being propagated vegetatively. Much confusion still exists in the classification of cvs. The main characters on which selection should be based are: (1) regular and prolific bearing every year to overcome the present tendency to biennial bearing; (2) fruit of good flavour, grade and keeping quality, including cold storage; (3) resistance to anthracnose and other diseases, and to insect pests, particularly fruit-flies; (4) early age to commence bearing; (5) dwarf habit to facilitate harvesting;

(6) cvs which fruit early and late in the season to extend the fruiting season; (7) cvs which will set a crop in the wetter tropics when rain may occur at flowering; this may possibly be obtained from Philippines and Indo-China cvs; (8) frost resistance where required.

Hybridization presents difficulties because: (1) it is laborious and expensive; (2) cvs are very heterozygous; (3) very small numbers of fruits mature after pollination; (4) polyembryony; (5) the long time taken for succeeding generations; (6) a lack of knowledge on many aspects of the crop.

A little hybridizing has been done in India and Florida without much success* and no commercial cvs have yet been produced by this means. 'In Florida . . . from nearly 13,000 flowers pollinated 45 seeds were obtained and none produced a tree that seemed promising.'

PRODUCTION

The mango has been an important item of diet in India since ancient times, and is by far the most important fruit crop of that country, occupying about 60 per cent of the total area under fruits. Akbar, the Mogul emperor (1556–1605), planted an orchard of 100,000 mango trees. There are now some 2 million acres of mangoes in India, giving an annual production of approximately 5 million tons of fruit. Mangoes are now grown in most tropical countries at the lower altitudes, and in Florida, Queensland, Egypt and Natal. Most of the fruits are consumed locally in the producing countries. Fresh mangoes from India are exported to Kuwait, Bahrein, Singapore, Malaya and elsewhere. India also exports approximately 1,000 tons of mango pickles annually, of which the largest quantity goes to the United Kingdom. Fresh mangoes have not become popular in temperate countries due to the difficulty of transporting them in good condition and the people in these countries have not acquired a taste for them; nor have canned mangoes gained popularity. Fresh mangoes are sent to the United Kingdom by air from the Kenya coast and experimental shipments have been sent in cold storage from the West Indies and elsewhere.

REFERENCES

BHARATH, S. (1958). Mango Propagation—A Modified System of Approach Grafting. *Trop. Agriculture, Trin.*, **35**, 190–194.

GANGOLLY, S. R., SINGH, R., KATYAL, S. L. and SINGH, D. (1957). *The Mango.* New Delhi: Ind. Counc. Agric. Resc.

HOSEIN, I. (1958). Mango Propagation by a "T" Graft Method. *Trop. Agriculture, Trin.*, **35**, 181–189.

MUKHERJEE, S. K. (1953a). Origin, Distribution and Phylogenic Affinity of the Species of *Mangifera* L. *J. Linn. Soc., London*, **55**, 65–83.

MUKHERJEE, S. K. (1953b). The Mango—Its Botany, Cultivation, Uses and Future Improvement, especially as Observed in India. *Econ. Bot.*, **7**, 130–162.

SINGH, L. B. (1960). *The Mango.* London: Leonard Hill.

* Some promising hybrids have now been obtained in India.

BOMBACACEAE

About 20 genera and 150 spp. of trees in the tropics round the world, closely related to the Malvaceae. Lvs alternate, simple or digitate, stipules deciduous. Fls hermaphrodite, usually large and showy; calyx 5-toothed, valvate; petals 5; stamens free or monadelphous, anthers 1-locular; ovary superior, 2–5-locular. Capsule loculicidally dehiscent or indehiscent. Seeds with little or no endosperm, often embedded in hairs from the wall of the fruit.

USEFUL PRODUCTS

Ceiba pentandra (q.v.) yields the kapok of commerce. Other interesting and useful plants are:

Adansonia digitata L., the baobab, which is widespread in the drier regions of tropical Africa, is often planted near villages in West Africa. This weird and massive tree is bottle-shaped and is remarkable for its enormous trunk in comparison with the small crown of foliage. The trunk is comparatively short, 13–17 m high, but is 10–14 m or more in girth, with short thick branches. A specimen with a girth of 37 m has been recorded. They live to a very great age and it is estimated that some may be 5,000 years old. The bark is unarmed. Lvs palmate with 5 sessile leaflets; fls large, 12·5–15 cm in diameter, white, with numerous monadelphous purple stamens; frs oblong, 15–20 cm long, pendulous on long stalks, woody, indehiscent, with large seeds embedded in dry acid pulp. The young leaves are used as a soup vegetable. The fruit pulp, which contains tartaric acid, is made into a drink and is also used as a food seasoner. The seed kernels are edible and contain 12–15 per cent oil. The inner bark yields a strong and durable fibre which is made into ropes. The bark is beaten to make cloth. Various parts of the tree are used in native medicines.

Several species of *Bombax* yield kapok, of which the most important are: *B. buonopozense* P. Beauv., the red-flowered silk-cotton tree of West Africa, a large tree up to 40 m high, with stout conical spines on the trunk and branches; the white floss from the fruits is of excellent quality. *B. malabaricum* DC., which extends from India to Australia, also yields fibre from its fruits, but the floss is not quite as resilient as true kapok.

Durio zibethinus Murr. is the famous durian of the Far East, which is widely cultivated throughout Malaysia. Attempts to establish it elsewhere have seldom been successful. The seeds quickly lose their viability. The trees grow up to 30 m tall; lvs oblong-acuminate with golden hairs on under-surface; fls cauliflorous, large, white or pink; frs ovoid, spiny, foetid, 12–25 cm in diameter, thick walls splitting into 5 valves; seeds large, covered with pulpy cream-coloured edible aril. The trees begin fruiting at about 7 years old. The fruits take about 3 months to develop. The fruit is not fully ripe until it drops from the tree when it gives out an abominable stench of over-ripe cheese, rotten onions, turpentine and bad drains. The custard-like pulp must be eaten within a short time as it quickly turns rancid and sour. Malays and other people in the Far East are very partial to it, as are some Europeans who have managed to overcome the odour. The fruit taste is difficult to describe, but it is sweet, aromatic, persistent and with a touch of garlic; it has also been described as 'French custard passed through a sewer'. Wild animals are very attracted by it, particularly elephants, tigers and monkeys. People in Malaya build shelters in wild durian trees in the forest and descend by ladders to pick up fruits as they drop, hoping to reach them before the elephants. It is believed to have aphrodisiac properties. The naturalist Wallace considered that 'it was worth a journey to the East, if only to taste of its fruit'. It gave the name to Corner's 'Durian Theory' of the evolution of angiosperms.

Ceiba Mill. (x = ?)

Nine species of large deciduous trees in tropical and subtropical America. *C. pentandra* (q.v.), the kapok, extends to West Africa and south-east Asia. *C. acuminata* (S. Wats.) Rose and *C. aesculifolia* (HBK.) Britton and E. G. Baker are used as a source of kapok in Mexico.

Ceiba pentandra (L.) Gaertn. (2n = 72–84) KAPOK

Syn. *Bombax pentandrum* L.; *Eriodendron anfractuosum* DC.

It is sometimes known as the white silk-cotton tree.

USES

The tree yields kapok, which is the floss derived from the inner capsule wall in which the seeds lie loose when ripe. Each hair or fibre is a single cell, 0·8–3 cm long, thin-walled, with a wide air-filled lumen. The hairs are lustrous due to a waxy coating and are very resilient, elastic, light, water-repellent and buoyant, the buoyancy being about 5 times that of cork. It is long lasting and is not attacked by fungi and pests such as rodents. Because of these properties, kapok is used for stuffing and insulating purposes. It is used in life-belts, life-jackets, life-buoys, mattresses, pillows, upholstery, saddles, sleeping-bags, surgical bandages, mackintosh linings, clothing for aviators and other protective clothing. Because of its lightness (it is 8 times as light as cotton) it requires less weight of kapok to fill a mattress than any other material. Kapok has low thermal conductivity and is one of the best known sound absorbers per unit of weight. In padded form it is used for thermal and acoustic insulation in aeroplanes, tanks, studios, hospitals, etc. Recently ways have been found of spinning kapok, so it can now be made into yarn and textiles.

Very young unripe pods are eaten in Java. The seeds are crushed, roasted and used in soups in West Africa. They contain 20–25 per cent of an edible oil which is used for culinary purposes, as a lubricant, and for soap manufacture. The press cake remaining is used as livestock food. Parts of the plant are used for native medicines. The soft light wood is utilized for dugout canoes, stools and carvings; the large plank buttresses may be made into doors, tables and platters. Large cuttings strike easily and are used as live fence posts. In many parts of the world some sacred significance is attached to the tree by local peoples.

SYSTEMATICS, ORIGIN AND DISTRIBUTION

Baker (1965) recognizes three varieties of *Ceiba pentandra*:

Var. *caribaea* (DC.) Bakh., which occurs wild in the American tropics and in evergreen, moist, semi-deciduous and gallery forests of West Africa. It is a gigantic tree, reaching 70 m in height and is the tallest tree in Africa. The trunk is unforked and spiny with large buttresses; branches horizontal, flowering irregularly, fls rose or cream-coloured, lvs narrow, frs rather short and broad, dehiscent, kapok grey to white. $2n = 80, 88$.

Var. *guineensis* (Schum. & Thonn.) H. G. Baker, which grows wild in savanna woodlands in West Africa. It is seldom more than 18 m in height; the trunks are spineless without buttresses and are often forked; branches strongly ascending, flowering annually, lvs broad, frs elongated and narrow at both ends, dehiscent, kapok grey. $2n = 72$.

Var. *pentandra* (syn. var. *indica* (DC.) Bakh.), the cultivated kapok of West Africa and Asia. Ecologically this variety shows a wide range of tolerance and can be grown in forest and savanna regions. It is a tree of moderate height, up to 30 m; the trunk is unbranched, usually spineless, with small or no buttresses; branches horizontal or ascending to varying degrees; lvs intermediate in breadth, flowering annually, fruit short or long, narrowed at both ends or banana-shaped, usually indehiscent, kapok usually white. $2n = 72–84$.

There seems little doubt that *C. pentandra* var. *caribaea* originated in the American tropics, as all the other species of *Ceiba* occur in this region. It ranges from southern Mexico to the southern boundary of the Amazon basin. In West Africa it is found wild from the Cape Verde peninsula in Senegal, southwards to Angola, and eastwards almost to the Great Rift valley. It seems reasonable to assume that fruits or seeds enveloped in kapok, which are buoyant and water-resistant, were carried by currents from the New World to Africa in remote times. Its extended range in Africa, the many uses to which it is put, and its religious significance, all point to ancient establishment in that continent, as does its resistance to the swollen shoot virus.

In Africa var. *guineensis* evolved in response to the savanna environment. Natural hybrids between var. *caribaea* and var. *guineensis* were then produced to give var. *pentandra*. A mutation for indehiscent fruits occurred and was taken into cultivation by man who propagated it vegetatively. Var. *pentandra* was better suited to man's use as it has easily climbable branches, deterrent spines are usually absent, it fruits regularly every year and, as the fruits are indehiscent, the kapok cannot escape and is easily collected. The indehiscent form cannot survive without the aid of man as it has no means of effective dispersal. Baker has produced artificial hybrids between var. *caribaea* and var. *guineensis* which are indistinguishable from var. *pentandra* and has also shown that the African and Asian cultivated forms are identical.

There is evidence that kapok had reached Java by the 10th century. Transportation from Africa to Asia must have been by man, as happened with a number of other crops. Baker suggests that Arab traders took kapok from West Africa across the continent and thence to India and the East. In south-east Asia kapok is distributed from western India, through Malaysia, Indo-China, and Indonesia to the Philippines, and as far as Samoa and Tahiti.

The evolution and history of kapok bears a close parallel with that postulated for cotton, *Gossypium* spp. (q.v.), except that cotton came in the reverse direction.

ECOLOGY

Kapok is a tropical tree and thrives best at elevations below 1,500 ft. It will grow under a wide range of conditions, but for high production requires abundant rainfall during the vegetative period and a drier period for flowering and fruiting, with an annual rainfall of 50–60 in. per annum. Fruits are not set at night temperature below 68°F. For best results it should be planted on good deep permeable soils (in Indonesia these are volcanic loams) with freedom from waterlogging. Exposed situations should be avoided as the tree is easily damaged by high winds.

STRUCTURE

A deciduous, fast-growing tree, 10–30 m in height, but reaching 70 m in var. *caribaea*. Very shallow rooted. Branches dimorphic. Trunk tapering, wide near ground; widely buttressed in some forms, with or without short conical sharp spines on trunk and branches. Crown thin; branches whorled, whorls usually of 3 branches, horizontal, giving pagoda-form, or ascending in some forms. Lvs alternate, glabrous, crowded at the ends of twigs; stipules small, lanceolate, falling early; petiole 8–20 cm long with pulvinus

---

Fig. 3. **Ceiba pentandra**: KAPOK. **A**, leafy shoot (× ⅓); **B**, flower in longitudinal section (× 1)—after Cobley (1956); **C**, fruit (× ⅓).

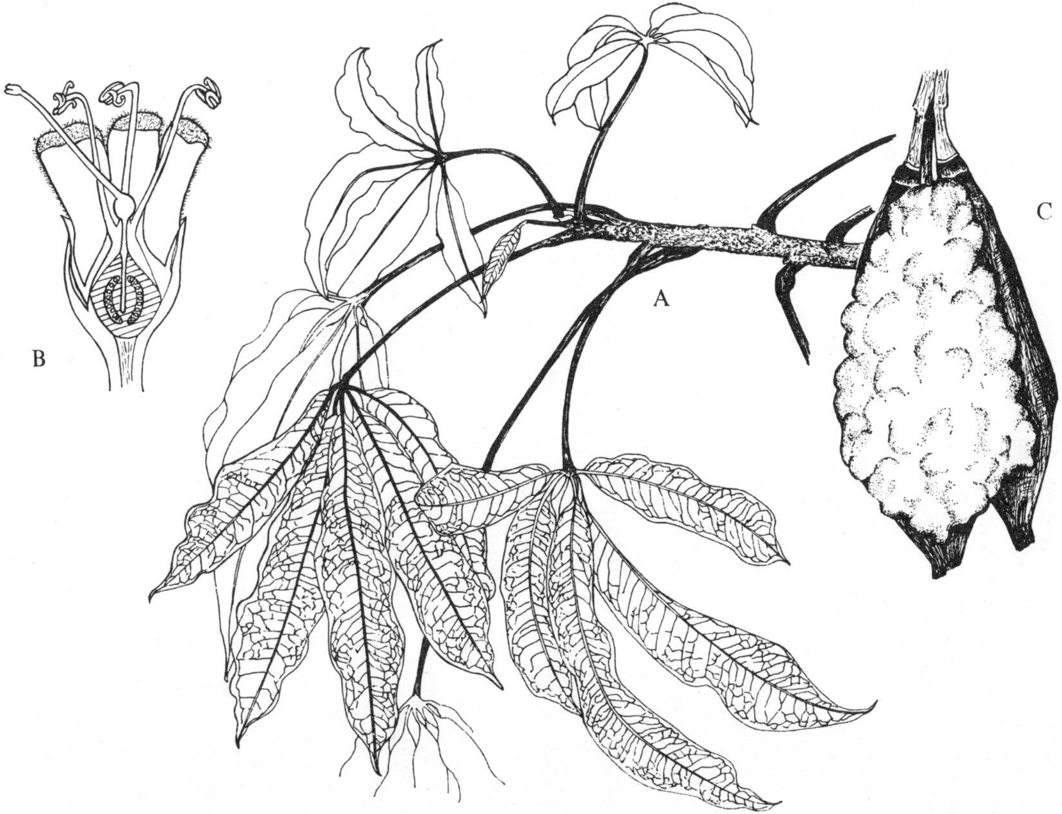
B
A
C

at both ends; lamina digitate with 5–11 leaflets, shortly stalked, lanceolate, acuminate, entire or slightly toothed, curved back and drooping, glaucous beneath, 8–16 × 2–4 cm. In areas with a dry season fls and frs produced after leaf fall. Fls in axillary fascicles, usually many-flowered; pedicels 2·5–4 cm long, articulate at top; calyx 1–1·5 cm long, 5-lobed, glabrous without; petals united at base, 2–3 cm long, obovate, usually dirty white in colour, with foetid milky smell, glabrous within, densely silky without; staminal column united at base, dividing into 5 branches, 3–5 cm long, each with 2–3 1-celled convoluted anthers; ovary 5-celled; style constricted at base with a dilation above top of staminal tube, then becoming oblique, obscurely 5-lobed. Frs ellipsoidal leathery pendulous capsules, usually tapering at both ends, 7·5–30 × 3–7·5 cm, turning brown when ripe, dehiscing by 5 valves or indehiscent. Seeds many, dark brown, obovoid, 4–6 mm in diameter, embedded in copious white, pale yellow or grey floss.

POLLINATION

The flowers open about 15 minutes after sunset and remain open until the petals fall off the next day. Seeds may be set by self-pollination when the stigmas and stamens of adjacent flowers come into contact. Flowers are also pollinated when bees visit the flowers early in the morning. However, Baker and Harris (1959) have shown that bats are the major agents of cross-pollination when they visit the flowers at dusk and during the night, mainly for nectar. At temperatures below 68°F pollination does not occur as the growth rate of the pollen tubes is then too slow for them to reach the ovary before the flowers fall. Setting of fruits is dependent upon fertilization of 20–120 ovules per ovary, the number varying with the cv. The size of pod and the amount of kapok is proportional to the number of seeds set. Fruit shedding also occurs independently of fertilization.

GERMINATION

Seed for sowing should be obtained from mature, full-sized pods from high-yielding trees. They may be planted either in nursery beds at 12 × 9 in. or by direct seeding in the field. Germination is fairly rapid.

CHEMICAL COMPOSITION

Kapok fibre has a cellulose content of about 64 per cent and a lignin content of about 13 per cent. The seeds contain 20–25 per cent oil, which is almost identical to cottonseed oil, and is used for the same purposes. The press cake contains about 26 per cent protein.

PROPAGATION

Kapok is usually propagated by seeds when grown on a plantation scale. If sown in a nursery, seedlings are transplanted in the field at 8–10 months old, the crown being removed, leaving about 4 ft of stem. It is also easily propagated by means of cuttings, 2–3 in. in diameter and 4–6 ft long, of

2–3-year-old wood. It has been stated that these should be from orthotropic branches.

HUSBANDRY

Seedlings and cuttings are usually planted at a spacing of about 20 ft apart in the field. The tree comes into bearing in the 3rd or 4th year, producing about 100 pods which give about 1 lb of cleaned floss. It is not in full bearing until the 7th–10th year, when it yields 330–400 pods per tree per year, which gives 3½–4 lb of kapok. Individual trees may yield up to 6–10 lb of kapok per year. The trees continue bearing for 60 years or more. The pod contains by weight approximately: 44 per cent husk, 32 per cent seeds, 17 per cent floss and 7 per cent placenta. The average weight per pod is about 1 oz. The pods are harvested when fully ripe and, in the dehiscent types, before they open. They are usually harvested by climbing the tree. The pods are then hulled and the kapok is dried in cage-like structures in the sun. The seeds are removed by beating with sticks or by machine. The quality of kapok is judged by its freedom from foreign matter and seeds, moisture content, colour, smell and lustre.

MAJOR DISEASES AND PESTS

Kapok appears to have few serious diseases or pests. In Ghana it is an alternate host of cocoa swollen shoot virus, but shows considerable resistance to it.

IMPROVEMENT

Little work seems to have been done on improving the crop. Selection of high yielding indehiscent types followed by vegetative clonal propagation would appear to be worth while.

PRODUCTION

Before World War II Indonesia was the largest producer of kapok and dominated the world market. Since then there has been a decline in kapok production due to destruction and neglect of the trees in Java during the Japanese occupation. Exports from the chief producing countries are now about 50 million lb, of which Thailand produces about half. Other exporting countries include Çambodia, Indonesia, East Africa, India and Pakistan. The United States is the biggest importer, taking more than half the world production.

REFERENCES

BAKER, H. G. (1965). The Evolution of the Cultivated Kapok Tree: A Probable West African Product. In *Ecology and Economic Development in Tropical Africa*, 185–216, ed. D. Brokensha, Berkeley: Calif.

BAKER, H. G. and HARRIS, B. J. (1959). Bat Pollination of the Kapok Tree, *Ceiba pentandra* (L.) Gaertn. (Bombacaceae). *J. W. Afr. Sc. Assoc.*, **5**, 1–9.

KIRBY, R. H. (1963). *Vegetable Fibres*. London: Leonard Hill.

CANNABIDACEAE

A small family in the order Urticales, sometimes included in the family Moraceae or Urticacae. It contains two genera only, *Humulus* with 3 spp., of which *Humulus lupulus* L. is the European hop, and the monocarpic *Cannabis* (q.v.)

Cannabis sativa L. (x = 10, 2n = 20) **HEMP**

USES

The plants provide three products, namely, fibre from the stems, oil from the seeds and narcotics from the leaves and flowers. The areas of production and methods of cultivation vary with the product required.

For the production of fibre it is grown mainly in temperate countries. The white bast fibres provide the true hemp of commerce. It is a soft fibre, valuable because of its length of 3–15 ft, its strength and durability. Its main use is as a substitute for flax in the manufacture of yarns and twines. It is also used for ropes, nets, sail-cloth, canvas, tarpaulins, etc. As a material for ropes it has largely been superseded by sisal. The short fibres or tow provide oakum used in caulking planks in ship-building, packing for pumps, etc.

For the production of oil the female plants are left after the male plants have been harvested for fibre. The seeds yield 30–35 per cent of a drying oil which is used as a substitute for linseed oil in paints and varnish; it is also used in soap manufacture. The seeds are used in bird and poultry feed.

Three types of narcotics are produced: (1) *bhang* (Hindustani) or *hashish* (Arabic) which is the dried leaves and flowering shoots of male and female plants, both cultivated and wild, and has a low resin content; (2) *ganja* which is the dried unfertilized female inflorescences of special cvs grown in India; the drug is official in the British Pharmacopoeia and is used medicinally as a sedative and hypnotic; (3) *charas* which is the crude resin collected by rubbing the tops of plants with the hands or beating them with a cloth; it is produced mainly in Yarkand in central Asia. In all these drugs the active principle is a resin from the glandular hairs on the leaves, stems and inflorescences. Little resin is produced when hemp is grown in temperate countries and hotter conditions are required. As a narcotic, it is consumed alone or as a beverage, but is more often used for smoking for euphoric purposes. Excessive smoking is harmful and may cause insanity. Cultivation and use as a narcotic is prohibited by law in most countries. It is also known as marijuana in the United States and Britain, where it is usually smoked illicitly in cigarettes.

ORIGIN AND DISTRIBUTION

C. sativa is a native of central Asia and is of very ancient cultivation in Asia and Europe. It is said to have reached China more than 4,500 years ago. It spread to the New World in post-Columbian times. It is now known in most tropical and temperate countries. It is naturalized in parts of India and elsewhere as a weed of waste land.

ECOLOGY

Hemp can be grown over a great range of altitudes, climates and soils. The best development of fibre is in mild humid climates of the temperate regions with temperatures of 60–80°F during the growing season, with adequate rainfall, particularly for germination and establishment, and on rich loamy soils. For the production of narcotic resin, hotter more tropical conditions are required. Sex expression is altered by environmental conditions, particularly the length of day, short days favouring the monoecious condition.

STRUCTURE

Robust, tall, erect, annual herb, 1–5 m high; all parts with viscid pubescence; dioecious, but occasionally monoecious, usually with male and female plants in roughly equal numbers. Stems angular, sometimes hollow, degree of branching depending on conditions of cultivation; when grown thickly for fibre production almost unbranched. Lvs opposite near base of stem, spirally arranged above; stipules small, pointed, persistent; petioles 4–6 cm long; lamina palmate with 3–11 leaflets, upper lvs often 1 leaflet only, leaflets sessile, narrowly lanceolate, serrate, tip acuminate, $6\text{–}10 \times 0{\cdot}3\text{–}1{\cdot}5$ cm, paler green beneath. Male fls in axillary and terminal panicles with few lvs; calyx with 5 deeply divided, widely spreading lobes, yellow, 5 mm long; petals absent; stamens 5 with slender filaments and long pendent anthers dehiscing by an apical pore. Male plants die soon after anthesis. Female inflorescences axillary and terminal, leafy; fls in pairs, each fl. enveloped by membraneous spathe-like, dark-green secondary bract with thick glandular hairs; calyx entire, thin, closely enveloping ovary; ovary sessile with solitary, pendulous ovule; styles 2-partite, 5 mm long, curving downwards and outwards. Female plants continue to live for 20–40 days after pollination until seeds are ripe. Fr. a smooth, shining achene, brownish in colour, covered by persistent calyx and enveloped by enlarged secondary bract. Seeds with fleshy endosperm and curved embryo; 100 seeds weigh 2 g.

POLLINATION

The flowers are wind-pollinated.

GERMINATION

The seed will germinate at low temperatures, but not below 1°C. Good seed should give a 90 per cent germination and, if properly stored, will remain viable for up to 2 years.

CHEMICAL COMPOSITION

Hemp fibre contains about 70 per cent cellulose. The seeds contain about 22 per cent protein and 32 per cent oil. The therapeutic and narcotic properties are contained in the resin, which contains a mixture of cannabinol and allied compounds.

PROPAGATION

Hemp is propagated by seed.

HUSBANDRY

For fibre production hemp is sown thickly with a seed-rate of 50–80 lb per acre either broadcast or in drills. Harvesting is done at the beginning of flowering, 4–5 months after sowing. The male plants produce the best fibre and are sometimes harvested first, the female plants sometimes being allowed to stand to set seed for oil production. The stems are cut near ground level and are retted either in water or by dew. After retting the stems are dried, broken and scutched. The yield of dried stems is 1½–3 tons per acre, which yield about 25 per cent fibre.

When grown solely for seed production a wider spacing is used and yields of 1,200–1,500 lb per acre are obtained.

For ganja production in India the crop is sown in rows 4 ft apart at a seed-rate of 5–6 lb per acre and plants in the rows are thinned when 8 in. high. The male plants are pulled out as soon as they can be recognized and before the pollen is shed. The unfertilized female plants remain and resin begins to form rapidly on the inflorescences, which are harvested when the flower stalks begin to turn yellow, about 5 months after sowing. The inflorescences are trodden and pressed into flat cakes; the average yield is about 250 lb per acre.

IMPROVEMENT

Monoecious cvs have been raised in Europe.

Fig. 4. **Cannabis sativa : HEMP. A,** flowering shoot of female plant (×½); **B,** male flower (×4); **C,** female flower (×6); **D,** fruit (×4); **E,** seed (×4).

A
B
C
D
E

PRODUCTION

The largest producer of hemp is the Soviet Union with an annual production of some 115,000 tons, compared with the estimated world production (excluding China) of 280,000 tons. The principal exporting countries are Italy and Yugoslavia and the principal importing countries are West Germany and France. Outside Europe, hemp is produced commercially in China, Japan, Chile and Peru. Increased production occurred in the United States as the result of a government-sponsored programme during World War II. Ganja is grown by a few licensed growers in Bengal, Mysore and Madras under supervision of the Excise Department, the drug being a monopoly of the Indian Government. Bhang is cultivated illegally to a small extent in many countries, *e.g.* by the Rastafarian sect in Jamaica in 1959–60.

REFERENCES

HAARER, A. E. (1953). Hemp (*Cannabis sativa*). *World Crops*, **5**, 445–448.

KIRBY, R. H. (1963). *Vegetable Fibres*. London: Leonard Hill.

CARICACEAE

A small, somewhat anomalous family with four genera, of which three are in tropical and subtropical America and one in Africa. Usually small trees and shrubs with terminal clusters of leaves and latex in all parts. Leaves spirally arranged, exstipulate. Often dioecious.

Carica L. (x = 9)

About 40 spp. in tropical and subtropical America. *C. papaya* (q.v.) is cultivated throughout the tropics for its edible fruits. *C. candamarcencis* Hook f., the Mountain Papaya, a native of the Andes, is cultivated at high altitudes in the tropics for its small, egg-shaped fruits, which are eaten stewed or made into preserves.

Carica papaya L. (2n = 18) PAPAYA, PAPAW, PAWPAW

USES

The ripe fresh fruits are eaten throughout the tropics for breakfast and dessert, and in fruit salads. They are used for making soft drinks, jam, ice-cream flavouring, crystallized fruit and are canned in syrup. Unripe fruits are cooked as a substitute for marrow and for apple sauce. Papain, prepared from the dried latex of immature fruits, is a proteolytic enzyme similar in action to pepsin, and is used in meat-tenderizing preparations, manufacture of chewing gum and in cosmetics, as a drug for digestive ailments, in the tanning industry for bating hides, for degumming natural silk and to give shrink-resistance to wool. Young leaves are sometimes eaten as spinach. In Java a sweetmeat is made from the flowers. The leaves and young fruits are used to tenderize meat. The seeds are used in some countries as a vermifuge, counter-irritant and abortifacient.

ORIGIN AND DISTRIBUTION

Carica papaya has never been found wild, but it is probable that it originated in southern Mexico and Costa Rica. It is closely related to *C. peltata* Hook. & Arn., which occurs in this area, and may have arisen by hybridization. Oviedo, who was Director of Mines in Hispaniola from 1513–1525, states that the seeds were brought from the coast beyond

Panama to Darien and from there it was carried to San Domingo and other islands of the West Indies. It was taken by the Spaniards to Manila in the mid-16th century and reached Malacca shortly afterwards. From there it was taken to India. It was reported in Zanzibar in the 18th century and had probably been taken there at an earlier date. It was seen in Uganda in 1874. It has now spread to all tropical and subtropical countries.

CULTIVARS

There are a number of named cvs, but these are difficult to maintain in dioecious plants. The hermaphrodite 'Solo', introduced into Hawaii from Barbados, is one of the best cvs; the fruits are pyriform, about 6×4 in., and weigh about 1 lb.

ECOLOGY

Papaya is a tropical plant and is grown in latitudes to 32°N and S; it is killed by frost. Near the equator it produces good crops up to about 5,000 ft. It requires full sun, but windbreaks should be provided. Low temperatures result in fruits of poor flavour. It thrives best in well-drained fertile soil with a pH of 6–6·5; it cannot stand waterlogging. In dry regions it is sometimes grown under irrigation.

STRUCTURE

A short-lived, quick-growing, soft-wooded tree, 2–10 m in height, usually unbranched, but branching may be induced by injury to apical meristem or by cutting back. Latex vessels in all parts.

STEM: Straight, cylindrical, with prominent leaf scars, tissue spongy, hollow, 10–30 cm in diameter.

LEAVES: Clustered near apex of trunk, large, spirally arranged; petiole 25–100 cm long, hollow, pale green or tinged purple; lamina 25–75 cm in diameter, orbicular, glabrous, palmately and deeply 7–11-lobed, lobes deeply and broadly toothed, pale green beneath with prominent veins.

FLOWERS: Plants usually dioecious, but hermaphrodite forms occur. Fls fragrant. Staminate fls in pendent axillary panicles, 25–75 cm long, sessile;

Fig. 5. **Carica papaya: PAPAYA. A,** top of female plant ($\times \frac{1}{8}$); **B,** male flower in longitudinal section ($\times 1$); **C,** female flower in longitudinal section ($\times 1$); **D,** hermaphrodite flower of *elongata* type ($\times 1$); **E,** hermaphrodite flower of *pentandria* type ($\times 1$); **F,** hermaphrodite flower with carpelloid stamens ($\times 1$); **G,** fruit in longitudinal section ($\times 1/5$); **H,** seed in longitudinal section ($\times 3$).

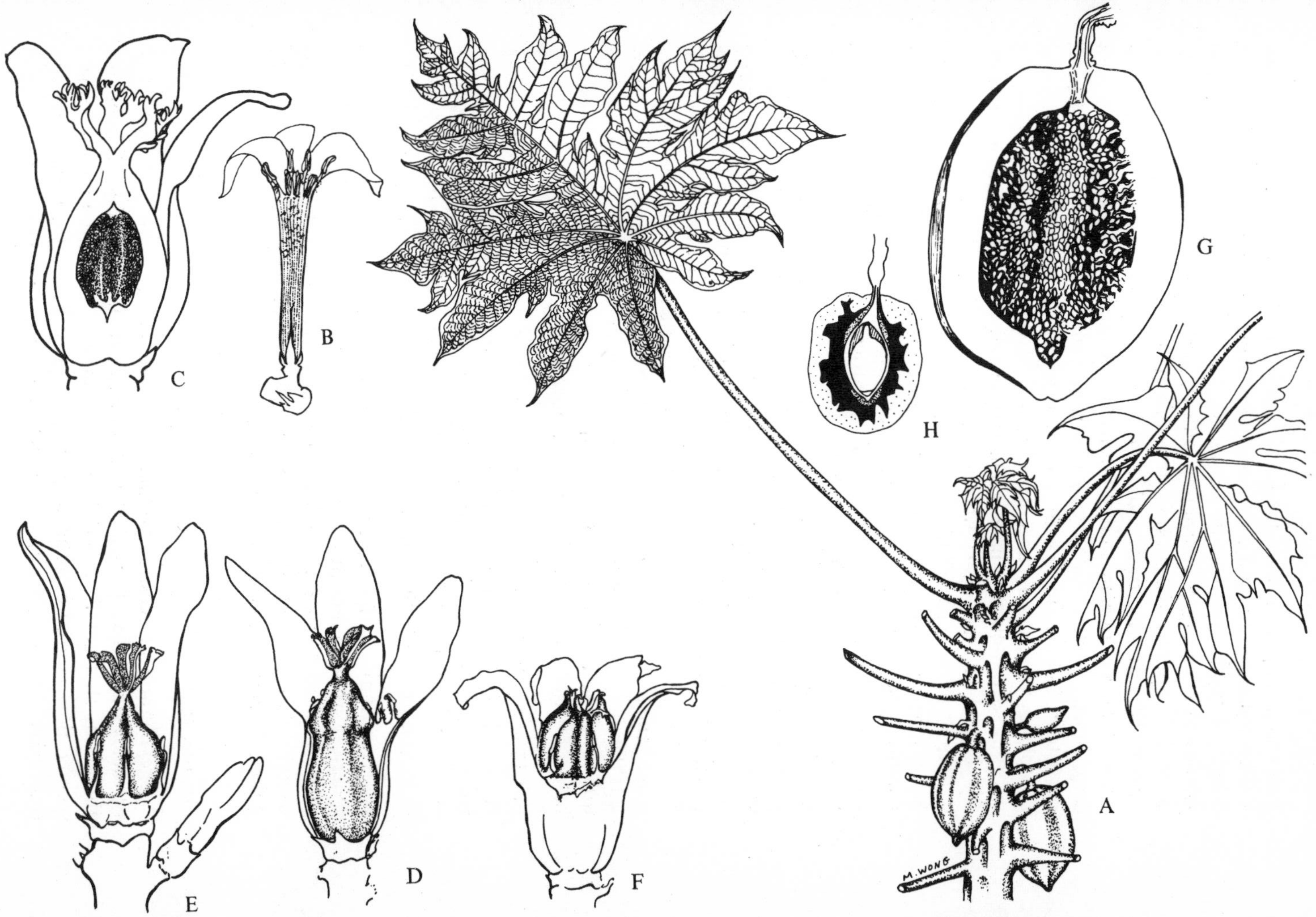
C
B
G
H
A
M. WONG
E
D
F

calyx cup-shaped, 1 mm long, 5-toothed, falling with corolla; corolla trumpet-shaped, about 2·5 cm long, with 5 spreading lobes about one-third length of tube, creamy white or yellow; stamens 10, inserted at throat of corolla in 2 whorls, 5 alternating with petals having filaments about twice as long as anthers and 5 opposite petals with shorter filaments; filaments and anthers woolly; anthers oblong, 2-celled, yellow; rudimentary pistil of some authors is an extension of floral axis. Pistillate fls 3·5–5 cm long, axillary, shortly-stalked, solitary or in few-flowered cymes; calyx cup-shaped, persistent, 3–4 mm long with 5 narrow teeth; corolla of 5 petals almost completely free; petals lanceolate, twisted, fleshy, yellow; ovary large, 2–3 cm long, ovoid-oblong, pale green, terminated by 5 sessile, deeply 5-cleft, fan-shaped stigmas; ovary with central cavity and numerous ovules.

Several other types of fls have been described. In the *elongata* type, fls hermaphrodite in short-peduncled clusters; petals united for one-third of length; stamens 10 in 2 series borne at throat of corolla; fully functional pistil elongate, developing into long cylindrical frs. In the *pentandria* type, fls hermaphrodite, corolla tube short, stamens 5 attached by long filaments near base of ovary and lying in furrows between lobes of ovary, which is more or less globose and develops into fr. which is 5-furrowed. Intermediate forms occur in which some or all (2–10) of stamens become carpelloid and produce ridged or irregular shaped frs. The proportion and type of fls produced can vary on the same tree, depending on the age of tree, season, etc. Female sterility is favoured by warm weather and such trees may become female fertile in cool months. Female trees do not undergo sex reversal; this only occurs in hermaphrodite and male trees. The dioecious condition has developed from the hermaphrodite condition. (See IMPROVEMENT below for the genetics of the different forms.)

FRUITS: Fleshy berry, 7–30 cm long, weighing up to 9 kg, ovoid-oblong to nearly spherical from pistillate fls; pyriform, cylindrical or grooved from hermaphrodite fls; skin thin, smooth, green, turning yellowish or orange when ripe; flesh yellow to reddish-orange, edible, with consistency of butter, and with mild and pleasant flavour; central cavity 5-angled.

SEEDS: Many, parietal, attached in 5 rows to interior wall of ovary; spherical, about 5 mm in diameter, black or greyish, wrinkled, enclosed in gelatinous sarcotesta formed from the outer integument; embryo median, straight, with ovoid, flattened cotyledons surrounded by fleshy endosperm. About 20 dried seeds per g.

POLLINATION

The method of natural pollination is not known with certainty. It is variously stated that the papaya is wind-pollinated, as the pollen is light and abundant, that small insects like thrips may assist, and that the sweet-scented flowers are nocturnal and are pollinated by moths. Isolated female trees have set fruit when they were 800 yards from the nearest male tree.

GERMINATION

Gelatinous envelope round seed is removed before planting; air-dried seeds will retain viability for 2–3 years. Germination takes 2–3 weeks and is epigeal.

CHEMICAL COMPOSITION

The edible portion of fresh fruit contains approximately: water 88 per cent; sugar 10 per cent; protein 0·5 per cent; fat 0·1 per cent; acids 0·1 per cent; ash 0·6 per cent; fibre 0·7 per cent. It is a rich source of vitamin A and has some vitamin C. The latex contains the enzymes papain and chymopapain, both of which have protein-digesting and milk-clotting properties.

PROPAGATION

Papayas are normally propagated by seed. These may be planted in wooden flats, which are placed in the sun while the cotyledons are still large and green; they are then transplanted into containers or polythene bags 3–4 weeks after germination; after a further 3–4 weeks (8–10 weeks after sowing), when they are 6–8 in. high, they are ready to plant in the orchard. Seed may also be sown in nursery beds at ½ in. depth at a spacing of 4 × 1 in.; 6,000 seedlings are required to plant one acre with 5–7 seedlings per hole. For papain production in East Africa seeds are frequently planted at stake with 10–30 seeds per hill; on germination these are thinned to about 5 plants which are allowed to grow for 6 months until the sex can be determined on flowering; ultimately one female plant is left per hill and one male plant for every 25–100 female plants. In planting hermaphrodite cvs such as 'Solo', it is customary to remove the female plants so that fruits of uniform size and shape are produced.

Papayas can be propagated vegetatively by cuttings or grafting, but this is not economic for commercial production.

HUSBANDRY

The usual spacing is 8–12 ft apart. It may be necessary to water the transplants and to shade with bracken or other material. After-cultivation is limited to weed control. Intercropping is sometimes practised and mulching is advantageous. Good responses have been obtained from organic and nitrogenous manures. In Hawaii a NPK 8-12-6 mixture at the rate of 2 lb per mature tree is recommended. Trees come into bearing at 9–14 months. Although trees may live for 25 years, yields decline with age. For papain production the productive life is usually 3 years, thereafter the trees get too high and latex yields decline.

For fresh fruit papayas are harvested when the first traces of yellow appear on the skin, after which they will ripen in 4–5 days. They should be cut with a sharp knife. Yields per tree vary from 30–150 fruits per annum, giving up to 15 tons of marketable fruit per acre. Shipments from Hawaii to the United States are treated with methyl bromide to kill Mediterranean fruit flies, *Ceratitis capitata* (Wied.). In cold storage unripe fruits should not be kept at below 50°F as this prevents successful ripening.

For papain production tapping commences when the unripe fruits are 4 in. or more in diameter and is done by making 3–4 vertical cuts about $\frac{1}{8}$ in. in depth. The cuts may be made by a razor blade inserted into rubber and mounted on a stick. The latex may be caught on a tray covered with cheese cloth mounted on a wire frame which is clamped on to the trunk at a convenient height. The latex coagulates and is scraped off with a wooden scraper. Non-ferrous containers may also be used to collect the latex. The coagulated latex must be dried quickly in the sun, but preferably in special ovens at a temperature of 130–140°F. Delay in drying causes oxidation and produces a bad colour; papain should be as white as possible. Fruits are re-tapped between the previous cuts at about weekly intervals. When the fruits are ripe they have little latex, but tapped fruits are edible on ripening. Latex which dries on the fruits may be scraped off, but is of a lower grade. Tappers should wear rubber gloves to protect their hands from the latex. A tree produces 50 per cent of its total production of papain in its first year of cropping, 30 per cent in its second year and 20 per cent in its third year, after which the crop is replanted. Yields of dry papain are 60–120 lb per acre per annum.

MAJOR DISEASES AND PESTS

The most serious disease in the West Indies is a mosaic virus which stunts the plants, causes yellow mottling and distortion of the leaves, bending down of petioles, followed by death of the tree. Diseased plants yield little or no crop. In some areas of Trinidad it is now almost impossible to grow papayas on account of the virus. This mosaic appears to be much more virulent than that in Hawaii and elsewhere. In Puerto Rico the vector is *Aphis spiraecola* and in Hawaii *Myzus persicae* (Sulz.). The vector of the bunchy-top virus is the leaf hopper, *Empoasca papayi.*

Pythium spp. cause a collar- and foot-rot when papayas are grown under waterlogged conditions. Anthracnose, *Colletotrichum gloeosporioides* Penz., causes spotting of ripe fruits.

Mites—*Tetranychus* spp., *Tenuipalpus bioculatus* McG. and *Hemitarsonemus latus* (Banks) are the most serious pests of papayas in Hawaii. Birds damage the fruits.

IMPROVEMENT

Sex in papaya is controlled by 5 pairs of genes which occur in 3 sex-determining complexes in the sixth chromosome. Because of tight linkages between the genes, the sex-determining complexes produce phenotypic results analogous to that which 3 alleles of a single gene with pleiotropic effects would produce. For practical convenience the complexes are generally designated as: M_1 dominant for maleness, M_2 dominant for hermaphroditism, and m recessive for femaleness. The combinations M_1M_1,

M_2M_1, and M_2M_2 are lethal and no viable seeds are produced. M_1m gives male trees, M_2m hermaphrodite trees and mm female trees. The crossing of dioecious trees (mm × M_1m) produces females and males in the ratio of 1 : 1. The selfing of hermaphrodite trees (M_2m) and the crossing of two hermaphrodites produce females and hermaphrodites in the ratio of 1 : 2. Hermaphrodites (M_2m) pollinated by males (M_1m) produce females, hermaphrodites and males in the ratio of 1 : 1 : 1. The size and shape of the fruits are determined by the parentage (see STRUCTURE above). By selfing hermaphrodites of known genetical constitution, e.g. 'Solo', uniform pyriform fruits are produced on the hermaphrodite progeny and uniform larger round fruits on the female progeny. By crossing mm × M_2m, half the progeny will be female with round fruits and half will be hermaphrodite with cylindrical fruits, and all the progeny will be fruitful.

It has been shown that yellow flower colour is dominant to white, purple stem colour is dominant to green, and yellow flesh colour is dominant to red.

One of the difficulties of the breeder is that there are no reliable characters to distinguish male, hermaphrodite and female trees until they flower. In dioecious cvs both male and female trees should be the progeny of the same parents which have the desired characters. Maintaining characters in hermaphrodite forms is easier. The characters required are: the yield of fruit or papain per tree and per acre; early and low-bearing with short internodes; uniformity of shape, texture and flavour, particularly for export markets; resistance to pests and diseases, particularly to mosaic.

PRODUCTION

Much of the papaya crop in the tropics is consumed locally, as it is difficult to transport the fruits satisfactorily over long distances. Fresh fruits are exported by air and in cold storage by sea from Hawaii to the United States, but little fresh fruit reaches other temperate countries. Papayas are now being canned and this market will probably increase.

The principal producer of papain before World War II was Ceylon, but since 1944, Tanganyika (now Tanzania) has become the leading producer, with maximum production of 306,485 cwt in 1948. Smaller quantities are produced in Uganda. The chief importer of papain is the United States. The market is very sensitive to overproduction.

REFERENCES

BECKER, S. (1958). The Production of Papain—An Agricultural Industry for Tropical America. *Econ. Bot.*, **12,** 62–79.

GREENWAY, P. J. *et al.* (1953). *The Papaw.* Tanganyika Dept. of Agric. Pamphlet **52.**

STOREY, W. B. *et al.* (1941). Papaya Production in the Hawaiian Islands. *Bull. Hawaii agric. Exp. Sta.*, **87.**

STOREY, W. B. (1958). Modification of Sex Expression in Papaya. *Hort. Adv.*, **2**, 49–60.

COMPOSITAE

One of the largest and most natural families of flowering plants with about 800 genera and 20,000 spp., mostly herbaceous annuals and perennials, widely distributed throughout the world. Lvs alternate or opposite, simple or divided, exstipulate. Fls hermaphrodite or unisexual, aggregated into small or large heads or capitula, surrounded by an involucre of free or connate bracts, with fls collected closely together on a receptacle, which is usually convex, each fl. subtended by a bract or scale, or naked. Fls of 2 types: (a) ligulate, zygomorphic, in which the 5 petals are united into a strap-shaped structure on one side of fl. (ray florets), (b) corolla tubular, gamopetalous, usually with 5 lobes (disc florets). Capitula may have all ray florets or all disc florets, but often with marginal ray florets, which may be sterile or lack stamens and act as an attraction for insects, and inner hermaphrodite disc florets in centre of capitulum. Calyx much modified and may be thread-like to form a pappus or may be dry and chaffy. Stamens usually 5, alternating with corolla lobes, epipetalous, usually with free filaments and syngenesious 2-locular anthers with longitudinal introrse dehiscence and forming a ring round style; usually protandrous with pollen being shed inside tube and is then forced out by developing style, which is usually unreceptive at first. Pollination is usually entomophilous. Ovary inferior, 1-locular, 1-ovuled; style mostly bifid. Fr. a sessile achene, sometimes beaked, and often crowned by pappus for dispersal. Seeds without endosperm, with straight embryo and plano-convex cotyledons. Some spp. are apomictic. The family contains a large number of weed spp., some of which now have a pantropical distribution.

USEFUL PRODUCTS

OIL-SEEDS: *Carthamus tinctorius* (q.v.), safflower, *Guizotia abyssinica* (q.v.), niger seed, and *Helianthus annuus* (q.v.), sunflower, are grown for their seeds which yield edible and drying oils.

VEGETABLES: *Lactuca sativa* (q.v.), lettuce, is widely cultivated as a salad crop. *Helianthus tuberosus* (q.v.), Jerusalem artichoke, is grown for its edible tubers. Other vegetables include:

Cichorium endivia L., endive, is an annual or biennial herb, which is probably a native of the Mediterranean region, but is sometimes considered as coming from India. It is grown as a salad plant, the leaves, often with curled margins, being eaten after blanching. It grows quite well in the tropics.

C. intybus L., chicory, a perennial herb which is native in Europe, is used as a salad plant and for greens; it was known to the Greeks and Romans. Its stout tap-root is roasted and ground and is used as a coffee substitute or for admixture with coffee. Chicory is also grown for forage. It does not grow well in the hot tropics.

Cynara scolymus L., globe artichoke, is a thistle-like perennial herb, native of the Mediterranean region. It is grown for its immature flower-heads, the fleshy base of the involucral bracts and receptacle being eaten after boiling. It does not do well in the tropical lowlands, but can be grown successfully in the highlands.

Tragopogon porrifolius L., salsify or oyster plant, is a native of Eurasia and is cultivated for its edible tap-root.

OTHER PRODUCTS: *Chrysanthemum cinerariaefolium* (q.v.) is the principal source of the insecticide pyrethrum. The dried capitula of *Anthemis nobilis* L. and *Matricaria chamomilla* L. produce chamomile, which is used medicinally. Some *Artemisia* spp. yield santonin which is used as a vermifuge, *A. cina* Berg. from Russia and *A. maritima* L. from Pakistan being the main source of the drug. The essential oil from the dried leaves of *A. absinthium* L., wormwood, is used to flavour absinthe. *A. dracunculus* L., tarragon, a native of western Asia, is grown for its pungent aromatic leaves which are used in vinegar and pickles.

Parthenium argentatum A. Gray, guayule of Mexico and the southern United States, has been utilized as a minor source of rubber, as have the roots of *Taraxacum kok-saghyz* Rodin, the Russian dandelion. *Mikania scandens* (L.) Willd. is sometimes grown as a cover-crop in the tropics.

ORNAMENTALS: Many composites are grown as garden plants, of which the commonest genera in the tropics are *Ageratum*, *Aster*, *Calendula*, *Centaurea*, *Chrysanthemum*, *Coreopsis*, *Cosmos*, *Dahlia*, *Gaillardia*, *Gerbera*, *Helianthus*, *Senecio*, *Tagetes*, *Tithonia* and *Zinnia*.

KEY TO PRINCIPAL CROPS

A. Florets all ligulate; capitula less than 1·5 cm in diameter .... *Lactuca sativa*

AA. Florets all tubular, much exserted; capitula 2·5–4 cm in diameter .... *Carthamus tinctorius*

AAA. Outer florets ligulate; inner florets tubular

B. Ray florets white; capitula 3–4 cm in diameter... *Chrysanthemum cinerariaefolium*

BB. Ray florets yellow

C. Capitula usually 15 or more cm in diameter ... *Helianthus annuus*

CC. Capitula 2–3 cm in diameter .... *Guizotia abyssinica*

Carthamus L. (x = 8, 12)

About 30 spp. distributed in Asia, Africa and the Mediterranean region. *C. tinctorius* (q.v.) is the only economic sp.

Carthamus tinctorius L. (2n = 24) SAFFLOWER

USES

The dried florets were the source of the red dye, safflower carmin, in Egypt, the Middle East and India. The colour is fugitive and has tended to be displaced by aniline dyes, but it is still used for dyeing cloth used on ceremonial occasions in India, where it is also used for colouring cakes and biscuits and for rouge. It should not be confused with saffron, which is obtained from the stigmas of *Crocus sativus* L. (Iridaceae). Safflower is now grown mainly as an oil-seed crop. In India the oil is used primarily for cooking, illumination and soap manufacture. Elsewhere it is used mainly in paints and varnish; because of its low linolenic acid content it has excellent colour-retention properties and no after-yellowing. The decorticated cake or meal has a high protein content and is used as a stock-feed; the undecorticated cake is used as manure. The tender shoots may be used as a pot-herb and the seeds are edible, usually being roasted first. It may also be used as fodder.

ORIGIN AND DISTRIBUTION

Safflower is only known in cultivation, with primary centres in Afghanistan and the Nile Valley and Ethiopia. It is supposed to have originated from *C. lunatus* L. which occurs wild over the entire range of the genus, or more probably from *C. oxyacantha* Bieb., which occurs as a weed from northern India to Turkey. *C. tinctorius* was cultivated in Egypt in very early times and spread throughout the Mediterranean region and eastwards to China. It was taken early by the Spaniards to Mexico. It was introduced experimentally as an oil crop into the United States in 1925, where it has been grown on a commercial scale since 1950, particularly in California. Trials have been made in Australia and South Africa.

CULTIVARS

Several races are recognized in India, varying in structure, particularly in the degree of spininess, and in oil- and dye-content. Spiny forms are considered better for oil and spineless forms for dye production. Cvs with high oil content have been selected in the United States.

CEOLOGY

Safflower is not suited to the low hot tropics. It is usually grown as a rain-fed crop. It shows considerable resistance to drought and wind.

Emerging plants need cool short days for root growth and the development of the rosette and higher temperatures and longer days for stem growth and flowering. High rainfall and humidity are harmful, encouraging disease, and a dry atmosphere is required during and after flowering for proper seed set and high oil content. It does best on deep, well-drained soils and cannot tolerate waterlogging.

STRUCTURE

A much-branched, glabrous, herbaceous annual, 0·5–1·5 m tall, with varying degrees of spininess. Tap-root long and stout. Early growth is slow; young plants have rosette habit, later branching from near base. Branches stiff, cylindrical, whitish in colour. Lvs spirally arranged, sessile, oblong to ovate-lanceolate, dark green, glossy, 10–15 × 2–4 cm, with spines along margin and at tip. Capitula terminal, 2·5–4 cm in diameter, with spreading outer leafy spiny bracts and inner triangular bracts, spine-tipped, forming a conical involucre, with small opening at tip, through which 30–90 florets protrude. Florets all tubular, hermaphrodite, usually orange-yellow in colour; corolla tube about 4 cm long with 5 pointed segments expanded above tube; staminal tube exserted, bright yellow, with 5 united anthers with introrse dehiscence; style pushes up through staminal tube, after which 2-lipped stigmas open. Achenes white or pale grey, shining, 4-angled, about 8 mm long, with 30–50 per cent hull; pappus absent.

POLLINATION

Cross-pollination by insects occurs, but some inbreds have been found to have less than 5 per cent out-crossing. Heavy rain at flowering adversely affects pollination.

GERMINATION

Practically no germination occurs at 36°F. The average time for germination at 41°, 48°, and 60°F was 16, 9, and 4 days respectively.

CHEMICAL COMPOSITION

The oil content of the seeds varies from 20–38 per cent; the thinner the hull the greater the oil content. The drying oil has a high linoleic acid content, about 75 per cent, and a very low linolenic acid content. The crude protein content of the expressed meal or cake varies from 20–55 per cent, depending on the amount of hull removed during processing. The florets contain 0·3–0·6 per cent of a scarlet red dye, carthamin, which is insoluble in water, and about 30 per cent of yellow pigment, which is soluble in water and which is removed by washing the dried florets.

PROPAGATION

Safflower is grown from seed.

HUSBANDRY

In India safflower is usually planted as a winter crop, often in mixed cultivation with cereals or pulses. It may be broadcast or planted in rows with a seed rate of 15–40 lb per acre. Control of weeds is important in the early stages of growth. In India the plants are topped as soon as the first buds appear to induce branching. The usual length of the growing season under favourable conditions is about 120 days. Good features of the crop are that it does not lodge or shatter and is not subject to bird damage. Spininess makes harvesting difficult. For dye production the fully opened florets are gathered every 2–3 days and yields of 80–120 lb per acre of dried florets are obtained. For oil production the plants are harvested when fully ripe and the seed is threshed and winnowed. Yields in India are 400–600 lb of seed per acre. Average yields in California are 1,700 lb per acre, but over 4,000 lb per acre have been obtained with irrigation. On suitable soils safflower is more productive than linseed.

MAJOR DISEASES AND PESTS

The most serious disease is rust, *Puccinia carthami* (Hutz.) Corda, which is widespread, and wilt, *Sclerotinia sclerotiorum* (Lib.) de Bary, causes damage in India. The worst pest in Europe and India is the larvae of *Acanthiophilus helianthi* (Rossi).

IMPROVEMENT

A few high yielding cvs have been developed in India and the United States.

PRODUCTION

In most countries where safflower is grown the crop is used locally. India exported relatively large quantities of seed in certain post-war years prior to 1956. Before 1942, India also exported the dye safflower carmine. Significant expansion of the crop has occurred in the United States since 1950 and production in 1963 was estimated at nearly 400,000 tons, of which about 225,000 tons of seed were exported, mainly to Japan. There is a small commercial production in Australia and Canada.

REFERENCES

KNOWLES, P. E. (1955). Safflower—Production, Processing and Utilization. *Econ. Bot.*, **9**, 273-299.

CROP RESEARCH DIVISION, U.S.D.A. (1959). Safflower. *U.S.D.A. Farmers' Bull.*, 2133.

Fig. 6. **Carthamus tinctorius**: SAFFLOWER. **A**, habit (×¼); **B**, capitulum (×1); **C**, capitulum in longitudinal section (×1); **D**, floret (×2); **E**, achene (×3).

D
E
B
A
C

Chrysanthemum L. (x = 9)

Over 100 spp. of annual and perennial herbs and subshrubs, mostly native of the Old World. Lvs alternate. Capitula terminal on long peduncles or in corymbose clusters; involucral bracts with membraneous tips or edges; receptacle naked; pappus scale-like or absent; ray florets pistillate and mostly fertile; disc florets hermaphrodite.

C. cinerariaefolium (q.v.), the Dalmatian insect flower, is the main source of the pyrethrum of commerce. *C. coccineum* Willd. (syn. C. *roseum* Adam.), the Persian insect flower, also yields pyrethrum, but its toxicity is less than the Dalmatian plants; it is also the ornamental pyrethrum of gardens. *C. marschallii* Aschers., the Caucasian insect flower, also yields pyrethrum, but is not grown commercially. The autumn-flowering chrysanthemums of gardens and florists are derived from *C. indicum* L., native of China and Japan, and *C. morifolium* Ramat. (syn. *C. sinense* Sabine), native of China. They have been cultivated in China since 500 B.C. and were introduced into Japan about 800 A.D., but they did not reach Europe until nearly the end of the 18th century.

Chrysanthemum cinerariaefolium (Trev.) Bocc. (2n = 18) PYRETHRUM

Syn. *Pyrethrum cinerariaefolium* Trev.

USES

The use of pyrethrum flowers for insecticidal purposes originated in Persia, *C. coccineum* being the species used. It was introduced into Europe early in the 19th century and into the United States about 1860. Later *C. cinerariaefolium* was found to be more effective and became the main source of pyrethrum. For this purpose the dried capitula were powdered, but kerosene extracts were made about 1920. The use was much extended in the 1930's, but it assumed great importance during World War II, when there was a considerable increase in demand. It was found to be highly effective against flies, fleas, lice and mosquitos, and to protect food and other produce. Aerosol sprays became standard equipment in some war areas. It was used in mosquito-repellent cream and in ointment for scabies. It is used in livestock sprays, in fog generators for warehouses, etc., and in mosquito coils for burning. Pyrethrum solutions are used for dipping dried fish and meats against beetle infection with *Dermestes* spp. and blow-flies, *Calliphora* spp.

Despite the post-war development of DDT and other chlorinated hydrocarbons and the organo-phosphorus insecticides, pyrethrum has maintained its position as an insecticide due to its effectiveness against a broad range of insects with little development of resistant strains, its rapid paralytic action or knock-down, its low toxicity to mammals and other warm-blooded animals, and its freedom from taint. It is non-inflammable and leaves no oily residue. For these reasons it is particularly valuable in the home and where there are foodstuffs. With the use of synergists such as sesame oil, piperonyl butoxide and others, the cost of application is reduced.

These substances are not themselves insecticidally active, but have the property of enhancing the toxicity of pyrethrins and so reducing the amount that is needed to achieve a given level of insecticidal activity.

ORIGIN AND DISTRIBUTION

C. cinerariaefolium occurs wild on the Dalmatian coast of Yugoslavia. It was introduced into Japan in 1881, who became the principal producer between World Wars I and II. In the early 1920's trials were made in Switzerland and France. Seed from Switzerland and Japan was grown in England from 1924 onwards, notably at the Rothamsted Experimental Station, which supplied seeds of the Harpenden strains to Kenya in 1929. In the same year Captain Gilbert Walker, who was the first planter to grow pyrethrum commercially in Kenya, obtained seed from Dalmatia which was grown on his farm at Nakuru. With the outbreak of World War II Kenya became the leading producer, a position which she still retains, and the Kenya strains have the highest known content of pyrethrins. With the increased demand during the war, production was extended to the highlands of Tanzania and to Kigezi in south-western Uganda. Pyrethrum has been tried in many countries. Production has now been extended to Ecuador and New Guinea.

ECOLOGY

Pyrethrum thrives best in areas of medium rainfall with 35–50 in. per year, evenly distributed, and on well-drained loams. In northern temperate countries such as Japan and Dalmatia flowering takes place over a period of 1–2 months only in the summer. In the tropics flowering and high content of pyrethrins are obtained only at high altitudes, *e.g.* in Kenya between 6,000–9,000 ft. Flowering is extended over a period of 9–10 months. In Kenya flowering starts in May and reaches a maximum during the period September to January; thereafter the yield of flowers decreases rapidly and the plants pass through a dormant phase of 2–3 months. Chilling is necessary to initiate flower buds. A mean maximum temperature of 75°F, if prolonged for a week or more, leads to the inhibition of flower production and the plants remain 'blind'. About 10 days at or below 60°F is necessary to stimulate flower bud development. Maximum flowering follows 3 months after the period of maximum cold. Yields of flowers and content of pyrethrins are inversely related to the mean maximum temperature. Kroll (1964) has shown that on farms in Kenya at 7,000 ft with a mean temperature of 61·4°F the average pyrethrins content is 1·44 per cent; at 7,500 ft with a mean temperature of 57·1°F the average pyrethrins content is 1·51 per cent; at 8,200 ft with mean temperature of 56·4°F the average pyrethrins content is 1·56 per cent. Rainfall increasing from 1–4 in. per month also increased pyrethrins content, but no further effect was obtained in excess of 4 in. per month.

STRUCTURE

Tufted perennial herb, 30–60 cm in height. Young and non-flowering plants have a close rosette habit. Roots numerous and fibrous. Greyish pubescence on stems, lvs and bracts. Lvs alternate, 10–30 cm long including long slender petiole, pinnate; pinnae narrow, 2–5 cm long, pinnatifid with linear acute segments. Capitula 3–4 cm in diameter, borne singly on long slender peduncles. Involucral bracts lanceolate, hyaline-tipped, 5–8 mm long. Peripheral ray florets about 18–22 in number; corolla white, about 1·5 cm long, oblanceolate, 2-veined and 2-toothed; unisexual and female; bilobed stigma projects through folded base of corolla. Disc florets closely packed on slightly convex receptacle, spirally arranged, hermaphrodite, yellow, about 5 mm long; calyx small, 5-toothed; corolla tubular with 5 reflexed short lobes; stamens 5, inserted at base of corolla, anthers connate with introrse dehiscence; ovary inferior, tapering towards base, 5-ridged, numerous oil glands and ducts in walls which secrete an oleoresin; style short; stigma hairy and unopened brushes pollen through staminal tube and 2 lobes then open outwards exposing stigmatic surface. Achenes about 4 mm long, tapering, 5-ribbed, pale brown, 1-seeded.

POLLINATION

Pyrethrum is self-sterile and must be cross-pollinated to produce viable seed. It is insect-pollinated, mainly by Coleoptera and Diptera.

GERMINATION

The germination is often rather poor due to the presence of unfertilized and non-viable seeds. Viability is lost if stored for long periods. Germination takes 10–15 days; 1 lb of seeds gives about 15,000 seedlings.

CHEMICAL COMPOSITION

The insecticidal toxicity of pyrethrum is due to the substances pyrethrin I, pyrethrin II, cinerin I and cinerin II, known collectively as pyrethrins. Pyrethrin I has the greatest toxicity. The amount of pyrethrins in the flowers varies with the country of origin, the altitude and the cvs grown, as does the proportion of the components. Japanese and Dalmatian pyrethrum contains about 1·0 per cent pyrethrins in the dried flowers with less pyrethrin I than pyrethrin II. Kenya pyrethrum contains a minimum of 1·3 per cent pyrethrins, and often considerably more, with a higher proportion of pyrethrin I. The liquid extract as now exported from Kenya contains 25–30 per cent of pyrethrins with the approximate composition of 10 parts of

---

Fig. 7. **Chrysanthemum cinerariaefolium: PYRETHRUM. A,** leaf (× ½); **B,** inflorescence (× ½); **C,** capitulum from above (× 1); **D,** ray floret (× 3); **E,** disc floret (× 6).

A
B
C
D
E

pyrethrin I, 7 parts pyrethrin II, and 3 each of cinerin I and II. About 90 per cent of the pyrethrins are contained in the ovary and developing achenes, which have many more oil glands than the rest of the flower. Fully opened capitula have more pyrethrins than buds and fertilized achenes more than unfertilized ones. The pyrethrins content is greater at the higher altitudes.

PROPAGATION

Much of the pyrethrum is grown from seed, which is sown thinly in nurseries, lightly covered with sifted soil, and kept moist under temporary shade. Seedlings are transplanted to the field when they are about 4 months old and 4–5 in. high.

The crop may also be planted from splits obtained by dividing mature plants. This permits the selection of high yielding clones with erect habit and large flowers and the crop comes into flower quicker and more evenly than with heterogeneous seedlings. Care must be taken to avoid taking splits from 'blind' plants. An indication of the production of a clone may be given by the number of dead flower stems at the end of the season. The number of splits per plant may be increased by earthing up the plants to encourage the growth of adventitious roots on the young stems. Due to the necessity of cross-pollination more than one clone should be planted or a small proportion of seedlings may be interplanted with the splits.

HUSBANDRY

In the Kenya highlands pyrethrum is often grown in rotation with wheat or maize and grass leys. Seedlings or splits are usually planted at 3×1 ft. On hillsides the rows should be on the contour and other soil conservation measures employed. It is important to obtain a good initial stand and blanks should be supplied as soon as possible. Adequate weed control is essential.

In East Africa application of 150–200 lb per acre of triple superphosphate at planting time is recommended and the effect lasts over 2–3 seasons. Nitrogen and potash have produced no increase in yields. Mulch between the rows increases yields, particularly in dry seasons, as does irrigation during dry spells in the growing season. Ridge planting increases yields and is recommended; it also helps to conserve the soil. Pyrethrins content is not influenced by manuring.

The first picking takes place about 4 months after planting and thereafter at intervals of 2–3 weeks during the flowering period which extends over 9–10 months of the year in Kenya. The flower-heads only are harvested and this is done when they are fully expanded and preferably when the outer 4–5 rows of disc florets are open. The pyrethrins content is not reduced in overblown flowers but, due to seed formation, the number of flowers per plant is reduced. The flowers are picked by hand, usually by women and children, and a skilful picker can harvest up to 60 lb of fresh flowers per day. About 200 lb per acre of dried flowers are produced during the 1st year, increasing to 800–1,000 lb per acre per annum for the second and third years at 8,000 ft and 300–400 lb per acre at 6,000 ft. Thereafter the yields decline and it is usual to plough in the pyrethrum after the 3rd year and to plant the land with some other crop in the rotation. Thus to maintain the acreage under pyrethrum it is necessary to replant one-third of the acreage each year. After the picking season is over the fields are cleaned up during the dry weather and the dry stalks of the previous season are cut back just before the rains break.

After harvesting the flowers are spread on trays and are dried in specially constructed driers in which there is an upward natural draught of hot air heated by passing over hot flues. The process takes 6–8 hours during which the moisture content is reduced from 75–83 per cent to 8–10 per cent. The fresh flowers should not be left in heaps where they would ferment, but should be spread thinly on the wire trays at a density of ¾ lb per sq. ft. The dried flowers are then sold for processing. In the early days of the Kenya industry pyrethrum was exported as baled dried flowers which contained a minimum of 1·3 per cent pyrethrins. It is now marketed as an extract containing 25–30 per cent pyrethrins.

MAJOR DISEASES AND PESTS

Few serious diseases and pests occur. In Kenya the true bud disease caused by *Ramularia bullunensis* Speg. is troublesome, as is *Aphelenchoides ritzemabosi* C. B. & T. The false bud disease also occurs and appears to be a physiological disorder which produces a shepherd's crook effect in the peduncle. The most serious pests are *Thrips tabaci* Lind. which damage the inflorescences and *T. nigropilosus* Uzel. which damage the leaves.

IMPROVEMENT

The most important character in the improvement of pyrethrum is the yield of pyrethrins per acre. Other characters include vigour of plant, upright growth to facilitate ease of cultivation and harvesting, heavy flower yield, heavy weight of individual flowers to reduce harvesting cost, and resistance to pests and diseases. The chilling requirements to initiate flower buds differ between individual plants and clones and it is necessary to test selections at the different altitudes and a different cv. may be required at 7,000 ft as compared with 9,000 ft near the equator. Incompatibility between some clones has been reported.

The first step is the selection of individual plants which may be multiplied vegetatively to produce clonal material. Interclonal hybrids are then produced and selections from them made in the same way. Back-crossing and line-breeding are also practised. Polycrosses produced by growing several clones together are also used as a source of breeding material. Clones are tested for three seasons, the normal cultivated life of the plant, and then for a further three seasons at various altitudes. Allowing for the time taken to increase planting material a minimum of 9 years is required before a new cv. can be issued.

In Kenya the first new cv., now known as 'C1', was a cross between two high-yielding clones (14 × 24). Although flower production was not more prolific than some unselected plants, due to the high pyrethrins content, the yield of active principles per acre was considerably increased. 'C1' was not suited to the lower altitudes, but the next issue 'C47', a hybrid with even higher toxic content, is better adapted to these altitudes. Cvs with over 2 per cent pyrethrins content have now been produced and still higher contents are in prospect. Cvs yielding up to 76 lb of pyrethrins per acre over three seasons are under trial.

PRODUCTION

Prior to World War I Dalmatia was the principal exporter of pyrethrum. Between World Wars I and II Japan produced almost all the world's requirements with an annual production reaching some 12,000 tons. The first exports from Kenya were made in the early 1930's and by 1940 production had reached 5,859 tons. An agreement was made with the Ministry of Supply early in the war, who undertook to buy for four years the output from 10,000 acres at a satisfactory price of £177 16s. 0d. per ton *c.i.f.* New York. By 1946 production in Kenya was 7,400 tons, after which there was a temporary decline. The pyrethrum was produced on European farms, but African farmers began growing the crop on a small scale in the 1950's and there are now over 30,000 African growers. Present production in Kenya is in the order of 12,000 tons of dried flowers per annum. Cultivation in Tanganyika began in 1939 and substantial quantities are now produced in the Southern Highlands and on Mount Kilamanjaro in Tanzania. A plantation in Kigezi in south-western Uganda grew pyrethrum during the period 1942–1945. The Kivu region of the Congo started producing pyrethrum during World War II and production continues. Since that time production has begun in Ecuador and New Guinea. It is also grown in Brazil and Japan and to a small extent in India. The United States is the largest importer of pyrethrum.

REFERENCES

KROLL, U. (1963). The Effects of Fertilizers, Manures, Irrigation and Ridging on the Yield of Pyrethrum. *E. Afr. agric. & for. J.*, **28**, 139–145.

KROLL, U. (1964). Effect of Mean Temperature on the Content of Pyrethrins of *Chrysanthemum* (*Pyrethrum*) *cinerariaefolium*. *Nature*, **202**, 1351–1352.

Pyrethrum Post, 1948–   . Published quarterly by the Pyrethrum Bureau, Kenya.

WALKER, G. (1950). Pyrethrum—the Pyrethrum Industry of Kenya, in *East African Agriculture*, 154–163, ed. J. K. Matheson and E. W. Boville. London: Oxford University Press.

Guizotia Cass. (x = 15)

A small genus of about 5 spp. of herbs in tropical Africa. *G. abyssinica* (q.v.) is grown as an oil-seed crop.

Guizotia abyssinica (L.f.) Cass. (2n = 30) NIGER SEED

USES

The seeds yield a yellow, edible, semi-drying oil with little odour and a pleasant nutty taste. It is used for culinary purposes, as an illuminant and for soap manufacture. It is used to a limited extent in paints, for which the Ethiopian seed is superior to the Indian as it has a higher linoleic acid content. The press cake is used for livestock feed and as manure. The seeds are fried and eaten in India and are used in chutneys and condiments.

ORIGIN AND DISTRIBUTION

G. abyssinica is of African origin and occurs sporadically from Ethiopia to Malawi. It was taken early to India. The only areas of extensive cultivation are in Ethiopia and India. It occurs as a casual in England.

ECOLOGY

Niger seed is grown as a rain-fed crop in areas of moderate rainfall seldom exceeding 40 in. and will produce a reasonable crop on rather poor soil.

STRUCTURE

A branched annual herb, 0·5–1·5 m tall. Young stems glandular-hairy. Lvs mostly opposite, but upper lvs sometimes alternate, sessile, clasping stems, lanceolate, serrate, somewhat scabrous, 5–20 × 1–3 cm, tip acuminate. Capitula 2–3 cm in diameter, borne on leafy shoots in lf. axils, 2–5 together. Outer involucral bracts 5, leaf-like, ovate, hairy, about 1 cm long; inner bracts lanceolate, chaffy. Peripheral ray florets about 8 in number, shortly ligulate, female, about 15 × 6 mm, several-nerved, 3–4-lobed. Disc florets tubular, about 1 cm long, 5-lobed, hermaphrodite; tips of 5 connate anthers exserted above corolla tube with bifid hairy stigma above. Corolla of both disc and ray florets yellow and hairy outside at base. Achenes 3·5–5 mm long, 3–4-angled, black, shiny, broadening upwards, pappus absent.

POLLINATION

Capitula are visited by bees and other insects.

GERMINATION

Seed can be stored for a year or more without deterioration.

CHEMICAL COMPOSITION

The seeds contain 30–50 per cent oil and about 20 per cent protein. Ethiopian oil contains about 70 per cent linoleic acid; Indian oil about 50 per cent with a corresponding increase in oleic acid. The press cake contains about 33 per cent protein.

PROPAGATION

The crop is grown from seed.

HUSBANDRY

Niger seed is often grown in mixed cultivation in India with finger millet, *Eleusine coracana* (L.) Gaertn., for which the land receives good preparation. When grown in pure stand little preparation is given and the seed is broadcast or planted in rows 12–14 in. apart with a seed rate of 4–10 lb. per acre. The plants flower about 3 months after sowing and are harvested after a further 6 weeks. The stems are cut near ground level and, after drying for a few days, the seed is threshed and winnowed. Pure stands yield 350–400 lb per acre of seed.

MAJOR DISEASES AND PESTS

No serious diseases and pests have been reported.

IMPROVEMENT

Very little work has been done in the improvement of the crop.

PRODUCTION

Ethiopia and India are the chief producers of niger seed, Ethiopia producing 100,000–200,000 tons per annum and India about 75,000 tons per annum. India has exported up to 20,000 tons (1952), but since 1955 her exports have been very small. Exports from Ethiopia have also declined and were 6,000 tons in 1962.

Fig. 8. **Guizotia abyssinica**: NIGER SEED. **A**, flowering shoot ($\times \frac{1}{2}$); **B**, lower leaf ($\times \frac{1}{2}$); **C**, young capitulum ($\times 1\frac{1}{2}$). **D**, capitulum in longitudinal section ($\times 3$); **E**, ray floret ($\times 4$); **F**, disc floret ($\times 4$); **G**, disc floret in longitudinal section ($\times 5$); **H**, achene ($\times 8$); **I**, achene in longitudinal section ($\times 8$).

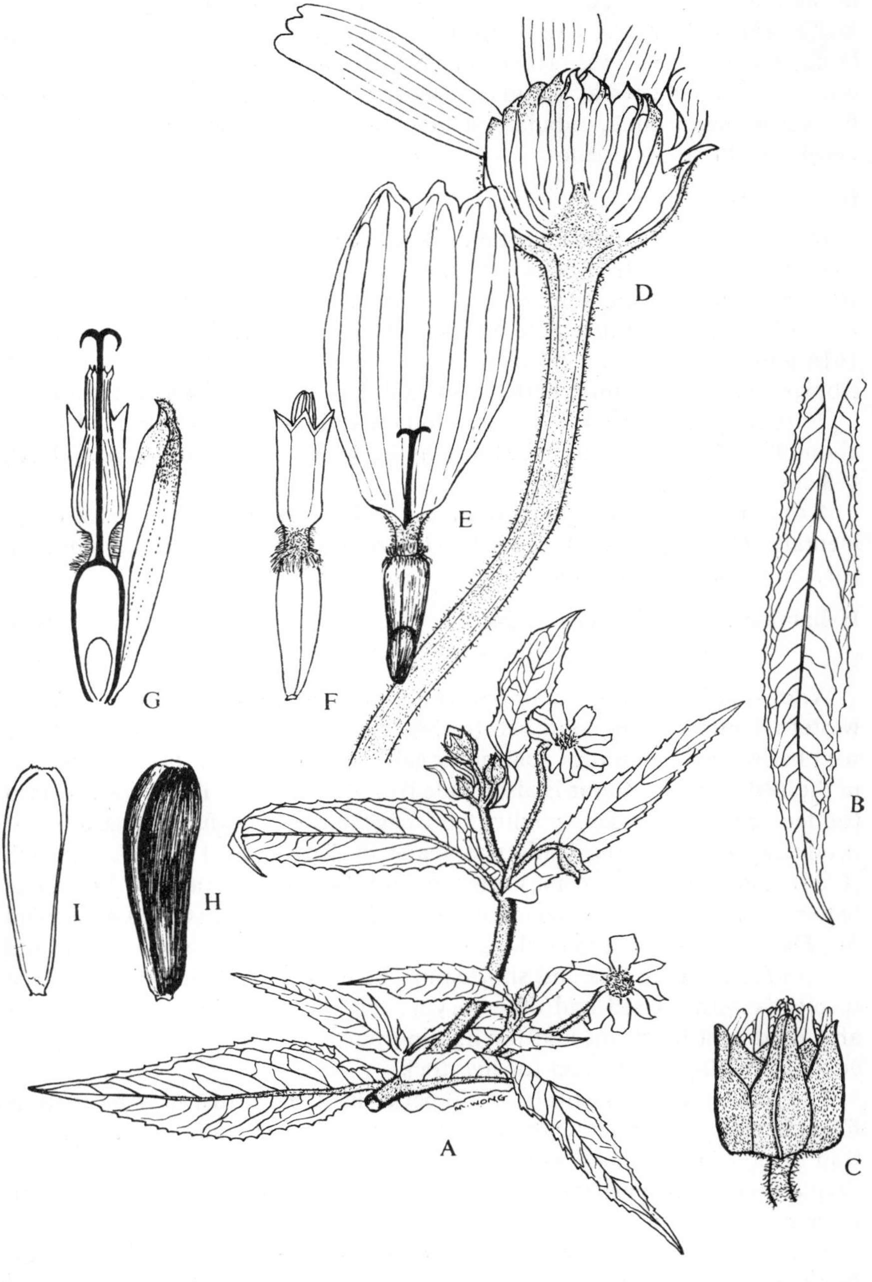

A
B
C
D
E
F
G
H
I
M. WONG

Helianthus L. (x = 17)

Over 100 spp. of tall annual and perennial herbs, with the greatest number in the United States, but extending through Mexico to South America and northwards to Canada. Lvs opposite or spirally arranged, simple; capitula large; bracts herbaceous; receptacle flat or conical with scales which partially enclose achenes at maturity; ray florets neuter, usually yellow; disc florets hermaphrodite, tubular, yellow, purple or brown; pappus of 1–4 deciduous bristles or scales.

USEFUL PRODUCTS

H. annuus (q.v.) is an important oil-seed crop.

H. tuberosus L., Jerusalem artichoke, is North American in origin and was cultivated by the Indians in the north-eastern United States in pre-Columbian times for its edible tubers. It was introduced into Europe before 1616 and has now been taken to most countries. It can be grown in the tropics. It is a perennial herb, 1–3 m tall. Its tubers, in which the carbohydrate is in the form of inulin, develop on stolons below ground. They are about 10–20 cm long and the skin may be white or red. Capitula yellow, 4–8 cm in diameter.

A number of spp., including interspecific hybrids, are good garden ornamentals. *H. decapetalus* L. and *H. rigidus* (Cass.) Desf. are the common garden perennial sunflowers.

Helianthus annuus L. (2n = 34) SUNFLOWER

USES

Wild sunflower was an important food plant of the Indians of the western United States long before the discovery of the New World. The seed kernels are eaten raw, roasted or salted, a habit which began in Russia in the 18th century. Flour is also made from the kernels. The seeds are used for feeding to livestock, poultry and cage birds. The first suggestion of producing oil from the seeds was made in Russia in 1779, but Burkill (1935) states that it was first grown as an oil crop in Bavaria in 1725. The better quality refined oil, which is pale yellow in colour, is used as a salad oil. The oil is also used in cooking and in the manufacture of margarine and compound cooking fats and shortening. Being a semi-drying oil, it is used in blends with linseed and other drying oils in paints and varnishes. It is also used as a lubricant and for lighting purposes. The decorticated press cake is a high-protein food for livestock.

The hulls, which constitute 35–50 per cent of the seeds, may be used as fillers in feed cakes and meals, as bedding for livestock, and in the preparation of polishing abrasives. The heads after threshing are fed to livestock. Sunflowers are also grown for fodder and silage for livestock and as a green manure crop.

H. annuus is commonly grown as an ornamental garden-plant, of which there are a number of colour forms, including ones with red and red-banded ray florets; also double forms and one with variegated foliage.

ORIGIN AND DISTRIBUTION

The cultivated sunflower, *H. annuus* var. *macrocarpus* (DC.) Ckll., is not known in a truly wild state. Heiser (1965) considers that it is possible that the original form was similar to *H. annuus* subsp. *jaegeri* Heiser, which is now found from Utah and Arizona to southern California. Earlier Heiser (1954) considered that it might have arisen from *H. annuus* subsp. *lenticularis* (Dougl.) Ckll., which occurs in western North America from southern Canada to northern Mexico. The wild *H. annuus* subsp. *annuus*, which occurs as a ruderal on dumps and vacant lots in central and eastern United States and Canada, is considered a hybrid between subsp. *lenticularis* and the cultivated sunflower. Heiser considers that introgression occurred between the weedy subspecies and other annual species of *Helianthus* and that men took forms to the middle west where they were adapted as weeds around village sites. The seeds were eaten by Indians and have been found on an archaeological site dated 2,000–3,000 years old. In this area the cultivated variety arose due to selection by man. No other annual species occur in the area and here on the periphery of distribution the large number of recessive genes which are present in var. *macrocarpus* would become manifest. Sunflower is the only important crop to have evolved within the present confines of the United States.

The earliest record of introduction into Europe is seeds brought back by the Spanish expedition to New Mexico in 1510 and planted in the Madrid Botanic Garden. L'Obel described the plant in 1575 and its origin was wrongly ascribed to Peru. It was introduced into Russia in the 18th century by Peter the Great and was used for chewing as well as ornament. Sunflowers have now been taken to most countries, both temperate and tropical, but the largest acreage is now in northern Caucasus, Ukraine and Volga river regions of Russia, the Balkans and Argentina.

CULTIVARS

Cvs vary greatly in: height of plant; time to maturity; number, diameter and colour of heads; size, shape, colour, oil and husk content of seeds; suitability for different environments. Cvs may be divided into the following types:

GIANT TYPES: 6–14 ft tall; generally late maturing; large heads 12–20 in. in diameter; seeds large, white or grey or with black stripes; oil content rather low; *e.g.* 'Mammoth Russian'.

SEMI-DWARF TYPES: 4½–6 ft tall; early maturing; heads 7–9 in. in diameter; seeds smaller, black, grey or striped; higher oil content; *e.g.* 'Pole Star', 'Jupiter'.

DWARF TYPES: 2–4½ ft; early maturing; flower heads 5½–6½ in. in diameter; seeds small; highest oil content; *e.g.* 'Advance', 'Sunrise'.

ECOLOGY

Sunflowers can be grown from the equator to as far north as 55°N. They can stand a little frost. They grow best at medium to high elevations in the tropics, but can also be grown in the lowlands. They are not suited to the wet tropics. Too heavy rain during the early stages of growth or cool wet weather during ripening causes attack by *Botrytis cinerea.* It is essential that the rain is evenly spaced during the growing season. Sunflowers have highly efficient root systems and can be grown in areas which are too dry for many crops and on rather poor soil. In South Africa reasonable yields have been obtained with 10 in. of rain by growing dwarf, early maturing cvs with small leaf area at relatively wide spacing. Giant types require less arid conditions. They can be grown on a wide range of soils, but the soil should be deep and well drained, and they cannot tolerate very acid or waterlogged soils.

STRUCTURE

A very variable, erect, hirsute, annual herb, 0·7–3·5 m tall. Tap-root strong to a depth of 3 m and with a large lateral spread of surface roots. Stems often unbranched, at first round, later thick, angular and woody; abundant xylem; pith often becoming hollow. Lower lvs cordate and opposite, soon becoming ovate and alternate in spiral with two-fifths phyllotaxis; petiole long, decussate; lamina with 3 main veins, sinuate-toothed, hispid with stiff appressed hairs on both sides, 10–30 × 5–20 cm, tip acute or acuminate. Capitula terminal, 10–40 cm in diameter, sometimes drooping; receptacle flat or dilated and convex; involucral bracts ovate or ovate-lanceolate, acuminate, ciliate, arranged in 3 rows. Outer ray florets neuter, with ligulate elliptic corolla, showy, yellow, strongly 2-nerved, deciduous, about 6 × 2 cm. Disc florets numerous, spirally arranged, hermaphrodite, about 2 cm long, subtended by bract; pappus scales 2, chaffy, deciduous; corolla tubular, dilated near base, somewhat hairy, 5-lobed; stamens 5; filaments flattened, free; anthers long, connate, ending in triangular appendage, often dark brown in colour; ovary inferior, pubescent, with single basal ovule; style slender; stigma 2-lobed. Achenes obovoid, compressed, slightly 4-angled, variable in size and colour, seldom less than 1 cm long, white, cream, brown, black, or white or grey with black stripes.

POLLINATION

The disc florets are protandrous. The pollen is shed in the anther tube, which is partly exserted from corolla. The developing stigma pushes up the

---

Fig. 9. **Helianthus annuus : SUNFLOWER. A,** flowering shoot (× ½); **B,** portion of capitulum in longitudinal section (× 1½); **C,** ray floret (× 1); **D,** disc floret in longitudinal section (× 4); **E,** achene (× 3); **F,** achene in longitudinal section (× 3).

A
B
C
D
E
F
M. WONG

pollen mass. Later the stigmas open outwards. The flowers are visited by bees and long-tongued flies for pollen and nectar and they bring about cross-pollination by crawling over the surface of the capitulum. Sunflowers are good honey plants. Most cvs are self-incompatible. Shortage of pollinating insects may occur in some areas, such as parts of East Africa, with inadequate pollination and many of the achenes may be empty of seeds.

GERMINATION

Care should be taken that only well-filled 'seeds' are planted. Dried seeds, suitably stored, will retain their viability for several years.

CHEMICAL COMPOSITION

The seeds usually contain 25–35 per cent of a semi-drying oil, but cvs have been bred in Russia with thin husks which are said to yield up to 50 per cent oil. The oil contains 44–72 per cent linoleic acid, the amount depending upon the conditions under which the crop is grown. Slow maturing seeds and crops grown in cool climates are richest in linoleic acid and with less oleic acid. The protein content of the seeds is 13–20 per cent and that of decorticated cake about 37 per cent and is of high biological value and digestibility. The stems and husks are rich in potash.

PROPAGATION

The crop is grown from seed. Sunflowers can be grafted onto the perennial Jerusalem artichoke, *Helianthus tuberosus*.

HUSBANDRY

For seed production the crop is usually grown at a spacing of 2–3 × ½–1 ft with a seed rate of 4–10 lb per acre. A fine tilth is desirable and the seed is sown at a depth of 1–1½ in. On most soils the response to fertilizers is small, but the application of phosphorus may be economic in the tropics. The crop is susceptible to boron deficiency. The time to maturity depends on the cv. and is usually 4–5 months, but cvs have been produced in Russia which will produce a crop in as little as 70 days. The heads are harvested when the involucral bracts turn yellow and the seeds become loose, but before shedding begins. The tall cvs are usually harvested by hand, but early dwarf types may be combine harvested. The heads are dried and are threshed either mechanically or by hand. The seeds receive further drying and for storage should not contain more than 12 per cent moisture. Average yields reported range from 800–1,400 lb of seed per acre, but over 3,000 lb per acre has been recorded. For fodder or silage the seed rate is about 35 lb per acre and the crop is harvested at the flowering or dough stage.

MAJOR DISEASES AND PESTS

Grey mould, *Botrytis cinerea* Pers. ex Fr., attacks young stems, leaves and ripening heads in cool wet weather. Rust, *Puccinia helianthi* Schw., and stem rot, *Sclerotinia sclerotiorum* (Lib.) de Bary also attack the crop. Broomrapes, *Orobanche* spp., are parasitic on the root. The sunflower moth, *Homoeosoma electellum* (Hulst), attacks the seeds. The crop is very subject to damage by birds and rodents.

IMPROVEMENT

Selection for increased oil content was begun in Russia in 1860 and since that time a large number of cvs have been produced. It is claimed that cvs with up to 50 per cent of oil have been produced. Other characters which have been bred for are: single large heads; dwarfness for mechanical harvesting; early maturity to extend range of the crop; large kernel forms with low hull content for chewing; resistance to rust and other diseases; resistance to *Orobanche* spp. which has been successfully achieved; resistance to sunflower moth which is obtained by crossing with wild species which have the 'armoured' layer present in the fruit coat; resistance to bird damage by breeding for peduncles with crook-neck so that capitula hang facing downwards.

Use has been made of self-incompatibility to produce hybrid F_1 seed, as with maize, to give hybrid vigour. The Canadian cv. 'Advance' is obtained in this way, but the cross must be repeated each year as F_2 seed breaks down.

PRODUCTION

As with many other crops, the main areas of production are far removed from the centre of origin. The largest producer is the Soviet Union with an annual crop of over 10 million acres which produces 4–5 million tons of seed. The second largest producer is Argentina, where large production only began in the mid 1930s to replace imported olive oil, but now has an annual crop of about 3 million acres. Rumania grows over 1 million acres per annum. Bulgaria, Hungary, Yugoslavia, Turkey, South Africa and Uruguay are the next largest producers. In the tropics Tanzania is the largest producer with 10,000–20,000 tons per annum and there is some small production in Kenya and Rhodesia. The Soviet Union and Bulgaria are the largest exporters of sunflower seed and oil and Western and Eastern Germany are the largest importers.

REFERENCES

BLACKMAN, G. E. (1951). The Sunflower. *World Crops*, **3**, 51–53.

HEISER, C. B. (1954). Variation and Subspeciation in the Common Sunflower, *Helianthus annuus*. *Am. Midland Naturalist*, **51**, 287–305.

HEISER, C. B. (1965). Sunflowers, Weeds and Cultivated Plants, in *The Genetics of Colonizing Species*, ed. H. G. Baker and G. L. Stebbins, 391–401. New York: Academic Press.

Lactuca L. (x = 8, 9, 17)

About 100 spp. of annual and perennial herbs with copious latex and panicles of small capitula, occurring throughout the world, but most abundant in the temperate regions of the Old World.

USEFUL PRODUCTS

L. sativa (q.v.) is lettuce. *L. indica* L., a native of Asia, is grown in China and Japan and has been introduced into Java and parts of Malaysia. It is grown for its leaves, which are usually eaten after cooking. The dried latex of the European *L. virosa* L. provides the drug lactucarium.

Lactuca sativa L. (2n = 18) LETTUCE

USES

Lettuce is the most widely cultivated salad crop. The leaves may also be boiled as spinach.

ORIGIN AND DISTRIBUTION

Most authorities consider that lettuce is a cultigen derived from *L. serriola* L., a prickly biennial ruderal of Europe, western Asia and northern Africa. Lundqvist (1960) considers, in the light of cytological and genetical evidence, however, that *L. sativa* probably originated by hybridization of other species, including *L. saligna* L., and that *L. serriola* arose from the same or subsequent hybridization and now exists as a camp-following weed. *L. sativa* and *L. serriola* are interfertile and natural hybrids occur.

The centre of origin of *L. sativa* appears to be the Middle East. The first records of lettuce as a vegetable is a long-leaved form depicted on Egyptian tombs dated 4500 B.C. Lettuce was grown by the ancient Greeks and Romans. The Moors developed many types. Lettuce reached China in the 7th century A.D. The first incontestable record of cabbage lettuce is by Fuchs in 1543. Lettuce is a comparatively recent introduction into the tropics and no truly tropical races have been evolved.

CULTIVARS

The following botanical varieties are recognized:

Cabbage or head lettuce, var. *capitata* L., which is the most widely grown. Rosettes very compact and cabbage-like with firm heart; lvs broad, almost orbicular, tender; midrib branching into small veins before reaching apex of lamina. In tropical lowlands cvs 'Mignonette' which often does not form hearts under these conditions, and 'Great Lakes' are recommended, and 'Iceberg' and 'New York' at the higher altitudes.

Fig. 10. **Lactuca sativa :** LETTUCE cv. 'Mignonette'. **A**, plant in vegetative phase (×½); **B**, plant in flower (×½); **C**, portion of inflorescence (×½); **D**, capitulum (×4); **E**, floret (×5); **F**, fruits (×4).

A
B
C
D
E
F
M WONG

Cos lettuce, var. *longifolia* Lam. Rosettes cylindrical or conical, upright; lvs obovate to oblong, rigid and rather coarse in texture; prominent midrib running almost to lamina apex which is rounded. Most cos cvs are self-folding, forming loose oblong hearts. Cvs recommended for the tropics are 'Little Gem' and 'Superb White'.

Leaf or curled lettuce, var. *crispa* L. Loose rosettes; lvs similar in texture and in appearance of midrib to cabbage lettuce, but differ in their inability to form hearts. A very heterogeneous group, including extremely curly and fringed cvs. Cv. 'Black Seeded Simpson' is recommended in Trinidad.

Asparagus or stem lettuce, var. *asparagina* Bailey (syn. var. *angustana* Irish). Young fleshy stems are used, usually cooked, and not the leaves, which are coarse and unpalatable. Basal lvs narrow, lanceolate, alternate, sometimes with pointed apex, not forming heads. This variety is grown almost exclusively in China.

ECOLOGY

Lettuce thrives better at the higher altitudes in the tropics; in the lowlands they often do not heart well and tend to bolt very quickly. They grow best in light, well-manured, well-drained soils with adequate watering.

STRUCTURE

An annual, glabrous, lactiferous herb forming a dense basal rosette and later tall, branched, flowering stems, 30–100 cm tall. Tap-root developing quickly, at first slender, later thickening, growing to a depth of about 1·5 m. Stem at first short with radical leaves spirally arranged. Shape and compactness of head and shape and size of lvs varying with cvs (see above). Rosette lvs 12–25 cm long, almost sessile; red anthocyanin pigment present in some cvs, *e.g.* 'Mignonette'; stem lvs becoming progressively smaller, ovate to orbicular, cordate, clasping stem. Capitula numerous in dense corymbose panicle with small sagittate scale-like bracts. Involucre 10–15 mm long; bracts lanceolate, blunt; receptacle naked. Fls few, all ligulate and hermaphrodite, pale yellow, exserted above involucre; stamens 5, anthers connate, short-tailed below and with short terminal appendages; stigma bifid. Achenes 3–4 mm long, narrowly obovate, compressed, 5–7-ribbed on each face, white, yellowish, grey or brown; tip constricted into a narrow beak, surmounted by pappus of 2 equal rows of soft white simple hairs.

POLLINATION

The buds grow rapidly in the 24 hours before anthesis, which takes place before the flowers open. All the flowers of the capitulum open at once and remain open for 2 hours only. During this time the unexpanded hairy stigma sweeps the pollen upwards and out of the anther column. The stigmas then fold back and the flowers are automatically self-pollinated. The flowers are visited by flies so a little out-crossing may sometimes occur.

GERMINATION

The optimum temperature for germination is 77°F, above which the percentage germination falls rapidly. Seeds tend to lose their viability rather rapidly in the tropics. Germination takes 4–5 days. The cotyledons are lanceolate and emerge in a horizontal position.

CHEMICAL COMPOSITION

Lettuce leaves contain approximately: water 94·3 per cent; protein 1·2 per cent; fat 0·2 per cent; carbohydrate 2·9 per cent; fibre 0·7 per cent; ash 0·7 per cent. They are rich in vitamins A and E.

PROPAGATION

Lettuce is grown from seed. 1 oz of seed produces about 3,000 plants.

HUSBANDRY

For home production small sowings should be made every two weeks in order to ensure a good succession. The seeds may be sown in drills 1 ft apart, thinned to 3 in. apart in the row at 2–3 weeks and with a final spacing of 1 ft. If transplanted the leaves should be at least 2 in. long and may be cut back in dry weather. Lettuce responds well to organic manures and to top dressings of fertilizers. The soils should never be allowed to dry out as the plants bolt quickly if growth is checked. For commercial production the seedlings are usually raised in nurseries or flats and are ready for transplanting in about 6 weeks. Lettuce is harvested about 3 months after sowing. If lettuce is obtained from an unknown source in the tropics it is advisable to wash the leaves in a dilute solution of potassium permanganate as a precaution against dysentery.

MAJOR DISEASES AND PESTS

Pythium spp. causes damping-off of seedlings and rotting of the leaves. A mosaic virus is reported from India. The lettuce looper, *Phytometra ni* Hubn., is seasonally common to the Greater Antilles. Slugs and snails can be troublesome.

IMPROVEMENT

Most lettuce cvs have been raised in temperate countries and there is a need to breed cvs suited to the low hot tropics.

PRODUCTION

Lettuce is grown for local consumption in the tropics. It is often grown by market gardeners near towns.

REFERENCE

LUNDQVIST, K. (1960). On the Origin of Cultivated Lettuce. *Hereditas*, **46**, 319–349.

CONVOLVULACEAE

About 45 genera and 1,000 spp., mainly twining annual and perennial herbs, but erect herbs and shrubs also occur, widely distributed throughout the world, particularly in the tropics. Latex often present. Lvs alternate, exstipulate, usually simple. Fls hermaphrodite, actinomorphic; corolla funnel-shaped with 5 lobes; stamens 5 inserted towards base of corolla, dehiscing longitudinally; ovary superior, usually 2-celled; fruit usually a capsule. *Cuscuta* spp., the dodders, are leafless, twining, total parasites, lacking chlorophyll.

USEFUL PRODUCTS

Ipomoea batatas (q.v.), the sweet potato, is the only important economic plant. Jalap, a resinous purgative, is obtained from the tubers of *Exogonium purga* (Hayne) Lindl. (syn. *Ipomoea purga* Hayne), a native twiner of eastern Mexico, now cultivated in Mexico, Jamaica and India. The seeds of *Rivea corymbosa* (L.) Hall. f, a native of Mexico, are used as a narcotic and intoxicant. A number of species are grown as ornamental climbers and include: *Ipomoea* spp. (q.v.) the rampant *Argyreia speciosa* Sweet from south-east Asia; *Porana paniculata* Roxb. from India. *Operculina* spp., from tropical America, have straw-coloured fruits with enlarged calyces and large capsules and are much appreciated for modern floral decorations; they are sometimes known as wooden-roses.

Ipomoea L. (x = 15)

A large genus of about 400 spp., mainly annual and perennial herbaceous twiners, but with a few erect shrubs, mostly in the tropics. Axillary flowers borne singly or in few-flowered cymes.

USEFUL PRODUCTS

I. batatas (q.v.) is the sweet potato.

I. aquatica Forsk. (syn. *I. reptans* Poir.) is an aquatic, floating or trailing, herbaceous perennial found throughout the tropics. It is often cultivated in Asia and is propagated by cuttings. The young terminal shoots and leaves are used as spinach. The vines are used as fodder for cattle and pigs and as fish food. In Malaya it is widely grown in fish ponds by the Chinese who feed it to their pigs; the pig manure is used to fertilize the fish ponds; thus fish, pork and spinach are provided.

I. eriocarpa R. Br. is used as a spinach in India and as a green fodder. *I. pes-tigridis* L. is grown for fodder in India and is fed to cattle both in the green state and as hay.

I. pes-caprae (L.) Sweet and *I. stolonifera* (Cyrill.) J. F. Gmel. are strand plants with a pantropical distribution and can be used for binding sand. The former is fed to pigs by Chinese in Malaya, but it taints the milk of cows.

A number of climbing species with handsome flowers are grown as ornamentals in the tropics. These include: *I. cairica* (L.) Sweet; *I. horsfalliae* Hook.; *I. leari* (Hook.) Paxt.; *I. nil* (L.) Roth; *I. purpurea* (L.) Roth (the common morning glory); *I. quamoclit* L. (syn. *Quamoclit pennata* (Desr.) Boj.); *I. tricolor* Cav. *I. alba* L. (syn. *I. bona-nox* L.; *Calonyction aculeatum* (L.) House), native of tropical America, is the moonflower, which is cultivated for its large, fragrant, night-blooming flowers.

Several species are used in native medicine throughout the tropics. The dried seeds of *I. nil* are official in the Indian Pharmacopoeia and are used as a purgative. In recent years morning glory seeds are being bought for chewing as a narcotic.

Ipomoea batatas (L.) Lam. (2n = 90) SWEET POTATO

USES

Sweet potatoes are grown throughout the tropics for their edible tubers, which are an important source of food in many countries. They are usually eaten boiled or baked, and may be candied with syrup or used as a puree. They are used for canning, dehydrating, flour manufacture, and as a source of starch, glucose, syrup and alcohol. They are also fed to livestock. In the tropics they are usually harvested as required, but they may be sliced and sun-dried for storage. In temperate countries they have to be cured for storage during the winter or they decay quickly. The tender tops and leaves are used as a pot-herb in Africa, Indonesia and the Philippines. The vines are widely used as a fodder for stock.

ORIGIN AND DISTRIBUTION

I. batatas is not known in a wild state. There seems little doubt that it originated in tropical America, but exactly where and what the parent species was or were is not known with certainty. In pre-Columbian time, sweet potatoes were grown in Mexico and parts of Central and South America and the West Indies. It had also reached Polynesia and New Zealand, but it was unknown at that time in Europe, Africa and Asia.

I. batatas is a hexaploid (2n = 90). It has been suggested that it, or its progenitor, was derived by amphidiploidy from a tetraploid (2n = 60) and a diploid (2n = 30) to produce a triploid (2n = 45), followed by subsequent doubling of the chromosomes to produce the hexaploid. *I. tiliacea* (Willd.) Choisy [syn. *I. fastigiata* (Roxb.) Sweet] has been suggested as one possible parent. *I. tiliacea* occurs wild in tropical America, Africa and Asia and bears a close resemblance to *I. batatas.* Its roots are sometimes tuberous and are occasionally eaten in South America. Nishiyama (1961) considers that *I. batatas* has probably arisen from *I. trifida* G. Don, which occurs wild in Mexico. He obtained 20 seedlings from seed collected from a Mexican plant and has since propagated these vegetatively in Japan. All this material is hexaploid (2n = 90) and resembles *I. batatas.* Although no edible tubers are produced, some have thickened roots. All the plants cross with *I. batatas.* F_1 hybrids show normal meiosis and are highly fertile in back-crosses to either parent. The possibility of *I. trifida* being a weedy species derived from the cultigen, *I. batatas,* should not be overlooked.

Whatever the origin of the sweet potato may be in tropical America, we are still left with the intriguing problem as to how the crop reached Polynesia in pre-Columbian times. It has often been suggested that man was responsible. Darlington (1963) even goes as far as to say that the fact that it is an allohexaploid 'leads to the outstanding contribution of botanical genetics to pre-history in demonstrating the spread of the sweet potato from Peru to Polynesia, a spread which proves the meeting of the Old and New World peoples assumed by Hornell and Heyerdahl in Easter Island some thousand or more years ago'.

I do not think it is necessary to invoke the aid of man in this transfer. Why should sweet potatoes be the only crop which man took from the New to the Old World before 1492? Sweet potatoes are invariably propagated by stem cuttings in the tropics and these would not survive the voyage unless planted and watered. They can also be grown from tubers, but these keep badly unless cured, and curing is not done in the tropics. Oviedo in 1526 stated that 'when the batatas are well cured they have been carried to Spain when the ship happens to make quick passage but more often they are lost on the voyage'. It is unlikely that they would survive the longer period in open canoes or rafts to Polynesia. As there is no tradition of growing *I. batatas* from seed, it is unlikely that seed would have been taken, and this is confirmed by Yen (1961a), who states that in South America and the Pacific 'in no case was purposeful propagation by seed recorded in practice or by local tradition'.

A number of species of *Ipomoea* have a wide distribution in the tropics and I suggested that *I. batatas* had been taken into cultivation independently in South America and Polynesia (Purseglove, 1963). I now consider this unlikely in view of *I. trifida* possibly being the progenitor of the cultivated species. It has been shown that sweet potato capsules float in water and that seeds, which have an almost impervious testa, germinate after immersion in sea water, and I have suggested that sweet potato capsules were carried by currents from the New World to Polynesia (Purseglove, 1965). Some cvs of *I. batatas* climb by twining; thus a fruiting plant could have floated across attached to a branch or log, both plant and support having been swept out to sea by a flood. Once the plant had become established in Polynesia, where most of their staple foods were root crops, it could easily have been taken into cultivation. Fryxell (1965) has shown that the hard seeds of *Gossypium tomentosum* Nutt. ex Seem float in sea water for at least a year with undiminished viability and he considers that the tetraploid ancestor of this species was carried by currents from Central America to Hawaii. Similarly he considers that *G. hirsutum* L. race *taitense* was carried by currents from South America to the Marquesas Islands and Samoa. 'Pacific Regattas' do not seem to be necessary, therefore, to carry *Gossypium* and *Ipomoea batatas* across the Pacific.

From Polynesia the Maoris took the sweet potato to New Zealand in the 13th or 14th century, where it was their staple food, as they grew no grain crops. Its name there was *kumara*, which is said to be the same name by which it was known in Peru, and this is used as evidence that man must have taken it. Evidence based on vernacular names is not always very reliable and I do not regard this as conclusive. That it was in ancient cultivation in Polynesia and by the Maoris can be judged from the religious ceremonial which surrounds its cultivation. The Maoris found a method of curing and storing the tubers during the winter and also a method of sprouting them to obtain planting material for each new season.

Columbus took the sweet potato to Spain in 1492 on the return from his first voyage to the New World, nearly 80 years before *Solanum tuberosum* (q.v.) reached Europe. The latter then took over the name of *Ipomoea batatas* which is derived from the Carib word *batata*. The Spaniards took the sweet potato to the Philippines from Mexico soon after their conquest of the latter country and the Portuguese introduced it early to their settlements in Africa and Asia. The Chinese obtained it from the Philippines in 1594 and it reached Japan in 1698, where it is now the second most important crop. Sweet potatoes were grown by Virginian colonists as early as 1648. By the mid-19th century the crop was widely grown from Zanzibar to Egypt.

In the East and parts of Africa sweet potatoes are gradually ousting yams because they give a quicker return with less work, and in some areas they are yielding place to cassava on account of higher yields with even less work involved.

CULTIVARS

A number of clonal cvs are known in the United States and elsewhere. These may be obtained from selections of existing clones, mutations, chance seedlings or deliberate hybridization. Most cvs are self-incompatible, but the self-fertile 'HES107' has been produced in Hawaii. In the United States modern cvs include 'Red Nancy' and 'Orlis' which are mutants and 'Kanda', 'Nemagold' and 'Allgold' which are seedling selections. Gooding (1964) found at least 88 distinct cvs in the West Indies, of which 'Black

Rock' and 'Red Nut' were well known, but are now being replaced as new cvs are bred. Cvs can be divided into those with dry mealy flesh on cooking and those with moist soft gelatinous flesh; the latter are erroneously called yams in the United States. Cvs vary in the colour of the skin, flesh and shape of the tubers, as well as vegetative characters, depth of rooting and time of maturity.

ECOLOGY

Sweet potatoes are grown from 40°N to 32°S. On the equator they are grown from sea level to 9,000 ft, but grow best where the average temperature is 75°F or more with a well-distributed annual rainfall of 30–50 in. and an abundance of sunshine. A frost-free growing period of 4–6 months is essential. In high rainfall areas they are often planted towards the end of the wet season. They will not persist in long periods of drought without irrigation. In East Africa sweet potatoes are often planted along swamp margins during the dry season in order to maintain adequate planting material and to provide out-of-season tubers.

The crop can be grown on a wide range of soils, but a well-drained, sandy loam with a clay subsoil is considered ideal. It cannot stand waterlogging and is usually grown on mounds or ridges. In south-western Uganda successful dry season crops were grown in pure peat of drained papyrus swamps.

I. batatas is a short-day plant and a photoperiod of 11 hours or less promotes flowering. Less flowering occurs at 12 hours and none at all at 13½ hours daylight. Above 30°N and S little flowering occurs. In Japan flowering is induced by grafting it on to other species of *Ipomoea*, including the moonflower, *I. alba* (q.v.). Flowering can also be induced by spraying with a dilute solution of 2–4, D.

STRUCTURE

A perennial herb, but treated as an annual in cultivation, with vine-like, trailing or twining stems, 1–5 m long, with latex in all its parts.

ROOTS: An extensive, fibrous, adventitious root system is produced from nodes of cutting; trailing stems in contact with soil also root at nodes. Tubers, about 10 per plant, develop in top 9 in. of soil by secondary thickening of some adventitious roots, both from those of original cutting and those from creeping stems. By earthing up behind, the plant can be continued indefinitely, but usually whole plant is lifted at harvest. Final structure of tuber is very complex with conducting tissue, parenchymatous storage cells, latex vessels, and an outer periderm which replaces ruptured

---

Fig. 11. **Ipomoea batatas**: SWEET POTATO. **A**, vegetative shoot (×½); **B**, 1–3, different leaf forms (×½); **C**, flower (×½); **D**, flower opened out (×½); **E**, tuber (×½).

B2
B1
B3
C
A
D
E

epidermis; secondary roots on tubers in 5–6 rows. Tubers fusiform to globular, smooth or ridged; periderm white, yellow, orange, red, purple or brown; flesh white, yellow, orange, reddish or purple.

STEMS: Prostrate or ascending, sometimes twining, thin, 3–10 mm in diameter, internodes 2–10 cm long, glabrous or pubescent when young, light green to purple in colour, angular or terete; bundles bi-collateral.

LEAVES: Very variable, even on same plant, depending on age; spirally arranged with phyllotaxy of 2/5, simple, exstipulate; petiole 5–30 cm long, grooved on upper surface, 2 small nectaries at base; lamina mostly ovate in outline, entire to deeply digitately-lobed, base usually cordate in first leaves, tip acute or obtuse, glabrous or with variable pubescence, 5–15 × 5–15 cm, green to purple in colour, sometimes with purple stain at base; veins palmate, green or purple beneath.

FLOWERS: Axillary, solitary or cymose; peduncle 3–15 cm long; bracteoles 2, small, lanceolate; calyx deeply 5-lobed, 1–1·5 cm long, cuspidate, sometimes ciliate; corolla funnel-shaped, 2·5–5 cm long, 2·5–4 cm in diameter, plicate, faintly 5-lobed, purplish in colour with deeper colour in throat and paler at margin; stamens 5, attached near base of corolla, of unequal length with longest same length or shorter than style, filaments white with glandular hairs, anthers white, pollen papillate; ovary, surrounded by lobed orange nectary, 2-celled, style filiform, white, glabrous, about 2 cm long, stigma white, capitate, 2-lobed.

FRUIT AND SEEDS: Glabrous or hirsute dehiscent capsule, 5–8 mm in diameter, containing up to 4 seeds, but usually only 1 or 2 develop. Seeds black, angular, glabrous, 3 mm long, testa very hard, micropyle invaginated, albuminous; cotyledons bilobed, transversely folded.

POLLINATION

Being a short-day plant, sweet potatoes seldom flower in temperate countries, unless it is induced (see above). In the tropics flowers and fruits are often produced, but the amount of flowering varies with the cv., environmental conditions and the season, The flowers open before dawn and close and wilt the same morning between 9–11 a.m., but remain open longer in cool, cloudy weather. The stigma is receptive in the bud from 6–8 p.m. and the anthers begin to dehisce from 11–12 p.m. Pollen can still germinate on the stigma some hours after wilting and remains viable for a day at ordinary temperatures. Natural cross-pollination occurs and is carried out by hymenopterous insects, particularly bees.

Almost all cvs are self-sterile and there is also complete or partial cross incompatibility between some of them. Seeds are only formed naturally when cross-compatible cvs are grown together. Incompatible pollen fails to grow on the stigma. The incompatibility is homomorphic and not connected with heterostyly. A prolific self-fertile seedling, 'HES107', was discovered in Hawaii. In Trinidad 'C26' has shown a small degree of self-compatibility.

GERMINATION

Under normal conditions, seeds germinate very irregularly, but can be induced by clipping the hard testa or treating with concentrated sulphuric acid for 45 minutes. Scarified seeds germinate in 1–2 days, when the radicle emerges and grows rapidly. The hypocotyl carries the cotyledons to the surface partially enclosed in the testa, which quickly falls away. The deeply bilobed cotyledons then expand and are carried apart by the growth of their petioles. The first foliage leaves then grow and by 21 days there are 3–4 of these, which are simple, ovate and cordate.

CHEMICAL COMPOSITION

Fresh sweet potatoes provide about 50 per cent more calories than Irish potatoes. The average composition is: water 70 per cent; protein 1·5–2 per cent; fat 0·2 per cent; carbohydrate 27 per cent; fibre 1 per cent. The sugar content is 3–6 per cent and varies with cv. and condition; it increases on storage and during cooking. Yellow- and orange-fleshed cvs are rich in vitamin A. The leaves contain: water 86 per cent; protein 3·2 per cent; fat 0·8 per cent; carbohydrate 8·5 per cent.

PROPAGATION

In the tropics sweet potatoes are almost invariably propagated by stem cuttings, 12–18 in. long, taken from the apical growth of mature plants. The bottom leaves are removed and the lower half of the cutting is inserted in the soil at an angle. In Uganda it is customary to use partially wilted cuttings. In India the central portion of the cutting is buried in the soil leaving a node exposed at either end.

In temperate countries propagation is by sprouts or slips obtained by planting small or medium-sized tubers close together in nursery beds or hotbeds. The sprouts reach a length of 9–12 in. in 4–6 weeks, when they are pulled out and transplanted in the field.

In breeding work the sweet potato is propagated by seed (see GERMINATION above). The seedlings are extremely variable and some of them may produce no tubers at all. Sweet potatoes can also be propagated by single-leaf cuttings.

HUSBANDRY

Sweet potatoes are usually planted on ridges or mounds. The ridges, about 18 in. high, are usually 3–4 ft apart with the cuttings 12 in. apart. Mounds, about 2 ft high, are usually 3–4 ft apart, with several cuttings planted on each mound.

The crop responds well to organic manures. The results from artificial fertilizer trials are conflicting and too much nitrogen may encourage vine growth at the expense of tubers. Jacob and Uexkull (1963) recommend 500–1,000 lb per acre of a 6 : 9 : 15 NPK mixture.

In the tropics the crop receives little cultivation after it is established. In the West Indies and the United States it is customary to turn back the vines from time to time to prevent rooting at the nodes in order to ensure a more even crop and fewer small tubers. The time taken to maturity is 3–6 months depending on the cv., and can be judged by the leaves turning yellow and beginning to drop or by cutting a tuber when the sap dries up rapidly without discoloration.

In Africa the tubers are harvested as required. Elsewhere the crop may be stored, but it stores poorly and may lose 10–15 per cent in weight in the first 2–3 weeks after harvesting. In temperate countries the tubers are cured before storage by keeping them at 85–95°F with a relative humidity of 85–90 per cent for 1–3 weeks and then gradually reducing the temperature to 50–55°F and maintaining a relative humidity of 80 per cent. Under these conditions the tubers will keep for 3–7 months.

Yields from 1–20 tons per acre have been reported, but a yield of 7–8 tons per acre may be regarded as satisfactory.

MAJOR DISEASES

Fungal diseases of sweet potatoes are not very serious in the tropics. Stem rot, *Fusarium oxysporum* Schlect. f. *batatas* (Wr.) Synder & Hansen is destructive in the United States, as is black rot, *Ceratocystis fimbriata* Ell. & Hals., which also occurs in the tropics. A soft rot during storage is caused by *Rhizopus* spp. Viruses are becoming increasingly important in East Africa, the West Indies and the United States; they include mosaic and internal cork disease.

MAJOR PESTS

The sweet potato weevil, *Cylas formicarius* (Fab.), is the major pest in most countries, the larvae feeding on the roots. The weevil, *Euscepes batatae* Waterhouse, is a serious pest in the drier parts of the West Indies. The moth, *Megastes grandalis* Gn., does much damage in Trinidad and South America. The eggs are laid in the axils of the leaves and the larvae burrow down the stem to the tubers. Rats, wild pigs and other vermin can cause serious damage in some countries.

IMPROVEMENT

Most sweet potato cvs are very heterozygous and, as has been shown above, they exhibit a very wide range of variability. In the tropics chance seedlings will occur and it is likely that some of them will be maintained, as peasant cultivators will seldom discard any plant which produces food. Mutations occur fairly frequently. Self-incompatibility and cross-incompatibility between some cvs add to the difficulty of breeding. However, when an improved cv. has been obtained, it can be maintained as a clone as the crop is propagated vegetatively and it can be multiplied quickly.

Sweet potato breeding is being done in the United States, Japan, Hawaii, New Zealand, Puerto Rico and Trinidad. In 1956 the Food Crop Unit was started at the Imperial College of Tropical Agriculture, now the University of the West Indies, in Trinidad and H. J. Gooding was appointed Plant Breeder. At that time yields in the West Indies were low, about ½–3 tons per acre. This was partly due to virus infection, but also because a large mixture of cvs were being grown, many of which were inferior.

The selection standards used are: (1) yield of at least 8 tons per acre; (2) early maturity; (3) sparse vine growth; (4) globular tubers without furrows to reduce waste and preferably red in colour; (5) shallow-rooted tubers to facilitate mechanical harvesting; (6) good keeping and storage qualities; (7) good cooking quality and palatability; (8) resistance to diseases, particularly viruses; (9) resistance to pests, *Megastes grandalis* in Trinidad and *Cylas formicarius* and *Euscepes batatae* elsewhere; (10) high content of vitamins A and C.

Gooding made a collection of cvs from the various West Indian islands, worked out the synonymy, and made selections from them. Seedlings were obtained from open-pollinated seeds and from controlled hybridization. Any cv. which shows promise is sent to the other islands for trials. A number of cvs have been raised which have given over 8 tons per acre in experimental planting and which have yielded well in commercial production (Gooding, 1964; Williams, 1964).

In hybridization it is necessary to bag the flowers before they open; glassine bags are very satisfactory for this purpose. It is not necessary to emasculate self-sterile cvs, but if there is any evidence of self-fertility the stamens should be removed. The bags should be replaced after pollination. The seeds are sown in pots and after two months are planted out in the field, where they will produce tubers in five months. Some seedlings produce no tubers whatsoever.

In progeny of the self-compatible cv. 'HES107' in Hawaii more than half the population of tuber-forming plants produced 1 lb or less total weight of tubers per plant, but two unique plants produced 75 tubers weighing 20·5 lb and 65 tubers weighing 27·5 lb respectively. Poole (1955) states that there is evidence of quantitative factor inheritance, that vigour is not depressed by inbreeding, and that short stem length and low weight of root yield are dominant by relatively few factors over long stem length and high root yield. Inheritance of a number of other characters has been worked out.

PRODUCTION

Sweet potatoes are grown throughout the tropics and the largest acreage is in Africa. World production has been estimated at approximately 110 million metric tons. Outside the tropics the largest producer is Japan with over 6 million tons, but substantial quantities are also produced in China, the United States (700,000 tons) and New Zealand.

REFERENCES

BURKILL, I. H. (1954). Aji and Batata as Group Names within the species *Ipomoea batatas*. *Ceiba*, **4**, 227–240.

COOLEY, J. S. (1951). The Sweet Potato—Its Origin and Primitive Storage Practices. *Econ. Bot.*, **5**, 378–386.

CROSBY, D. G. (1964). The Organic Constituents of Food. iii. Sweet Potato. *J. Food Sci.*, **29**, 287–293.

ELMER, O. H. (1960). Sweet Potatoes and their Diseases. *Bull. Kans. agric. Exp. Sta.*, 426.

FRYXELL, P. A. (1965). Stages in the Evolution of *Gossypium* L. *Adv. Frontiers Pl. Sci.*, **10**, 31–56.

GOODING, H. J. (1964). Some Aspects of the Methods and Results of Sweet Potato Selection. *Emp. J. exp. Agric.*, **32**, 279–289.

MACDONALD, A. S. (1963). Sweet Potatoes, with particular Reference to the Tropics. *Field Crop Abstr.*, **16**, 219–225.

NISHIYAMA, I. (1961). The Origin of the Sweet Potato. *Tenth Pacific Science Congress, Hawaii*, 119–128.

POOLE, C. F. (1955). Sweet Potato Genetic Studies. *Hawaii Agric. Exp. Sta. Bull.*, **27**.

PURSEGLOVE, J. W. (1963). Some Problems of the Origin and Distribution of Tropical Crops. *Genetica Agraria*, **17**, 105–122.

PURSEGLOVE, J. W. (1965). The Spread of Tropical Crops, in *The Genetics of Colonizing Species*, ed. H. G. Baker and G. L. Stebbins: New York: Academic Press.

WILLIAMS, D. B. (1964). Improvement of the Sweet Potato by Breeding at the University of the West Indies. *J. Agric. Soc. Trinidad and Tobago*, **899**, 419–429.

YEN, D. E. (1961a). Evolution of the Sweet Potato (*Ipomoea batatas* (L.) Lam.). *Nature, Lond.*, **191**, 93–94.

YEN, D. E. (1961b). Sweet-Potato Variation and its relation to Human Migration in the Pacific. *Tenth Pacific Science Congress, Hawaii*, 93–117.

CRUCIFERAE

A very natural family of about 300 genera and 3,000 spp., mainly herbaceous, and with the majority in temperate regions. Lvs exstipulate, usually alternate and simple. Fls usually hermaphrodite, regular, hypogynous; sepals 4 in 2 whorls; petals 4 in 1 whorl, cruciform, usually clawed with spreading limb; stamens typically 6 and tetradynamous with 2 outer short stamens and inner whorl of 4 with longer filaments; anthers usually 2-locular with introrse longitudinal dehiscence; ovary superior, usually of 2 united carpels; 1-locular, with parietal placentation, divided by spurious membranous septum; short style with 2-lobed stigma. Fr. a pod-like capsule, if longer than broad a siliqua, when as long as broad a silicula; dehiscing by 2 valves (sides) separating from below upwards leaving septum intact; rarely indehiscent and joined as in a lomentum. Seeds with no endosperm and curved embryo.

USEFUL PRODUCTS

The following species provide vegetables, oils, condiments and stock feed. Most of them originated in and are more suited to the temperate regions, but many of them are grown in the tropics, even at low altitudes, so that mention must be made of them.

KEY TO THE CROP GENERA

A. Fruit a silicula (as broad as long)
- B. Robust perennial herb; lvs oblong, up to 60 cm, not lobed .... *Armoracia*
- BB. Slender annual herb; basal lvs pinnatifid, much smaller .... *Lepidium*

AA. Fruit a siliqua (longer than broad)
- B. Indehiscent at maturity .... *Raphanus*
- BB. Dehiscent at maturity
  - C. Aquatic herb .... *Nasturtium*
  - CC. Not aquatic .... *Brassica*

Armoracia rusticana Gaertn. (x = 8; 2n = 32) HORSE-RADISH

Syn. *Cochlearia armoracia* L.; *Armoracia lapathifolia* Gilib.

A stout glabrous perennial herb, native of south-eastern Europe, with long fleshy cylindrical roots, large oblong serrate lvs and small white fls. The grated roots are used as a condiment, particularly with roast beef. The pungent taste is due to the glucoside sinigrin which is broken down in water by enzyme action. It is sometimes grown at higher altitudes in the tropics. The roots of *Moringa oleifera* Lam. (Moringaceae), horse-radish tree, a native of India, are sometimes used as a substitute; this small tree is now grown throughout the tropics and its tender pods are used in culinary preparations.

Brassica L. (x = 8, 9, 10, 11)

About 40 spp. of annual or biennial, rarely perennial herbs, native of the north temperate parts of the Old World, especially in the Mediterranean region. All have tap-roots, which may be fleshy; stems erect or ascending, glabrous or with simple hairs; lower basal lvs often pinnatifid with large terminal lobe (lyrate); inflorescence an ebracteate raceme; sepals erect, inner pair often saccate at base; petals long-clawed, usually yellow; stamens 6; stigma usually capitate; fruit a siliqua with convex valves, tipped by an indehiscent and usually seedless beak; seeds in a single row in each loculus, spherical; cotyledons obcordate, emarginate.

KEY TO THE CROP SPECIES

A. Plants glaucous; lvs thick, glabrous at maturity; fls large, 1·2–1·5 cm long.
  B. Inflorescence elongated and open at anthesis, 10–25 cm long; young radical lvs glabrous .... *B. oleracea*
  BB. Inflorescence short at anthesis, not more than 5 cm long, with fls clustered at top; young radical lvs with scattered hairs.
    C. Roots slender, not tuberous .... *B. napus*
    CC. Roots tuberous .... *B. napobrassica*
AA. Plants grass-green; lvs thin, often with scattered hairs; fls less than 1·2 cm long.
  B. Beak of ripe siliqua conical or slender, not as long as rest of fr.
    C. Siliqua at maturity wide-spreading.
      D. Roots tuberous .... *B. rapa*
      DD. Roots not tuberous.
        E. Leaves lobed, those on stem not clasping, margin notched .... *B. juncea*
        EE. Leaves lobed, those on stem clasping ... *B. campestris*
        EEE. Leaves not lobed .... *B. chinensis*

CC. Siliqua at maturity closely appressed to rachis *B. nigra*
BB. Beak of siliqua flat, same length or longer than rest of fr. .... *B. alba*

Brassica alba (L.) Rabenh. (2n = 24) WHITE MUSTARD

Syn. *B. hirta* Moench., *Sinapis alba* L.

This is the mustard of 'mustard and cress', the young seedlings being eaten as salad. The seeds, which are yellowish in colour and white within, contain 30 per cent oil. They are ground with the seeds of *B. nigra* and starch to produce table mustard. It is also grown as a fodder crop and as green manure. It is not much grown in the tropics, but it can be grown in boxes and cut 6 days after sowing for salad. *B. alba* is a native of the Mediterranean region. An annual hairy herb, 30–80 cm high; fls yellow; petals about 1 cm long; frs hairy, 3-veined, 2·5–4 cm long. Self-sterile. The seeds contain the glucoside sinalbin which is hydrolysed in the presence of water by the enzyme myrosin to give the pungent flavour.

Brassica campestris L. (2n = 20) FIELD MUSTARD

An important oil seed crop in India, the oil being used for cooking. Two main varieties are recognized: var *sarson* Prain (Indian colza or sarsan) and var. *toria* Duthie & Fuller (Indian rape or toria). A slender erect annual 0·3–1 m tall with broad-based, stem-clasping lvs, which are somewhat hairy and glaucous, small yellow fls, and frs 2·5–5 cm long. Self-sterile. It is grown in pure stand with seed broadcast at 4–5 lb per acre or may be grown mixed with cereals. Yields are 7–11 cwt per acre.

Brassica chinensis L. (2n = 20) CHINESE CABBAGE, PAK-CHOI

The leaves are eaten cooked as a vegetable or as salad. It originated in eastern Asia and is grown extensively in China and Japan. It has now spread to Malaysia, Indonesia, and the West Indies. It is one of the easiest and most productive vegetables in the tropics for all elevations. A biennial herb, grown as an annual; basal lvs broad, shining, 20–50 cm long, with thickened white petioles, not forming a compact head; fls pale yellow about 1 cm long; frs slender 3–6 cm long. Usually planted 1 ft apart and harvested after 2–3 months. The var. *pekinensis* (Rupr.) Sun., pe-tsai, is used as salad, the blanched heart being used; it resembles a giant cos lettuce.

Brassica juncea (L.) Czern. & Coss. (2n = 36) INDIAN MUSTARD

This species is possibly of African origin, but was taken early to Asia. It is extensively cultivated from eastern Europe to China and in Africa. It is an important oil seed in India, where it is known as *rai*. It is one of the most pungent of the cultivated mustards. The seeds contain 35 per cent oil, which is edible and is used for cooking and as a substitute for olive oil. The leaves are used as a pot-herb, but they must be boiled twice. They contain the glucoside sinigrin. An erect, much-branched annual, to 1 m tall, basal lvs stalked, up to 20 cm long, with very large ovate terminal segment; fls pale yellow, 7–9 cm long; frs 3–5 cm long. Self-fertile. In India the seed is broadcast at the rate of 4–5 lb per acre and yields about 11 cwt per acre.

Brassica napobrassica (L.) Mill. (2n = 38) RUTABAGA, SWEDE

Syn. *B. napus* L. var. *napobrassica* (L.) Rchb.

Swedes are European in origin and are now grown there for feeding to livestock. They are also eaten as a vegetable, the tubers being sliced and boiled. They are seldom grown in the tropics. A glaucous biennial with tuberized stem-base and tap-root, which is usually yellow fleshed, and either violet, bronze, white or yellow outside. It can be distinguished from the turnip by the short 'neck' formed by the tuberized base of the epicotyl.

Brassica napus L. (2n = 38) RAPE

Rape was in ancient cultivation in the Mediterranean region. It is grown in Europe as green fodder for livestock and for its seeds from which rape or colza oil is extracted, for which purpose it is also grown in Japan. The residual rape-seed cake is fed to livestock. Rape oil is edible; it is used for greasing loaves of bread before baking. It is also used as an illuminant and lubricant and for soap manufacture. Rape is not much grown in the tropics. A much branched, erect annual or biennial; lvs glaucous; lower lvs lyrate-pinnatifid, stalked; upper leaves lanceolate, sessile, clasping stem; fls pale yellow, 1·2–1·5 cm long; frs 5–11 cm long. Self-fertile.

Brassica nigra (L.) Koch (2n = 16) BLACK MUSTARD

A native of Eurasia and long cultivated in Europe. This was the first species to provide table mustard for use as a condiment, but the seeds are now usually mixed and ground with those of *B. alba*, which are not so pungent. The seeds yield 28 per cent of a fixed oil, which is used in medicine and soap-making. The seeds also contain about 1 per cent of a volatile oil which is used as a counter-irritant when greatly diluted. The crop is little grown in the tropics. An annual branched herb up to 1 m high; lvs all stalked, up to 16 cm long; fls bright yellow, about 8 mm long; frs about 2 mm long are held erect and appressed to rachis; seeds 1 mm or more in diameter, minutely pitted, dark brown without, yellow within. Self-sterile. The seeds contain the glucoside sinigrin from which, by the action of the enzyme myrosin in the presence of water, is produced the volatile oil which is responsible for the pungent aroma and flavour.

Fig. 12. **Brassica chinensis**: CHINESE CABBAGE. **A,** flowering plant (× ¼); **B,** flower from side (× 3); **C,** flower from above (× 3); **D,** flower in longitudinal section (× 3); **E,** fruits (× ½); **F,** fruit (× 1).

E
F
C
D
B
A
M. Wong

Brassica oleracea L. (2n = 18) WILD CABBAGE

The wild cabbage is indigenous to the Mediterranean region, south-western Europe and southern England, where it grows on maritime cliffs. It was taken into cultivation at least 4,500 years ago and gave rise to many cultivated races, which agree with the wild species in fls, fr. and seeds, but vary widely in vegetative morphology and in the form of the immature inflorescence. These races include some of the world's most important vegetables and are now widely cultivated almost everywhere. On flowering the inflorescence lengthens so that the buds overtop the open fls; sepals erect; petals 1·2–1·5 cm long, about twice as long as sepals, lemon yellow; all 6 stamens erect; fr. 5–10 cm long, cylindrical with a short tapering, usually seedless beak; seeds 8–16 in each cell, dark grey-brown, 2–4 mm in diameter.

KEY TO THE CULTIVATED VARIETIES (RACES)

A. Stem of first year elongated
  B. Stem branched and leafy .... var. *acephala*
  BB. Stem unbranched; axillary buds arrested and swollen .... var. *gemmifera*
AA. Stem of first year short
  B. Stem tuberous .... var. *gongylodes*
  BB. Stem not tuberous
    C. Terminal bud arrested, compact, much swollen. Inflorescence not developed in first year........ var. *capitata*
    CC. Inflorescence partly developed in first year
      E. Inflorescence forming a compact mass of thickened colourless peduncles, bracts and undeveloped fls .... var. *botrytis*
      EE. Young inflorescence not compacted, green in colour .... var. *italica*

Var. acephala DC. BORECOLE, COLLARD, KALE

Erect branched herbs, which do not form heads like cabbages, and are closest to the wild species. Thousand-headed kale, which branches early and produces much foliage, and marrow-stemmed kale with thick succulent stems, are grown for feeding to livestock. Curled kale with crinkled leaves and other finer-leaved cvs are used as boiled green vegetables. Collards, in which the smooth leaves form a rosette at the apex of the stem, can stand higher temperatures than kale.

Var. botrytis L. BROCCOLI, CAULIFLOWER

The edible portion is the white curd-like mass of a close aggregation of abortive flowers on thick hypertrophied branches produced at the top of a short stem. Broccoli are hardier and take longer to mature. In addition to being boiled as a vegetable or cooked with a cheese sauce, cauliflowers are made into soup and are also pickled. During growth the leaves are often bent over the heads to keep the latter white. They require more care, richer soil and better cultivation than cabbages. The Indian cv. 'Early Patna' does reasonably well in the lowland tropics.

Var. capitata L. CABBAGE

The cabbage is of very ancient cultivation and has been grown in Europe since at least 2500 B.C. It was introduced into England by the Romans and is now grown throughout the world, including the lowland tropics. The plant is a biennial and has a very short stem surmounted by a mass of thick overlapping leaves forming a compact head, which may be pointed or round, green or red, smooth or crinkled. Savoy cabbages have dark green crinkled leaves. Cabbages may be eaten raw or cooked. They may be pickled, for which purpose red cabbages are usually used. Sauerkraut, a favourite food in Germany and Russia, is a form of human silage in which sliced cabbage is fermented in its own juice together with salt. Large 'Drumhead' cvs are grown for feeding to stock. The recommended cvs for the lowlands in Trinidad are 'Jersey Wakefield' and 'Charleston Wakefield' with pointed heads, 'Copenhagen Market' and 'Succession' with round heads.

Var. gemmifera Zenk. BRUSSELS SPROUTS

Biennial with simple erect stem to 1 m high, with axillary buds forming compact heads or sprouts, about 3 cm in diameter, which are eaten after boiling. It is a cool season crop and does not do well in the lowland tropics.

Var. gongylodes L. KOHLRABI

Syn. *B. caulorapa* Pasq.

A biennial in which secondary thickening of the short stem produces the spherical edible portion, 5–10 cm in diameter, which is eaten cooked; it may be green or purple. Glaucous lvs have long slender petioles. It does reasonably well in the lowland tropics. The cvs recommended in Trinidad are 'Early Purple Vienna' and 'Early White Vienna'.

Var. italica Plenck SPROUTING BROCCOLI

Similar to the cauliflower, but terminal and axillary green heads of buds are produced, not so compact as the cauliflower, and are harvested just before the flowers open. It withstands hot and drier conditions better than the cauliflower. Cv. 'Texas 107' is recommended in Trinidad.

Brassica rapa L. (2n = 20) TURNIP

Syn. *B. campestris* L. var. *rapifera* Metz.

Turnips have been cultivated in Europe for over 4,000 years. It is a native of central and southern Europe, and has now spread all over the world. Turnips were taken to Mexico in 1586 and to Virginia in 1610. They can be grown in most parts of the tropics. A biennial with swollen tuberous white-fleshed tap-root; 'neck' lacking (cf. swede); basal lvs grass-green, stalked, lyrate-pinnatifid, bristly; fls bright yellow, sepals spreading, petals 6–10 mm long, about 1½ times as long as sepals, 2 outer stamens curved outwards at base and much shorter than inner stamens; fr. 4–6·5 cm long with long tapering beak; seeds blackish or reddish-brown, 1·5–2 mm in diameter. The swollen root is eaten cooked and the tops may be used as spinach. Cv. 'Purple Top White Globe' is recommended in Trinidad. The seed is sown thinly in drills, 1 ft apart, and the seedlings are thinned to 6 in. apart. The roots are ready for harvesting in 8–10 weeks.

Lepidium sativum L. (x = 8; 2n = 16, 32) GARDEN CRESS

A native of the Levant and also occurs wild in Abyssinia. It has been cultivated since very early times in Europe and is now cultivated as a salad plant all over the world, the young seedlings being eaten. An erect glabrous annual, 15–30 cm tall; basal lvs long-stalked, lyrate, with toothed obovate lobes; stem leaves once or twice pinnate; uppermost lvs sessile, linear; entire; fls minute, white, protogynous; fr. a silicula, 5–6 × 3–4 mm. The seed should be sown thickly in shallow boxes, the seedlings, harvested in the cotyledon stage, being ready for cutting 10 days after sowing.

Nasturtium officinale R. Br. (x = 8; 2n = 32) WATERCRESS

An aquatic perennial herb widely distributed in a wild state in Great Britain, south and central Europe, and western Asia, and now introduced into many parts of the world; it has become a serious river-weed in New Zealand. The tips of the leafy stems are eaten as salad and are often used to garnish steaks and other meats; it may also be cooked as a vegetable. Stems hollow, angular, glabrous, 10–60 cm. long; procumbent and rooting at nodes below, then ascending or floating; lvs lyrate-pinnate with 3–9 leaflets; fls 4–6 mm in diameter, white; fr. a siliqua, 13–18 mm long; seeds in 2 distinct rows in each cell. Watercress can be grown from seed, but is usually planted from cuttings placed about 6 in. apart. It is best grown in clear running water. When established, the more the tops are gathered the better, as this induces branching. The nearly sterile hybrid (2n = 48) of *N. officinale* (2n = 32) × *N. microphyllum* (Boenn.) Rchb. (2n = 64) is also grown for salad.

Raphanus sativus L. (x = 9; 2n = 18) RADISH

The radish, which probably originated in western Asia, was cultivated in ancient Egypt, Assyria, Greece and Rome. It has now spread throughout the world. It is grown for its young tender tuberous roots which are usually eaten raw and have a pungent flavour. The leaves may also be eaten as salad. An annual bristly herb, 20–100 cm high; tap-root swollen, white or red, round, cylindrical or tapering, flesh white; lvs lyrate-pinnatifid; fls white to lilac, small; fr. inflated, indehiscent; 3–7 cm long and up to 1·5 cm in diameter with 6–12 seeds and long conical beak; seeds about 3 mm in diameter. Seed should be sown thinly in beds or drills and the roots are ready for pulling in 3–5 weeks; if left too long they become tough. The cvs recommended in Trinidad are 'Scarlet Globe', 'Sparkler' and 'White Icicle'.

Var. *longipinnatus* Bailey provides Chinese and Japanese radishes, which are extensively grown in eastern Asia. The large cylindrical white roots may attain a weight of 5 lb, and some Japanese cvs are said to weigh up to 40–50 lb. They are eaten raw and have a mild flavour, or they can be sliced and cooked. Radical lvs are long and narrow, up to 60 cm in length, with 8–12 pairs of pinnae. Cvs 'Hong Kong Summer' and 'Sutton's Chinese White' have been introduced into Trinidad.

R. caudatus L., (2n = 18) rat-tailed radish, is cultivated in India and eastern Asia for its long slender pods, up to 30 cm or more in length, which are used unripe as a vegetable.

ECOLOGY

Most of the cruciferous crop plants originated in temperate countries and grow best in cool moist climates and at higher altitudes in the tropics. They are not really suited to the low humid tropics. In Trinidad (lat. 10°N.) with a minimum monthly temperature ranging from 67–72°F and maximum temperatures of 86–91°F fair yields of cabbages and cauliflowers are obtained with irrigation in the dry season which is the coolest time of the year. In Puerto Rico (lat. 18°N.) the best crops are produced at altitudes above 3,000 ft. The best vegetables I have ever grown anywhere were at Kabale in south-western Uganda (lat. 1°S.) at an altitude of 6,200 ft, with a mean maximum temperature of 75°F, a mean minimum temperature of 50°F, a mean relative humidity of 91 per cent at 8.30 a.m. and 56 per cent at 2.30 p.m., and an average rainfall of 35 in. per annum. The ideal soil for Brassicas is a rich sandy loam.

POLLINATION

Crops of this family are cross-pollinated by insects, bees being largely responsible. Some of the crops are self-sterile. Some of them will not flower in the lowland tropics and few of them set seed under these conditions. Seed can be produced at high altitudes in the tropics and this was done in East Africa during World War II. Seed is also produced in India. It is essential that cvs grown for seed should be separated by at least ¼ mile to avoid cross-pollination.

GERMINATION

The varieties of *Brassica oleracea* intercross and produce useless hybrids, so that a reliable source of seed is essential. Under good storage conditions, the seeds retain their viability for several months. They usually give a rapid and good germination, which is epigeal.

CHEMICAL COMPOSITION

Representative values of 100 g of edible portion are given in the table below.

Crop	Water g.	Protein g.	Fat g.	Carbo-hydrates g.	Fibre g.	Ash g.	Vit. A i.u.	Ascorbic Acid mg.
Broccoli	87·3	4·5	0·6	6·4	1·6	1·2	560	94
Brussels sprouts	83·3	5·2	0·3	9·9	1·9	1·3	145	82
Cabbage	91·4	1·7	0·2	6·1	1·0	0·6	30	43
Chinese cabbage	91·0	1·7	0·3	5·4	0·6	1·6	2800	32
Cauliflower	89·4	2·8	0·4	6·5	1·0	0·9	10	82
Kale	85·4	4·5	0·7	7·5	1·3	1·9	2015	125
Kohlrabi	91·0	2·0	0·1	6·1	1·0	0·8	tr.	60
Mustard (*Brassica juncea*)	90·6	2·6	0·4	4·8	1·0	1·6	610	62
Radish	93·2	0·9	0·1	5·0	0·7	0·8	tr.	28
Rape (leaves)	83·3	2·9	1·7	11·2	1·8	0·9	1340	120
Turnip	92·5	0·8	0·2	5·7	0·8	0·8	tr.	28
Watercress	92·2	2·8	0·4	3·3	1·1	1·3	1105	44

From W. L. Woot-Tseun and M. Flores (1961). *Food Composition Table for Use in Latin America.*

PROPAGATION

With the exception of watercress which is usually planted from cuttings, all the cruciferous crops are grown from seed. Broccoli, brussels sprouts, cabbages, cauliflowers and kohlrabi are usually sown in nurseries or boxes and transplanted to the field; the others are usually sown direct and thinned as required.

HUSBANDRY

The land should be well cultivated before planting. The Brassica crops respond well to animal and other organic manures, and 5–20 tons per acre is applied in Trinidad market gardens. Fertilizers are also applied at the rate of 4–5 cwt per acre of a 4 : 7 : 5 NPK mixture and a top dressing of nitrogen may be given later. Cruciferous crops should be rotated with crops of other families. In Trinidad most of the vegetables grown for the local market are grown with irrigation during the dry season on land which is cropped with rice in the wet season. The usual spacing for cabbages and cauliflowers is $3 \times 1\frac{1}{2}$ ft and they take 3–4 months from planting to harvesting. Further details of cultivation are given above under the individual crops. For successful vegetable production a high standard of husbandry is essential.

MAJOR DISEASES AND PESTS

The most serious diseases of cole crops in the tropics are: blackleg, *Phoma lingam* (Tode ex Fr.) Desm.; black rot, *Xanthomonas campestris* (Pamm.) Dows; downy mildew, *Peronospora parasitica* (Pers. ex Fr.) Fr.; club root, *Plasmodiophora brassicae* Woron.; cabbage mosaic virus. The root-knot nematode, *Meloidogyne* spp., can cause serious damage.

Major pests are: cabbage white butterfly, *Ascia monuste* (L.); cabbage bud-worm, *Hellula phidilealis* Wlk.; diamond-back moth, *Plutella maculipennis* (Curt.); snails and slugs can cause much damage.

IMPROVEMENT

The breeding of cruciferous vegetable crops has received considerable attention in temperate countries and a very wide range of cvs are available. Very little breeding work has been attempted in the tropics, except in Hawaii and Puerto Rico. Elsewhere trials have been made of introduced cvs, most of which have been bred for temperate conditions. Cvs which will thrive in the low hot humid tropics are urgently required.

REFERENCES

CAMPBELL, J. S. and GOODING, H. J. (1962). Recent Developments in the Production of Food Crops in Trinidad. *Trop. Agriculture, Trin.*, **39**, 261–270.

CHILDERS, N. F. *et al.* (1950). *Vegetable Gardening in the Tropics.* Fed Exp. Sta. Puerto Rico Circ. 32.

HERKLOTS, G. A. C. (1972). *Vegetables in South-east Asia.* London: George Allen & Unwin.

NIEUWHOF, M. (1969). *Cole Crops.* London: Leonard Hill.

CUCURBITACEAE

About 90 genera and 750 spp. almost equally divided between the Old and New World tropics, with 7 genera common to both hemispheres. A few spp. extend into the temperate regions, but all are frost-tender. The cultivated spp. are tendril-climbing or prostrate annual, or occasionally perennial rapid-growing herbs. Stems mostly soft, glabrous, hairy or prickly; bundles bi-collateral. Lvs spirally arranged with a phyllotaxis of $\frac{2}{5}$, usually simple, broad and often deeply cut; tendrils lateral, simple or branched, and may twist to right or left at different points along their axes. Fls axillary, solitary or in inflorescences, unisexual, monoecious or dioecious, actinomorphic; calyx adnate to ovary, 5-lobed; corolla polypetalous or sympetalous of 5 petals; stamens usually 3, mostly syngenesious with contorted anthers, 1 anther always 1-locular, others 2-locular; ovary inferior, placentas often 3, parietal, ovules many, stigmas thick. Fr. typically a fleshy berry-like structure with a hard rind (pepo). Seeds without endosperm, usually flattened.

USEFUL PRODUCTS

The cultivated species dealt with below belong to the genera *Benincasa*, *Citrullus*, *Cucumis*, *Cucurbita*, *Lagenaria*, *Luffa*, *Momordica*, *Sechium* and *Trichosanthes*. The following species of less importance are also grown in the tropics:

Cyclanthera pedata Schrad. var. *edulis* Schrad., a strong-smelling vigorous glabrous annual vine, native of Mexico, is sometimes cultivated for its ovoid yellowish edible frs, about 5 cm long.

Sicana odorifera (Vell.) Naud., a glabrous perennial with solitary fls and orange-crimson to purple-black frs, fragrant, 30–60 cm long, is cultivated in parts of South America for its ornamental frs, which are sometimes eaten.

GENERAL REFERENCES

JEFFREY, C. (1962). Notes on Cucurbitaceae, including a Proposed New Classification of the Family. *Kew Bull.*, **15**, 337–371.

WHITAKER, T. W. and DAVIS, G. N. (1962). *Cucurbits*. London: Leonard Hill.

KEY TO THE PRINCIPAL CULTIVATED GENERA

A. Fr. a fleshy many-seeded pepo with hard firm rind
  B. Mature fr. covered with white wax .... *Benincasa*
  BB. Mature fr. not covered with white wax
    C. Staminate fls in racemes .... *Luffa*
    CC. Staminate fls solitary or fasciculate
      D. Fls white .... *Lagenaria*
      DD. Fls lemon yellow to deep orange
        E. Lvs pinnatifid .... *Citrullus*
        EE. Lvs deeply or shallowly lobed, not pinnatifid
          F. Corolla gamopetalous, 5-partite to middle *Cucurbita*
          FF. Corolla rotate, deeply 5-partite to, or almost to base .... *Cucumis*
AA. Fr. fleshy, but not a pepo
  B. Fr. pyriform, 8–20 cm long, with a single large seed *Sechium*
  BB. Fr. filiform, tuberculate, 5–25 cm long, many seeded.. *Momordica*
  BBB. Fr. long-tapering, very slender, 30–150 cm long, many seeded .... *Trichosanthes*

Benincasa Savi (x = 12)

Two annual spp. in tropical Asia, one of which, *B. hispida* (q.v.) is cultivated.

Benincasa hispida (Thunb.) Cogn. (2n = 24) WAX OR WHITE GOURD

Syn. = *B. cerifera* Savi

USES

The young and mature fruits are boiled as a vegetable. The ripe fruits are cut into pieces and candied with sugar. The seeds are fried and eaten. The young leaves and flower buds are used as spinach.

ORIGIN AND DISTRIBUTION

B. hispida occurs wild in Java. It is now cultivated throughout tropical Asia and has been introduced into many tropical countries, but is little grown outside Asia.

ECOLOGY

The wax gourd is better suited to the medium dry areas of the lowland tropics and grows better in northern than in southern Malaya.

STRUCTURE

Robust, annual, hispid, climbing herb to several m. Stems stout, light green with scattered rough hairs; tendrils 2–3-fid. Lvs large, petiole 10–20 cm. long, lamina 10–25 × 10–20 cm, 5–11 angled or lobed, dentate, base cordate. Fls monoecious, solitary in lf. axils, large, yellow, 6–12 cm diam.; calyx with 5 lobes; corolla rotate, petals 5 almost free; male fls peduncle 5–15 cm long; female fls short stalked, ovary densely hairy, 2–4 cm long; stigmas 3, curved, bilobed. Fr. large, 20–35 × 15–20 cm, nearly spherical to long-oblong, at first hairy usually becoming glabrous, dark green, covered with white, easily removable wax; flesh white, in middle spongy with numerous seeds. Seeds flat, smooth, with narrow base, whitish or pale brown, 1–1·5 × 0·5–0·7 cm.

CHEMICAL COMPOSITION

The fruits contain: water 96 per cent; protein 0·4 per cent; fat 0·1 per cent; carbohydrate 3·2 per cent; mineral matter 0·3 per cent. The seeds yield a pale yellow oil.

HUSBANDRY

Wax gourds are grown throughout the East for local consumption, but seldom on a large scale. They are usually planted on mounds and the fruits are harvested after 4–5 months. The heavy fruits need strong supports either by stout poles or by trailing over roofs or trees.

Citrullus Schrad. ex Eckl. & Zeyh. (x = 11)

Four species, native to tropical and subtropical Africa, of which one is also native to tropical Asia. Climbing or trailing with long runners and foetid musky odour. The pinnately-lobed lvs distinguish the genus from *Cucumis* and *Cucurbita.*

USEFUL PRODUCTS

C. lanatus (q.v.) is the watermelon.

C. colocynthis (L.) Schrad., which grows wild from India to tropical Africa, produces the drug colocynth, official in the British Pharmacopoeia, and is the dried pulp of unripe fully grown fruits, the size of cricket balls. The violent purging action is due to a bitter alkaloid and a resin.

Citrullus lanatus (Thunb.) Mansf. (2n = 22) WATERMELON

Syn. *C. vulgaris* Schrad.; *Colocynthis citrullus* (L.) O. Ktze.

USES

The sweet juicy pulp of the ripe fruit is eaten fresh. It is a valuable alternative to drinking water in desert areas. The dried parched seeds are chewed,

---

Fig. 13. **Benincasa hispida :** WAX GOURD. **A**, leafy shoot (× ⅓); **B**, male flower (× ½); **C**, male flower in longitudinal section (× ½); **D**, female flower in longitudinal section (× ½); **E**, fruit (× ⅓); **F**, fruit in longitudinal section (× ⅙).

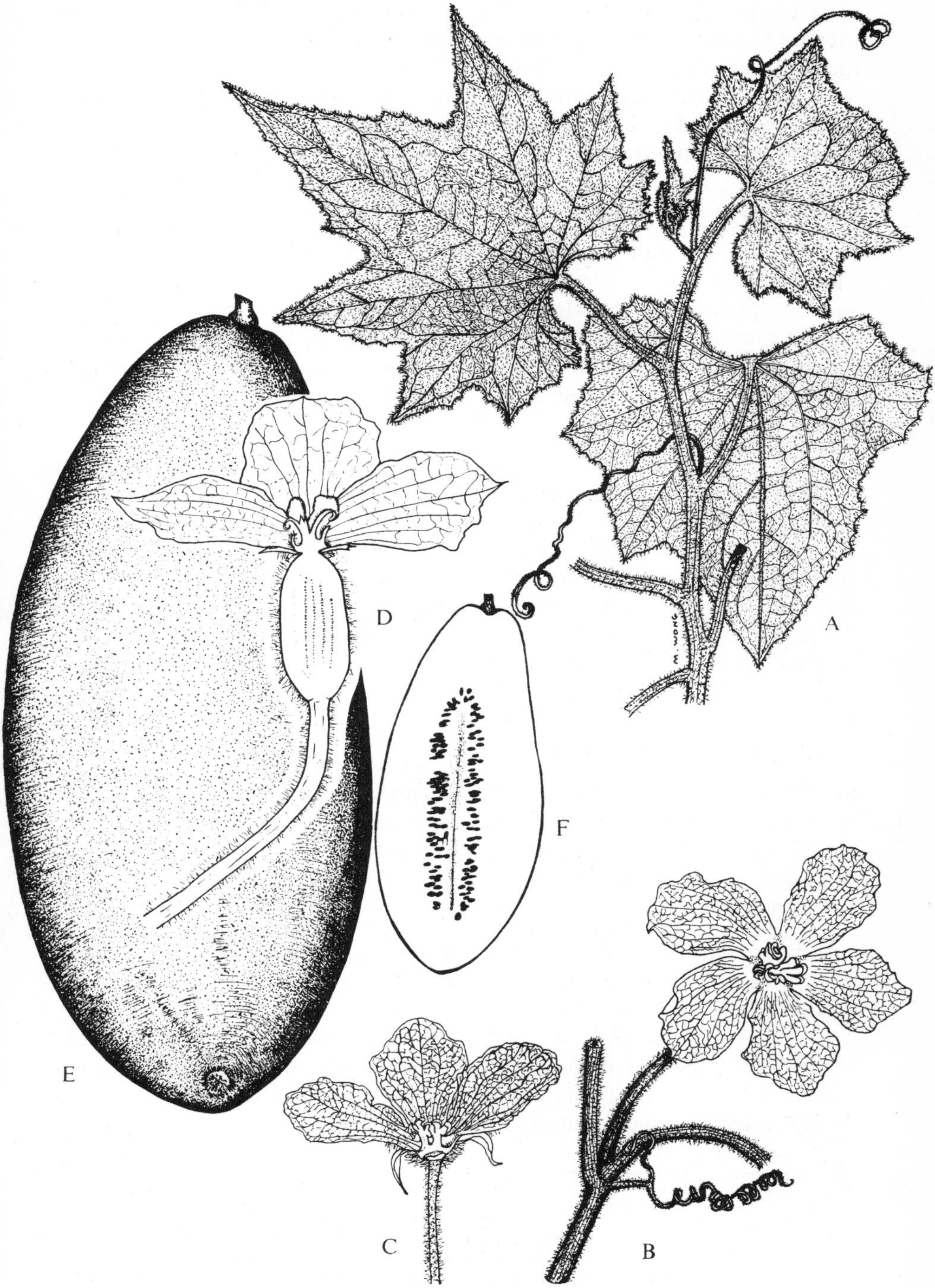
A
D
F
E
C
B

particularly in southern China. They are sometimes ground and baked into bread. Oil extracted from the seeds is used in cooking and as an illuminant; the seed-cake can be used as livestock feed.

C. lanatus var. *fistulosus* (Stocks) Duthie & Fuller ($2n = 24$) is grown in India for its small round fruits, the size of small turnips, which are cooked as a vegetable and are also made into preserves and pickles.

ORIGIN AND DISTRIBUTION

C. lanatus is a native of tropical and subtropical Africa, where a form with small, somewhat bitter fruits is spontaneous in some areas. It is of ancient cultivation in the Mediterranean region and was grown in Egypt in remote times. It reached India in prehistoric times, but was not taken to China until the 10th or 11th century A.D. It was taken to the New World in post-Columbian times and is now widely spread throughout the tropics.

CULTIVARS

There is great variation within the species ranging from small hard bitter inedible fruits to large succulent sweet fruits. The cvs vary in vigour, earliness, and productivity; shape, colour and markings of fruits; thickness and texture of rind; colour, texture, flavour and sugar-content of flesh; size, colour and number of seeds. There are a large number of named cvs in the United States and elsewhere; those recommended in Trinidad are 'Dixie Queen' and 'Kleckley Sweet'.

ECOLOGY

Watermelons are grown throughout the tropics and subtropics, but do best in the hot drier areas with an abundance of sunshine. They are killed by frost. They grow best on fertile sandy soil, particularly on sandy river banks. They are fairly drought-resistant and will not stand waterlogging.

STRUCTURE

Slender, hairy, monoecious annual, often sprawling over the ground. Root system very extensive and superficial. Stems rather thin, angular, grooved, 1·5–5 m long, with long white hairs. Tendrils 2–3-fid. Lvs 5–20 × 2–12 cm, 3–4 pairs pinnate lobes, lobes again divided and toothed with broad apices; petiole 1–10 cm long, shorter than lamina. Fls unisexual, solitary, axillary, usually more male fls than female; calyx 5-lobed; corolla deeply 5-partite, pale yellow, 2·5–3 cm diam; petals 1–1·5 cm long;

Fig. 14. **Citrullus lanatus**: WATERMELON. **A**, flowering shoot (× ⅓); **B**, male flower (× ½); **C**, male flower in longitudinal section (× 1½); **D**, female flower in longitudinal section (× 1½); **E**, fruit (× ¼); **F**, fruit in longitudinal section (× ⅛).

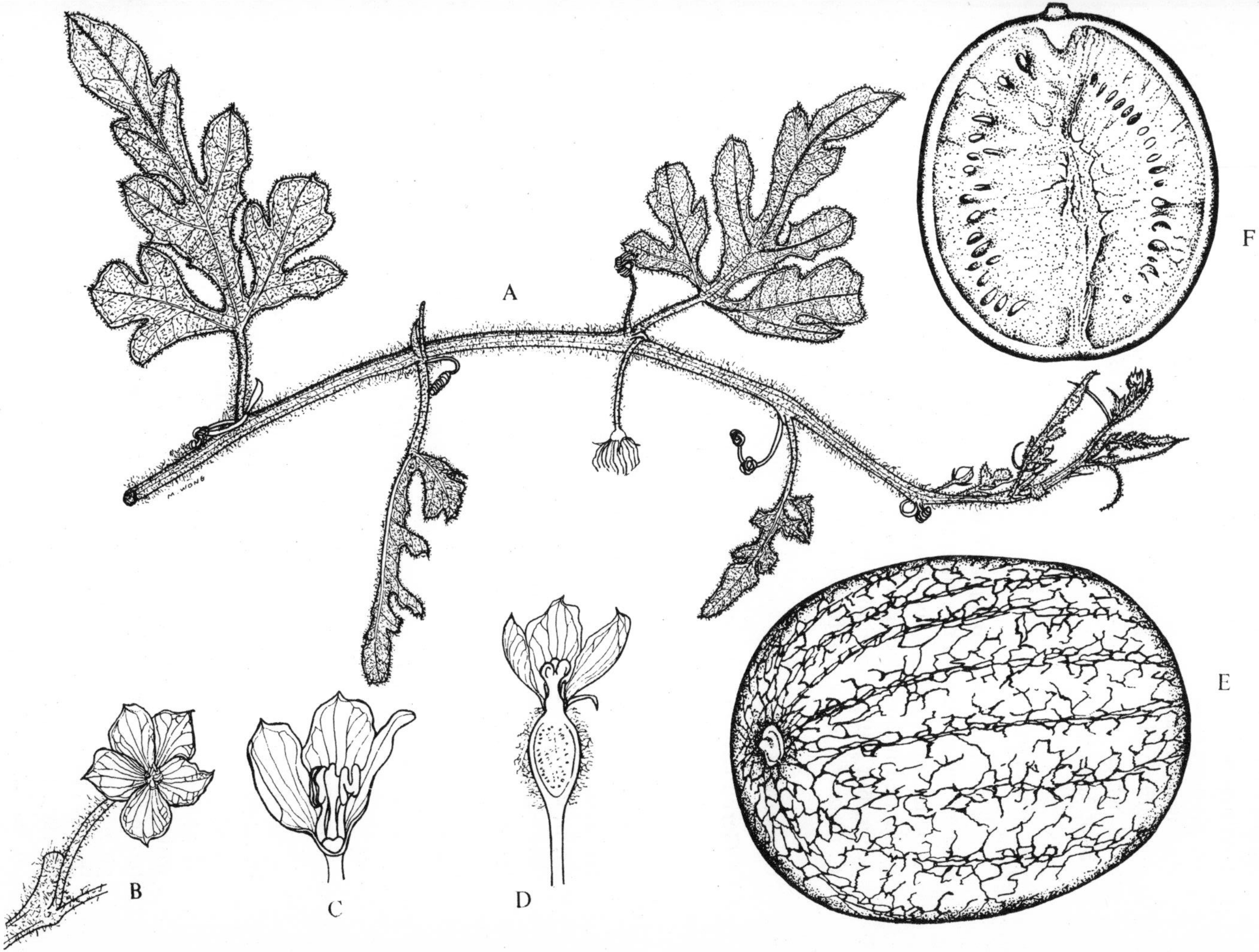
A
B
C
D
E
F
M. WONG

stamens 3–5 with short free filaments, anthers almost free; female fl. with 3 staminodes, ovary ovoid with woolly hairs, style 0·5 cm long with 3 lobes. Fr. globose or oblong, varying greatly in size up to 60 cm or more long; rind mostly glabrous, green or cream, striped or mottled green, hard, but not durable; flesh red, green, yellow or whitish, usually sweet, many-seeded. Seed white, black, reddish, yellow, flat, smooth, 0·6–1·5 × 0·5–0·7 cm; about 15 seeds per g.

POLLINATION

The watermelon is pollinated by insects, particularly honey-bees. Hand-pollination assists fruit setting. It has been shown that pollen germination is better and pollen-tube growth faster when the pistillate flowers are pollinated immediately after anthesis. Perfect self-fertile flowers sometimes occur, but self-pollination in these is rare unless aided by bees or by hand. *C. lanatus* is not cross-fertile with other genera or species.

GERMINATION

Germination is epigeal.

CHEMICAL COMPOSITION

The edible portion, which constitutes 60 per cent of the whole fruit, contains: water 93·4 per cent; protein 0·5 per cent; fat 0·1 per cent; carbohydrate 5·3 per cent; fibre 0·2 per cent; ash 0·5 per cent; vitamin A 70 mcg. The seeds contain 20–45 per cent of a yellowish edible semi-drying oil and 30–40 per cent protein; they are rich in the enzyme urease.

PROPAGATION

Grown from seed which is usually planted *in situ*.

HUSBANDRY

Watermelons are usually planted on mounds 8–10 ft apart, 6–8 seeds being sown in each mound at a depth of 1–2 in., later being thinned to 1–2 plants per mound. The amount of seed required to plant 1 acre is 1–2 lb. The young fruits are usually thinned to two per plant or per mound. A little straw is usually placed under the developing fruits. The fruits are ready for harvesting 4–5 months after sowing and ripe fruits can be recognized by giving a dull thud when tapped and the withering of the tendrils. In harvesting, as much peduncle as possible should be left on the fruit and it should be cut with a sharp knife. Watermelons require careful handling as they are easily damaged and they cannot be stored for more than 2–3 weeks. A good crop should give 500 marketable fruits per acre with an average weight of 20 lb per fruit.

MAJOR DISEASES AND PESTS

The most serious diseases are: Fusarium wilt, *Fusarium oxysporum* Schlecht. ex Fr. f. *niveum* (E.F.Sm.) Snyder & Hansen; anthracnose, *Colletotrichum orbiculare* (Berk. & Mont.) Arx; downy mildew, *Pseudoperonospora cubensis* (Berk. & Curt.) Rostov.; powdery mildew, *Erysiphe cichoracearum* DC.; watermelon mosaic virus.

The most serious pests are: root-knot nematode, *Meloidogyne* spp.; cucumber beetles, *Diabrotica* spp. and *Acalymma trivittata* Mann.; *Anasa scorbutica* (F.); *Pycnoderes quadrimaculatus* Guerin; *Aphis gossypii* Glover; *Margaronia* spp. The last four are major pests of cucurbits in the West Indies.

IMPROVEMENT

Breeders in the United States have paid attention to the following characters: vigour, earliness and yield; rind as thin as possible and yet not damaged by handling; deep-red crisp flesh with high sugar content and freedom from stringiness; as few seeds as possible, darker coloured seeds preferred; resistance to *Fusarium* wilt and anthracnose. Cvs are very heterozygous. Inbreeding does not seem to affect vigour.

PRODUCTION

Watermelons are grown throughout the drier areas of the tropics for local consumption. They are grown in southern Europe and the central and southern United States for local and northern markets.

REFERENCES

BEATTIE, J. H. and DOOLITTLE, S. P. (1951). Water-melons. *U.S. Dep. Agric. Fmrs' Bull.*, **1394**.

PARRIS, G. K. (1949). Water-melon Breeding. *Econ. Bot.*, **3**, 193–212.

Cucumis L. (x = 7, 12)

About 40 spp. of tendril-climbing or trailing annuals and perennials, mostly native of tropical Africa, monoecious or rarely dioecious. Lvs entire or palmately lobed; tendrils simple; staminate fls usually in clusters; corolla rotate, deeply 5-lobed, yellow; stamens 3, free; pistil with 3–5 placentas and an equal number of stigmas, many ovuled; frs usually fleshy, indehiscent, globular to elongated.

USEFUL PRODUCTS

Two spp. are widely grown: *C. melo* (q.v.) for fresh fruits and *C. sativus* (q.v.) as a salad vegetable. *C. anguria* (q.v.) is occasionally grown for pickling and as a vegetable.

KEY TO THE CULTIVATED SPECIES

A. Fruit mostly with blunt prickles, spiny, or muricate
  B. Lvs deeply-lobed with rounded sinuses, frs short ... *C. anguria*
  BB. Lvs not deeply-lobed; if lobed shallowly lobed with acute sinuses; frs usually long .... *C. sativus*
AA. Fruit smooth or somewhat scaly or netted, glabrous or somewhat pubescent .... *C. melo*

Cucumis anguria L. (2n = 24) WEST INDIAN GHERKIN

USES

The fruits are mostly used in pickles. They are also eaten as a cooked vegetable and are used in curries. They should not be confused with small cucumbers which are now mainly used for gherkins.

ORIGIN AND DISTRIBUTION

Meeuse (1958) has shown that *C. anguria* is a cultigen descended from a non-bitter mutant of the African wild species, *C. longipes* Hook., which normally has bitter fruits. The 2 spp. cross readily and the F_1 and F_2 generations are highly fertile. It was probably introduced to the New World through the early slave trade, and in isolation and with a certain amount of selection the present cultigen has been developed.

STRUCTURE

Slender, trailing, monoecious annual herb with stiff hairs throughout. Stem angled with small simple tendrils. Lvs long-petioled; lamina, 4–9 cm long, deeply 3–5-lobed with rounded sinuses. Fls small, yellow, 3–10 mm in diameter, staminate fls in clusters, pistillate fls solitary on long slender hirsute peduncles. Fr. oval to oblong, 4–5 cm long, covered with long sharp glistening hairs on warty pimples; rind pale green turning to ivory on ripening; flesh greenish. Seeds many, small, white, smooth, 3–5 mm long.

PRODUCTION

C. anguria is grown in Brazil and occasionally in the West Indies.

REFERENCE

MEEUSE, A. D. J. (1958). The possible origin of *Cucumis anguria* L. *Blumea, Suppl.* **4**, 196–204.

Fig. 15. **Cucumis anguria**: WEST INDIAN GHERKIN. **A**, flowering shoot ($\times \frac{1}{2}$); **B**, male flower in longitudinal section ($\times 10$); **C**, female flower in longitudinal section ($\times 10$); **D**, young fruit ($\times \frac{1}{4}$).

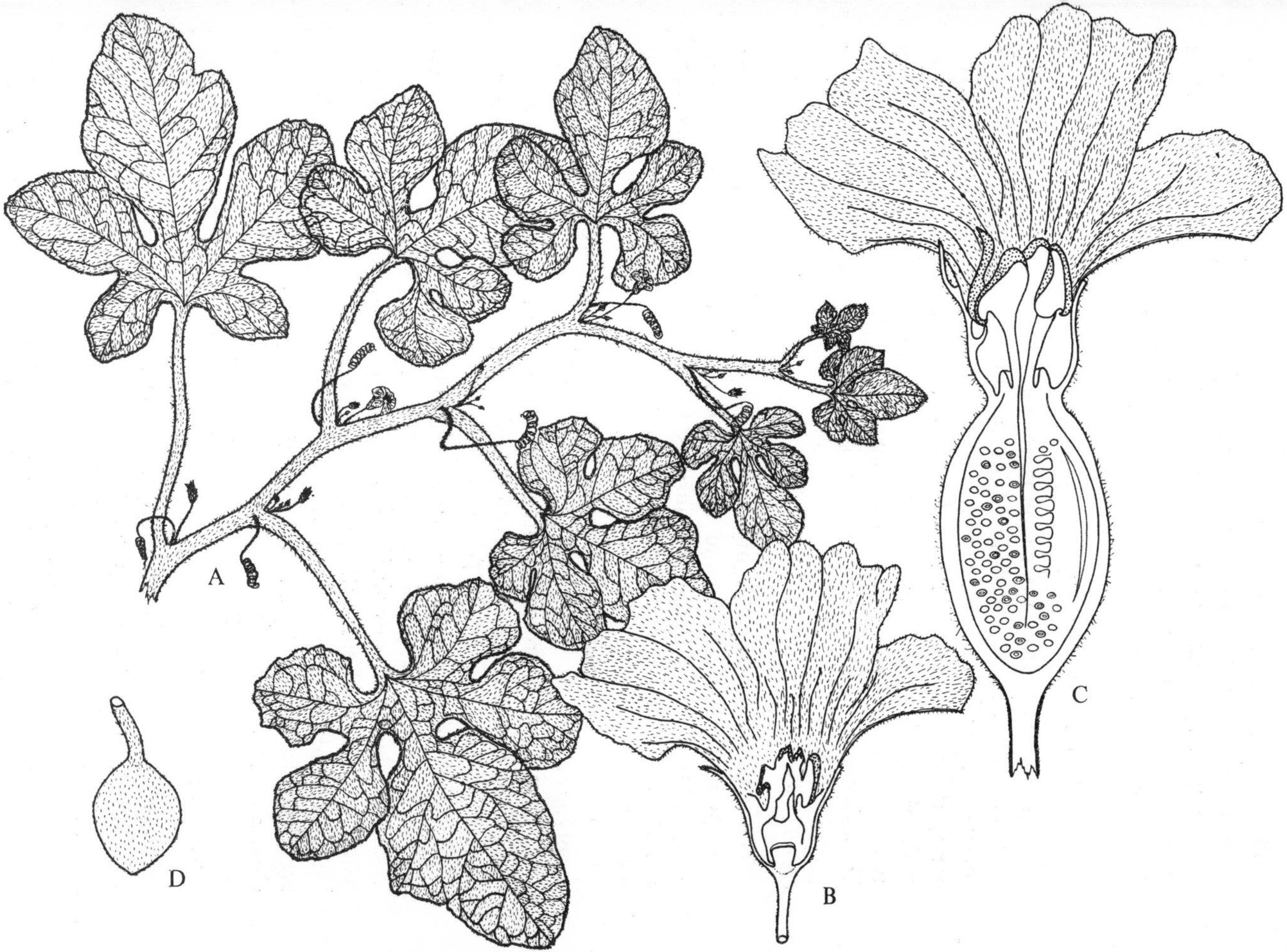
A
B
C
D

Cucumis melo L. (2n = 24) MELON

USES

The fresh flesh of the fruit is eaten out of the rind after removal of the seeds as dessert, often with a little sugar and sometimes with a little powdered ginger. The fruits of some cvs are used for preserves and as vegetables. Some are grown as curiosities. The seeds are edible and are chewed and they yield an edible oil.

ORIGIN AND DISTRIBUTION

The place of origin of the melon is not known with certainty, but as the wild species of *Cucumis* occur in Africa, it is likely that it originated in that continent. It does not appear to have been known to the ancient Egyptians, nor to the Greeks and seems to have reached Europe towards the decline of the Roman Empire. It was introduced into Asia at a comparatively late date and well-developed secondary centres of variation now occur in India, China, Persia and southern Russia. It is now world-wide.

CULTIVARS

The classification of this highly polymorphic species is in a confused state. A number of species and varieties have been erected from time to time, but this hardly seems to be justified as all the forms hybridize readily and there are many intermediate forms. The most commonly cultivated types are:

(1) Cantaloupe melon of Europe with thick, scaly, rough rind, often deeply grooved.

(2) Musk-melon, grown mainly in the United States, with smaller fruits, rinds finely netted to nearly smooth, and very shallow ribs. The cv. 'Smith's Perfect' is recommended in Trinidad.

(3) Casaba or winter melon with large fruits, maturing late and of good storage quality; rind usually smooth, often striped or splashed and yellow in colour; flesh firm with little musky odour or flavour. Cv. 'Honey Dew' with ivory skin and green flesh is included here.

(4) A number of types, often with elongate fruits resembling cucumbers, are grown in India, China and Japan and are used as vegetables, e.g. the Conomon of Japan.

ECOLOGY

Melons are now grown from the tropics to the temperate regions. They are killed by frost. The crop requires plenty of sunshine and heat and does

---

Fig. 16. **Cucumis melo :** MELON. **A,** flowering shoot (×½); **B,** male flower in longitudinal section (×3); **C,** female flower in longitudinal section (×3); **D,** fruit (×⅓); **E,** fruit in longitudinal section (×⅓).

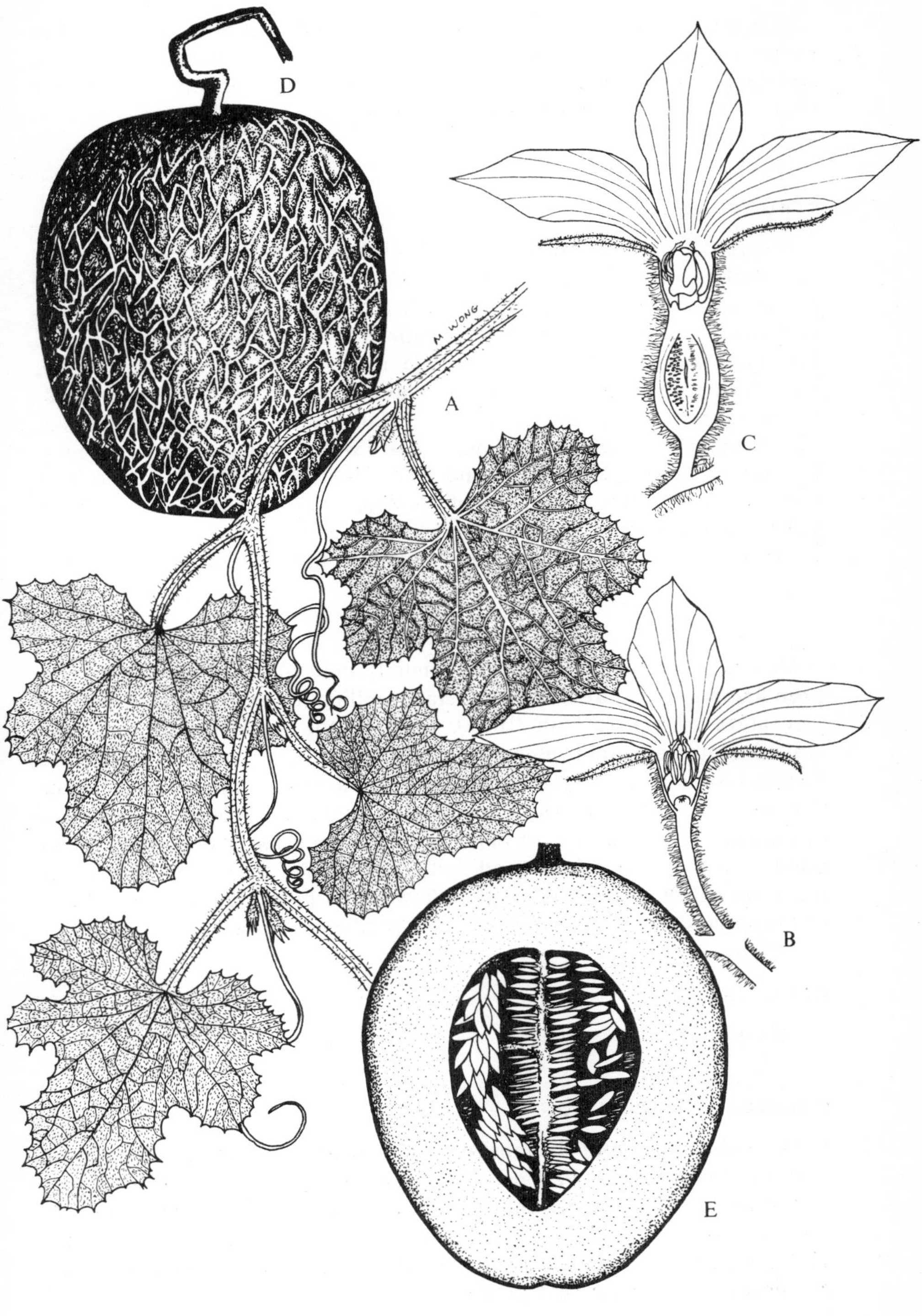
D
M WONG
A
C
B
E

well in the hot dry tropics, or during the dry season. It does not do well in the humid tropics or in a damp atmosphere when foliage diseases are serious and the fruits are of poor quality. It grows best on rich loamy soils, as near neutral as possible, and will not tolerate high acidity. It is often grown under irrigation.

STRUCTURE

A variable, trailing, softly hairy annual. Vines are monoecious or andromonoecious. Root system large and superficial. Stems ridged or striate. Lvs orbicular or ovate to reniform, angled or shallowly 5–7-lobed, 8–15 cm in diameter, dentate, base cordate; petiole 4–10 cm long; tendrils simple. Fls staminate and clustered, pistillate and solitary, or hermaphrodite, 1·2–3·0 cm in diameter, yellow, on short stout pedicels; calyx 5-lobed, 6–8 mm long; corolla deeply 5-partite, petals round, 2 cm long; stamens 3, free, connectives of anthers prolonged; pistil with 3–5 placentas and stigmas. Fr. very variable in size, shape and rind, globular or oblong, smooth or furrowed, rind glabrous and smooth to rough and reticulate, pale to deep yellow, yellow-brown, or green, flesh yellow, pink or green, many seeded. Seeds whitish or buff, flat, smooth, 5–15 mm long. About 30 seeds per g.

POLLINATION

Melons are pollinated by insects, mainly honey-bees. The male flowers open first and are more numerous. Hand-pollination assists setting, but the percentage set ranges from 10–50. Most of the musk-melon cvs in the United States are andromonoecious, *i.e.* with male and hermaphrodite flowers on the same plant, and the proportion of self-pollination is higher than with the British cvs which are mainly monoecious. The andromonoecious condition is regarded as more primitive from the evolutionary viewpoint, and the monoecious condition was achieved by a single dominant mutation which changed the perfect flower to a pistillate one. All sections of the species are cross-fertile, but interspecific crosses are sterile.

GERMINATION

This is epigeal.

CHEMICAL COMPOSITION

The edible portion of the musk-melon, which constitutes 45–80 per cent of fruit, contains; water 92·1 per cent; protein 0·5 per cent; fat 0·3 per cent; carbohydrate 6·2 per cent; fibre 0·5 per cent; ash 0·4 per cent; vitamin A 350 mcg. 'Honey Dew' melons contain: water 87·4 per cent; carbohydrate 11·1 per cent. The edible seed kernel contains approximately 46 per cent of a yellow oil and 36 per cent protein.

PROPAGATION

Melons are grown from seed which is planted at a depth of ½–1½ in. with a seed-rate of about 2 lb per acre.

HUSBANDRY

Melons may be planted in pits, on the flat, on hills or ridges at a distance of 4–5 ft apart. A plant should be permitted to carry about 4 fruits. Melons respond well to organic manures which should be applied at 10–15 tons per acre. Artificial fertilizers may be applied at the rate of 500–600 lb per acre of a 4 : 16 : 4 NPK mixture. Yields per acre are about 8,000 lb. They are usually left on the plant until fully ripe, unless required for distant markets, and they are ready for harvest 3–4 months after planting. Ripeness can be ascertained in some cvs which show an abscission crack where the fruit is attached to the peduncle.

MAJOR DISEASES AND PESTS

The most serious diseases and pests are the same as *Citrullus lanatus* (q.v.). The form of Fusarium wilt is *F. oxysporum* f. *melonis* (Leach & Currence) Syn. & Han. It is also attacked by the musk-melon rust, *Macrosporium cucumerinum* Ell. & Ev. Melons are also attacked by cucumber and melon mosaic viruses. In Malaya and other parts of the tropics a fly, *Bactrocera cucurbitae*, causes young fruits to fall by tunnelling in the pedicel.

IMPROVEMENT

Many improved cvs have been raised in the United States. Attention has been paid to high density of the fruit which is usually associated with thick flesh and a small seed cavity. Resistance has been obtained to *Fusarium* wilt, powdery and downy mildews, and melon aphid. Little or no work has been done on improvement for tropical conditions.

PRODUCTION

Melons are grown for local consumption in the hot dry tropics, but they are more extensively grown in warm temperate countries.

REFERENCES

BEATTIE, J. H. and DOOLITTLE, S. P. (1951). Musk-melons. *U.S. Dep. Agric. Fmrs' Bull.*, **1468**.

DAVIS, G. N., WHITAKER, T. W. and BOHN, G. W. (1953). Production of Musk-melons in California. *Calif. Univ. Agric. Exp. St. Circ.*, **429**.

Cucumis sativus L. (2n = 14) **CUCUMBER**

USES

The fruits are eaten as a salad vegetable before they are fully mature and are usually peeled. In the East they are often eaten as a cooked vegetable. The young fruits, usually of small-fruited cvs, are pickled as gherkins; the smallest for mixed pickles and small to medium sized fruits for dill pickles. The seed kernels are occasionally eaten and yield an edible oil. The young leaves are eaten as salad or cooked as spinach in Indonesia and Malaya.

ORIGIN AND DISTRIBUTION

It has been suggested that the cultigen, *C. sativus*, originated in northern India, where the related *C. hardwickii* Royle occurs wild, although this might be a 'weedy' form of *C. sativus* which has escaped from cultivation. The cucumber with 7 pairs of chromosomes and several distinct morphological features stands apart from other species with 12 pairs of chromosomes which are indigenous to tropical Africa. It was in Egypt at the time of the XIIth Dynasty and was known to the Greeks and the Romans. It was in China by the 6th century A.D. It has now spread throughout the world.

CULTIVARS

Numerous cvs have been developed in many parts of the world differing in size and shape of fruits; thickness, spininess and colour of rind, ranging from whitish-green to dark-green, others turning yellow or rusty-brown when mature. The cvs can be divided into the following classes:

(1) Field cucumbers with white or black spines. The cv. 'Burpee Hybrid' is recommended in Trinidad.

(2) English or forcing cucumber with long fruits up to 90 cm long which are nearly spineless. They set fruits parthenocarpically.

(3) Sikkim cucumber of India with reddish-brown fruits.

(4) Pickling cucumbers with small fruits used for production of gherkins.

In India cvs are divided into hot weather cvs with small egg-shaped, dark-green fruits and rainy season cvs with much longer fruits. Locally acclimatized cvs in the tropics do better than those introduced from temperate regions.

ECOLOGY

Cucumbers require a warm climate, but not as hot as for melons. In cool temperate countries they are usually grown under glass. The optimum temperature for growth is about 85°F and the optimum night temperature 65–70°F. They need a fair amount of water without too high an atmospheric humidity which facilitates disease, particularly downy mildew. They cannot stand waterlogging. Although they can be grown on a variety of soil, they do best on a well-manured sandy loam with a pH of 6·5–7·5.

STRUCTURE

Trailing or climbing monoecious annual herbs with stiff bristly hairs. Root system extensive and largely superficial. Stems strongly 4-angled; tendrils unbranched. Lvs long-petioled, triangular ovate, rough, 3–5-angled or shallowly lobed with acute sinuses, dentate, palmately 5–7-nerved, 7–20 cm long, base deeply cordate, apex acuminate; petiole 5–15 cm long. Fls 3–4 cm in diameter, males predominating, borne in axillary clusters on slender pedicels, females usually solitary, axillary, on stout peduncles; calyx 5–10 mm long, with 5 narrow lobes; corolla to 2 cm long, bell-shaped, yellow, deeply 5-partite, hairy, wrinkled; stamens 3, filaments free, ending in thickened connective with anthers on outer face; female fls with 3 united inferior carpels, simple style and 3 thick stigmas. Fr. pendulous, variable in shape and size (see CULTIVARS above), nearly globular to oblong and elongated with scattered spinous tubercles and warts, particularly when young; flesh pale-green with characteristic cucumber odour, many seeded except in parthenocarpic cvs, seeds flat, white, 8–10 × 3–5 mm. 50 seeds per g.

POLLINATION

Bees are the main pollinating agents. The female flowers develop later than the more numerous male flowers and have a ring of nectiferous tissue surrounding the style. An abundance of light tends to increase the number of staminate flowers; reduction in light increases the number of pistillate fls. High temperatures and long days tend to keep the plants in the staminate phase. English or forcing cucumbers are parthenocarpic and seedless; if they are pollinated they produce seeds which spoil them for the market. In the case of non-parthenocarpic cvs hand pollination assists fruit setting.

GERMINATION

The best and quickest germination, which is epigeal, is obtained at 80°F.

CHEMICAL COMPOSITION

The edible portion, which is about 80 per cent of the fruit contains: water 95·0 per cent; protein 0·7 per cent; fat 0·1 per cent; carbohydrate 3·4 per cent; fibre 0·4 per cent; ash 0·4 per cent. The seed kernels contain approximately 42 per cent oil and 42 per cent protein.

PROPAGATION

By seed.

HUSBANDRY

Cucumbers are planted on hills, 3–4 ft apart, with several seeds per hill and thinned to 2–3 plants; or in rows 4–5 ft apart thinned to 1 ft between plants. The seed rate is about 2 lb per acre. The crop responds well to organic manures at the rate of 10–15 tons per acre and mixed fertilizers can be applied at the rate of 600 lb per acre of a 4 : 16 : 4 NPK mixture. The plants are sometimes staked and the tip of the main stem may be nipped off to encourage branching. In California 15 in. of irrigation water is recommended. Cucumbers are harvested before they are fully mature, unless required for seed, and picking usually begins about 2 months after sowing and thereafter every few days. Yields are of the order of 4,000–6,000 lb per acre. Pickling cucumbers are picked when they are 2–6 in. long.

MAJOR DISEASES AND PESTS

These are the same as given for *Citrullus lanatus* (q.v.) and *Cucumis melo* (q.v.).

IMPROVEMENT

Breeding has been largely confined to temperate countries and cvs have been obtained which are resistant to downy mildew and mosaic.

PRODUCTION

Cucumbers are grown in the tropics for local sale and consumption.

REFERENCE

DAVIS, G. N. and HALL, B. J. (1958). Cucumber Production in California. *Calif. Univ., Agr. Exp. Sta. Ext. Serv. Manual*, **24**.

Cucurbita L. (x = 10)

About 25 spp., all New World in origin, many of them xerophytic and indigenous to the arid areas of northern Mexico and south-western United States. Monoecious annual or perennial scandent herbs; stems long-running or short and bushy, more or less prickly, angled or furrowed, often rooting at nodes; tendrils branched; lvs simple, alternate, shallow to deeply lobed; fls large, solitary, yellow; calyx and corolla campanulate, latter 5-partite to middle of length; male fls long-pedunculate, stamens 3, anthers connivent into long twisted body; pistillate fls short-pedunculate, ovary oblong or discoid, inferior, unilocular with 3–5 placentas; style thick, stigmas each 2-lobed; fr. a pepo, seeds numerous.

There is considerable confusion in most tropical countries as to the identity of the cultivated spp. and this is made worse by the fact that the American and English names, squash, pumpkin and marrow are often

---

Fig. 17. **Cucumis sativus**: CUCUMBER. **A**, flowering shoot (×⅓); **B**, male flower in longitudinal section (×3); **C**, female flower in longitudinal section (×2); **D**, fruit (×½); **E**, fruit in transverse section (×½).

C
A
B
D
E

applied indiscriminately to their fruits. In this account the diagnostic features and origin and distribution of the 4 annual spp. grown for their edible fruits are given, but, as their uses, husbandry, etc., are much the same, the spp. are then grouped together.

USEFUL PRODUCTS

The immature fruits are eaten as a fresh vegetable, either stewed, boiled or fried. Mature fruits are used for baking, making jam and for pies. They are also canned, mainly for pie-stock. They are used for feeding to livestock. In some countries, particularly Asia, the seeds, which yield an edible oil, are roasted and eaten. The leaves and flowers are cooked and eaten as a pot-herb in Asia and Africa; they may be dried and stored.

The culinary names may be defined as follows:

(1) Summer squash: immature fruits of *C. pepo* used as a table vegetable.

(2) Winter squash: mature fruits of *C. maxima*, *C. mixta*, *C. moschata* and *C. pepo* used as a table vegetable, for baking, in pies and for making jam; also as a livestock feed. The flesh is usually fine-grained and mild-flavoured and is thus suitable for baking. The fruits can be stored for 6 months or more without deterioration.

(3) Pumpkins: the mature fruits of *C. maxima*, *C. mixta*, *C. moschata* and *C. pepo* used in pies and as forage; the flesh is somewhat coarse and strongly flavoured and is not usually served as a table vegetable.

(4) Cushaw: mature fruits of certain cvs of *C. mixta* used for baking and for forage.

(5) English vegetable marrow: fruits of *C. pepo*, which are boiled immature as a vegetable, and in a mature state for jam, and for storage and cooking in the winter.

C. ficifolia Bouché, Malabar gourd or fig-leaf gourd, is a perennial sp., which has been cultivated since ancient times in the highlands of Mexico, Central and South America. The seeds are edible and the flesh is candied and also fermented to produce an alcoholic beverage. Remains of it have been found in a pre-maize, pre-ceramic horizon at Huaca Prieta, Peru, dated 4000–3000 B.C. It is a monoecious vigorous perennial with circular, usually lobed leaves, up to 25 cm in diameter, and globular or cylindrical fruits, 15–20 cm long, green with white stripes and blotches; the seeds are usually black.

The small fruits of *C. pepo* L. var. *ovifera* (L.) Alef., which are hard-shelled, bitter and inedible, in many shapes, sizes and colours, are grown for curiosity and ornament.

KEY TO THE ANNUAL EDIBLE SPECIES

A. Stems soft, round, moderately bristly; peduncle soft, terete, enlarged by soft cork .... *C. maxima*

AA. Stems hard, angular; peduncle basically angular, grooved.
- B. Stems and lvs with spiculate bristles; peduncle hard, sharply angular, grooved........................ *C. pepo*
- BB. Stems and lvs lacking bristles; peduncle hard, smoothly grooved, flared at fr. attachment .... *C. moschata*
- BBB. Stems and lvs lacking bristles; peduncle hard, greatly enlarged in diameter by hard cork, not flared at fruit attachment .... *C. mixta*

Cucurbita maxima Duch. ex Lam. (2n = 40) PUMPKIN, WINTER SQUASH

ORIGIN AND DISTRIBUTION

Seeds of *C. maxima* have been excavated in Peru and dated 1200 A.D., but no remains have been found in Mexico and Central America. From the little evidence available it appears that it was domesticated in South America and was not taken elsewhere until post-Columbian times. A related sp., *C. andreana* Naud., occurs wild in Argentina and Bolivia and hybridizes readily with *C. maxima*; it may be the prototype or a recent degenerate 'weedy' offshoot of *C. maxima. C. maxima* has now spread to most of the world.

STRUCTURE

Annual monoecious long-running vine or rarely bush, only slightly harsh to touch; stems soft, round in cross-section. Lvs not rigid, moderately upright, usually reniform, serrate, not lobed or only shallowly so, cordate with very deep sinus, occasionally with white blotches. Fls acute or obtuse in bud; sepal lobes linear; corolla bright yellow; androecium short, thick, columnar; stigmas small, yellow, smooth; mature peduncle soft, round in cross-section, becoming irregularly thickened with soft cork in most cvs, not enlarged at fr. attachment. Fr. variable, soft- or hard-shelled, dull or brightly coloured; flesh fine-grained, various shades of yellow; seeds plump, usually smooth, white or pale brown, not separating easily and cleanly from pulp; funicular attachment acute, asymmetrical; seed margin smooth, obtuse.

Two vars sometimes recognized:

(1) var. *maxima*—winter squashes, e.g., 'Mammoth', 'Hubbard', 'Delicious', 'Buttercup'.

(2) var. *turbaniformis* Alef.—turban squashes with ovary protruding considerably from receptacle.

Cucurbita moschata (Duch. ex Lam.) Duch. ex Poir (2n = 40)

PUMPKIN, WINTER SQUASH

ORIGIN AND DISTRIBUTION

There is good archaeological proof that *C. moschata* was widely distributed in both North and South America and occurs at early levels in both

Mexico (5000 B.C.) and Peru (3000 B.C.). It is generally thought to have been domesticated first in Central America or Mexico so that an early movement from north to south is likely. As it will tolerate hotter conditions than the other cultivated spp. of *Cucurbita*, it is the most widely grown throughout the tropics of both hemispheres today.

STRUCTURE

Annual, monoecious, long-running vine, soft pilose, not harsh; stems moderately hard, round or smoothly 5-angled. Lvs large, shallowly lobed; occasionally with white blotches, to 20 × 30 cm, petiole 12–30 cm long. Fls acuminate in bud; sepals large, flat, often foliaceous lobes to 5 cm long; corolla yellow to yellow-orange, 10–12 cm long; androecium long and slender, 4 cm long, columnar; stigmas large, 2 cm long, bright orange or green; mature peduncle hard, smoothly 5-angled without cork development, often greatly enlarged at fr. attachment. Fr. variable, not hard-shelled, dull in colour; flesh yellow to dark orange, fine- to coarse-grained with gelatinous fibres; seeds plump and fat, dingy white to dark brown, separating from pulp readily and cleanly; funicular attachment obtuse, slightly asymmetrical; seed margin scalloped, irregular in outline, often darker coloured than rest of testa.

Cvs include 'Butternut', 'Kentucky Field', 'Sugar', 'Winter Crookneck'.

Cucurbita mixta Pang. (2n = 40) PUMPKIN, WINTER SQUASH

Originally included in *C. moschata.* It was described from material collected in Mexico and Central America. It is separated from other spp. of *Cucurbita* by sterility barriers which are effective enough to maintain its identity. It is intolerant of cool temperatures.

ORIGIN AND DISTRIBUTION

C. mixta is not of such ancient cultivation as *C. maxima* and *C. pepo.* Material from Mexico has been dated 100–760 A.D. It seems to have been widely distributed in northern Mexico and the south-western United States in pre-Columbian times.

STRUCTURE

Annual monoecious vine, pilose not harsh; stems hard, 5-angled. Lvs large, cordate, shallowly to moderately lobed, usually with white blotches. Fl. acuminate in bud; sepals long, linear; corolla yellow to orange-yellow; androecium long, slender, columnar; stigmas large, bright orange to

---

Fig. 18. **Cucurbita moschata : PUMPKIN. A,** flowering shoot (× ⅓); **B,** male flower in longitudinal section (× ½); **C,** female flower in longitudinal section (× ½); **D,** young fruit (× ½).

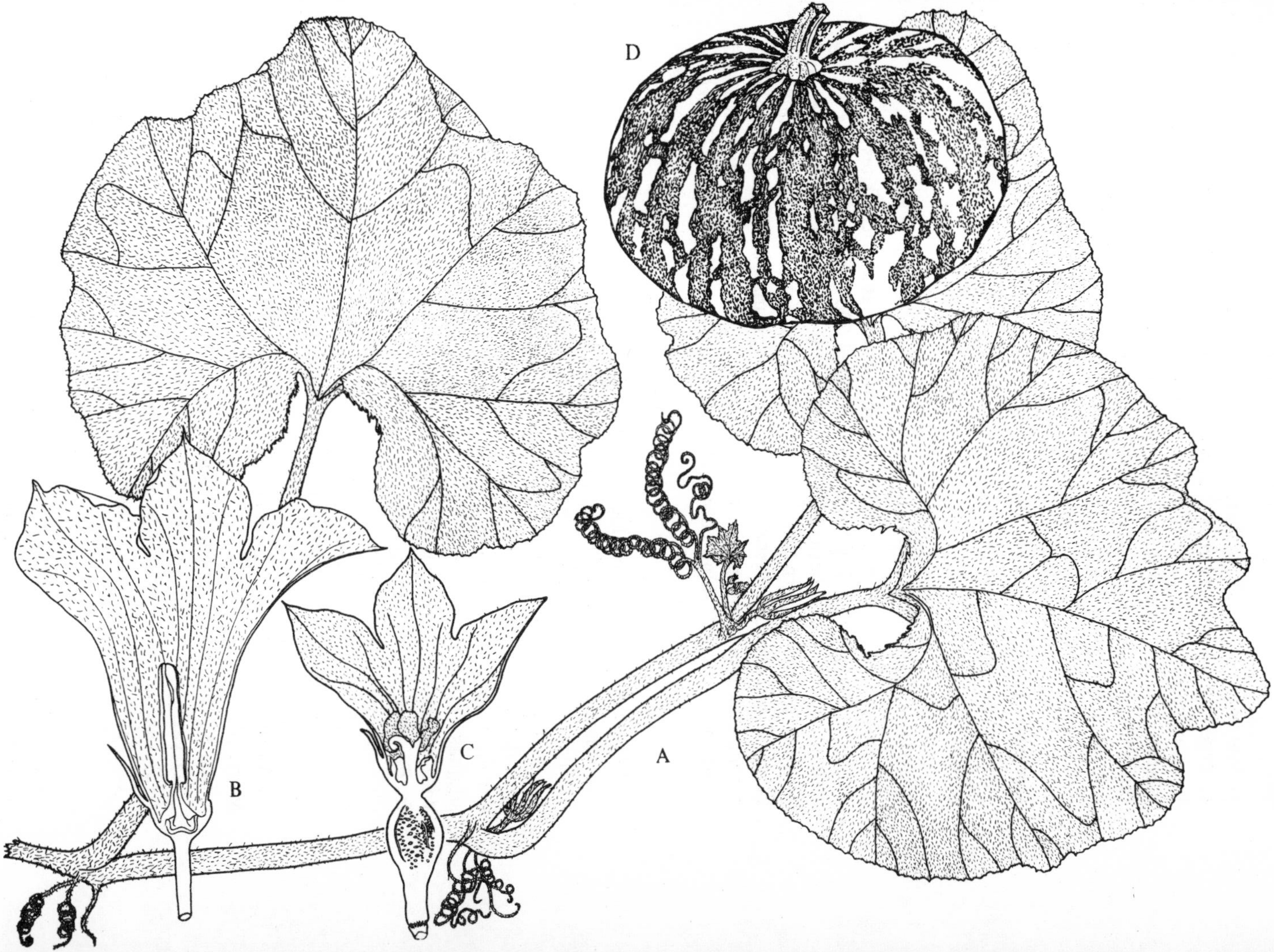
D
A
C
B

yellow or green, rough; mature peduncle hard, basically 5-angled, but often enlarged by hard, warty cork, but not flared at fruit attachment. Fr. variable, hard- or soft-shelled, usually dull in colour; flesh white to pale tan or yellow, coarse-grained, with soft, but not gelatinous fibres; seeds plump, white or tan, separating readily and cleanly from pulp; funicular attachment obtuse, not or only slightly asymmetrical; seed margin barely scalloped, acute.

Cvs include 'Cushaw', 'Japanese Pie'.

Cucurbita pepo L. (2n = 40) MARROW

It also provides summer squash, winter squash, pumpkin, and ornamental gourd.

ORIGIN AND DISTRIBUTION

C. pepo has been recorded from many archaeological sites in North America, the oldest being from Mexico dated 7000–5500 B.C. It was widely distributed over northern Mexico, the south-western United States and possibly on the eastern seaboard of the latter country in pre-Columbian times. *C. texana* Gray grows wild in Texas and may be either the putative parent or a weedy offspring. *C. pepo* has now been widely distributed and some of its cvs will stand cooler climates than the other spp. of *Cucurbita.*

STRUCTURE

A large coarse polymorphic sp., variable in morphological and reproductive characters. Annual herb, monoecious, long-running or occasionally bushy, with spiculate bristles, harsh to touch; stem hard, frequently 5-angled. Lvs cordate, serrate, deeply lobed with narrow, acute sinuses, with or without white blotches. Fls acuminate in bud; sepals short, awl-shaped; corolla bright yellow to orange-yellow; androecium short, thick, conical; stigmas small, yellow, smooth; mature peduncle hard, sharply 5-angled without cork development, not or only slightly enlarged at attachment to fr. Fr. very variable in size, shape and colour, hard- or soft-shelled, often dull coloured; flesh white to dark-yellow, coarse-grained; seeds large or small, tan to dingy white, separating readily and cleanly from pulp; funicular attachment obtuse, symmetrical; seed margin smooth, obtuse; testa lacking in some cvs.

Cvs include 'Sugar', 'Connecticut Field', 'Yellow Crookneck', 'English Marrow'.

Some authorities recognize 3 edible vars:

(1) var. *pepo*—Field pumpkins
(2) var. *medullosa* Alef.—Vegetable marrows
(3) var. *melopepo* Alef.—Bush squash and pumpkins; summer squashes.

NOTE

Hybrids can be obtained only with difficulty from most species combinations of cultivated *Cucurbita* and no naturally-occurring interspecific hybrid has ever been found. *C. maxima* and *C. pepo* are more closely related to *C. moschata* than to each other, *C. moschata* occupying a central position. *C. mixta* has the closest affinity to *C. pepo*.

ECOLOGY

Although tropical in origin, the modern cvs are adapted to a wide range of conditions. The perennial *Cucurbita ficifolia* grows at higher altitudes in the tropics than the other species, is tolerant of a cool climate, and is a short-day plant. The annual species are insensitive to day length. They require relatively warm temperatures; *C. maxima* and *C. pepo* are more tolerant of cool temperatures than *C. moschata* and *C. mixta*. None of them does well in the ever-wet tropics, preferring dry areas or areas of medium rainfall. They are often grown in the dry season. They can be grown on most soil types, provided these are well drained and of good fertility.

POLLINATION

The number of staminate flowers produced always exceeds that of the pistillate ones. Pollination is effected by insects, mainly bees, and they are naturally cross-pollinated. Hand-pollination assists fruit setting.

GERMINATION

The seeds take about 1 week to germinate. Germination is epigeal. The withdrawal of the cotyledons from the seed is aided by the development of a lateral outgrowth or peg from the lower surface of the hypocotyl.

CHEMICAL COMPOSITION

The edible portion, which is about 70 per cent in mature fruits, has the following approximate composition: water 90 per cent; protein 1 per cent; fat 0·2 per cent; carbohydrate 8 per cent; fibre 0·5 per cent. The immature fruits contain more water and less nutrients, but have less waste. The seed kernels contain 40–50 per cent oil and 30 per cent protein.

PROPAGATION

These crops are grown from seed, but they can be grown from cuttings if required as they root at the nodes.

HUSBANDRY

They are usually planted in hills, 4–5 ft apart for bush and small-vine cvs, 8–12 ft apart for long-running cvs. Several seeds are planted per hill and later thinned to 1–2 plants. The seed rate is 3–4 lb per acre. They respond well to ample dressings of organic manure, and artificial fertilizers may be applied at the rate of 600–800 lb per acre of a 5 : 10 : 10 NPK mixture. In peasant agriculture in Africa they are often interplanted with other crops such as maize, or near houses where they receive the household refuse. Those grown for immature fruits, which are harvested before the rind begins to harden, produce the first usable fruits 7–8 weeks after planting and continue bearing for several weeks and yield about 3–5 tons per acre. Those grown for mature fruits take 3–4 months before harvesting and yield 10 tons or more per acre. For storage they should be kept at 75–85°F for two weeks to harden the shell and they then store best at 50–55°F with low humidity.

MAJOR DISEASES AND PESTS

These are the same as *Citrullus lanatus* (q.v.) and *Cucumis melo* (q.v.), but attacks are usually not so serious.

IMPROVEMENT

Some work has been done in the United States and elsewhere on breeding of *C. maxima* and *C. pepo*, but very little work has been done on *C. moschata* and *C. mixta* which are more suited to tropical conditions. Cvs within each sp. cross readily, so they should be hand-pollinated and bagged or grown in isolated plots for seed production. *C. maxima* × *C. pepo* does not produce fertile seeds. *C. maxima* × *C. moschata* occasionally sets fruit, producing a few viable seeds and the F_1 hybrids are self-fertile.

PRODUCTION

The *Cucurbita* crops are little grown in the humid tropics, but are widely grown in the drier areas, particularly in Africa for local consumption. In the United States and Europe they are grown commercially for immature and mature fruits and for canning.

REFERENCES

THOMPSON, R. C., DOOLITTLE, S. P. and CAFFREY, D. F. (1955). Growing Pumpkins and Squashes. *U.S. Dep. Agric. Fmrs' Bull.*, **2086**.

WHITAKER, T. W. and BOHN, G. W. (1950). The Taxonomy, Genetics, Production and Uses of the Cultivated Species of *Cucurbita*. *Econ. Bot.*, **4**, 52–81.

Lagenaria Ser. (x = 11)

Usually considered a monotypic genus now spread throughout the tropical and warmer regions of the world, but Willis (1966) gives 6 spp.

Lagenaria siceraria (Molina) Standl. (2n = 22) BOTTLE GOURD

Syn. *L. vulgaris* Ser.; *L. leucantha* (Duch.) Rusby

Also known as the white-flowered gourd and the calabash gourd, but should not be confused with the true calabash, which is the fruit of *Cresentia cujete* L., a small West Indian tree of the family Bignoniaceae.

USES

Mainly cultivated for the dry hard shells of the fruits, which are of many shapes, and which are used for various domestic utensils such as bowls, bottles, ladles, milk-pots, churns, spoons, work-baskets and containers of many types; they are also used for floats for fishing nets, pipes, musical instruments, etc. They are sometimes carved. They were used before the invention of pottery in many regions. The fruits of most cvs are too bitter for food, but some are eaten in the same way as pumpkins and marrows; they are also used in curries. The young shoots and leaves of some cvs are occasionally used as a pot-herb. The seeds of others are used as a masticatory and the oil from the seeds for culinary purposes.

ORIGIN AND DISTRIBUTION

L. siceraria is probably the only cultigen which was common both to the Old and New Worlds since very remote times; the spread of sweet potatoes (q.v.) from Peru to Polynesia and coconuts (q.v.) from Polynesia to Panama, although pre-Columbian, is probably more recent. No other crops were common to both hemispheres before 1492. It is probable that the bottle gourd originated in Africa where it occurs subspontaneously, as it also does in India. It has been shown that gourds are capable of floating in sea water for up to 224 days with no decrease in the viability of the seeds and it seems probable that they drifted from tropical Africa to the coast of Brazil. The archaeological evidence in the Old World is not nearly as abundant as that from the New World, but remains have been found in an Egyptian tomb dated 3500–3300 B.C. It must have reached Asia at a very early date and China not later than the 1st century A.D. It was taken to New Zealand by the Maoris about the 12th century. It has been found in Mexico in horizons dated 7000–5500 B.C. and at Huaca Prieta in Peru dated 4000–3000 B.C. It must have been one of the most ancient crops cultivated by man in the tropics and certainly the most widely spread.

CULTIVARS

Many cvs are known differing in shape and palatability of the fruits.

ECOLOGY

It is chiefly suited to semi-dry areas, but is found throughout all tropical and subtropical environments.

STRUCTURE

Annual, monoecious, long-running or climbing, musky-scented herb; stem robust, longitudinally furrowed, pilose with jointed, gland-tipped hairs; tendrils usually bifid. Lvs with petiole 5–30 cm long with 2 pore-like glands at junction with lamina, which is widely ovate, with cordate base with broad sinuses, dentate, lobes absent or 3–7-lobed or undulate, soft, white hairy beneath, 10–30 cm broad. Fls axillary, solitary, short-lived; calyx tube campanulate with 5 narrow spreading lobes, 1–1·5 cm long; petals 5, free, white, obovate, woolly, 2·5–5 × 2–4 cm; male fls on long peduncles 5–25 cm long; stamens with 3 free filaments, anthers white, lightly-adhering; female fls on short, robust peduncles, 2–5 cm long; lengthening after fertilization; staminodes absent, ovary with white, woolly hairs, 2·5–3 cm long, placentae 3, stigmas 3, thick, bilobed. Fr. very variable in shape and size, 10–100 cm or more long, with hard, durable rind; flattened, globular, bottle- or club-shaped, sometimes crook-necked or coiled, many-seeded. Seeds compressed, ridged to 2 cm long, white or tan.

POLLINATION

The flowers are pollinated by insects, mainly bees.

CHEMICAL COMPOSITION

In edible cvs the edible portion, which constitutes up to 80 per cent of the fruit, contains: water 90·7 per cent; protein 0·7 per cent; fat 0·2 per cent; carbohydrate 6·3 per cent; fibre 1·5 per cent; ash 0·6 per cent. The seed kernels contain 45 per cent oil.

HUSBANDRY

They are usually cultivated near houses on mounds of well-manured soil. Several seeds are planted per mound, the strongest seedling being retained. They require strong supports so that the fruits may be suspended. It is possible to change the shape of the developing fruit by constricting it with string. The vines begin fruiting 3 months after planting. For gourds the fruits are harvested when dead ripe; they are opened by cutting across the neck; the flesh is bored out as completely as possible; the rest is allowed to decay and is then emptied out. They are treated with ashes and dried by hanging over the smoke of fires. For vegetables the fruits are gathered while still young and tender. In India each plant is said to yield 10–15 fruits weighing 1–3 lb each.

PRODUCTION

They are extensively grown for gourds in Africa and other parts of the tropics, but the use is declining with the increase in pottery, tin and enamel utensils. They are widely grown in India as a vegetable.

Fig. 19. **Lagenaria siceraria: BOTTLE GOURD. A,** flowering shoot (× ½); **B,** male flower in longitudinal section (× 1); **C,** female flower in longitudinal section (× 1); **D,** fruit (× ½).

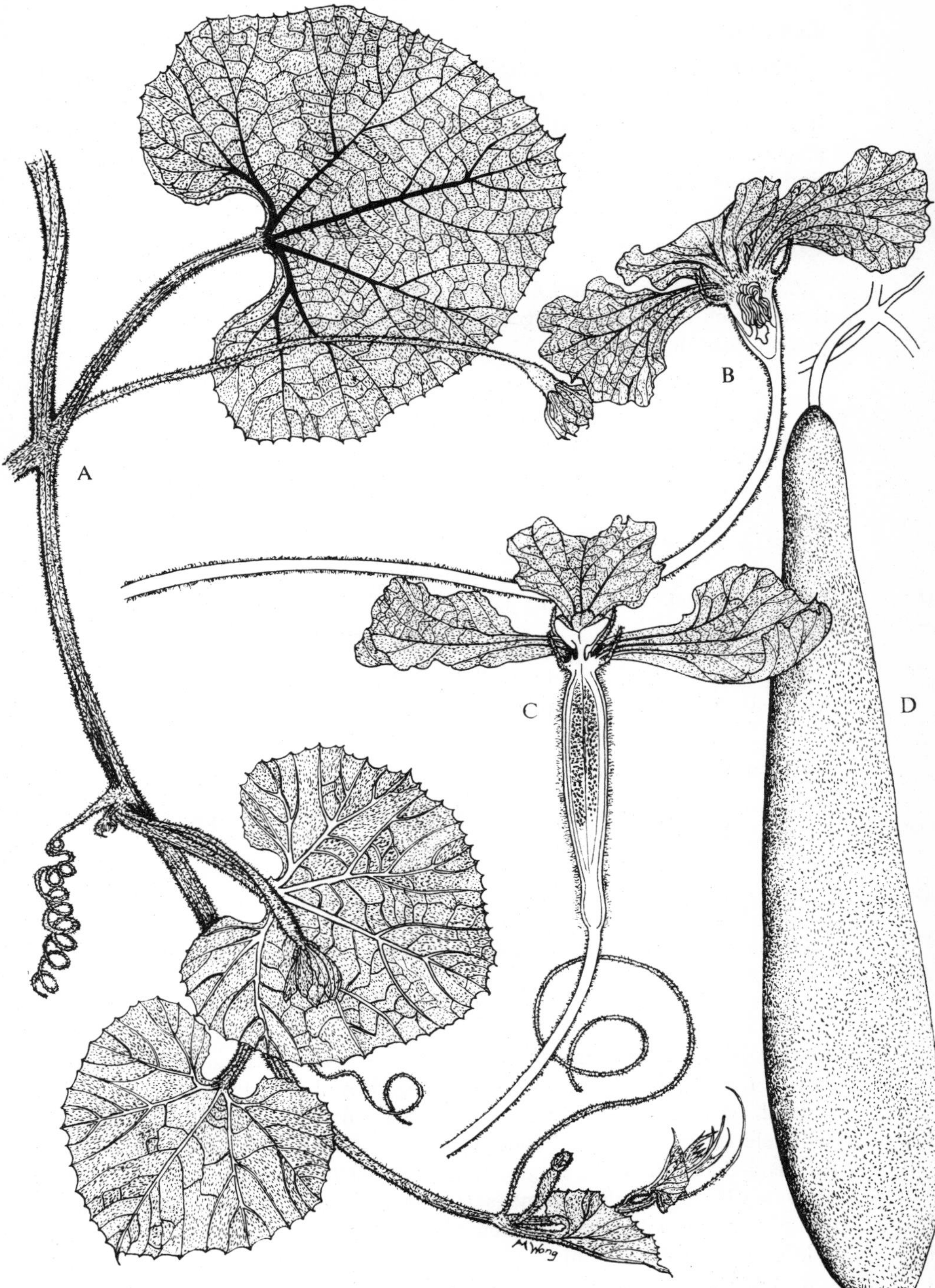
A
B
C
D
M Wong

Luffa Mill. (x = 13)

About 8 spp. in tropical Asia. Monoecious annual vines; tendrils branched; lvs nearly glabrous, 5–7-lobed; fls yellow, showy; petals 5, free; staminate fls in racemes; pistillate fls solitary; anthers free; pistil with 3 placentae and many ovules; stigmas 3, bilobed; fruit oblong or cylindrical, rind becoming dry at maturity, interior fibrous.

USEFUL PRODUCTS

L. acutangula (q.v.) and *L. cylindrica* (q.v.) are grown for their immature fruits which are eaten as vegetables; the ripe fruit of the latter yields commercial loofah sponges.

KEY TO 2 CULTIVATED SPECIES

A. Ovary ribbed and angled; club-shaped and not tomentose; lvs not deeply lobed .... *L. acutangula*
AA. Ovary not ribbed, usually cylindrical and tomentose; lvs mostly deeply lobed .... *L. cylindrica*

Luffa acutangula (L.) Roxb. (2n = 26) ANGLED LOOFAH

USES

It is cultivated throughout India and the East for its young tender fruits which are used as a vegetable.

ORIGIN AND DISTRIBUTION

L. acutangula probably originated in India where it is found wild in the north-west of the country.

ECOLOGY

Unlike most cucurbits it grows well in the low humid tropics.

STRUCTURE

Stout monoecious climber, foetid when bruised. Stem acutely 5-angled; hairy tendrils 3-fid or more. Lvs 5–7-angled or shallowly lobed, pale green beneath, scabrous, 10–25 × 10–25 cm. Male fls in racemes of several fls; peduncle 15–35 cm long; female fl. solitary, borne in same lf. axis as male fls. Fls 4–5 cm diameter, fragrant; calyx 5-partite glandular; petals yellow, obovate, 2–2·5 cm long, stamens 3; ovary inferior, filiform, with 10 longitudinal ribs on which are swollen glands, style short, stigmas 3. Fls open in later afternoon or evening. Fr. club-shaped, crowned by enlarged sepals and style, angled, 10-ribbed, many-seeded, 15–50 × 5–10 cm. Seeds black, pitted, flattened, 1–1·3 × 0·7–0·9 cm.

HUSBANDRY

Like all cucurbits loofahs grow and fruit better in rich soil. They are planted on hills 3 ft apart, or in rows 5–6 ft apart and plants 8–12 in. in row. They may be allowed to trail or are staked. Harvesting of immature tender fruits begins 2 months after sowing. On maturity the fruits become bitter and inedible. Each plant yields 15–20 fruits.

Luffa cylindrica (L.) M. J. Roem. (2n = 26) SMOOTH LOOFAH

Syn. *L. aegyptiaca* Mill.

Also known as sponge gourd, dish-cloth gourd, and vegetable sponge.

USES

In India and the East the young tender fruits of the non-bitter types are eaten fresh like cucumbers or cooked as a vegetable or used in soups. It was the early Portuguese explorers who discovered how to use the loofah sponge, which is the fibro-vascular network of the ripe fruit. Before World War II 60 per cent of the loofahs imported into the United States were used in filters in marine steam engines and in diesel engines. They are also used as bath sponges and for cleaning purposes and in the manufacture of pot-holders, table mats, door and bath mats, insoles, sandals and gloves. Because of their shock and sound absorbing properties they were used in steel helmets and armoured vehicles of the U.S. Army. The seeds yield an edible oil. In Japan the sap from the stem is used in toilet preparations. It is also used in native medicines in the East.

ORIGIN AND DISTRIBUTION

L. cylindrica is of ancient cultivation in the Old World and appears to have been domesticated in tropical Asia, possibly in India. It appears to have reached China about 600 A.D. It is now cultivated or grows as an escape in practically all the tropical regions of the world. Because of the loss of the Japanese crop during World War II and its importance as filters, attempts were made to produce it in many countries in the New World and Africa.

CULTIVARS

Cvs are recognized which produce the best vegetables and those which produce the best sponges.

ECOLOGY

Although tropical in origin the best loofahs are grown in Japan. Too heavy a rainfall during flowering and fruiting is harmful. The soil should be moderately rich and a medium well-drained loam gives the best results.

STRUCTURE

A vigorous annual climber. Stem 5-angled; tendrils 2–3-fid or more. Lvs broadly ovate to reniform, dark-green, rather deeply 5–7-lobed, dentate, scabrous, 6–25 × 8–27 cm, apex acute, base cordate; petiole hispid, 5–10 cm long. Fls yellow, 5–10 cm in diameter; male fls in axillary 4–20-flowered inflorescences; stamens 5, free; female fls solitary in same leaf axils as males; ovary cylindrical, smooth, pubescent; stigmas 3. Fls open in early morning. Fr. nearly cylindrical, normally with light furrows or stripes, but not ribbed, 30–60 cm long, surmounted by stout style and calyx. Seeds black, flat, smooth, faintly winged, 10–15 mm long.

POLLINATION

Loofahs are pollinated by insects.

GERMINATION

The cotyledons are oval in shape and about 5 cm long, germination being epigeal.

CHEMICAL COMPOSITION

The edible portion of tender fruit contains: water 93 per cent; protein 1·2 per cent; fat 0·2 per cent; carbohydrate 3·1 per cent; fibre 2·0 per cent; ash 0·5 per cent. Seed kernels, 51 per cent of weight of seeds, contain about 46 per cent of oil and 40 per cent protein.

PROPAGATION

Loofahs are propagated by seeds obtained from ripe fruits.

HUSBANDRY

Loofahs are usually planted 3–4 ft apart. The best sponges are obtained by training the vines over trellises. The side branches are pruned to encourage the growth of the main stem and early male flowers and sometimes the early female fls are removed. The number of fruits per vine is limited to 20–25. They are harvested when fully mature, as indicated by yellowing of the base and apex, about 4–5 months after planting. Under optimum conditions, 24,000 fruits are obtained per acre. The fruits are processed by immersing them in tanks of running water until the outer walls disintegrate.

Fig. 20. **A, Luffa cylindrica :** SMOOTH LOOFAH. **A1,** flowering shoot (× ½); **A2,** female flower in longitudinal section (× 1); **A3,** male flower in longitudinal section (× 1). **B, Luffa acutangula :** ANGLED LOOFAH. **B1,** fruits (× ½); **B2,** fruit in longitudinal section (× ½).

A2
B2
M. WONG
A3
A1
B1

They are then washed to remove the seeds and pulp, and bleached and dried in the sun. The cultivation as vegetables is the same as *L. acutangula*.

MAJOR PESTS AND DISEASES

The plants are attacked by downy and powdery mildews (see *Citrullus lanatus*). Fruit flies, *Dacus* spp., are the most serious pest.

IMPROVEMENT

Little work has been done on this aspect of the crop.

PRODUCTION

Japan is the principal producer of loofahs, which are superior to those grown elsewhere. Brazil also produces substantial quantities. When the Japanese supplies were cut off during World War II, efforts to produce loofahs in Africa, Central America and the West Indies were not very successful, as the product generally was of rather poor quality.

REFERENCE

PORTERFIELD, W. M. (1955). Loofah—The Sponge Gourd. *Econ. Bot.*, **9**, 211–223.

Momordica L.

About 40 spp. of annual or perennial climbing herbs in tropical Africa and Asia; monoecious or dioecious. The young tender fruits of some species are used as vegetables. These include: *M. balsamina* L. (balsam apple), *M. charantia* (q.v.), *M. cochinchinensis* Spreng., and *M. dioica* Roxb. ex Willd. The first two spp. are sometimes grown as ornamentals. A number of spp. are used in native medicines in Africa and Asia.

Momordica charantia L. — BITTER GOURD

Also known as bitter cucumber or balsam pear.

USES

The young bitter fruits are cooked and eaten as a vegetable in India and the Far East. The bitterness is reduced by steeping the peeled fruit in salt water before cooking. It is also used as an ingredient of curries and the fruit is pickled. The seed mass of the ripe fruit is used as a condiment in India. The tender shoots and leaves are used as spinach. *M. charantia* is used in native medicines.

Fig. 21. **Momordica charantia**: BITTER GOURD. **A,** leafy shoot (×½); **B,** male flower in longitudinal section (×2); **C,** female flower in longitudinal section (×2); **D,** fruit (×½); **E,** fruit in longitudinal section (×½); **F,** fruit in transverse section (×⅔); **G,** seed (×2); **H,** seed in longitudinal section (×2).

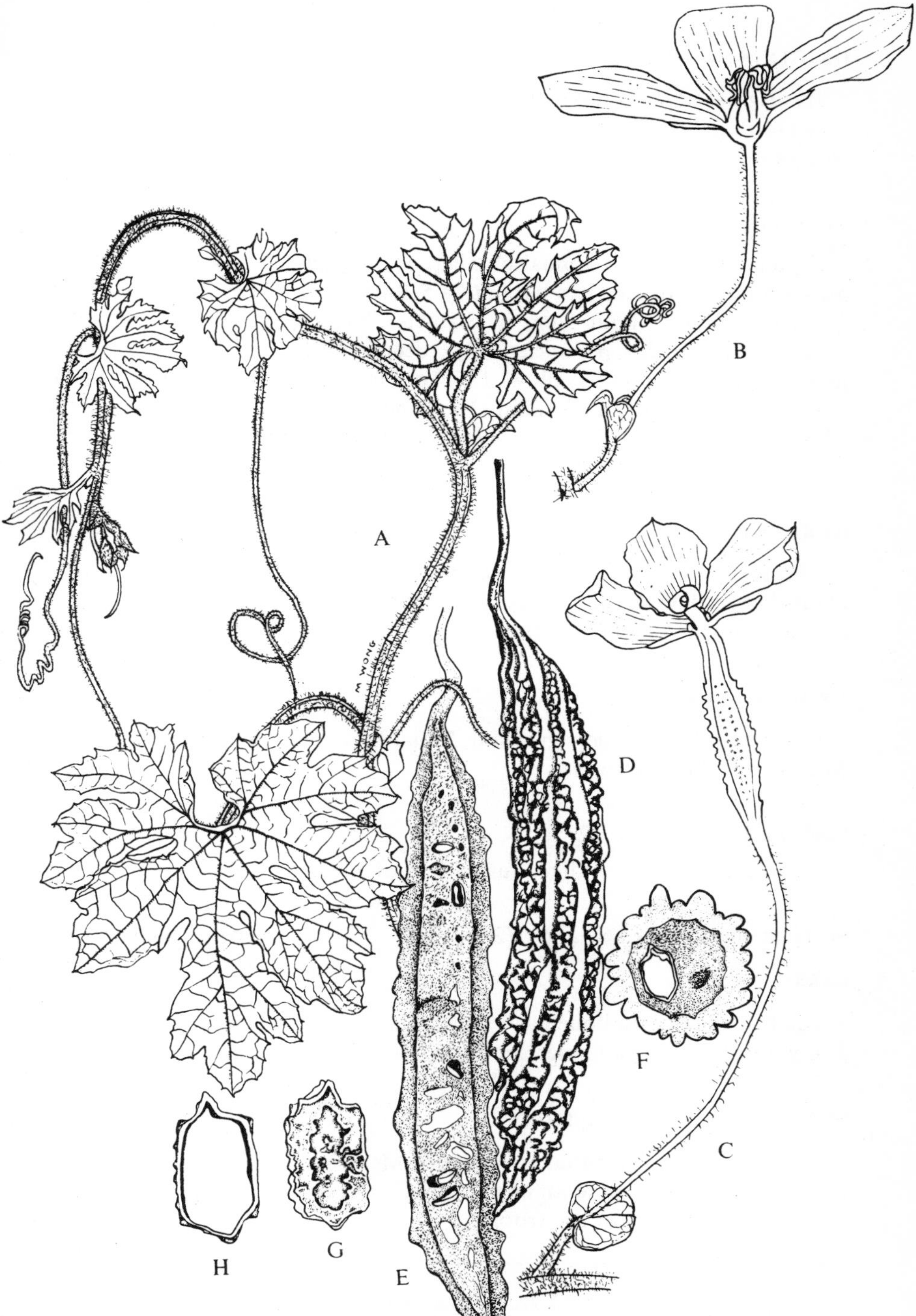
A
B
C
D
E
F
G
H
M WONG

ORIGIN AND DISTRIBUTION

Its original home is unknown, except that it was in the Old World. It was taken to Brazil early in the slave trade and is now widespread throughout the tropics.

STRUCTURE

Slender monoecious annual climber. Stem 5-angled and furrowed; tendrils simple or forked. Lvs palmately 5–9-lobed, 5–17 cm in diameter. Fls axillary, solitary, about 3 cm in diameter, prominent sessile bract at or near base of slender peduncle; calyx deeply 5-fid; corolla rotate, parted nearly to base; petals 5, yellow, 1·5–2 cm long; stamens 3; filaments free, anthers united; stigmas 3, bifid. Fr. pendulous, 5–25 cm long, fusiform, ribbed with numerous tubercles. Seeds numerous, 1–1·5 cm long, brownish, with scarlet aril.

HUSBANDRY

The normal spacing is 2 × 1 ft and the plant is usually given supports. The plants begin to flower 30–35 days after planting and the first fruits are ready for gathering 15–20 days later.

PRODUCTION

Bitter gourds are widely grown in India, Indonesia and by the Chinese in Malaya and Singapore.

Sechium P. Br. (x = 12)

A monotypic genus indigenous to central Mexico.

Sechium edule (Jacq.) Swartz (2n = 24) CHOYOTE; CHRISTOPHINE

USES

The fruits are eaten boiled as a vegetable, as are the large tuberous roots. The young leaves and tender shoots are sometimes eaten as spinach.

ORIGIN AND DISTRIBUTION

The choyote is indigenous to southern Mexico and Central America and was a common vegetable among the Aztecs prior to the Conquest. It has now spread throughout the tropics.

Fig. 22. **Sechium edule :** CHOYOTE. **A,** flowering shoot ($\times \frac{1}{2}$); **B,** male flower ($\times 5$); **C,** female flower ($\times 5$); **D,** fruit ($\times \frac{1}{2}$); **E,** fruit in longitudinal section ($\times \frac{1}{2}$).

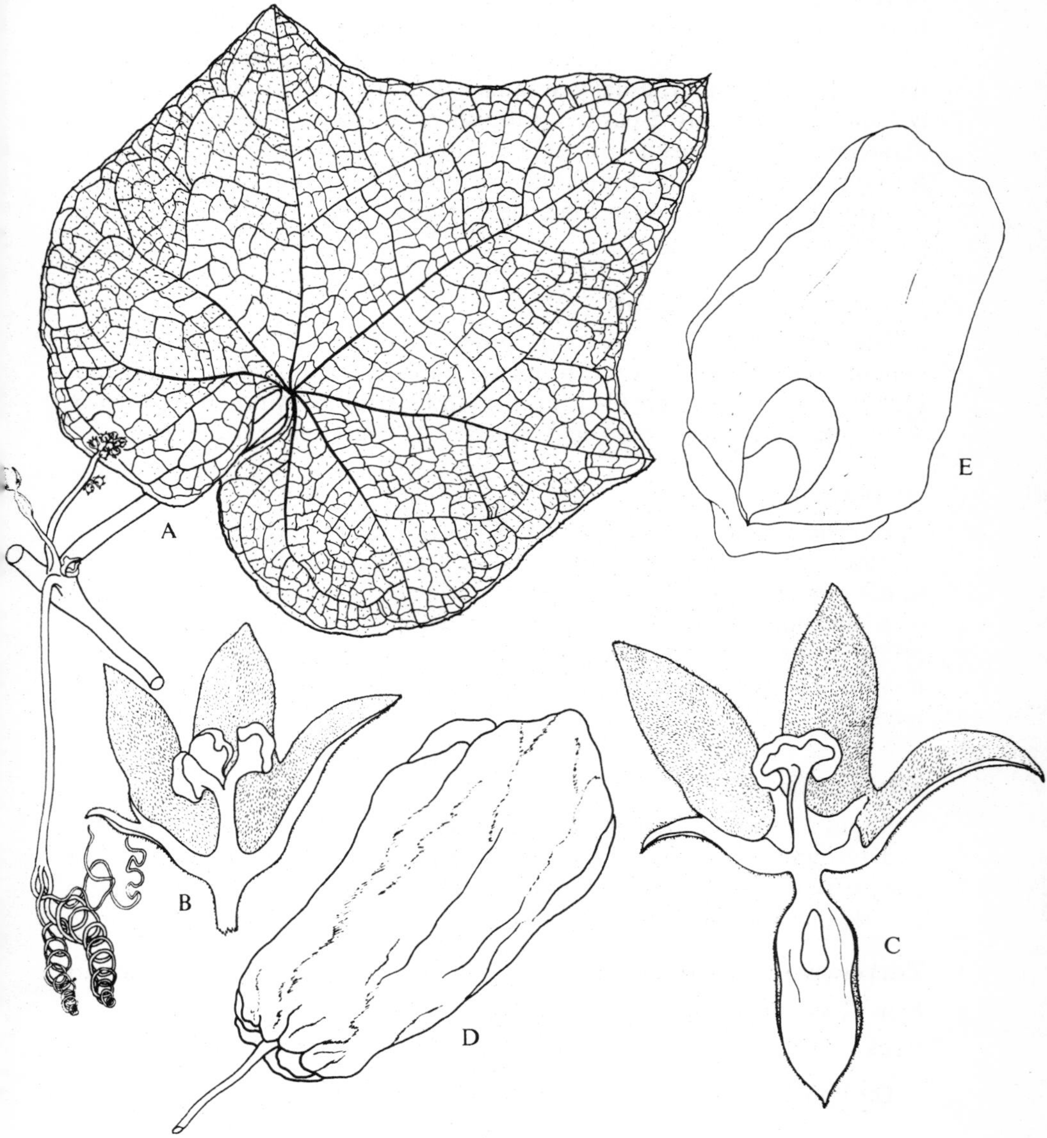
A
B
C
D
E

STRUCTURE

A monoecious, robust, climbing or sprawling, sparsely hairy, herbaceous, perennial vine, up to 12 m long, with tuberous root. Stems longitudinally furrowed; tendrils large, branched. Lvs broadly ovate, angled or faintly lobed, scabrous, 10–25 cm in diameter, base cordate, apex apiculate; petiole 3–15 cm long. Fls small, axillary; male fls in small peduncled clusters; female flowers solitary, usually in same axil as male; calyx deeply 5-partite; corolla rotate, deeply 5-partite, greenish or cream in colour; stamens 5, filaments connate; pistil with inferior ovary, connate style and stigmas forming a small head. Frs fleshy, pyriform with longitudinal furrows, whitish or green, 10–20 cm long, flesh whitish, single-seeded. Seed large, flat, white, 3–5 cm long.

CHEMICAL COMPOSITION

The edible portion, which constitutes about 80 per cent of the fruit, contains: water 89·8 per cent; protein 0·9 per cent; fat 0·2 per cent; carbohydrate 7·7 per cent; fibre 0·4 per cent; ash 1·0 per cent; vitamin A 650 mcg. The roots contain 79·0 per cent water and 17·8 per cent carbohydrate.

HUSBANDRY

The choyote grows best at altitudes above 1,000 ft in the tropics in areas of moderate rainfall. The whole mature fruit is planted horizontally and thinly covered with soil. They are usually spaced 4–6 ft apart along a fence or trellis, but if allowed to sprawl will need much more room. Harvesting of the fruits begins 3–5 months after planting and continues for many months. They are picked when ½–1 lb in weight and can be stored for several weeks.

Trichosanthes L.

About 15 spp. of annual and perennial herbs, some of them with large underground tubers, in the tropics of Asia, Polynesia and Australia. The unripe fruits of *T. cucumerina* (q.v.) are used as a vegetable. Several species are used in local medicines.

Trichosanthes cucumerina L. SNAKE GOURD

Syn. *T. anguina* L.

USES

The immature fruit is boiled and eaten. The ripe fruit is fibrous and bitter.

Fig. 23. **Trichosanthes cucumerina**: SNAKE GOURD.
A, flowering shoot (×⅓); **B**, male flower in longitudinal section (×1); **C**, female flower in longitudinal section (×1); **D**, fruit (×⅛); **E**, fruit in longitudinal section (×⅛); **F**, seed (×1); **G**, seed in longitudinal section (×1).

B
D
E
A
F
G
C
M. WONG

ORIGIN AND DISTRIBUTION

T. cucumerina occurs wild from India to Australia and a cultivated form of it has long been grown in India and the Far East. It is now grown occasionally in the West Indies.

STRUCTURE

Monoecious climbing annual herb. Stem slender, 5-angled, furrowed; tendrils branched. Lvs angular or 5–7-lobed, dentate, 10–25 cm long. Fls axillary, white, about 5 cm in diameter; male fls in racemes, female fls solitary; calyx 5-partite; corolla rotate, 5-partite, with long white hairs inside; stamens 3; ovary 1-celled, long, hairy, with many ovules. Frs very slender, long-tapering 30–150 cm long, greenish-white. Seeds thick, brownish, sculptured, 1–1·5 cm long.

HUSBANDRY

The crop usually receives little attention, except that it is often grown up some support. A weight is usually hung to the end of the growing fruit to keep it straight.

EUPHORBIACEAE

A large, difficult and not very natural family with about 280 genera and 8,000 spp. of herbs, shrubs and trees of very diverse habit, widely distributed throughout the world. White latex sometimes present, which may be irritant. Lvs usually alternate and stipulate, sometimes reduced. Monoecious or dioecious. Fls small, male or female; sepals generally 3–5; petals usually absent; stamens 1 to numerous; anthers usually 2-celled; ovary superior, usually 3-celled; frs typically a 3-lobed capsule.

USEFUL PRODUCTS

FRUITS: *Antidesma bunius* (L.) Spreng. (bignay), which occurs wild from India to Australia, is cultivated in Malaysia and elsewhere for its fruits which are made into preserves. *Baccaurea motleyana* Muell.-Arg. (rambai) and *B. sapida* Muell.-Arg. have edible fruits and are cultivated in south-east Asia. *Phyllanthus acidus* (L.) Skeels (Otaheite gooseberry), a native of Madagascar and India, and *P. emblica* L. (emblic), native of tropical Asia, are cultivated for their acid fruits which are eaten cooked; their branches resemble pinnate lvs.

OILS: *Aleurites* spp. (q.v.) are a source of drying oils. *Ricinus communis* (q.v.) yields castor oil.

ROOT CROPS: *Manihot esculenta* (q.v.), cassava, is a very important tropical root-crop.

RUBBER: *Hevea brasiliensis* (q.v.), Para rubber, produces most of the world's natural rubber. *Manihot* spp. (q.v.) have also been used as a source of rubber.

OTHER PRODUCTS: *Croton tiglium* L., native of south-east Asia; is cultivated in India and Ceylon for its seeds which yield croton oil, one of the world's most powerful purgatives. *Euphorbia antisyphilitica* Zucc., indigenous to Mexico and Central America, is the source of candelilla wax.

ORNAMENTALS: The following are commonly cultivated in the tropics as ornamental shrubs: *Acalypha hispida* Burm. f.; *A. wilkesiana* Muell.-Arg.; *Codiaeum variegatum* (L.) Blume (garden crotons); *Euphorbia milii* Ch. des Moulins (Christ's thorn); *E. pulcherrima* Willd. ex Klotzsch (poinsettia);

E. tirucalli L. (milk bush), which is also used as hedges for kraals in East Africa; *Jatropha* spp. A number of succulents, *e.g. Euphorbia* spp., are grown as pot plants.

Aleurites Forst. (x = 11)

A small genus of trees with 6 spp. in tropical and subtropical eastern Asia and Malaysia. The seeds yield drying oils. *A. montana* (q.v.) can be grown at higher elevations in the tropics for tung oil production; *A. fordii* Hemsl., a native of the cooler parts of west and central China, is not suited to the tropics, where it has been tried in many countries. China is the main producer of tung oil from *A. fordii,* but plantings have also been made in south-eastern United States. *A. trisperma* Blanco is a native of the Philippines, where it is used locally for oil production. *A. moluccana* (L.) Willd., the candlenut, is native in Malaysia and is now widely distributed in the tropics. It yields an inferior oil which contains no eleostearic acid.

Aleurites montana (Lour.) Wils. (2n = 22) TUNG, MU-TREE

USES

Tung oil is very quick drying with good waterproofing properties. It is used in varnish and paint manufacture and in making linoleum, oilcloth and in insulating compounds. The cake after expressing the oil cannot be used as stock feed, but it is a good fertilizer.

ORIGIN AND DISTRIBUTION

A. montana is native of subtropical parts of China. Experimental plantings have been made in many tropical countries and it is grown commercially as a plantation crop in Nyasaland, now Malawi, where it was first introduced in 1931.

ECOLOGY

A. montana requires not less than 40 in. of rain during the growing season and a cool dry season of 4 months during which the trees shed their leaves and remain dormant. It will not thrive under lowland tropical conditions. In Malawi it is grown at altitudes of 2,000–5,000 ft. It will tolerate poor shallow stony soils, but naturally yields better on deeper, more fertile soils. Soils should be slightly acid with good drainage; it is very susceptible to accumulations of ash and waterlogging.

STRUCTURE

A deciduous tree to 20 m in height. White latex in all its parts. Two types of trees are recognized in Malawi: (1) A-type clones in which the main

---

Fig. 24. **Aleurites montana: TUNG. A, flowering shoot ($\times \frac{1}{2}$); B, male flower ($\times 1\frac{1}{2}$).**

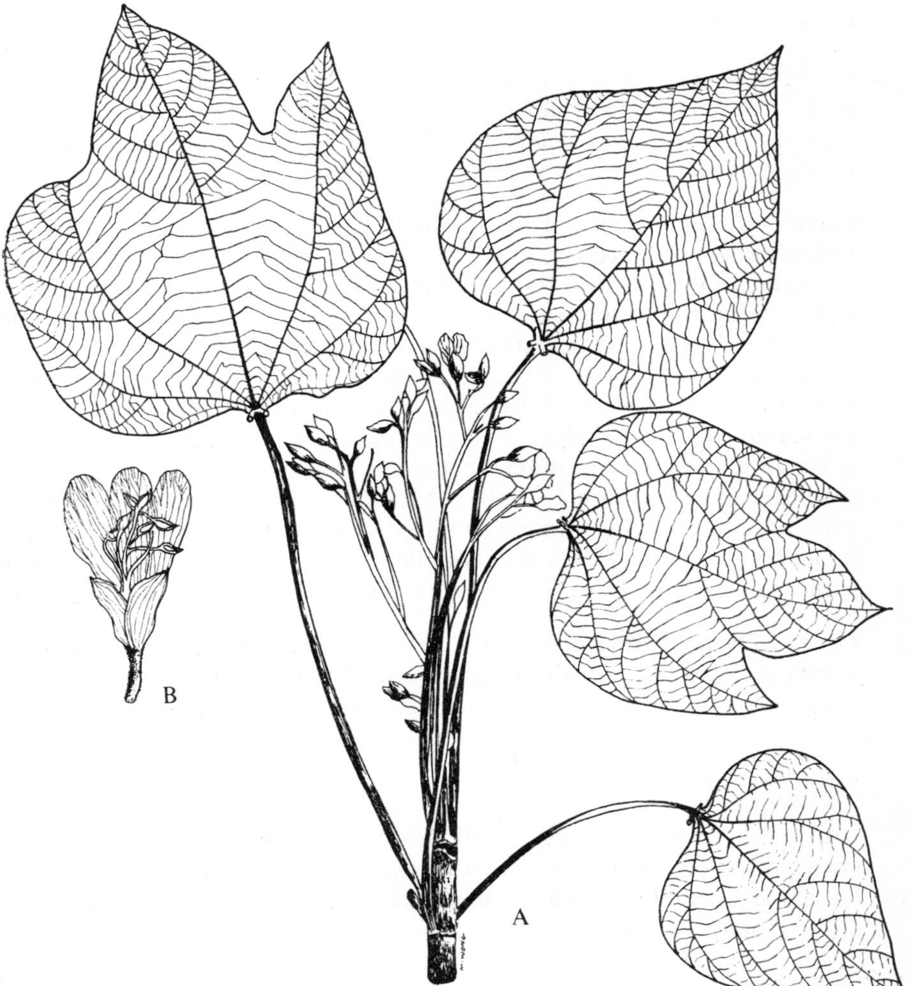
B
A

trunk increases steadily in height as it develops succeeding tiers of primary branches, one above the other at some distance apart; takes 3–5 years to come into bearing; (2) B-type clones in which the main trunk loses its identity after producing first or second whorl of branches; secondaries arise at short intervals; trees low, dense and bushy in habit; a precocious bearer, coming into bearing in third year and giving high yields in first 6 years. Both types are relatively surface rooting.

LEAVES: Spirally arranged; petioles longer than lamina; lamina ovate, entire or 3–5-lobed, 5–30 cm long and nearly as wide, 5–7 veins from cordate or truncate base, apex acuminate; 2 stalked concave green nectaries at junction of petiole and lamina.

FLOWERS: Unisexual and trees usually bear both male and female fls. Inflorescences of three types: (1) Male which emerge at tips of branches simultaneously with new lvs, corymbose, large with about 200 male fls; trees with this type are predominantly male and bear little fruit. (2) Female in racemes at end of short leafy shoots of new season's growth; smaller and denser than male; about 20 fls. (3) Mixed of 2 types: (a) on predominantly male trees with a few female fls borne at end of mainly male inflorescences; (b) mainly on 'bearer' trees with predominance of female fls and a few males in a subsidiary position. In both male and female fls, males being smaller, calyx splits into 2–3 lobes; petals 5, longer than calyx, 2 or more cm long, white with base in old fls tinged pink. Male fls with 8–20 stamens in 3 whorls, outermost whorl alternating with 5 glands on disc. Female fls with 3–5-celled ovary with 1 ovule in each; style with 2 thick branches.

FRUIT: Drupe, flattened spherical, 4–6 cm in diameter, 3–5-angled, thick woody pericarp. A few fruits dehisce on tree, but most fall without doing so. Seeds oval, about 3 cm long, with rough testa; endosperm white.

POLLINATION

Flowers are protogynous, female fls opening before male on same tree, cross-pollinated, entomophilous. Predominantly male trees have a longer blossoming period than bearers, are usually the first to begin flowering, and a few should be planted as pollinators.

GERMINATION

Fresh seeds usually give 75–90 per cent germination, which is reduced on storage unless stored stratified in moist sand in a cool place. On germination the radicle emerges and, when hypocotyl loop appears above surface, seedlings should be transplanted from germination beds to the nursery. Hypocotyl on straightening then pulls cotyledons above surface. Approximately 140 seeds per lb.

CHEMICAL COMPOSITION

The proportion of air-dried seeds in air-dried fruits is on average 43 per cent; air-dried seeds contain 4–6 per cent moisture and an average of 39 per cent oil, giving 30 per cent oil in factory expression. Air-dried seeds contain about 63 per cent kernel, which has oil content of 62 per cent. The superior quality of tung oil is due to the high content of eleostearic acid. The press-cake is poisonous.

PROPAGATION

Tung may be grown from seeds, but unselected seedling material is variable and contains approximately 50 per cent of unproductive, predominantly male trees. In Malawi selected clones are now budded on to seedling rootstocks, and the method is essentially the same as described for Para rubber, *Hevea brasiliensis* (q.v.). Budding may be done in the field or in the nursery. Male trees may be topworked.

HUSBANDRY

Land preparation and planting are similar to rubber; spacing of mature tung is usually 25–30 ft depending on vigour of clone. Closer initial spacing may be used, *e.g.* 10 ft, and later thinned. Windbreaks are desirable. Intercropping with maize or soya beans may be practised in the early years. Tung responds to nitrogenous fertilizers and to tung cake. The ripe fruits are allowed to fall on the ground. After harvesting the hulling is done by hand or mechanically. The seeds are then air-dried, after which they are shelled mechanically and the kernels, with some shell (which facilitates oil extraction) are ground and then pressed to extract the oil. The overall yield in Malawi in 1959 was 600 lb of oil per acre, but with improved planting material and management this could be substantially increased, when 1 ton per acre of air-dried seed may be obtained.

MAJOR DISEASES AND PESTS

The two most serious diseases in Malawi are a die-back caused by *Botryosphaeria ribis* Gr. & Dug. and a root rot caused by *Armillaria mellea* (Vahl ex Fr.) Kummer. Although attacked by various insect pests, none is serious.

IMPROVEMENT

The main aim is increase in yield of good quality oil per acre. High yielding clones, which can be vegetatively propagated, have been selected from seedling material in Malawi and have given 100 per cent increase in yield over unselected seedling material. Hybridization of high yielding clones has been done to produce legitimate seedling families and selections from these are being made. Hybrids between *A. montana* and *A. fordii* have been produced but showed little merit.

PRODUCTION

Aleurites montana provides a small proportion of the tung oil exported from China. Since its introduction into Malawi in 1931 there was a steady increase in planting, which reached a peak of 21,500 acres in 1954.

REFERENCES

FOSTER, L. J. (1962). Recent Technical Advances in the Cultivation of the Tung Oil Tree, *Aleurites montana*, in Nyasaland. *Trop. Agriculture, Trin.*, **39**, 169–187.

GOLDBLATT, L. A. (1959). The Tung Industry. ii. Processing and Utilization. *Econ. Bot.*, **13**, 343–364.

HILL, J. (1965). *Aleurites montana:* The Relative Value of some Malawi Selections. *Trop. Agriculture, Trin.*, **42**, 311–321.

HILL, J. and SPURLING, A. J. (1966). A note on the Classification of Montana Tung (*Aleurites montana*). *Trop. Agriculture, Trin.*, **43**, 19–24.

POTTER, G. F. (1959). The Domestic Tung Industry. I. Production and Improvement of the Tung Tree. *Econ. Bot.*, **13**, 328–342.

WEBSTER, C. C. (1950). The Improvement of Yield in the Tung Oil Tree (*Aleurites montana*). *Trop. Agriculture, Trin.*, **27**, 179–220.

WEBSTER, C. C., WIEHE, P. O. and SMEE, C. (1950). *The Cultivation of the Tung Oil Tree (Aleurites montana) in Nyasaland.* Zomba: Govt. Printer.

Hevea Aubl. (x = 9)

The genus *Hevea* is native to South America, where it is known from the Amazon valley, the upper Orinoco valley, the Guianas and the Matto Grosso region of Brazil; it is typical of the Amazonian 'hylaea'. *Hevea* exhibits much morphological variability and a wide range of ecological sites. Nine spp. are now recognized. Some of the subspecific variation is biologically stabilized and with a definite geographical range; some is a mere response to environmental conditions. No barriers to hybridization occurs between spp.; all may be crossed artificially and some natural crosses occur. It is probably a young genus geologically and in a state of evolutionary flux.

The type sp., *H. guianensis*, was described by Aublet from French Guiana in 1775. Spruce collected 8 more spp. from Rio Negro area of Brazil in 1849–1864. Mueller published the first synopsis of genus in 1873–1874; he recognised 11 spp. and transferred *Siphonia brasiliensis* HBK. to the genus *Hevea*. Further spp. were later erected, but Ducke after many years of field observation recognized 12 spp., with numerous vars. and forms. Schultes (1956), who made extensive collections in the Amazon region, now considers there are only 9 spp.

They range from large forest trees to little more than shrubs. Most spp. are native of the tropical rain forest with a wide range of ecological sites, from deeply flooded alluvial land and acid bogs to high, well-drained upland and rocky slopes and hill-tops.

All spp. contain latex in all their parts. Lvs spirally arranged or sub-opposite at the ends of shoots, trifoliate; leaflets entire, pinnate-veined; long petioles with nectaries at apex; monoecious with male and female fls in the same inflorescence. Fls small in panicled cymes; calyx 5-toothed or 5-lobed; petals absent; disc of 5 free or united glands; stamens 5–10, filaments united into a column and anthers sessile; ovary 3-celled with 1 ovule per cell; stigma nearly sessile, 2-lobed. Frs large woody capsules, usually dehiscing explosively into 2-valved cocci.

H. brasiliensis (q.v.) produces some 99 per cent of the world's natural rubber, its yield and quality of latex being superior to all other spp.

The other spp. of *Hevea*, of which some are tapped in a wild state and some may be important for breeding, are as follows:

H. benthamiana Muell.-Arg.: Occurs only north of Amazon River in north-western part of Amazon and upper Orinoco basins. Overlaps with *H. brasiliensis* only in a small area west of Manaos, where natural hybrids between the two spp. occur. Trees may reach a height of 90 ft, but are usually smaller; leaves with golden-brown indumentum on upper surface. Grows in low alluvial flooded areas and bogs. Lower in yield than *H. brasiliensis*; pure white latex yields good quality rubber which is obtained from wild trees. It is resistant to *Dothidella ulei.*

H. camporum Ducke: Native of open savannas in the headwaters of the Madeira River. Related to *H. pauciflora*, but a dwarf sp.

H. guianensis Aubl. and its var. *lutea* (Spruce ex Benth.) Ducke & R. E. Schultes: Very variable and found throughout range of genus. Tree to 100 ft or more; leaflets erect. Prefers well-drained upland and occurs up to 6,000 ft. Yellowish latex yields inferior rubber for which wild trees are tapped.

H. microphylla Ule: Endemic in uppermost Rio Negro basin in Brazil, Colombia and Venezuela. Very distinct sp.; tree to 60 ft; fruits do not dehisce explosively. Grows in low-lying ground, often permanently flooded. White watery latex almost completely lacks rubber.

H. nitida Mart. ex Muell.-Arg.: Occurs throughout most of the Amazon valley and uppermost Orinoco. Usually medium-sized tree; upper surface of leaflets very shiny. Usually grows in forests on sandy soil. Thin white latex acts as an anti-coagulent with that of other spp. It is resistant to *Dothidella ulei.* The var. *toxicodendroides* (R. E. Schultes & Vinton) R. E. Schultes, known only from isolated rocky mountains in East Colombia, seldom exceeds 8 ft in height and is xerophytic; it has a relatively high percentage of rubber in latex.

H. pauciflora (Spruce ex Benth.) Muell.-Arg.: Occurs in Rio Negro and upper Orinoco basins and in the Guianas. Medium-sized tree; leaflets large, membranous; seeds very large. Grows on rocky hillsides and high well-drained river-banks. White latex has low rubber and high resin content. The var. *coriacea* Ducke has wider and somewhat disrupted range of *H.*

nitida; leaflets smaller than sp. and leathery; latex whitish to tawny yellow.

H. rigidifolia (Spruce ex Benth.) Muell.-Arg.: Endemic to uppermost Rio Negro basin of Brazil, Colombia and Venezuela. Medium-sized tree, 60 ft; leaves thick, coriaceous. Grows on high well-drained soils. Cream coloured latex poor in rubber and high in resin.

H. spruceana (Benth.) Muell.-Arg. [(syn. *H. discolor* (Spruce ex Benth.) Muell.-Arg.]: Very abundant in lower Amazon basin. Tree with trunk enlarged at base; under surface of leaflets velvety; flowers brownish; capsules and seeds largest in genus. Grows on low and deeply flooded river banks. Watery latex almost devoid of rubber. Hybrids with *H. brasiliensis* can be used as rootstocks for latter and show some promise for resistance to root disease in the Americas.

NOTE

The related *Micranda minor* Benth. which is widespread and abundant in the Amazon and upper Orinoco basin, is a large tree to 110 ft, with abundant pure-white latex which yields a rubber of high quality, but cannot be repeatedly tapped.

Hevea brasiliensis (Willd. ex Adr. de Juss.) Muell.-Arg. (2n = 36)

PARA RUBBER

USES

Prior to the discovery of the New World, the Indians of Central and South America used the latex of certain plants for making balls, bottles, crude footwear and for waterproofing fabric. Columbus on his second voyage, 1493–1496, records the use of latex, of which one of the main sources must have been *Castilla elastica* Cerv. (q.v.). La Condamine reported the use of rubber in torches and various articles, such as bottles and syringes, in Ecuador in 1736. This was obtained from a local tree called *heve*, probably *Castilla ulei* Warb. The vernacular name was later used by Aublet to give the generic name *Hevea*.

The first use of *Hevea* spp. by the Indians of the north-western Amazon was probably gathering the seeds for food; they are sometimes eaten ceremonially and may also be used in time of famine. The seeds of the related *Micranda* spp. and *Vaupesia cataractarum* R. E. Schultes are used in a similar manner. The seeds are toxic to man until the cyanic poisons are removed by long soaking or by boiling. During boiling the oil may be extracted and may be used for illumination. It has been suggested that the Indians cultivated the trees for these purposes and also extended the species beyond their natural range, with the result that hybridization and introgression of *H. brasiliensis* occurred, particularly with *H. pauciflora*. Schultes (1956) refutes these hypotheses.

After the discovery of the use of the latex of *Hevea* spp. by the Indians for the manufacture of various articles, bits of rubber were brought back to Europe and aroused considerable curiosity. The French gave it the name of caoutchouc, a vernacular name of the product. In 1770 the English chemist Priestly discovered accidentally that it would rub out pencil marks, hence the name india-rubber, and small cubes were put on the market at 3s. each. Some attribute this discovery to Nairne, who started selling cubes of it in Cornhill in 1770. Gradually other uses were found for the product. Small exports were made from Brazil in the form of bottles and also articles such as shoes, balls and toy figures. In 1763 French chemists found that it could be dissolved in turpentine; later ether, and then rectified petroleum were found to be better solvents. Rubber tubing was first made in 1791.

In 1820 Hancock began manufacturing goods of masticated rubber in England, extending this to France in 1828 and to the United States in 1832. In 1823 MacIntosh made waterproof raincoats (mackintoshes) by coating fabric with rubber dissolved in naphtha; unfortunately they became tacky in the sun. Then came the all-important discovery of vulcanization by Goodyear and/or Hancock in 1839, in which rubber was heated with sulphur. Unlike the raw product, which became sticky when hot and brittle when cold, vulcanized rubber retained its physical properties unchanged from 0°–100°C; compounding with active and inert ingredients and formation into useful shapes was now possible. This led to a sudden and large expansion in the demand for raw rubber with extensive exploitation of the wild trees in Brazil, until by 1865 the demand exceeded the supply, which in turn led to the founding of rubber plantations in the East (see below). Rubber was also collected from other plants, notably *Ficus elastica* Roxb. (Moraceae) in India, *Castilla elastica* Cerv. (Moraceae) and *Manihot glaziovii* Muell.-Arg. (Euphorbiaceae) in tropical America, and later *Funtumia elastica* (Preuss) Stapf (Apocynaceae) and *Landolphia* spp. (Apocynaceae) in Africa, the first four of which were also planted. Thus until 1900 no single plant was established as the principal source of rubber.

The use of waterproof clothing in the American Civil War brought about the first rubber boom. A further impetus was given when Dunlop rediscovered the pneumatic tyre in 1888 and with the development of the motor-car after Daimler's invention of the internal combustion engine in 1885. Today some 70 per cent of the total rubber consumption is in the manufacture of tyres and tubes and other items associated with automotive transport.

It has been estimated that some 50,000 different products are made from rubber directly or indirectly. About 6 per cent of the world's rubber is used for footwear, boots, shoes, soles and heels, and 4 per cent for wire and cable insulation. Miscellaneous manufactured articles include rubberized fabrics, shock absorbers, washers and gaskets, transmission and conveyor belting, hoses, sports goods, household and hospital supplies, contraceptive appliances, in paints, etc. Sponge rubber from foamed latex is used in upholstery, mattresses, etc. Vulcanite or ebonite, a hard high-sulphurized rubber, is used in electrical and radio engineering industries and for protective lining in chemical plants. Rubber powder with bitumen is used for road surfacing.

The loss of Para rubber from Malaya and Indonesia due to the Japanese occupation in World War II led to the use of many other sources of natural rubber, including the Russian dandelion (*Taraxacum kok-saghyz* Rodin: Compositae), the Mexican guayule (*Parthenium argentatum* A. Gray: Compositae) and *Cryptostegia* spp. (Asclepiadaceae). It also led to the rapid developments in synthetic rubber production. The increasing use of plastics adds further competition to natural rubber.

ORIGIN AND DISTRIBUTION

Hevea brasiliensis grows wild in the tropical rain forests of the Amazon basin. With the exception of a small area west of Manaos, it is apparently confined to areas south of the Amazon river, to the Paraná and Matto Grosso regions of Brazil and Bolivia and to the Madre de Dios in Peru.

As has been shown above, there was considerable confusion in the early sources of rubber, as the latex of several tree species yielded this product. Local Indians knew of its use and small quantities reached Europe from the Para area early in the 19th century. It was the discovery of vulcanization in 1839 which led to the greatly increased demand and the extensive exploitation of the wild *Hevea* trees both in the lower and upper Amazon. Spruce describes how the entire population set itself to harvest rubber. The tappers, or *seringueiros* as they were called, hacked the trees with small axes which caused callus growth and often spoiled the trees for further use.

In 1824 Hancock had suggested growing rubber in plantations, but no notice was taken of this. In 1870 Sir Clements Markham of the India Office, who had been responsible for collecting planting material of *Cinchona* (q.v.) in South America, suggested that rubber should be obtained from tropical America and plantations established in Asia in order to maintain the world's supply, which was in danger of extinction. At his request James Collins reviewed rubber-producing plants and his *Caoutchouc of Commerce* was published in 1872. He had access to Spruce's collections and recommended obtaining *Hevea* in Brazil. Through Markham's efforts and those of Sir Joseph Hooker, Director of the Royal Botanic Gardens, Kew, and President of the Royal Society, a few seeds were collected by Farris in Brazil and sent to Kew in 1873. Twelve plants were raised and sent to Calcutta, but failed in Sikkim. A second consignment of seeds sent from Brazil to India in 1875 failed to germinate.

Markham then sent Robert Cross to Panama to obtain seeds of *Castilla elastica*, to Para where he obtained 1,000 plants of *Hevea* and to Ceara to get planting material of *Manihot glaziovii*. Cross returned to Kew on 22nd November, 1876, but none of his *Hevea* plants were to reach the East.

In the meantime Hooker commissioned H. A. Wickham (later Sir Henry), who was residing at Santarem at the time, to collect seeds of *Hevea*. This he did with great promptitude; he collected some 70,000 seeds in an area between the Tapajoz and Madeira Rivers of the central Amazon basin for which he received £10 per 100 seeds. It is often said that it was an extremely lucky chance that almost the entire rubber plantations of the East were derived from these seeds, as the quality of the material has never been surpassed. Schultes (personal communication) informs me, however, that *H. brasiliensis* in western Amazonia is better, and that the strain from Territorio de Acre in Brazil and Beni in Bolivia is far superior. It should be remembered that at that time there was no real knowledge of what species or even genus of plants was the best source of rubber; Wickham was not a botanist and must have depended upon native knowledge for the gathering of the seeds; the seed has short viability and must be gathered shortly after shedding, the bulk of which takes place over a comparatively short period. By chance his seeds were collected in an area where *H. brasiliensis* was the only species of *Hevea* available; elsewhere *H. guianensis* is commoner and *H. benthamiana* as common and there are five other spp. of *Hevea* which Wickham might have collected and which give inferior rubber. Furthermore, most species are very variable and interspecific crosses are common. It was indeed fortunate that the East was stocked from this source and that Cross's plants, collected nearer the mouth of the Amazon, where inferior strains occur, never reached the East. Other collections of Para rubber and other species of *Hevea* were taken to Java later, but none have been found as good as the Wickham material.

Wickham chartered a cargoless ship, S.S. *Amazonas*, which was lying in the Amazon at the time, thus avoiding delay which might well have resulted in all the seeds losing their viability. It is usually stated that Wickham smuggled the seed out of Brazil, but this is not the case as the export of *Hevea* seeds from Brazil was not prohibited at that time and an official Brazilian statement has shown that they were brought out with the goodwill and co-operation of the government. (*Borracha no Brasil*, Rio de Janeiro, 1913). The seeds arrived at Liverpool and were taken by a special train to Kew where they arrived on 14th June, 1876. They were promptly planted; less than 4 per cent germinated and some 2,800 young plants were raised. Meanwhile the Government of India, who had financed the search, had concluded that its success with rubber was uncertain and so the Colonial Office arranged for the bulk of the seedlings to be sent to Ceylon. Consequently on 9th August, 1876, 1,919 seedlings in 38 Wardian cases (miniature greenhouses) were dispatched there and planted at Henaratgoda. Two days later 50 plants were sent to Singapore; it has been stated that these died on the Singapore quay due to delay in payment of freight; another report states that 5 plants survived in poor condition but subsequently died. In the same year a few plants were sent to Buitenzorg in Java.

The following year, on 10th June, 1877, Kew sent a further 22 plants of Wickham's collection to the Singapore Botanic Gardens. Of these, 9 plants were planted in the Gardens, 9 were planted in Sir Hugh Low's residency garden at Kuala Kangsar in Perak, Malaya, one may have been taken to Malacca and there were 3 losses. Murton, superintendent of the Singapore Botanic Gardens, attempted to increase his trees by vegetative propagation, but found that only cuttings from young seedlings rooted satisfactorily. The trees in Singapore began fruiting in 1881 and Cantley, who succeeded

Murton, multiplied the stock, but, due to the lack of funds, they were inadequately looked after. When H. N. Ridley arrived in Singapore in 1888 as the first scientific director of the Botanic Gardens there were 9 trees of the original introduction, 21 5-year-old trees and 1,000 younger trees, mainly seedlings. The Kuala Kangsar trees also fruited and Swettenham (later Sir Frank) planted some 200 seedlings in 1884. The large introduction into Ceylon flourished and seeded and the seed crop there in 1888 was 20,000 seeds. Singapore's seed supply was augmented from Ceylon, again from the same Wickham stock. During the growth of the industry over 7 million seeds, as well as many seedlings, were distributed from the Singapore Botanic Gardens alone. Java made additional introductions of seed from the Wickham stock, but in 1896 and in subsequent years obtained further seeds from Brazil, including *H. spruceana* and *H. guianensis*. New introductions from Brazil were not distributed as none were as good as the introductions from Wickham's collection.

With great perseverance, little encouragement or help and slender financial resources, Ridley set to work in Singapore. He showed that *Hevea brasiliensis* was superior to all other rubber-producing plants then available. Within 4 months of arriving he reported that very little interest was being shown in rubber in Malaya, although he forecast an increase in demand. He discovered the excision method of tapping, the opening up of the same cut which increased the flow of latex (wound response), the best time to tap, and the regeneration of the bark, which could then be re-tapped (see below). He devised methods of coagulating the latex and studied tree growth, yields, pests and diseases. Simultaneously, Trimen, and later Willis, were carrying on work in Ceylon and research was also being done in Java.

Having shown that rubber was a commercial proposition, Ridley then tried to persuade the reluctant planters to give the new crop a trial, but at first he met with little success. Coffee was then the important crop in Malaya and for his efforts he was known as 'Mad Ridley'. In 1893 he distributed plants and seeds to residents and district officers of the Federated Malay States to plant near their homes. Low at Kuala Kangsar had also distributed some seeds. The first plantation was made on Ridley's advice by a Malacca-born Chinese, Tan Chay Yan, in 1898. It was fortunate for Malaya that, when rubber-planting began in earnest (coffee prices had slumped and the plants were attacked by pests and diseases while there was an increased demand and price for rubber), there was a good supply of planting material at the Singapore Botanic Gardens and Ridley with the latest 'know-how' to advise and help the planters. Ridley stayed long enough in Singapore to see the great rubber boom of 1910 and Malaya reach its important position in world rubber production which it has never lost. Thus rubber, the most recently domesticated economic crop, has developed into a large-scale industry during the present century. Its centre of production is on the opposite side of the world from where it originated, Asia now producing 92 per cent of the world's natural rubber. Ridley, 'the father of Malaya's rubber industry', died in his 101st year on 24th October, 1956.

Para rubber was introduced into various tropical African countries early in the 20th century. Thus Uganda received a seedling from Kew in 1901 and obtained seeds from Ceylon in 1903. By 1912 nearly 1,500 acres had been planted, which had increased to nearly 15,000 acres by 1920, but the area has now declined. Nigeria is now the largest producer in Africa and this continent supplies 7 per cent of the world's rubber. The Firestone Tyre and Rubber Company began plantations in Liberia in 1924. The Ford Motor Company made the first attempt to establish *Hevea* as a plantation crop within its natural range at Fordlandia on the Tapajos River in Brazil in 1928. This was a failure due to attacks of South American leaf blight, *Dothidella ulei,* which limits severely the economic production of Para rubber in the New World. Fortunately the dreaded disease has not reached the Old World. Brazil, which was the only rubber-exporting country up to 1870, a product of wild trees, now provides only one per cent of the world's total and imports rubber from Malaya. The Goodyear Plantation Company made plantings in the Philippines in 1928, and these estates were a repository for selected planting materials from the East. Subsequently the United States Department of Agriculture has continued research, collection of new planting material and development programmes in South and Central America.

ECOLOGY

Wild Para rubber grows in the tropical evergreen rain forest of the Amazon basin, often in periodically flooded areas, but larger trees are found on the well-drained plateaux. Most of the planted rubber is grown between 15°N and 10°S where the climax vegetation is lowland tropical forest and where the climate is hot, humid and equable, with temperatures ranging from 74–95°F and a well-distributed rainfall of 75–100 in. or more per annum. Areas with wide temperature ranges and pronounced dry seasons are not ideally suited to rubber, although it is grown under these conditions in southern India and Vietnam.

Wickham had collected his seeds on an undulating plateau, but Cross had obtained his from a swampy area. On account of the latter the trees sent to Singapore in 1877 were transferred to the edge of a swamp; those at Kuala Kangsar were planted on moist, well-drained soils which were more suited to *Hevea.* Although the crop will tolerate a wide range of soils with pH from 4–8, the optimum range is pH 5–6 and it does best on deep, well-drained loams. Shallow, badly-drained or peaty soils should be avoided and lime is deleterious.

Much of the rubber in Malaysia is grown on land formerly covered with lowland Dipterocarp forest. It is important in clearing the forest to maintain a ground cover, and this is done by planting leguminous cover crops. Early plantings had usually been clean weeded which caused erosion and loss of organic matter.

STRUCTURE

A quick-growing tree, rarely exceeding 25 m in height in plantations, but wild trees over 40 m have been recorded. Copious latex in all parts, usually white, sometimes tinged yellow.

ROOTS: Well-developed tap-root, 2·5 m long at 3 years, with laterals 7–10 m long.

TRUNK, BARK AND BRANCHES: Trunks of seedling trees taper from base and are conical in shape; trunks of buddings are cylindrical. Both seedlings and buddings have periodicity of growth. During the resting stage whorls of scale lvs occur round terminal bud. Terminal buds of main stems of saplings grow rapidly producing long internodes towards the end of which lvs are clustered. Lvs thus appear in tiers. In trees which are old enough, lvs shed, terminal buds of branches grow rapidly and trees are temporarily bare of lvs, a condition known as 'wintering'; new lvs are then produced at proximal end and inflorescences in axil of scale lvs and lower foliage lvs; new leaves then harden off. Fruits borne on proximal part of long shoots. In Malaya there is usually a 'wintering' of all trees after the dry weather at the beginning of year and often a second complete or partial leaf-change in August or September. At other times when there is a short dry spell, trees may change their lvs on one or two branches only.

Bark consists of outer corky layers and greenish cork meristem; then the hard bark which is orange-brown in colour and with large numbers of stone cells which are more numerous towards the periphery, parenchyma and disorganized sieve tubes and a few latex vessels; then comes the soft bark consisting mainly of vertical rows of sieve tubes with some medullary rays and latex vessels which are more numerous nearest the cambium. Thickness of bark and proportion of tissues varies with different clones, and age of tree, whether virgin or renewed bark, whether seedling or budded, and in case of seedling with height from ground. Bark thickness varies from 6·5–15 mm, but is usually 10–11 mm. Complete renewal of bark after tapping takes 7–8 years.

Latex vessels, which are modified sieve tubes, are formed from the cambium as cells, which fuse and the cross-walls disintegrate. They run in concentric cylinders in a counter-clockwise direction at an angle of about 3·5° to the vertical; clockwise spirals have now been found in some clones. The vessels are laterally interconnected within each ring, but connections are disrupted as trunk expands. The number of vessels per ring and number of rings vary with age and thickness of bark and with the clone. Oldest, outermost vessels contribute little or no latex on tapping virgin bark as they are broken during increase in girth; in renewed bark the number of functional vessels is increased (also see TAPPING below).

LEAVES: Spirally arranged, trifoliate, glabrous, with deciduous stipules and 3 extra-floral nectaries at junction of leaflets with petiole, which only secrete nectar on the new flush during flowering. Oldest lvs of flush are larger with longer petioles than those produced at end of flush. Petiole 2–70 cm long, but usually about 15 cm. Leaflets short-stalked, elliptic or obovate, entire, base acute, apex acuminate, dark green above, paler glaucous beneath, veins pinnate with about 20 pairs; lamina 4–50 × 1·5–15 cm, usually about 15 × 5 cm; petiolule 0·5–2·5 cm. Young lvs purple-bronze, becoming green on hardening and turning orange-brown, red or yellow before falling.

FLOWERS: Borne in many-flowered, axillary, shortly pubescent panicles on basal part of new flush; fls small, scented, unisexual, shortly-stalked, with larger female fls at terminal ends of main and lateral branches and more numerous smaller male fls, with 60–80 males to each female fl.; calyx yellow, bell-shaped, with 5 narrow triangular lobes; petals absent, male fls 5 mm long with 10 sessile anthers set in 2 superposed circles of 5 each on central slender column; female fls 8 mm long with green disc at base and 3-celled, shortly pubescent ovary and 3 white short sessile stigmas. Flowering takes place over a period of about 2 weeks with some male fls opening first, lasting for 1 day and then dropping, followed by female fls which remain open for 3–5 days; remainder of male fls then open.

FRUITS: Only a small proportion of female fls set fruit and of these 30–50 per cent fall off after a month and more fall off later. Fruit wall attains full size in about 3 months, when embryo still microscopic; ripens 5–6 months after fertilization. Mature fruit a large, compressed, 3-lobed capsule, 3–5 cm in diameter, with 1 seed per carpel. Mesocarp thin, coriaceous; endocarp woody. Dehisces explosively and noisily with endocarp breaking into 6 pieces and seeds are thrown a distance of about 15 yards.

SEEDS: Large, ovoid, slightly compressed, shiny, 2–3·5 × 1·5–3 cm, testa grey or pale brown with irregular dark brown dots, lines and blotches. The testa being derived from the female parent and the seed shape being determined by the pressures of the capsule, it is possible to identify the female parent of any seed by its markings and shape and is the most reliable method of identifying 'clonal' seed. Endosperm white in viable seeds, turning yellow in older seeds. Weight of seed 2–4 g.

POLLINATION

Entomophilous. No pollen grains of *Hevea* have been found in spore traps over the flowering period. In Malaya 21 spp. of insects have been seen to visit the flowers, but midges and thrips predominate. In Puerto Rico pollination is by midges of the genera *Atrichopogon*, *Dasyhelea* and *Forcipomyia* and thrips (*Frankliniella* spp.). In Brazil midges involved also include the genera *Culicoides* and *Stilobezzia*. Midges are only active around sunrise and sunset; thrips operate throughout the day. As the main pollinators do not fly far and wild *Hevea* in Brazil is widely scattered with about 2 trees per acre, it is likely that much of it is self-pollinated and that

Wickham's collection may have included a number of inbred lines (see BREEDING below).

Although it is frequently stated that many clones are self-incompatible, this is not so, although it has been found, *e.g.* 'PIL. B84.' Out of 2,744 clones tested for male sterility in Malaya only 11 were suspected to be completely male sterile and 26 to be partially male sterile; the rest were found to be normal. The majority of clones set more seed when the opportunity for crossing occurs.

In hand-pollination all the flowers are removed from the inflorescence except for a few ripe, but unopened female flowers; staminal column from male parent is removed with clean forceps and inserted over stigma of female flower, which is then sealed with a plug of cotton wool dipped in latex. Relatively heavy losses of young fruits occur and pollination success is about 4–5 per cent. The flowers and developing fruits may be dusted with fungicides and insecticides, more particularly against *Oidium* attack. As fruits dehisce explosively, maturing fruits must be protected by a net bag. Clones do not always flower at the same time, but pollen may be stored for 20 days over 30 per cent H_2SO_4 at 6°C (relative humidity of 70 per cent). It can be transported in vials in thermos jars with ice.

GERMINATION

Seeds are viable for a short time only, so must be planted as soon after harvesting as possible. Viable seeds germinate in 3–25 days (also see PROPAGATION below). Germination is hypogeal.

CHEMICAL COMPOSITION

Fresh latex consists of a colloidal suspension of rubber particles in an aqueous serum. The content of rubber hydrocarbon varies from 25–40 per cent, with an average of about 30 per cent. It is cis-1,4-polyisoprene with a formula of $(C_5H_8)_n$. It is manufactured in the plant from carbohydrates and its function in the plant is not known. Although it is an energy-rich compound, there does not seem to be any means of readily releasing this energy within the plant. Non-rubber constituents present in the latex include proteins, resins, sugars, glucosides, tannins, alkaloids and mineral salts.

The kernels, which constitute 50–60 per cent of the seed, contain 40–50 per cent of a semi-drying oil, which may be used for soap-making. The cake after oil extraction contains about 30 per cent protein. Seeds contain a cyanogenetic glucoside, linamarin, and an enzyme, linase, which hydrolyses the glucoside to hydrocyanic acid.

Fig. 25. **Hevea brasiliensis :** PARA RUBBER. **A,** shoot with dehiscing fruit (×½); **B,** inflorescence (×½); **C,** male flower cut open (×3); **D,** female flower in longitudinal section (×3); **E1–2,** fruits (×½); **F,** seed (×½).

A
B
C
D
F
E1
E2

PROPAGATION

SEED: All the original plantings and a fair amount of the world's rubber was from unselected seeds from the original Wickham introduction. It shows considerable variability in yield, growth, branching and bark characters. In the present highly competitive market it is essential that these plantings should be replaced by the better planting material which is now available (see IMPROVEMENT below).

In addition to seedling rubber, it is also necessary to plant seeds to obtain stocks for budding. Seeds may be planted at stake, usually two or more seeds per hole, and the seedlings are then selectively thinned. This is still extensively practised for budding in the field. Alternatively the seeds may be planted in nurseries. Seeds are viable for a short time only—up to a month—and so must be planted soon after harvesting. They can be stored for a few weeks more in damp powdered charcoal or damp sawdust in perforated polythene bags and this method is used when transporting seeds. Commercial 'clonal' seeds are stored in cold storage at about 4°C; this often gives reduced, but tolerable germination.

For nursery planting seeds are germinated in shaded beds of friable soil, sand or coir dust, the seeds being pressed into the surface of the beds, touching each other, shaded with sacking and watered daily. Germinated seeds are removed daily after the fifth day and germination is more or less complete in 3 weeks. The germinated seeds are transplanted into shaded nurseries at spacings depending on the type of plant required (see below).

For seedling rubber the germinated seeds are planted in the nursery at 1 × 1 ft or 18 × 8 in., giving approximately 40,000 plants per acre, of which 25,000–30,000 should reach the standard size for pulling at 10–15 months. The entire plant is dug or pulled out of the ground, the stem cut back to 18–24 in. of brown bark, the tap root pruned to 18–24 in. and the side roots to 4 in. The seedling stumps are then tied into bundles, transported to the field and planted. It is also possible to plant the germinated seeds in baskets or perforated polythene bags. 'Twin' seedlings may be produced by decapitating the hypocotyl and then splitting it from top to bottom leaving each half attached to a cotyledon.

CLONAL SEED: This is an unfortunate term as such seeds are the sexual progeny of budded clones and seed garden progeny is probably a better name. The seeds may be obtained from monoclone plantings of self-fertile clones or by planting a mixture of a small number of clones, which are known to produce high-yielding families. This is usually done in special seed gardens, isolated from other rubber, and on flat land to facilitate harvesting of the seeds. In a polyclone garden the progeny can only have two parents or may be the result of self-pollination; as the male parent is not known the seed is illegitimate. It is usually desirable to obtain as much cross-pollination as possible, as selfing may reduce vigour. The clones used as seed parents should produce 20 lb or more of dry rubber per tree per annum. The seed harvested rarely exceeds 3 lb per tree per annum. If at some future date good parents can be found which are male- or self-sterile, seeds of guaranteed parentage could then be produced (see Cocoa).

As clonal seedlings are more variable than budded rubber and their average yield is less, it is desirable to select for vigour in the nursery and the field, and later for yield in the early years of tapping, cutting out until the desired spacing is obtained. Seedlings are cheaper to produce than buddings and can be tapped a year earlier than brown buddings.

Legitimate seed may be produced by hand pollination between selected clones, but it is much more expensive to produce and the yield of seed is less.

VEGETATIVE PROPAGATION: Soon after the introduction of rubber to the Singapore Botanic Gardens in 1877, Murton showed that cuttings from young seedlings rooted easily, but this became progressively more difficult as the trees got older and by the time their yield potential had been assessed rooted cuttings could not be obtained. However, in 1960, by using modern techniques of mist propagation, it was found possible to root leafy cuttings of certain clones (see Cocoa). Unfortunately trees derived from cuttings are blown over more easily and are difficult to bud successfully. Quite early, rubber was successfully marcotted, but the marcots were more susceptible than seedlings to termites and to wind damage. All forms of cleft and similar graftings were unsuccessful, as exuding latex prevented union.

The first successful bud-grafting of *Hevea* was made in Java by van Helten in 1916 and Maas in Sumatra in 1917. This was modified by Forkert, who showed that the best results were obtained by taking the bud slip with a sliver of wood, which could be removed with the minimum distortion of the bark without latex exudation. It is essential that rootstock and scion should be in an active state of growth and that their cambial tissues should be closely appressed and tied in place. Open-pollinated seed from vigorous, strong-rooted, high-yielding parents should be used for producing stocks. There does not appear to be any incompatibility between normal stocks and scions or other effects. For many years it was customary to bud onto seedling stocks at 12–18 months, either in the field or nursery. This is now known as brown budding. Since 1960 this is being superseded by Hurov's method of green budding of seedlings 4–6 months old (see below). Crown budding or 3-component budding is used in the American tropics to obtain crowns resistant to *Dothidella ulei*. An alternative to the latter is high-stump budding at 6–8 ft.

BUDWOOD NURSERIES: These are established by planting germinated seeds in rows 3 ft apart with 2 seeds, 6 in. apart, every 2 ft along the rows. The stronger seedling of the two is budded at 1 year old with the desired clone, or the smaller seedling if the larger one fails, giving a final stand of 3 × 2 ft. Such nurseries are usually maintained for about 7 years during which time 5 crops of budwood are taken; for brown budding the first buds are taken at 15–18 months, when 1 plant yields 2–2½ yards of budwood, of which 1 yard yields 10 buddings. For green budwood nurseries see below.

BUDDED STUMPS: Stocks are grown in nurseries from germinated seeds planted 1 ft apart in rows alternately 1 ft and 2 ft apart and are budded at 12–18 months as near the ground as possible. Two parallel cuts are made and the latex that exudes is removed before cutting the panel at the top and the bud is inserted after removing the sliver of wood taken with the bud, care being taken not to touch the surfaces which come in contact. The bark panel on the stock is pulled over the bud, and the central portion of a palm frond is applied and bound tightly with hessian twine. A rubber leaf is tied to hang over the graft to protect it from rain and sun. The patch is inspected after about 3 weeks, the green colour of the patch indicating that it has taken. The flap may be cut off at this time. It is inspected again after a further 2 weeks, after which the stock may be topped to force the bud to grow. The budded stump may then be lifted or left for a week and then lifted, during which time the bud dormancy is broken. The cut end should be dipped in hot wax to seal it. Lateral roots are pruned to 2–6 in. long. Budded stumps may be established in large polythene bags before transfer to field.

STUMPED BUDDINGS: Germinated seeds are sown in the nursery at 2 × 1 ft, subsequently selectively thinned to 2 × 2 ft. Brown budding is the same as above. The stock is cut back after a successful take and the scions are grown on in the nursery for a further 12–18 months. They are then cut back to 6–7 ft above the union and pulled as above for transplanting but leaving more root. Such plants can be brought into tapping 18 months earlier than budded stumps, but this method is not much used as stumped buddings are more costly to produce and transport and losses are higher due to shock at transplanting.

GREEN BUDDING: Green budwood is usually obtained by cutting back plants in the nursery to force many shoots and by stripping green dormant buds from the non-leafy parts of new terminal growth flushes; it can also be obtained from young scions or vigorous flushes on mature trees. The budwood must be protected against drying out and may be stored in damp sawdust for up to 3 days. The stocks may be grown in polythene bags in the nurseries. Young green bark is taken off the stock of the same size as the bud slip, 2½ in. long, leaving a small flap at the top. The bud slip is applied and bound with transparent plastic strip, 12 in. long × ¾ in. wide. The best age for green budding is 3–6 months; if very young stocks are used the starch reserves may be depleted and die-back of the scion occurs, although application of fertilizers may help. Die-back also occurs if unfavourable dryish weather follows the cutting back. Thus the best time to bud is when the stocks are big enough and when favourable weather can be expected to follow. Three weeks after budding the plastic strip may be removed and if the bud is still green the stock is cut back. The buddings are transplanted to the field when one or more whorls of leaves have hardened off, usually about 3 months after budding. A special green budwood nursery may be made by planting rootstocks at 3 × 3 ft and budding with the required clone. The scions are allowed to grow for 6 months and the tops removed at 3 ft. They are pruned subsequently to form bushes with 3–4 main branches. After cutting back new green budwood is obtained in 2–3 weeks and yields 30–40 buds per plant at each cutting.

FIELD BUDDING: Seedling stocks planted at stake in the field may be green budded at 5–6 months, but this should not be done during or when dry weather is expected. Brown budding may also be done at stake on plants about 1 year old. For field budding two or more seeds are planted at stake; if more than two these are selectively thinned to two. The most vigorous seedling is budded and, if this fails, the second seedling is then available.

AFTER CARE: All planting material, green or brown buddings or seedlings are pruned to restrict development to one scion stem without branches to about 10 ft in height. In brown buddings snagging may be necessary by removing the remains of the seedling stock at the level of the scion union.

CROWN BUDDING: This is used in the New World where *Dothidella ulei* is serious. A three-component tree is produced in which high-yielding, but susceptible oriental clones are budded on seedling rootstocks and this is grown, controlling the disease with fungicide. It is then crown budded at a height of 6–8 ft to produce a disease resistant crown. The latter, being low-yielding, tends to lower the yield at the proximity of the cut to the upper union and is rather subject to wind damage.

HUSBANDRY

PLANTING: If rubber is planted in areas of primary forest, all commercial timber is usually extracted; the remainder of the trees are felled, and the stumps are removed along the lines or contours of the future planting. Mechanical clearing of the forest is possible on big estates. In the early plantings all plant debris was burnt, followed by clean weeding, resulting in severe leaching, loss of soil structure, and soil erosion. This was followed by no-burn and minimum disturbance with regeneration of natural cover which was controlled by periodic slashing. Nowadays it is usual to burn and to establish cover crops (see below). In replanting old rubber land, the old trees may be removed mechanically or they may be poisoned and, if cut down, the stump surfaces are creosoted.

High densities of planting give the highest yield per acre, but the trees take longer to reach tappable size and give lower yields per tree and per tapper. Optimum density must be a compromise between yield per tree and yield per tapper. Currently final tapping stands of 100–120 trees per acre are recommended and are obtained by thinning from original plantings of 150–180 buddings per acre or 200–240 seedlings per acre according to the degree of nursery selection. Half the thinning is done between 3 years old and the beginning of tapping on a basis of girth, and the remainder during the first 3 years of tapping on the basis of yield. Planting distances may be 15 × 16 ft for buddings and 13 × 14 ft for seedlings, or 30 × 8 ft for buddings and 30 × 6 ft for seedlings. The wider spacing between rows increases the susceptibility to wind damage. Square or rectangular planting cannot be done on steeply sloping land where contour terraces are necessary.

In some countries the rubber is planted at wider spacings with permanent intercrops such as coffee or cocoa. On most smallholdings intercropping is usually practised for the first 3–5 years, but it is recommended that annuals should not be nearer the rubber than 4 ft and bananas 6 ft, and that cultivation should cease after 3½ years. Herbicidal weed control is extensively practised, particularly against grasses, especially *Imperata cylindrica* Beauv. (*lalang*) which has a very deleterious effect upon rubber and also provides a fire hazard.

COVER CROPS: On poor soils or when replanting rubber on any but the best soils, leguminous cover crops are strongly advocated. A good cover crop should have the following characteristics: (1) should be perennial and easily multiplied and established, preferably by seed; (2) rapid and vigorous growth is desirable to cover the soil and suppress weeds; (3) should not be competitive with the rubber; (4) should not require too frequent control to prevent it scrambling over or shading the trees; (5) should tolerate pruning or slashing if required and not die off when cut back; (6) should be able to establish itself on poor soil; (7) should have abundant leaf and provide rich litter which rots rapidly; (8) should fix nitrogen and provide this nutrient for the rubber; (9) should grow well in the early stages in full sunlight, but should also be shade tolerant and persist when the tree canopy closes over; (10) should be resistant to drought; (11) should be resistant to pests and diseases, particularly those which might attack the rubber; (12) should not form products which are toxic to the rubber; (13) should be easily eradicated when required.

In Malaya a mixture of *Calopogonium mucunoides* Desv. which grows quickly, gives a good early cover in sun and lasts 18 months, *Centrosema pubescens* Benth. which makes slow early growth but persists under shade, and *Pueraria phaseoloides* (Roxb.) Benth. which is the most vigorous of them when well established with little overhead shade, is recommended in the proportion of 2 : 2 : 1. The seeds should be treated with concentrated sulphuric acid or mechanically scarified before sowing to speed up germination and they are usually inoculated with the correct *Rhizobium*. They are sown in drills with rock phosphate at a seed rate of 5–7 lb per acre. It is necessary to keep them free of weeds during the early growth. Other legumes used as cover crops in rubber are the shrubby *Moghania macrophylla* (Willd.) O. Ktze. (syn. *Flemingia congesta* Roxb.), which can withstand repeated brush cutting, and the climbing *Psophocarpus palustris* Desv. Large, coarse grasses have a deleterious effect on rubber, but the short *Axonopus compressus* (Swartz) Beauv. and *Paspalum conjugatum* Bergius have been found to be beneficial.

MANURING: In Malaya, rubber on the coastal alluvial clay soils seldom responds to fertilizer application with the exception of nitrogen. Inland loam and clay loam soils respond mainly to nitrogen and phosphorus. On the inland sandy soils responses to potassium are found, as well as to nitrogen and phosphorus. Magnesium deficiency is widespread, also in Indonesia, Ceylon and Nigeria, and produces a chlorosis in the interveinal areas of the leaf and the yellowing is usually contiguous with the leaf margin. It is corrected by the application of magnesium limestone or kieserite. Manganese deficiency may also occur in Malaya and produces pale green leaves with midrib and veins dark green; it may be corrected by the application of manganese sulphate. Potassium deficiency with marginal and tip chlorosis, followed by a marginal necrosis, is fairly common in Malaya, and also occurs in Vietnam, the Ivory Coast and Nigeria.

Nurseries normally receive a NPK mixture and P is commonly applied in the planting holes. The value of fairly generous fertilizer application to ensure rapid growth of immature rubber, particularly during the first two years after planting or budding, has been well established. In the absence of leguminous cover crops this should be continued up to tapping as there is a positive correlation between the growth made during this period and the yield during the first 6–7 years of tapping; fertilizing also ensures the maximum growth which is important in bringing trees to a tappable size as quickly as possible. Mature rubber usually gives modest but profitable yield responses to fertilizers.

Where there are vigorous pure leguminous creeping covers N fertilizers may be discontinued from the third to the fifth or sixth year after planting. Rock phosphate ploughed in at the rate of 4½ cwt per acre before sowing the legumes markedly increased their rate of establishment compared with broadcast dressings of 1–2 cwt per annum.

Various formulations using ammonium sulphate, Christmas Island rock phosphate, potassium chloride and kieserite (26 per cent MgO) are recommended in Malaya, depending on the soil type and the age of the trees. It is considered sound practice on the inland soil types to apply 2–3 lb of this NPKMg fertilizer per mature tree annually up to 10 years before replanting. Foliar analysis is now being used in Malaya as a guide to fertilizer requirements.

TAPPING: Ridley's invention of excision tapping at the Singapore Botanic Gardens in 1889 and its subsequent development was a major contribution to the economic exploitation of Para rubber. Previous to this, the incision method by making a series of cuts with a cutlass often resulted in injury and gave poor yields. In the excision method a thin paring of bark is made from a sloping cut with a knife with a V-shaped cutting edge, leaving a grooved channel along which the latex can flow. Ridley showed that draining out the latex stimulates the production of more and that on re-opening of the cut the flow of latex was increased to a certain maximum. He also showed that the bark was renewed and could be tapped again. After testing various patterns of tapping he finally established the double fishbone method. Nowadays most rubber is tapped on a single half spiral cut.

The latex vessels occur in the phloem and the number of functional latex vessels increases as the cambium is approached; consequently it is necessary for the cut to be as near the cambium as possible without wounding it. The latex vessels occur in a counter-clockwise spiral at about 3·5° from the vertical (see STRUCTURE above); thus by making the tapping cut from top left to bottom right at an angle of 25–30° more latex vessels are severed and gives 7–8 per cent more yield than the reverse; the angle also permits the latex to run down without overflowing. The thickness of the bark varies with the age of the tree and the clone. There is more corky tissue in seedling trees and virgin bark and in the tapering trunks of the seedling there are more latex vessels nearer the base; in budded trees with their cylindrical trunks the number is roughly constant at all heights for bark of the same age. The latex runs along the cut and then down a vertical guide line, where a metal spout driven into the tree channels the latex into a cup, usually of glass or earthenware, held in a simple wire loop whose ends grip the trunk. Tapping is done as early in the day as possible; since the turgor pressure in the latex vessels is greatest at this time. The latex is obtained from an elongated heart-shaped area around the tapping panel with most of it from the area below the cut. The amount of bark removed at each tapping should be as little as possible to re-open the vessels, about $\frac{1}{16}$ in., which on alternate daily tapping uses up $\frac{7}{8}$ in. per month or 10 in. per year.

The first tapping begins when a proportion of the stand has reached a certain girth in 5–7 years. In buddings the first tapping is made when 70 per cent of the trees have reached a girth of 20 in. at the height of opening of 50–60 in. and is usually done on a half spiral at 30° on alternate days. The second panel is opened at the same height on the opposite side of the trunk. In this system 10 years is allowed for bark renewal before tapping again. On seedling trees the first tapping is made when 70 per cent of the trees have reached 20 in. girth at 30 in. high or 22 in. girth at 20 in. high. The first cut is made at 20–30 in. from ground level at 25°, usually on a half spiral and the trees are tapped every three days. The second panel is opened with at least 20 in. girth at 40–50 in. high, renewed bark being tapped after 13 years or so.

An international notation for tapping systems gives the length of cut as a fraction of the circumference, *e.g.* S/2 is a half spiral, S/1 is a full spiral, and the frequency of tapping, *e.g.* d/1 daily tapping, d/2 alternate days. The usual system on low panels is S/2.d/2 and is referred to as 100 per cent. S/2.d/1 would be 200 per cent; S/2.d/3 would be 67 per cent. If the trees are rested for a period this can be added as a fraction *e.g.* 6m/9 in which the tree is tapped for 6 months and rested for 3 months; thus S/2.d/2.6m/9 is 67 per cent intensity.

Continuous tapping on S/2.d/2 100 per cent retards girth increment severely and canopy development to a lesser extent in some high-yielding clones and is thought to be responsible for trunk snap by wind. In this case it may be advisable to give a rest period and S/2.d/2.8m/12 nearly doubled the girth increment in clone 'RRIM 501,' gave greater yields per tapper, but reduced yield per acre to 70 per cent. In brown bast susceptible material (see below) a reduced rate of exploitation is desirable, *e.g.* S/2.d/3 67 per cent. In clone 'Glen 1' this reduced yield per acre to 89 per cent but increased the yield per tapper to 130 per cent. On average over a number of clones S/2.d/3 67 per cent gave 83 per cent of the yield of tapping S/2.d/2 100 per cent. In older trees high panels may be opened at 100 in. but a ladder has to be used. High panels are normally only opened when double-cut tapping is practised, usually when lower bark of second renewal is being exploited.

Tapping begins at first light about 6 a.m. in Malaya. In Thailand it is customary to tap at night from 2 a.m. onwards. Task size varies greatly with the age and condition of the rubber, topography, etc. Between 6 a.m. and 9 a.m. a tapper can tap 400–500 trees for normal tapping, and tasks of up to 600 trees are sometimes given. With double-cut, ladder-tapping the task is about 200 trees. The tapper removes the coagulated latex from the previous cut and makes the new cut either upwards or downwards; tapping the high cut upwards gives rather higher yields than tapping downwards. The tapper returns to collect the latex at 11 a.m. and carries it in pails to a central point for bulking and transport to the factory. No tapping can be done during rain or if the panel is wet.

Yield stimulation can be done by the application of growth substance analogues such as 2,4-D and 2,4,5-T. They are usually applied at one per cent active ingredient in an oil carrier, usually 5 parts palm oil and 3 parts petroleum grease. They are normally applied to lightly-scraped bark 2–3 in. below the tapping cut at 6-monthly intervals. These compounds postpone the sealing of the cut ends of the vessels, thus prolonging the flow of the latex after tapping and the latex is withdrawn from a larger area of bark than in unstimulated trees and hence the area in which latex is regenerated in the vessels is also increased. They should not be used on young rubber as the extra yield is at the expense of girth increment, and in addition to increasing the risk of wind snap, they also restrict the future yield capacity of the tree. They are not recommended on thin renewed bark of old trees or trees on poor soils. The best yields are obtained by tapping virgin and first renewal bark on an S/2.d/2 100 per cent system with yield stimulation on the renewed bark (S/2.d/3 67 per cent for clones susceptible to brown bast). Yields drop markedly in bark of second renewal and a double-cut tapping with 2S/2.d/3 133 per cent without stimulation or 2S/2.d/4 100 per cent with half yearly stimulation of the high cut only are usually recommended. Yield stimulants are also used when the decline in yield occurs as the tapping cut approaches the junction of bark of different ages and on trees to be cut out within 6 years.

When rubber is to be replanted as yield declines with age due to the cumulative effect of tapping or for replacement with higher yielding material, intensive or slaughter tapping is done for 3 years before cutting out. Such trees may be tapped with 2 high V-cuts and 2 low half-spiral cuts tapped every 4th day (intensity 200 per cent) with the application of yield stimulants.

PROCESSING: Cleanliness of utensils and all equipment during all stages of processing is essential from tapping to the final product if high-grade rubber is to be produced. With clones in which latex coagulates easily, or when the latex is not delivered to the factory quickly, 10–25 per cent ammonia is added as an anti-coagulant. On arrival at the factory the latex is bulked and strained. An estimate of the dry rubber content of the latex, usually 30–35 per cent, is made by specific gravity measurements, usually with special hydrometers. In the preparation of smoked sheet rubber the latex is diluted with water to 12–15 per cent dry rubber content and is then re-strained using a finer screen. It is poured into the coagulating tanks and coagulated by adding acetic or formic acid, mixing thoroughly and quickly. One part of 4 per cent formic acid is required to coagulate 100 parts of latex diluted to 12 per cent rubber content; if ammonia has been added as an anti-coagulant more will be required. Froth formed on the surface is skimmed off as this would cause bubbles in the coagulated rubber. The rubber coagulates into a thick curd and the clear serum is run off. By using separators in the tank, thick sponge-like sheets are produced and these are passed 6–8 times between rollers of decreasing distances apart to produce a sheet not more than 2·5 mm in thickness and then through grooved rollers to produce a corrugated or ridged surface. The sheets are dried by hanging in a smoke house for 4 days at about 50°C. The final product should be a uniform golden-brown, semi-transparent sheet, without opaque spots or blemishes. The rubber is graded and packed in bales, 19 × 19 × 24 in., for export.

Crepe rubber is produced by passing coagulated sheet rubber through special shearing rollers running at unequal speeds. It may be bleached by adding sodium sulphite to the latex. Drying is done in special houses with controlled heat but not smoked. The lower-grade brown crepes are made from cup lump and other scrap.

Rubber may also be shipped as concentrated latex with a rubber content of 60–70 per cent, usually obtained by centrifuging. It can also be obtained by creaming with alginates or evaporation under reduced pressure. To prevent coagulation the ammonia content is adjusted to approximately 0·7 per cent for shipment.

Much effort is now being directed towards making natural rubber more competitive with synthetic by processing it into comminuted, or crumb form which can be dried rapidly and compacted into small, easily handled blocks. A scheme has also been developed under which both new and old forms of dry rubber produced in Malaya can be graded and sold to technical specification in a few uniform grades, instead of being marketed in a large number of grades based on visual characteristics such as colour, freedom from bubbles, mould, etc., which may not affect manufacturing properties.

YIELDS: These vary enormously according to the planting material used and the standard of husbandry. Average yields of unselected seedling rubber in India are about 300 lb dry rubber per acre per annum, and in

Malaya 400–500 lb per acre. Average yields of the best budded clones of the RRIM 500 series in commercial production in Malaya are about 1,300 lb, while in one trial clone 'RRIM 600' yielded at the rate of 3,885 lb per acre in its 9th year of tapping, having given an average of 2,178 lb over the 9 years. Modern clones yield approximately 600 lb per acre in 1st year of tapping, 1,000 lb in the 2nd year, 1,300 lb in the 3rd year, 1,400 lb in the 4th year, and 1,500 lb in the 5th year. In trials on estates RRIM 600 clones averaged 2,200 lb in the 5th year of tapping. There are quite considerable commercial areas of RRIM 600 series clones in Malaya that are yielding over 2,000 lb of dry rubber per acre.

MAJOR DISEASES

South American leaf blight, *Dothidella ulei* P. Henn.,* is the most serious disease of rubber, but is at present confined to South and Central America and Trinidad. It is the major limiting factor to rubber production in the New World, making plantation production almost impossible in many areas. The high-yielding Far Eastern clones are extremely susceptible. In the asexual stage on young leaves olive green spots appear on the under-surface and usually result in leaf shedding. If the leaves do not fall a sexual stage is reached on older leaves with small raised black perithecia usually on the upper surface and often bordering holes in the lamina where diseased tissue has died out. It also attacks young stems and fruits. Infected trees are defoliated at every new flush, resulting in die-back and eventually death of the tree. The best method of control is by breeding resistant clones; resistance is usually obtained from *Hevea benthamiana*, but they are usually much lower yielding than the Far Eastern clones. Crown budding of Far Eastern clones with resistant crowns has also been tried (see PROPAGATION above).

Powdery mildew, *Oidium heveae* Steinm., is restricted to Asia and Africa, but is only really serious in Ceylon, particularly at the higher altitudes of 1,000' ft. Circular white colonies occur on lower surface of leaves and cause leaf-fall during refoliation. In Ceylon regular treatment with sulphur is necessary.

Black stripe and leaf blight, *Phytophthora palmivora* (Butl.) Butl., attacks the tapping panel just above the cut (black stripe) and vertical black lines are seen in the wood on scraping away sunken discoloured areas of bark. It is treated with fungicide weekly (not copper) and tapping knives are disinfected. It causes a rapidly spread necrosis killing leaves, branches and pods. Leaf blight is only serious in India and parts of Central America.

Mouldy rot, *Ceratocystis fimbriata* Ell. & Halst., only infects freshly tapped bark, giving greyish mould just above cut followed by wounds and killing of the cambium. It is controlled by stopping or reducing tapping intensity and applying fungicide.

Pink disease, *Corticium salmonicolor* Berk. & Br., causes pinkish mycelial encrustation over the surface of the bark, followed by wilt, die-back, cankers and latex bleeding. It is most serious on 3–4-year-old trees, which can be sprayed with copper.

---

* The correct name is now *Microcyclus ulei* (P. Henn.) v. Arx.

White root rot, *Fomes lignosus* Kl., is the most serious root disease in Malaya and Indonesia. Rhizomorphs white, flattened, sparsely branched. Sporophores orange brown with white margin. Leaves go yellow and die; die-back of branches occurs followed by death of trees; when these symptoms occur it is too late to save the tree. It is essential to control from planting (see PLANTING above), followed by regular foliar and collar inspection; dead roots and infected tissue are removed and collars painted with protectant such as Fomac.

Red root rot, *Ganoderma pseudoferreum* (Wakefield) Over. & Steinm. Rhizomorphs reddish skin round root surface; sporophores dark red-brown. It spreads more slowly than white root rot; controlled by removal and isolation of infected material.

Brown bast: Initial symptom is increased yield of watery latex, followed by a drying up of part or whole of the tapping cut. Affected parts are darker coloured and cankerous wood growths occur. Some clones are more susceptible than others. The true cause is not known, but it is often considered to be a physiological condition due to overtapping. Total or partial dryness can also result from more than one cause, including root disease, necrosis extending from branch and fork wounds, sodium arsenite damage, etc. The affected bark should be excised and, if healthy yielding bark is found lower down, tapping can be continued, but at reduced intensity.

MAJOR PESTS

Rubber is relatively little affected by pests. Termites (*Coptotermes* spp.), and cockchafers (*Holotrichia* spp.) can be troublesome locally, while elephants, wild pigs, deer, porcupines, rats and squirrels may cause damage to bark and trunk, particularly when young. The giant snail, *Achatina fulica* Fer., causes damage in Indonesia.

IMPROVEMENT

AIMS: 'The large-scale production of latex by *Hevea* is forced on it by man and is not normal for the tree. . . . The biological formation of rubber within the plant is, on the other hand, a normal process which continues as long as the cells are active. . . . The latex remains relatively static and no appreciable movement occurs until there is an injury comparable to the onset of tapping. . . . The development of high-yielding rubber trees therefore involves the intensification of a normally low rate of rubber formation into a high rate of rubber renewal' (Polhamus, 1962). Sustained yield is the main aim of rubber improvement and planting material is required which (1) comes into production as quickly as possible, (2) opens with a production as high as possible, (3) shows the most rapid increase in production with age, (4) continues this yield increment as long as possible and (5) has a long tapping life.

Factors other than yield which are considered in selection and breeding are: (1) Vigour of growth, particularly during the early years as this reduces the period of expensive weeding and upkeep when there is no compensatory yield. A dense crown is required which will shade out weeds; abundant foliage promotes rapid growth but this may lead to trunk snap. (2) Resistance to storm damage particularly trunk snap in young trees which is very prevalent in some clones. This is due mainly to the disparity between a heavy dense crown and the girth of the trunk, the latter being reduced by early and heavy tapping and high yields. Brittleness of wood and branching at an acute angle also affect storm damage. Undue susceptibility to trunk snap can probably be prevented by delayed opening of tapping, periodic tapping, pruning or a combination of these. (3) Bark quality—desirable features are reasonable thickness, abundance of latex vessels, absence of excessive amount of hard stone cells in outer layer, rapid rather than prolonged latex flow, and rapid renewal of even bark. (4) Latex quality—it should be free flowing and not liable to premature coagulation in the cup; the resin content should not be more than 6 per cent, otherwise the sheets are soft and slow drying; for crepe manufacture hardness is also required and the colour should preferably be white; for latex concentrate it should have mechanical stability and be white in colour. (5) Disease resistance—in the New World resistance to South American leaf blight (*Dothidella ulei*) is essential if rubber is to be grown economically. In all areas, and particularly in Malaya, non-susceptibility to brown bast is most desirable. No disease is a limiting factor to *Hevea* cultivation in Malaya, but the more susceptible material to the various diseases should be excluded, *e.g.* clones vary considerably in their resistance to pink disease. Resistance to *Phytophthora*, leaf-fall and die-back have been demonstrated in Costa Rica. (6) Suitability for particular local conditions and purposes. Clonal material should therefore be tested over as wide an area as possible.

There is an element of risk involved in large-scale planting of insufficiently tested material. The grower has the choice between materials which yield at a moderate rate under continuous conventional tapping and others whose potentially higher annual yield may have to be reduced to the same level by reduced tapping frequency to combat brown bast or to encourage girthing, although they may give much higher yields in later life when the critical phases are past.

METHODS AND RESULTS: As was shown above (see ORIGIN AND DISTRIBUTION) almost all the rubber in the Far East originated from Wickham's historic collection on the Tapajos River in Brazil in 1876. It was also shown (see POLLINATION) that it is probable that the parent trees of this seed were inbred and might represent genetically pure lines. Planted in close proximity in the East cross-pollination took place resulting in hybrid vigour. In unselected seedling populations from this introduction large variability in yield was found. The yield frequency curve showed a marked skew distribution with a small number of the best trees contributing a considerable proportion of the total yield—4 per cent of yield obtained from only 1 per cent of crop. The high coefficient of variability provides valuable material for selection.

High-yielding mother trees were selected in 1910–1915 in Java and Sumatra and planted in gardens to produce illegitimate seed. Refinements in the technique of budding about 1918 permitted clonal planting. Only a few of the early selections proved to be genetically optimal producers, the remainder owing their superiority to chance favourable environmental conditions of soil or space, etc., and it was then realized that the true value of a seedling tree could only be determined with certainty by the yield of buddings made from it. On the Prang Besar Estate in Malaya after World War I attempts were made on a commercial basis to utilize the advances in selection initiated in Java and Sumatra. Over one million trees were kept under observation and 618 of these were selected for budding in trials. Only a few performed well and only 2 or 3, e.g. 'PB 86', were planted over a considerable period.

The method developed for breeding rubber is the controlled hand-pollination (for method see POLLINATION above) between selected high-yielding clones. The legitimate seedlings are germinated in rat-proof enclosures and planted in budwood nurseries. Buds from these are then budded on seedling rootstocks (see PROPAGATION above) and the clonal buddings are tested against a proved clone. They are then selected for yield and secondary characteristics and then further tested in large-scale experiments on different soil types. The final recommendation of a clone for use on a commercial scale is usually 20 years after the date of the original hand-pollination. Clonal buddings are also planted in seed gardens for the production of 'clonal' seed (see PROPAGATION above). It is essential to rogue any off-type trees; the only sure method of identifying buddings is by comparing the shape and pattern of the seeds with those of a standard clonal seed collection (see SEEDS above). Preselection methods by determining significant correlations between anatomical or physiological features of young plants and their final performance would permit the best seedlings to be selected from a larger original population of legitimate seedlings and the search for such characters continues.

The progress in rubber breeding is illustrated by a brief summary of the work at the Rubber Research Institute of Malaya (R.R.I.M.). The yields are given as the average over the first 5 years of tapping in lb dry rubber per tree per annum. The yield from unselected seedlings was 4·3 lb. Selections were made in seedling population on estates in Malaya and in seedling material from Ceylon, Java, etc. and clonal buddings made; the best selections, *e.g.* 'PIL.B84' gave 9 lb. During the period 1928–1931 hand-crossing, using mainly Pilmoor clones, produced 1,530 seedlings from which 974 new clones were made and tested in small-scale trials. From these 30 clones were selected, RRIM 500–529, and tested in large-scale trials. The average yield of 17 of these clones was 15 lb. 'RRIM 513' was the best of the series. Work was discontinued during the economic depression 1932–1936, but from 1937–1942 until the Japanese occupation, hand-crosses were made using 17 of the RRIM 500 series and 15 primary clones, which included some of the best parents in South-east Asia. 'TJIR 1' from Indonesia was the outstanding parent and was a parent of 1,068 of the 1,981 seedlings produced. 39 clones were selected for large-scale testing and were given the numbers RRIM 600–638. The average yield of dry rubber of the first 5 years of tapping of 24 of these clones gave 18·8 lb per tree per annum. Outstanding clones in this series were RRIM 600, 605 and 623. In one trial 'RRIM 600' yielded at the rate of 3,885 lb per acre in its 9th year of tapping (1964). From the above yields it can be seen that the slope of the curve of increase in yield is flattening out with subsequent generations and it seems probable that the ceiling of yield improvement of the Wickham material may be being reached.

From 1947–1958 further crossing was made between 45 parent clones to produce 4,830 seedlings in 110 seedling families. The parents included the best clones of the RRIM 500 and 600 series and 17 primary clones, including 'TJIR 1' and 10 primary clones not previously used from Java, Sumatra, Malaya, Vietnam and South America. Selection from these forms the basis of RRIM 700 series.

Since 1959 many other crosses have been made, including wide-crosses, back-crosses, selfing and sib-crosses of both primary and secondary clones and 4 clones from Peru.

The R.R.I.M. recommend clones for planting based on the following classification: Class 1—suitable for large-scale planting, their performance having been confirmed from commercial areas; Class II—promising clones but not yet fully tested, which are recommended for moderate scale planting of not more than one-third of the total replanting programme and should contain at least 3 clones; Class IIIA—experimental planting of up to 25 acres per clone, which are usually being tested on a large scale but only preliminary information is available; Class IIIB—planting of small blocks of one task only, mainly to ensure that a source of budwood is available if some of these clones prove successful. From 1947–1964 over one million acres of rubber were planted or replanted in Malaya, of which 30 per cent was under 'clonal' seedlings and the remainder was budded.

In the Western Hemisphere selection has been predominantly for disease resistance, more particularly to *Dothidella ulei.* Jungle selections of *Hevea brasiliensis* for resistance have been made; these have been crossed with high-yielding clones from the East and these progenies have been back-crossed to the Eastern clones in an attempt to combine resistance and yield. Interspecific crosses between high-yielding Eastern clones and *H. benthamiana* have also been made and these show outstanding resistance to leaf blight and are reasonably promising in regard to yield. Blight-resistant clones from Brazil have been exchanged for high-yielding clones from Malaya with quarantine propagation at Coconut Grove, Florida and inspection and re-packaging at Kew, England.

The Rubber Research Institute of Malaya opened a Unit at the University of the West Indies, Trinidad, in 1961 to study the cytogenetics of *Dothidella ulei* and work is now being done on the epidemiology and pathogenic strains of the disease and the assessment of the degree of host resistance, including *H. brasiliensis* and *H. benthamiana* crosses.

A very small proportion of clones, less than 0·5 per cent, are suspected to be male sterile. Should any of these prove to pass on a higher yield potential they could then be used as the female parents in producing 'clonal' seed in seed gardens of known parentage. Inbreeding of self-fertile high-yielding clones and subsequent crossing may result in high-yielding clones showing hybrid vigour.

PRODUCTION

At the beginning of the 20th century the world's supply of Para rubber came almost entirely from wild trees in Brazil and the world's consumption of rubber from *Hevea* and other natural sources was 52,500 tons. The annual consumption roughly doubled during each decade. Although rubber was synthesized in the 19th century, the manufacture of synthetic rubber and its use on a large scale did not occur until World War II. Production of synthetic rubber outside the Soviet Union first overtook that of natural rubber in 1961 when over 2 million tons of both synthetic and natural rubber were produced. Natural rubber entering world trade comes almost entirely from *Hevea brasiliensis.* With rising production costs and competition from synthetic rubber the emphasis now is on replanting with the greatly improved planting material rather than expansion in planted area.

Malaya, with some 4 million acres under rubber, is now the biggest producer, and produced 752,000 tons in 1962, over 35 per cent of the world's total of natural rubber. Malayan production first surpassed that of Indonesia in 1959, which with 4½ million acres, was relegated to second place in the quantity produced. Production in Malaya is expected to reach 1 million tons or more by 1968. Other important producing countries in Asia, which produces 92 per cent of the world's rubber, given in order of production are: Thailand, Ceylon, South Vietnam and Sarawak. The percentage value of rubber to other domestic exports is approximately: Malaya 50 per cent, Indonesia 45 per cent, Thailand 25 per cent, South Vietnam 70 per cent and Sarawak 25 per cent. Africa produces 7 per cent (1962) of the world's supply of natural rubber, with Nigeria the biggest producer and still expanding, followed by Liberia and the Congo. Brazil now produces only 1 per cent of the world's total.

Most temperate countries import natural rubber, but the countries importing the greatest quantity given in decreasing amounts are United States, Soviet Union, Eastern Germany, United Kingdom, Japan, Western Germany and France. The annual consumption in the United States is now about 20 lb per caput compared with 1 lb per caput for all the other countries combined.

REFERENCES

Annual Reports, *Journal* (publ. quarterly), *Planters' Bulletins* (publ. bi-monthly) and *Research Archives of Rubber Research Institute of Malaya, Kuala Lumpur.*

Annual Reports and Quarterly Journal of Rubber Research Institute, Ceylon.

BOUYCHOU, J. G. (1954). *Manuel du planteur d'Hevea.* Tome 1, Paris: Institut Français du Caoutouch.

DJIKMAN, M. J. (1951). Hevea: *Thirty Years of Research in the Far East.* Coral Gables: Univ. Miami Press.

EDGAR, A. T. (1960). *Manual of Rubber Planting (Malaya).* Kuala Lumpur: Incorp. Soc. of Planters.

HILTON, R. N. (1959). *Maladies of* Hevea *in Malaya.* Kuala Lumpur: Rub. Res. Inst., Malaya.

POLHAMUS, L. G. (1962). *Rubber.* London: Leonard Hill.

PURSEGLOVE, J. W. (1959). History and Functions of Botanic Gardens with Special Reference to Singapore. *Gdns'. Bull., Singapore,* **17**, 125–154.

RAO, B. S. (1965). *Pests of* Hevea *Plantations in Malaya.* Kuala Lumpur: Rub. Res. Inst., Malaysia.

ROSS, J. M. (1964). Summary of Breeding carried out at the R.R.I.M. during the Period 1928–1963. *Res. Archives of Rub. Res. Inst. Malaya,* Document 28.

RUBBER RESEARCH INSTITUTE OF MALAYA (1961). *Proceedings of the Natural Rubber Research Conference, Kuala Lumpur, 1960.* Kuala Lumpur: R.R.I.M.

SCHULTES, R. E. (1956). The Amazon Indian and Evolution of *Hevea* and related genera. *J. Arnold Arbor.,* **37**, 123–147.

SHORROCKS, V. M. (1964). *Mineral Deficiencies in* Hevea *and Associated Cover Plants.* Rub. Res. Inst., Malaysia.

WYCHERLEY, P. R. (1964). The Cultivation and Improvement of the Plantation Rubber Crop. *Res. Archives of Rub. Res. Inst. Malaya,* Document 29.

Manihot Mill. (x = 9)

100–200 spp., depending on definition and viewpoint, confined to the Western Hemisphere from Arizona to Argentina. Mainly shrubs and subshrubs, but trees occur. Latex in all parts. Lvs alternate, entire or palmately lobed. Monoecious with male and female fls in same inflorescence.

The root tubers of *M. esculenta* (q.v.) are an important source of starchy food throughout the tropics. Several spp. yield rubber. *M. glaziovii* Muell.-Arg., Ceara rubber, was widely distributed by Kew from 1877 onwards from Cross's collection in Brazil in 1876 for trial as a plantation crop for rubber production. Quite extensive plantings were made in East Africa and by 1912 there were 1,500 acres in Uganda and over 100,000 acres in German East Africa. Small tree, 10–20 m, fast growing. Yield low; latex coagulates immediately on exposure; cannot be tapped continuously: rubber of good quality, but resin content rather high. It was abandoned in favour of *Hevea brasiliensis* (q.v.). It is used for temporary shade for cocoa in West Africa.

M. dichotoma Ule, Jequie Manicoba rubber, and *M. piauhyense* Ule, also from Brazil, were tried for rubber production, but were unsatisfactory.

Manihot esculenta Crantz (2n = 36) CASSAVA, MANIOC, TAPIOCA

Syn. *M. utilissima* Pohl, *M. aipi* Pohl, *M. dulcis* Pax, *M. palmata* Muell.-Arg.

USES

Root tubers are an important food and carbohydrate source in many parts of the lowland tropics. Care is required due to the presence of hydrocyanic glucoside (see SYSTEMATICS and CHEMICAL COMPOSITION below). Sweet cassavas may be eaten raw after peeling; in bitter cassava HCN is destroyed by washing and cooking; peeled roots of both may be boiled or roasted. Fresh tubers do not keep long but may be sliced and dried in the sun, with or without parboiling, the latter enhances the keeping quality, and may be stored for several months. In West Africa a thick paste, *fufu*, is made by peeling and cutting up boiled roots and pounding them in a wooden mortar. Throughout Africa the cultivation of cassava was encouraged as a famine reserve and against the ravages of migratory African locust (*Locusta migratoria migratorioides* R. & F.) to which it is almost immune. Bitter cassava is planted in areas where wild pigs, baboons, porcupines and other game are a serious menace and is often one of the few crops which can be grown in such areas.

Starch is prepared by grating or grinding washed peeled tubers and washing out the starch by squeezing in repeated changes of water. Tapioca is made by gently heating washed and clean starch on hot iron plates which partly cooks it and causes agglutination into small round pellets. Cassava flour is made by grinding the sun-dried slices or chips. In north-eastern South America a coarse meal, *farinha*, is prepared by grating washed peeled tubers and squeezing the mass in long, sleeve-like baskets, *tipiti*, which expresses the juices; the compressed pulp is then toasted over a low fire. The latex and extracted juice may be concentrated by boiling to produce *cassureep* which is used in sauces and is a constituent of the West Indian pepper-pot. In West Africa, *garri*, a meal like *farinha*, is made, but the pulp is fermented for a rather longer period. Cassava starch is used for food and also in the manufacture of adhesives and cosmetics and in sizing textiles, laundering and paper making. Tapioca is used for puddings, biscuits and confectionery.

The leaves are used as a pot-herb, particularly in Africa. Beer and other alcoholic drinks may be made from cassava. Sweet cassava may be fed to livestock. In Trinidad it is used as a nurse-crop for cocoa.

SYSTEMATICS, ORIGIN AND DISTRIBUTION

Many cvs or clones are known and some of the variation is noted under STRUCTURE below. Some authorities have given different specific names to bitter and sweet cassavas, the former being *M. esculenta* (syn. *M. utilissima*) and the latter *M. palmata* (syn. *M. dulcis*). This distinction is not justified as the two merge into each other and the toxicity of a clone varies from place to place. For practical purposes, however, it is usual to divide the cvs into: (1) sweet cassavas with low HCN content with HCN confined to phelloderm of tubers; (2) bitter cassavas with high HCN content with HCN generally distributed throughout tubers, including core. Cvs with firm yellow flesh usually contain more HCN than those with softer whiter flesh. They can also be classified into: (1) short-season cvs which mature as early as 6 months after planting and cannot be left in ground longer than 9–11 months without serious deterioration; these are often sweet cassavas; (2) long-season cvs which take at least a year or more to mature a crop and some of them may be left in the ground 3–4 years without serious deterioration; these tend to be bitter cassavas.

M. esculenta is not known in a wild state. There are two geographical centres of speciation of the genus, one in western and southern Mexico and parts of Guatemala and the other in north-eastern Brazil. Cvs of *M. esculenta* are found in both these regions and, by the time of Columbus, cassava had reached its present limit of cultivation in the New World extending from 25°N to 25°S. There is some evidence that cvs have hybridized with local native spp. in both centres of speciation to form a number of complexes and that some wild weedy spp., *e.g. M. saxicola* Lanj., are derived from the cultivated complex. From the available evidence, Rogers (1963) postulates that the cultivated *M. esculenta* could have arisen in both centres. There is evidence that the crop was grown in Peru some 4,000 years ago and in Mexico 2,000 years ago. Throughout the American tropics the sweet cassavas are more widely distributed than the bitter cassavas; the former occurred in the older civilizations, they have received more human selection, and are usually secondary foods. Bitter cassavas were probably cultivated later and where grown usually provide the dominant staple.

Cassava appears to have been taken by the Portuguese to São Tomé and Fernando Po in the Gulf of Benin and to Warri and the mouth of the Congo River on the mainland during the last half of the 16th century. Elsewhere on the mainland of West Africa it was of little importance prior to the 19th century and its spread and use increased in the 20th century. It was not known north of River Niger before 1914.

Cassava was taken from Brazil to Réunion in 1736 and thence to Madagascar. It was recorded in Zanzibar in 1799. It seems likely that cassava was either absent or unimportant in most of East Africa before 1850, except along the coast and in the vicinity of Lake Tanganyika to which it had been brought from the west. Speke states that there was no cassava on the north-western shore of Lake Victoria in 1862, but Stanley found it growing in Uganda in 1878. Much of the spread during the present century in Africa is due to the suzerain powers encouraging its cultivation as a famine reserve and against locust attack. Africa now has more cassava than all the rest of the world.

Rumph in 1692 did not record cassava in Malaysia. It was taken from Mauritius to Ceylon in 1786 and reached Calcutta in 1794. It is probable that it was taken at an earlier date to the Philippines from Mexico. It has now spread to all tropical areas.

ECOLOGY

Cassava is essentially a lowland tropical crop, but it can be grown up to 5,000 ft on the equator. It cannot stand cold or frost. Cassava is grown in areas with rainfall of 20–200 in. per annum. Except at planting, it can stand prolonged periods of drought, when it sheds its leaves, but comes away again quickly with the rain. It is therefore valuable in regions with low and uncertain rainfall. It grows best on sandy or sandy loam soils of reasonable fertility, but it can be grown on almost all soil types provided they are not waterlogged, too shallow or too stony. Cassava will produce an economic crop on exhausted soils unsuitable for other production and consequently is often the last crop taken in the rotation in shifting cultivation. It is exhaustive of potash. Too high fertility may result in excessive vegetative growth at the expense of tuber and starch formation.

STRUCTURE

A short-lived shrub, 1–5 m in height, with latex in all its parts.

ROOTS: Tubers develop as swellings on adventitious roots a short distance from stem by a process of secondary thickening. Number, shape, size, angle at which they penetrate the ground, colour of outer cork and internal tissues vary greatly. Usually 5–10 tubers per plant, cylindrical or tapering, 15–100 cm long, 3–15 cm across, occasionally branched. Consist of: (1) outer skin or periderm; cork layers may be rough or smooth, white, light to dark brown, pink or red; (2) thin rind or cortex, usually white, but may be tinged pink or brown; (3) core or pith mainly consisting of parenchyma rich in starch with few xylem bundles and latex tubes; usually white but may be yellow or tinged red; this is the edible portion. In addition to tubers, fibrous adventitious roots develop horizontally and vertically. Old tubers become lignified.

STEMS: Vary greatly in height and branching habit, branching near base in some cvs, high up in others; angle of branches with main stem determines whether erect or spreading forms. Branches usually glabrous, slender, with lvs towards apex and with prominent nubbly lf. scars below; colour of periderm and lf. scars ranging from silver-grey to greenish yellow, pale or dark green, pale reddish or dark brown, or streaked with purple.

Fig. 26. **Manihot esculenta**: CASSAVA. **A**, young shoot with leaf (× ¼); **B**, leaf base with stipules (× ½); **C**, flowering shoot (× ½); **D**, female flower (× 1); **E**, male flower in longitudinal section (× 3); **F**, female flower in longitudinal section (× 3); **G**, fruit (× ½); **H**, seed (× ½); **I**, tubers (× 1/10).

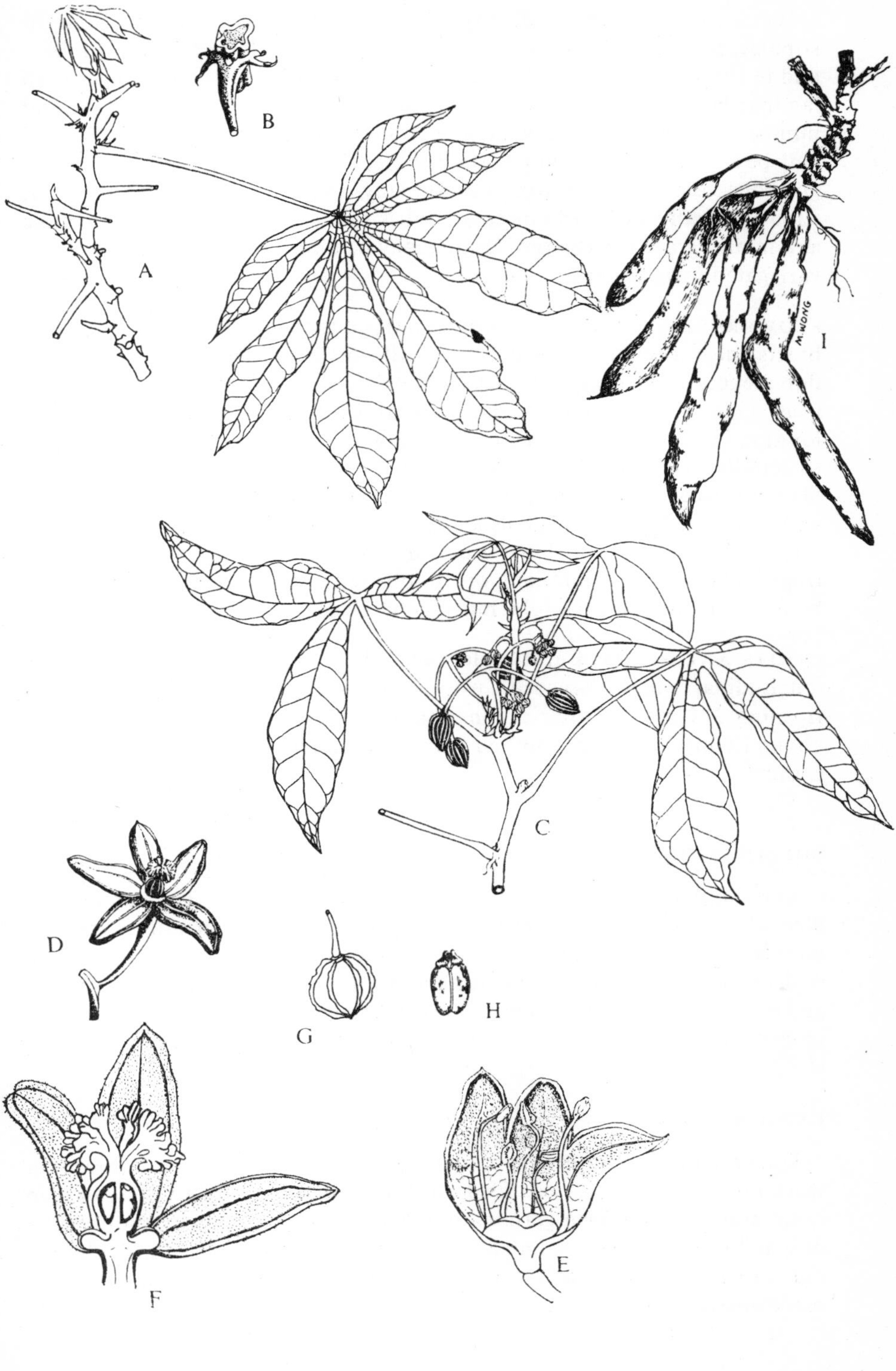
A
B
I
M.WONG
C
D
G
H
F
E

LEAVES: Spirally arranged, phyllotaxis 2/5. Variable in size, colour of stipules, petioles, midribs and lamina, in number of lobes, depth of lobing and in shape and width of lobes. Petiole 5–30 cm long, usually longer than lamina; lamina deeply palmately divided with 3–9 lobes, usually 5–7, and number may vary on same plant, lobes obovate-lanceolate, entire, 4–20 × 1–6 cm, base narrowed to subpeltate lf. base, tip acuminate, glabrous above, glaucous and pale beneath, occasionally slightly hairy beneath veins; stipules usually with 3–5 lanceolate lobes to 1 cm long, deciduous; petiole and midrib green to deep red; lamina may be tinged red; a green and yellow variegated form occurs.

FLOWERS: Borne in axillary racemes near ends of branches, 3–10 cm long; bracts linear, deciduous; fls unisexual with male and female fls in same inflorescence, latter near base. Sepals 5, pale yellow or tinged with red, glabrous without, puberulous within; apetalous. Male fls pedicel 0·5–1·0 cm long; calyx 3–8 mm long, campanulate with 5 triangular lobes extending to middle of calyx; stamens 10 in 2 whorls, alternately short and long, filaments free, anthers small; basal nectiferous disc, orange in colour, fleshy with 5 double lobes. Female fls usually larger than males, pedicel 1–2·5 cm long; calyx 5-partite to base, 1 cm or more long; ovary 3-carpellary, 6-ridged, glabrous, 3–4 mm long on 10-lobed disc, style connate surmounted by 3 lobed stigmas, each lobe much subdivided.

FRUITS AND SEEDS: Capsule globose, 1·5 cm long, glabrous with narrow longitudinal wings, 3-seeded; woody endocarp splitting explosively on ripening 3–5 months after pollination and ejecting seeds. Seeds ellipsoid, about 12 mm long, grey mottled with dark blotches and with pronounced caruncle.

POLLINATION

Entomophilous; cross-pollinated; protogynous, male fls opening 7–8 days after females. In artificial pollination fls enclosed in muslin bag on morning of anthesis and pollination made same afternoon; muslin bag replaced by bag of mosquito netting a few days later to let in light and air and to collect seed on dehiscence. The percentage success of hand-pollination in crossing cvs is 0–56 per cent, with an average of 14 per cent.

GERMINATION

Under natural conditions germination protracted and erratic. In breeding work micropylar end of seed filed until white embryo is just visible to give rapid and fairly uniform germination. Seeds placed on moist sand in petri dish at 30°C and are potted up as soon as radicle appears. Seedlings planted out into propagating plots when 6 in. high and in 5–6 months they have made enough growth to take at least 6 cuttings per plant.

CHEMICAL COMPOSITION

This varies, but a fairly typical percentage composition of the edible portion of fresh tuber which comprises about 80 per cent of whole tuber is: water 62; carbohydrate 35; protein 1·0; fat 0·3; mineral matter 1·0. Tubers relatively rich in calcium and vitamin C. In starch extraction fresh tubers yield 20–25 per cent starch. The leaves are rich in protein and vitamin A and may contain HCN.

The toxic principle is hydrocyanic acid which is liberated when the enzyme linase acts upon the glucoside linamarin. In sweet cassava HCN confined to rind; in bitter cassava HCN more widely distributed and must be destroyed by boiling, roasting, expression or fermentation. Activity of enzyme kept in check when tuber is growing, but liberation of HCN proceeds actively when tubers dug and wilt; stale tubers are therefore more poisonous. The amount of HCN varies from 10–370 mg per kg of fresh tuber; less than 50 mg is considered innocuous, 50–100 mg is moderately poisonous, over 100 mg is highly poisonous. The toxicity is markedly influenced by the local soil and climatic conditions and changes in toxicity can occur when moving cvs from one country to another.

PROPAGATION

All economic planting is done with stem cuttings, 9–12 in. long, preferably taken from the middle of the stems. Great care is necessary to avoid planting cuttings infected with viruses. Cuttings from the base of stems are more likely to be infected with mosaic virus (see below). Propagation from seeds is confined to breeding work. Stems may be grafted and this is used in testing for resistance to virus diseases.

HUSBANDRY

Cassava may be planted in pure stand, but it is often interplanted with annual food crops; it is often the last crop taken off the land before the resting period. A deep thorough cultivation allows maximum tuber development, but the crop does not usually receive much care and attention. Spacing is usually 3–5 ft between plants. The cuttings are often inserted for about half their length into the soil at an angle, but they may be planted vertically or horizontally. Sometimes 2–3 cuttings may be planted close together. They may be planted on the flat, on ridges, or on mounds. Weeding should be done during the early stages and the plants are usually earthed up 2–3 months after planting to encourage tuber production. As soon as the branches interlace cultivation should cease as much damage would result from breaking the brittle branches. Cassava is seldom manured. Short-season cvs are harvested 6–10 months after planting; long-season cvs are often left for longer periods of 2 years or more. The tubers are usually dug individually as required, as they begin to rot within 48 hours after being taken from the ground. Yields will vary with the cv. grown, whether short- or long-season, the time to harvesting, the care taken in cultivation, and the amount of virus disease. In India yields vary from 1–12 tons per acre. Under intensive cultivation yields of 20–26 tons per acre have been recorded. The great value of cassava in peasant agriculture is that a crop can be grown and reaped for little labour and it can be kept in the ground until required, more particularly for times of food shortage.

MAJOR DISEASES AND PESTS

The only serious diseases of cassava are the virus diseases given below.

Mosaic virus was first recorded in East Africa in the 1890's but was not reported as causing serious damage until the 1920's. Yields were seriously reduced in Madagascar in 1925 and then in parts of East, Central and West Africa. Whitish or yellowish chlorosis of young lvs occurs, together with lf. distortion and reduction in size of lf. A bad outbreak in susceptible clones causes great reduction in yield. The vectors are white flies (*Bemisia* spp.) and the virus can only be transmitted in the young lvs; if there are a sufficient number of vectors the symptoms appear in 2–3 weeks in susceptible cvs. In resistant cvs the virus moves rapidly down the plant and is then restricted to the basal portion of the stem where it can do no harm. Pruning, plucking lvs and unfavourable conditions of growth cause increased infection. Planting of infected cuttings spreads the disease and the damage is more serious than natural infection by the vector; losses may be as high as 95 per cent in non-tolerant cvs.

Brown streak virus is serious in the coastal regions of East Africa. Low temperatures are inimical to infected plants and they usually die, although this is not always the case. Resistance is tested by growing the cvs at high altitudes. The symptoms are chlorosis of lvs, but only on mature lvs and with no distortion of lamina as in mosaic, brown streaks in cortical tissue of stem which later coalesce to form patches with necrotic lesions on lf. scars, followed by shrinking and death of internodal tissue, and discoloured areas in roots. *Bemisia* spp. are probably vectors and it can also be transmitted mechanically and by planting infected cuttings.

Fomes lignosus Klotzsch may cause rotting of tubers.

Scale insects such as *Aonidomytilus albus* Ckll. and red spider, *Tetranychus telarius* L. are reported as causing damage. Wild pigs and other mammals, from rats to hippopotami, may cause extensive damage particularly to sweet cassavas.

IMPROVEMENT

AIMS: (1) Resistance to virus diseases. Throughout tropical Africa resistance to mosaic is of prime importance and in East Africa resistance to brown streak as well. (2) High yield of tubers with high starch content. (3) Good quality tubers—those used directly as food should be palatable and with low fibre content; for starch production high fibre content is unimportant. (4) HCN content—sweet cassavas with low HCN content are preferred by majority of growers, but bitter cassavas may be necessary in areas infested with pigs and other game. (5) Time to maturity—early maturing cvs are desirable in some areas and are usually more palatable, whilst in other areas cvs may be required that may remain long in the ground without deterioration as a famine reserve. (6) Enhanced protein content—desirable from nutritional point of view.

METHODS AND RESULTS: Important work of breeding cassavas resistant to mosaic and later to brown streak was begun by Dr Storey at Amani in Tanganyika (now Tanzania) in the 1930's. Two lines of attack were adopted:

A short-term policy involving crosses of the most resistant clones of *M. esculenta.* Cvs were introduced from most tropical countries, including Brazil and parts of Africa, but most of the introductions were found to be very susceptible to mosaic. Resistance was found mainly in some of the local cvs and by crossing these considerably more resistance was obtained in the hybrids, but no immune clones were obtained. No plant withstood infection from either disease when subjected to infection by grafting. It takes approximately 3 years to produce each successive generation. (For methods used see POLLINATION and GERMINATION above.)

A long-term policy involving the study of resistance in other *Manihot* spp. and the production of interspecific hybrids (cf. sugar-cane breeding for mosaic resistance). The following spp. were used and results obtained:

(1) *M. glaziovii,* Ceara rubber (see above). The F_1 plants with cassava were very vigorous and grew into tree-like forms resembling the *M. glaziovii* parent and were highly resistant to mosaic and brown streak, but the roots were woody. The F_1 plants were of low fertility and certain crosses were sterile. By back-crossing to cassava, when better fertility was obtained, forms were produced which bore good yields of palatable tubers and were resistant to both viruses. In third back-crosses two clones were produced (46106/27 and 4763/16) which yielded 27 and 35 tons of fresh tubers per acre after 20 months in the field. Third and fourth back-crosses have now been issued to cultivators.

(2) *M. dichotoma,* (see above) crossed with cassava; the F_1's were good seed producers, but did not propagate readily from stem cuttings; they were completely immune to the virus; tubers were lacking. Further back-crosses were made and also intercrosses with other hybrids. A fourth back-cross of *M. dichotoma* to cassava crossed with a third back-cross of *M. glaziovii* to cassava possesses high resistance to the viruses and yields are promising.

(3) *M. saxicola* Lanj. is rich in protein and has been used in an attempt to increase the protein content of the tubers. Unfortunately it has a high HCN content and is susceptible to mosaic.

(4) *M. melanobasis* Muell.-Arg., a low-growing plant with tubers rich in protein and HCN, was crossed with cassava; back-crosses to cassava gave outstanding yields and were fairly resistant to brown streak but not to mosaic.

(5) *M. catingae* Ule. F_1 hybrids with cassava were tree-like, vigorous and with woody root systems, but with resistance to the viruses. These have now

been crossed with resistant clones of *M. glaziovii* and *M. melanobasis* origin with cassava and improved tubers obtained without loss of resistance.

PRODUCTION

World acreage has been estimated at approximately 17 million acres, more than half of which is grown in Africa, and world production at about 62 million tons. Very little cassava and its products enter into international trade. Pre-war exports of cassava products were approximately 325,000 tons and in the post-war period have been reduced to about half this figure. The main exporting countries are Indonesia, Madagascar and Brazil and the principal importing countries are U.S.A. and European countries. The increase in acreage of cassava during the 20th century must be greater than almost any other crop.

REFERENCES

JENNINGS, D. L. (1957). Further Studies in Breeding Cassava for Virus Resistance. *E. Afr. agric. J.*, **22**, 213–219.

JENNINGS, D. L. (1960). Observations on Virus Diseases of Cassava in Resistant and Susceptible Varieties. *Emp. J. exp. Agric.*, **28**, 23–34, 261–270.

JONES, W. O. (1959). *Manioc in Africa.* Stanford Univ. Press.

NICHOLS, R. F. W. (1947). Breeding Cassava for Virus Resistance. *E. Afr. agric. J.* **12**, 184–194.

ROGERS, D. J. (1963). Studies of *Manihot esculenta* Crantz and related species. *Bull. Torrey Bot. Club*, **90**, 43–54.

Ricinus L. (x = 10)

A monotypic genus.

Ricinus communis L. (2n = 20) CASTOR

USES

The ancient Egyptians used the oil as an illuminant. Up to the beginning of the 20th century the main use of castor oil in the western world was in medicine, mainly as a purgative. Now the bulk of the crop is utilized in industry. It is water resistant and is used for coating fabrics and other protective coverings. It is used in the manufacture of high-grade lubricants, soap and printing inks, in textile dyeing and for preserving leather. The dehydrated oil is an excellent drying agent which compares favourably with tung oil and is used in paints and varnishes. The hydrogenated oil is used in the manufacture of waxes, polishes, carbon paper, candles and crayons. Fractions of the oil are used in the production of 'Rilson', a polyamide nylon-type fibre, and large quantities of oil are used for this purpose in France and Brazil. They are also used for plastics, ointments and cosmetics. The pomice or residue after crushing is used as a fertilizer; it is toxic due to the presence of ricin, but a method of detoxicating the meal has now been found so that it can safely be fed to livestock. The stems can be made into paper and wall-board. It has long been grown as an ornamental plant and red-leaved, red-fruited forms occur. The seeds are poisonous.

ORIGIN AND DISTRIBUTION

Ricinus communis originated in Africa and grows wild in East and North Africa, the Yemen and the Near and Middle East. It was cultivated in ancient Egypt as long ago as 4000 B.C. It was taken at an early date to India and beyond, and was recorded in China in the Tang period, A.D. 618–906. It was introduced into the New World shortly after Columbus. The castor plant is now naturalized in many tropical and subtropical countries.

ECOLOGY

Castor requires a warm climate and is killed by frost. At least 140–180 day growing season is required before the first killing frost. It can be grown over a wide altitude range in the tropics and with both low and medium rainfall. Heavy rainfall and waterlogging should be avoided. At sustained temperatures above 100°F seed may fail to set. The best soils for its cultivation are rich well-drained sandy or clayey loams. In the U.S.A. it is grown with irrigation.

STRUCTURE

Annual herb or short-lived perennial, 1–7 m high, glabrous, very variable. In the wild state it reaches the size of a small tree.

ROOTS: Well-developed tap-root with prominent laterals, which produce a surface mat of feeding roots.

STEMS: Green or reddish, becoming hollow with age, with well-marked nodes and prominent leaf scars. A single stem is first produced which terminates in an inflorescence at 6th–10th node in dwarf early cvs, at 8th–16th node in later-maturing cvs, and at 40th or more nodes in tall and wild plants. As the panicle develops, 2–3 sympodial branches grow out, one from each node immediately below it; these end in inflorescences and one or more sympodial branches grow out from the nodes immediately below them and the process is continued. Thus, the development along each axis is sequential and a plant will have inflorescences at various stages of development. Degree of branching varies considerably.

LEAVES: Spirally arranged with a phyllotaxis of 2/5, peltate; stipules 1–3 cm long, united, sheathing bud, deciduous; petiole pale green or reddish, round, 8–50 cm long, with 2 nectiferous glands at junction to lamina, 2 glands on either side at base and 1 or more glands on upper surface towards base; lamina orbicular, 10–75 cm in diameter, palmately 5–11-partite for about half length; segments ovate or lanceolate, acuminate, serrate, dark green or reddish above, paler green beneath.

FLOWERS: Borne in terminal, many-flowered panicles, 10–40 cm long, unisexual, with male fls at the base and female fls on the top 30–50 per cent of inflorescence; some commercial hybrids entirely female. Male fls in 3–16 flowered cymes; pedicels 0·5–1·5 cm long; sepals 3–5, ovate, spreading, green, 5–7 mm long; petals absent; stamens numerous, 5–10 mm long, with much branched filaments, each branch ending in a small, spherical, pale yellow anther. Female fls in 1–7 flowered cymes; pedicels 4–5 mm long; sepals 3–5, green, 3–5 mm long, connate and bursting irregularly, quickly falling; petals absent; ovary superior, 3-celled, 1 ovule per cell; ovary wall covered with fleshy soft green spines each terminating in a transparent bristle which breaks off as fruit develops; style very short; stigmas 3, deeply bifid, fleshy, red, papillate, long persistent.

FRUITS: Globose capsule with elongated pedicel, 3-lobed, 1·5–2·5 cm in diameter, usually spiney; unripe frs green, but red in some cvs, turning brown on ripening; woody remains of pistil at apex. Dehisce by woody pericarp splitting along dorsal suture; in wild plants and some cvs splitting is violent ejecting seeds; most modern cvs are indehiscent and seed is held in capsule for several weeks.

SEEDS: Ovoid, compressed dorsally, tick-like, shining, pale grey or pale buff to almost black with darker mottling; yellowish-white caruncle at base; very variable in size, 0·5–1·5 cm long; 450–5,000 seeds per lb. Testa brittle, about 20 per cent weight of seed; endosperm copious, white; embryo small; cotyledons thin and papery.

POLLINATION

Plants examined by me in Trinidad are markedly protogynous; most of the female flowers have set seed and the fruit are developing before the male flowers open on the same inflorescence. The anthers burst explosively on drying or when touched scattering copious pollen. After shedding pollen male flowers soon absciss. It is usually stated that castor is mainly wind pollinated. At the same time as the flowers are opening the glands on the young leaves on the sympodial branches below the inflorescence exude copious nectar, so it seems probable that insects play some part in pollination.

GERMINATION

Sound seed will retain their viability for 2–3 years. The brittle testa is easily damaged and crushed seeds may cause clogging in mechanical

Fig. 27. **Ricinus communis**: CASTOR. **A**, young shoot, inflorescence and leaves (×¼); **B**, male flower in longitudinal section (×2); **C**, female flower in longitudinal section (×2); **D**, developing fruits and male flowers (×½); **E**, mature fruits (×⅓); **F**, dehiscing carpel (×½); **G**, seed-front view (×1); **H**, seed-side view (×1).

D
C
B
G
H
E
F
A
M. WONG

planters. It is advisable to treat the seeds to prevent subsequent damping off. More even germination may be obtained by pouring boiling water on the seeds and leaving them to soak for 24 hours. Some castor seeds show dormancy, but this may be broken by removing the caruncle and cutting a small hole in the testa. Germination is epigeal and emergence takes 7–10 days; sometimes longer.

CHEMICAL COMPOSITION

Seeds of commercial cvs contain 40–55 per cent of a non-drying oil. Large-seeded cvs have less oil than smaller-seeded cvs. The oil contains 80–90 per cent of ricinoleic acid, which is unique among vegetable oils in having the hydroxyl group near a double bond. On dehydration a drying oil is produced, which does not turn yellow on drying or baking; hence its value in paints and varnishes. The seeds contain a toxic protein, ricin, which acts as a blood coagulant. Eating a single seed may cause serious illness and more may be fatal. They contain a powerful allergen, a protein polysaccharide. Neither the ricin nor the allergen is carried over into the oil if it is properly extracted, but remains in the meal. A method has now been found to detoxicate the meal. The leaves contain ricin, but in much smaller quantities than in the seeds; they should not be fed to livestock.

PROPAGATION

By seed.

HUSBANDRY

In India castor is grown almost entirely as a dryland crop and is often rotated with finger millet (*Eleusine coracana*) or in mixed cultivation with other crops. In the United States it is usually grown with irrigation. A thorough deep preparation of the seed-bed is essential. With dwarf cvs in the United States with mechanized production, castor seeds are planted at a depth of 1½–3 in. in rows 38–40 in. apart, with a spacing of 8–10 in. within the rows and with a seed rate of 12–14 lb per acre. In India, with taller cvs, the spacing is usually 3 × 2 ft. In Africa 2–4 seeds are planted per hole, later thinned to one plant when 8–12 in. high; tall unbranching cvs are topped at 3 ft to encourage branching. Castor should be cultivated regularly to control weeds until it is 2–3 ft high, but cultivation should be shallow because of the surface roots. It is an exhaustive crop. In the United States 40–120 lb of nitrogen per acre in split applications is customarily applied. Dwarf, non-shattering cvs have been developed in the United States which are harvested mechanically, but it is usual to wait until all the capsules are dry and the plants have been defoliated by frost or by a chemical defoliant. The African peasant farmer often prefers dehiscing types so that there is no need to thresh the seed. Immature capsules should not be picked as the seeds in them have a lower oil content. Care should be taken to avoid scarifying or damaging the testa as oil exudes and the seeds turn rancid. In the tropics seeds are also harvested from wild and naturalized plants. In India yields of up to 900 lb of seeds per acre are obtained in pure stand, with an average yield of 400–500 lb per acre. In the United States yields of 1,000–2,500 lb per acre are obtained.

MAJOR DISEASES AND PESTS

Diseases are more serious in regions of high rainfall. In the United States the most serious diseases are leaf spots caused by *Alternaria ricini* (Yoshii) Hansford and *Xanthomonas ricinicola* (Elliott) Dowson, which cause partial to complete defoliation and also attack the capsules. In India the caterpillars of *Arhaea janata* L. attack the leaves and *Dichocrocis punctiferalis* Guen. the capsules.

IMPROVEMENT

Most of the work on breeding has been done in the United States where dwarf, non-shattering hybrid and inbred cvs have been produced. 'Baker 296', a dwarf inbred 3–4 ft high, is the main cv. grown in Texas. Hybrid seed has been produced in a similar manner to hybrid maize, e.g. 'Pacific Hybrid 6' which is grown in Arizona and California; this and other similar cvs flower at the 7th–9th node, average $7\frac{1}{2}$ ft in height and are predominantly female; it is thus necessary to plant a pollinator with them which is usually one of the parents used in developing the hybrid. It is necessary to obtain fresh hybrid seed each year. Medium duration selections (165 days) have been made in India by crossing high-yielding, long-duration types with short-duration cvs from the United States and South Africa and gave 11–20 per cent higher yields than local cvs.

PRODUCTION

The demand for castor seed and oil has increased greatly in recent years. By 1960 world production had reached over 500,000 tons. Brazil is the largest producer with over 200,000 tons, followed by India with about 100,000 tons and Thailand with 40,000 tons. Since World War II there has been a substantial increase in production in the United States. Other exporters include Ecuador, South Africa, Ethiopia, and Tanzania. Prior to 1945 seed was obtained from wild plants in Tanzania, since when the crop has been planted and in 1960 this country exported 18,400 tons. The three largest importers of castor seed and oil are the United States, France and the United Kingdom.

REFERENCES

ALYADURAL, S. G. *et al.* (1963). Evolving short duration Castor Strains. *Indian Oilseeds J.*, 7, 237–242.

BRIGHAM, R. D. and SPEARS, B. R. (1960). Castorbeans in Texas. *Bull. Tex. agric. Exp. Sta.*, 954.

CROP RESEARCH DIVISION, AGRIC. RESC. SERV. (1960). Castorbean Production. *U.S.D.A. Farmers' Bull.*, 2041.

WEIBEL, R. O. (1948). The Castor-oil Plant in the United States. *Econ. Bot.*, **2**, 273–283.

LAURACEAE

About 47 genera and 1,900 spp. of trees and shrubs, except for the twining herbaceous, parasitic *Cassytha* spp., mainly in the tropics and subtropics. Aromatic oil glands in all parts. Lvs simple, usually alternate or spirally arranged, leathery and evergreen; stipules absent. Fls mostly bisexual, regular, small, green or yellow, sepals usually 6 in 2 whorls; petals absent; stamens typically 12 in 4 whorls, the innermost often reduced to staminodes, anthers dehiscing by 2 or 4 small upturned flaps; ovary superior of 1 carpel. Frs usually 1-seeded fleshy berries.

USEFUL PRODUCTS

The genus *Cinnamomum* (q.v.) yields spices and essential oils. *Persea americana* (q.v.) is the avocado pear. Sassafras, a flavouring material, is obtained from the bark of *Sassafras albidum* Nees in eastern North America. *Laurus nobilis* L., a native of Asia Minor, is the sweet bay and the laurel of antiquity; its leaves are used for culinary purposes. Several spp. produce valuable timber, notably *Ocotea rodioei* (Schomb.) Mez, greenheart of British Guiana.

Cinnamomum Blume (x = 12)

About 110 spp. of evergreen trees and shrubs in Asia, Australia and Fiji. Bark often aromatic. Lvs mostly opposite and usually 3-nerved from base. Fls small in axillary and terminal panicles; sepals 6; stamens usually 9 with 3 staminodes. Fr. small berry in cup-like perianth.

USEFUL PRODUCTS

SPICES AND ESSENTIAL OILS: *C. zeylanicum* (q.v.) produces the true or Ceylon cinnamon. Cassia or Chinese cinnamon is the bark of *C. cassia* (Nees) Nees ex Blume, a native of Burma and cultivated in southern China; it also produces cassia buds, which are the dried unripe fruits and are used as spice; cassia oil is obtained from the leaves. Indian cassia comes from *C. tamala* (Buch.-Ham.) Nees & Eberm., the leaves of which are also used as a spice; it is a native of India. Padang cassia, with smooth bark and no cork, is *Cinnamomum burmanni* (Nees) Blume, an Indonesian tree, the bark of which is exported to the United States and with a considerable market in the East. Inferior substitutes for cinnamon and cassia are Oliver's bark from *C. oliveri* Bailey in Australia and Mossoia bark from *C. mossoia* Schewe in New Guinea.

Camphor, an important essential oil used in industry, is obtained by distilling the wood of *C. camphora* (L.) Nees & Eberm. It is used in the manufacture of celluloid, in disinfectants and chemical preparations, and has a wide range of medicinal uses. Safrole, produced from the residual oil after camphor extraction, is used in soap and perfume manufacture. *C. camphora* is a native of China, Japan and Taiwan. It has been introduced into many tropical countries, where it does best at higher altitudes. Camphor is a handsome evergreen tree, 20–30 m in height; stout dormant buds are produced; lvs small, ovate, whitish beneath, 3–10 × 2–5 cm, spirally arranged, with 1 longitudinal vein (midrib); fls small, 4 mm in diameter, pale yellow. Frs black and fleshy when ripe, about 1 cm in diameter. Taiwan is the principal producer. There is a little commercial production in the United States, mainly in Florida. A large proportion of the world's supply of camphor is now produced synthetically from pinene, a turpentine derivative, and from coal tar.

Cinnamomum zeylanicum Breyn (2n = 24) CINNAMON

USES

The bark, exported as quills, is used as a spice or condiment, for flavouring cakes and sweets, in curry powder, and in incense, dentifrices and perfumes. Cinnamon bark oil is distilled from the chips and bark of inferior quality; it is used for flavouring confectionery and liqueurs and in pharmaceutical and dental preparations and soaps. Cinnamon leaf oil is distilled from dried green leaves and is used in perfumes and flavourings and also in the synthesis of vanillin.

ORIGIN AND DISTRIBUTION

Cinnamon occurs wild in Ceylon and south-western India to 6,000 ft, but is commoner in the lowlands. It is doubtful whether cinnamon and cassia (see above) barks were used by ancient Egyptians in embalming, as is stated in some works. They appear to have reached Egypt and Europe by 5th century B.C. and were known to Herodotus, being brought by Sabaeans from India or Ceylon to southern Arabia, which became the centre of the spice trade; the traders concealing the true source of these spices. Brown (1956) considers it unlikely that the cinnamon of antiquity was derived from *C. zeylanicum.* The Portuguese occupied Ceylon in 1536 largely for the sake of cinnamon trade, of which they maintained the monopoly, until they

Fig. 28. **Cinnamomum zeylanicum:** CINNAMON. **A,** flowering shoot ($\times \frac{1}{2}$); **B,** flower in longitudinal section ($\times 10$); **C,** fruits ($\times \frac{1}{2}$); **D,** fruit in longitudinal section ($\times 1$).

A
B
C
D

were ousted by the Dutch in 1656, who in turn were ousted by the British in 1796 and who carried on the monopoly until 1833. The Dutch began commercial cultivation in Ceylon in 1770. Cinnamon was introduced into Java in 1825 and has since been cultivated in southern India, the Seychelles, Madagascar and Brazil and tried in other tropical countries, but Ceylon produces the best quality and the bulk of the world's supply.

ECOLOGY

Soil and other environmental conditions affect quality. Wild trees are confined to tropical evergreen rain forest. The best cultivated cinnamon is grown at low altitudes in Ceylon on poor white sands, with an average temperature of 85°F and 85–100 in. of rainfall per annum. It is naturalized in the Seychelles and Madagascar, the seeds being dispersed by birds.

STRUCTURE

An evergreen tree, 8–17 m high in wild state. Bark and lvs strongly aromatic. Lvs stiff, evergreen; petiole 1–2 cm long, grooved on upper surface; lamina ovate or elliptic, 5–17 × 3–10 cm, strongly 3-veined from near base, lateral veins to $\frac{3}{4}$ or more length of lf., lf. reddish when young, turning dark green above with paler veins and pale glaucous beneath. Fls in lax axillary and terminal panicles at ends of twigs; peduncle creamy white, 5–7 cm long; individual fls very small, about 3 mm diameter, with foetid smell, each subtended by a small ovate hairy bract; sepals 6, campanulate, pubescent; stamens 9 in 3 whorls with glands at base, filaments hairy, anthers 4-celled opening by 4 small valves, staminodes 3; ovary superior, 1-celled. Fr. fleshy berry, black, 1-seeded, ovoid, 1·5–2 cm long, with enlarged calyx at base. In Ceylon trees flower in January and fruits ripen 6 months later.

POLLINATION

Cinnamon is probably pollinated by insects, especially flies.

GERMINATION

Seeds soon lose their viability and should be sown fresh after the removal of the pulp. Germination takes 2–3 weeks.

CHEMICAL COMPOSITION

The principal aromatic substance in cinnamon bark is cinnamic aldehyde, but it also contains eugenol and other substances. Cinnamon bark oil contains about 60 per cent cinnamic aldehyde and 10 per cent eugenol. Chips yield 0·5–2 per cent oil, although Singhalese distillers may get as low as 0·2 per cent. Leaves dried for 3 days yield 0·5–2 per cent leaf oil; this contains 65–95 per cent eugenol and usually under 3 per cent cinnamic aldehyde.

PROPAGATION

Cinnamon is usually propagated by seed. It is often necessary to net or bag the trees to protect the fruits from birds. The fruits are kept in heaps until the pulp rots; the seeds are then washed and dried in the shade. They may be sown thickly in nurseries with light shade, which is removed when the seedlings are 6 in. high. After 4 months clumps of seedlings are transplanted into baskets and after a further 4–5 months they are planted in the field. Seeds may also be planted at stake with a number of seeds per hole. Cinnamon can be propagated by cuttings of young 3-leaved shoots, by layering shoots and by the division of old rootstocks; by using the last method the time to harvest is reduced.

HUSBANDRY AND PREPARATION

In planting up forest land it is advisable to leave tall trees at 50 ft intervals to provide light shade. The usual spacing between planting holes is 6–8 ft. The crop is weeded until the weeds are shaded out. The plants are coppiced after 2 years, the stems being cut within a few inches of the ground and the cut surfaces covered with earth. This encourages the formation of shoots, of which 4–6 are allowed to grow for a further 2 years before harvesting; they are kept straight by pruning. The stems are cut when they are 6–10 ft high and ½–2 in. in diameter. The stool is then pruned and regenerates.

The stems are cut during the rains to facilitate peeling. Leaves and twigs are trimmed off and the bark is removed by making two longitudinal cuts, and half the circumference of bark is removed as an entire piece. After fermenting for 24 hours in covered heaps, the epidermis, cork and green cortex are removed by scraping. On drying the scraped bark contracts into pipe or quill form. These are then packed one inside the other to form a compound quill, 3½ ft long. These are dried further in the shade and rolled by hand daily to make them firm and compact. Dried quills are pale brown in colour. The best quality cinnamon is obtained from thin bark from shoots in the centre of the bush and from the middle portion of the shoots.

The yield from the first cutting after 4–5 years is 50–60 lb of quills per acre per annum, increasing to 150–200 lb in subsequent cuttings and thereafter declining after 10 years.

The quills are graded into 5 qualities according to the thickness of the bark, appearance, colour and aroma. Commercial bark should not be more than 0·5 cm thick and the thinner the bark the better the grade. Broken quills are exported as quillings and the inner bark of twigs and twisted shoots as featherings; they are used mainly for grinding or for the distillation of cinnamon oil. Trimmings from cut shoots before peeling, shavings of outer and inner bark which cannot be separated and pieces of thick outer bark are exported as chips, which are used mainly for distillation.

MAJOR DISEASES AND PESTS

No serious diseases and pests have been recorded. Stripe canker, *Phytophthora cinnamomi* Rands, causes damage in young trees.

IMPROVEMENT

Little or no work has been done on the improvement of the crop. Various types are recognized in Ceylon and some are known to be inferior. Selection, together with the vegetative propagation of clonal material, would obviously be advantageous.

PRODUCTION

The Portuguese, in the 16th century, extracted a tribute of 250,000 lb of cinnamon bark annually from the Singhalese king. Production at the beginning of the 19th century was about 500,000 lb annually. Ceylon still produces the best quality and greatest quantity of cinnamon and now exports 50,000–60,000 cwt annually, of which nearly half is imported by Mexico. West Germany, the United States and Great Britain are the next largest importers. The Seychelles export about 25,000 cwt of cinnamon annually and also considerable quantities of cinnamon leaf oil. Indonesia exports approximately 150,000 cwt of cassia annually, substantial quantities of which go to the United States.

REFERENCES

BROWN, E. G. (1955–1956). Cinnamon and Cassia; Sources, Production and Trade. *Col. Pl. Anim. Prod.*, **5**, 257–280 and **6**, 96–116.

SAMARAWIRA, I. ST. E. (1964). Cinnamon. *World Crops*, **16**, 45–49.

Persea Mill. (x = 12)

Hutchinson (1964) states that there are 10 spp., all of which are trees, occurring mainly in tropical America; also known from Asia, and with 1 sp. (*P. indica* Spreng.) in the Canary Islands. *P. americana* (q.v.) is widely cultivated in the tropics and subtropics for its edible fruits.

Persea americana Mill. (2n = 24) AVOCADO

Syn. *P. gratissima* Gaertn.

USES

Avocado or alligator pear is one of the finest salad fruits, although some people regard it as an acquired taste. The fresh, smooth, buttery pulp is eaten and it is the most nutritious of all fruits. It is an important item of diet in Central America. It was little known outside tropical America before the 20th century. Its consumption is increasing in the United States and the United Kingdom. It is usually served as half fruits with lemon juice, vinegar or Worcester sauce, and salt and pepper. It is used in salads. The pulp, which may be preserved by freezing, is used as a sandwich filling or spread, and in ice creams and milk shakes. It does not cook satisfactorily. Avocado oil is used in cosmetics.

ORIGIN AND DISTRIBUTION

P. americana originated in Central America and the early Spanish explorers record its cultivation from Mexico to Peru, but it was not in the West Indies at that time. It was introduced into Jamaica about 1650 and subsequently spread throughout the region. It was taken to southern Spain in 1601. The spread to the Old World tropics was much later than most other New World crops. It was in Mauritius in 1780. Most of its spread in Asia was in the mid-19th century. It was reported in Zanzibar in 1892. Avocados are now grown in most tropical and subtropical countries, including South Africa and Australia. It was first recorded in Florida in 1833 and in California in 1856, but commercial production in these states did not begin until 1900.

SYSTEMATICS

Only one species is now usually recognized and there are 3 ecological races:

1. MEXICAN: Native of the highlands of Mexico, where wild progenitors have been found. It is the only race with anise-scented leaves. The fruits are small, usually 8 oz or less, ripening 6–8 months after flowering; with thin smooth skin, comparatively large seeds, which are often loose in the cavity; the cotyledons are smooth; the pulp has the highest oil content of all races with up to 30 per cent. It is the hardiest of the races, being the most resistant to cold, but is of little commercial importance, except for its hybrids with the other races. Some authorities consider it a distinct botanical variety—*P. americana* var *drymifolia* Mez = *P. drymifolia* Cham. & Schlecht.

2. GUATEMALAN: Native of the highlands of Central America; the wild prototype has been found in Guatemala. The fruits are large, usually 1–2 lb in weight, ripening in 9–12 months after flowering. They are borne on long stalks, the skin is thick, brittle and hard, often warty, the seeds are usually small and held tightly in the cavity; the seed coats are adherent. The pulp has a medium oil content of 8–15 per cent. The Guatemalan race is less resistant to cold than the Mexican.

3. WEST INDIAN: Native of the lowlands of Central America, but it did not spread to the West Indies until after the discovery of the New World. The leaves are lighter in colour than the other races. The fruits are large, but the range in size is less than the Guatemalan race. The fruits ripen 6–9 months after flowering. They are borne on short stalks; the skin is smooth and leathery, but not as thick as the Guatemalan. The seeds are large and are often loose in the cavity; the cotyledons are rough. The pulp has the lowest oil content of 3–10 per cent. It is the tenderest of the races and is best suited to the hot low tropics.

Inter-racial hybrids occur and are important economically.

CULTIVARS

Over 700 cvs have been tried in the United States during the present century, but only a few have shown promise. Three of the most important cvs are:

'FUERTE': A Mexican × Guatemalan hybrid. The fruits are pyriform, weigh 8–16 oz, with 18–26 per cent of oil. It is resistant to cold. It provides about 75 per cent of the Californian crop. It was introduced into Trinidad, but is now seldom seen there.

'LULA': Seedling from a Guatemalan race; the male parent is unknown, but was probably West Indian. The fruits are pyriform, weigh 14–25 oz, with 12–15 per cent oil. The skin is smooth and light green. It grows rapidly and is precocious and productive; it is suited to the tropics. It is the leading cv. in Florida and one of the two most grown cvs in Trinidad, where it fruits late in the year in September.

'POLLOCK': West Indian race. The large fruits are oblong to pyriform, weigh 30–50 oz with 3–5 per cent oil. It has a vigorous spreading habit and is well suited to the lowland tropics. It is extensively grown in Florida and is one of the two most important cvs in Trinidad, where it fruits early in the year in June.

ECOLOGY

The races show varying resistance to cold. Avocados can be grown on a wide range of soil types, but they are extremely sensitive to poor drainage and cannot stand waterlogging; they are intolerant of saline conditions. They are grown with irrigation in California. The wood is soft and brittle and liable to wind damage, so that windbreaks are desirable.

STRUCTURE

An evergreen tree to 20 m high, but budded trees are usually shorter. Shallow-rooted and there are no visible root hairs. Growth is in flushes. Lvs spirally arranged, variable in shape and size; petioles 1·5–5 cm long; lamina elliptic, ovate-oblong or obovate-oblong, 5–30 × 3–15 cm, glaucous beneath, pinnatinerved. Fls in axillary many-flowered panicles crowded at the ends of the branches; peduncles and pedicels yellowish-green, pubescent; fls bisexual, fragrant, subtended by lanceolate deciduous brown hairy bracts, 4 mm long; calyx of 6 sepals in 2 whorls, 1·0–1·5 cm in diameter, yellowish, densely tomentose within and without; petals absent; stamens 9 in 3 whorls; those of inner whorl are longer and each has 2 orange nectaries at the base,

---

Fig. 29. **Persea americana**: AVOCADO. **A**, leafy shoot (× ½); **B**, flower from above (× 6); **C**, flower in longitudinal section (× 4); **D**, stamen of inner whorl with nectaries (× 6); **E**, fruit in longitudinal section (× ½).

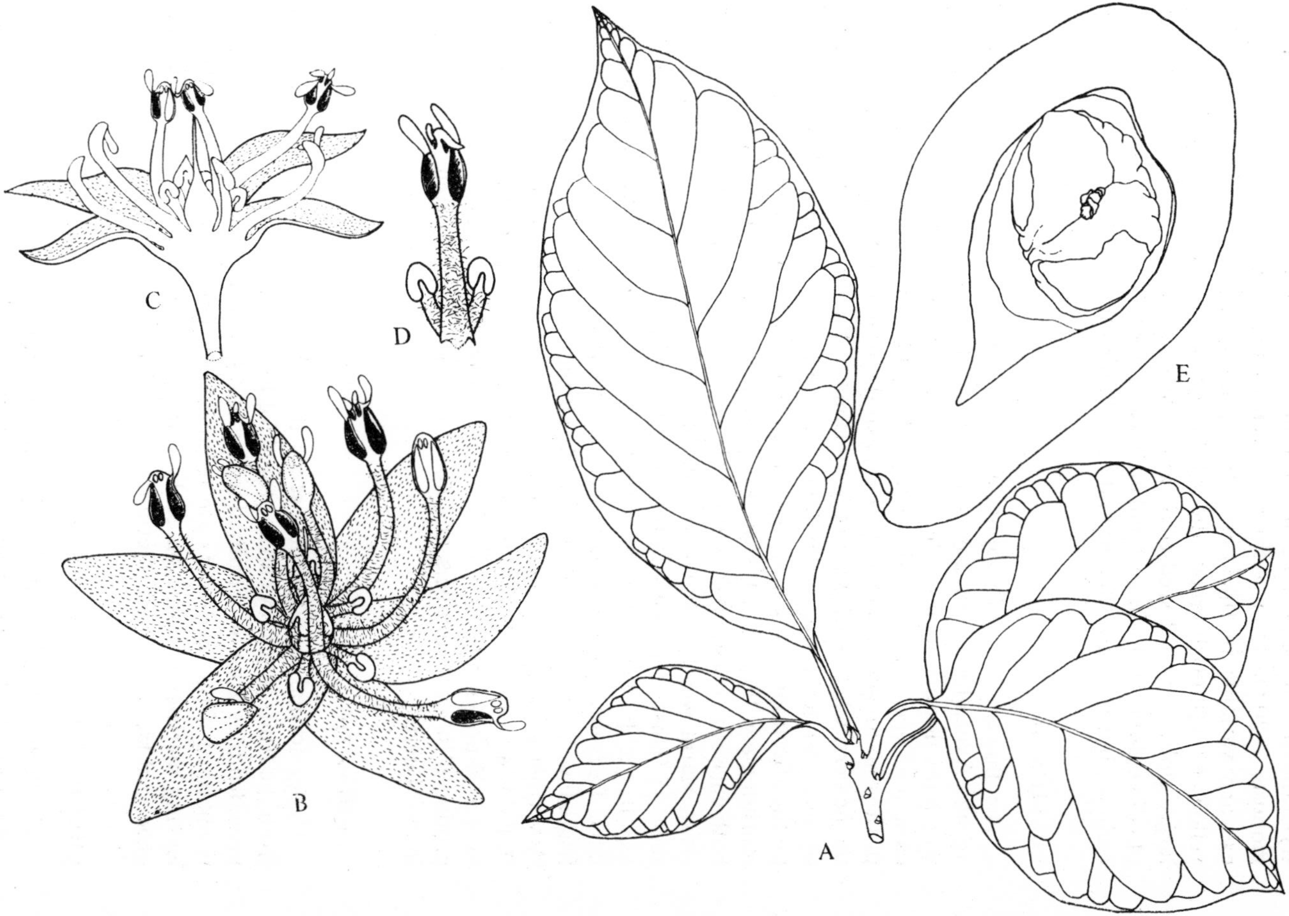
C
D
E
B
A

staminodes 3 in innermost whorl; filaments hirsute, anthers dehisce by 4 small valves or flaps hinged at upper end; ovary 1-celled with slender hirsute style and simple stigma. Fr. large fleshy berry, single-seeded, pyriform or globose, 7–20 cm long, yellowish-green to maroon and purple in colour; exocarp variable in thickness and texture (see races above); mesocarp yellow or yellowish-green and of butter-like consistency, edible. Seeds large, globose with 2 seed-coats and 2 large fleshy cotyledons, whitish or pink in colour, enclosing a small embryo. Sometimes frs develop without an embryo, but they are smaller and are cylindrical in shape.

POLLINATION

The flowers are entomophilous, bees being the principal pollinating agents. Avocados are protogynous and exhibit synchronous dichogamy in which the flowers open on the first day when the pistil is receptive and then close and open again on the second day when the pollen is shed and the pistil is no longer receptive. Cvs vary in the time of opening and closing of the flowers. Class A opens on the morning of the first day, then closes, and opens again in the afternoon of the following day, *e.g.* 'Lula'. Class B opens on the afternoon of the first day, and the second opening is on the following morning, *e.g.* 'Pollock'. It would thus appear necessary to interplant cvs of both classes to ensure pollination. The periodicity is most marked in bright, clear, warm weather, but is less marked on cool, cloudy days. Some cvs such as 'Hass' are self-fruitful. 'Fuerte', which exhibits the periodicity, usually yields well in monoclonal blocks in California. I know of two isolated trees in south-western Uganda, separated by several miles, which fruited well, but this was at a high altitude where the weather is cool. Nevertheless, it does seem desirable in hot regions to interplant both types. 'Lula' is an effective pollinator for 'Pollock'. The cv. 'Collinson' produces no viable pollen and so must have pollen from another cv. The flowers are produced in great numbers and about one fruit sets for each 5,000 flowers in cv. 'Fuerte'.

GERMINATION

Under normal conditions the seeds are viable for 2–3 weeks after removal from the fruits, but they can be stored for longer periods in dry peat at 42°F provided they are not permitted to dry out. Quicker and better germination is obtained if the seed-coats are removed. By splitting the cotyledons and leaving meristemic tissue on each section, several clonal seedlings can be produced from one seed. Germination is hypogeal.

CHEMICAL COMPOSITION

The flesh contains: water 65–80 per cent; sugar about 1 per cent; protein 1–4 per cent; oil 3–30 per cent; it is rich in vitamin B and good in vitamins A and E. The oil, which is similar in composition to olive oil, is highly digestible. Avocados have the highest energy value of any fruit; in California they average 1,000 calories per lb, with a maximum of 1,375 calories per lb.

PROPAGATION

Avocados can be easily grown from seed, but as cvs are very heterozygous, they should be propagated vegetatively. Mexican rootstocks have the greatest resistance to cold, but are incompatible with West Indian scions; Guatemalan and hybrid scions are successful on stocks of all races. Guatemalan and West Indian races produce the most vigorous seedlings. The seeds may be planted in nurseries, 14–18 in. apart in rows 2 ft apart, and should not be covered with more than ½ in. of soil. Stocks, 2–4 months old, are side (veneer) grafted with terminal tips 2–2½ in. long; this is the usual practice in Florida. Somewhat older stocks may be shield-budded which is the usual custom in California. Budded and grafted plants are transplanted to the orchard at 9–16 months old; in California the period is more usually 21–30 months. Leafy cuttings may be rooted, but not well enough for economical propagation. Poor seedling trees and unsatisfactory cvs can be top-worked using cleft grafts.

HUSBANDRY

The normal planting distance is 20–40 ft, depending on the vigour of the cvs and whether they are of upright or spreading growth. They are sometimes planted at $35 \times 17\frac{1}{2}$ ft with the removal later of alternate trees when growth makes this necessary. If there is any danger of waterlogging they should be planted on mounds. Transplants are planted with a ball of earth round the roots. Windbreaks should be provided. Orchards may be clean cultivated, but this should be shallow because of the surface roots, or they may be intercropped with pigeon peas or other crops when young; cover-crops may be planted or grass permitted to grow; weeds may be controlled by herbicides.

Little pruning is usually done except to train the trees and cut out dead wood. Suckers from stock should be removed. In upright cvs such as 'Pollock' the central shoot is pruned to develop a spreading habit; in spreading cvs such as 'Fuerte' branches may be thinned or shortened. Old tall trees may be headed back to facilitate harvesting. Manuring is similar to that used in citrus (q.v.). Zinc and copper deficiencies may occur and are corrected by foliar sprays.

Seedling avocados begin bearing at 5–6 years; vegetative-propagated plants usually earlier, but it is usual to remove the fruits until 3–4 years old. Avocados have a marked tendency to biennial bearing, which is partly controlled by genetic factors.

Mature fruits can remain on the tree for some time before harvesting. Maturity can be tested by the oil content or by picking a few fruits to test ripening. Immature fruits will not soften and ripen properly; mature fruits will ripen in 3–6 days at 80°F and in 30–40 days at 40°F. The fruits should be clipped from the tree. For storage and transport temperatures of 38–45°F are used, depending on the cv. but they cannot be stored for long periods. Yields are very variable with 100–500 or more fruits per tree. Good orchards in California give 6,000–12,000 lb of fruit per acre per annum.

MAJOR DISEASES AND PESTS

The most serious disease of avocados in the New World is root rot caused by *Phytophthora cinnamomi* Rands, which may kill the tree. Other serious diseases are: Cercospora spot, *Cercospora purpurea* Cooke; black spot or anthracnose, *Colletotrichum gloeosporioides* Penz.; and scab, *Sphaceloma perseae* Jenkins, all of which attack the leaves and fruits. Scale insects, mealy-bugs and mites may cause damage. In Puerto Rico the sugar-cane root weevil, *Diaprepes abbreviatus* (L.), may cause serious damage.

IMPROVEMENT

Many new clonal cvs have been selected from seedling trees in California, Florida, the West Indies, Central America, Hawaii and the Philippines. Hand-hybridization has not produced any good cvs. Selection is based on the following characters: suitability for the local environment, particularly hardiness to cold or tolerance to heat; regular and heavy bearing; time of bearing so as to spread the crop; ease of propagation; low spreading crown to facilitate harvesting; medium size fruits of uniform shape required by particular market; smooth dark green fruits usually preferred; small seeds which are tight in cavity; quantity and quality of pulp with freedom from fibres; resistance to diseases and pests.

PRODUCTION

Avocados are an important food crop in Central America. Elsewhere in the tropics they are grown for home consumption and the local markets. Commercial production is confined to California, Florida, Cuba, Argentina, Brazil, South Africa, Australia and Hawaii. In temperate countries the avocado is still regarded as a luxury fruit, but it is likely to gain in popularity as more people become acquainted with it. Production in the United States increased from 130 tons in 1924 to approximately 21,000 tons in 1948 and 40,000 tons in 1958.

REFERENCES

HODGSON, R. W. (1950). The Avocado—A Gift from the Middle Americas. *Econ. Bot.*, **4**, 253–293.

HUME, E. P. (1951). Growing Avocados in Puerto Rico. *Fed. Exp. Sta. Puerto Rico Circ. 33.*

RUEHLE, G. D. (1958). The Florida Avocado Industry. *Bull. Fla Univ. Agric. Exp. Sta. 602.*

LEGUMINOSAE

One of the three largest families of flowering plants with some 690 genera and about 18,000 spp. of herbs, shrubs, trees and climbers; the other large families being Compositae and Orchidaceae. Hutchinson (1959, 1964) and others make it an order, Leguminales, and divide it into three families, Caesalpiniaceae, Mimosaceae and Fabaceae (syn. Papilionaceae). In this treatment I retain the family Leguminosae and divide it into three subfamilies; Caesalpinioideae, which Hutchinson considers to be the most primitive and more closely related to the primitive Rosaceae; Mimosoideae; and Papilionoideae (syn. Papilionatae, Faboideae, Lotoideae), the climax group. The first two subfamilies are mainly tropical and are of little importance agriculturally, while the Papilionoideae, which is the largest of the subfamilies with about 12,000 spp., is widely distributed in both tropical and temperate regions, and provides a large number of crop plants.

Despite the great variety of forms, Leguminosae constitutes a very natural and easily recognizable family, which resolves itself, according to the structure of the flower, into three very natural subfamilies. Lvs usually alternate and compound, pinnate or trifoliate. Fls mostly hermaphrodite and usually with 5 sepals and 5 petals. Ovary is superior with a single carpel, cavity and style. Fruit is a pod and is usually a legume. A legume has been defined as 'a fruit formed by a single carpel and dehiscent by both ventral and dorsal sutures so as to separate into two valves'. By this definition not all leguminous fruits are legumes. The pod may be round or flat or winged, thick or thin, straight or coiled, short or long, papery or leathery, woody or fleshy, splitting open or indehiscent, in some cases breaking into 1-seeded joints (lomentum); in a few spp. fr. may be 1-seeded, fleshy and indehiscent. Endosperm usually absent.

NITROGEN FIXATION

Although the value of legumes in improving and sustaining soil fertility has been known since ancient times, it was not until towards the end of the 19th century that it was found out that legumes added nitrogen to the soil and the way in which this was brought about. Many species of legumes have nodules on their roots containing bacteria which have the power of fixing atmospheric nitrogen, some of which is then available to the host plant and the soil nitrogen is increased by sloughed, disintegrating nodules; in return the bacteria are supplied with carbohydrates by the host. This property has made legumes of great importance in agriculture; they provide protein-rich food for man and stock; they have an important place in crop rotations; they are used in admixture with grasses in leys and pastures; they are used as cover crops and green manures. This has been well authenticated in temperate countries, but comparatively little information is available as to the value of legumes in the maintenance of soil fertility in the tropics.

The bacteria, *Rhizobium* spp., which are normally free-living in the soil when they do not fix nitrogen, are attracted to the roots of legumes from the seedling stage onwards. They enter through the root hairs and pass into the cortex. Here they cause the cells to divide. These cells are tetraploid and produce the nodules which appear on the surface of the roots. The size and shape of the nodules varies considerably, but their form is constant for each species of legume. The growth and efficiency of the nodules is influenced by the carbon/nitrogen ratio of the plant and by the presence in the soil of phosphate, calcium, magnesium, molybdenum and boron. The presence and number of nodules is no guide to their effectiveness. If they are ineffective the bacteria may be parasitic on the host for nitrogen and the legumes must obtain their nitrogen from the soil, in which case they deplete the nitrogen reserves quicker than cereals and grasses. Effective nodules contain red leghaemoglobin which can be seen on cutting them across. Ineffective nodules are usually small, hard, spherical and a greenish colour inside.

In the tropics legumes have the ability to produce nodules on more acid soils and soils more deficient in phosphorus, calcium and other nutrients than in temperate countries. In the temperate regions nodulation is poor in seasons characterized by short days and low light intensity; whether this is applicable in the tropics is not known. It is generally considered that at the high temperatures in the tropics significant excretion of nitrogen by the roots does not take place.

A number of strains of *Rhizobium* are recognized, six of which have been raised to specific rank. Each strain is able to infect the root system and produce nodules on any of a group of related legumes, but not on other groups. The cowpea strain of *R. japonicum* is widespread in the tropics and can bring about nodulation in a large number of genera of tropical legumes. It is probably the ancestral and typical organism and is the least specialized of the strains. When introducing legumes into new areas in the tropics it is advisable to inoculate them with the correct strain of *Rhizobium* if it is not of the cowpea type. Soya beans, for example, have their own specific strain of the bacteria. Very few species of the Caesalpinioideae produce nodules, whereas nodulation is fairly general in the Mimosoideae and Papilionoideae.

More critical work is urgently needed on nitrogen fixation by tropical legumes. Webster and Wilson (1966) dealing with cover crops in the tropics state that 'it has not been definitely established that leguminous covers generally fix nitrogen under field conditions, but evidence from pot experiments, and from soil and leaf analyses, makes it highly probable that they do'. They also say that 'growing annual legumes as green manures in rotation with other crops is only likely to be of value when sufficient soil moisture is available . . . (that) the beneficial effects are short lived . . . (and that) it is not likely to be practicable to maintain fertility by green manuring alone'.

REFERENCES

COMMITTEE DIVISION TROPICAL PASTURES, C.S.I.R.O., AUSTRALIA (1962). A Review of Nitrogen in the Tropics with particular reference to Pastures. *Commw. Bur. Pastures and Field Crops, Bull. 46.*

DESOUZA, D. I. A. (1966). Nodulation of Indigenous Trinidad Legumes. *Trop. Agriculture, Trin.*, **43**, 265–267.

WEBSTER, C. C. and WILSON, P. N. (1966). *Agriculture in the Tropics.* London: Longmans.

WHYTE, R. O., NILSSON-LEISSNER, G. and TRUMBLE, H. C. (1953). *Legumes in Agriculture.* Rome: FAO Agric. Studies 21.

KEY TO THE SUBFAMILIES

A. Fls actinomorphic; petals valvate in bud .... *Mimosoideae*
AA. Fls zygomorphic; petals imbricate in bud
B. Aestivation imbricate-ascending with posterior (uppermost) petal innermost.................... *Caesalpinioideae*
BB. Aestivation imbricate-descending with posterior (uppermost) petal outermost and 2 anterior petals forming a keel .... *Papilionoideae*

CAESALPINIOIDEAE

This subfamily has 152 genera and nearly 2,800 spp. of trees and shrubs, rarely herbs, mostly tropical and subtropical and most numerous in tropical America. Lvs nearly always alternate, pinnate or bipinnate; stipules paired, mostly deciduous; stipels mostly absent. Fls zygomorphic, often showy, usually hermaphrodite; sepals 5 or 4 by union of 2 upper sepals, mostly free, sometimes much reduced when 2 bracteoles which are large and calyx-like cover bud; petals 5 or fewer with upper petal innermost in bud; stamens 10 or fewer, free to variously connate, dehiscing lengthwise or by terminal pore; ovary superior, 1-locular, 1-many ovules, style simple. Fr. a legume or indehiscent and drupaceous. Seeds sometimes arillate, rarely with endosperm.

USEFUL PRODUCTS

Tamarindus indica (q.v.) is grown throughout the tropics for its edible fruits. *Cassia* spp. (q.v.) are the source of the drug senna. Other species yielding useful products include:

Haematoxylon campechianum L., logwood, yields the purplish-red dye haematoxylon, which with iron salts gives a very permanent black. It is native to tropical America and the West Indies. It is a small thorny tree, about 8 m high, with small pinnate lvs arising from swollen joints; pinnae 2–4 pairs, wedge-shaped and usually deeply notched; fls very small, yellow, in many-flowered axillary racemes. The Spaniards first found it in Honduras before 1525. The sapwood is removed and the heartwood is exported as logs or as extracted dye. Jamaica has long been a centre of production and much of the logwood is obtained from wild or self-sown trees which have to be removed on clearing the land.

Caesalpinia sappan L., sappanwood, is a native of south-east Asia. Its heart-wood gives a red dye, which, under the name of bresil, was imported into Europe during the Middle Ages. The Portuguese discovered a similar dye in *C. echinata* Lam., brazilwood, in South America. They gave it the same name as the eastern product and from this the country of Brazil got its name. The pods of *C. coriaria* (Jacq.) Willd., divi-divi, and *C. spinosa* (Mol.) Ktze., tara, which are native in tropical America, are used commercially as a source of tannin. *C. pulcherrima* (L.) Sw., Pride of Barbados, which is probably American in origin, is widely cultivated as an ornamental shrub throughout the tropics. Forms with yellow, red and yellow, and pink flowers occur.

Copals are hard resins of recent, semi-fossil, or fossil origin which are used in paints, varnishes, inks, plastics, sizing, adhesives, fireworks and many other products. They may be obtained by wounding the trees when the resin exudes and hardens or in a fossil state from the ground. The main sources of African copal are *Copaifera* spp. and *Daniellia* spp. *Hymenaea courbaril* L., locust tree, is the chief source of South American copal and is also a valuable timber tree.

Copaifera officinalis L., a Venezuelan tree, yields copaiba balsam, which is used in medicine and industry.

A number of species of Caesalpinioideae produce valuable timbers, of which the best known are probably *Peltogyne paniculata* Benth., purpleheart, and *Mora excelsa* Benth., mora, both of which are found in South America.

The subfamily has a number of very attractive ornamental trees and shrubs of which the most widely cultivated are *Amherstia nobilis* Wall. from Burma, *Bauhinia* spp., *Cassia* spp., *Delonix regia* (Boj ex Hook.) Raf. (flamboyant) which was discovered in Madagascar in 1824, and *Peltophorum pterocarpum* (DC.) Heyne (yellow flame) which is a native of Malaysia.

The genera *Brachystegia*, *Isoberlinia* and *Julbernardia* are dominant over wide areas of woodland in tropical Africa.

Cassia L. (x = 6, 7, 8)

A large genus of about 600 spp. of trees, shrubs and herbs throughout the tropics, but most numerous in the dry open forests of tropical America. Lvs simply paripinnate, petiole glands often present. Fls hermaphrodite, mostly yellow, in axillary or terminal racemes or panicles; sepals 5; petals 5, subequal, uppermost inside the others; stamens 5–10, filaments thick of unequal length, anthers dehiscing by terminal pore, some anthers may be abortive. Pods flat or cylindrical, sometimes 4-angled or winged, splitting or indehiscent, many seeded, often with partitions between seeds.

USEFUL PRODUCTS

C. angustifolia (q.v.) and *C. senna* (q.v.) produce the senna of commerce. A number of other species are used as purgatives throughout the tropics and in other native medicines. The pods of *C. fistula* L., Indian laburnum, were exported to Europe for the pulp round the seeds which is used as a purgative. This tree, up to 6 m high, with hanging racemes of yellow fls up to 5 cm in diameter, has been widely distributed throughout the tropics as a garden ornamental and is the most handsome of all the Cassias. It is a native of India. *C. alata* L., a shrub of the New World tropics, is used in the treatment of ringworm.

Several species have been tried as green manure crops and *C. didymobotrya* Fresen., an East African shrub, is still used for this purpose in Asia. *C. siamea* Lam., a rapidly growing, short-lived tree from Asia, is extensively grown in tropical Africa for firewood and poles. The seeds of *C. occidentalis* L., native of tropical America, are used as a substitute for coffee in its area of origin and in West Africa. A number of species are grown throughout the tropics as ornamental trees.

Cassia angustifolia Vahl (2n = 28) INDIAN OR TINNEVELLY SENNA

The leaves and pods of *C. angustifolia* and *C. senna* (q.v.) are official in the British and United States Pharmacopoeias and are used as laxatives. *C. angustifolia* is indigenous to Somaliland and Arabia and is now cultivated in southern India for senna. It is a small shrub 60–75 cm high. Lvs paripinnate with 3–7 pairs of elliptic acuminate leaflets, 2·5–5·0 × 0·5–1·5 cm. Fls in terminal racemes, yellow. Pods flattened, 4–7 × 2 cm. Seeds dark brown. The purgative properties are due to emodin and allied glucosides.

The scarified seeds are usually broadcast at the rate of 15 lb per acre and the crop is usually rain-fed or grown with light irrigation. It is treated as an annual. The first flush of flower stalks is cut to induce lateral branching. The bluish-green leaflets are stripped by hand at 3–5 months. A second stripping is made after about a month and the plants are then allowed to flower and set seeds. The leaves are dried in the shade for 7–10 days and become yellow green in colour. The pods are dried and beaten to remove the seeds.

As a dry-land crop 300 lb of cured leaf and 75–150 lb of pods are obtained per acre; as a wet-land crop 750–1,250 lb of leaf and 150 lb of pods per acre are obtained. Before World War II India exported about 3,000 tons of senna annually. Since then the production has declined somewhat. Much of the senna is exported to the United Kingdom and the United States.

Cassia auriculata L. (2n = 14, 16, 28) AVARAM

C. auriculata, also known as the tanner's Cassia, is a native of India, where it is also cultivated for its bark, which is an important source of tanning material. It is also used to rehabilitate barren tracts and as a green manure. It is a stout shrub. Lvs paripinnate with 7–9 pairs of elliptic leaflets with mucronate tips and large stipules. Flowers yellow.

The crop is usually sown thickly and thinned during the 1st year. The shrubs reach a height of 3 m after 2 years and 5 m after 4 years. Harvesting begins when the shrubs are 2–3 years old. The old branches and twigs are cut and the bark is stripped off and dried. Coppiced bushes can be harvested annually. The stripped bark constitutes about 20 per cent by weight and a yield of about 1,400 lb of green bark per acre is obtained. The dried bark contains about 18 per cent tannin. At one time India produced up to 50,000 tons annually, but imported wattle bark (q.v.) from East and South Africa now tends to take the place of avaram in Indian tanneries.

Cassia senna L. (2n = ?) ALEXANDRIAN SENNA

Syn. *C. acutifolia* Del.

C. senna is a native shrub of Egypt, the Sudan and the Sahara region. The leaves and pods are picked from wild and cultivated plants in a similar manner to *C. angustifolia* (q.v.) and are exported as senna. Lvs paripinnate with 3–9 pairs of lanceolate leaflets, 1·5–3·5 × 0·5–1·2 cm; fls yellow in erect racemes; pods flattened, 4–6 × 1·8–2·5 cm. In the Sudan much of the senna exported is from the Northern Province, where the crop is grown on poor sandy soils and survives as a perennial with an occasional irrigation. Yields of up to 1,120 lb may be obtained after 70 days. The Sudan exports about 500 tons per year, the United States taking most of the leaves and United Kingdom the pods.

Tamarindus L. (x = 12)

A monotypic genus.

Tamarindus indica L. (2n = 24) TAMARIND

USES

The rather tart, brown pulp round the seeds is eaten fresh; mixed with sugar it makes a sweetmeat (tamarind balls); it is used for seasoning other food and in curries, preserves, chutneys and sauces; a refreshing acid drink and sherbet are made with it. The seeds may be eaten after the removal of the testa and roasting or boiling. They are sometimes made into flour. In India they are used as a source of carbohydrate for sizing cloth, paper and jute products and for vegetable gum in the food processing industry. The flowers and leaves are used in salads, curries and soups. The leaves, if allowed to lie wet on cloth, cause the latter to rot. Many parts of the tree are used in native medicines in Africa and Asia. The over-ripe fruits can be used to clean copper and brass. The tree is a useful ornamental and shade tree for avenues and gardens. It makes excellent charcoal.

ORIGIN AND DISTRIBUTION

Tamarinds grow wild in the drier savannas of tropical Africa. It appears to have been introduced into India at an early date and has now been widely distributed throughout the tropics.

ECOLOGY

The tree is particularly well adapted to semi-arid tropical regions, but can be grown in monsoon regions provided the soil is well drained. It does not do well in ever-wet climates. It will tolerate poor soil. In Africa it is frequently seen growing on or beside termite mounds. In Africa, also, it is often associated with the baobob, *Adansonia digitata* L. (q.v.).

STRUCTURE

A semi-evergreen, spreading tree to 20 m high with a compact rounded crown with drooping branches which reach to within a few feet of the ground. Trunk stout, often becoming twisted, with grey scaly bark. Lvs alternate, paripinnate, 7–15 cm long, with pulvinus at base of petiole; leaflets 10–20 pairs, opposite, entire, almost sessile, oblong, 1–2·5 × 0·5–1 cm, pale green beneath, with unequal rounded base and rounded or emarginate apex. Inflorescences in small terminal drooping racemes, 5–10 cm long. Fls 2–2·5 cm in diameter, zygomorphic: bracteoles 2, reddish, boat-shaped, about 8 mm long, enclosing bud but falling before fls open; sepals 4, reflexed, ovate, cream-coloured, 1–1·5 cm long; petals 3, borne on top of fl., obovate, pale yellow streaked with red, 1–1·5 cm long; fertile stamens 3 alternating with staminodes, filaments 1 cm long, united for about half their length, curling upwards under petals, anthers transverse, reddish-brown, dehiscing longitudinally; pistil oblique, curving upwards, green, sparsely hairy, longer than stamens, with small clavate stigma. Pods usually curved, oblong, 5–10 × 2 cm, constricted, scurfy brown with brittle shell, 1–10-seeded, indehiscent. Seeds obovate, flattened, brown, about 1 cm long, joined to each other by tough fibres running through brown sticky pulp which surrounds seeds.

POLLINATION

This is probably entomophilous.

CHEMICAL COMPOSITION

The pulp, which constitutes about 40 per cent of the pod, contains about: water 20·6 per cent; protein 3·1 per cent; fat 0·4 per cent; carbohydrate 70·8 per cent; fibre 3·0 per cent; ash 2·1 per cent. The tartness of the pulp is due to tartaric acid, of which it contains about 10 per cent. The carbohydrate is mainly in the form of sugars. The seeds contain: starch 63 per cent; protein 16 per cent; semi-drying oil 5·5 per cent.

PROPAGATION

Tamarinds are normally grown from seeds, but they can be budded or propagated by cuttings.

HUSBANDRY

Tamarinds are seldom, if ever grown in plantations and usually receive little attention.

MAJOR DISEASES AND PESTS

Pods in Trinidad are riddled with insects, which often eat all the pulp.

IMPROVEMENT

Little or no work has been done. The pulp is usually better developed and juicier in the Indian races than in those from Africa.

PRODUCTION

Tamarinds are grown throughout the tropics for local consumption. The dried pulp is exported from India to Europe and America for the preparation of chutneys and meat sauces.

MIMOSOIDEAE

This subfamily has 56 genera and about 2,800 spp. of trees and shrubs, very rarely herbs, mainly confined to the tropics and subtropics and are more numerous in the southern hemisphere. Lvs usually bipinnate, rarely once pinnate, sometimes reduced to phyllodes; stipules present, sometimes spine-like. Fls actinomorphic, small, usually sessile and massed in cylindrical spikes or globose heads; sepals usually 5, mostly valvate and united to form a toothed or lobed calyx; petals same number as sepals, valvate, free or connate; stamens often numerous, free or monadelphous; anthers small, versatile, often with apical gland, dehiscing longitudinally; ovary 1-locular, superior, style usually filiform, stigma small and terminal. Fr. dehiscent or indehiscent, sometimes a lomentum.

Fig. 30. **Tamarindus indica** : TAMARIND. **A,** flowering shoot (× ½); **B,** flower (× 1); **C,** flower with sepal removed (× 1); **D,** fruits (× ½); **E,** seed (× 1).

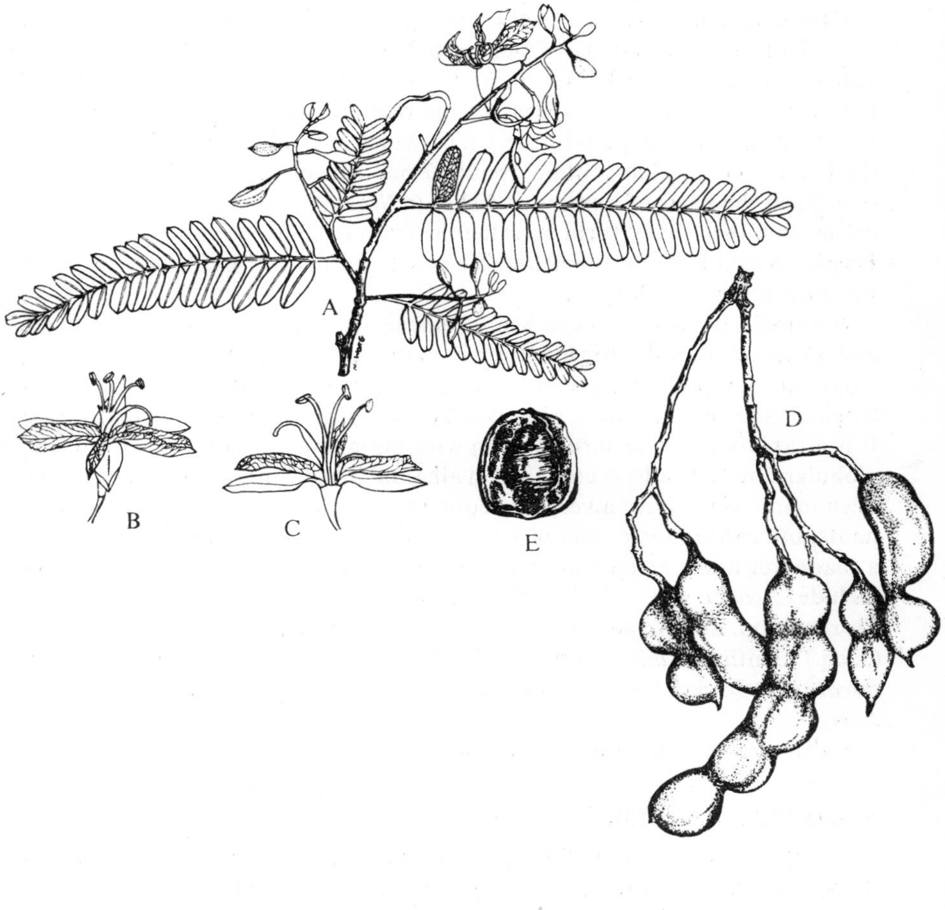
A
B
C
D
E

USEFUL PRODUCTS

Acacia mearnsii (q.v.) is the source of wattle bark.

Inga laurina (Sw.) Willd., sackysac, is a native tree of the West Indies and Central America and has edible fruits, which are often sold in local markets. Lvs pinnate, usually with 2 pairs of leaflets; fls small, white, fragrant, in spikes 10–12 cm long; pods oblong, flat, 5–15 × 2–3·5 cm, with sweet, white, slightly aromatic pulp which is the edible portion. This and other species of *Inga* are extensively planted as shade trees for coffee in the New World.

Parkia filicoidea Welw. ex Oliv., African locust bean, is the commonest tree of the park savannas of northern Nigeria and they are owned individually. It is a tree to 30 m or more tall with bipinnate lvs, red club-shaped flower heads about 5 cm in diameter and numerous pods, 15–45 × 2 cm. The yellow, dry, powdery pulp of the fruits, which is rich in carbohydrates, and the black seeds, which contain about 17 per cent of fat and are rich in protein, form the basis of numerous dishes in West Africa. The pods are fed to cattle and the leaves form a valuable manure. The pods of *P. speciosa* Hassk., which have an objectionable smell and taste of garlic, are used for flavouring foods in Malaya.

A number of species of the Mimosoideae are important as browse plants and a source of fodder for livestock in many parts of the tropics; the leaves, twigs, and/or pods are eaten. *Leucaena glauca* (L.) Benth., a native of tropical America, has now been widely distributed throughout the tropics. It is a rapidly-growing shrub to 4 m with bipinnate leaves and small, white, globular flower-heads. It contains an alkaloid, mimosine, but selections have been made which have a very low content. Mimosine is non-toxic to ruminants, but causes horses and pigs to shed their hair. *L. glauca* is also used as a shade tree and a green manure. Other plants used for browsing and fodder include *Acacia* spp., *Albizia* spp., *Pithecellobium dulce* (Roxb.) Benth., *Prosopis* spp. and *Samanea saman* (Jacq.) Merr. (syn. *Pithecellobium saman* (Jacq.) Benth.). *Albizia* spp. are also used for shade trees, of which *A. falcata* (L.) Backer (syn. *A. moluccana* Miq.) is widely used in south-eastern Asia.

Calliandra spp. are good ornamental shrubs or small trees.

Acacia Mill. (x = 13)

A large genus of about 900 spp. of trees and shrubs, very rarely herbs, in the tropics and subtropics, with the greatest number in Africa and Australia, in areas with dry climates. Often armed with prickles or spines. Lvs bipinnate with numerous small leaflets; in many Australian spp. lvs reduced to phyllodes, in which pinnae and leaflets are absent and the petiole is flattened into a leaf-like organ. Stipules often modified into a pair of thorns at base of petiole, which are sometimes much enlarged, hollow and ant-inhabited. Fls small, yellow or white, in axillary cylindrical spikes or globose heads, with numerous free stamens. Pods variable, flat, cylindrical, straight or curled, usually dehiscent.

USEFUL PRODUCTS

A. mearnsii (q.v.) is grown for tannin production. Species of *Acacia* in natural vegetation, especially in arid regions, are important as reserve fodder; the leaves and nutritious pods are browsed or the trees are lopped. They are often cultivated for this purpose, also as shade trees or soil improvers, and for binding sand dunes or controlling soil erosion. Some species contain cyanogenetic glucosides and are toxic.

The following species yield useful products:

A. catechu (L.f.) Willd., cutch, is a small thorny tree which is a native of India. It has yellow flowers in axillary spikes. The manufacture of cutch is a very ancient industry in India. It is a greyish resinous substance obtained by boiling chips of heartwood in water and then evaporating the liquid. The paler form, known as catechu, is used for chewing with betel leaf, *Piper betle*(q.v.), as a masticatory and gives the characteristic red colour of spittle with lime. Cutch is used as a brown dye and to a lesser extent as tannin.

A. farnesiana (L.) Willd., cassie flower, is a small armed shrub to 4 m high. It is probably a native of tropical America and has now been widely distributed throughout the tropics, often becoming naturalized. It is grown for its bright golden-yellow, fragrant flowers. Cassie perfume is obtained from the flowers and is extracted by hot fat. The centre of the industry is Cannes in southern France.

A. nilotica (L.) Willd. ex Del. (syn. *A. arabica* (Lam.) Willd.), babul, occurs wild in tropical Africa and extends to India. It is a spiny tree to 14 m with yellow flowers in globose heads. It is planted in India for its bark, which contains 12 per cent tannin and is extensively used in local tanning. It also yields babul gum which is similar to gum arabic.

A. senegal (L.) Willd., Sudan gum arabic, is widespread in the dry savannas of tropical Africa and extends to the Red Sea and eastern India. It is a spiny tree to 12 m tall with fragrant creamy-white flowers borne in spikes, 5–10 cm long. Gum arabic is obtained mainly from wild and semi-cultivated trees in the Sudan where it grows in areas with a rainfall as low as 10–14 in. per year. Incisions are made in the bark and the tears of gum which exude are collected about one month later. The gum is colourless, odourless and tasteless; it is slowly and completely soluble in water. It is used in the textile, mucilage, paste, polish and confectionery industries, in lithographic work, as a glaze in painting, and as an emulsifying and demulcent agent in pharmacy. The Sudan has exported gum arabic to Europe for over 2,000 years.

A. seyal Del. is the source of shittim wood, and is widespread in the drier parts of tropical Africa, and extending to Egypt. It is a small thorny tree to 12 m high with yellow flowers in globose heads. It yields an edible gum of good quality, but is inferior to that of *A. senegal.* It is used as a browse plant and fodder for livestock. The Ark of the Covenant, as well as the furniture of the Tabernacle, were made from shittim wood. It was also used for the coffins of Egyptian kings because of its 'incorruptible' nature.

Acacia mearnsii De Wild. (2n = 26) BLACK WATTLE

Syn. *A. decurrens* (Wendl.) Willd. var. *mollis* Lindl.

USES

Wattle bark, sometimes known as mimosa bark, contains 30–45 per cent tannin, with an average of 35–39 per cent. It is now the most widely used tanning material in the world, this use developing largely during the present century. It is extensively used in the tanning of hides and skins in the production of many different classes of leather. It is now usually exported as a solid extract. Tannin is used to a small extent in the treatment of boiler water, in dye production, and in mining. The timber is light and tough and is used for fuel, charcoal, poles and pit-props. In Ceylon black wattle is used as a green manure and windbreak for tea.

SYSTEMATICS

In the past there has been considerable confusion as to the correct scientific name for black wattle. It has often been wrongly named *A. decurrens* (Wendl.) Willd. and *A. mollissima* Willd. The name has now been finalized by Brenan and Melville (1960) as *Acacia mearnsii* De Wild. It has been confused with two other Australian species, namely:

A. decurrens (Wendl.) Willd., green wattle, which has linear leaflets, 5–12 mm long (1·5–4 mm in *A. mearnsii*), and very angular young stems with little tomentum. It ripens its seeds in 5–6 months, whereas *A. mearnsii* takes 14 months. *A. decurrens* is as high yielding and has as high a tannin content as *A. mearnsii*, but the tannin is more highly coloured, producing more highly coloured leather and so is less preferred. It is hardier than *A. mearnsii*, is more resistant to cold and insect attack and grows better on poor sites.

A. dealbata Link, silver wattle, is similar to *A. mearnsii* in leaflet size, but can be distinguished by the grey pubescence of the young shoots (golden-yellow in *A. mearnsii*), silvery-grey colour of the mature bark, absence of glands at the insertion of each pair of pinnae, and almost entire absence of constrictions between seeds in the pod. The tannin content of the bark is only 20 per cent and gives a much darker colour in tanning. Its sale is prohibited by law in Natal.

A third and very distinctive species has been introduced into many parts of the tropics from Australia, namely: *A. pycnantha* Benth., golden wattle, which does not have bipinnate or feathery leaves, but lanceolate phyllodes. The bark yields over 40 per cent tannin, but the small size of tree and slow rate of growth render it uneconomic for large-scale production.

ORIGIN AND DISTRIBUTION

Black wattle is a native of eastern Australia and Tasmania, where the exports of bark between 1850–1900 caused the depletion of natural trees in accessible areas. It was introduced into Natal in South Africa in 1864 as a quick-growing tree for shelter belts and fuel. The first exports of tan bark from Natal were made in 1886. Black wattle was introduced into the Kenya Highlands in 1903 to provide fuel for the railway, but later developed into an important tan-bark industry. It was taken to India in 1840, but it is not grown to any considerable extent, as most of the Indian sub-continent is unsuitable for its growth. There has been extensive planting in Rhodesia and Tanzania since World War II. Although it has been introduced into Uganda (Kigezi district), Madagascar, Java, the Philippines and North and South America, there has been little commercial development in these regions, but plantations have been started in Brazil.

ECOLOGY

In Natal the crop grows best in the 'mist-belt' between 2,000–4,500 ft, with a rainfall of 35–40 in. per annum, and on deep, well-drained, red or chocolate lateritic soils. In Kenya and elsewhere on the equator the crop is grown at altitudes of 6,000–8,000 ft. Black wattle cannot tolerate extremes of heat or cold. It can be grown on a wide range of soil types and will tolerate poor soils with little humus provided that they are friable and of good depth. Adequate weed control and thinning are essential in the early stages.

STRUCTURE

A quick-growing unarmed tree to 15–20 m high with rounded crown. Dense surface feeding mat of roots. All parts except fls pubescent or puberulous. Young shoots ridged, angled, with golden-yellow tomentum. Lvs feathery, bipinnate; petiole decurrent, 1·5–2·5 cm long, with a gland above; rachis 4–12 cm long with numerous raised glands on upper surface; pinnae 12–21 pairs, 3–4 cm long; leaflets minute, 20–70 pairs, linear-oblong, 1·5–4 × 0·5–0·7 mm. Fls minute, pale yellow, fragrant, in globose heads, 5–8 mm in diameter, on axillary panicles at the ends of branches; peduncles 2–6 mm long; calyx campanulate; petals 5, free, valvate, about 2 mm long; stamens numerous, exserted, anthers minute, versatile, pale yellow; style filiform. Pods dark brown when ripe with golden-brown or grey pubescence, constricted and contracted between seeds (moniliform), 3–10 × 0·5–0·8 cm, with 3–12 joints, dehiscent along one suture. Seeds black, smooth, elliptic, compressed, about 5 × 3·5 mm; with conspicuous caruncle; areole 3·5 × 2 mm.

POLLINATION

The fragrant flowers are visited by bees which are the main pollinating agents. Cross-pollination takes place freely, but the flowers are also self-fertile.

GERMINATION

The pods take 14 months to mature from flowering. There are approximately 40,000 seeds per lb. The seeds have very hard seed-coats, which causes poor and delayed germination unless pre-treated. This is usually done by putting the seeds in hot water at 90–100°C and leaving them to cool and soak overnight. The water is then poured off and the seeds are washed in two or three changes of clean water and dried in the shade. Seeds treated in this way will give over 90 per cent germination and can be kept for considerable periods without impairing germination. Vast numbers of seeds germinate when felled plantations are burnt.

CHEMICAL COMPOSITION

Black wattle bark, air-dried with 12·5 per cent moisture, contains on the average 34–39 per cent of catechol tannin, an uncrystallized colloidal substance with astringent properties. It forms insoluble compounds with gelatine-yielding tissue, thus enabling it to convert hides and skins into leather, giving increased weight, thickness, durability and resistance to water and decay. Tan liquors contain 15–25 per cent tannin depending on the speed of tanning required. Approximately 1 lb of pure tannin (100 per cent) is required for every 2 lb of air-dry sole leather, less being used for lighter leathers. Solid wattle extract contains a legal minimum of 60 per cent tannin, sometimes as much as 63 per cent, and 20 per cent soluble non-tannins. Its production saves space and freight and permits tanners to use more uniform material of known tannin content. Wattle tannin is considered superior to quebracho, which is obtained from the wood of two South American trees, *Schinopsis balansae* Engl. and *S. lorentzii* (Griseb.) Engl. of the family Anacardiaceae (q.v.). Wattle tannin gives a better colour to the leather and does not precipitate in acid solution, thus penetrating the hides faster.

PROPAGATION

Commercial plantations are always grown from seed (see GERMINATION above). Attempts at vegetative propagation from cuttings have been unsuccessful, but a fair degree of success has been obtained with marcots on young branches.

HUSBANDRY

A fine seedbed is required for planting. The seeds are sown *in situ* in rows 6–12 ft apart, either in continuous lines with a seed rate of 2–3 lb per acre

---

Fig. 31. **Acacia mearnsii : BLACK WATTLE. A,** flowering shoot (×½); **B,** leaf rachis showing glands (×1); **C,** portion of inflorescence (×10); **D,** flower in longitudinal section (×15); **E,** pods (×½); **F,** portion of pod with seed (×2);

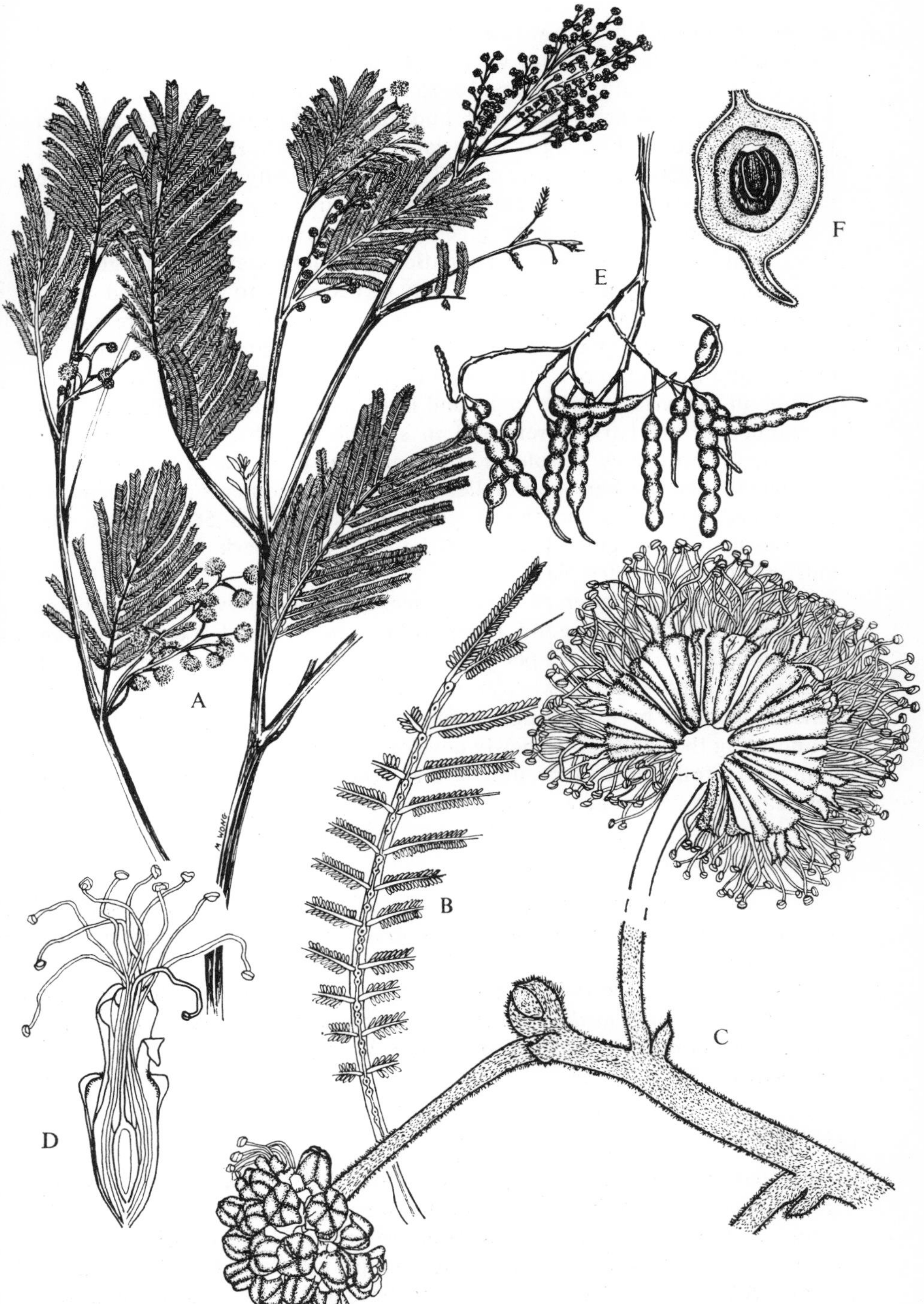
A
B
C
D
E
F
M WONG

or spot sowing at 3 ft or so with a seed rate of 1–1½ lb per acre. It is essential to keep the plantation free of weeds until the canopy is established and to thin the seedlings until they are 10–15 ft high. The first thinning is made when the seedlings are 3–4 in. high and they are thinned to 3–4 ft apart in the rows. Further thinning is done later. Some authorities recommend a final stand of 250–300 trees per acre. The crop is usually harvested when the trees are 6–8 years old, but it may be deferred until they are 12 years old. Felled plantations are usually regenerated by burning brushwood over them, which causes prolific germination of fallen seeds. Some growers favour non-burning, in which case it may be necessary to sow treated seeds.

Harvesting is done during the rains when the bark strips more easily. The thickest heaviest bark with the highest tannin content is at the base of the tree and this is removed from standing trees by ring barking at 4 ft and pulling off the bark in strips to ground level. The tree is then felled and the remainder of the bark is removed in suitable lengths. If the factory is close-by the bark may be sold or transported as green bark. If it is not, it is usual to dry the bark over fallen logs with the outer side uppermost. This takes 8–14 days and the weight is reduced by 38–58 per cent. On average 1·8 tons of green bark gives 1 ton of dried bark. The dried bark must be stored in a dry place.

Yields of 4–6 tons of green bark per acre are obtained when the trees are 7–8 years old. In Natal good 10-year-old trees averaged 7 tons of green bark and 33 tons of timber per acre. The dried bark may be chopped and pressed for export, but nowadays much of it is exported as wattle extract. This is obtained by extracting it from the chopped bark with water and evaporating the solution. In newly developed areas it is considered that the minimum profitable factory unit is one producing 5,000 tons of extract per year from 25,000 acres of wattle.

MAJOR DISEASES AND PESTS

Few serious pathogens occur in Africa. Gummosis, a physiological disease, in which gum is produced from cracks in the bark, can render the bark valueless.

The major insect pests in Natal are: *Acanthopsyche junodi* Heyl., wattle bagworm; *Achaea lienardi* Boisd., wattle looper caterpillar; *Jassus cederanus* Naudé, froghopper; *Lygidolon laevigatum* Reut., jassid. No serious pests are recorded from Kenya.

IMPROVEMENT

Populations are very heterogeneous. Now that vegetative propagation can be done by marcotting it should be possible to select for higher bark yield, higher tannin content, resistance to frost, pests and diseases and general suitability for marginal areas. Selfing has been found to reduce vigour. All the Australian tan wattles are diploid ($2n = 26$) and interspecific crosses can be made between *A. mearnsii* and *A. decurrens* (Wendl.) Willd.,

A. dealbata Link, *A. pycnantha* Benth. and *A. baileyana* F. Muell. Hybrids with *A. decurrens* suggest the possibility of combining the desirable characters of both species.

PRODUCTION

South Africa is the major producer and exports over 150,000 tons of bark and extract (two-thirds of the world's total) per annum from over 500,000 acres. Kenya is the next largest producer and exports about 25,000 tons per annum. Production is extending in Tanzania, where the Colonial Development Corporation has put down 33,000 acres at Njombe. Africans also produce bark for this factory. Black wattle is also grown in Rhodesia. As with so many other tropical crops, the main areas of production are far removed from the centre of origin. Australia, the home of black wattle, now imports wattle extract from Africa.

The United Kingdom is the largest importer in Europe. The United States is taking increasing quantities of wattle extract because of decreasing supplies of quebracho from South America due to over-exploitation and of locally produced chestnut extract due to disease. Although India is richly endowed with indigenous tanning material, she now imports wattle extract from Africa.

In addition to supplying tannin, black wattle yields large quantities of timber which is used for fuel, charcoal, mine-props and hard-board.

REFERENCES

BRENAN, J. P. M. (1959). *Leguminosae: Subfamily Mimosoideae* in *Flora of Tropical East Africa.* London: Crown Agents.

BRENAN, J. P. M. and MELVILLE, R. (1960). The Latin Name of Black Wattle. *Kew Bull.* **14**, 37–39.

HOWES, F. N. (1953). *Vegetable Tanning Materials.* London: Butterworth.

WIMBUSH, S. H. (1941). A Comparison of Wattle Growing in Natal and in Kenya. *E. Afr. agric. J.*, **6**, 121–126, 220–224.

PAPILIONOIDEAE

According to the International Rules of Botanical Nomenclature it would appear that the correct name for this subfamily is either Faboideae or Lotoideae. It is sometimes designated Papilionatae.

About 480 genera and 12,000 spp. of trees, shrubs, herbs and climbers, generally distributed throughout the world, with the more primitive woody genera mostly in the tropics and the more advanced herbaceous genera commoner in the temperate regions. Due to the very distinctive structure of the flower, members of this subfamily are very homogeneous and easy to recognize.

Lvs usually alternate and mostly compound, pinnate, trifoliate or digitate; stipulate; stipels often present at base of individual leaflets. Fls zygomorphic and typically papilionaceous; mostly hermaphrodite; calyx tubular and usually 5-toothed; petals 5, imbricate with descending aestivation; upper (adaxial) petal exterior, usually largest, forming standard (vexillum); 2 lateral petals more or less parallel with each other forming wings (alae): and lowest 2 petals interior, usually joined by lower margins, to form keel (carina), which encloses stamens and ovary. Stamens usually 10, monadelphous (all united by filaments) or diadelphous with 9 united by filaments and with upper or vexillary stamen free; rarely all stamens free; mostly all perfect; anthers 2-locular, usually dehiscing lengthwise by slits. Ovary superior, of 1 carpel, usually 1-locular, sometimes with false septa; ovules 1-many on ventral suture. Fr. usually a legume or pod, splitting along dorsal or ventral sutures or both; sometimes indehiscent; occasionally jointed and breaking into 1-seeded segments. Seeds usually without endosperm.

USEFUL PRODUCTS

PULSE CROPS: A pulse may be defined as the dried edible seed of a cultivated legume. Pulse crops are of very ancient cultivation in both the Old and the New World. They are next in importance to cereals as sources of human food and contain more protein than any other plant product. Animal protein is still a rarity in the diet of vast numbers of the poorer people in the tropics and pulses often provide the chief, and in some cases the only source of protein. Before the introduction of the potato, pulses constituted a great part of the food of the poorer classes in Europe. Most of the many different kinds of beans and peas grown throughout the tropics are harvested when the seeds are ripe and the pods are dry. Their low moisture content and hard testa permit storage over long periods. In addition to providing dry pulses, many of the crops are grown for their green edible pods and unripe seeds and the leaves and shoots of some of them are used as pot-herbs. Very little work had been done on the breeding and selection of the tropical pulse crops, although the local cultivators must have carried out some selection for palatability and adaptation to their local environments over the many centuries in which they have been grown.

The most widely grown pulse crops in the tropics are: common bean, *Phaseolus vulgaris* (q.v.); Lima bean, *P. lunatus* (q.v.); cowpea, *Vigna unguiculata* (q.v.); pigeon pea, *Cajanus cajan* (q.v.). In India three of the smaller-seeded pulses are of great importance, namely: chick pea, *Cicer arietinum* (q.v.); green gram, *Phaseolus aureus* (q.v.); black gram, *P. mungo* (q.v.). Other pulse crops of less importance in the tropics as a whole are: *Canavalia* spp. (q.v.); *Cyamopsis tetragonoloba* (q.v.); *Dolichos uniflorus* (q.v.); *Lablab niger* (q.v.); *Phaseolus aconitifolius* (q.v.); *P. acutifolius* (q.v.); *P. calcaratus* (q.v.); *Psophocarpus tetragonolobus* (q.v.). Pulse crops grown on the edge of the tropics as winter crops or at high altitudes include:

Lathyrus sativus (q.v.); *Lens esculenta* (q.v.); *Phaseolus coccineus* (q.v.); *Pisum sativum* (q.v.); *Vicia faba* (q.v.). Yam beans, *Pachyrrhizus* spp. (q.v.), are grown mainly for their edible tubers.*

OIL SEEDS: Groundnut, *Arachis hypogaea* (q.v.), is one of the most important annual oil seeds. Soya bean, *Glycine max* (q.v.), which is also a pulse crop, is not much cultivated in the tropics, but is of importance in the Far East and the United States.

DYES: Indigo is obtained from *Indigofera* spp., of which the most important are: *I. arrecta* Hochst. ex A. Rich., *I. sumatrana* Gaertn. and *I. tinctoria* L. The crop is still cultivated in India, but its importance has declined due to synthetic dye production. There were 1,688,900 acres under cultivation in India in 1896; by 1956 this had declined to 10,600 acres. The species now mainly grown is *I. arrecta*, a native of tropical Africa. Indigo is still used for local dyeing in tropical Africa and elsewhere, but seldom enters into international trade. Dyes are also obtained from camwood from *Baphia nitida* Afzel ex Lodd., barwood from *Pterocarpus erinaceus* Poir. and *P. soyauxii* Taub., all of which are West African trees, and red sanderswood from *P. santalinus* L.f. from the East Indies.

OTHER PRODUCTS: Sunn hemp, *Crotalaria juncea* (q.v.) is an important fibre crop in India. *Derris* spp. (q.v.) and *Lonchocarpus* spp. are a source of rotenone and are used as insecticides and as fish poisons. Tonka bean, *Dipteryx* spp. (q.v.), is used for flavouring. Balsam of Tolu obtained from *Myroxylon balsamum* (L.) Harms. and balsam of Peru from *M. pereirae* (Royle) Klotzsch are used medicinally and as fixatives for perfumes. Liquorice is obtained from the roots of *Glycyrrhiza glabra* L. and gum-tragacanth from *Astragalus gummifer* Labill., both from western Asia, but they are not truly tropical. Fenugreek, *Trigonella foenum-graecum* L., is a native of southern Europe and Asia; its seeds are used in curries in India; it is also used in medicine and in making artificial maple flavouring. Jumbie beads are the very poisonous red and black seeds of *Abrus precatorius* L., a pan-tropical species; they are used for necklaces and rosaries and in India for goldsmiths' weights. Ambatch, one of the lightest known woods, obtained from *Aeschynomene elaphroxylon* (Guill. & Perr.) Taub. of tropical Africa, is used locally as floats for fishing nets, harpoons and rafts; it is a constituent of the Nile sudd.

COVER CROPS, GREEN MANURES AND FODDERS: Because of the probable fixation of nitrogen in their root nodules (see above) legumes are widely grown in the tropics as cover crops and green manures. Cover crops are often planted under tree crops. For the characteristics of a good cover see rubber (p. 160). Green manures are planted in rotation with annual crops and are ploughed into the soil. Because of their high protein content legumes are widely planted in pure stand or in admixture with grasses for grazing and fodder for livestock. Many of the legumes used for these purposes have very hard impervious testas and are pre-treated before

* The yam bean of tropical Africa, *Sphenostylis stenocarpa* Harms, is cultivated for both its tubers and its seeds.

sowing to ensure a quick and even germination. This is done by soaking the seeds in hot water, which is allowed to cool and stand overnight, or by putting them in concentrated sulphuric acid for 15 minutes, after which they are carefully washed, or by mechanical scarification.

The legumes most widely used for cover crops, green manures and fodders are:

Calopogonium mucunoides Desv., a native of tropical America, has been widely introduced throughout the tropics and has become naturalized in some areas. It is a vigorous, creeping, short-lived, hairy perennial, which has a tendency to climb by twining. Lvs trifoliate; leaflets ovate, 3–10 × 2–7 cm; inflorescence axillary; fls small, blue, 7–10 mm long; pods narrow, 3–4 × 0·5 cm, with dense brown hairs; seeds square, flattened, pale brown, 3–4 mm in diameter. The seed rate in pure stand is 5–8 lb per acre. Its value lies in its rapid initial growth, producing a dense mat of foliage, 1–2 ft high, at 5 months. It is rather unpalatable to stock because of its hairiness.

Centrosema plumieri (Turp.) Benth., a native of tropical America, has been widely introduced. It is a quick-growing, deep-rooted, drought-resistant, trailing or twining, glabrous perennial. Lvs trifoliate; petiole 3–10 cm long; leaflets ovate, acute, 5–12 × 3–10 cm; inflorescence raceme of 2–6 fls; bracteoles about 1·5 cm long, twice as long as calyx; fls white or pale yellow, sometimes suffused red or purple, inverted with standard at base, about 5 cm in diameter; pods 8–15 × 1 cm with beak, 1–2·5 cm long, and 2 longitudinal ridges on either side of margin; seeds deep brown, 7 × 5 mm. It is sown at a seed rate of 7–10 lb per acre and can be used as forage as well as a cover crop.

Centrosema pubescens Benth. is also a native of tropical America and has been widely introduced throughout the tropics. It is recommended for planting in rubber in Malaya and the East and is usually grown in admixture with *Calopogonium mucunoides* (q.v.) and *Pueraria phaseoloides* (q.v.). Although it is rather slow growing at first, it later forms an excellent cover and will persist under shade. It is similar in appearance to *C. plumieri* from which it can be distinguished by its pubescent lvs, smaller bracteoles which are 8 mm long and equal in length or shorter than calyx, and narrower pod. Seeds olive green with red markings, 5 × 4 mm. It is sown at a seed rate of 5–7 lb per acre. It is also grown for fodder and in pasture mixtures with grasses in Nigeria and elsewhere.

Pueraria phaseoloides (Roxb.) Benth., tropical kudzu, is one of the most important and widely planted cover and green manure crops of the tropics and often becomes naturalized. It is indigenous to, and occurs wild throughout the lowlands of Malaysia. It is a very vigorous, densely pubescent, slender, climbing or trailing, perennial herb with procumbent stems rooting at the

---

Fig. 32. **A, Calopogonium mucunoides. A1,** leaf (× ½); **A2,** fruits (× ½); **A3,** seed (× 1). **B, Centrosema pubescens. B1,** flowering shoot (× ½); **B2,** flower in longitudinal section (× 1); **B3,** pod (× ½); **B4,** seed (× 4).

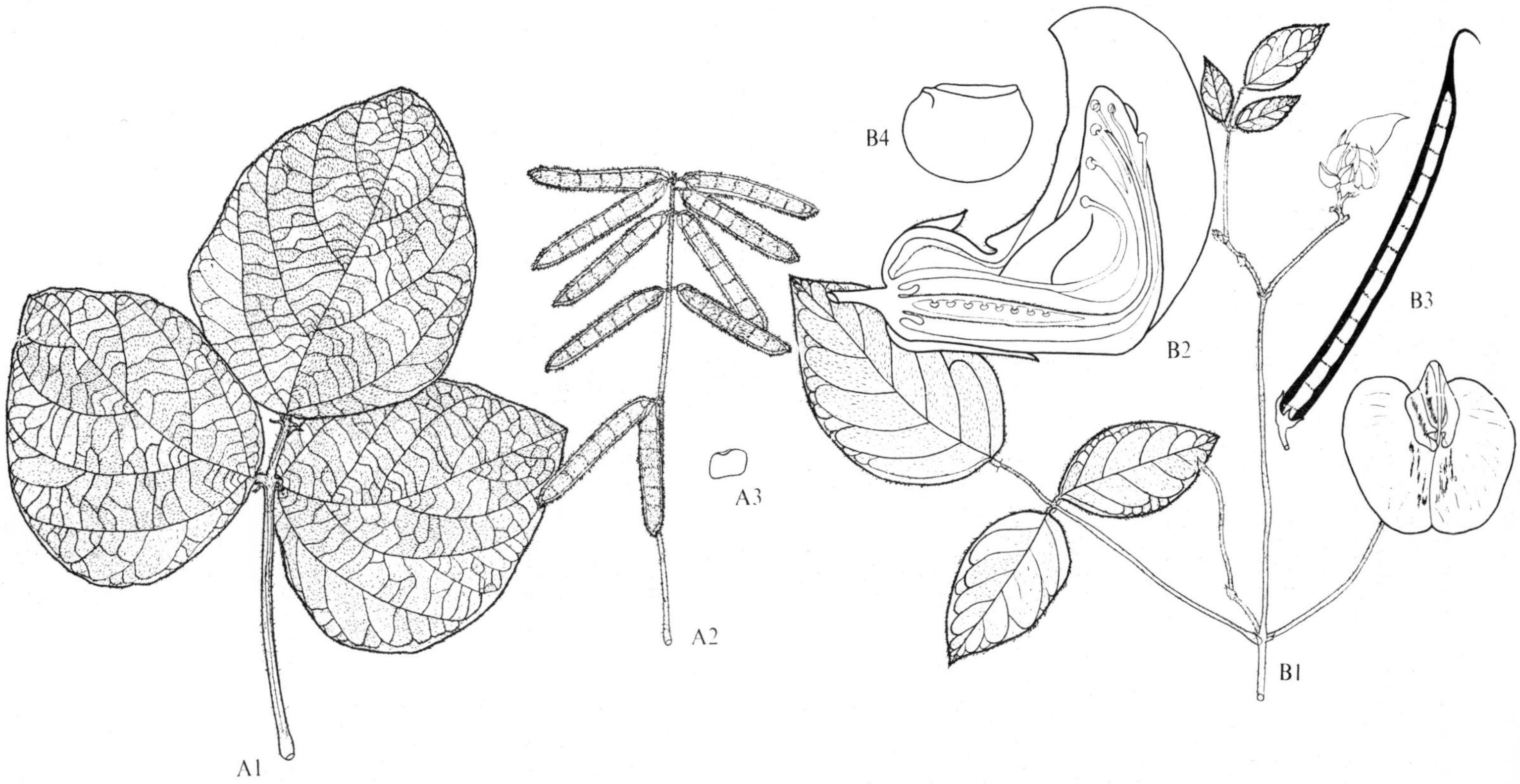
A1
A2
A3
B1
B2
B3
B4

nodes and with tuberous roots. Lvs trifoliate, petiole 2·5–15 cm long; leaflets ovate, 5–12 × 4–10 cm; inflorescence axillary raceme, 10–20 cm long, with glandular swollen nodes; fls mauve, pink or blue and white, 1·5 cm long; pods linear, flattened or rounded about 8 × 0·6 cm, beaked, brown-hairy at first, often smooth when ripe; seeds small, oblong, dark in colour, about 3 mm long. It is planted at a seed rate of 5 lb per acre. It is suited to all tropical areas with fair to high rainfall and moderate to high temperatures. It quickly forms a dense thick cover which suppresses weeds and prevents colonization by grasses and persists well, but it will not tolerate dense shade. Tropical kudzu is now attracting considerable attention as a fodder and pasture plant throughout the tropics, often in mixtures with grasses.

Pueraria thunbergiana (Sieb. & Zucc.) Benth. is the well-known kudzu of Japan and China, which has now been spread widely in the subtropics and has been extensively planted in the south-eastern United States. It does not do as well as *P. phaseoloides* in the tropics.

Stizolobium aterrimum Piper & Tracy (syn. *Mucuna pruriens* DC. var. *utilis* Wall.), the Bengal bean, is probably a native of tropical Asia and has been widely distributed throughout the tropics. It is a strong, vigorous, twining, annual herb. Lvs trifoliate; leaflets 15 cm or more in length; inflorescence pendent raceme, 45–75 cm long, many-flowered; fls dark purple, 3–4 cm long, pods sickle-shaped, about 10 cm long, black when mature with short white hairs, prominent median ridge, 4–5-seeded; seeds oblong, black, very shiny, 1·0–1·2 cm long, with prominent white hilum. It is grown as a green manure crop in Brazil and elsewhere and for green fodder, silage and hay. The seed rate is 30–40 lb per acre. The seeds are eaten in times of famine but only after boiling with repeated changes of water and the removal of the testa. The seeds and pods, whole or ground, can be fed to livestock with no ill effects.

Stizolobium deeringianum Bort., Florida velvet bean, first attracted attention when it was introduced into Florida about 1876. It is better suited to colder climates than *S. aterrimum* and is now widely distributed in the subtropics.

In addition to their use as pulses, many of the pulse crops can be grown and used as green manures, cover crops and fodders.

Other crops grown for these purposes include: *Aeschynomene americana* L., joint vetch, native of tropical America; *Alysicarpus vaginalis* (L.) DC., alyce clover, from tropical Asia; *Clitoria laurifolia* Poir. and *C. ternatea* L., butterfly peas, from tropical America; *Crotalaria* spp. (q.v.); *Desmodium* spp., tick clovers or beggar weeds, widespread in the tropics; *Moghania*

Fig. 33. **A, Pueraria phaseoloides. A1,** leaf (× ½); **A2,** flower in longitudinal section (× 4); **A3,** pod (× ½); **A4,** seed (× 1). **B, Stizolobium aterrimum. B1,** leaf (× ½); **B2,** flower in longitudinal section (× 1); **B3,** pod (× ½); **B4,** seed (× ½).

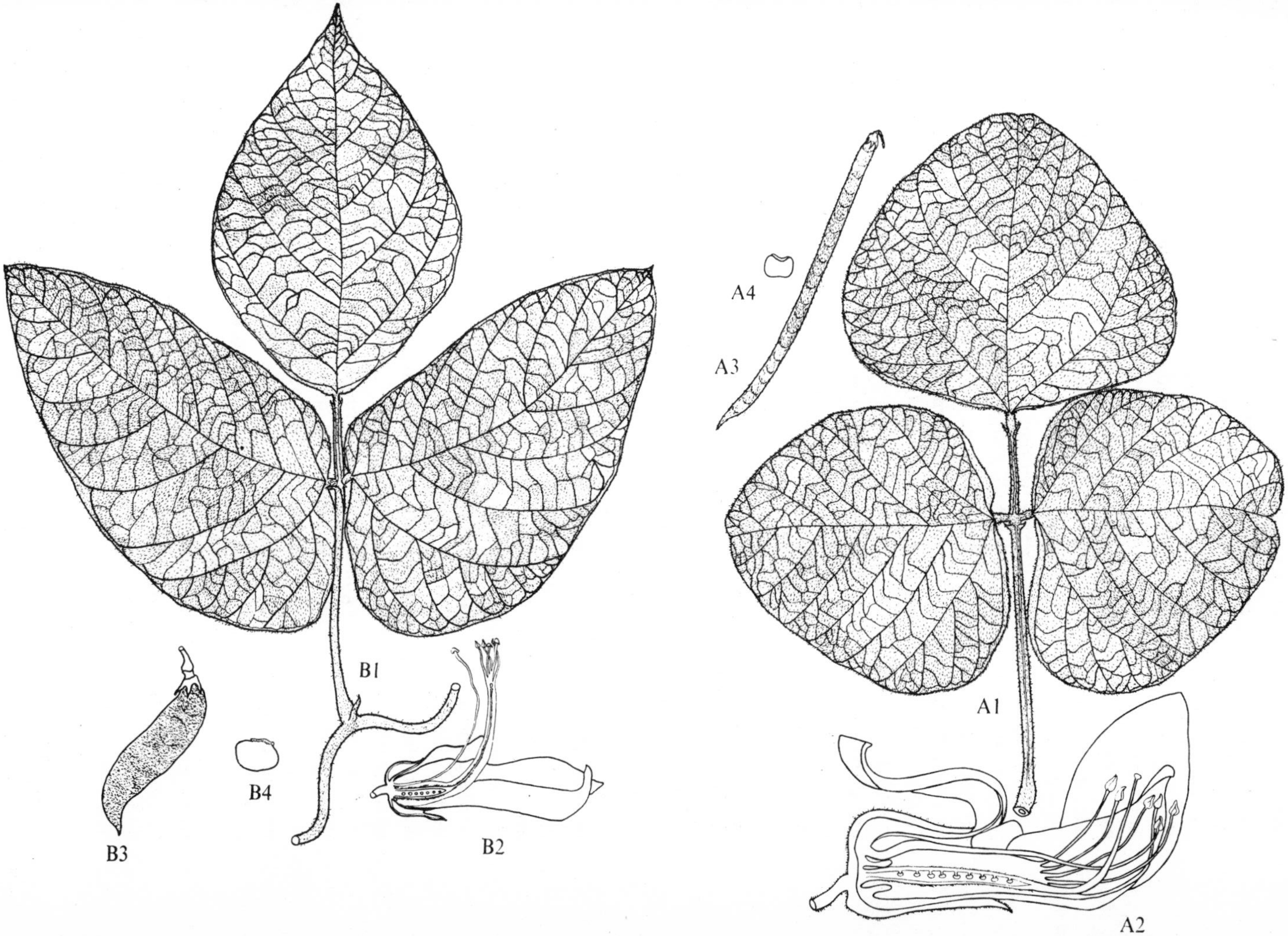
A1
A2
A3
A4
B1
B2
B3
B4

macrophylla (Willd.) O. Ktze. (syn. *Flemingia congesta* Roxb.), wild hops, from tropical Asia; *Indigofera* spp., indigo, widespread in the tropics—*I. spicata* Forsk. (syn. *I. hendecaphylla* Jacq.) is somewhat toxic to stock; *Lespedeza* spp. from Asia and Australia; *Lupinus* spp., lupins—*L. termis* Forsk., Egyptian lupin, is grown in the Sudan on flooded lands too hard and too salty for other crops, its mildly poisonous, bitter seeds may be eaten after boiling and straining; *Medicago sativa* L., lucerne or alfalfa, of Mediterranean origin, is now widely cultivated as a fodder in the temperate regions and the subtropics and at higher altitudes in the tropics; *Sesbania* spp. throughout the tropics; *Stylosanthes erecta* P. Beauv. (syn. *S. guineensis* Schum. & Thonn.) from West Africa and *S. guianensis* (Aubl.) Sw., Brazilian lucerne from tropical America, are becoming increasingly important as pasture legumes in the tropics; *Tephrosia* spp., particularly *T. candida* (Roxb.) DC. from India and *T. vogelii* Hook f. from tropical Africa, which are toxic to stock and are also used as fish poisons; *Trifolium* spp., clovers, are grown at the higher altitides; *Trigonella foenum-graecum* (q.v.), fenugreek, which is grown in India as a green fodder and soil renovator; *Zornia* spp.

SHADE TREES: Although it is traditional to provide shade trees for cocoa, coffee and tea in some countries, the benefits obtained may be open to question and there has been a considerable controversy on this subject during recent years. This matter is discussed under cocoa (pp. 587–588), coffee (pp. 471–472) and tea (p. 606). As is pointed out by Webster and Wilson (1966), under good environmental conditions, 'provided nutrient uptake is not limiting, photosynthetic activity, net assimilation rate, growth and yields increase with increasing light intensity up to, or closely approaching full daylight. On the other hand where nutrient uptake is below optimum, shade may be helpful by reducing light intensity and thus limiting the level of photosynthetic activity to a level that can be sustained by the available nutrient supply'. If the crops are grown under favourable conditions of soil and climate, the increase in yields from fertilizers, particularly nitrogen, is much greater in full sunlight than under shade trees. In areas of marginal rainfall for the crops, shade trees which will compete with the crop for soil moisture cannot be used. Temporary shade is usually regarded as essential for young cocoa. In the case of young coffee mulching can have the desired effect. 'In the absence of both shade and mulching, young coffee and cocoa plants tend to make rapid initial vegetative growth, to bear one or two relatively heavy crops and then to suffer from unthriftiness and die-back.' Because of the probable fixation of atmospheric nitrogen and the high nitrogen content of the leaves, flowers and fruits, many of the best shade trees are found in the Leguminosae. These include:

Gliricidia sepium (Jacq.) Walp. (syn. *G. maculata* Steud.), Nicaraguan cocoa shade, which is a native of tropical America, is one of the most widely used shade trees for bananas, coffee and young cocoa. It can be planted from large poles and can be topped to produce material for green manuring.

A number of species of *Erythrina* are used as shade trees. These include: *E. glauca* Willd., swamp immortelle and *E. poeppigiana* (Walp.) O. F. Cook (syn. *E. micropteryx* Poepp.), mountain immortelle, which are the best shade for cocoa in the West Indies and tropical America; *E. indica* Lam. and *E. subumbrans* (Hassk.) Merr. (syn. *E. lithosperma* Miq.) in Asia; *E. abyssinica* Lam. in East Africa. These species may be propagated by large cuttings, as well as seeds. They are often grown as fences.

Temporary shade is provided by *Cajanus cajan* (q.v.), *Sesbania* spp. and *Tephrosia* spp.

ORNAMENTALS: Among the most handsome tropical woody climbers are: *Mucuna bennettii* F. Muell., a native of New Guinea, with large trusses of flame-coloured flowers; *M. rostrata* Benth. from tropical America with orange flowers; *Strongylodon macrobotrys* A. Gray with long hanging racemes of jade-green flowers; all are of comparatively recent cultivation. More widely known are: *Clitoria ternatea* L., butterfly pea; *Clianthus dampieri* Cunn., glory pea; and trees belonging to the genera *Erythrina* and *Millettia*.

REFERENCES

FAO (1959). *Tabulated Information on Tropical and Subtropical Grain Legumes.* Rome: FAO.

GROSZMANN, N. M. (1956). Pulse Crops. Beans and Peas. *Queensland Agr. Dept. Adv. Lft.*, 245, Revised.

HUTCHINSON, J. (1964). *The Genera of Flowering Plants.* Vol. 1. Oxford University Press.

RUBBER RESEARCH INSTITUTE, MALAYA (1954). Establishing a Legume Cover. *Planters' Bull. R.R.I.*, **14**, 86–94.

—— (1955). Further Legume Covers. *Planters' Bull. R.R.I.*, **17**, 31–38.

—— (1958). Establishment and Maintenance of Legume Covers. *Planters' Bull. R.R.I.*, **39**, 129–133.

THOMPSTONE, E. and SAWYER, A. M. (1914). The Peas and Beans of Burma. *Dept. Agric. Burma, Bull.*, **12**, of 1914.

WEBSTER, C. C. and WILSON, P. N. (1966). *Agriculture in the Tropics.* London: Longmans.

WHYTE, R. O., MILSSON-LEISSNER, G. and TRUMBLE, H. E. (1953). *Legumes in Agriculture.* Rome: FAO.

WILLIAMS, W. A. (1967). The role of the Leguminosae in Pasture and Soil Improvement in the Neotropics. *Trop. Agriculture, Trin.*, **44**, 103–115.

KEY TO GENERA OF TROPICAL PULSE CROPS AND LEGUMINOUS OIL SEEDS

A. Fruits ripening underground
- B. Lvs pinnate; leaflets 4 .... *Arachis*
- BB. Lvs trifoliate .... *Voandzeia*

AA. Fruits ripening above ground
- B. Lvs pinnate with terminal leaflet present .... *Cicer*
- BB. Lvs even-pinnate; terminal leaflet represented by tendril, or bristle, or absent ....
  - C. Wings of corolla free or nearly so from keel ..... *Lathyrus*
  - CC. Wings of corolla adherent to keel for half or more of their length
    - D. Style bearded in tuft or ring at apex; calyx lobes short and broad .... *Vicia*
    - DD. Style bearded down one side
      - E. Calyx lobes leafy; pods several-seeded .... *Pisum*
      - EE. Calyx lobes long-subulate; pods 1–2-seeded. *Lens*
- BBB. Lvs trifoliate
  - C. Stamens monadelphous; fruits short, erect, 4-ribbed .... *Cyamopsis*
  - CC. Vexillary stamen free from near base upwards
    - D. Style bearded or pilose at or towards stigma
      - E. Keel of corolla coiled.......................... *Phaseolus*
      - EE. Keel of corolla bent inwards at right angles, beaked .... *Dolichos*
      - EEE. Keel of corolla arched or curved; stigma prominent
        - F. Stigma strongly oblique or introrse; roots not tuberous .... *Vigna*
        - FF. Stigma on inner face of style, subglobose; roots tuberous .... *Pachyrrhizus*
    - DD. Style beardless on stigmatic part
      - E. Long-running or twining stems .... *Lablab*
      - EE. Stems not twining, but erect or spreading
        - F. Leaflets ovate; fls white or purple .... *Glycine*
        - FF. Leaflets lanceolate-oblong; fls yellow or orange .... *Cajanus*
  - CCC. Vexillary stamen united in upper part with others, free below
    - D. Pods square, 4-winged .... *Psophocarpus*
    - DD. Pods flattened, not 4-winged .... *Canavalia*

Arachis L. (x = 10)

A small genus of about 19 spp. of tropical and subtropical herbs in South America from River Amazon through Brazil, Bolivia, Uruguay and northern

Argentina to about 35°S. The cultigen *A. hypogaea* (q.v.) is not known in a wild state; the other species are wild and perennial, forming an important part of the herbage which is extensively grazed. All species are geocarpic, ripening their fruits underground.

Arachis hypogaea L. (2n = 40) GROUNDNUT, PEANUT, MONKEYNUT

USES

Large quantities of groundnuts are consumed locally in the areas of production. The world trade depends largely on the European demand for groundnut oil, the extraction of which began in Marseilles over 100 years ago. The non-drying oil is used as a substitute for olive oil as a salad and cooking oil. It is used in the manufacture of margarine and inferior quality oil for soap, and as a lubricant. High quality oil is used in the pharmaceutical industry. A considerable quantity of oil in India is used in the manufacture of vegetable ghee by hydrogenation. The cake after expression of the oil is a high-protein livestock food. The best quality cake may be ground into flour for human consumption.

The nuts are eaten raw or after roasting, for which the Virginia types are preferred as they contain less oil than the Spanish types, which are better for oil extraction. More than half the edible peanut stocks of the United States are used for peanut butter, which is made by grinding roasted blanched kernels after the removal of the testa and the germ. The nuts are used in confectionery and in curries. A favourite dish in West Africa is 'groundnut chop'. Ardil, a synthetic textile fibre, is manufactured from the protein. The green haulms make excellent fodder and hay. In the United States the crop is often fed to hogs with the hogs doing the harvesting.

ORIGIN AND DISTRIBUTION

Arachis hypogaea originated in South America and was probably domesticated in the Gran Chaco area, including the valleys of Paraguay and Paraná Rivers. They became widely distributed throughout the continent and have been found in excavations in coastal Peru dated about 800 B.C. and must have been brought from the east. The first record in North America is from a cave in Mexico dated about the time of Christ. Groundnuts appear to have been widely distributed in the West Indies in pre-Columbian times, but there is no record of them in the present United States before that time. In the 16th century the Portuguese took them from Brazil to West Africa and the Spaniards took them across the Pacific to the Philippines from where they spread to China, Japan, Malaysia, India and as far as Madagascar. They are now found in all tropical and subtropical countries.

CULTIVARS

Although some authorities have established botanical varieties, it is best regarded as a single variable species in which distinctive cvs are recognized. Bunting (1955, 1958) separates the cvs on clearly identifiable characters of agronomic significance, of which the most important are: (1) branch form: alternate or sequential (see STRUCTURE below); (2) habit: erect bunch, spreading bunch or running; (3) size and shape of pod, beaked or keeled, constricted or not constricted; (4) number of seeds per pod and colour of testa after storage. The best time for identification is at harvest, combined with colour observations before sowing the next crop as the testa may appear pink on harvesting and the colour deepens during storage.

Cvs can be divided into the following sections:

1. VIRGINIA: These are alternate branched forms which have: (1) true runners and spreading bunch forms; (2) lateral branches frequently exceed main axis in length, especially in prostrate forms; (3) main axis is vegetative only; (4) nodes of lateral branches of all orders occur in alternating pairs of two vegetative and two reproductive branches; (5) long season forms of more or less indeterminate growth subject to environmental restrictions and may be weakly perennial; (6) plants are dark green in colour; (7) seeds exhibit a marked dormancy, only germinating after 30–360 days rest period; (8) seeds typically 2 per pod; (9) testa colour of mature seed invariably deep russet brown; (10) moderately resistant to *Cercospora* leaf-spot.

2. SPANISH-VALENCIA. These are sequential branched forms which have: (1) erect bunch forms; (2) ascending lateral branches do not overtop main axis; (3) nodes of main axis above the primary laterals are reproductive followed by sterile axils; (4) nodes of primary laterals usually continuously reproductive for the first 6 nodes or so followed by sterile axils; (5) strictly annual short-season forms maturing in 90–110 days; (6) plants are lighter green in colour; (7) no seed dormancy; (8) pods 2–6-seeded; (9) exhibit wide range of kernel size and testa colour, but when brown are light brown or bronze; (10) highly susceptible to *Cercospora* leaf-spot.

This section may be further subdivided:

(a) *Spanish*—pods mostly 2-seeded

(b) *Valencia*—pods typically more than 2-seeded, usually 3–4-seeded, but may get up to 6. The stems are thicker than the Spanish types and have a light reddish or purple tinge.

In the past there has been a tendency to confuse spreading bunch of Virginia with erect bunch types of Spanish-Valencia, but the form of branching is quite distinct, and the former are longer, denser, more bushy and more densely leafy. They are longer lived, taking 120–140 days to maturity.

There are relatively few cvs in the United States and many cvs are found in Africa which are not grown in North America. Bunting (1958) gives a key to the cvs. Most of the commercial crop of West Africa is of the 'Matevere' group, belonging to the Virginia section.

ECOLOGY

Groundnuts are grown to 40°N and S of the equator. It is a warm-season crop and is killed by frost. Most of the crop is produced in areas with 40 in.

or more annual rainfall and there should be at least 20 in. in the growing season. Dry weather is required for ripening and harvesting. The most suitable soils are well-drained, loose, friable, sandy loams, well supplied with calcium and with moderate amounts of organic matter. It can be grown on heavier soils, but this makes harvesting more difficult as the soil adheres to the nuts and it may also stain them. Soils which crust or cake are unsuitable because of the difficulty of peg penetration. Well-aerated soil with good drainage is essential as the crop cannot tolerate waterlogging.

STRUCTURE

An erect or trailing, sparsely-hairy, annual herb, 15–60 cm high.

ROOTS: Many of the wild species have tuberous roots, but not *A. hypogaea*, which has a well developed tap-root with many lateral roots. Adventitious roots develop from the hypocotyl and aerial branches, particularly in prostrate forms. There are no root hairs as there is no true epidermis, absorption taking place 8–10 mm behind root-cap. Nodules are usually present, but inoculation with the correct strain of *Rhizobium* has given varied and inconsistent results.

STEMS: Branching dimorphic with monopodial vegetative branches and reduced reproductive branches. The main stem or central axis develops from the terminal bud of the epicotyl. The first two lateral monopodial branches arise from buds in the axils of the cotyledons. A monopodial vegetative branch then develops at each node above on the main axis, usually for the first 3–5 nodes. The lateral branches may also produce secondary monopodia. The reduced reproductive branches are also produced singly at nodes on the monopodial branches. Even when the leaf appears to subtend a single flower, there is always a reduced leaf in the axil, showing that the flower is borne on a branch. Where several flowers are borne in succession on such a branch, each flower is subtended by a reduced leaf or cataphyll, although the flowers may appear superficially to arise in clusters on these strongly reduced branches. The arrangement of the monopodial vegetative branches and reduced reproductive branches is of two distinct types:

(1) Alternate branched forms which occur in the true runners (prostrate) and spreading bunch (upright) forms of the Virginia group of cvs. In these all axils of the main stem after bearing the last monopodium are sterile, thus there are no reproductive branches on the main stem. On the lateral monopodia the first 2 axils produce secondary monopodia, followed by a reproductive branch at each of the next 2 nodes and this process is repeated up the stem with alternate pairs of vegetative and reproductive branches before the branch terminates in a series of sterile axils. Sometimes in higher alternations the pair of reproductive branches may fail to develop so that a pair of sterile axils appear between pairs of vegetative branches. Alternate branched forms are the more primitive.

(2) Sequential branched forms which occur in the true erect bunch forms of the Spanish-Valencia group of cvs. The central axis produces a few erect or ascending monopodial branches from the lower nodes, followed by reproductive branches from the next nodes before ending in a series of sterile axils. The lateral monopodia may produce 0–2 vegetative branches, then fruiting branches at successive nodes, followed by sterile axils, thus repeating the structure of the main stem. Vegetative branching is seldom resumed once reproductive branches are produced.

Young stems angular at first with solid pith; this later breaks down and the old stems are hollow and cylindrical.

LEAVES: Spirally arranged with phyllotaxy of 2/5, pinnate with 2 opposite pairs of obovate leaflets, mucronate, 3–7 × 2–3 cm, entire; stipules prominent, linear-pointed, about 2·5–3·5 cm long, adnate to petiole for less than half their length; petiole 3–7 cm long; pulvini at point where stipules become free and at base of each leaflet cause characteristic sleep movements in which the petiole folds downwards and the leaflets fold upwards until they touch.

FLOWERS: Borne on compressed spikes in the axils of foliage leaves and never at the same node as vegetative branches; 1 to several fls. Fls are most abundant at lowest nodes. Fls sessile, 5–7 cm long; calyx tube elongates as bud develops, 4–6 cm long, very slender and often mistaken for pedicel; calyx lobes 5 with 4 fused into a superior lip above standard and 5th linear tooth below keel, 5 mm long; petals and staminal column adnate at base and inserted on top of calyx tube; petals 5, yellow; standard about 1·5 × 1·5 cm; stamens 10, monadelphous, filaments fused for more than half their length, 2 stamens (opposite standard) sterile, 4 stamens with oblong, bilocular, introrse anthers alternating with 4 stamens with smaller globose anthers; pistil with ovary of single sessile carpel with 2–6 ovules, surmounted by long filiform style which elongates with calyx tube and ends in a club-shaped stigma among anthers.

FRUITS: Immediately after fertilization fruit first appears at the end of pointed stalk-like structure (carpophore) known as the peg, which elongates by means of an intercalary meristem at base of sessile ovary. Cells at tip of ovary become lignified and push base of stigma to one side forming a protective cap as peg enters soil. Peg is positively geotropic, but not negatively phototropic. The peg elongates and grows downwards penetrating the soil to a depth of 2–7 cm, when it loses its geotropism and the tip turns to the horizontal position and the ovary swells rapidly. Young peg is conical in shape and the tip retains this shape until the maximum penetration of the soil is reached. The time taken for peg to reach the soil depends on the

---

Fig. 34. **Arachis hypogaea**: GROUNDNUT. **A**, leafy shoot ($\times \frac{1}{2}$); **B**, flower ($\times \frac{1}{2}$); **C**, flower in longitudinal section ($\times 3$); **D**, developing fruit ($\times \frac{1}{2}$); **E**, base of plant showing flower, developing fruits and roots with nodules ($\times \frac{1}{2}$).

C
A
D
B
E

initial distance from the soil, but if this is more than 15 cm it usually fails to reach the ground and the tip dies. Fruit enlargement begins at the base. If the apical seed aborts the terminal section swells no further. Mature fruit is a structurally dehiscent, but functionally indehiscent legume, oblong, 1–8 × 0·5–2 cm, and may contain 1–6 seeds. The dry pericarp of the mature fruit is reticulate with 10 longitudinal ridges. The reticulations are due to mechanical tissue below the veins in the hardened mesocarp. The endocarp consists of parenchymatous tissue surrounding the developing seeds which serves as storage tissue. It collapses as the pod matures eventually forming the thin papery lining. Failure of fertilization of an ovule or early embryo abortion may produce 1-seeded fruits.

SEEDS: Mature seeds elongated and are cylindrical or ovoid, 1–2 × 0·5–1 cm. Cvs vary in size, shape and in colour of testa of seeds; colour ranging from white, pink, red, purple and shades of brown. Seeds have a thin testa and contain no endosperm. They have two massive cotyledons, an epicotyl with 3 buds of which the terminal bud has 4 foliage leaves and the two cotyledonary laterals have 1–2 leaves, the hypocotyl, and a large radicle. 400–1,300 seeds per lb.

POLLINATION

The first flowers appear 4–6 weeks after planting and the maximum number of flowers are produced after a further 4–6 weeks. Only one flower on each inflorescence opens on one day and there is an interval of one to several days between the opening of successive flowers. The flower bud is 6–10 mm long 24 hours before anthesis, after which the bud elongates slowly during the day and rapidly during the night. Anthesis and pollination takes place at sunrise the following day at the time of petal expansion, self-pollination taking place within the closed keel. The stamens with oblong anthers elongate and dehisce first, after which filaments of the stamens with globose anthers elongate to the level of the oblong anthers and dehisce. The flowers wither 5–6 hours after opening, after which the calyx tube is shed, leaving only the ovary and base of the style. There is a marked tendency to cleistogamy. Bunting (1955) records that in Africa the bulk of the commercially useful crop in erect bunch types is produced from flowers formed below the surface of the soil from the lower nodes of the primary branches. This may be due to the lower nodes being covered by soil during after-cultivation. Insects, including bees, visit the flowers, but out-crossing must be rare as cvs show genetic uniformity in progeny rows.

GERMINATION

Virginia cvs have a seed dormancy of varying periods, but germination can be accelerated by moistening the seeds with the macerated kernels of non-dormant Spanish-Valencia cvs. On planting, the radicle emerges and grows rapidly and may reach a length of 10–14 cm in 4–5 days. Germination is neither epigeal nor hypogeal, as the hypocotyl carries the cotyledons to the surface of the soil where they remain. The length of the hypocotyl is dependent on the depth of planting and, if the seeds are buried deeply, may reach a length of 10–12 cm. The terminal bud develops quickly followed by the two lateral cotyledonary buds. The main axis exhibits little inhibitory growth on the lateral branches.

CHEMICAL COMPOSITION

The seeds are rich in non-drying oil and protein, the content of which varies with the cv. Virginia types contain 38–47 per cent oil and Spanish types 47–50 per cent oil. The average composition of shelled nuts is approximately: water 5·4 per cent; protein 30·4 per cent; fat 47·7 per cent; carbohydrate 11·7 per cent; fibre 2·5 per cent; ash 2·3 per cent. Decorticated groundnut cake contains about: water 10·3 per cent; protein 46·8 per cent; fat 7·5 per cent; carbohydrate 23·2 per cent; fibre 6·4 per cent; ash 5·8 per cent. The oil contains about 53 per cent of oleic acid and 25 per cent linoleic acid. The principal proteins are arachin and conarachin. Groundnuts are rich in vitamins B and E.

PROPAGATION

All commercial production is grown from seed. Groundnuts can be propagated vegetatively from cuttings. Cuttings from erect, sequential-branched forms, both from the main and lateral stems produce plants similar in branching and flowering habit to the parent plant. If cuttings are taken from the main axis of alternate-branched forms, no flowers are produced on the main axis, but if cuttings are taken from lateral branches, they continue to behave as laterals and produce inflorescence on all branches including the initial cutting.

HUSBANDRY

ESTABLISHMENT: Groundnuts are often grown in rotation with cotton, tobacco, maize and other cereals and are usually grown late in the rotation. In peasant cultivation they are often interplanted with other crops. The seedbed should be thoroughly prepared. They may be planted on the flat or on ridges. Narrow ridges should be avoided with bunch types as this may make pegging difficult. Ridging facilitates lifting, but it reduces the number of plants per acre.

The spacing and seed rate is related to the growth habit, but usually the closer the spacing the higher the yields and it helps to prevent too much damage from rosette virus. It also suppresses late flowering, or at least renders it ineffective, thus producing more compact fruiting and more even maturity and quality. Stands of 100,000 plants per acre are recommended, but it is impossible to exceed 70,000 plants per acre with mechanical planting. For bunch types planted by hand a spacing of 1 ft × 6 in. is recommended with a seed rate of 60–80 lb per acre of shelled nuts, or 2 ft × 4 in. for mechanical cultivation. For spreading types or runners the usual spacing is 2 × 1 ft with a seed rate of 40 lb per acre. When planted by hand it is advisable to plant two seeds per hole as this gives a better stand and allows wider inter-row spaces. Seed for sowing is better stored in the shell, thus maintaining viability; hand-shelling of the seed is preferable as it causes less damage. The seeds are often treated with mercurial or thiram dressing before sowing, especially machine-shelled seeds which are often damaged. In the Sudan it was found that the addition of dieldrin reduced termite damage. Inoculation with the correct strain of *Rhizobium* has given inconsistent results. Chemical seed treatment renders artificial inoculation useless. It is important to sow as early in the season as possible if high yields are to be obtained. The crop is not aggressive in the early stages and weeding should be done during early growth. If weeds get out of hand the crop is difficult to harvest. Hoeing and mechanical cultivation must stop after peg clongation at about 8 weeks as this disturbs the pegs and also facilitates the spread of rosette.

MANURING: Although groundnuts are usually grown on sandy soils of poor fertility the response to fertilizers is often low. The crop is very unpredictable in its response to fertilizers. These are usually best applied to other crops in the rotation. Nodulation is only effective after three weeks and nitrogen gives a response in some areas. Sulphur and calcium increase the amount of nodulation. Groundnuts can utilize phosphorus at a lower level than most crops and there is little response to phosphorus in many areas. However, high economic returns and long residual effects have been obtained from single superphosphate applied at the rate of 11 lb P_2O_5 per acre in northern Nigeria. This may be due to the sulphur and calcium content, as large increases were also produced by gypsum. Responses to ammonium sulphate may also be due to the sulphur content. Groundnuts make a heavy drain on soil potash, yet response to potash application is uncommon, particularly in Africa. The crop also makes a heavy drain on calcium. This is usually applied in the United States and the foliage is also often dusted with gypsum at the time of flowering. In Sierra Leone, Piggott (1960) states that the only fertilizers improving yield were calcium, potassium and magnesium, the last two elements having no effect alone, but only in the presence of calcium, when a combination of all three gave the highest yields. Calcium was effective in improving the shelling percentage by reducing the number of 'pops' or empty pods. Placement of the fertilizers is important. In northern Nigeria pelleted superphosphate can be placed within ½ in. of the seeds at planting without damage. Nutrients are absorbed through the peg and developing fruits as well as the roots.

HARVESTING: The time to maturity is 3½ months or so for bunch types and 5 months or more for spreading types. The crop is often harvested by hand. The plants are pulled up and turned over in the sun to dry before stripping the nuts. Drying on tripods gives greater viability of seeds. It is necessary to harvest at the optimum time to combine the maximum yield with the minimum losses from shedding and, in the case of erect bunch types, from sprouting. Losses are heavy if the soils get hard.

The extent of mechanization depends upon the local circumstances and the cost of labour. In Senegal complete mechanization was four times as expensive as manual cultivation and three times as expensive as animal cultivation. The crop is becoming increasingly mechanized in the United States. The mechanical harvester produced by the National Institute of Agricultural Engineering at Silsoe in England shows promise.

Yields are very variable and usually range between 400–1,500 lb of unshelled nuts per acre, but yields of over 2,000 lb per acre have been obtained. The shelling percentage is 80 per cent for bunch cvs and 60–75 per cent for spreading cvs. It is advisable that the nuts should be stored in the shell if possible as quality and viability deteriorate rapidly after shelling.

MAJOR DISEASES

Cercospora arachidicola Hori and *C. personata* (Berk. & Curt.) Ellis & Everh. cause a serious leaf spot which is characterized by dark spots surrounded by a yellow ring and is commonest on the lower leaves. It may cause defoliation. The disease is widespread and is serious at times in most countries.

Sclerotium rolfsii Sacc. produces a wilt and is widespread. It causes the death of branches or the entire plant, with reddish fruiting bodies on the stem at ground level. *Rhizoctonia bataticola* (Taub.) Butl. also causes a wilt, as does *Pseudomonas solanacearum* (E. F. Sm.) Dows.

Aspergillus flavus Link. attacks stored seeds and has also been found on living plants. It produces aflatoxin which is known to cause fatalities in turkeys, possibly from carcinoma of the liver. Other birds and mammals are similarly affected.

Rosette virus is the most serious disease in Africa, particularly in the wetter parts. The vector is *Aphis laburni* Kalt. The whole plant is severely stunted; the younger leaves are chlorotic and mottled, with successive leaves becoming smaller, curled, distorted and yellow. Spacing has an important effect on the incidence of the disease, close spacing assisting in keeping down the general level of infection.

MAJOR PESTS

In the United States the worst pests are the caterpillars of *Anticarsia gemmatalis* (Hubn.) and the beetle *Pontomorus leucoloma.* In India the major pests are caterpillars of *Amsacta albistriga* Walk. and *Stomopteryx nerteria* Meyr. and the beetle *Sphenoptera perotetti* G. Thrips and aphids damage the crop in many countries and a number of pests damage stored groundnuts.

IMPROVEMENT

Improvement of groundnut cvs is difficult as, for all practical purposes, the crop is 100 per cent inbred, crossing is difficult and time-consuming, and only a short time is available when it can be done. Furthermore individual plants produce so few seeds that the recovery of improved types in a small segregating population is improbable. This can be overcome by vegetative propagation by cuttings and can be used for testing the progeny. Some authorities consider that pure lines are unstable, but this is probably due to accidental seed mixture. Bunting (1955) found that, for all practical purposes, cvs are stable after vigorous rogueing and filtering through progeny rows. Chance outcrossing may occur, but this is considered rare. *Arachis hypogaea* is a tetraploid and chromosome irregularities might produce off-types.

Improvement consists of: collection and trial of local and introduced cvs; selection in mixed cvs; purification of cvs by rogueing at harvest, followed by sorting the seeds for colour before sowing—cvs can be purified in two seasons by this means; hybridization which probably offers the best chance of success. Emasculation is done in the evening and hand-pollination is carried out the next morning.

Selection aims will include adaptation to the local environment, earliness, drought resistance, shelling percentage, high oil and protein content, resistance to diseases and pests, and suitability for mechanical harvesting. Selection of local and introduced cvs in the Congo is said to have given yield increases of 33 per cent. Resistance to *Cercospora* leaf spot and rosette has been found in long-season cvs in Tanzania. Generally speaking, however, breeding results have been disappointing. Despite many years' work in the United States, very little progress has been made and average yields there have shown little increase. Some progress is reported from breeding and selection work in Madras and Mysore for Indian conditions. It is claimed that cvs have been produced which give 20 per cent more yield than the normal crop with a higher shelling percentage of 73 per cent and a higher oil content of 51·2 per cent.

PRODUCTION

Groundnuts are the second largest source of vegetable oil, the largest being soya beans. There has been a great increase in production during the last 60 years and in 1962–1963 a record crop of nearly 14 million tons was produced. India is by far the largest producer with an annual production of about 4·5 million tons from approximately 16 million acres, but most of it is now consumed locally and only a small part of the crop enters world trade compared with pre-war years. China is the second largest producer with about 2·5 million tons per annum, but now exports only negligible quantities. In many tropical countries groundnuts are grown for domestic consumption. Nigeria is the largest exporter of groundnuts with annual exports of about 0·5 million tons; most of the crop is produced in the

Northern Region. Other countries exporting substantial quantities are Senegal, the Sudan, Niger and Gambia. The United States produces about 0·8 million tons annually which is used internally. Other substantial producers in the New World are Brazil and Argentina. In all the major producing countries, with the exception of the United States, production is predominantly for oil extraction. Most of the exports of groundnuts and oil are taken to Europe, France taking nearly half the total quantity and the United Kingdom about one-sixth. The quantity of groundnuts entering world trade is nearly 1·5 million tons per annum.

In 1946 the British Government proposed a bold plan to grow 3,210,000 acres of groundnuts using mechanical production in East Africa in order to alleviate the world's chronic shortage of fats. It was hoped to produce 800,000 tons of groundnuts per annum. Such was the ill-fated East African groundnut scheme. The Overseas Food Corporation was founded to supervise and carry out the project. It was proposed to clear the first 150,000 acres in 1947 and that the total acreage would be established by 1951. The first area chosen was Kongwa in central Tanganyika and it was hoped that eventually there would be 107 mechanized units each of 30,000 acres. However, the bush at Kongwa proved highly intractable and by the end of 1947 less than 10,000 acres had been cleared and £4,250,000 had been spent. The following year clearing also began at Urambo also in central Tanganyika. By 1949 some 50,000 acres had been cleared at Kongwa, of which 25,000 acres were under groundnuts and 20,000 acres of sunflower; at Urambo there were 500 acres of groundnuts and 2,700 acres of sunflower. By 1951 some 65,000 acres were under crops. The scheme was later abandoned after over £36,000,000 had been spent. As stated in a 1951 White Paper 'the original aims of the scheme have proved incapable of fulfilment . . . the groundnut is not a plant which lends itself readily to mass methods over vast acreages'. Poor irregular rainfall with a prevalence of drought years led to abandonment at Kongwa, while disease was a serious problem at Urambo. The debacle is an object lesson for those who would force an abrupt change without sufficient experimentation, particularly in Africa. It is somewhat ironic that at this time there were large stocks of groundnuts at Kano in Northern Nigeria stored in vast pyramids, which were awaiting export because of the shortage of rolling stock on the Nigerian railway.

REFERENCES

ARANT, F. S. *et al.* (1951). *The Peanut—the Unpredictable Legume.* Washington: Nat. Fertilizer Assoc.

BOVILL, E. W. (1950). The Tanganyika Groundnut Scheme, in *East African Agriculture.* ed. by Matheson, J. K. and Bovill, E. W., 114–124. London: Oxford Univ. Press.

BUNTING, A. H. (1955). A Classification of Cultivated Groundnuts. *Emp. J. exp. Agric.*, **23**, 158–170.

BUNTING, A. H. (1958). A Further Note on the Classification of Cultivated Groundnuts. *Emp. J. exp. Agric.*, **26**, 254–258.

ORAM, P. A. (1958). Recent Developments in Groundnut Production with special reference to Africa. *Field Crop Abstr.*, **11**, 1–6, 75–84.

PIGGOTT, C. J. (1960). The Effect of Fertilizers on Yield and Quality of Groundnuts in Sierra Leone. *Emp. J. exp. Agric.*, **28**, 58–64.

WOOD, A. (1950). *The Groundnut Affair*. London: Bodley Head.

Cajanus DC. (x = 11)

It is often stated that the genus is monotypic, but in addition to the cultigen *C. cajan* (q.v.), there is a wild sp. in Africa.

Cajanus cajan (L.) Millsp. (2n = 22) PIGEON PEA

Syn. *C. indicus* Spreng.

Other common names include red gram, Congo pea, no-eye pea.

USES

The young green seeds are eaten as a vegetable in many countries and are canned in Puerto Rico and Trinidad. The ripe dry seeds are boiled and eaten as a pulse. In India these are split and made into dhal, which may be prepared either by a dry or wet method. In the dry method the dry seeds are placed in the sun for 3–4 days and are then split in a mill and this process is repeated 3–4 times. In the wet method the seeds are soaked in water for 6–10 hours, mixed with red earth overnight, then dried in the sun, after which the red earth is removed by sieving and the seeds are finally split into dhal in a handmill. The split dhal is then cleaned by repeated winnowing and sieving to remove the hulls and broken pieces and is then treated with castor or sesame oil to preserve its quality, prevent insect attack and give it an attractive appearance. The yield of dhal is about 66 per cent by the dry method and 80 per cent by the wet method. Dhal is graded according to the method of preparation, that produced by the dry method fetching the better price, and also by the percentage of broken bits, unripe and shrivelled seeds and the boldness of the grain. The dried husks, seeds and broken dhal are used as cattle feed in India. The green pods are sometimes used as a vegetable.

The tops of the plants with fruits provide excellent fodder and are also made into hay and silage. Pigeon peas may be planted alone or in pastures as browse plants. They are also planted as green manures and cover crops, and are used as temporary shade in young cocoa and other crops, as windbreaks and for anti-erosion work. The dried stalks are used for firewood, thatching and baskets in India.

ORIGIN AND DISTRIBUTION

The pigeon pea is probably a native of Africa, where it is sometimes found wild or naturalized. Seeds have been found in Egyptian tombs of the

XIIth Dynasty and it was cultivated there before 2000 B.C., by which time Egypt had established trade relations with tropical Africa to the south and Syria to the east. Pigeon peas were cultivated in Madagascar from very early times. It appears that they were taken to India in prehistoric times, and this region now constitutes a centre of diversity with the greatest number of cvs. The crop was taken to the New World in early post-Columbian days, but it did not reach the Pacific until later, being taken to Guam in 1772. Pigeon peas are now widely spread throughout the tropics and subtropics.

CULTIVARS

Two botanical varieties have been recognized:

var. *flavus* DC.—earlier maturing, shorter plants with yellow standards, and green glabrous pods, which are light coloured when ripe, and are usually 3-seeded. These are the *tur* cvs of India, where they are extensively cultivated in the Peninsula.

var. *bicolor* DC.—perennial, late-maturing, large, bushy plants, with dorsal side of standard red or purple or streaked with these colours, and hairy pods blotched with maroon or dark coloured, with 4–5 seeds, which are darker coloured or speckled when ripe. These are the *arhar* cvs of India which are more extensively cultivated in the north of the country.

These distinctions are open to doubt as the varieties cross readily, while many of the characters such as colour and shape of the flowers, pods and seeds follow simple Mendelian inheritance and are distributed in both varieties. Many cvs are recognized with over 100 in India alone. Most cvs are diploid, but tetraploids and hexaploids have been found in India.

ECOLOGY

Pigeon peas show a wide adaptability in regard to climate and soils. They are drought resistant with a deep root system which permits good growth under semi-arid conditions with under 25 in. of rain per year. They are less suitable for the very wet tropics. Most cvs are very sensitive to frost. They can be grown on almost all soil types, provided the soil is not markedly deficient in lime and they will not tolerate waterlogging. Most cvs, notably the tall, late-maturing ones, exhibit a photoperiodic effect, being short-day plants, and this affects the time of maturity according to the date of sowing and also the height of the plant and therefore the ease of harvesting.

STRUCTURE

A woody, short-lived, perennial shrub, 1–4 m tall, sometimes grown as an annual. A pronounced deep tap-root with longer laterals in the spreading than in the erect cvs. The number, position (1st branch at 6th to 16th node) and angle (30° in erect cvs to 60° in spreading cvs) of the lateral branches varies with cv. Young stems angled and hairy. Lvs spirally arranged with a phyllotaxis of 2/5, trifoliate, with a tendency to be deciduous; petiole grooved above, 2–8 cm long; stipules small, ovate, hairy, about 4 mm long;

stipels small; pulvinus at base of petiole and at base of leaflets; leaflets lanceolate to narrow elliptic, acute at both ends, entire, hairy on both surfaces, greyish beneath with minute yellow resinous glands; longer-stalked terminal leaflet usually larger than short-stalked laterals, 6–15 × 2–6 cm. Inflorescences small, terminal and/or axillary racemes, 4–12 cm long, with several fls. Flowering extends over several months. Fls about 2·5 cm long; calyx lobes 4, upper 2 lobes being united; standard orbicular, auricled, yellow, or dorsal side red or purple or yellow veined with red or purple; wings and keel yellow, of equal length; keel incurved at apex, obtuse; stamens with free vexillary stamen and rest connate of unequal length; anthers uniform, small, oblong, yellow, dorsifixed; ovary and base of style hairy; stigma knob-shaped. Fruit a flattened pod with diagonal depressions between seeds, 2–8-seeded, 4–10 × 0·6–1·5 cm, beaked, usually hairy, green or dark maroon or blotched with maroon; do not shatter in the field. Seeds vary in size, shape and colour, usually round or oval, about 8 mm in diameter, white, greyish, red, brown, purplish or speckled, with small white hilum. 100 seeds weigh 11–13 g.

POLLINATION

Pigeon peas are self- and cross-compatible. The stamen filaments elongate in the bud and pollen is shed the day before the flowers open. However, the flowers are visited by bees and other insects and about 20 per cent (5–40 per cent) cross-pollination can occur. The majority of the flowers open between 11 a.m.–3 p.m. and often remain open for 6 hours only. Rain at flowering reduces fertilization. For hybridization it is essential to emasculate before 9 a.m. on the day before the flowers open and they may be hand-pollinated at the same time.

GERMINATION

The seeds usually give a good germination, which is hypogeal.

CHEMICAL COMPOSITION

The raw unripe seeds contain about: water 67·4 per cent; protein 7·0 per cent; fat 0·6 per cent; carbohydrate 20·2 per cent; fibre 3·5 per cent; ash 1·3 per cent. The green seeds constitute about 45 per cent of the weight of the whole pod. Dry ripe pigeon peas contain about: water 10·1 per cent; protein 19·2 per cent; fat 1·5 per cent; carbohydrate 57·3 per cent; fibre 8·1 per cent; ash 3·8 per cent. The composition of Indian dhal (split seeds) is: water 15·2 per cent; protein 22·3 per cent; fat 1·7 per cent; carbohydrate 57·2 per cent; ash 3·6 per cent.

Fig. 35. **Cajanus cajan :** PIGEON PEA. **A**, shoot with flowers and fruits (× ½); **B**, flower in longitudinal section (× 2½); **C**, seeds (× 1½).

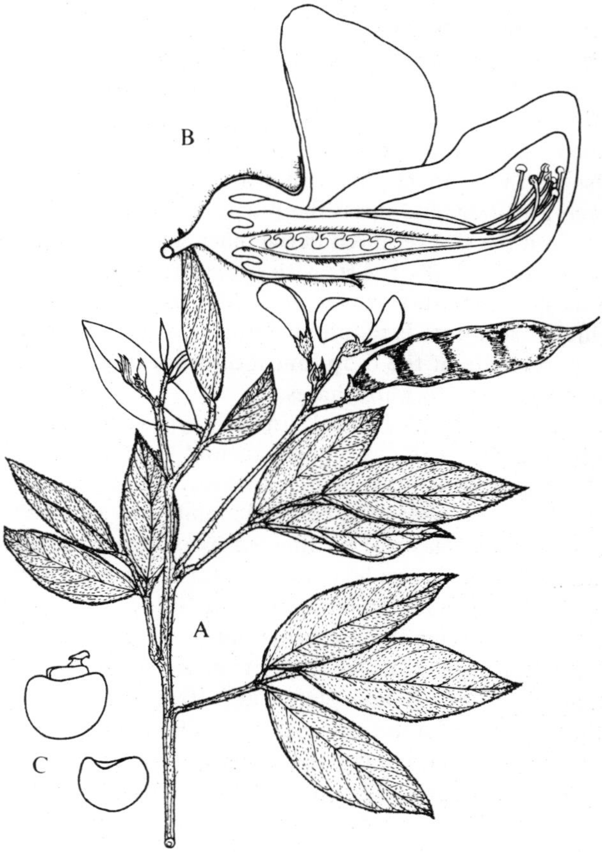
B
A
C

PROPAGATION

The crop is grown from seeds planted *in situ*. Seedlings are difficult to transplant, but it is possible if it is done carefully with soil round the roots. Pigeon peas can be propagated by cuttings if required.

HUSBANDRY

In India and Africa the crop is often grown in mixed cultivation with sorghum, finger millet, bulrush millet, maize, sesame and other crops. The seed is either broadcast or with one row of pigeon peas to every 3–5 rows of the main crop. The main crop is harvested first after which the pigeon peas continue to grow and are harvested later. Growth is slow at first but, once established, the crop requires little attention. When grown in pure stand the spacing will depend upon the purpose for which the crop is grown, the cv. used, and the size required in the mature plant. In pure stand the spacing used varies from 3–6 × 1–4 ft, with a seed rate of 10–20 lb per acre. When grown as green manure, cover or fodder crop, they are usually broadcast and a higher seed rate is required. Limited work has been done on the fertilizer requirements of the crop, but it appears to give little response, except on some occasions to phosphates.

Podding begins in 12–14 weeks in early types and they require 5–6 months to reach maturity; late cvs require around 9–12 months to reach maturity. The tall, late-maturity cvs, being short-day plants, have a limited cropping season, which in Trinidad is early November to late February. The dwarf early maturity cvs may not be affected by photoperiod and can yield all the year round. The crop can be continued for 3–4 years or may be ratooned, as is done when grown for green manure or fodder. As a pulse crop the yields usually drop off after the 1st year and it is best treated as an annual. In India and Burma it is usual to harvest the whole plants, which are then dried and threshed. A few plants are often grown near homesteads for green peas. In pure stand the yield per acre of green pods varies between 1,000–4,000 lb, but over 8,000 lb has been obtained, and dried seeds between 500–1,000 lb per acre, but up to 1,800 lb has been recorded. In mixed cultivation yields of 200–800 lb are obtained. The conversion ratio from fresh green pods to dried peas is about 3·3.

MAJOR DISEASES AND PESTS

Fusarium udum Butl. causes a wilt and is the most serious disease of pigeon peas in India. It is soil-borne and attacks all parts of the plant and causes blackening in the tissues of the root and base of the main stem.

Physalospora cajanae is serious in the West Indies and causes a collar and stem canker, particularly when the crop is grown as a perennial. The primary symptoms are small, grey, scutiform lesions with dark edges on the stems and branches. Usually the plant is girdled and dies. A similar disease has been ascribed to *Diplodia cajani* and *Phoma* sp.

The most serious pests in India are the caterpillars of *Heliothis armigera* Hubn. and *Exelastis atomosa* W., and the fly *Agromyza obtusa* M. In the

West Indies the larvae of a pyralid moth, *Elasmopalpus rubedinellus* (Zell.) feeds on the developing seeds and can cause considerable damage. An allied borer, *Ancylostomia stercorea* (Zell.) is occasionally common and *Heliothis virescens* (F.) is also a major pest which feeds on the developing seeds rendering them unfit for sale. Infection by borer takes place at the time of flowering. Bruchids are serious storage pests in Asia.

IMPROVEMENT

Classification, selection and hybridization of cvs is being done in India, Hawaii, Puerto Rico and Trinidad. The aims of the breeders are: (1) high yield per acre of green pods or dried seeds; (2) ease of picking which is obtained in dwarf plants with compact head of pods clustered towards ends of branches; (3) day neutral plants in order to spread production; (4) early-maturing plants which will give a crop in the off-season when market prices are high; (5) determinate bearing strains to limit the period of harvesting; (6) cvs which bear heavily for a long period for 'back garden' use; (7) resistance to diseases and pests, particularly *Fusarium* wilt and stem canker; (8) good colour of cooked seed for canning.

Hybrid vigour is shown in crosses between pure lines, but this is difficult to utilize because of technical difficulties and the production of hybrid seed would be costly. One plant of an Hawaiian hybrid produced 1,430 pods with 1,150 g seeds (shelling percentage 73 per cent) and an average of 4·6 seeds per pod. At the University of the West Indies in Trinidad crosses are being made between the higher-yielding late cvs which tend to crop only during a particular period of the year and determinate, early-maturing cvs.

PRODUCTION

Statistics are difficult to obtain due to local consumption, mixed cropping and pigeon peas are not widely exported. India is the largest producer where pigeon peas are the second most important pulse with 5–8 million acres planted annually and a production of over 1·5 million tons of dried seed. In Puerto Rico green pigeon peas for the fresh market and canning are the basis of a $1-million industry. A canning industry is now being built up in Trinidad. Pigeon peas are also grown in Africa and South America.

REFERENCES

CAMPBELL, J. S. and GOODING, H. J. (1962). Recent Developments in the Production of Food Crops in Trinidad. *Trop. Agriculture, Trin.*, **39**, 261–270.

GOODING, H. J. (1962). The Agronomic Aspects of Pigeon Peas. *Field Crop Abstr.*, **15**, 1–5.

HENDERSON, T. H. (1965). *Some Aspects of Pigeon Pea Farming in Trinidad.* Univ. West Indies, Dept. of Agric. Econ. Occas. Series No. 3.

MATTA, D. N. and DAVE, B. E. (1931). Studies in *Cajanus indicus. Mem Agric. Dept. India, Bot. Ser.*, **19**, 1–25.

Canavalia DC. (x = 11)

About 40 spp. in the tropics and subtropics of both hemispheres. Annual or perennial herbs, usually twining or trailing. Lvs trifoliate; petiole with large pulvinus at base; stipules caducous. Inflorescence axillary raceme, many-flowered, with 2 or more fls arising from prominent pedicellar glands. Fls violet, rose or white; calyx with 2 upper and 3 smaller lower lobes; standard large, reflexed; stamens monadelphous, but vexillary stamen free at base; anthers all alike; style beardless. Pods oblong or linear with wing or rib near upper suture, 2-valved.

USEFUL PRODUCTS

C. ensiformis (q.v.), *C. gladiata* (q.v.) and *C. plagiosperma* (q.v.) are grown as pulse crops and green manures. *C. maritima* (Aubl.) Thou. (syn. *C. rosea* (Sw.) DC.) is a pantropical strand plant and is useful as a sand-binder.

KEY TO THE CULTIVATED SPECIES

A. Plants usually bushy and erect; pods more than 10 times as long as broad; hilum less than half length of seed
- B. Seeds white .... *C. ensiformis*
- BB. Seeds cinnamon coloured .... *C. plagiosperma*

AA. Plants twining; pods less than 10 times as long as broad; hilum nearly as long as seed .... *C. gladiata*

Canavalia ensiformis (L.) DC. (2n = 22) JACK BEAN, HORSE BEAN

USES

It is grown as a green manure. It is also used as a fodder, but livestock eat it with reluctance and it is more palatable when dry. The young pods and immature seeds are used as a vegetable.

ORIGIN AND DISTRIBUTION

C. ensiformis has been recovered from archaeological sites in Mexico dated 3000 B.C. It is a native of Central America and the West Indies and has now been widely introduced throughout the tropics.

ECOLOGY

It is hardy, deep-rooted and drought-resistant and will tolerate shade.

Fig. 36. **Canavalia ensiformis**: JACK BEAN. **A**, leaf (×½); **B**, flower (×1); **C**, flower in longitudinal section (×2); **D**, pod (×½); **E**, seed (×½).

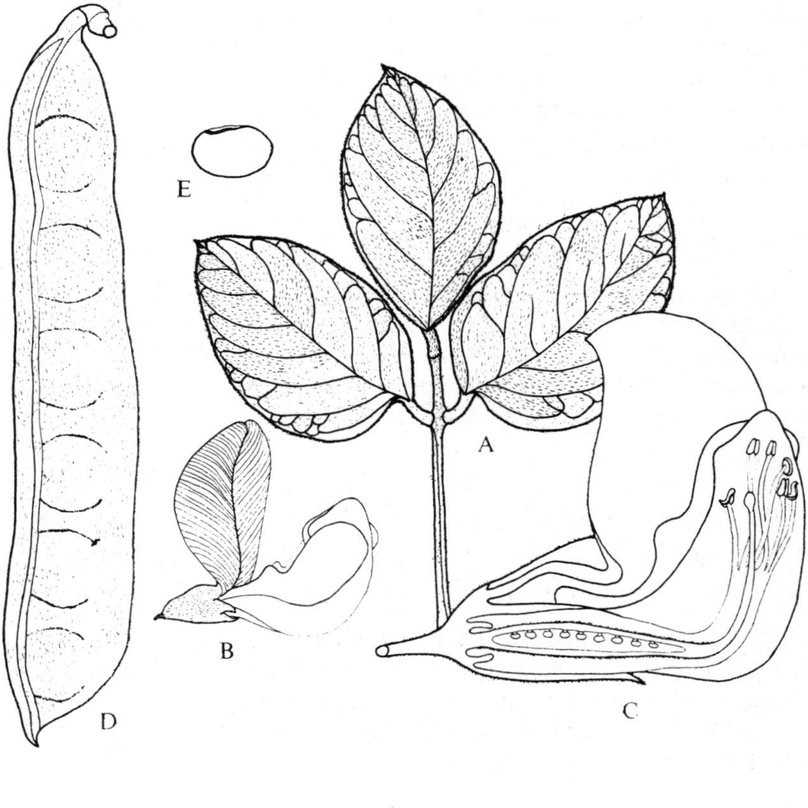
E
A
B
D
C

STRUCTURE

A bushy erect annual herb, 1–2 m high; tips of branches tend to twine, particularly in shade. Lvs trifoliate, shortly hairy; petiole usually longer than leaflets, grooved above, stout, with large pulvinus at base and at base of each leaflet; stipels present; leaflets elliptic to ovate, rounded at base and apex, 6–15 × 5–10 cm, terminal leaflet long-stalked, side leaflets short-stalked with unequal base, entire. Inflorescence curved raceme with 10–50 fls borne in groups of 3–5 on swollen pedicellar glands. Fls 2–2·5 cm long; calyx with 2 upper and 3 smaller lower lobes; corolla rose or violet, gradually fading towards base; stamens 10 united for most of length; anthers globose, dorsifixed, deep yellow; pistil beardless with clavate terminal stigma. Pods stout, straw-coloured when ripe, pendent, 20–30 × 2–2·5 cm, ribbed near upper suture; 8–20-seeded; usually only the 2–3 lowest fls of each inflorescence produce pods. Seeds white, about 2 × 1·3 cm, somewhat flattened; hilum less than half length of seed, 8 mm long, pale brown with orange margin.

POLLINATION

Bagged flowers set pods and seeds. Hand-crossing is difficult as the emasculated flowers tend to drop, but it can be done and should be confined to the lowest 2–3 flowers of an inflorescence. Under natural conditions the flowers are visited by bees and 20 per cent or more cross-pollination occurs.

GERMINATION

Germination is epigeal.

CHEMICAL COMPOSITION

The immature fruits and beans contain about: water 75·2 per cent; protein 6·9 per cent; fat 0·5 per cent; carbohydrate 13·3 per cent; fibre 3·3 per cent; ash 0·8 per cent. The ripe dried beans contain: water 11·0 per cent; protein 23·4 per cent; fat 1·2 per cent; carbohydrate 55·3 per cent; fibre 4·9 per cent; ash 4·2 per cent.

PROPAGATION

Jack beans are grown from seeds.

HUSBANDRY

It is established as a green manure crop by broadcasting at a seed rate of 40–60 lb per acre. The yield of green matter is 16–20 tons per acre and the dried beans 1,200 lb per acre.

Canavalia gladiata (Jacq.) DC. (2n = 22, 44) SWORD BEAN

USES

It is used as a green manure, cover crop, and for forage. The young pods and beans are extensively used as a vegetable in tropical Asia, but to a lesser extent elsewhere. The ripe seeds should be eaten with caution as they may be somewhat poisonous; they should be boiled in salt and the water changed.

ORIGIN AND DISTRIBUTION

C. gladiata is of Old World origin and is probably derived from *C. virosa* Wight & Arn., which grows wild in tropical Asia and Africa. It is widely cultivated in the East, particularly in India, and has now spread throughout the tropics. It has become naturalized in some areas.

STRUCTURE

A large perennial climber. The vegetative parts and flowers are similar to *C. ensiformis*, but the petioles are shorter than the leaflets, the leaflets are larger, 10–18 × 6–14 cm, glabrescent, and are acuminate with a short point at apex; fls are larger, 3·5 cm long, and are white or pinkish; the pods are more curved and with more strongly developed ridges, 15–25 × 3·5–5 cm, 8–16-seeded; the seeds are larger, typically claret coloured, 2·5–3·5 cm long, with a dark brown hilum, 2–2·5 cm long, extending almost the entire length of the seed.

POLLINATION

As *C. ensiformis.*

CHEMICAL COMPOSITION

The unripe fresh beans contain about: water 88·6 per cent; protein 2·7 per cent; fat 0·2 per cent; carbohydrate 6·4 per cent; fibre 1·5 per cent; ash 0·6 per cent.

HUSBANDRY

When grown as a vegetable it is usually grown near houses and allowed to trail on walls and trees.

Canavalia plagiosperma Piper

This species is very similar to *C. ensiformis*, both in leaves and flowers, but the seeds are larger and are cinnamon coloured, 2·7 × 1·7 cm, with a dark brown hilum 1·0 cm long. It probably originated in tropical South America and seeds of it have been found in pre-ceramic levels at Huaca Prieta in Peru dated about 2500 B.C. A form in Trinidad, which is probably this species, crosses with *C. ensiformis*, when the cinnamon colour of the seeds is dominant over the white testa of *C. ensiformis*, segregating in the F_2's to 9 cinnamon, 3 cinnamon and white, 3 white speckled cinnamon and 1 white.

REFERENCES

PIPER, C. V. (1925). The American Species of *Canavalia* and *Wenderothia Contr. U.S. Nat. Herb.*, **20**, 555–585.

SAUER, J. D. (1964). Revision of *Canavalia*. *Brittonia*, **16**, 106–181.

Cicer L. (x = 7, 8)

About 14 spp. of small annual and perennial herbs in western Asia. *C. arietinum* (q.v.) is grown as a pulse crop.

Cicer arietinum L. (2n = 16) CHICK PEA, GRAM

USES

Gram is the most important pulse of India, where the whole dried seeds are eaten cooked or boiled or in the form of dhal, which is prepared by splitting the seeds in a mill and separating the husk. The seeds are sprinkled with water and heaped overnight to soften the husk and are then dried before milling. The yield of dhal is about 80 per cent of the weight of the whole grain. Flour (baisin) is made by grinding the seeds and is one of the chief ingredients with ghee and sugar of many forms of Indian confectionery. Green pods and tender shoots are used as a vegetable. The dry stems and leaves after threshing and the husks and broken pieces from the dhal making are fed to livestock. An acrid liquid from the glandular hairs is collected by spreading a cloth over the crop at night, which absorbs the exudation with the dew. It contains about 94 per cent malic acid and 6 per cent oxalic acid and is used medicinally and as vinegar.

ORIGIN AND DISTRIBUTION

C. arietinum is not known in a wild state, but is found as an escape in Mesopotamia and Palestine. It appears to have originated in western Asia and to have spread at a very early date to India and Europe. The crop was known to the ancient Egyptians, Hebrews and Greeks. It has been introduced in recent times to tropical Africa, Central and South America, and Australia, but is of little importance in these areas. During World War II imports to East Africa from India were restricted and the local Indian community stimulated production by African cultivators.

CULTIVARS

The bulk of the crop in India is still grown from unselected cvs.

ECOLOGY

Gram is very drought-resistant and requires a cool dry climate and light well-aerated soils. In India it is grown as a winter crop and it must have cold or cool nights with dew for successful cultivation. It cannot tolerate heavy rains and so is unsuited to the wet hot tropics where it often fails to flower. It grows best on heavy clay soils and with a rough seedbed.

STRUCTURE

An erect or spreading, much-branched, annual herb, 25–50 cm tall; all parts covered with clavate glandular hairs. Tap-root well developed; lateral roots numerous with large nodules; root system more extensive in late-maturing, spreading cvs than in early, erect cvs. Lvs imparipinnate, about 5 cm long, yellowish-green to dark bluish-green; stipules ovate, about 8 mm long, notched; leaflets ovate, elliptic or obovate, serrate, $0{\cdot}8$–$2 \times 0{\cdot}5$–$1{\cdot}5$ cm. Fls axillary on jointed peduncle, 2·4–4 cm long, usually solitary, small, drooping in bud, extending as fl. opens and then deflexed as fade and in fr.; calyx united with 5 teeth, corolla about 1 cm in length, white, greenish, pink or blue, pink fls fading to blue with age; standard broad and clawed, wings free, keel incurved; stamens 10, vexillary stamen free, others connate; anthers uniform; ovary sessile; style incurved, filiform, not bearded; stigma terminal. Pods swollen, oblong, 2–3×1–2 cm, 1–2-seeded, beak obliquely placed, sometimes sterile. Seeds angular, 0·5–1·0 cm in diameter, with pointed beak and small hilum; testa smooth, wrinkled or rough, ranging in colour from white, yellow, red, brown to nearly black. Average weight of 100 seeds 17–27 g.

POLLINATION

Lowest flowers open first with usually only one open on each branch. They usually open on two successive days, but close at night. Flowering continues on each plant for more than a month. Anthers grouped above stigma before flowers open, so usually self-pollinated. Flowers are visited by bees and occasional natural cross-pollination occurs. On cloudy and on wet days little pollination takes place and empty pods result.

GERMINATION

Germination is hypogeal. The first two leaves are scale-like and the later leaves are pinnate.

CHEMICAL COMPOSITION

Analysis of the whole dried seed gives approximately: water 9·8 per cent; protein 17·1 per cent; fat 5·3 per cent; carbohydrate 61·2 per cent; fibre 3·9 per cent; ash 2·7 per cent.

PROPAGATION

Gram is grown from seed.

HUSBANDRY

Gram is usually grown in India as a cold-weather crop, either in admixture with cereals and other crops or in pure stand, when the seed is either broadcast or planted in rows about 10 in. apart with a seed rate of 30–40 lb per acre. The crop is sometimes grown under irrigation in India. It is harvested 4–6 months after sowing and is cut, dried and threshed. Yields range from 400–1,600 lb of dried pulse per acre, with an average of about 600 lb. White or light coloured, bold seeds command the highest price.

MAJOR DISEASES AND PESTS

Gram blight caused by *Mycosphaerella rabiei* Kov. (syn. *Ascochyta rabiei* (Pass.) Lab.) is the most serious disease in India. It is seed-borne and causes brown spots on the leaves, stems and pods and eventually the death of the plant. It is also attacked by a rust, *Uromyces ciceris-arietini* (Grogn.) Jacz. & Boyer, and wilt caused by *Rhizoctonia bataticola* (Taub.) Butl. and *Fusarium orthoceras* Appel & Wr. var. *ciceria* Padwick.

The most serious pest is the gram caterpillar, *Heliothis armigera* Hubn. and it is very susceptible to store pests, particularly pulse beetles, *Bruchus* spp.

IMPROVEMENT

Attempts have been made in India to collect and classify cvs and isolate improved cvs, but much of the crop is still grown from unselected mixed cvs. White, bold-seeded 'Kabuli' gram, which is late-maturing and low-yielding has been crossed with brown-seeded, high-yielding cvs. The hybrids give medium-sized seeds with fair yields. It is hoped to produce early-maturing, pale and large-seeded cvs with resistance to blight and wilt.

PRODUCTION

The areas of maximum production are in India and the Middle East. Most of the crop is consumed locally. India grows over 20 million acres annually with a production of 4–5 million tons. The greatest acreage is in northern India in the Upper Ganges basin and the Central Provinces.

Fig. 37. **Cicer arietinum :** CHICK PEA. **A,** leafy shoots (×½); **B,** leaf (×1); **C,** flower (×3); **D,** stamens and pistil (×3); **E,** pods (×½); **F,** seeds (×2).

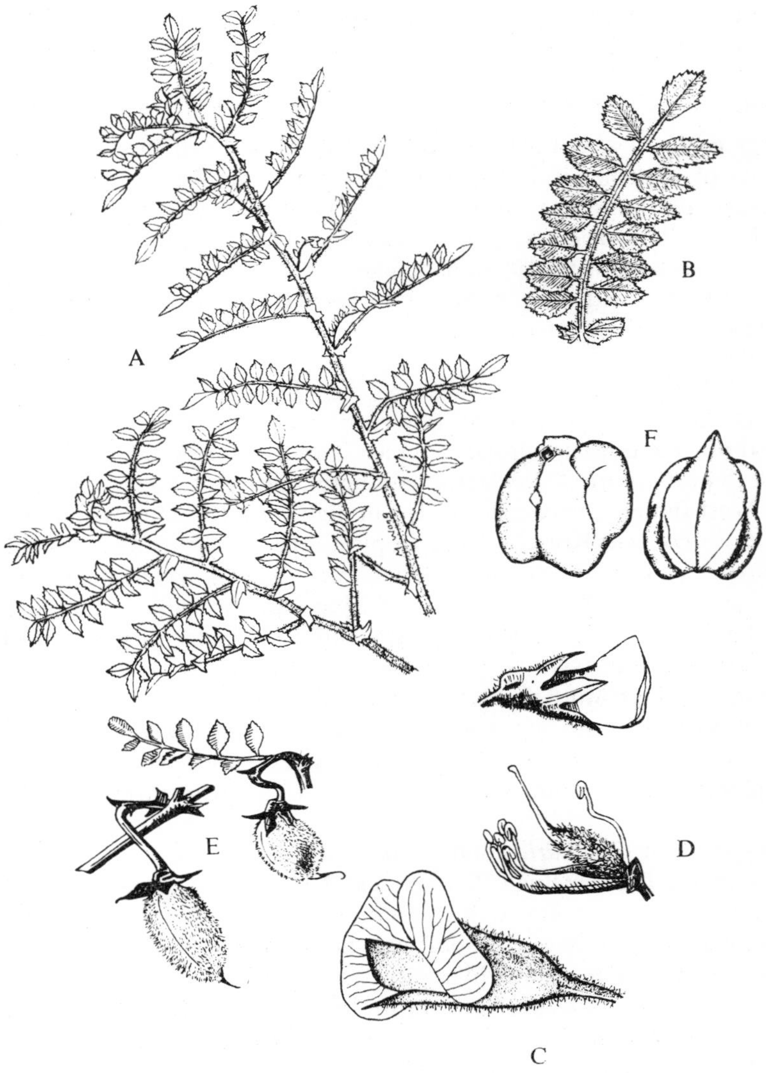

A
B
F
E
D
C

REFERENCES

HOWARD, A., HOWARD, G. L. C. and KAHN, A. R. (1915). Some varieties of Indian Gram (*Cicer arietinum* L.). *Mem. Dept. Agri. India, Bot. Ser.*, 7, 213–285.

SHAW, F. J. F. and KHAN, A. R. (1931). Studies in Indian Pulses: (2) Some Varieties of Indian Gram (*Cicer arietinum* Linn.). *Mem. Dept. Agri. India, Bot. Ser.*, **19**, 27–47.

Crotalaria L. (x = 7, 8)

A large genus of about 550 spp. of shrubs and herbs of the tropics and subtropics of both hemispheres, with the greatest number in Africa. Lvs alternate, simple (unifoliate) or digitately 3-, 5- or 7-foliate; stipulate; fls in terminal or axillary racemes, often showy, mostly yellow, standard prominent, keel beaked; stamens 10, monadelphous, 5 with small versatile anthers alternating with 5 with long basifixed anthers; fr. inflated pod, 2-valved; seeds loose and rattling at maturity in dry pod.

USEFUL PRODUCTS

C. juncea (q.v.) is an important fibre crop and green manure.

Several other species are grown as green manures and fodders, of which the most important are: *C. anagyroides* Kunth., native of tropical America and the West Indies, an erect shrub, 1–2 m high, with trifoliate lvs and yellow fls; *C. intermedia* Kotschy from tropical Africa with trifoliate lvs and pale yellow fls; *C. mucronata* Desv. (syn. *C. striata* DC.), widely distributed throughout the Old World tropics, an erect undershrub, 1–2 m high, with yellowish silky hairs, trifoliate lvs and yellow fls, purple-striped; *C. zanzibarica* Benth. (syn. *C. usaramoensis* Bak. f.) from East Africa, a short-lived, quick-growing shrub, widely grown as a cover crop and green manure in Asia. All these spp. are considered non-toxic. Several *Crotalaria* spp. are used locally for bast and tie fibres. A few spp. make good ornamental plants, *e.g.* *C. retusa* L., which is now widespread throughout the tropics.

Many species of *Crotalaria* contain alkaloids and are more or less toxic to farm animals. These include: *C. burkeana* Benth. which causes crotalism (*styfsiekte*) in cattle in South Africa, resulting in acute inflammation of the horn-forming membranes of the hoof; also in South Africa, *C. dura* Wood & Evans causes pulmonary and liver diseases in horses and cirrhosis of the liver in cattle; *C. retusa* L. has been shown to cause cirrhosis of the liver in humans and cattle in Jamaica; *C. maypurensis* HBK. is suspected of losses in cattle from the same cause in Guyana.

Crotalaria juncea L. (2n = 16) SANN OR SUNN HEMP

USES

Sunn hemp is second in importance to jute, *Corchorus capsularis* (q.v.), as a bast fibre in India. It has greater tensile strength and is more durable under exposure than jute. In the United Kingdom it is used as a substitute for hemp, *Cannabis sativa* (q.v.), but it is not as strong as the latter. It is essentially a cordage fibre and is used in the manufacture of twine and cord, marine cordage, fishing nets, matting, sacking, rope soles for shoes and sandals, etc. It was used extensively during World War II for camouflage nets. It is made into cigarette and tissue paper.

Crotalaria juncea is one of the most widely grown green manures throughout the tropics. It is grown in rotation with rice, maize, tobacco, cotton and other crops and in sugar cane, pineapples (Hawaii), coffee (Brazil) and in orchard crops. Dried stalks and hay are used as stock fodder. Reports on the toxicity of the seeds vary, but they are fed to pigs in Rhodesia and horses in the Soviet Union.

ORIGIN AND DISTRIBUTION

C. juncea is never found truly wild and probably originated in India where it has been cultivated from ancient times. It has now spread to most tropical countries as a green manure.

ECOLOGY

Sunn hemp is the fastest growing of the *Crotalaria* spp. and is very effective in smothering weeds. It is hardy and drought-resistant and grows on almost all soil types. For fibre production light, loamy, well-drained soils are preferred; on low-lying, clay soils it makes vigorous growth, but the fibre is coarser and the yields are lower. It is said that vegetative growth is favoured by long days, but seed setting is then poor.

STRUCTURE

An erect shrubby annual, 1–3 m in height, with vegetative parts covered with short downy hairs. There is a long, strong tap-root and numerous well-developed lateral roots; nodules are freely produced and are much-branched and lobed, up to 2·5 cm in diameter. Stems ribbed. Lvs simple, spirally arranged with a phyllotaxy of 2/5; stipules minute, pointed; petiole short, about 5 mm long with pulvinus; lamina linear-elliptic to oblong, entire, 4–12 × 0·5–2 cm. Inflorescence terminal open raceme to 25 cm in length with minute linear bracts. Fls showy; pedicels short with 2 small lateral pointed appendages near base of calyx; sepals 5, hairy, shortly united at base, lobes pointed with 3 lower sepals united at tips to form a boat-like structure under keel and separating in fr., with 2 upper sepals similarly united; corolla deep yellow; standard erect, about 2·5 cm in diameter, rounded, sometimes streaked purple on dorsal surface; wings shorter, keel twisted, very acute; stamens 10, almost free to base, 5 with short filaments and long narrow anthers and 5 with long filaments and small rounded anthers; style, long, bent, with row of hairs along one surface and terminal stigma with tuft of hairs. Fr. a softly-hairy inflated pod, about 3 × 1 cm, grooved along upper surface and with short pointed beak, pod light brown when ripe, several seeded. Seeds small, flattened, kidney-shaped, dark grey to black, loose at maturity and rattling in pod. 15,000 seeds per lb.

POLLINATION

The lowest flowers in the inflorescence open first and the flowers remain open for 2 days. In the bud the stamens with the linear anthers are longest and dehisce in the bud. The filaments of stamens with ovate anthers then elongate pushing the liberated pollen to the orifice of the keel. The stigma lies at the orifice above the pollen mass. The wings and keel are articulated by a ball-and-socket joint. Large bees such as *Xylocopa* spp. and *Megactile* spp. alight on the wings and set the piston mechanism in operation, when the stigma followed by a mass of pollen is forced against the abdomen of the insect. The ovate anthers dehisce at the tip of the keel at the end of the first day or the following morning and the pollen from them is liberated by further visits of the bees. Extensive cross-pollination occurs and self-pollination only takes place after the stigmatic surface has been stimulated by the bee or done mechanically. Bagged flowers do not set seeds.

GERMINATION

Germination is usually rapid and uniform.

CHEMICAL COMPOSITION

The seeds contain approximately: water 8·6 per cent; protein 34·6 per cent; fat 4·3 per cent; carbohydrate 41·1 per cent; fibre 8·1 per cent; ash 3·3 per cent.

PROPAGATION

The crop is grown from seed.

HUSBANDRY

Both for fibre and as a green manure the crop is usually broadcast on a well-prepared seedbed. For fibre production it is sown thickly to prevent branching with a seed rate of 60–80 lb; for green manure the seed rate is about 35–40 lb per acre, but as little as 12–15 lb per acre are sometimes sown, particularly when grown for seed. No further cultivation is done after sowing. For green manure the crop is ploughed in after 2–2½ months when the plants begin to flower as it decomposes more rapidly at this stage. The

Fig. 38. **Crotalaria juncea**: SUNN HEMP. **A**, shoot with flowers and fruits (× ½); **B**, flower in longitudinal section (× 2); **C**, seed (× 4).

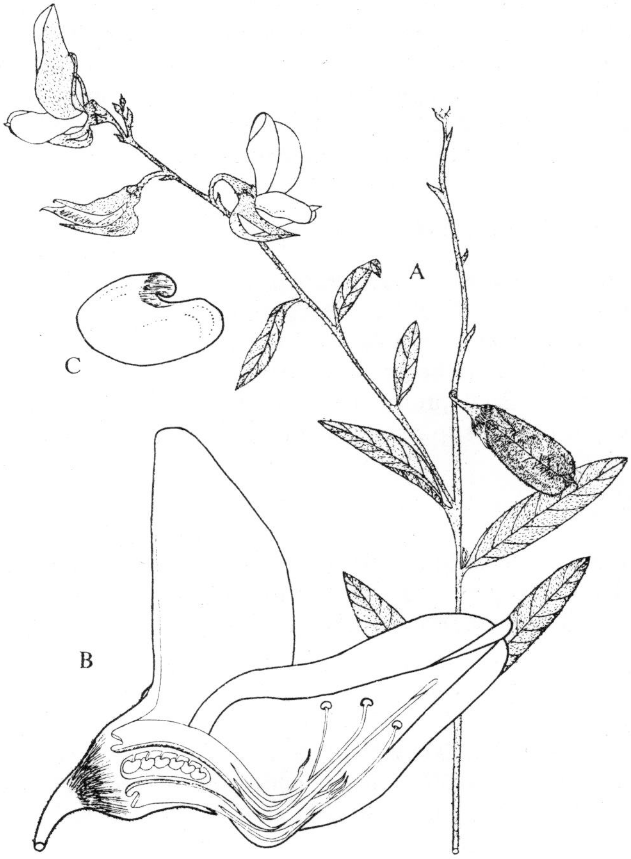
A
B
C

yield of green matter is 8–12 tons per acre, which on decomposition may add 60–100 lb of nitrogen to the soil.

As a fibre crop, which may be sometimes irrigated, it is usually cut or pulled when the pods are forming at 3–3½ months. Plots required for seed take 4–4½ months giving a yield of 400–600 lb per acre. The stems used for fibre are tied into bundles of about 100 each, the tops and roots are cut off, and they are left in the field for a few days until the leaves have dropped off. The bundles are then retted in clean water, stagnant or slow-flowing, for about a week, after which the bark is peeled off and the fibres are separated by beating and washing. Average yields vary from 300–500 lb per acre of dry fibre, which is about 8 per cent of the weight of the dried stems. The commercial fibre is 3–5 ft long, whitish, grey or yellow in colour, and quality is judged by the length, fineness, colour, uniformity and freedom from extraneous matter. The lignified phloem sclerenchyma cells composing the fibre have rounded ends and a rather thick lumen.

MAJOR DISEASES AND PESTS

Wilt, *Fusarium udum* Butl. var. *crotalariae* Padwick, and anthracnose, *Colletotrichum curvatum* Briant & Martyn, are two common and serious diseases in India. The caterpillars of the sunn hemp moth, *Utetheisa pulchella* L., and the stem borer, *Laspeyresia pseudonectis* M. are the two most serious pests. Seed production may be hampered by pod borers. Unlike most *Crotalaria* spp., sunn hemp is attacked by root nematodes.

IMPROVEMENT

No systematic classification of cvs has been attempted, but some selection of improved types suited to particular localities, with attention paid to early maturity, fibre yield and resistance to disease, has been made in India.

PRODUCTION

India grows about 800,000 acres of sunn hemp per year with an annual production of 80,000–100,000 tons, of which 20–30 per cent is exported, mainly to the United Kingdom, Belgium and the United States.

REFERENCES

HOWARD, A. and G. L. C. (1919). Studies in the Pollination of Indian Crops. *Mem. Agri. Dept. India, Bot. Ser.*, **10**, 195–220.

KIRBY, R. H. (1963). *Vegetable Fibres.* London: Leonard Hill.

SINGH, B. N. and S. N. (1936). Analysis of *Crotalaria juncea* with special reference to its use in Green Manure and Fibre Production. *J. Amer. Soc. Agron.*, **28**, 216–227.

Cyamopsis tetragonoloba (L.) Taub. (x = 7; 2n = 14) CLUSTER BEAN

Syn. *C. psoralioides* DC.

USES

Cyamopsis is a small genus of 3 spp. in tropical Africa and Asia, of which *C. tetragonoloba* is the cluster bean or guar. The young tender pods are eaten as a vegetable and the seeds as cattle feed in India and Pakistan, where it is also used as fodder and green manure. The seed flour has exceptionally high viscosities at low concentrations, possessing 5–8 times the thickening power of ordinary starch. The principal mucilaginous substance is manno-galacton. It is used to improve the strength of certain grades of paper, for backing postage stamps, in textile sizing and as a stabilizer and thickener in food products such as ice-cream, bakery mixes and salad dressings. *Sesbania aculeata* Pers. has recently been suggested as a source of guar gum.

ORIGIN AND ECOLOGY

The plant is not known in a wild state and is probably indigenous to India. It has now been introduced into the drier tropics as a fodder and green manure, and is grown for gum production in India and the south-western United States. The crop is hardy and very drought-resistant and grows well on alluvial and sandy loams.

STRUCTURE

It is a robust bushy annual, 1–3 m tall, and tall and dwarf cvs are recognized. Branches stiff, usually with white hairs, erect, angled, and grooved. Lvs alternate, trifoliate; leaflets ovate, sparsely serrate. Fls small in dense axillary racemes; standard about 8 mm long, white, wings pinkish; stamens 10, monadelphous; anthers uniform, ovate; style incurved; stigma terminal, capitate. Pods linear, in stiff erect clusters, compressed with double ridge on dorsal side and single ridge below, 4–10 cm long, with marked beak, and 5–12 oval seeds, about 5 mm long, varying in colour from white or grey to black. 100 seeds weigh about 6 g.

HUSBANDRY

In India it is usually grown in mixed cultivation. In pure crop the seed is usually broadcast with seed rate of 10–20 lb per acre when grown for seed and of 30–40 lb per acre for green manure or fodder. The crop comes into bearing in 3–3½ months after sowing. Yields of about 10,000 lb of green fodder and 600–800 lb of seed are obtained under rain-fed conditions and yields of nearly double this under irrigation. The dry seeds contain about 33·3 per cent protein and 40 per cent carbohydrate. In northern India it is usually grown as a summer crop.

Derris Lour. (x = 11, 12, 13)

About 70 spp., mainly lianes, rarely shrubs and trees, throughout the tropics, but with the largest number of species in south-eastern Asia. Lvs imparipinnate without stipels or tendrils. Extracts from the root of some species are used as insecticides and fish poisons, *D. elliptica* (q.v.) being the main cultivated species, but *D. ferruginea* Benth., a native of India, and *D. malaccensis* Prain from Malaysia are also used.

Derris elliptica Benth. (2n = 22, 24) DERRIS, TUBA ROOT

USES

The powdered root is of ancient use in south-eastern Asia as a hair-wash against lice and as a fish poison. By 1877 it was cultivated in Singapore for local use as an insecticide for vegetables. In the early 1900's it was exported to Europe and the United States for the same purpose and was often used in conjunction with other insecticides. It is used as: (1) dusting powders with suitable carriers such as talc or kaolin; (2) an aqueous suspension for spraying with a dispersal agent and water; (3) extracts of powder in an organic solvent such as turpentine.

For use as a fish poison the pounded roots are soaked in water. The fish become stupefied at first, making violent efforts to escape, during which time they are speared. They then become sluggish, turn on their backs and float on the surface, when they can be taken out of the water by hand. The fish may be eaten with impunity. I well remember taking part in a large fish hunt on a wide river in Sarawak. Several hundred people were assembled in canoes before dawn. The silence was broken by the rhythmic beating of the tuba root (derris). At a signal from the headman by the firing of a gun at dawn the tuba and liquid were thrown into the river. Within a short time the fish were making frantic attempts to escape and then followed a glorious free-for-all and the fish were speared. The hunt went on for several hours and continued down river as the current took the tuba. Large quantities of fish were caught.

ORIGIN AND DISTRIBUTION

Derris elliptica occurs wild from eastern India to New Guinea. It is not found wild in southern Malaya where the plants seldom flower. It has been distributed to most tropical countries for trial.

ECOLOGY

The plant grows in the forests and secondary jungle. It is grown in areas with 90–130 in. of rain and average temperatures of 85°F. It will tolerate a wide range of soil conditions, but does best on rich friable loams; stony and swampy ground is unsuitable. It is a short-day plant.

STRUCTURE

A perennial woody evergreen climber to over 16 m in length; in cultivation it is grown as a sprawling bush. Roots up to 1 cm in diameter and to over 2 m in length, dark reddish-brown. The rotenoids (see below) occur as discrete globules of resin in parenchymatous cells of the root. Trunk about 1 m high and 10 cm in diameter with long whippy branches at first rusty-tomentose. Lvs alternate, imparipinnate, stalked, with 9–13 opposite leaflets, exstipellate, elliptic, 6–15 × 3–7 cm. Fls borne in axillary racemes, 10–20 cm long, with 5–20 small pinkish fls, about 1·5 cm long; calyx with very short teeth; petals glabrous; vexillary stamen free at base but connate from middle with others in a closed tube, stamens of equal length, anthers versatile. Pods flattened, strap-shaped, brownish, indehiscent, about 5 × 2 cm, 1–4-seeded, upper suture narrowly winged. Seeds flattened.

CHEMICAL COMPOSITION

The roots contain rotenoids of which the most important and most toxic is rotenone ($C_{23}H_{22}O_6$); other toxic constituents are deguelin, tephrosin and toxicarol. Derris insecticides should not contain less than 3 per cent rotenone on a dry-weight basis and 15 per cent total ether extractives. Rotenone first appears when the roots are about 6 weeks old and showing suberization; roots about 5 mm in diameter are considered to have the highest content.

PROPAGATION

Derris is propagated vegetatively by well-ripened stem cuttings, 8–12 in. long and about ¾ in. in diameter with 3–12 nodes. The cuttings are planted at a slant in nursery beds of open rooting material, 6 × 2 in. apart, with the bud at least 1 in. above the ground. They are ready for transplanting after 6 weeks. Where it is necessary to conserve planting material single-node, soft-wood cuttings, treated with growth hormone, can be used.

HUSBANDRY

The rooted cuttings are planted in the field at a distance of 3 × 2–3 ft. They are harvested after 2 years' growth when the roots contain the maximum concentration of rotenoids, after which it declines. About 95 per cent of the roots are in the top 18 in. of soil. The roots are dried rapidly in the sun or in ovens at 130°F until the moisture content is reduced to about 10 per cent. They must be stored in a cool dry place as the roots deteriorate if moist or if exposed to sunlight and air.

MAJOR DISEASES AND PESTS

No serious diseases or pests have been reported.

IMPROVEMENT

Wild *Derris elliptica* in Java contains about 0·5 per cent rotenone; this has been increased to 8–11 per cent by selection and cultivation, and some cvs now contain 13 per cent rotenone on a dry-weight basis. In Malaya the highly toxic 'Changi No. 3' has been crossed with the high-yielding 'Sarawak Creeping' to combine these two desirable characteristics.

PRODUCTION

The first exports to the Western World were made in the early 1900's. The main producing countries are Malaya, Sarawak, Indonesia and the Philippines. The United States imported 6½ million lb of derris in 1940. *Lonchocarpus nicou* (Aubl.) DC. from Guyana, *L. urucu* Killip & Smith from Brazil and *L. utilis* Killip & Smith from Peru are also important sources of rotenone and in 1946 the United States imported 11 million lb, chiefly from Peru and Brazil.

REFERENCE

HOLMAN, H. J. (1940). *A Survey of Insecticidal Material of Vegetable Origin*. London: Imperial Institute.

Dipteryx Schreb.

About 8 spp. of trees in tropical America. The seeds of some species contain coumarin and are used commercially. The main source is from *D. odorata* (q.v.), but those of *D. oppositifolia* (Aubl.) Willd. are also used. Those of *D. puncata* Blake and *D. rosea* Spruce are also said to contain coumarin. Frs drupaceous.

Dipteryx odorata (Aubl.) Willd. TONKA BEAN

USES

The cured beans, which have an odour of new-mown hay, are used chiefly for flavouring and scenting tobacco, but are also used to perfume and flavour other products such as confectionery, liqueurs, soap, and in perfumery. They are used as a substitute for vanilla in ice-cream and other products. The tree also produces kino from incisions in the bark. The timber is very hard, tough and durable.

ORIGIN AND DISTRIBUTION

D. odorata is indigenous to South America and occurs in the forests of Colombia, Venezuela, the Guianas and Brazil, being most abundant along

Fig. 39. **Derris elliptica**: DERRIS. **A**, leaf (×½); **B**, inflorescence (×½); **C**, flower (×2); **D**, flower in longitudinal section (×5); **E**, pods (×½); **F**, seed (×2).

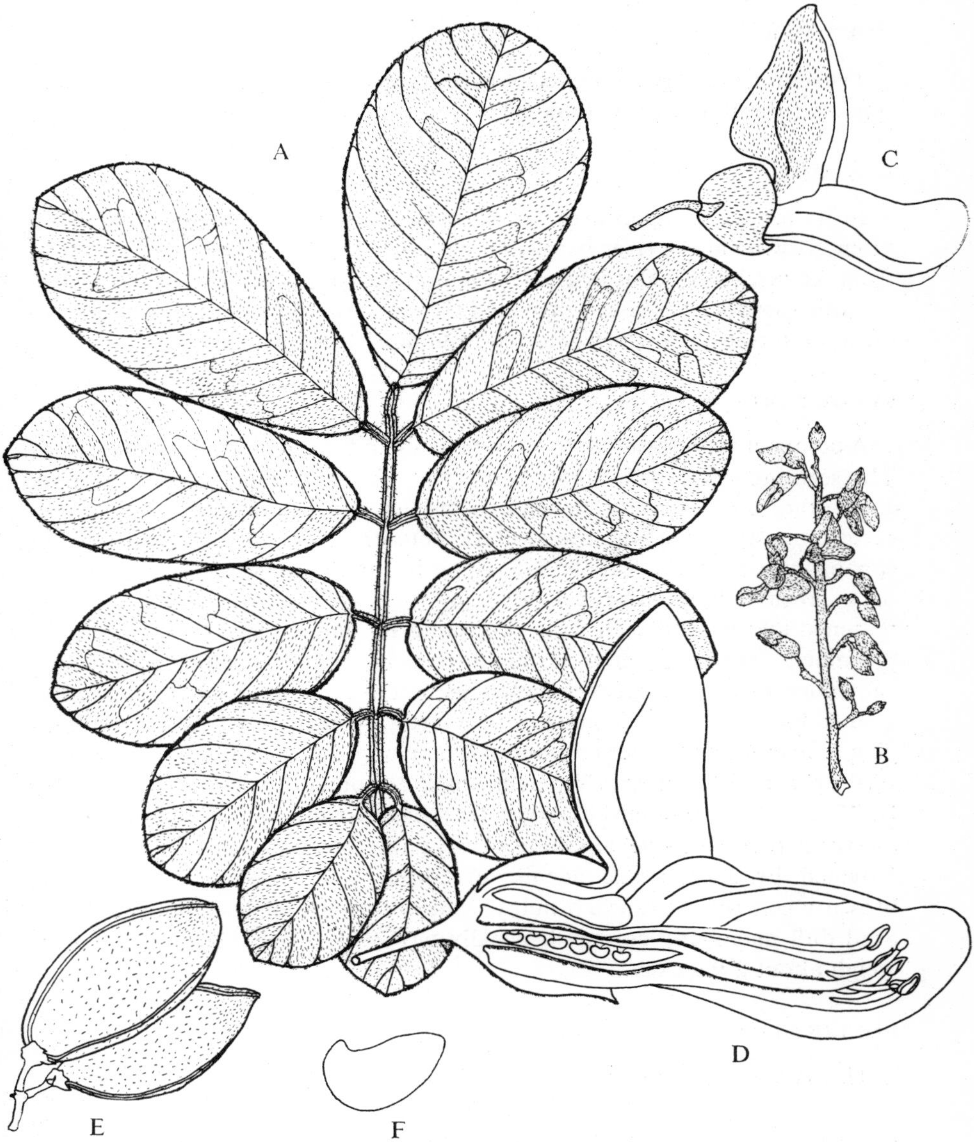
A
C
B
D
E
F

the tributaries of the Orinoco. It has been introduced into Trinidad and other West Indian islands.

CULTIVARS

Two commercial types are recognized, namely, the Angostura–Venezuelan type and the Brazilian or Para type.

ECOLOGY

In its natural environment, tonka bean is a tree of the tropical forest, often growing along river banks. Under cultivation in Trinidad it is grown from sea level to 1,000 ft in areas with a rainfall of 60–110 in. It will survive on soils too poor for cocoa provided they are well drained, but grows best on more fertile soils which are rich in humus. It is a calcifuge.

STRUCTURE

A compact evergreen tree to 40 m high and a trunk up to 1 m in diameter. The seedling develops a vigorous tap-root, but this descends to less than 1 m; other anchor roots descend to greater depths and there is a dense surface mass of feeding roots. Bark smooth, pale grey. Lvs alternate, subabruptly pinnate; leaflets 3–6, opposite or alternate, leathery, glossy, dark green, elliptic, unequal-sided, up to 15 × 8 cm, exstipellate, rachis flattened and winged with wing projecting beyond leaflets. Inflorescence a prominent, many-flowered terminal panicle. Fls rose-violet in colour, subtended by an ovate, early-deciduous bract, 5 mm long, and enclosed by 2 pink bracteoles which quickly turn brown and drop; calyx rose-pink, glandular-punctate, tube short about 4 mm long, 2 posterior sepals petaloid, 10–12 mm long, 3 anterior sepals in small 3-toothed lip; corolla rose-pink, 10–12 mm long; standard emarginate; stamens connate into a sheath split above; anthers versatile, alternate ones smaller or abortive; ovary stipitate, 1-ovuled. Fr. drupaceous, indehiscent, elliptic, 7–10 × 3–6 cm; pale yellow-brown when ripe; mesocarp pulpy; endocarp hard enclosing single seed. Seed dull mahogany coloured, usually wrinkled, 3–5 × 1–2 cm. The seeds weigh about 2·9 g each.

POLLINATION

This is entomophilous.

GERMINATION

When the whole fruit is planted germination begins after about 4 weeks and it is advisable to assist the cotyledons and shoot to free themselves

---

Fig. 40. **Dipteryx odorata :** TONKA BEAN. **A,** shoot with fruit ($\times \frac{1}{2}$); **B,** flower in longitudinal section ($\times 5$); **C,** fruit in longitudinal section ($\times \frac{1}{2}$).

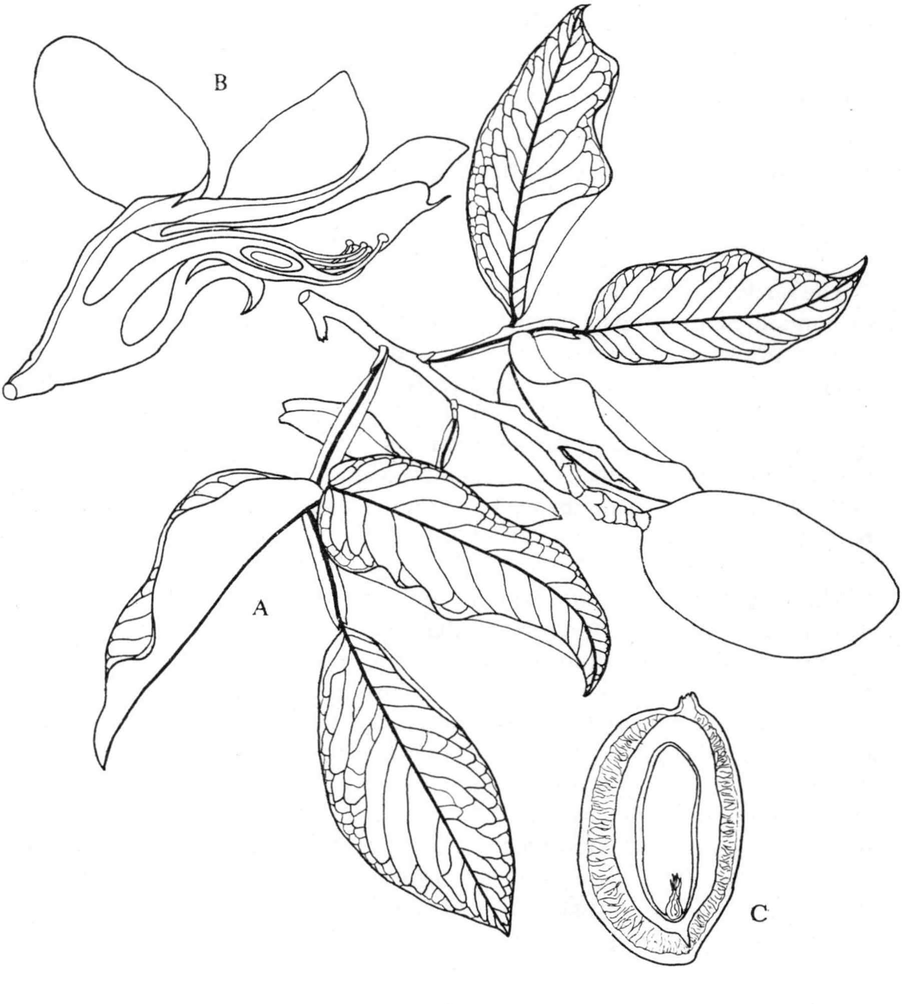
B
A
C

from the shell. When seeds, after removal of the endocarp, are planted, germination takes 1–2 weeks.

CHEMICAL COMPOSITION

The characteristic odour and flavour is due to coumarin, which is liberated during fermentation and drying from a glucoside by enzyme action. The marketed product contains about 8 per cent moisture, 2–3 per cent coumarin and 25 per cent of a solid fat, tonka butter.

PROPAGATION

The crop is usually grown from seed, but it can be propagated vegetatively by budding, by cuttings as in cocoa (q.v.), and by marcotting.

HUSBANDRY

The seeds may be planted at stake or seedlings may be raised in nurseries or containers and transplanted at about 8 weeks old. Temporary shade is usually provided by bananas, cassava and pigeon peas, but this should be removed when the trees are 6 ft high to prevent them becoming spindly. In pure stand the usual initial spacing is 10 × 10 ft with subsequent thinning when 10 years old and at further intervals. The trees are topped at about 6 ft to induce branching. Trees which receive little attention begin bearing at 7–10 years, but with improved cultivation and management a crop may be obtained in the 2nd year. The trees tend to produce a good crop every 2nd or 3rd year.

The fallen fruits are gathered and in Trinidad harvesting takes place in March–May. The fruits are dried in the sun for a few days, after which they are cracked along the suture by a hammer to release the single seeds, which are then shade dried. The best quality beans are long, bold, and firm, with a deep colour and finely wrinkled. The average yield is 1–2 lb of dried beans per tree, but there are records of individual trees producing up to 50 lb. The beans are then sold for curing. This consists of soaking the beans for several days in barrels of strong rum or alcohol with an alcohol percentage of 45–65 per cent. At higher percentages the coumarin is dissolved out of the seeds, but, as the surplus rum is used for subsequent curing, any dissolved coumarin is usually deposited on the seeds thus cured. Properly cured beans should be pliable and show a heavy crystalline coumarin deposit on the testa.

MAJOR DISEASES AND PESTS

Tonka bean is remarkably free from pests and diseases, except for bats which pick the fruits and carry them away. After eating the pulpy flesh they drop the seeds.

IMPROVEMENT

Selection of high-yielding, regular bearers, followed by vegetative propagation would be an obvious advantage. Dwarf forms are known and could be used as rootstocks.

PRODUCTION

The principal production is from wild trees in Venezuela, and to a lesser extent in Brazil and Colombia. The crop has also been planted in Venezuela and the West Indies. During the 1930's the crop was highly remunerative and in 1936 was fetching about 8s. 6d. per lb in London. At this time Venezuela was sending beans to Trinidad to be cured and in 1936 Trinidad exported nearly 1 million lb of cured beans. Venezuela now cures its own beans. Since World War II prices have slumped and production has decreased due to competition from synthetic coumarin and vanillin.

REFERENCE

POUND, F. J. (1938). History and Cultivation of the Tonka Bean (*Dipteryx odorata*) with Analyses of Trinidad, Venezuelan and Brazilian Samples. *Trop. Agriculture, Trin.*, **15**, 4–9, 28–32.

Dolichos Lam. (x = 11, 12)

About 100 spp. of annual and perennial herbs, usually twining, occasionally subshrubs, of the tropics and subtropics, with the greatest number of species in the Old World. Lvs pinnately trifoliate, stipellate. Fls in axillary racemes and clusters at node-like thickenings of peduncle; keel bent inwards at right-angles (cf. curved in *Vigna*; coiled in *Phaseolus*); stamens diadelphous. Pod large, flat, usually curved and beaked. *D. uniflorus* (q.v.) is a pulse crop. The correct name of *D. lablab* is now *Lablab niger* (q.v.).

Dolichos uniflorus Lam. (2n = 24) HORSEGRAM

Syn. *D. biflorus* auct. non Linn.

USES

Horsegram is the poor man's pulse crop in southern India where the seeds are parched and then eaten after boiling or frying, either whole or as a meal. The seeds are an important food for cattle and horses and are usually fed after boiling. The stems, leaves and husks are used as fodder. In Burma the dry seeds are boiled, pounded with salt and fermented to produce a sauce similar to soya sauce from *Glycine max.* (q.v.). Horsegram is also grown as a green manure crop.

ORIGIN AND DISTRIBUTION

D. uniflorus occurs wild in the Old World tropics. The main area of cultivation is in India.

ECOLOGY

It is grown as a dry-land crop in areas of moderate rainfall not exceeding 35 in. In areas of higher rainfall it is usually sown after the rains have ceased. It is hardy and drought-resistant and is grown on all soil types, even on poor soils with very little rain.

STRUCTURE

A low-growing, slender, sub-erect, annual herb with slightly twining, downy stems and branches, 0·3–0·5 m high. Lvs trifoliate; leaflets pilose, entire, membranous, broadly ovate, acute, 2·5–5 cm long, laterals oblique; stipules ovate-lanceolate, about 1 cm long; stipels minute. Fls 1–3 together in lf-axils about 1 cm long, each on a short peduncle; calyx downy with lanceolate teeth; corolla pale yellow; standard ovate-emarginate, longer than wings, hardly reflexed; keel narrow, obtuse; stamens diadelphous; anthers uniform, basifixed; style filiform, persistent, glabrous. Pod linear, recurved, beaked, downy, dehiscent, 5–7-seeded; valves tough, chartaceous. Seeds small, rhomboidal, flattened, 3–6 mm long, light red, brown, black or mottled, with shining testa and small inconspicuous hilum.

HUSBANDRY

The seeds are sown broadcast or in rows with a seed rate of about 40 lb per acre and the crop receives little attention. For harvesting as pulse the duration of the crop is 4–6 months and yields of 150–300 lb are obtained, but up to 600 lb has been recorded. When grown as fodder it is harvested after about 6 weeks and yields 2–5 tons of green fodder.

MAJOR DISEASES AND PESTS

The most serious diseases in India are a root-rot caused by *Rhizoctonia* sp. and anthracnose, *Glomerella lindemuthianum* (Sacc. & Magn.) Shear. The gram caterpillar, *Azazia rubricans* Boisd. is destructive to the crop.

PRODUCTION

The only country growing large acreages of horsegram is India.

Glycine Willd. (x = 10)

About 10 spp. of twining, climbing or procumbent perennial herbs, and rarely erect, annual herbs, of tropical and warm temperate regions of the Old World. Lvs mostly trifoliate; fls small, white to purplish, in short axillary racemes; stamens usually monadelphous; pods narrow, usually flattened.

USEFUL PRODUCTS

G. max (q.v.) is an important food and forage crop. *G. javanica* L., which is wild and widespread in tropical Africa and Asia, is becoming more widely used throughout the tropics as a fodder and mixed with grasses in pastures. It is a variable, twining, pubescent, perennial herb with racemes of small white or purple fls. It is very palatable to stock and seeds profusely.

Glycine max (L.) Merr. (2n = 40) SOYA BEAN, SOYBEAN

Syn. *G. soja* Sieb. & Zucc.; *G. hispida* (Moench) Maxim.; *Soja max* (L.) Piper

USES

Soya beans are one of the world's most important sources of oil and protein. They are an important food crop in eastern Asia, where the unripe seeds are eaten as a vegetable and the dried seeds are eaten whole, split or sprouted; they are processed to give soya milk, which is a valuable protein supplement in infant feeding; they are made into curds and cheese. Soya sauce is made from mature fermented beans and is one of the principal flavourings and sauces of eastern Asia; it is also one of the principal ingredients of Worcester and other western sauces. It is made by mixing boiled beans with wheat flour and salt and fermenting them in trays by means of the fungus, *Aspergillus oryzae*, for up to a week. The fermented beans are then submerged in brine in jars, which are left exposed to the sun for many months to extract the flavour. In Indonesia boiled beans are fermented by *Aspergillus* and made into cakes (tempe).

The seeds yield an edible, semi-drying oil which is extensively used in the Far East as food. There has been a vast increase in the use of soya bean oil in the United States and Europe since World War II. The bulk of the oil in the United States is used for edible purposes, particularly as a salad oil and for the manufacture of margarine and shortening. The oil is used industrially in the manufacture of paints, linoleum, oilcloth, printing inks, soap, insecticides, disinfectants, etc. 'Lecthin', phospholipides obtained as a by-product of the oil industry, is used as a wetting and stabilizing agent in the food, cosmetic, pharmaceutical, leather, paint, plastic, soap and detergent industries.

Soya meal, the residue after the extraction of the oil is a very rich protein feeding stuff for livestock for which there is an increasing demand. The meal and soya bean protein are used in the manufacture of synthetic fibre (artificial wool), adhesives, textile sizes, waterproofing, fire-fighting foam, etc. Soya flour prepared from the whole beans produces full-fat flour with about 20 per cent oil; that from mechanically-expressed meal gives low-fat flour with 5–6 per cent oil; that prepared from solvent-extracted meal gives defatted flour with about 1 per cent oil. The flour is used in bakery and other food products; as additives and extenders to cereal flour and meat products, in health foods, etc.

Soya beans are an important pasture and fodder crop; they are made into hay and silage; they are also used as a green manure and cover crop.

ORIGIN AND DISTRIBUTION

The cultigen, *G. max*, is thought to be derived from *G. ussuriensis* Regel & Maack, a slender, prostrate, twining legume, which is found wild throughout eastern Asia, possibly in hybridization with *G. tomentosa* Benth., which grows wild in southern China. *G. gracilis* Skvortzor is considered an intermediate semi-cultivated species between the wild *G. ussuriensis* and *G. max*. Soya beans have been cultivated in China since before the times of written records, of which the first is in 2838 B.C. They have also been an important food crop in Manchuria, Korea and Japan since the earliest times. It was introduced into the Jardin des Plants, Paris, in 1740, and the Royal Botanic Gardens, Kew, in 1790, but little interest was shown in the crop in Europe until the first shipments of soya beans were made in 1908. Soya beans were first taken to the United States in 1804, but there was little commercial production until the 20th century, first mainly as a forage and pasture crop, but with increasing use of seed since 1930 and the most rapid increase since 1942, in response to the war-time demand for edible oils and fats.

The crop has long been cultivated in Indonesia. Soya beans have been introduced into most tropical countries during the 20th century. Attempts to popularize and extend the cultivation of the crop in Africa, India and the West Indies have met with little success, as the dried beans are difficult to cook, they are not found palatable by indigenous people, while there are other pulses and oil seeds adapted to local conditions and which satisfy the dietetic habits of the people.

CULTIVARS

A great number of cvs are recognized in the Far East which vary in: time to maturity; height and plant type; size, colour, oil and protein content of the seed; the uses to which they are put. For oil production yellow seeds are usually richest in oil and give meal and flour with an attractive colour. For immature seeds for use as a vegetable large yellow- or green-seeded forms are preferred. For hay and fodder cvs are usually brown- or black-seeded and the plants are often twining. Black-seeded cvs are usually the richest in protein.

Prior to 1908 there were not more than 8 cvs in the United States. By 1922 over 800 cvs had been introduced, of which 33 were valuable agriculturally. In 1931 collectors returned from Asia with nearly 3,000 introductions. By 1947 over 10,000 seed lots had been introduced by the United States Agricultural Department. Comparatively few cvs are in extensive production out of this vast amount of material. New cvs are also obtained by selection and hybridization.

ECOLOGY

Glycine max is a subtropical plant, but cultivation now extends from the tropics to 52°N. The general climatic requirements approximate to those of maize and the greatest development in the United States has been in the corn belt. They are grown mainly in areas where the summer is hot and rather damp. They will not withstand excessive heat or severe winters, but are less susceptible to frost than cowpeas. They can be grown on a wide range of soil types, but thrive best on sandy or clay loams and alluvial soils of good fertility. The strain of *Rhizobium japonicum*, the nitrogen-fixing bacteria in the root nodules, is specific to the soya bean. Inoculation is desirable if the crop is taken into a new area. If nodules are absent or ineffective, the crop is exhausting of soil nitrogen.

The soya bean is a short-day plant. All cvs flower more quickly with 14–16 hours of darkness. Some cvs will not produce flowers unless there is 10 hours or more of darkness. Most cvs have a narrow range in which they will mature properly and produce satisfactory yields. Cvs taken to areas with longer days may not mature at all. Those taken to areas with shorter days flower early, giving low yields and inferior quality seed. The period to maturity varies from 75–200 days.

The time taken to maturity is of extreme importance in adaptation to a particular latitude. Nine maturity groups are recognized in the United States, each with a narrow range of latitude, ranging from short, very early-maturing cvs adapted to the short summers and long days of southern Canada and the northern States (Group O and I), to tall, late-maturing cvs of the Gulf coast (Group VIII). With a new introduction it is important to test for adaptation to local conditions.

STRUCTURE

Most cvs are erect, bushy, pubescent annuals, 20–180 cm tall, with tawny or grey hairs on stem, lvs, calyx and pod. Some of the more primitive cvs are prostrate and twining, a tendency which is increased with shade. Two types of growth occur; cvs with determinate growth in which the terminal bud develops into a terminal inflorescence; cvs with indeterminate growth in which the stem does not terminate in an inflorescence.

ROOTS: Radicle develops into a tap-root, which may extend to 150 cm, but most roots are in top 30–60 cm of soil; nodules when present are small, spherical, sometimes lobed.

STEMS: Number of axillary buds on the well-developed main stem which grow out to form lateral branches depends on cv. and density of planting; 1–3 laterals usually develop. Buds in axils of cotyledons and primary lvs do not normally develop unless tip damaged.

LEAVES: Alternate, trifoliate, rarely 5-foliate; petiole long, narrow, cylindrical; stipules small, free, lanceolate; stipels minute; leaflets ovate to lanceolate, usually pale green in colour, variably pubescent, 3–10 × 2–6 cm, base rounded, apex acute or obtuse; lateral leaflets often slightly oblique. Most cvs drop lvs when pods begin to mature.

FLOWERS: Borne on short, clustered, axillary racemes of 3–15 fls; up to 30 fls recorded in terminal raceme of determinate types; 20–80 per cent of fls drop off without forming pods. Fls small; bracteoles 2, ovate, acute; calyx hairy, persistent, united for half length and with 2 upper and 3 lower lobes, anterior tooth longest; corolla white or lilac; standard about 5 mm long, hardly reflexed, ovate, emarginate; wings narrow, obovate; keel shorter than wings, not fused along upper surface; stamens monadelphous or vexillary stamen free at base and occasionally entirely free; anthers uniform, globose; ovary hairy, sessile, few-ovuled; style curved, glabrous, with capitate stigma.

FRUITS: Pods borne in clusters on short stalks in groups of 3–15, hairy, pale yellow, grey, tawny or black; slightly curved and usually somewhat compressed, 3–7 × 1 cm, dehiscent, 1–5-seeded but 2–3-seeded in most cvs.

SEEDS: More or less globose in most grain cvs, flattened in some vegetable and hay cvs; testa straw, yellow, green, brown or black or blotched and mottled in combinations of these colours; hilum small; cotyledons yellow or green, usually the former, size very variable. Weight of 100 seeds ranging from 5–40 g, but most grain cvs 10–20 g.

POLLINATION

Soya beans are completely self-fertile and are normally self-pollinated. The flowers open early in the morning and the pollen is shed just before or at the time of opening and is shed directly onto the stigma. The flowers are visited by bees and other insects, so that cross-pollination can take place, but this is usually considerably less than 1 per cent.

GERMINATION

The seedlings usually appear above ground in 5–7 days. Germination is epigeal, the bent epicotyl first emerging and by straightening and elongating pulls the cotyledons from the soil.

CHEMICAL COMPOSITION

The composition of the mature soya bean varies with the cv. and the soil and climatic conditions. Black-seeded cvs are richest in protein and have a low oil content; yellow seeded cvs have a higher oil content and are low in protein. The composition of the dried seed is as follows: water 5·0–9·4 per

Fig. 41. **Glycine max**: SOYA BEAN. **A**, leaf (×½); **B**, flower from below (×7); **C**, flower in longitudinal section (×7); **D**, pod (×½); **E**, seed (×½).

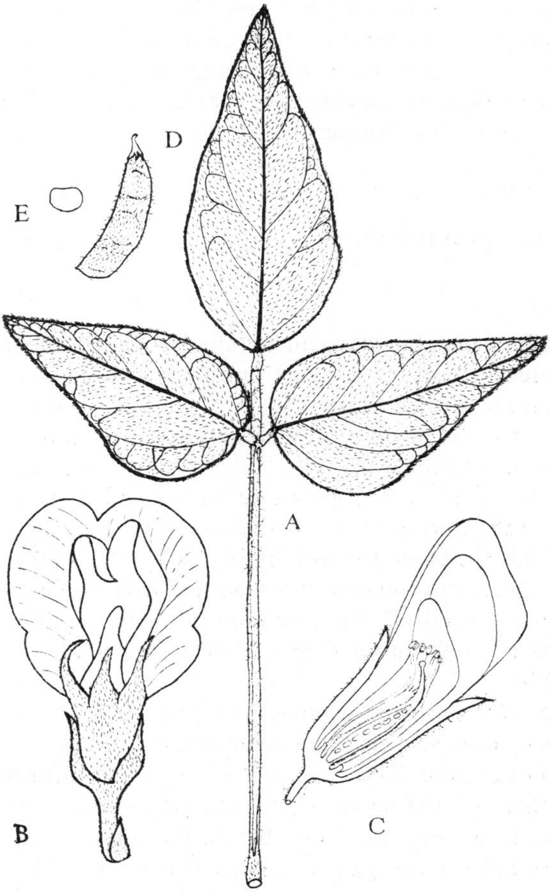
D
E
A
B
C

cent; protein 29·6–50·3 per cent; fat 13·5–24·2 per cent; carbohydrate 14·0–23·9 per cent; fibre 2·8–6·3 per cent; ash 3·3–6·4 per cent. Soya bean meal, solvent-extracted, contains about: water 9·6 per cent; protein 45·7 per cent; fat 1·3 per cent; carbohydrate 31·4 per cent; fibre 5·9 per cent; ash 6·1 per cent.

Mature beans contain a higher percentage of protein than any other pulse and most other foodstuffs. The semi-drying oil is midway in character between linseed and cottonseed oils, containing about 51 per cent linoleic acid, 30 per cent oleic acid and 6·5 per cent linolenic acid. The seeds contain appreciable quantities of vitamin B complex.

PROPAGATION

The crop is grown from seed.

HUSBANDRY

The ease with which the crop can be fully mechanized is one of the reasons for the vast increase in the production of soya beans in the United States since 1942. In the United States the crop is usually grown in rotation with maize; in Manchuria in rotation with sorghum and millet; in some parts of the Far East in rotation with rice. It is often interplanted with maize and other crops. It requires a well-prepared, firm seedbed. The spacing depends on the cv. and the purpose for which the crop is required. For seed production in the United States it is usually planted at 1½–4 ft between rows and 1 in. between plants in the row at a seed rate of 40–60 lb per acre. Spacing in the Far East is usually 2–3 ft between rows with 4–5 in. between hills and with 2 plants per hill with a seed-rate of 35–45 lb per acre. The planting depth is 1–2 in.

Soya beans respond to phosphate and potash fertilizers and to calcium on lime-deficient soils. The usual recommendation is 30–60 lb P_2O_5 and 50–75 lb K_2O per acre. The crop is intercultivated until flowering begins, after which there should be sufficient shade to control weed growth. When grown for seed, harvesting is done before the pods shatter and the plants are cut at ground level or pulled. They are dried for some days and then threshed. Much of the crop in the United States is combine harvested. Yields of up to 1,500–2,400 lb per acre of dried seeds are obtained in the United States, the average yield for the whole country increasing from 660 lb per acre in 1924 to 1,290 lb per acre for the 5 years average of 1948–1952 and with further subsequent increases. In Manchuria and the Far East yields of 900–1,200 lb per acre are obtained and in Java 540 lb per acre. The beans should be stored at a moisture content of 12 per cent or less.

When grown for forage or hay in the United States, soya beans are often mixed with maize, sorghum or Sudan grass. Yields of 9–10 tons per acre of green matter are obtained. For hay the crop is cut when the pods are well formed, but before the leaves are shed and yields 1–5 tons per acre of hay. For green manure soya beans are often mixed with cowpeas.

MAJOR DISEASES

The crop is relatively free from serious epidemics, particularly in the United States, but the incidence of diseases is now increasing. The commonest diseases are: bacterial blight, *Pseudomonas glycinea* Coerper; bacterial pustule of leaves, *Xanthomonas phaseoli* (E. F. Smith) Dows. var. *sojensis* Hedges; pod and stem blight, *Diaporthe phaseolorum* (Cook & Ellis) Sacc. var. *batatis* (Hart. & Field) Wehmeyer; all of which occur in north-eastern Asia and the United States. Other diseases include: frogeye, *Cercospora daizu* Miura (syn. *C. sojina* Hara); wildfire, *Pseudomonas tabacum* (Wolf & Foster) Stapp; downy mildew, *Peronospora manshurica*. Two viruses attack the plant: soybean mosaic virus and yellow bean mosaic virus. Five species of root-knot nematode, *Meloidogyne* spp., and a cyst nematode, *Heterodera* sp., attack the roots.

MAJOR PESTS

Soya beans are comparatively free from insect pests of which the most serious are: soybean pod borer, *Laspeyresia glycinivorella*, in north-eastern Asia; pod borer, *Etiella zinckerella*, in Indonesia; blister beetles, *Epicauta* spp., and velvet bean caterpillar, *Anticarsia gemmatalis* (Hubn.) in the United States.

IMPROVEMENT

METHODS: These include:

(1) *Introduction of cvs:* Many cvs have been introduced into the United States from the Far East (see CULTIVARS above). These are grown and grouped according to their probable adaptation and use and distributed for further testing. Their photoperiod is important in determining the areas in which they can be grown successfully.

(2) *Selection:* This includes the purification and selection of pure lines from introduced and local cvs.

(3) *Hybridization* in order to combine the superior features of the best cvs; back-crossing is used in the transfer of genes for disease resistance and high oil content. Artificial crossing is tedious because of the smallness of the floral organs. Emasculation and pollination is done on the day before the flowers open and the best results are obtained between 3–7 p.m. Hybridization is now widely employed in the United States.

(4) *Polyploids and irradiation:* Tetraploids have been induced by colchicine treatment. They have shorter thicker stems, longer internodes, shorter broader leaves, larger seeds and pollen, and are later maturing than diploids, to which they are inferior and are of little commercial value. Neutron irradiation produced 228 mutant plants in a population of 4,200. Most of the mutants were curiosities, but some characters such as greater vigour, shatter resistance, changes in oil and protein content, and changes in time to maturity may be of use in breeding.

The United States Regional Soybean Laboratory at Urbana, Illinois, has done much to co-ordinate breeding programmes in the various states with spectacular results.

AIMS AND RESULTS: These include:

(1) *Increased yields:* Yield is determined by the size and number of seeds per plant, number of pods per node, number of seeds per pod, and the percentage of abortive seeds. Average yields in the United States have more than doubled during the past 30 years, largely as a result of improved cvs.

(2) *Maturity:* Time from sowing to harvest is affected by photoperiod and is extremely important for adaptation to a particular latitude. It is correlated with the height of the plant; tall, late-maturing types being dominant. It is essential to test the cvs under local conditions. Cvs have now been developed to cover a wide range of latitudes.

(3) *Freedom from lodging and shattering:* Cvs have been developed which will stand erect without lodging and which will hold the seed when ripe without shattering. This is particularly important for modern combine harvesting.

(4) *Testa colour:* This is unimportant for fodder or green manuring. Yellow seeds are preferred for oil and they usually have a higher oil content. Dark seeds give an undesirable colour to flour and meal.

(5) *Oil content:* This is influenced by the environment and the season. By hybridization cvs with a higher oil content have been developed. In the United States the old standard oil cvs with an oil content of 15–18 per cent have now been almost entirely replaced with cvs having 19–22 per cent. The drying properties of the oil may be important for special purposes. The iodine values range from 99·6–143·2.

(6) *Protein content:* This varies inversely with the oil content. A high protein content may be required for nutritional purposes and for meal for livestock.

(7) *Disease resistance:* This is now receiving attention in the United States and resistance has been obtained to bacterial blight, bacterial pustule, frogeye, and downy mildew.

(8) *Vegetable soya beans:* Numerous introductions have been made into the United States of soya beans suitable for use of the immature seeds as vegetables, but little breeding work has been done.

(9) *Forage-type soya beans:* Very slender-stemmed cvs were selected in the United States prior to 1920 for use as forage and green manures, but the acreage is now declining.

PRODUCTION

The cultivation of soya bean as an oilseed crop is a comparatively recent development, and the crop is now the most important source of vegetable oil. The production of high protein-content meal as a livestock feeding stuff is gaining in importance at the expense of oil, both in the United States and for export to Western Europe. The United States has been the largest producer of soya beans since 1954 and its output has been steadily expanding; it now produces 60 per cent of the world output. The staggering increase in the United States is shown by the following statistics. The acreage has risen from a negligible figure in 1900 to $1\frac{3}{4}$ million acres in 1924, which was mainly for fodder, to $15\frac{1}{2}$ million acres in 1953, and the present acreage is about 30 million acres (1966). Increased mechanization and over 100 per cent increase in yields have been responsible for this remarkable increase. The United States now produces about 20 million tons per annum, of which about 75 per cent is used for local consumption and the remainder exported; Japan, the Netherlands, Spain and Canada being the largest importers of beans and oil. Nearly all the oil used locally in the United States is used for salad oil and in the manufacture of compound cooking fats and margarine, each commodity taking about one-third, or 0·5–0·6 million tons of oil.

The total world acreage is about 70 million acres with a production of over 30 million tons. The second largest producer is China and Manchuria. Other countries producing over 0·1 million tons are Canada, Brazil, Indonesia, Japan, North and South Korea, and the Soviet Union. Nigeria and Tanzania export small quantities.

REFERENCES

MARKLEY, K. S. Ed. (1950). *Soybeans and Soybeans Products*, 2 vols. New York: Interscience Pub.

POEHLMAN, J. M. (1959). *Breeding Field Crops*. New York: Holt.

Lablab Adans. (x = 11, 12)

The genus as now constituted contains 2 spp.

Lablab niger Medik. (2n = 22, 24) HYACINTH BEAN

Syn. *Dolichos lablab* L.; *Lablab vulgaris* Savi.

Also known as bovanist bean, lablab bean, seim bean, lubia bean (Sudan), Indian bean, and Egyptian bean.

USES

The young pods and tender beans are popular vegetables in India and are also used elsewhere in the tropics for this purpose. The ripe and dried seeds are consumed as a split pulse in India. The beans are also sprouted, soaked in water, shelled, boiled and smashed into a paste, which is fried with spices, etc. The haulms, either green or as hay or silage, are used as livestock fodder and the dried seeds are also fed to livestock. The growing crop is also grazed and has the valuable property of remaining green in the dry season. It is also grown as a green manure and cover crop. It is the green manure which is grown in rotation with cotton and sorghum in the Sudan Gezira, and is the last crop in the rotation, after which the land is bare fallowed for a season.

ORIGIN AND DISTRIBUTION

Lablab niger is probably of Asian origin and has been cultivated in India since earliest times. *Var. lignosus* is found wild in some areas in India. The crop was taken to Africa. It has since been distributed to many tropical countries and has become naturalized in some areas.

VARIETIES

Two botanical varieties are recognized in India and are sometimes considered as distinct species. They are:

var. *lablab.* (var. *typicus* Prain.) which is a short-lived, perennial, twining herb, usually treated as an annual. The pods are longer and more tapering, with the long axis of the seeds parallel to the suture. It is grown in India as a garden crop, mainly for the green pods.

var. *lignosus* (L.) Prain (syn. *Dolichos lignosus* L.) a longer-lived, semi-erect, bushy perennial, but is usually treated as an annual. It is sometimes known as the Australian pea. The pods are shorter and more abruptly truncated and the long axis of the seeds is at right angles to the suture. The plants have a characteristic, strong, unpleasant smell. This variety is grown as a field crop in Asia, mainly for the ripe seeds and fodder.

The varieties are confused and doubtfully distinct. They cross naturally and artifically. There is considerable variation in both varieties in the colour of the flowers and seeds, as well as the size of pod and habit of the plant. A tall climbing cv. occurs in Trinidad with wide short pods and with the long axis of the seeds at right angles to the suture.

ECOLOGY

Lablab niger is grown as a dry-land crop. The field crop is hardy and drought-resistant and can be grown in areas with a low rainfall of 25–35 in. It can tolerate poor soils, provided they are well drained. In India and Burma it is often grown on sandy river banks exposed when the monsoon floods subside. The garden cvs require better conditions. The crop can be grown from sea-level to 7,000 ft in Asia. It is photoperiodic and both long- and short-day cvs are said to occur. Short-day cvs in India take 6–47 weeks to flower according to the sowing date.

STRUCTURE

An herbaceous perennial herb, often grown as an annual, 1·5–6 m tall; usually twining, but bushy forms also occur; very variable in colour of stems, foliage, fls, pods and seeds, and in shape of pods. Lvs alternate, trifoliate; petiole slender, laterally compressed, narrowly grooved above, swollen pulvinus at base; stipules and stipels small, triangular to lanceolate; petiolules with thickened pulvini; leaflets ovate, lateral leaflets oblique, entire, 5–15 × 4–15 cm, sub-glabrous or softly hairy, acute or acuminate. Inflorescences axillary, erect, long-stalked, stiff racemes, to 30 cm or more in length, but much shorter in some cvs. Fls arising 1–5 together from prominent tubercular thickenings on peduncle; pedicels short; bracteoles 2–5 mm long; calyx 4-lobed, upper 2 sepals connate; corolla white or purple; standard reflexed, orbicular, emarginate, 1·5 cm long; wings obliquely obovate; keel incurved at a right angle; stamens 10, vexillary stamen free, rest connate; anthers uniform, basifixed; ovary sessile with several ovules; style incurved with soft hairs towards top of upper surface; stigma terminal, glabrous, capitate. Pods variable, obliquely oblong, often curved and flattened, 5–15 × 1–5 cm, curved beak with persistent style; tuberculate along margin, 3–6-seeded. Seeds variable in size and colour, rounded, slightly compressed, white, cream, buff, reddish, brown or black, or variously speckled; hilum white, prominent, raised, extending one-third distance round seeds. Weight of 100 seeds 25–50 g.

POLLINATION

The flowers are visited by insects so some cross-pollination probably occurs. Pollination and seed-setting is reduced in cold weather.

GERMINATION

Germination, which is epigeal, takes about 5 days. The cotyledons are not leaf-like and the first pair of leaves are opposite, simple and cordate.

CHEMICAL COMPOSITION

The immature pods contain about: water 82·4 per cent; protein 4·5 per cent; fat 0·1 per cent; carbohydrate 10·0 per cent; fibre 2·0 per cent; ash 1·0 per cent. The approximate composition of the dry pulse is: water 9·6 per cent; protein 24·9 per cent; fat 0·8 per cent; carbohydrate 60·1 per cent; fibre 1·4 per cent; ash 3·2 per cent.

PROPAGATION

The crop is grown from seed.

HUSBANDRY

The garden cvs in India are given heavy manuring, frequent irrigation and require supports for climbing. They are sown in pits, 6–10 seeds per pit in July or August and are thinned to 4 vines after 1 month. They flower in November and the green pods are harvested from December to March. They are sometimes kept for a 2nd year and begin to yield again in July. The field cvs are frequently intersown with a cereal crop such as *Eleusine coracana* and are rain-fed, receiving very little attention or care. They are planted in June and July and harvested from October to March. The seed rate in pure stand is 50–60 lb per acre. The yield in mixed cultivation is about 400 lb of dried seeds per acre and in pure stand up to 1,300 lb per acre. As green manure or fodder, the crop remains green during the dry season. The yield of green fodder is 5–10 tons per acre.

MAJOR DISEASES AND PESTS

Xanthomonas phaseoli (E. F. Smith) Dows. can cause a severe defoliation in humid weather. A mosaic virus attacks the crop in India, where the most serious pests are a pod borer, *Adisura atkinsoni* Moore, and a stink bug, *Coptosoma cribraria* F.

IMPROVEMENT

Little work has been done on improving the crop. In India a hybrid has been produced between a garden cv. and a field cv., combining the good pod quality of the former with the hardiness of the latter. In Trinidad the 'Waby' bean, named after the selector, has fleshy pale green stringless pods, and is useful as a salad bean. Numerous trials at the Gezira Research Farm in the Sudan for more than 25 years produced no strain good enough to replace the main unselected crop as originally grown at the beginning of the scheme.

PRODUCTION

Hyacinth beans are widely grown in India and south-eastern Asia and in Egypt and the Sudan. They are an important cover crop in the Sudan. Although introduced elsewhere they are of comparatively little importance. No statistics of production are available.

REFERENCE

SCHAAFFHAUSEN, R. V. (1963). *Dolichos lablab* or Hyacinth Bean. *Econ. Bot.*, **17**, 146–153.

Lathyrus L. (x = 7)

About 130 spp. of annual and perennial herbs, mostly of the temperate regions of the northern hemisphere, with a few species in South America and on tropical mountains. Often climbing by means of tendrils; stems winged or strongly angled; lvs even-pinnate, often with terminal tendril; fls solitary or in axillary racemes. Differs from *Vicia* in flattened style, bearded down inner surface and wings nearly or quite free from keel.

A few spp. are grown for food or forage, *e.g. L. sativa* (q.v.). Some are ornamentals. The sweet pea, *L. odoratus* L., and everlasting pea, *L. latifolius* L., can be grown at higher elevations in the tropics.

Fig. 42. **Lablab niger**: HYACINTH BEAN. **A,** leaf (×½); **B,** flower in longitudinal section (×3); **C,** pods (×½); **D,** seeds (×½).

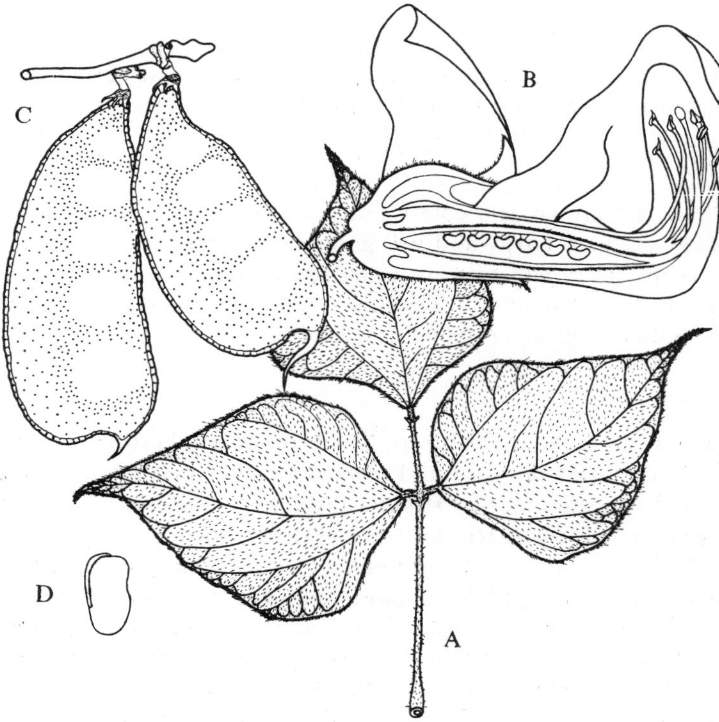

B
C
D
A

Lathyrus sativus L. (2n = 14) GRASS PEA, CHICKLING PEA

USES

Grass pea is the cheapest pulse in India. It occurs as a weed in the barley fields. It is grown for fodder and the seeds are eaten by the poorer classes in India and the Middle East, particularly in times of famine. They are boiled and eaten and made into chapaties, paste balls and curries. The seeds are fed to livestock. The leaves are eaten as a pot-herb. The seeds eaten over a long period can cause lathyrism, a paralysis of the lower limbs in man, but it is possible that this is due to admixture with seeds of *Vicia sativa* L., which occurs as a weed in the grass pea crop. Large white seeds are best for human food and it is advisable to parch and boil carefully.

ORIGIN AND DISTRIBUTION

Lathyrus sativus is a native of southern Europe and western Asia. It is grown in India, but has not been widely distributed in other regions.

ECOLOGY

Grass peas are grown as a cold-weather crop in India and the northern Sudan. They are very hardy and will germinate and grow on land too dry for other crops. They will tolerate waterlogging and a wide range of soil conditions, including poor soil. They are often grown as a catch-crop in the rice fields, both as a grain crop and as a fodder.

STRUCTURE

A procumbent, slender, glabrous, much-branched, herbaceous annual, with a well-developed tap-root and winged stems. Lvs alternate, pinnate, ending in 1–3 delicate tendrils; stipules prominent, leaf-like, ovate, entire, attached to petiole for ¼ length; leaflets 2–4 entire, linear-lanceolate, sessile, 5–7·5 × 1–1·3 cm (frequently a third perfect leaflet is opposed by a tendril). Fls axillary, solitary, about 1·5 cm long; long-peduncled; calyx teeth longer than tube; corolla blue or purple with white keel; stamens diadelphous; anthers uniform; stigma spatulate. Pods oblong, retuse, flattened, 2·5–4 cm long, dorsally 2-winged, 3–5-seeded. Seeds wedge-shaped, angled, white, brown, grey or mottled. 100 seeds weigh about 6 g. Germination hypogeal.

CHEMICAL COMPOSITION

The seeds contain about: water 10·0 per cent; protein 28·2 per cent; fat 0·6 per cent; carbohydrate 58·2 per cent; mineral matter 3·0 per cent.

HUSBANDRY

The crop is usually broadcast with a seed rate of 35–40 lb per acre. It matures in 4–4½ months. Yields of 900–1,000 lb per acre of dry seeds are obtained and 1,200–1,400 lb of hay.

MAJOR DISEASES

The following diseases are reported as causing damage in India: mildews, *Oidium erysiphoides* Fr. and *Peronospora lathyripalustris* Gaumann; rusts, *Uromyces pisi* (Pers.) Wint. and *U. fabae* (Pers.) de Bary; wilt, *Fusarium orthoceras* App. & Wr. var. *lathyri*.

PRODUCTION

Annual production in India is about 0·5 million tons from 4 million acres.

Lens Mill. (x = 7)

About 6 spp. of low, erect or subscandent herbs with pinnate lvs in the Mediterranean region and western Asia. *L. esculenta* (q.v.) is the only cultivated species.

Lens esculenta Moench (2n = 14) LENTIL

Syn. *Lens culinarus* Medik.; *Ervum lens* L.

USES

Lentils are a nutritious pulse. The split seeds (dhal) are used in soups. The percentage recovery of the dhal is 65–80 per cent. Flour from the ground seed is used mixed with cereals in cakes and is also used as invalid and infant food. The husks, bran and dried haulms provide fodder for livestock. The young pods are used as a vegetable in India. The red pottage for which Esau sold his birthright was made of lentils.

ORIGIN AND DISTRIBUTION

Lentils are of very ancient cultivation in Egypt, southern Europe and western Asia, and from these areas spread northwards in Europe, eastwards to India and through much of China, and southwards to Ethiopia. They have now been taken to most subtropical and warm temperate regions and to high altitudes in the tropics. They are now cultivated in Chile and Argentina.

CULTIVARS

The cvs can be divided into two distinct types which are sometimes given subspecific rank:

(1) Large-seed cvs (subsp. *macrospermae* (Baumg.) Barul.) with large pods, generally flat; seeds large, 6–9 mm in diameter, flattened; cotyledons generally yellow, sometimes orange; fls large, white, rarely blue. Mediterranean, Africa and Asia Minor.

(2) Small-seeded cvs (subsp. *microspermae* Barul.) with small convex pods; seeds 3–6 mm, convex; fls small, violet blue to white or pink. Chiefly south-western and western Asia.

ECOLOGY

Lentils are grown as a winter crop in India and are particularly important in northern India and Pakistan. They are grown from sea level to 11,000 ft. They can be grown on a wide range of soils from light loams to black cotton soils and will tolerate moderate alkalinity. They are not suited to the hot wet tropics.

STRUCTURE

An erect, much-branched, softly hairy, light-green, annual herb, with slender stems, 25–40 cm tall. Lvs pinnate, usually ending in a tendril or bristle; stipules linear; stipels absent; leaflets opposite or alternate, 4–7 pairs, sessile, lanceolate, entire, about 1·3 cm long. Inflorescences axillary, 1–4-flowered with slender peduncle; fls small, up to 8 mm long; calyx with 5 narrow lobes; corolla pale blue, white or pink; stamens diadelphous; anthers uniform; style inflexed, bearded along inner surface. Pods oblong, compressed, smooth, broad, rarely more than 1·3 cm long, 1–2-seeded. Seeds bi-convex, small, lens-shaped, grey to light red speckled with black; hilum minute. 100 seeds weigh about 2 g. Germination hypogeal.

POLLINATION

Although usually self-fertilized, cross-pollination can occur.

CHEMICAL COMPOSITION

The whole seed contains about: water 11·2 per cent; protein 25·0 per cent; fat 1·0 per cent; carbohydrate 55·8 per cent; fibre 3·7 per cent; ash 3·3 per cent.

HUSBANDRY

In India the crop is grown in mixed cultivation, sometimes in a standing crop of rice. When grown in pure stand it is either broadcast or planted in rows about 9 in. apart with a seed rate of about 50 lb per acre. The crop matures in about 3½ months. The yields range from 400–600 lb per acre in dry cultivation, but up to 1,500 lb can be obtained with irrigation.

MAJOR DISEASES

The crop is attacked by rust, *Uromyces fabae* (Pers.) de Bary, and a wilt *Fusarium orthoceras* App. & Wr. var. *lentis* Vas. & Srin.

PRODUCTION

The world acreage is estimated at about 3·5 million acres. India is the largest producer with about 1·5 million acres annually, producing about 0·3 million tons. Pakistan is the next largest producer, followed by Ethiopia, Syria, Turkey and Spain.

REFERENCE

BARULINA, H. (1930). Lentils in the U.S.S.R. and other Countries. *Bull. Appl. Bot. Leningrad, Suppl.* **40**.

Pachyrrhizus Rich. ex DC. (x = 11)

Six spp. of perennial herbs with trifoliate lvs and tuberous roots, usually climbing, native of tropical America. *P. erosus* (q.v.) and *P. tuberosus* (q.v.) are cultivated, mainly for their edible tubers. *P. ahipa* (Wedd.) Perodi, a non-climbing species, is cultivated in Bolivia for its small fusiform tubers.

KEY TO THE 2 PRINCIPAL CULTIVATED SPECIES

A. Leaflets palmately lobed or either prominently or obscurely dentate; pods 7·5–13 cm long .... *P. erosus*
AA. Leaflets entire; pods usually 13–30 cm long........... *P. tuberosus*

Pachyrrhizus erosus (L.) Urban (2n = 22) YAM BEAN

Syn. *P. angulatus* Rich. ex DC., *P. bulbosus* (L.) Kurz, *Dolichus erosus* L.

USES

The watery tubers are eaten raw or cooked. The young pods are sometimes eaten like French beans. Starch may be obtained from old tubers. The mature seeds and leaves contain a toxic substance. The roots and mature seeds contain rotenone and may be useful as an insecticide and fish poison (see *Derris*).

ORIGIN AND DISTRIBUTION

P. erosus occurs wild in Mexico and northern Central America. It was cultivated in Central America in pre-Columbian days. It was taken early by the Spaniards from Mexico to the Philippines and had reached Amboina before the end of the 17th century. It is now widely known in south-eastern Asia and is extensively grown by Chinese market gardeners in Singapore and elsewhere throughout the region. It is also cultivated in India and Hawaii. It has become naturalized in southern China and Thailand.

ECOLOGY

Yam beans can be grown successfully in the hot wet tropics. A well-tilled, loose, sandy soil is preferred.

STRUCTURE

An herbaceous twining hirsute herb, 2–6 m tall, climbing or trailing. Tubers solitary or several, simple or lobed, frequently turnip-shaped, with light brown skin and white flesh, to 30 cm in diameter. Stems with tawny hairs. Lvs trifoliate, alternate; stipules linear-lanceolate, 5–10 mm long;

petiole 3–18 cm long; stipels linear to 8 mm long; leaflets ovate or rhomboidal, obscurely or coarsely dentate or 5-lobed, 3–18 × 4–20 cm. Fls in many-flowered axillary racemes, 5–70 cm long, with 1–5 fls borne in dense cluster on short pedicels at each node of peduncle; fls 1·5–2·0 cm long; calyx irregularly 4-lobed; corolla violet or white; standard broad, suborbicular, emarginate, auriculate, about 2 cm long; keel connate at base, recurved, about 2 cm long; stamens diadelphous, anthers uniform; ovary subsessile; style ciliate, recurved; stigma subglobose on ventral surface just below apex. Pods 7·5–14 × 1·1–1·8 cm, flattened, finely strigose, almost smooth at maturity, constricted, 4–12-seeded. Seeds almost square, 5–10 mm in diameter, flattened, yellow, brown or red. Weight of 100 seeds about 20 g.

CHEMICAL COMPOSITION

The edible portion, which comprises 90 per cent of the fresh tuber, contains about: water 87·1 per cent; protein 1·2 per cent; fat 0·1 per cent; carbohydrate 10·6 per cent; fibre 0·7 per cent; ash 0·3 per cent.

HUSBANDRY

Yam beans are grown from seed at a spacing of 3 × 1 ft or on hills 3–4 ft apart with 3–5 seeds per hill. The plants are sometimes staked. If grown for tubers the inflorescences should be removed early and pod and seed formation prevented. The tubers should be harvested after 4–8 months after which they become too fibrous.

Pachyrrhizus tuberosus (Lam.) Spreng. (2n = 22) YAM BEAN, POTATO BEAN

USES

This is the yam bean of South America and parts of the West Indies. It is used in the same way as *P. erosus* (q.v.), but the young pods are not eaten as a vegetable as they have irritant hairs.

ORIGIN AND DISTRIBUTION

P. tuberosus grows in western South America, where it appears to be native in the headwaters of the Amazon River. It is introduced or cultivated in Venezuela, Trinidad, Jamaica, Puerto Rico and China. It was widely distributed by the Royal Botanic Gardens, Kew, in 1889, but does not appear to have been established outside the New World to any extent.

Fig. 43. **Pachyrrhizus erosus**: YAM BEAN. **A**, leaf (× ½); **B**, inflorescence (× ½); **C**, flower from below (× ½); **D**, flower in longitudinal section (× 4); **E**, pod (× ½); **F**, seed (× 2); **G**, young tuber (× ½).

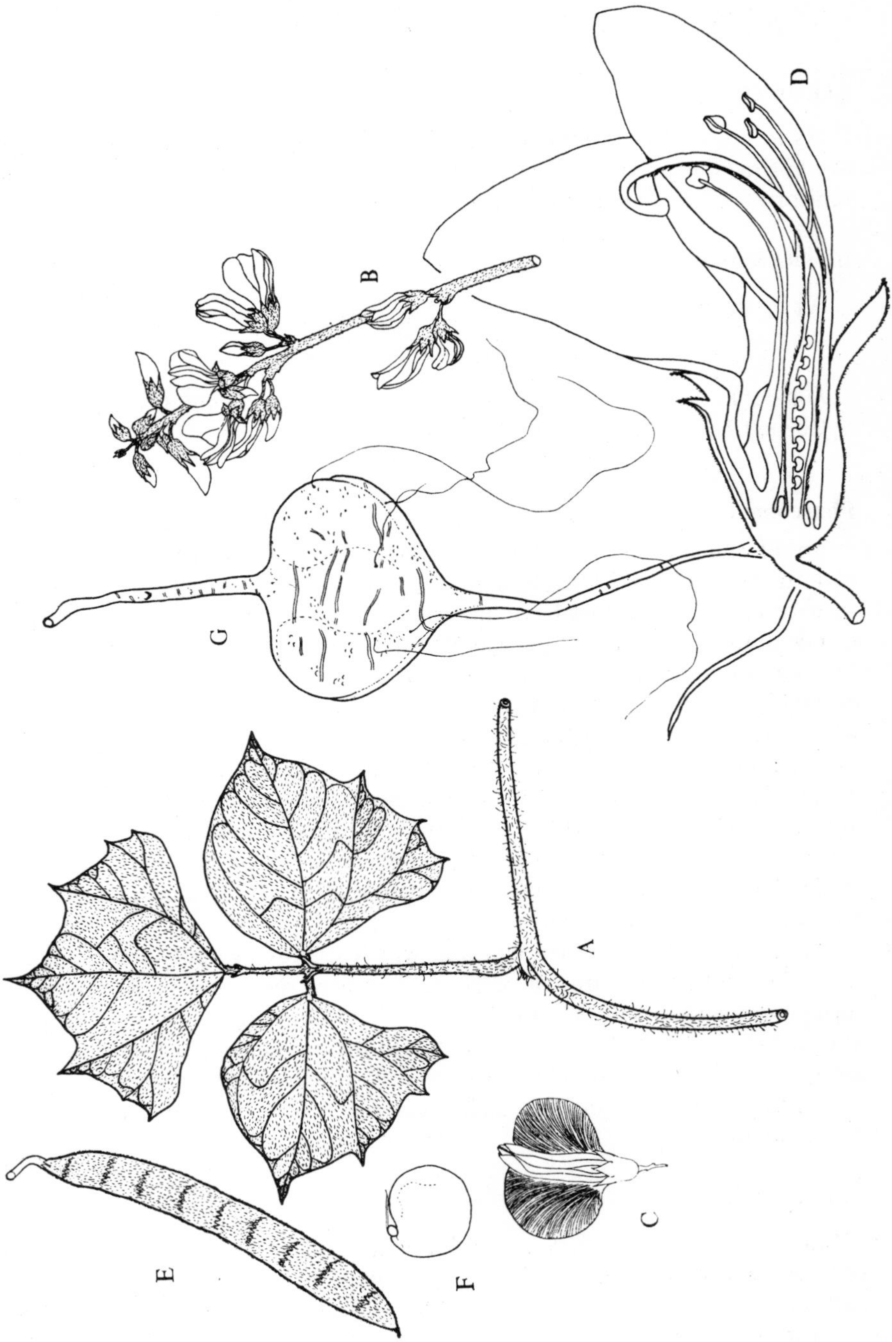
A
B
C
D
E
F
G

ECOLOGY

As *P. erosus.*

STRUCTURE

As *P. erosus,* but the tubers grow larger, the leaflets tend to be large and are entire, the pods are longer and have reddish hairs and the seeds are reniform and are 11–14 mm long.

HUSBANDRY

As *P. erosus.*

REFERENCE

CLAUSEN, R. T. (1944). *A Botanical Study of Yam Beans* (Pachyrrhizus). Cornell Univ. Agric. Exp. Stat., Mem. 264.

Phaseolus L. (x = 11, 12)

About 150 spp. of annuals and perennials throughout the warm regions of both hemispheres. Usually twining herbs, rarely woody at base; some forms erect. Some species have tuberous roots. Lvs pinnately trifoliate, very rarely unifoliate; stipules persistent; stipels small; leaflets usually entire. Fls few to numerous on axillary peduncles; bracts 2, often caducous; bracteoles sometimes persistent; calyx campanulate, 5-toothed; corolla white, yellow, red or purple; standard reflexed; wings same length or longer than standard; keel spirally coiled, which is the distinctive mark of the genus; stamens 10, diadelphous, with free vexillary stamen, of equal length; anthers uniform; style filiform, twisted, bearded on inner curve; stigma oblique. Pod straight or curved, sub-terete or compressed, several to many seeded, dehiscent in wild spp. Seeds variable in size, shape and colour. Germination epigeal or hypogeal.

Morphological and other features which distinguish the domesticated from the wild species are: increase in seed size; reduction in fleshiness of root system and loss of perennialism; reduction of parchment layers of pod and in shattering of pods and violent seed dissemination; reduction in the amount of glucoside phaseolunatin in seeds which is hydrolysed by enzyme action to produce hydrocyanic acid; decrease in impermeability of seeds to water intake.

USEFUL PRODUCTS

Pulses obtained from *Phaseolus* spp. are one of the most important sources of protein in the diet of many tropical people and supplement the carbohydrate staple foods of rice, maize and other cereals. Beans are rich in the amino-acids lysine and tryptophane and these complement the amino-acid zein found in maize, so that food with protein of a high biological value is achieved. In addition to supplying dried pulse, the immature pods and seeds of most of the beans are eaten as vegetables.

The most important and mostly widely cultivated species of *Phaseolus* are *P. vulgaris* (q.v.) and *P. lunatus* (q.v.). *P. aureus* (q.v.) and *P. mungo* (q.v.) are important pulse crops in India. *P. aconitifolius* (q.v.), *P. acutifolius* (q.v.), *P. angularis* (q.v.) and *P. calcaratus* (q.v.) are pulse crops of minor importance. *P. coccineus* (q.v.) is an important summer vegetable in Europe, but can only be grown at higher altitudes in the tropics. A number of species are grown as green manure and cover crops and also for fodder. These include *P. aconitifolius*, *P. aureus*, *P. calcaratus*, *P. lathyroides* L., *P. lunatus*, *P. metcalfei*, Woot. & Standl., and *P. mungo*.

REFERENCES

ALLARD, H. A. and ZAMERER, W. J. (1944). Responses of Beans (*Phaseolus*) and Other Legumes to Length of Day. *Bull. U.S. Dep. Agric. Tech.*, 867.

HARDENBERG, E. V. (1927). *Bean Culture.* New York: Macmillan.

KAPLAN, L. (1965). Archeology and Domestication in American *Phaseolus* (Beans). *Econ. Bot.*, **19**, 359–368.

PIPER, C. V. and MORSE, W. J. (1914). Five Oriental Species of Beans. *Bull. U.S. Dep. Agric.*, 119.

SHAW, E. J. F. and KAHN, A. R. (1931). Studies in Indian Pulses. *Dept. Agric. Ind. Mem. Bot. Ser.*, **19**.

KEY TO PULSES, *Phaseolus* spp.
Based on Hector (1936)

A. Leaflets deeply divided with 3–5 narrow lobes ..... *P. aconitifolius*
AA. Leaflets entire or very occasionally with 2–3 shallow broad lobes
- B. Fls white, cream, violet or red
  - C. Seeds oblong or globular, usually not more than 1·5 cm long, usually less, not broad nor prominently flattened
    - D. Calyx bracts small, inconspicuous, much shorter than calyx........................ *P. acutifolius*
    - DD. Calyx bracts broad, prominent, equalling or exceeding calyx .... *P. vulgaris*
  - CC. Seeds larger and broader, usually more than 1·5 cm long, mostly nearly or as wide as long, often very flat
    - D. Fls large and showy, usually more than 1·3 cm long, scarlet, occasionally white .... *P. coccineus*
    - DD. Fls small, 1·2 cm long or less, white or cream .... *P. lunatus*

BB. Fls clear yellow; beans of oriental origin
C. Plants and pods hairy; seeds usually dull
D. Pods with long hairs; seeds oblong, usually blackish; hilum concave .... *P. mungo*
DD. Pods with short hairs; seeds nearly or quite globular; hilum not concave........... *P. aureus*
CC. Plants and pods glabrous or nearly so; seeds smooth and shiny
D. Pods constricted between seeds; hilum not concave .... *P. angularis*
DD. Pods not constricted between seeds; hilum concave .... *P. calcaratus*

Phaseolus aconitifolius Jacq. (2n = 22) MAT OR MOTH BEAN

Syn. *P. trilobus* Ait.; *Vigna aconitifolius* (Jacq.) Verdc.

USES

In India the green pods are eaten as a vegetable and the ripe seeds, whole or split, are eaten cooked. It is grown for forage and hay for livestock and is a useful green manure.

ORIGIN AND DISTRIBUTION

P. aconitifolius is a native of India, Pakistan and Burma where it grows wild and is also cultivated. It spread in cultivation to Ceylon and China. It was first introduced into the United States in 1902 and is grown for fodder in Texas and California.

ECOLOGY

It is grown as a hot-season crop in India from sea-level to 4,000 ft. It requires high uniform temperatures and is very drought-resistant. It keeps alive with little water and does best with a well-distributed rainfall of 30 in. per annum. Heavy rain is harmful. It can be grown on many soil types, but dry light sandy soils are particularly suitable. It is a short-day plant.

STRUCTURE

A slender, trailing, hairy, annual herb, 10–30 cm tall. Easily distinguished from other cultivated *Phaseolus* spp. by its divided leaflets. Up to 12 trailing primary branches, 60–130 cm long, are produced from short erect main stem and about 25 secondary branches, 60 cm long, and 20 tertiary branches, 30 cm long. Lvs alternate, trifoliate; petiole grooved, 5–10 cm long; stipules peltate, 1·2 cm long, with lanceolate lobes; stipels small; leaflets 5–8 cm long, with slightly larger terminal leaflet with 5 acuminate lobes and lateral leaflet with 4 lobes. Inflorescences axillary; peduncle 5–10 cm long; racemes capitate with several small yellow fls, about 9 mm long. Pods small, 2·5–5 × 0·5 cm, nearly cylindrical, brown with short stiff hairs, short curved beak, 4–9-seeded. Seeds small, rectangular, 5 mm long, yellow to brown or mottled black; hilum linear, white. 100 seeds weigh about 1 g.

POLLINATION

The flowers are normally self-fertilized.

GERMINATION

Germination, which is epigeal, is rapid at 80°F.

CHEMICAL COMPOSITION

The mature seeds contain approximately: water 9·3 per cent; protein 23·0 per cent; fat 0·7 per cent; carbohydrate 59·0 per cent; fibre 4·0 per cent; ash 4·0 per cent. The composition of hay made from the whole plant is: water 5·6 per cent; protein 17·4 per cent; fat 2·5 per cent; carbohydrate 43·5 per cent; fibre 16·7 per cent; ash 14·3 per cent.

HUSBANDRY

In India the crop is usually interplanted with cereals. The spacing recommended in pure stand is 2½–3 ft between rows with a seed rate of 3–4 lb per acre planted at a depth of 1–1½ in. A good seedbed is required. Yields in the United States are 15–20 tons per acre of green matter, 3–4 tons per acre of hay, and 1,100–1,600 lb per acre of seed. In Madras the crop is rotated as a green manure with cotton.

PRODUCTION

The crop is grown in India for food. In the south-western United States it is grown for pasture, fodder and green manure.

REFERENCE

KENNEDY, P. B. and MADSON, B. A. (1925). The Mat Bean, *Phaseolus aconitifolius*. *Bull. Calif. agric. Exp. Sta.*, 396.

Phaseolus acutifolius Gray var. **latifolius** Freem. (2n = 22) TEPARY BEAN

USES

Tepary beans are used mainly as dry shelled beans. They have been tried as a hay and cover crop in the United States.

ORIGIN AND DISTRIBUTION

P. acutifolius occurs wild in Arizona and north-western Mexico. It was taken into cultivation in Mexico some 5,000 years ago and archaeological remains of this date have been found at the Tehuacan caves, but were later replaced by *P. vulgaris* (q.v.) at this site. They are now cultivated in Arizona and north-western Mexico. Tepary beans have been introduced into various

African territories and they may be useful as a catch-crop and where a rapid food supply is needed in areas of poor rainfall. As yet they have not been much grown outside their homeland. The present distribution of the tepary bean in the New World is a much contracted relic one, having been largely replaced by the common bean.

CULTIVARS

Freeman (1918) isolated and grew 47 cvs in Arizona, differing in colour, shape and size of bean. The variety cultivated is var. *latifolius*.

ECOLOGY

Tepary beans are particularly suited to arid regions and can withstand heat and a dry atmosphere. They will produce a crop when other beans would fail. They are very susceptible to waterlogging and frost and are not suited to the wet tropics. They are short-day plants.

STRUCTURE

A sub-erect annual to 25 cm high; bushy on poor land, otherwise recumbent, spreading or twining. First pair of leaves simple, about 5·5 × 3·5 cm, being narrower than *P. lunatus* and *P. vulgaris*, with truncate base (cordate in other 2 spp.); petioles shorter, 4–5 mm long. Mature leaves trifoliate, usually glabrous; petioles 2–10 cm long; stipules lanceolate, appressed to stem; leaflets ovate, 4–8 × 2–5 cm, acute, entire. Inflorescences axillary with 2–5 white or pale lilac fls; calyx 4-lobed, 3–4 mm long; standard half-reflexed, broad, emarginate, 8–10 mm long; wings 10–15 mm long; keel narrow with 2–3 turns to spiral. Pods compressed, 5–9 × 0·8–1·3 cm, 2–7-seeded (average 5-seeded), rimmed on margins, sharp beaked, hairy when young. Seeds roundish to oblong, about 8 × 6 mm; average weight 0·15 g (0·23 g in *P. vulgaris*; 0·5 g in *P. lunatus*), not glossy, white, yellow, brown or deep violet, either self-coloured or variously flecked.

POLLINATION

Not known with certainty, but presumably self-pollinated.

GERMINATION

The seeds absorb water very easily. In moist soil the testa wrinkles within 5 minutes; in water in 3 minutes. Germination is epigeal.

CHEMICAL COMPOSITION

The dried beans contain about: water 9·5 per cent; protein 22·2 per cent; fat 1·4 per cent; carbohydrate 59·3 per cent; fibre 3·4 per cent; ash 4·2 per cent.

HUSBANDRY

The seed rate is 25–30 lb per acre when broadcast and 10–15 lb per acre when planted in rows 3 ft apart and with plants 6 in. apart in the rows. The seeds are planted at a depth of 2–4 in. They are sometimes planted on hills 18 in. apart with 3–4 seeds per hill. Tepary beans will give a crop in as short a time as 2 months. In dry farming yields are 450–700 lb per acre, but under irrigation 800–1,500 lb per acre are obtained. When grown for fodder 5,000–9,500 lb per acre of oven-dry hay is produced and for this purpose it is sown at a seed rate of 60 lb per acre.

PRODUCTION

Growing of tepary beans is mainly confined to Mexico and Arizona.

REFERENCE

FREEMAN, G. F. (1918). Southwestern Beans and Teparies. *Univ. Arizona Agric. Exp. Stat. Bull.*, 68. Revised.

Phaseolus angularis (Willd.) Wight* (2n = 22) ADZUKI BEAN

USES

In Japan and China the dried pulse is used as human food, either cooked whole or made into meal which is used in soups, cakes and confections. It is also grown for herbage.

ORIGIN AND DISTRIBUTION

P. angularis is probably a native of Japan, but has long been established in China and Sarawak. It has been introduced into the southern United States, South America and the Congo.

ECOLOGY

In the Congo it is said to tolerate high temperatures. It is fairly drought-resistant and is susceptible to waterlogging. The climatic requirements are similar to soya beans (q.v.). In Japan it is grown in rotation with rice. It is a short-day plant.

STRUCTURE

A summer annual, usually bushy and erect, 25–75 cm tall; some of the later-maturing cvs are slightly vining and some are prostrate. Lvs trifoliate; petiole long; leaflets ovate, usually entire, 5–9 cm long. Inflorescences axillary, short, of 6–12 clustered bright yellow fls on short pedicels. Pods cylindrical, 6–12 × 0·5 cm, commonly straw-coloured, but blackish and brown forms occur, 5–12-seeded, somewhat constricted between seeds. Seeds oblong, seldom flattened, about 8 × 4 mm, maroon, straw, brown or black in colour, with long white hilum. 100 seeds weigh 10–20 g.

* Now usually referred to as *Vigna angularis* (Willd.) Ohwi & Ohashi.

POLLINATION

Flowers are self-fertile when bagged, but cross-pollination between cvs is frequent.

GERMINATION

Germination is hypogeal. The seeds retain their viability for over 2 years.

CHEMICAL COMPOSITION

The dry pulse contains about: protein 21–23 per cent; fats 0·3 per cent; carbohydrate 65 per cent.

HUSBANDRY

The crop is usually planted at a spacing of 3 × 1 ft with a seed rate of 20–25 lb per acre, but a closer spacing of 1 × 1 ft is given in the Congo. The time to maturity is 3–5 months. Yields range from 400–1,000 lb per acre.

PRODUCTION

Adzuki bean is mainly grown in Japan and China. In Japan it is the second most important pulse crop to soya beans.

REFERENCE

PIPER, C. V. and MORSE, W. J. (1914). Five Oriental Species of Beans. *Bull. U.S. Dep. Agric.*, 119.

Phaseolus aureus Roxb. (2n = 22) GREEN OR GOLDEN GRAM, MUNG

Syn. *V. radiata* (L.) Wilczek var. *radiata*

USES

P. aureus is an important crop in India, where it is 'esteemed as the most wholesome among the pulses, free from the heaviness and tendency to flatulence' which is associated with other pulses. The dried beans are boiled and are eaten whole or after splitting into dhal. They are parched and ground into flour after removal of the testa, the flour being used in various Indian and Chinese dishes. The green pods are eaten as a vegetable. In China and the United States it is used for bean sprouts. The beans are soaked overnight, drained and placed in containers in a dark room. They are sprinkled with warm water every few hours and the sprouts are ready in about a week. One pound of dry beans gives 6–8 lb of sprouts. The haulms are used as fodder and the husks and split beans are a useful livestock food. The crop is also grown for hay, green manure and as a cover crop.

SYSTEMATICS

P. aureus is much confused in the older literature with *P. mungo* L. (q.v.) and *P. radiatus* L. *P. aureus* can be distinguished from *P. mungo* by the spreading or reflexed pods with short hairs, the globose seeds, and flat hilum. In *P. mungo* the pods are erect or sub-erect with long hairs, the seeds are larger, oblong and smooth and the hilum is concave.

ORIGIN AND DISTRIBUTION

The crop is of ancient cultivation in India and the plant is not found in a wild state. It is probably derived from *P. radiatus* L., which occurs wild throughout India and Burma, and which is occasionally cultivated. *P. aureus* was an early introduction into southern China, Indo-China and Java. It has been introduced in comparatively recent times into East and Central Africa, the West Indies and the United States.

CULTIVARS

Various cvs are recognized differing in habit, height, period to maturity, colour of pods, and size and colour of seeds. The cvs can be divided into two main types depending on the colour of the seeds: (1) golden gram with yellow seeds; not a prolific seed producer and with a tendency to shatter; used mainly for pasture, hay, silage and as a cover crop, but good pulse cvs occur in India; (2) green gram with dark or bright green seeds, the latter being used for sprout production; the cvs seed more prolifically and the pods ripen more uniformly and have less tendency to shatter; they are more commonly planted for the production of pulse.

ECOLOGY

The crop is grown in India from sea-level to 6,000 ft, usually as a dry-land crop following rice. It does best on a good loam with a well-distributed rainfall of 30–35 in. per year. It is drought-resistant and is susceptible to waterlogging. Both short- and long-day cvs are found in India.

STRUCTURE

An erect or sub-erect, deep-rooted, much-branched, rather hairy, annual herb, 0·5–1·3 m tall, sometimes with a slight tendency to twining at the tips. Lvs alternate, trifoliate, dark or medium green; petiole long; stipules ovate; leaflets ovate, 1·5–12 × 2–10 cm. Inflorescence an axillary raceme; peduncle 2–13 cm long, with 10–20 fls clustered at top; calyx bracts ovate, as long as calyx; corolla yellow, standard 1–1·7 cm in diameter. Pods grey or brownish when mature, long and slender, 5–10 × 0·4–0·6 cm, spreading, reflexed, with short hairs, 10–15-seeded. Seeds small, globular, usually green, but sometimes yellow or blackish; hilum round, white, flat; testa with fine wavy ridges. 100 seeds weigh 3–4 g.

POLLINATION

The flowers are fully self-fertile when bagged and are almost entirely self-pollinated. The flowers are produced 6–8 weeks after sowing. Pollen is shed in the bud the evening before the flowers open, which takes place the following morning and they fade the same afternoon. Up to 42 per cent cleistogamy occurs. There is poor seed set if rain occurs during flowering.

GERMINATION

Germination is epigeal.

CHEMICAL COMPOSITION

The dry seeds contain about: water 9·7 per cent; protein 23·6 per cent; fat 1·2 per cent; carbohydrate 58·2 per cent; fibre 3·3 per cent; ash 4·0 per cent.

HUSBANDRY

The crop requires a good tilth. It is sown broadcast or in rows with a seed rate of 12–15 lb per acre. In the United States it is planted in rows 1½–3 ft apart with a seed rate of 5–8 lb per acre. The crop matures in 80–120 days. Yields of 400–500 lb of dried beans per acre are obtained, but over 1,000 lb per acre has been recorded in the United States. The pods have a tendency to shatter. Yields of air-dried hay in the United States are 1–2·5 tons per acre for golden gram, but less for green gram.

MAJOR DISEASES AND PESTS

P. aureus is not susceptible in the United States to many of the diseases which attack the common bean, soya bean and cowpea. It is attacked by nematodes and root rot caused by *Sclerotium rolfsii* Sacc.

IMPROVEMENT

A little sorting out of cvs has been done in India and the United States.

PRODUCTION

Large acreages are grown in India for local consumption. During World War II when sources from the Orient were cut off there was a large increase in production in the United States with a maximum of 110,000 acres in 1945, mainly for sprout production, but this acreage has now been much reduced. About 25 million lb of beans are used annually in the United States.

Fig. 44. **Phaseolus aureus:** GREEN GRAM. **A,** flowering shoot (×½); **B,** flower in longitudinal section (×3); **C,** pod (×½); **D,** seed (×3).

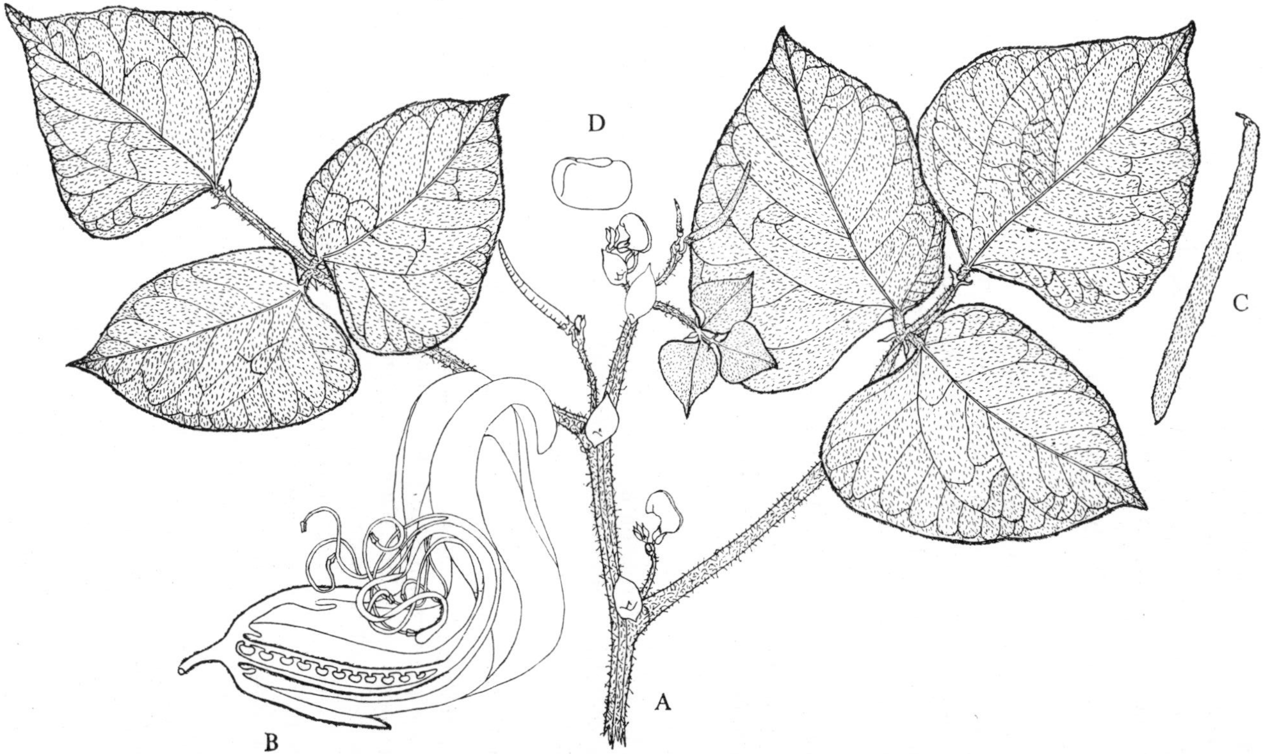
D
C
A
B

REFERENCES

BOSE, R. D. (1932). Studies in Indian Pulses: 4. Mung or Green Gram (*Phaseolus radiatus* L.). *Indian. J. agric. Sci.*, **2**, 607–624.

LIGON, L. L. (1945). Mung Bean. A Legume for Seed and Forage Production. *Bull. Okla. agric. Exp. Sta.*, 284.

Phaseolus calcaratus Roxb.* (2n = 22) RICE BEAN

USES

The dried pulse is used for human consumption in India, Burma, Malaysia, China, Fiji, Mauritius and the Philippines. The beans are usually boiled and eaten with or instead of rice. The young pods and leaves are used as a vegetable. The whole plant is used as fodder. It has been tried as a green manure and cover crop.

ORIGIN AND DISTRIBUTION

P. calcaratus occurs wild from the Himalayas and central China to Malaysia and has spread in cultivation in other parts of tropical Asia.

ECOLOGY

The wild forms grow on open sites and roadsides. The cvs can tolerate high temperatures and they are moderately drought-resistant. It is a short-day plant.

STRUCTURE

A short-lived, erect to sub-erect or twining annual, 1·5–3 m tall. Stems grooved with short white hairs. Lvs trifoliate; petiole 5–10 cm long; stipules conspicuous, ovate-lanceolate, attached below middle; stipels linear-lanceolate; leaflets ovate, 5–10 × 2·5–6 cm, usually entire, occasionally faintly 3-lobed. Inflorescence an erect axillary raceme; peduncle 7·5–20 cm long, with 5–20 fls. Fls 2–3 together at nodes; bracteoles linear-lanceolate, twice as long as calyx; calyx 4 mm long with short deltoid teeth; corolla yellow, much contorted; standard 1·5–2 cm in diameter. Pods long and slender, sub-terete, glabrous, shattering, 6–12 × 0·5 cm, 8–12-seeded. Seeds oblong with rounded ends, about 8 mm long, yellow, red, brown, black or speckled; hilum straight, concave, white. 100 seeds weigh 8–12 g.

POLLINATION

The flowers are self-fertile.

GERMINATION

This is hypogeal.

* Now usually referred to as *Vigna umbellata* (Thunb.) Ohwi & Ohashi; syn. *V. calcarata* (Roxb.) Kurz.

CHEMICAL COMPOSITION

The pulse contains about: water 10·5 per cent; protein 21·7 per cent; fat 0·6 per cent; carbohydrate 58·1 per cent; fibre 5·2 per cent; ash 3·9 per cent.

HUSBANDRY

In Burma the crop is usually grown in rotation with rice. It is usually broadcast at a seed rate of 60–80 lb per acre. The time to maturity is as little as 2 months. Yields of 200 lb or more per acre are obtained.

PRODUCTION

The crop is cultivated to a limited extent in south-eastern Asia.

Phaseolus coccineus L. (2n = 22) SCARLET RUNNER BEAN

Syn. *P. multiflorus* Willd.

USES

In temperate countries it is usually grown for its tender pods which are sliced and cooked. In Central America the green and dry seeds are eaten. It is occasionally grown as an ornamental. The fleshy tubers are boiled and eaten in Central America.

ORIGIN AND DISTRIBUTION

This bean occurs wild at altitudes of about 6,000 ft in the uplands of Chiapas and Guatemala in Central America, and perennial forms were domesticated and are still cultivated in this area. It is cultivated to a more limited extent in the Central Plateau of northern Mexico and in Costa Rica, Panama and Colombia. Beans dated 7000–5000 B.C. found in Mexico were probably from wild plants. Those at Tehuacan dated 200 B.C. are domesticates and were probably traded in from the humid uplands. They have now been widely distributed in temperate countries.

ECOLOGY

The scarlet runner bean is a plant of the humid uplands in the tropics. It does not set fruit in the tropical lowlands. It is a long-day plant and this, together with the fact that it is less sensitive than most *Phaseolus* spp. to cool summers, explains why it is so successful in Britain. It is killed by frost and consequently is grown as an annual in temperate countries.

STRUCTURE

Twining perennial to 4 m or more in height with thickened tuberous roots. Lvs trifoliate; leaflets ovate, 7·5–12·5 cm long. Fls 1·8–2·5 cm long, several to many on long axillary peduncles, usually bright scarlet, occasionally white. Pods 10–30 cm long. Seeds 1·8–2·5 × 1·2–1·6 cm, broad-oblong, not tapering at ends, convexly flattened, dark purple with red markings, rarely white. Germination hypogeal.

HUSBANDRY

In Central America they are often interplanted with maize. Sprouts from the tuberous roots often take over the maize fields for the first year or two of the fallow period. They are also planted in plots adjacent to houses and are treated as perennials with a life span of from two to several years. In temperate countries they are grown as annuals and are given long supports.

PRODUCTION

Outside Central America scarlet runner beans are seldom found in the tropics, although they are occasionally grown by Europeans at the higher altitudes.

REFERENCE

KAPLAN, L. (1965). Archeology and Domestication in American *Phaseolus* (Beans). *Econ. Bot.*, **19**, 353–368.

Phaseolus lunatus L. (2n = 22) LIMA OR SIEVA BEAN

Syn. *P. limensis* Macf.; *P. inamoenus* L.

It is also known as butter bean, Madagascar bean and Burma bean.

USES

P. lunatus is grown for dried shelled beans. Care is required in preparation as some cvs contain dangerous quantities of hydrocyanic acid. This is dissipated by boiling and changing the cooking water. The green shelled beans are cooked as a vegetable and the young pods and leaves are sometimes used for this purpose. The green beans are used for canning and freezing in the United States.

SYSTEMATICS

Some authorities separate it into two species namely:

P. limensis (syn. *P. lunatus* f. *macrocarpus*), the lima bean, which was considered perennial and has large seeds which are usually white.

P. lunatus, the sieva bean, which is annual and has smaller flattened or plump seeds of various colours.

However, perennial and annual forms, climbing and erect forms, all seed colours, sizes and shapes occur in both groups, and, as the species are completely interfertile, it is better to follow Mackie (1943) and regard it as a single species, *P. lunatus*.

ORIGIN AND DISTRIBUTION

Mackie (1943) considers that the centre of origin is in the Guatemala region of Central America, where wild endemic forms occur and that it was taken along the Indian trade routes in three directions, namely: the Hopi or northern branch through Mexico to the southern United States (small limas); the Inca or southern branch through Central America to Peru (large white limas); the Carib or West Indian branch across the Gulf of Mexico to the West Indies and thence to Brazil (tropical, perennial cvs which are short-day plants and the beans have the highest content of hydrocyanic acid).

However, since that time large lima beans have been found in excavations in Peru, dated 6000–5000 B.C., while the earliest record in Mexico of the small lima bean is 500–300 B.C. Furthermore, wild types of *P. lunatus* are found in other parts of Central America and in the Andes from Peru to Argentina. Thus, as Heiser (1965) points out, it seems probable that there was separate domestication in Central and South America from conspecific geographic races. Kaplan (1965) considers that the big lima of Peru may have first been domesticated east of the Andean highlands in warm humid lands and that the small limas of Mexico may have arisen in the Pacific coastal foothills of Mexico.

In post-Columbian times lima beans have been widely distributed, particularly in the tropics. Carib-type beans were taken by the Spanish galleons across the Pacific to the Philippines and from there spread to Asia. The slave trade took them from Brazil to Africa. All modern large lima beans trace back to Peru and they were taken at an early date to Madagascar. It is now the main pulse crop in the wet forest regions of tropical Africa, and they are widely grown in Burma. Lima beans have escaped from cultivation and maintain themselves in a wild state in many tropical countries.

CULTIVARS

The cvs can be divided into small annual bush limas and tall climbing pole limas, which usually require a support. Bush lima cvs recommended for the tropics are 'Fordhook 242' and 'Burpee Bush'. In the United States 'Fordhook 242', a potato type with large thick seeds, and 'Henderson', a baby lima with small, thin seeds, are the principal cvs grown in the United States for processing and marketing of fresh beans. Pole limas, recommended for the tropics are 'King of the Garden', which is the most popular home-garden cv. in the United States, 'Carolina', which is the sieva or butter bean of the southern United States, and 'Florida Speckled Butter'. For the production of dry lima beans in California 'Ventura', the standard lima with large thin beans, and 'Wilbar' and 'Western', the dry baby limas with small thin seeds are the principal cvs grown; they are all climbing types but are allowed to run on the ground.

ECOLOGY

Lima beans are grown from sea level to 8,000 ft in the tropics. They require a growing season free from frost and dry weather for maturing the seeds. They will tolerate wetter weather during the growing season than *P. vulgaris*. Some cvs fail to set seed during very hot weather and seed set is reduced above 80°F. Sieva or the small lima group are more resistant to hot arid conditions than the large lima group. Pole cvs require a longer season than bush cvs. They are fairly drought-resistant and need well-drained, well-aerated soil; they are sensitive to high acidity and a pH of 6–7 is preferred. Some cvs and wild plants of the Carib type are short-day plants; the other groups are day neutral. Lima beans are not as tolerant of adverse conditions as many of the other cultivated species of *Phaseolus*.

STRUCTURE

Wild plants and pole cvs are twining, perennial herbs, 2–4 m tall, with enlarged rootstocks; annual and small bush forms, 30–90 cm high, have been developed in cultivation. Lvs trifoliate; petiole 8–17 cm long; stipules minute, basifixed, broadly triangular; stipels minute, sessile; leaflets ovate, acuminate, usually shortly hairy below, 5–12 × 3–9 cm, lateral leaflets oblique. Inflorescence an axillary raceme to 15 cm long, many-flowered, with 2–4 fls at each node; bracts small; calyx small, campanulate with very short teeth; corolla small; standard usually pale green, occasionally violet, 0·7–1 cm broad; wings white; keel prolonged into a complete spiral; stamens 10, diadelphous with free vexillary stamen; style coiled. Pods oblong generally recurved, 5–12 × 1·5–2·5 cm, 2–4-seeded. Seeds very variable in size, shape and colour, 1–3 cm long; ranging from flat-seeded types to rounded seeds of potato types; white, cream, red, purple, brown or black, self-coloured or mottled; hilum white with translucent lines radiating from it to outer edge of testa. Weight of 100 seeds 45–200 g.

POLLINATION

Lima beans are normally self-fertilized, but natural cross-pollination can and does occur. Up to 18 per cent cross-pollination has been recorded in Burma.

GERMINATION

Germination, which is epigeal, does not take place at a soil temperature below 60°F.

Fig. 45. **Phaseolus lunatus: LIMA BEAN. A**, leaf (× ½); **B**, flower (×4); **C**, flower in longitudinal section (×4); **D**, pod (× ½); **E**, seed (× ½).

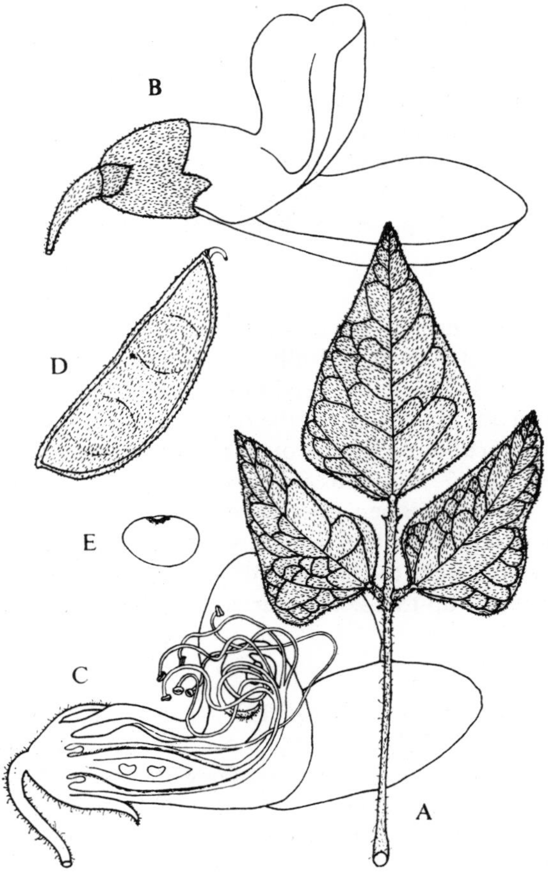
B
D
E
C
A

CHEMICAL COMPOSITION

The dried pulse contains about: water 12·6 per cent; protein 20·7 per cent; fat 1·3 per cent; carbohydrate 57·3 per cent; fibre 4·3 per cent; ash 3·8 per cent. The green beans contain about: water 66·5 per cent; protein 7·5 per cent; fat 0·8 per cent; carbohydrate 22·0 per cent; fibre 1·5 per cent; ash 1·7 per cent.

The mature beans contain the glucoside phaseolutanin, which gives them their characteristic flavour. Under damp conditions or when the tissues are broken down by chewing or grinding, an enzyme present in the seeds causes the liberation of hydrocyanic acid. The enzyme is destroyed by heat during boiling. Large lima beans of the Inca type and small limas of the Hopi type contain 25–55 p.p.m. HCN which is far below the tolerance set by law in the United States of 100 p.p.m. HCN. Carib type beans may contain dangerous quantities of HCN and a wild lima from Puerto Rico gave 997 p.p.m. HCN. The HCN content is greater in coloured beans; white-seeded types are relatively free, but over 100 p.p.m. HCN has been recorded in a few cases.

HUSBANDRY

In the United States bush cvs are sown at 28–36 × 2–8 in. and pole cvs in hills 3–4 ft apart with 3–4 seeds per hill for training on poles. The seeds are sown at a depth of 1–2 in. with a seed rate of 120–150 lb per acre for the large-seeded cvs and 50–70 per acre for the small-seeded cvs. Early maturing cvs are harvested from about 100 days onwards. Large lima cvs of Peru and Madagascar take up to 7–9 months and give a crop over a long period. Green beans are harvested when they have attained full size but before turning colour; they can be kept up to 11 days at a temperature of 28–32°F. Dry beans are harvested when the majority of pods are fully mature and have turned yellow. Yields of up to 1,200 lb per acre of dry beans are obtained.

MAJOR DISEASES AND PESTS

Downy mildew, *Phytophthora phaseoli* Thaxt., and pod blight, *Diaporthe phaseolorum* (Cook & Ellis) Sacc., are the most destructive diseases of lima beans and do not attack *Phaseolus vulgaris*. They are attacked by root-knot nematodes, *Meloidogyne* spp. Pod borers and leaf hoppers are a limiting factor to production in parts of the tropics. For other diseases and pests see *Phaseolus vulgaris*.

IMPROVEMENT

Selection of cvs has been done in the United States, particularly in California, for early maturity, vigour, heat-resistance, etc. Some success has been obtained in resistance to root-knot nematodes and certain diseases. The following characters are dominant: twining habit over bush; coloured testa over white; mottled testa over self-colour; flat thin seeds over fat potato types; coloured corolla over white.

PRODUCTION

Lima beans are an important crop in the United States for canning and freezing and for dry bean production. White butter beans are exported from Madagascar. They are grown as pulse crop in some tropical countries, particularly in parts of Africa and in Burma.

REFERENCES

HEISER, C. B. (1965). Cultivated Plants and Cultural Diffusion in Nuclear America. *Amer. Anthropologist*, **67**, 930–949.

KAPLAN, L. (1965). Archeology and Domestication in American *Phaseolus* (Beans). *Econ. Bot.*, **19**, 358–368.

MACKIE, W. W. (1943). Origin, Dispersal and Variability of the Lima Bean, *Phaseolus lunatus*. *Hilgardia*, **15**, 1–29.

Phaseolus mungo L. (2n = 22, 24) BLACK GRAM, URD

Syn. *Vigna mungo* (L.) Hepper

In the West Indies known as woolly pyrol.

USES

Black gram is one of the most highly prized pulses of India, more particularly in the vegetarian diet of high caste Hindus. It is boiled and eaten whole or after splitting into dhal. It is parched and ground into flour which is made into balls with spices, eaten as porridge, or baked into bread and biscuits. The green pods are eaten as vegetables. It is used as a green manure and cover crop and as a short-lived forage. The hulls and straw are used as cattle feed. The forage and hay are rather hairy and inferior to *P. aureus* (q.v.).

SYSTEMATICS

See *P. aureus* above.

ORIGIN AND DISTRIBUTION

P. mungo is of very ancient cultivation in India and is not known in a wild state. It probably originated from *P. trinervius* Heyne or *P. sublobatus* Roxb. which occur wild in India. It has been introduced in recent times elsewhere in the tropics, mainly by Indian immigrants, and in some countries as a green manure.

CULTIVARS

Various cvs are recognized in India and may be divided into early-maturing cvs with rather larger black seeds and late-maturing cvs with smaller olive-green seeds. A late-maturing cv. with a dense mass of herbage is grown as a green manure in the West Indies under the name of woolly pyrol.

ECOLOGY

Black gram is grown in India from sea level to 6,000 ft. It is drought-resistant and is grown in areas with a rainfall of not more than 35 in. per year. It is not suitable for the wet tropics and in areas with heavy rainfall it is grown in the season following the cessation of the rains. It is particularly suited to clay soils and is often grown on black cotton soils in India.

STRUCTURE

An erect or sub-erect, diffusely branched herbaceous annual, 20–80 cm high, occasionally trailing, with reddish-brown hairs. Lvs trifoliate; petioles long; stipules ovate; stipels falcate; leaflets ovate to lanceolate, entire, 5–10 cm long. Inflorescence axillary and may have 2–3 branches; fls 5–6 clustered at top of short hairy peduncle which elongates in fruit; bracteoles longer than calyx; calyx lobes linear; corolla pale yellow; standard 12–16 mm wide; keel spirally coiled with horn-like appendage; stamens diadelphous; style spirally twisted. Pods erect or sub-erect, buff to dark brown when mature, rounded, 4–7 × 0·6 cm, very hairy, 6–10-seeded, with short hooked beak. Seeds oblong with square ends, 4 mm long, mostly black, but green forms occur; testa smooth without ridges; hilum white, concave. 100 seeds weigh about 4 g.

POLLINATION

Flowering commences after 6 weeks. The flowers are self-pollinated and no case of cross-pollination has been recorded, adjacent cvs maintaining their purity. Pollination takes place in the bud in the evening. The flowers open in the following morning and fade in the afternoon. Up to 42 per cent cleistogamy has been recorded.

GERMINATION

This is epigeal.

CHEMICAL COMPOSITION

The dried pulse contains approximately: water 9·7 per cent; protein 23·4 per cent; fat 1·0 per cent; carbohydrate 57·3 per cent; fibre 3·8 per cent; ash 4·8 per cent.

HUSBANDRY

In India black gram is grown both as a summer and winter crop. It is often grown in rotation with rice and sometimes in mixed cultivation. It is

---

Fig. 46. **Phaseolus mungo :** BLACK GRAM. **A,** leaf (× ½); **B,** flower in longitudinal section (× 3); **C,** pods (× ¾); **D,** seed (× 1).

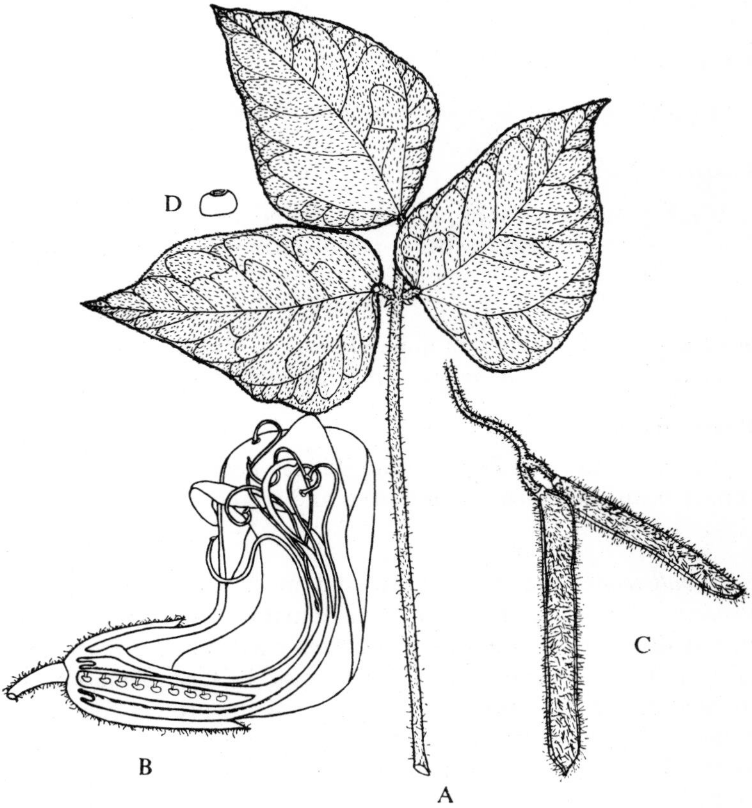
D
B
A
C

usually planted on a rough seedbed as too much tilth encourages vegetative growth at the expense of seed. It is sown either broadcast or in rows 10 in. apart with a seed rate of 10–12 lb per acre. The period to maturity is 80–120 days and yields of 400–500 lb per acre of dried seed are obtained.

MAJOR DISEASES AND PESTS

In India the crop is attacked by: downy mildew, *Erysiphe polygoni* DC.; rust, *Uromyces appendiculatus* (Pers.) Unger; leaf spot, *Cercospora cruenta* and a hairy caterpillar, *Diacrisia obliqua*.

IMPROVEMENT

A little sorting of cvs has been done in India.

PRODUCTION

Black gram is only important in India for local consumption, where large acreages are grown, being particularly important in Mysore.

REFERENCE

BOSE, R. D. (1932). Studies in Indian Pulses, 5. Urd or Black Gram (*Phaseolus mungo* Linn. var. *roxburghii* Prain). *Indian. J. agric. Sci.*, **2**, 625–637.

Phaseolus vulgaris L. (2n = 22) COMMON BEAN

It is also known as French bean, kidney bean, haricot bean, salad bean, runner bean, snap bean, string bean and frijoles.

USES

P. vulgaris is the best known and most widely cultivated species of *Phaseolus*. They are grown for their immature edible pods and for the dry ripe seeds and to a lesser extent for green-shelled beans. The leaves are used as a pot-herb in some parts of the tropics. In Latin America and parts of tropical Africa they furnish a large part of the protein food of the inhabitants, being grown mainly for the dried pulse. They are little grown in India, where the people prefer their own, better-known pulses. In Europe, the United States and other temperate countries they are grown mainly for the green immature pods which are eaten as a vegetable and are also canned and frozen. The whole dried beans are also cooked with tomato sauce and canned and are usually known as baked beans. The straw is used as forage.

ORIGIN AND DISTRIBUTION

Phaseolus vulgaris is of New World origin. The earliest remains of the common bean that are clearly domesticated is an uncharred pod valve with a carbon^{14} date of 4975 B.C. ±200 from the Coxcatlan cave in the Tehuacan valley in Mexico and is contemporary with maize. The evidence for domestication is the breadth of the pod and the absence of a well-developed parchment layer and dehiscence mechanism, being similar to those found in higher horizons. Because of the varietal composition of archaeological *P. vulgaris* in Mexico and elsewhere and the degree of endemism, Kaplan (1965) suggests multiple domestication within Middle America from an ancestral species that was widespread and polymorphic. The older *P. vulgaris* material from Peru is about 2,200 years old and coincides with the introduction of maize. The common bean could have been introduced into coastal Peru from Central America, although, as Heiser (1965) points out, it may have an independent domestication from the closely related *P. arborigeus* Burkhart, which occurs wild in the area. The Aztec kings received tribute of about 5,000 tons of beans annually, together with maize, cocoa, chillies, cotton and other produce.

The common bean was taken to Europe in the 16th century by the Spaniards and Portuguese and it reached England in 1594. They also took it to Africa and other parts of the Old World. *P. vulgaris* is now widely cultivated in many parts of the tropics and subtropics and throughout the temperate regions.

CULTIVARS

Many hundreds of cvs are recorded and new cvs are introduced annually They can be divided into:

(1) dwarf or bush cvs, which do not require supports and are early maturing.

(2) climbing or pole cvs, which require supports, take longer to mature and have a longer bearing season.

These are further subdivided in the United States according to their uses and the types are given below, together with the names of cvs which have yielded well in Trinidad and with a few cvs from elsewhere which are recommended for the tropics:

(i) Snap beans grown for pods, which are harvested before they are fully grown and while they are still tender and the seeds are still small. Flat or oval pods are preferred for the fresh market; round-podded cvs with white seeds are preferred for canning; flat-podded cvs are used for slicing or French beans. Three types of snap beans are recognized:

(a) wax or yellow-podded bush cvs, *e.g.* 'Cherokee Wax', 'Sure Crop', 'Yellow Waxpod';

(b) green-podded bush cvs, *e.g.* 'Bountiful' and 'Canadian Wonder Select' with flat pods; 'Contender' with long oval pods, 'Top Cross' with round pods;

(c) pole cvs, *e.g.* 'Kentucky Wonder', 'Tropic Wonder'.

(ii) Green-shell beans used in the green-shelled condition. These are of comparatively little importance.

(iii) Dry-shell or field beans which are grown for the dry ripe seeds. These are extensively grown in Latin America and tropical Africa, where there are many cvs showing a great variation in the colour of the testa.

Four types of dry-shell beans are recognized:

(a) medium field beans with seeds 1–1·2 cm long, *e.g.* 'Pinto' with pinkish-buff testa with small brown spots, extensively grown in the United States;

(b) pea or navy beans with seeds 8 mm or less in length, *e.g.* 'Small White' which is extensively grown for canning in California;

(c) red kidney beans with seeds 1·5 cm or more in length, which are very important in Latin America;

(d) marrow beans with seeds 1–1·5 cm long, *e.g.* 'Yellow Eye'.

ECOLOGY

The common bean is not suited to the ever-wet tropics such as Malaya, but does well in areas of medium rainfall from the tropics to the temperate regions. They grow best in Trinidad when planted towards the end of the rains. Excessive rain causes flower drop and increases the incidence of disease. Some rain is required for the critical flowering and setting period. Dry weather is required for harvesting dry-shell beans. They are killed by frost. They can be grown on most soil types from light sands to heavy clays and also on peat soils. They are sensitive to high concentrations of manganese, aluminium and boron. In testing for photoperiod in the United States all bush cvs were found to be day-neutral, half the cvs of the pole type were short-day plants and half were day-neutral.

STRUCTURE

A highly polymorphic species showing considerable variation in habit, vegetative characters, flower colour, and size, shape and colour of pods and seeds. A twining or erect annual herb.

ROOTS: The pronounced tap-root grows rapidly to a depth of 1 m and there are extensive lateral roots mainly confined to the top 6 in. of soil. Nodules irregular, knobbly.

STEMS: In twining cvs stems 2–3 m tall with 11–16 or 28–30 elongated nodes; in erect bushy cvs stems 20–60 cm tall with determinate growth and a main axis of 4–8 nodes terminating in an inflorescence. Sub-erect and spreading forms occur, together with intermediates between pole and bush types which develop weak runners.

LEAVES: Alternate, trifoliate, often somewhat hairy; petiole long, grooved above, marked pulvinus at base; stipules small, ovate, acute, basifixed; stipels minute, lanceolate; leaflets ovate, entire, acuminate, 8–15 × 5–10 cm, lateral leaflets asymmetrical.

Fig. 47. **Phaseolus vulgaris:** COMMON BEAN. **A,** leaf (×½); **B,** flower (×3); **C,** flower with corolla removed (×4); **D,** young pod (×1); **E,** seed (×2).

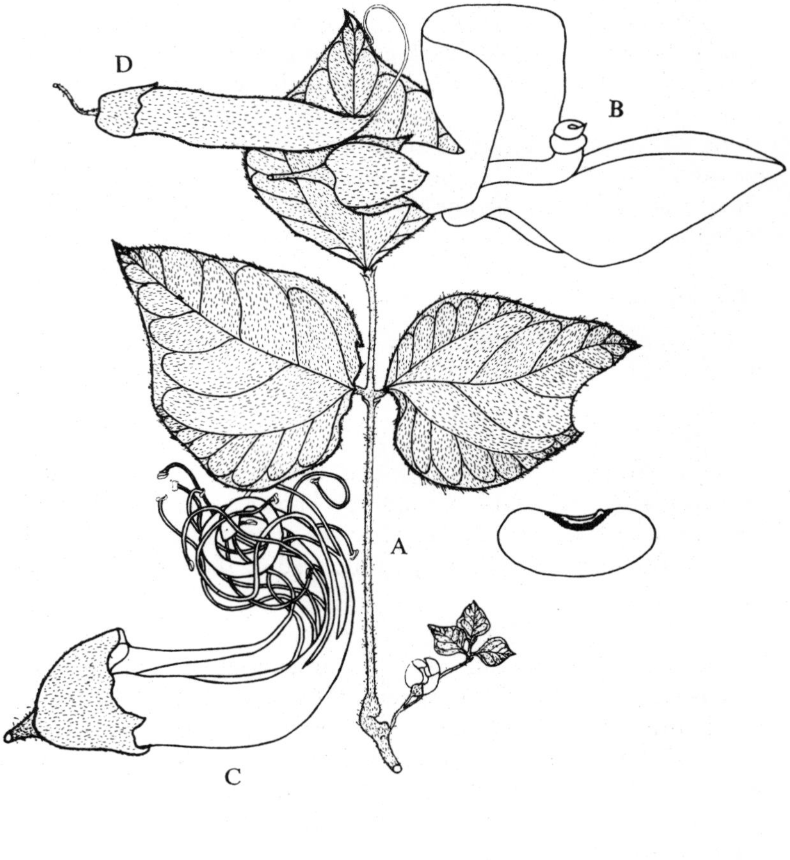
D
B
A
C

FLOWERS: Borne on axillary, few-flowered, lax racemes, usually shorter than lvs, with fls at or near apex of peduncle; pedicels short, 5–8 mm long; calyx bracts leafy, ovate, as long as or exceeding length of calyx, up to 6 mm long; calyx 3–4 mm long, with 1 upper and 3 lower teeth; corolla white, yellowish, pink or violet; standard 9–12 mm in diameter, emarginate; claws of wings 5–6 mm long; keel spirally twisted; vexillary stamen free, rest connate; anthers small, globose, uniform, basifixed; style long, twisted, hollow; stigma capitate.

FRUITS AND SEEDS: Pods slender, narrower than in *P. lunatus*, 8–20 × 1–1·5 cm, usually 4–6-seeded, but up to 12 seeds in some cvs, usually glabrous, straight or slightly curved, edges rounded or convex, beak prominent, yellow in wax cvs, light to dark green in green-podded cvs, sometimes with pink or purple splashes. Seeds very variable in colour, shape and size; white, yellow, greenish, buff, pink, red, purple, brown or black; self-coloured, mottled, blotched or striped; some eyed with different colour round hilum; 0·7–1·6 cm long; kidney-shaped, oblong or globular, ratio of length, breadth and thickness very variable, often somewhat compressed; hilum usually white; endosperm absent. Weight of 100 seeds 20–60 g.

POLLINATION

P. vulgaris is self-fertilized, pollination taking place at the time the flower opens.

GERMINATION

Germination, which is epigeal, is usually good and rapid. The seeds remain viable for 2 years. A condition known as bald head occurs in which little or no growth takes place after the appearance of the cotyledons and is caused by mechanical injury to the seed or the apical meristem.

CHEMICAL COMPOSITION

The average composition of the dried pulse is approximately: water 11·0 per cent; protein 22·0 per cent; fat 1·6 per cent; carbohydrate 57·8 per cent; fibre 4·0 per cent; ash 3·6 per cent. Green snap beans contain about: water 85·2 per cent; protein 6·1 per cent; fat 0·2 per cent; carbohydrate 6·3 per cent; fibre 1·4 per cent; ash 0·8 per cent.

PROPAGATION

The crop is grown from seed.

HUSBANDRY

In tropical Africa the crop is seldom grown alone and is usually interplanted with crops such as maize, sweet potatoes, cotton and coffee; a mixture of cvs is normally sown. In pure stand the bush cvs give the highest yields at a spacing of 1–1½ × 1 ft, but 2½–3 ft × 4–6 in. makes weeding easier.

Pole beans are usually planted on hills 3–4 ft apart with 4–6 seeds per hill, later thinned to 3–4 plants; they may also be planted in rows at a spacing of 3–4 × ½–1 ft. The seed rate varies with the seed size and the spacing, but is usually 20–50 lb per acre. In the United States 70 lb per acre is used for bush cvs and 20–30 lb per acre for pole cvs. The depth of planting is 1–3 in.

In peasant cultivation in Africa the crop is seldom manured. In the United States 300–500 lb of nitrogenous and phosphatic fertilizers is usually applied, and may or may not contain potash. In Queensland a response is seldom given to potash, but nitrogen has always proved beneficial; the normal application is 6–10 cwt per acre of a 4 : 15 : 2 NPK mixture. Fertilizers are best applied in bands 6 in. wide and 2 in. below the seeds. The crop is sometimes grown with irrigation in some arid subtropical regions.

Snap beans are harvested before the pods are fully grown and while the seeds are still small and do not cause the pods to bulge. Picking begins 2–4 weeks after the first flowers and 7–8 weeks after sowing in early cvs. They should be picked every 3–4 days and the number of pickings is greater in pole cvs than in bush cvs. The average yield of green or snap beans is about 2 tons per acre, but considerably higher yields are recorded. Dry beans are harvested as soon as a large percentage of the pods are fully mature and have turned yellow. Some cvs have a tendency to shatter. The whole plants are usually pulled, after which they are dried and threshed. In Africa the yield of dried pulse is about 500–1,000 lb per acre. The average yield in the United States is about 1,200 lb per acre.

MAJOR DISEASES

Anthracnose, *Colletotrichum lindemuthianum* (Sacc. & Magn.) Bri. & Cav., is one of the most destructive diseases of *Phaseolus vulgaris* and is now almost worldwide. Elongated dark red cankers occur on the stems and leaf veins and sunken spots with pink centres and darker borders on the pods. Cold wet weather favours attack. It may be carried by diseased seeds so it is essential to use disease-free seeds, preferably from an arid area such as California.

Fusarium root rot, *Fusarium oxysporum* Schlecht. ex Fr.f. *phaseoli* Kendr., is a serious disease in the New World, England and Australia. The first symptom is a red discoloration of the tap-root which later turns brown and the roots become dry and papery.

Rust, *Uromyces phaseoli* Arth., has a worldwide distribution. It produces reddish uredospores and later dark brown teleutospores, mainly on the leaves.

Mosaic of three different strains is widespread and is transmitted by aphids and by seed. The virus causes a ruffling, crinkling and yellow mottling of the leaves.

MAJOR PESTS

Many species of aphids and leaf hoppers are pests of beans, as well as numerous Lepidopterous caterpillars and Chrysomelid beetles. Bean beetle,

Acanthoscelides obtectus and cowpea beetles, *Callosobruchus* spp. attack the growing seeds and also the dried seeds; they are widespread throughout the tropics, including Africa. The most serious pests of the dried pulse are bean weevils, *Bruchus* spp. The Mexican bean beetle, *Epilachna varivesta* Muls., is the most serious bean pest in the United States; it also attacks *Phaseolus lunatus*. The bean fly, *Melanagromyza phaseoli* (Coq.), is widely distributed in tropical Africa, Asia and Australia. The eggs are laid on the leaves and the larvae bore through the stem, where they pupate at ground level, causing the plants to fall and die.

IMPROVEMENT

Improved strains of the cvs are constantly being produced in temperate countries, attention being paid to yield, improved habit, time to maturity, disease-resistance, etc., as well as the purpose for which the product will be utilized. Very little work has been done on improvement in Africa and elsewhere in the tropics.

PRODUCTION

The common bean is the most important pulse crop throughout tropical America and many parts of tropical Africa, being mainly grown on a field scale for the production of dried pulse. It is of little importance in India and most of tropical Asia, where indigenous pulses are preferred. Dry haricot beans are imported into the United Kingdom from Japan, Chile, the United States, Ethiopia, Mozambique, East Africa and Malawi. In the United States about 1·5 million acres are grown for dry bean production and under 0·5 million acres for snap beans and processing. It is ranked fourth among the frozen vegetables in the United States. In the United States fresh beans are on the market every month of the year, being produced in the southern states in the winter, the northern states in the summer, and in the intermediate states in the spring and autumn. The production of common beans in Latin America is more than twice that of the United States and the whole crop is consumed locally and some beans are imported.

REFERENCES

HARTER, U. and ZAUMEYER, W. J. (1944). A Monographic Study of Bean Diseases and Methods of Control. *U.S. Dept. Agric. Tech. Bull.*, 868.

HEISER, C. B. (1965). Cultivated Plants and Cultural Diffusion in Nuclear America. *Amer. Anthropologist*, **67**, 930–949.

KAPLAN, L. (1965). Archeology and Domestication in American *Phaseolus* (Beans). *Econ. Bot.*, **19**, 358–368.

Pisum L. (x = 7)

Six spp. of spreading or tendril-climbing herbs in the Mediterranean basin and western Asia, of which *P. sativum* (q.v.) is cultivated.

Pisum sativum L. *sens ampl.* (2n = 14) PEA

USES

The fresh green seeds are cooked and eaten as a vegetable; they are also canned and frozen and are the leading frozen vegetable of the United States. The pods of some cvs are eaten as well as the seeds. The ripe dried seeds, either whole, split or as flour, are used for human and stock food. Although dried peas were known from very early days in Europe, green peas were not used until the 16th century. The leaves are used as a pot-herb in Burma and parts of Africa. The plants and haulms are used for forage, hay, silage and also as a green manure. Gregor Mendel in 1858–1866 made his epoch-making experiments on *P. sativum.*

SYSTEMATICS

Two species or subspecies are sometimes recognized, but these are of doubtful validity as they are completely cross-fertile. The two species are:

1. *P. arvense* L. (syn. *P. sativum* subsp. *arvense* (L.) Poir.), the field pea, which is very hardy and is usually grown as a sprawling plant on a field scale for the dried seeds. It has stipules with a red spot and the flowers are coloured, usually purplish. The pods and seeds are small, and the latter are angular and brownish or grey in colour, often with blotches. It is sometimes grown for forage.

2. *P. sativum* L. (syn. *P. hortense* Aschers. & Graebn.; *P. sativum* subsp. *hortense* Poir.), the garden pea, which is more robust, but is less hardy, and is usually grown for green peas or pods. It has stipules without a red spot and the flowers are white. It has larger pods and seeds and the latter are round, softer, smooth or wrinkled and contain more sugar. It is subdivided by some authorities into: var. *macrocarpon* Ser., the edible-podded or sugar peas which lack the stiff papery inner parchment in the pod; and the var. *humile* Poir., the early dwarf pea, in which the pods have a stiff lining membrane.

ORIGIN AND DISTRIBUTION

Pisum sativum is not known in a wild state, but is very closely related or conspecific with *P. arvense* which occurs in a wild state in Georgia in Russia. It is sometimes considered that the cvs were domesticated from *P. elatius* Steven, which occurs as a weed of cultivated land in Europe and eastwards towards central Asia, but it may well be that this is a weedy form developed from the original parent. It is possible that *P. sativum* may have arisen by hybridization between *P. elatius* and *P. arvense*, followed by back-crossing.

The first cultivation of peas appears to have been in south-western Asia and they reached the Greeks via the Black Sea. The Latin and Germanic tribes got them from the Greeks. They spread to India and China by way of the Himalayas and Tibet. Peas reached the mountain regions of Ethiopia and east and central Africa before the advent of the Europeans and were already a well-established and important food crop in south-western Uganda and Rwanda by 1861. It is possible that they were introduced into Ethiopia during the Abyssinian invasion of Arabia in the 4th or 6th centuries.

CULTIVARS

A very great number of cvs have been recorded in temperate countries. In Kigezi in south-western Uganda the local field peas give higher yields than introduced cvs when grown under indigenous methods of cultivation (see below), although the latter give the highest yields with better cultural treatment. Local cvs in Kigezi include 'Mitali' with light-coloured seeds and 'Miseriseri' with dark green seeds; both are small-seeded. At lower altitudes in the tropics the best results are obtained by the edible-podded cvs 'Melting Sugar', which can also be used for green peas, 'Tall Grey Sugar' and 'Dwarf Grey Sugar'. At higher altitudes the garden peas 'Alderman', 'Telephone', 'Thomas Laxton', and 'Daisy' have given good results.

ECOLOGY

Peas require a cool, relatively humid climate with temperatures of 55–65°F. Consequently they seldom yield well in the tropics below 4,000 ft, or when grown in cool weather, as in India, where they are grown as a winter crop. Hot dry weather interferes with seed settlings. They require a reasonable level of soil fertility and a pH of 5·5–6·5. They cannot tolerate very acid soils or waterlogging.

STRUCTURE

A short-lived, climbing, glabrous, usually glaucous annual, 30–150 cm tall; dwarf, semi-dwarf and tall cvs occur depending on the number and length of the nodes. Tap-root well developed with many slender laterals. Stem weak, round and slender and tall cvs need sticks for climbing up. Lvs pinnate with 1–3 pairs of leaflets and a terminal branched tendril; stipules large and leaf-like, ovate, denticulate in the lower half, 2–6·5 cm long, rather larger than leaflets; stipels absent; leaflets ovate or elliptic, 2–6 in number, entire or with undulate margin, 1·5–5·5 × 1–3 cm; fls axillary, solitary or in 2–3-flowered racemes; bracts very small; calyx oblique with swelling on one side, lobes unequal, about 1 cm long; corolla usually white, but pink and purple to some field cvs; standard obovate or almost orbicular, 1·2–1·6 cm long; wings falcate-oblong, longer than keel, to which they adhere for about half their length; keel short, incurved, obtuse; stamens diadelphous with free vexillary stamen above grooved

---

Fig. 48. **Pisum sativum:** PEA. **A**, flowering shoot (× ½); **B**, flower (× 1); **C**, flower in longitudinal section (× 2); **D**, young pod (× 1); **E**, young pod opened (× 1).

A
B
C
D
E
M. WONG

staminal tube; filaments broad; anthers uniform; style sickle-shaped, flattened, bearded on inner surface; stigma minute, terminal. Pods tumid or compressed, shortly stalked, straight or curved 4–15 × 1·5–2·5 cm, 2–10-seeded, 2-valved, dehiscent by both sutures in most cvs; parchment-like endocarp in most cvs, but absent in edible-podded types. Seeds globose or angled, smooth or wrinkled, exalbuminous, green, grey or brownish, mottled in some cvs. Weight of 100 seeds 15–25 g.

POLLINATION

Pollen is shed in the bud and the flowers are almost entirely self-fertilized.

GERMINATION

Germination is hypogeal and the cotyledons which remain in the seeds are at an angle of about 120° from each other. The epicotyl appears above ground as a bent hook, later straightening. The first two foliage leaves are simple.

CHEMICAL COMPOSITION

The whole dry mature seeds contain about: water 10·6 per cent; protein 22·5 per cent; fat 1·0 per cent; carbohydrate 58·5 per cent; fibre 4·4 per cent; ash 3·0 per cent. Fresh green peas contain about: water 74·3 per cent; protein 6·7 per cent; fat 0·4 per cent; carbohydrate 15·5 per cent; fibre 2·2 per cent; ash 0·9 per cent.

HUSBANDRY

For the garden or the fresh pea market the seed is usually sown in drills 2–3 ft apart, with 2–3 in. apart between seeds, at a depth of 1–2 in., with a seed rate of 60–100 lb per acre. For processing or dried peas they are usually sown in rows 7 in. apart with a seed rate of 180–300 lb per acre. The crop responds well to organic manures and mixed NPK. For green peas the pods should be well filled, but the seeds should not be mature. Yield of pods is 50–150 bushels per acre and shelled peas about one ton per acre. The yield in Europe of dry peas is about 2,000 lb per acre with 25 cwt of haulms or straw. For hay field peas are often grown mixed with cereals.

In south-western Uganda and Rwanda field peas provide the staple pulse of the Bakiga and Banyarwanda tribes, being grown mainly between 6,000–8,500 ft. In this area peas are the first crop taken after a grass fallow or land opened up after grazing. Dwarf field cvs are broadcast in the grass; the plot is then rough-hoed once only producing a very rough seedbed. No weeding or further cultivation is done and yet fair crops are obtained for the minimum of work. The plants are cut when mature and are dried and threshed.

MAJOR DISEASES AND PESTS

Powdery mildew, *Erysiphe polygoni* DC., is the most widely spread and most serious disease of peas. Wilt, caused by two races of *Fusarium oxysporum* Schlecht. ex Fr. f. *pisi* is serious in North America.

The most serious pests are the pea aphis, *Macrosiphum onobrychis* (Boy.), in North America, and pea weevil, *Bruchus pisorum* (L.), which has a wide distribution.

PRODUCTION

Peas are an important garden and field crop throughout the temperate regions, where they are grown for fresh peas, and which are also canned or frozen; also for dry peas. In the tropics they are only grown at high altitudes or during the cool season, where in some areas they are grown as a pulse, and elsewhere in gardens for local use.

Psophocarpus Neck. ex DC.

Five spp. of twining, tuberous-rooted herbs in tropical Africa, Asia and Madagascar, with trifoliate lvs and 4-sided, 4-winged pods.

P. tetragonolobus (q.v.) is cultivated in parts of the tropics. *P. palustris* Desv. (syn. *P. longipedunculatus* Haask., *P. palmettorum* Guill. & Perr.) grows wild throughout tropical Africa and is occasionally cultivated for its edible pods, as on coral rag in Zanzibar.

Psophocarpus tetragonolobus (L.) DC. GOA BEAN

It is also known as asparagus pea, winged bean, four-angled bean, Manila bean, and princess pea. It should not be confused with *Lotus tetragonolobus* L. (syn. *Tetragonolobus purpureus* Moench.), also with 4-sided, 4-winged pods, which grows wild in the Mediterranean region and which was introduced into England under the name of asparagus pea or winged pea, and is occasionally grown for its young pods which are used as a vegetable.

USES

The Goa bean is grown mainly for its immature pods which are eaten cut up and cooked in the manner of French beans. The tuberous roots, which are smaller than *Pachyrrhizus* (q.v.), are eaten raw or cooked in some areas, notably in Burma. The ripe seeds are eaten after parching in Java. The young leaves, shoots and flowers may also be eaten as a vegetable. It has been suggested as a green manure, cover crop, fodder crop and a restorative fallow crop, because of its exceptional nodulation (see below).

ORIGIN AND DISTRIBUTION

The origin is uncertain, but it probably originated in tropical Asia, where it is now mainly cultivated. Burkill (1935) considers that it originated on the

African side of the Indian Ocean, possibly in Madagascar or Mauritius. However, it is not cultivated on the African mainland. It seems to have been long known to the Melanesians and is fairly extensively cultivated in New Guinea. It was first recorded by Rumpf in the Moluccas in the second half of the 17th century, who considers that it was brought there from elsewhere, possibly from Java. It has been introduced into the West Indies.

ECOLOGY

Goa beans thrive best in hot wet climates, although they appear to do better in the dry season in Trinidad. Loamy soils are regarded as the most suitable and the crop cannot tolerate waterlogging. In Burma it is often grown under irrigation. Thompstone and Sawyer (1914) and Masefield (1957, 1961) have reported on its exceptional capacity for forming very large and numerous root nodules. Masefield (1961) recorded the fresh weight of nodules from one plant at the age of 109 days of 585·1 g of fresh nodules, with an average of 23·12 g for all plants and with an average of 627 nodules per plant. The largest single nodule weighed 0·6 g with an extreme diameter of 1·2 cm. The crude protein percentage on dry matter of the shoots was 25·62 per cent. *Phaseolus vulgaris* grown under the same conditions gave an average of 1·55 g of fresh nodules per plant and *Pisum sativum* 0·41 g. Sugar-cane in Burma is said to give 50 per cent higher yields when following *Psophocarpus tetragonolobus.*

STRUCTURE

A twining glabrous perennial herb, usually grown as an annual. Roots rather numerous with the main long laterals running horizontally at a shallow depth, later becoming thick and tuberous. Stems produced annually, 2–3 m long, ridged. Lvs trifoliate; petiole long, deeply grooved on upper surface with large pulvinus at base; stipules 2-parted, small, lanceolate, inserted above base; stipels small, lanceolate, sessile; leaflets broadly ovate, entire, acute, glaucous beneath, 8–15 × 4–12 cm. Inflorescence an axillary raceme, up to 15 cm long with 2–10 fls; bracteoles ovate, persistent; calyx to 1·5 cm long with large cup and shorter ovate teeth of which upper 2 connate; corolla large, much exserted; standard broad, much reflexed, deeply emarginate, auricled at base, pale green at back, white or pale blue within, 2·5–4·0 cm in diameter; wings irregularly obovate; keel incurved, obtuse; vexillary stamen free at base, connate with others from middle; anthers uniform; ovary shortly stipitate; ovules many; style long, incurved, bearded lengthwise; stigma terminal, globose with dense hairs around and below it. Pods 15–30 × 2·5–3·5 cm, more or less square, with 4 longitudinal jagged ridges, 2-valved. Seeds 8–17, embedded in tissue of pod, nearly

---

Fig. 49. **Psophocarpus tetragonolobus :** GOA BEAN. **A,** leaf (×½); **B,** flower from below (×½); **C,** flower in longitudinal section (×2); **D,** pod (×½); **E,** seed (×1).

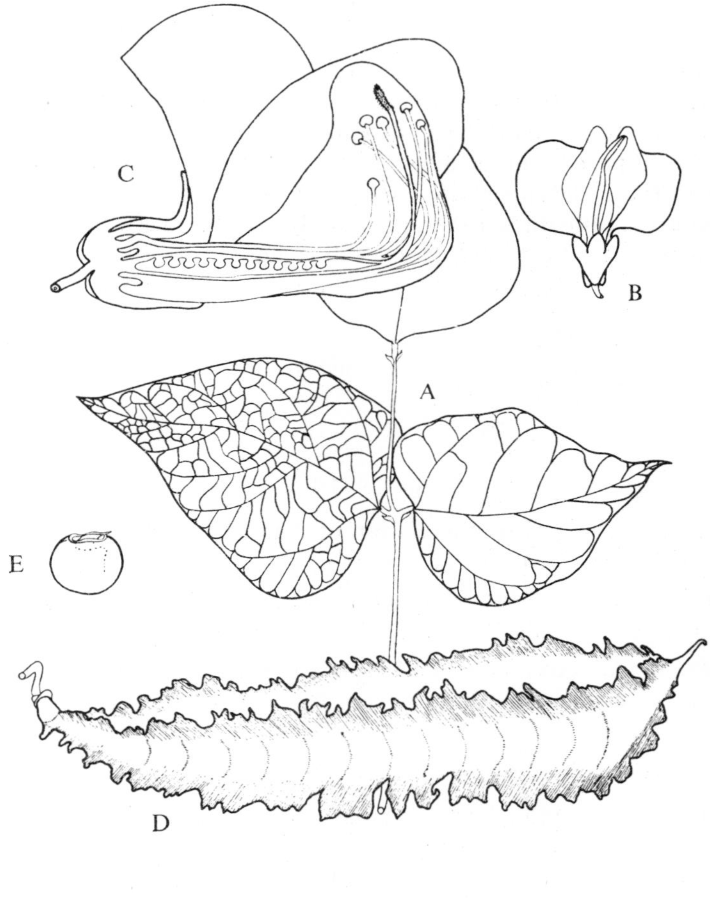
C
B
A
E
D

globular, up to 1 cm long, white, yellow, brown or black, smooth and shining. 100 seeds weigh about 30 g.

CHEMICAL COMPOSITION

The immature green pods contain about: water 91·8 per cent; protein 1·9 per cent; fat 0·2 per cent; carbohydrate 3·1 per cent; fibre 2·6 per cent; ash 0·4 per cent. The dry ripe seeds contain about: protein 37 per cent; fat 15 per cent; carbohydrate 28 per cent.

HUSBANDRY

It is usually planted at a spacing of 4 × 2 ft with strong stakes or a trellis for support. It is rather slow to come into bearing, but continues cropping for a year or so. In Burma the best seed for planting comes from the Shan States. For root production in Burma it is often planted 3–6 in. apart and is allowed to trail. The tubers are harvested 7–8 months after sowing and medium tubers, a little thicker than the thumb, are regarded with the greatest favour. The seeds take a little longer to ripen.

PRODUCTION

Cultivation of the crop is mainly confined to south-eastern Asia, where it is grown for local consumption.

REFERENCES

MASEFIELD, G. B. (1957). The Nodulation of Annual Leguminous Crops in Malaya. *Emp. J. exp. Agric.*, **25**, 139.

MASEFIELD, G. B. (1961). Root Nodulation and Agricultural Potential of the Leguminous Genus *Psophocarpus*. *Trop. Agriculture, Trin.*, **38**, 225–229.

THOMPSTONE, E. and SAWYER, A. M. (1912). The Peas and Beans of Burma. *Bull. agric. Dept. Burma, 12.*

Vicia L. (x = 5, 6, 7)

About 120 spp. of annual or perennial herbs, mostly tendril-climbing, widely distributed in the temperate regions, but with the greatest numbers in the Old World. Lvs even-pinnate; fls axillary, solitary or racemose; calyx lobes short and broad; wings adherent to keel, stamens usually diadelphous; anthers uniform; style slender; stigma terminal; pods flat, mostly several- to many-seeded; germination hypogeal.

V. faba (q.v.) is grown in temperate regions. Several species are grown for forage or green manure, notably *V. sativa* L., common vetch or tares. It is grown as a summer annual in cooler climates and as a winter annual in warmer climates, but is not suited to the tropics, except possibly at high altitudes.

Vicia faba L. (2n = 12, 14) BROAD BEAN

Syn. *Faba vulgaris* Moench.

It is also known as the horse bean, field bean, tick bean and Windsor bean.

USES

Broad beans are grown as garden cvs for the green shell beans and as a field crop for dried beans which are used as food for man and livestock. They are also grown for fodder and hay.

SYSTEMATICS

Some authorities divide it into 2 subspecies:

(1) subsp. *paucijuga* Mur., comprising several Indian cvs.

(2) subsp. *eu-faba*, which occurs both in Europe and Asia and is subdivided into 3 varieties: (a) var. *minor* Beck, tick peas or pigeon beans, with small rounded bolster-shaped seeds; (b) var. *equina* Pers., horse beans, with medium-sized, bolster-shaped seeds, with an average length of 1·5 cm; (c) var. *major* Harz., the broad bean used for culinary purposes, with large broad flat seeds, with average length of 2·5 cm, and with longer pods.

ORIGIN AND DISTRIBUTION

Vicia faba is one of the oldest of cultivated plants. It originated in the Mediterranean region or south-western Asia, and is closely related to *V. pliniana* Trabut, which grows wild in Algeria. Cvs with seeds 1 cm long were widely cultivated in prehistoric times, and have been found in the remains of lake-dwellings in England and Switzerland. They were grown by the ancient Egyptians, Hebrews, Greeks and Romans. They were an early introduction into western Asia and reached China in remote times. It was the only edible bean known in Europe in pre-Columbian times and was introduced into the New World after 1492.

CULTIVARS

A large number of cvs are recognized, but field beans are often very mixed.

ECOLOGY

Broad beans are a temperate crop. Even in the United States the summers are often too hot and the winters too cold for it. It can be grown towards the edge of tropics as a winter crop, as is done in the northern Sudan and Burma, or at high altitudes, as in Latin America. Experimental plots yielded well at 6,000–8,000 ft in south-western Uganda. It is not suited to the low hot tropics where it may flower well but usually produces no pods. It requires good fertile soils with a plentiful supply of lime and adequate and sustained water supply. It does well on heavy clay soils.

STRUCTURE

An erect, robust, leafy, glabrous, annual herb, 30–180 cm tall. Well-developed tap-root to 1 m or more with strong laterals; smaller roots with large clusters of small lobed nodules. Stems stout, hollow, square, winged at angles, with 1–7 branches from axils of cotyledons and basal nodes, otherwise unbranched. Lvs pinnate, often ending in small point (rudimentary tendril); stipules half-sagitate, dentate, often with an extra-floral nectary; leaflets 2–6, sub-opposite or alternate, ovate to elliptic, somewhat glaucous, mucronate, 5–10 cm long. Inflorescences short axillary racemes, 1–6-flowered. Fls fragrant, 2·5–3·7 cm long; calyx campanulate, oblique, teeth triangular; standard white, often faintly streaked with black on dorsal side, sub-erect, spatulate, longer than wings; wings often with purplish blotch; keel oblique, adnate to wings; stamens short, diadelphous; anthers uniform, usually dark coloured; style short, bent, with tuft of hairs near tip; stigma terminal. Pods stout, sub-cylindrical or flattened, beaked, 5–10 cm long in field cvs, up to 30 cm long in garden cvs, fleshy with white velvety lining when young, tough and hard at maturity. Seeds very variable in shape and size, strongly compressed to nearly globular, white, green, buff, brown, purple or black, 1–2·6 cm long; hilum prominent. Weight of 100 seeds 40–180 g.

POLLINATION

Anthers dehisce early and pollen is shed in the keel, so that self-pollination can occur. Nevertheless, the flowers are much visited by bees and cross-pollination can take place, as is shown by the considerable mixture of field beans in Europe.

GERMINATION

Germination is hypogeal.

CHEMICAL COMPOSITION

The dried beans contain about: water 14·3 per cent; protein 25·4 per cent; fat 1·5 per cent; carbohydrate 48·5 per cent; fibre 7·1 per cent; ash 3·2 per cent.

PROPAGATION

The crop is grown from seed.

HUSBANDRY

In the tropics broad beans are grown as a cool-weather winter crop or at high altitudes. It is usually grown in rows 1–2 ft apart with a seed rate of 120–150 lb per acre. In England yields of 1,800–2,400 lb per acre of dried beans are obtained and 25–30 cwt of haulms or straw. In the Sudan yields of ½ ton per acre of beans are considered satisfactory, but yields of over 1 ton per acre have been recorded.

MAJOR DISEASES AND PESTS

The most serious diseases of broad beans in the Sudan are: chocolate spot, probably caused by *Botrytis fabae* Sardina or *B. cinerea* Pers. ex Fr.; rust, *Uromyces fabae* (Pers.) de Bary; and mildews caused by *Leveillula taurica* (Lev.) Arn. and *Erysiphe polygoni* DC. The most serious pest is the bean aphis, *Aphis fabae* Scop.

PRODUCTION

Broad beans are of some importance for local consumption in the Sudan and Egypt, and in Latin America, notably in Mexico and Brazil. They are mainly grown, however, in temperate countries.

Vigna Savi (x = 10, 11, 12)

About 150 spp. of climbing or prostrate, rarely erect herbs, in the tropics of both hemispheres, with the greatest number in the Old World. Linnaeus included the Old World species in genus *Dolichos*. Lvs pinnately trifoliate, stipulate and stipellate. Fls whitish, yellowish or purple, mostly in alternate pairs on the ends of long peduncles (cf. open raceme of *Phaseolus*); often falling soon after opening; standard broad and large; keel incurved and arched upwards (cf. coiled in *Phaseolus*); stamens diadelphous; stigma lateral (cf. terminal in *Dolichos*). Pods slender, usually long and subterete, 2-valved. Seeds various, small, not broad and flat.

USEFUL PRODUCTS

The most important economic species is *V. unguiculata* aggreg (q.v.).

V. vexillata (L.) Benth., a twining or prostrate herb with tuberous roots and pink or purplish flowers, which turn yellow on fading, is widespread in tropical Africa. The roots are eaten in the same way as sweet potatoes (q.v.) in the Sudan and Ethiopia.

V. hosei (Craib) Backer, a native of Sarawak and the East Indies, is grown as a green manure and cover crop in south-eastern Asia and Queensland.

Vigna unguiculata (L.) Walp. *aggreg.* (2n = 22, 24) COWPEA

For synonymy and views see below.

Cvs grown for the dry seeds are also known as black-eye pea, black-eye bean, southern bean, China pea, Kaffir pea and marble pea. Cvs grown for their long immature pods are variously known as yard-long bean, asparagus bean, Bodi bean and snake bean.

USES

The dried seeds are an important pulse crop in the tropics and subtropics, particularly in Africa. The dried pulse may be ground into meal which is used in a number of ways (see Dalziel, 1937). The fresh seeds and immature pods are eaten and they may be frozen or canned as is sometimes done in the United States. The young shoots and leaves are eaten as spinach and provide one of the most widely used pot-herbs in tropical Africa; they are often dried and stored for dry-season use. The value as a pot-herb is due to the fact that cowpeas do not mature in a definite period, but continue producing new leaves if cut back regularly from an early stage. This characteristic also makes it a useful fodder plant for hay, silage or pasture; it is also used as a green manure and cover crop. The seeds are sometimes used as a coffee substitute. A form is grown in Northern Nigeria for a strong fibre which is obtained from the peduncles.

SYSTEMATICS

There is considerable confusion in the synonymy and names, partly due to the number of distinctive forms. The wild cowpea, indigenous to tropical Africa, is *V. unguiculata* (L.) Walp. (syn. *V. baoulensis* A. Chev.). Some authorities consider that all cvs belong to this species as they all cross readily with it producing fertile hybrids. According to this view, *V. sinensis* and *V. sesquipedalis* become synonyms of *V. unguiculata*.

Others recognize three distinct species, namely:

(1) *Vigna unguiculata* (L.) Walp. (syn. *Dolichos unguiculatus* L., *Phaseolus cylindricus* L., *Dolichos catjang* Burm., *Vigna cylindrica* (L.) Skeels, *V. catjang* Burm. Walp., *V. sinensis* var. *cylindricus*). This is the catjang cowpea, which is the most primitive of the cvs. This is cultivated in Africa, but is commoner and has more cvs in Asia.

(2) *Vigna sinensis* (L.) Savi ex Hassk. (syn. *Dolichos sinensis* L.), the common cowpea, with most cvs in Africa, where the crop is more highly specialized.

(3) *Vigna sesquipedalis* (L.) Fruw. (syn. *Dolichos sesquipedalis* L., *Vigna sinensis* var. *sesquipedalis* (L.) Koern.). This is the asparagus pea or yard-long bean, mostly grown for its immature pods, and is most widely cultivated in the Far East, but it is also recorded from Africa.

If it is desired to separate the cultivated forms from the wild *V. unguiculata*, the former should then bear the name of *V. sinensis*.

KEY TO THE CULTIVATED FORMS OR SPECIES

A. Spreading, sub-erect or erect annuals, 15–80 cm high; pods less than 30 cm long, hard and firm, not inflated when young

---

Fig. 50. **Vigna sesquipedalis**: YARD-LONG BEAN. **A**, leaf and pod (×½); **B**, flower in longitudinal section (×3); **C**, seed (×1½).

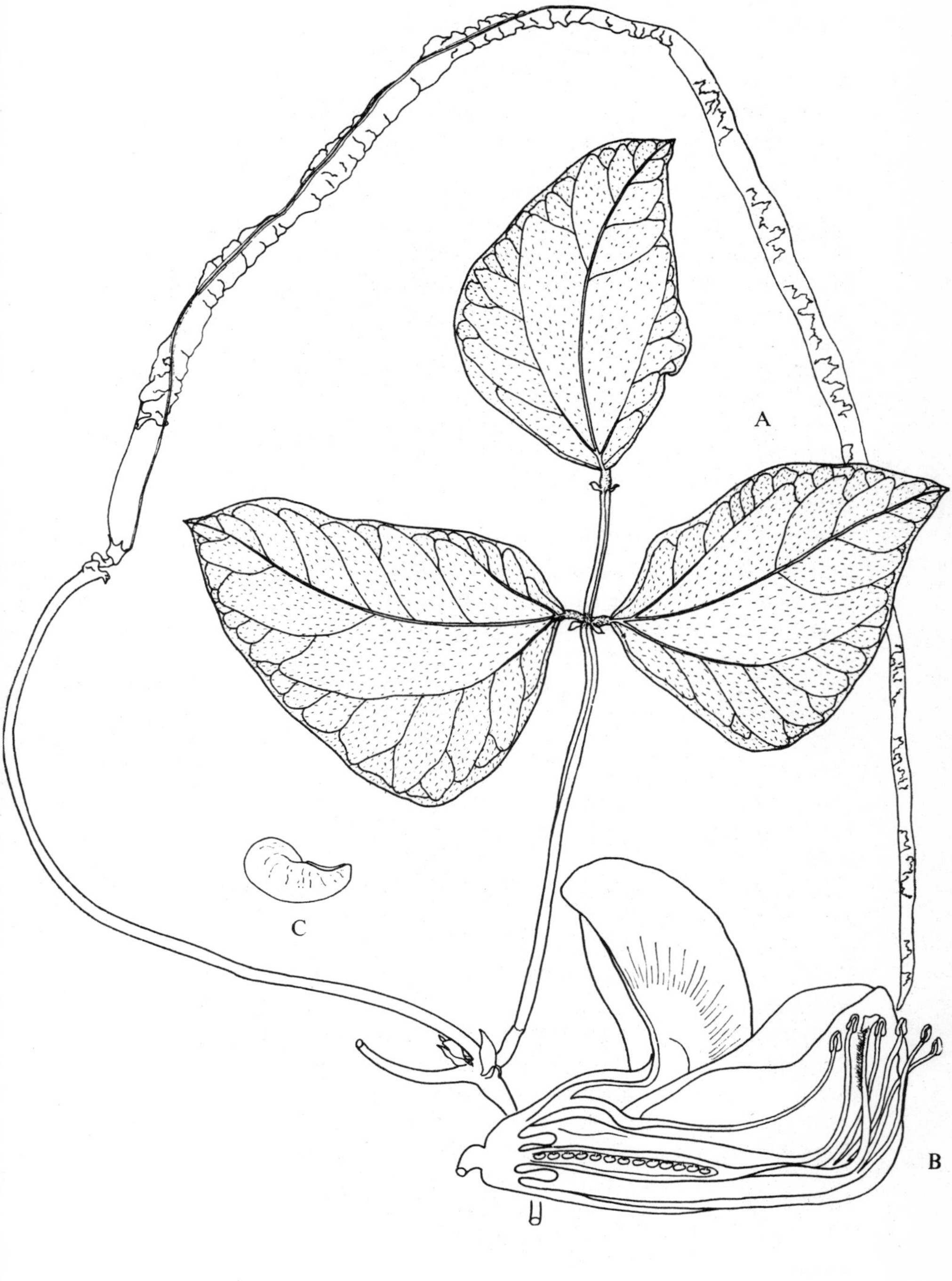
A
B
C

B. Pods 7·5–12 cm long, erect or ascending; seeds usually 3–6 mm long .... *V. unguiculata*

BB. Pods 10–30 cm long; pendent even when young; seeds usually 6–10 mm long .... *V. sinensis*

AA. Climbing annuals, 2–4 m high; pods 30–100 cm long, pendent, more or less inflated and flabby when young; seeds usually 8–12 mm long .... *V. sesquipedalis*

ORIGIN AND DISTRIBUTION

Cowpeas are of ancient cultivation in Africa and Asia and were known in India in sanskritic times. The early Greeks and Romans grew it, when it was known as phaseolus. As the wild *V. unguiculata* is widespread in tropical Africa, it seems reasonable to assume that the crop was domesticated in this region and that it spread from there in remote times through Egypt or Arabia to Asia and the Mediterranean. It was introduced by the Spaniards into the West Indies in the 16th century and was taken to the United States about 1700. Cowpeas are now widely distributed throughout the tropics and subtropics.

CULTIVARS

Brittingham (1946) has classified 67 cvs in the southern United States into 13 horticultural groups, based mainly on differences in flowers, pods and seeds. This can be simplified as follows:

(1) Asparagus bean group, *e.g.* 'Yard-long'. In Trinidad the cv. 'Long White', imported from Hong Kong, has given good results. In Hong Kong cvs can be divided into green-podded forms, $1\frac{1}{2}$–3 ft long, and 'white' podded forms with very pale green pods, up to $1\frac{1}{2}$ ft in length.

(2) Catjang group, *e.g.* 'Cream Lady', which does well in Puerto Rico.

(3) Cowpea group, which can be subdivided into:

(a) 'Crowder' cvs with seeds crowded in pods and may be black, speckled, brown or brown-eyed in colour. 'Brown Crowder' is a good cv. in Puerto Rico.

(b) 'Black eye' cvs in which the seeds are not crowded in the pods and are white with a black-eye pattern round the hilum. These are extensively grown in California and do well in Puerto Rico.

(c) 'Cream' cvs in which the seeds are not crowded in the pods and are cream in colour.

(d) Other cvs in which the spacing of the seeds is intermediate between 'Crowder' and 'Black-eye', *e.g.* 'Purple Hull' cvs with mature pods deep purple and seeds with a buff or maroon eye.

(e) Forage cvs, *e.g.* 'New Era', which has recently been introduced into West Africa, where it has been found useful for dried beans.

ECOLOGY

Cowpeas can be grown under a wide range of conditions. They are sensitive to cold and are killed by frost. They will tolerate heat and relatively dry conditions and can be grown with less rainfall and under more adverse conditions than *Phaseolus vulgaris* and *P. lunatus*. The asparagus beans can tolerate and require a higher rainfall than the common cowpeas. Cowpeas can be grown successfully on a great variety of soils, provided they are well drained. They are sometimes grown on very poor acid soils as a soil-improver. *Vigna sinensis* is said to be a short-day plant and *V. sesquipedalis* day-neutral.

STRUCTURE

Erect, prostrate or climbing, glabrous or glabrate, annual herbs. Tap-root stout with numerous spreading laterals in surface soil; nodules large, globular, about the size of small peas, often collected in groups. Stems very variable, erect or sub-erect in the catjang, usually procumbent in the cowpea, and twining anti-clockwise to a height of 2–4 m in the asparagus bean; may be tinged purple. Lvs trifoliate; petiole stout, 5–15 cm long, grooved; stipels inconspicuous, ovate to lanceolate; leaflets ovate-rhomboid, usually entire, but sometimes slightly lobed, apex acute, 6·5–16 × 4–11 cm, lateral leaflets oblique. Inflorescences axillary with few fls crowded near tip in alternate pairs on thickened nodes, usually 2–4-flowered; peduncles stout, grooved, often exceeding length of leaf, 2·5–15 cm long, with cushion-like nectary between each pair of fls; pedicels very short; calyx campanulate with long or short triangular teeth of which the upper 2 are usually connate and longer than the rest; corolla dirty white, sordidly yellow or violet, much exserted; standard 2–3 cm in diameter; keel truncate; stamens diadelphous, anthers uniform; ovary sessile, many-ovuled; style bent, bearded on inner curve immediately below oblique stigma. Pods variable (see SYSTEMATICS above), linear, semi-terete, 8–100 × 0·8–1 cm, 8–20-seeded. Seeds very variable in size, shape and colour, 2–12 mm long, globular to kidney-shaped, smooth or wrinkled, white, green, buff, red, brown or black and variously speckled, mottled, blotched or eyed; hilum white surrounded by a dark ring. Weight of 100 seeds 10–25 g.

POLLINATION

The flowers open early in the morning, close before noon and fall the same day. The extra-floral nectaries at base of the corolla attract ants, flies and bees, but a heavy insect is required to depress the wings and expose the stamens and stigma. The pollen is sticky and heavy. The degree of cross-pollination varies considerably in different areas. In California with its dry climate cowpeas are considered almost entirely self-pollinated; in humid areas in the United States considerable cross-pollination occurs, as has also been reported from Nigeria.

GERMINATION

Germination is epigeal and the germination percentage is usually high. The seeds remain viable for several years. The first pair of true leaves are simple and opposite, but exhibit considerable variation in size and shape.

CHEMICAL COMPOSITION

The dry pulse contains approximately: water 11·0 per cent; protein 23·4 per cent; fat 1·3 per cent; carbohydrate 56·8 per cent; fibre 3·9 per cent; ash 3·6 per cent. The young green pods contain about: water 86·2 per cent; protein 3·4 per cent; fat 0·3 per cent; carbohydrate 7·4 per cent; fibre 1·8 per cent; ash 0·9 per cent.

PROPAGATION

Cowpeas are grown from seed.

HUSBANDRY

In tropical Africa cowpeas are grown mixed with other crops or in pure stands. The seed is broadcast and the plants are later thinned, the thinnings being used for pot-herbs. The seed rate is 20–30 lb per acre. In the southern United States cowpeas are usually grown in rotation with cotton and maize. When grown for dry seed production a spacing of 2½–3 ft × 2–3 in. is used with a seed rate of 15–25 lb per acre. If grown for forage they are broadcast or drilled in rows 6–8 in. apart with a seed rate of about 90 lb per acre. For forage production in the United States they are usually grown mixed with Sudan grass, sorghum or maize. A wider spacing is required for procumbent spreading cvs than for erect and sub-erect cvs. Twining asparagus beans require supports. In the United States 600–1,000 lb per acre of a 4 : 8 : 8 NPK mixture is usually applied in bands 2 in. below the seeds. Cowpeas are usually grown as a rain-fed crop, but are sometimes irrigated in California.

Early maturing cvs produce a crop in about 3 months and green pods may be available for picking after 50 days. Late cvs may take up to 5 months to mature. The pods tend to ripen unevenly. For hay the crop is cut when most of the pods are well developed. In the tropics yields of 400–600 lb of dry beans per acre are obtained; in the United States yields of dry beans are 1,000–1,500 lb per acre, with up to 2,500 lb per acre of black-eye beans in California. The yield of hay is about 2 tons per acre.

MAJOR DISEASES AND PESTS

In California the most destructive diseases are cowpea wilt, *Fusarium oxysporum* Schlecht. ex Fr. f. *tracheiphilum* (E. F. Sm.) Synder & Hansen, and charcoal rot, *Sclerotium bataticola* Taub. *Colletotrichum lindemuthianum* (Sacc. & Morg.) Bri. & Cav. attacks cowpeas under rainy conditions

---

Fig. 51. **Vigna sinensis**: COWPEA. **A,** shoot with leaf and inflorescence (× ½); **B,** flower in longitudinal section (× 3); **C,** parts of corolla and calyx (× 1); **C1,** standard; **C2,** wings; **C3,** keel; **C4,** calyx; **D,** pod (× ½); **E,** seed (× 3).

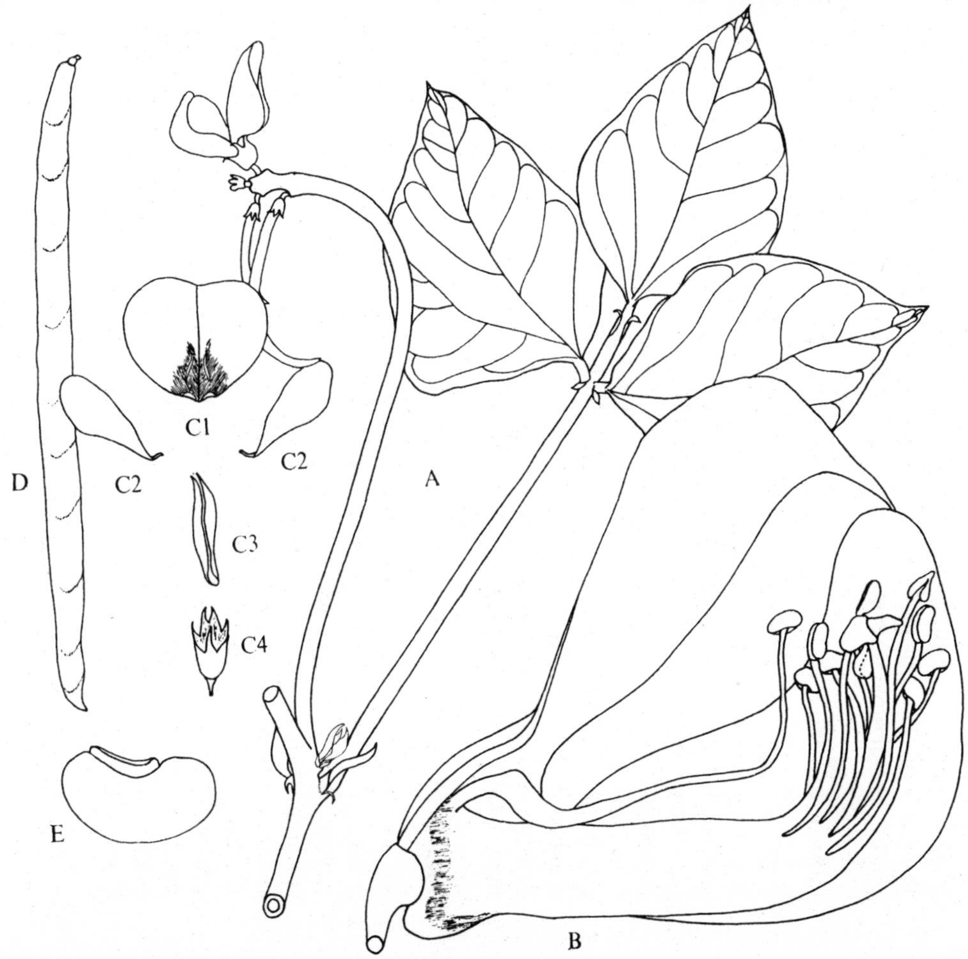
A
B
C1
C2
C2
C3
C4
D
E

in many parts of the world. Cowpea mosaic occurs in the West Indies, the vector of which is the bean-leaf beetle, *Cerotoma ruficornis* (Oliv.); the virus is also seed-borne. Cowpeas can be severely damaged by root-knot nematodes, *Meloidogyne* spp. and the string nematode, *Belonlaimus gracilis*. *Striga gesnerioides* (Willd.) Vatke (Scrophulariaceae) is a root parasite on cowpeas in tropical Africa.

The major pests are the same as those of *Phaseolus vulgaris* (q.v.). The most serious pest in the south-eastern United States is the cowpea curculio, *Chalcodermus aeneus* Boh. Cowpeas are very subject to store pests such as *Bruchus* spp. The boring larva of the fly, *Melanagromyza phaseoli* Coq., causes considerable damage to seedlings in India.

IMPROVEMENT

Much selection and hybridization has been carried out in the United States for size and shape of seed, colour of testa, vine shape, size, maturity and yield. Elsewhere little work has been done. Hybrids between 'Californian Blackeye' and 'Iron' cowpea, with back-crossing to the former, has produced resistance to nematodes, cowpea wilt and charcoal rot. A hybrid from this cross, which has been selected for several generations, is an excellent dual-purpose cv., giving high yields of good-quality, medium-sized seed, with the vigour and size of the 'Iron' parent, with resistance to diseases and drought and gives good crops on heavy soils. It is, however, late maturing and the large vines make mechanical harvesting difficult.

PRODUCTION

Cowpeas are the second most important pulse crop in tropical Africa, the most important being *Phaseolus vulgaris*, but the latter is less tolerant of adverse conditions. The seed cvs are grown in tropical Asia, but they are of less importance than some of the indigenous pulses. Asparagus beans are an important vegetable crop in south-eastern Asia, particularly with Chinese market gardeners. They are also grown in the West Indies. Cowpeas are grown to a limited extent in tropical America, more particularly in Venezuela. They are of some importance in the United States, where they are grown in the south-eastern states for green pods and green seeds, while California is the main producer of dry black-eye peas.

REFERENCES

BRITTINGHAM, W. H. (1946). A Key to the Horticultural Groups of Varieties of the Southern Pea, *Vigna sinensis*. *Amer. Soc. Hort. Sci. Proc.*, **48**, 478–480.

LORZ, A. P. *et al.* (1955). Production of Southern Peas (Cowpeas) in Florida. *Bull. Fla agric. Exp. Sta.*, 557.

MACKIE, W. W. (1946). Blackeye Beans in California. *Calif. Univ. Agric. Ext. Bull.*, 696.

Voandzeia Thou. (x = 11)

A monotypic genus, native of Africa, scarcely differing from *Vigna* except for the subterranean fruits.

NOTE: *Kerstingiella geocarpa* Harms, Kersting's groundnut, is cultivated in West Africa for its edible seeds. It is a herb with prostrate rooting branches fruiting below ground. It can be distinguished from *Voandzeia* by its deeply divided calyx with narrow lobes and the glabrous style. The axillary flowers are subsessile and the fruits, about 2 cm long, and usually 2-seeded, are buried by a carpophore in a similar manner to *Arachis hypogaea* (q.v.), which is different from the way in which *Voandzeia* buries its fruits (see below). *Kerstingiella* is also monotypic.

Voandzeia subterranea (L.). Thou. (2n = 22) BAMBARA GROUNDNUT

USES

Bambara groundnuts are grown for their edible seeds which are used as a nutritious pulse and not as an oil-seed. The fresh seeds are eaten in an unripe state or the pulse after soaking and boiling, as the dried seeds are very hard. The dried seeds are sometimes roasted and ground into flour.

ORIGIN AND DISTRIBUTION

V. subterranea is found wild in West Africa, but rarely so. It has been cultivated throughout tropical Africa for many centuries. It was taken at an early date to Madagascar, probably by Arabs. It had reached Brazil and Surinam early in the 17th century and was later taken to the Philippines and Indonesia.

CULTIVARS

Several cvs are recognized in Africa differing in the shape of the leaves and the size, hardness and the colour of the seeds. The greatest variation is found in Togo and Zambia.

ECOLOGY

Bambara groundnut will grow on poor soils in hot climates. It is grown in hot dry regions which are marginal for other pulses and groundnuts, giving higher yields than the latter on worn-out soils.

STRUCTURE

An annual herb with short creeping much-branched stems, rooting at nodes, and with very short internodes thus giving a bunchy appearance to the plant. Lvs pinnately trifoliate; petiole long, erect, grooved, thickened at base; stipellate; petiolules with marked pulvini; leaflets lanceolate to narrowly elliptic, glabrous; larger terminal leaflet, 5–10 × 1·5–3 cm, apex rounded, emarginate. Inflorescences axillary, 1–3-flowered, but usually 2-flowered; peduncle short, thick, 0·6–1·2 cm long; pedicels short; fls small; calyx glabrous with upper 2 lobes connate, lobes short and broad; corolla pale yellow; standard orbicular, about 9×7 mm; wings oblong-obovate, same length as keel, which is almost straight; stamen diadelphous, uniform, with free vexillary stamen; style short, bent, hairy along inner surface; stigma small, lateral. Tip of peduncle has a swollen glandular tip with a brush of hairs behind; after fertilization it bends towards soil and then excavates a tunnel drawing in the developing pods. The pods then develop underground. Frs rounded, wrinkled when mature, about 2 cm in diameter, usually 1-seeded, but occasionally 2-seeded. Seeds round, smooth, hard, varying in size, up to 1·5 cm broad, whitish, red, black or variously mottled or blotched; hilum white, sometimes surrounded by black eye. 600–900 seeds per lb.

POLLINATION

The flowers are self-pollinated. The petals are often undeveloped or fail to open in the bud and the flowers are cleistogamous.

CHEMICAL COMPOSITION

The ripe seeds contain: protein 16–21 per cent; fat 4·5–6·5 per cent; carbohydrate 50–60 per cent; thus providing a completely balanced food.

PROPAGATION

The crop is grown from seed.

HUSBANDRY

Bambara groundnuts are sown either in mixed cultivation with other crops or in pure stand at a spacing of about 12×8 in. with a seed rate of about 35 lb of shelled nuts per acre. The rows are usually earthed up and in some areas are lightly covered with soil to promote fruit production. The crop matures in 4 months from sowing. It is harvested by pulling up the entire plants to which the pods adhere by tough wiry stalks. The normal yield is about 500 lb of dried seeds per acre, but over 1,000 lb per acre has been recorded. The proportion of unshelled to shelled seed is approximately 3 : 2.

MAJOR PESTS AND DISEASES

The crop appears to be remarkably free from pests and diseases.

Fig. 52. **Voandzeia subterranea :** BAMBARA GROUNDNUT. **A**, plant with flowers and fruits ($\times \frac{1}{2}$); **B**, flower in longitudinal section ($\times 6$); **C**, seed ($\times \frac{1}{2}$).

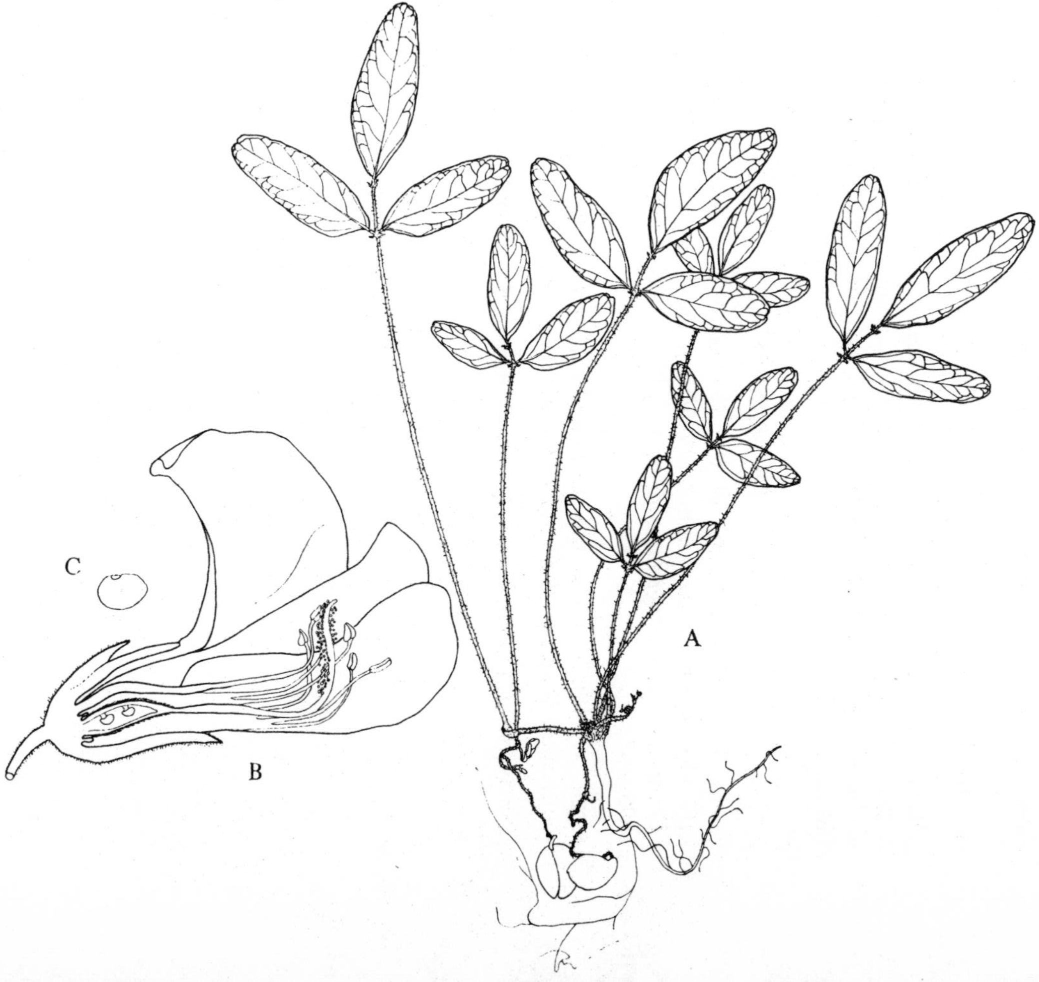
C
A
B

IMPROVEMENT

Little or no work has been done on the improvement of the crop, except some sorting and testing of cvs in Zambia.

PRODUCTION

Bambara groundnuts do not enter into world trade. They are cultivated in the drier regions of tropical Africa on poor soils, mainly for local production, but seldom on a large scale. The most extensive production is in Zambia.

REFERENCES

HEPPER, F. N. (1963). Plants of the 1957–58 West African Expedition. II. The Bambara Groundnut (*Voandzeia subterranea*) and Kersting's Groundnut (*Kerstingiella geocarpa*) wild in West Africa. *Kew Bull.*, **16**, 395–407.

RASSEL, A. (1960). Le voandzu, *Voandzeia subterranea* Thouars, et sa culture au Kwango. *Bull. agric. Congo Belge*, **51,** 1–26.

STANTON, W. R. (1966). *Grain legumes in Africa*. Rome: FAO.

MALVACEAE

About 50 genera and 1,000 spp. of herbs and shrubs, rarely small trees, with a world-wide distribution. The stems are often fibrous and the plants usually have a stellate indumentum. Lvs alternate, simple, usually palmately veined; stipules present. Fls actinomorphic, usually hermaphrodite; sepals usually 5, valvate, more or less united, usually with an involucre of bracteoles (epicalyx); petals 5, free, contorted or imbricate, often adnate to base of staminal column; stamens numerous, hypogynous, monadelphous, anthers 1-celled; ovary superior, syncarpous, usually 2 or more-celled; style 1, often branched above. Fr. usually dry, separating into 1-seeded indehiscent schizocarps or capsule dehiscing loculicidally. Seeds with some endosperm; cotyledons usually folded.

USEFUL PRODUCTS

Gossypium spp. (q.v.) yield cotton of commerce. The stems of many species yield fibres, some of which are grown commercially, *e.g. Hibiscus* spp. (q.v.) and *Urena lobata* (q.v.); others are used locally and some have been tested experimentally, *e.g. Malachra* spp., *Pavonia* spp., *Sida* spp., *Thespesia* spp. and *Wissadula* spp. *Abutilon theophrasti* Medic. (syn. *A. avicennae* Gaertn.), China jute or Indian mallow, is grown in China and the Soviet Union as a fibre crop. The fruits of some *Hibiscus* spp. (q.v.) are edible.

KEY TO ECONOMIC GENERA

A. Carpels with hooked spines . . . . *Urena*
AA. Carpels without hooked spines
  B. Epicalyx absent or when present consisting of 5 or more free or partially connate narrow bracteoles . . . *Hibiscus*
  BB. Epicalyx of 3 broad foliaceous more or less cordate bracteoles, usually deeply toothed . . . . *Gossypium*

Gossypium L. (x = 13) COTTON

About 30 spp. of annual subshrubs, perennial shrubs or small trees distributed in the tropical and subtropical regions of Africa, Asia, Australia and America. Whole plant irregularly dotted with black oil glands. Branches terete or slightly angled, tomentose, usually with monopodial vegetative branches and sympodial fruiting branches. Bracteoles 3, usually foliar and persistent; calyx cup-shaped; petals 5, imbricate; stamens numerous, lower part of filaments united into a tube, upper part free with unicellular anthers; styles clavate or furrowed, rarely divided at tip; ovary 3–5-locular. Fr. a dry, brittle, loculicidally dehiscent capsule. Seeds covered with one or two coats of unicellular hairs, or in some wild spp. almost naked.

Hutchinson, Silow and Stephens (1947) considered the genus from the evolutionary standpoint, taking into consideration the cytological, genetical, geographical and archaeological evidence then available. They regrouped all the wild and cultivated cottons of the world into 20 spp. belonging to 8 sections of the genus. They recognized 4 spp. in cultivation, the diploid Old World cottons *G. arboreum* (q.v.) and *G. herbaceum* (q.v.) and the tetraploid New World cottons *G. barbadense* (q.v.) and *G. hirsutum* (q.v.). Since then several new species have been discovered. Saunders (1961) lists 23 spp., while Fryxell (1965) states that the number of species is 32.

Crosses between species in the different sections are difficult to make and in some cases are impossible; the F_1 hybrids are usually completely sterile. Crosses between species in each section are usually possible and the F_1 hybrids are reasonably fertile; nevertheless there is a breakdown of genotypic balance in subsequent generations with the elimination of intergrading types by natural selection, so that the individual species tend to retain their identity. Fryxell (1965) has demonstrated that within each section there occur pairs of sibling species with close genetic and morphological similarity which are interfertile and have an adjacent or overlapping distribution. With the exception of the wild diploid New World species, each section of the genus has a different genome.

The wild lintless diploid species are perennial xerophytic shrubs or small trees occurring in arid regions of the tropics and subtropics, usually with a very limited distribution. They are mostly found growing on desert fringes, dry river beds and rocky hillsides. Due to man's influence, the cultivated linted species have changed from the stationary, almost relic state of the lintless species to an agressive colonizing phase with vast planted populations which has spread them into more mesophytic habitats and varied climatic conditions, thus giving enormous selection advantages and producing a great variety of forms. This was assisted by the development of the annual habit. With the exception of *G. herbaceum* race *africanum* and certain races of New World tetraploid cottons, which occur wild in littoral habitats, the various races of the cultivated cottons are never truly wild, but usually occur as escapes from cultivation. Their origin and distribution are intimately connected with their utilization by man.

The cultivated cottons have two types of unicellular hairs which are outgrowths from the epidermis of the seed: (a) fuzz in which the hairs are circular in cross section, at first thin walled, but cellulose is then deposited within the hair until the central cavity or lumen is almost obliterated;

these hairs are attached firmly to the seed and cannot be spun; (b) lint in which the hairs grow to a greater length and the deposition of cellulose is reduced giving a lumen or cavity so that the hairs collapse on drying to form a ribbon; the cellulose is deposited spirally and changes in the direction of deposition cause the ribbon to twist, giving characteristic convolutions or cling which permits spinning; this lint is easily detached from the seed. The mutation of the gene controlling the rate of secondary thickening of some of the seed-coat hairs occurred in the wild *G. herbaceum* race *africanum* in southern Africa and man later took cotton into cultivation to obtain the lint for spinning. The wild diploid cottons are lintless and only have fuzz. In some of the cultivated cottons the fuzz hairs are absent or much reduced, giving naked or tufted seeds when the lint is removed.

CLASSIFICATION OF GOSSYPIUM SPECIES

A. WILD LINTLESS SPECIES

Section I: **Sturtiana** (2n = 26) C GENOME

About 9 spp. in Australia, which include *G. sturtii* F. Muell., *G. robinsonii* F. Muell. and *G. australe* F. Muell.

Section II: **Erioxyla** (2n = 26) D GENOME

Three spp. with *G. aridum* (Rose & Standl.) Skovsted in western Mexico, *G. armourianum* Kearn. on San Marcos Island in the Gulf of California, and *G. harknessii* Brandeg. in Baja California and adjacent islands.

Section III: **Klotzschiana** (2n = 26) D GENOME

Two spp. with *G. klotzschianum* Anders. var. *klotzschianum* endemic in Galapagos Islands, *G. klotzschianum* var. *davidsonii* (Kell.) Hutch. in Baja California and adjacent islands, and *G. raimondii* Ulb. in northern Peru.

Section IV: **Thurberana** (2n = 26) D GENOME

Four spp. with *G. thurberi* Tod. from Arizona to Mexico, and *G. gossypioides* (Ulb.) Standl., *G. trilobum* (DC.) Kearn. and *G. lobatum* Gentry in Mexico.

Section V: **Anomala** (2n = 26) B GENOME

Three African spp. with *G. anomalum* Waw. & Pey., the most widespread of the wild spp. extending from the southern borders of the Sahara to south-west Africa, *G. triphyllum* Hochr. in south-west Africa, and *G. barbosanum* Phill. & Clem. endemic in Cape Verde Islands.

Section VI: **Stocksiana** (2n = 26) E GENOME

Five spp. centring in the Arabian peninsula with *G. stocksii* Mast. in Sind and south-east Arabia, *G. somalense* (Gürke) Hutch. in Somaliland,

G. areysianum (Deflers) Hutch. and *G. incanum* (Schwartz) Hillcoat in Aden Protectorate, and *G. longicalyx* Hutch. & Lee in East Africa and the Sudan.

B. OLD WORLD LINTED COTTONS

Section VII: **Herbacea** (2n = 26) A GENOME

Two spp. with a number of cultivated races in Asia and Africa.

1. *G. herbaceum* L. Perennial and annual shrubs or subshrubs with rounded capsules which are usually smooth with few oil glands. 5 races recognized:

(1) *africanum* (Watt) Hutch. & Ghose, a bushy perennial which occurs wild in southern Africa and is the most primitive of the Old World linted cottons. The mutation for lint production is believed to have occurred in this race and it is considered the common ancestor of the diploid cultivated cottons.

(2) *acerifolium* (Guill. & Perr.) Chev., perennial and most primitive of the cultivated herbaceums, having been domesticated either in Ethiopia or southern Arabia. It was probably the 'wool-bearing' tree of Theophrastus at Tylos on the Persian Gulf in 350 B.C. The Moslems spread it in Africa across to West Africa. This race is now of little commercial importance.

(3) *persicum* Hutch., a typical annual form developed to meet the limitations of the cold winters in Persia. It was widely spread round the Mediterranean by the Moslem invasion and was the first cotton cultivated in the Nile delta. This race has now spread as far as Afghanistan and Russian Turkistan.

(4) *kuljianum* Hutch., a small annual subshrub and now extends the furthest north of the cultivated cottons. It is a short-term race selected for areas of short, hot summers and long, cold winters and provides the very early cottons of Russia and western China.

(5) *wightianum* (Tod.) Hutch., large annual shrubs producing some of the better staples of the present Asiatic cottons. It was developed in western India from annual open-bolled forms introduced from Persia early in the 19th century.

2. *G. arboreum* L. Much-branched perennial shrubs to 2 m high or annual subshrubs 0·5–1·5 m high; capsules tapering, profusely pitted with prominent oil glands. 6 races recognized:

(1) *indicum* Silow, in its most primitive perennial forms such as the 'Rozi' cottons of western India, it is more closely related to *G. herbaceum* than are other races of *G. arboreum*. This race was taken to the East African coast in the 16th century and later spread to Madagascar, occurring as an escape from cultivation in some areas. The lint of the perennial forms

---

Fig. 53. **Gossypium herbaceum race wightianum:** ASIATIC COTTON. **A**, sympodial branch (×½); **B**, opening boll (×½); **C**, opened boll (×½). After Sastri (1956)—*Wealth of India*, Vol. 4, p. 176.

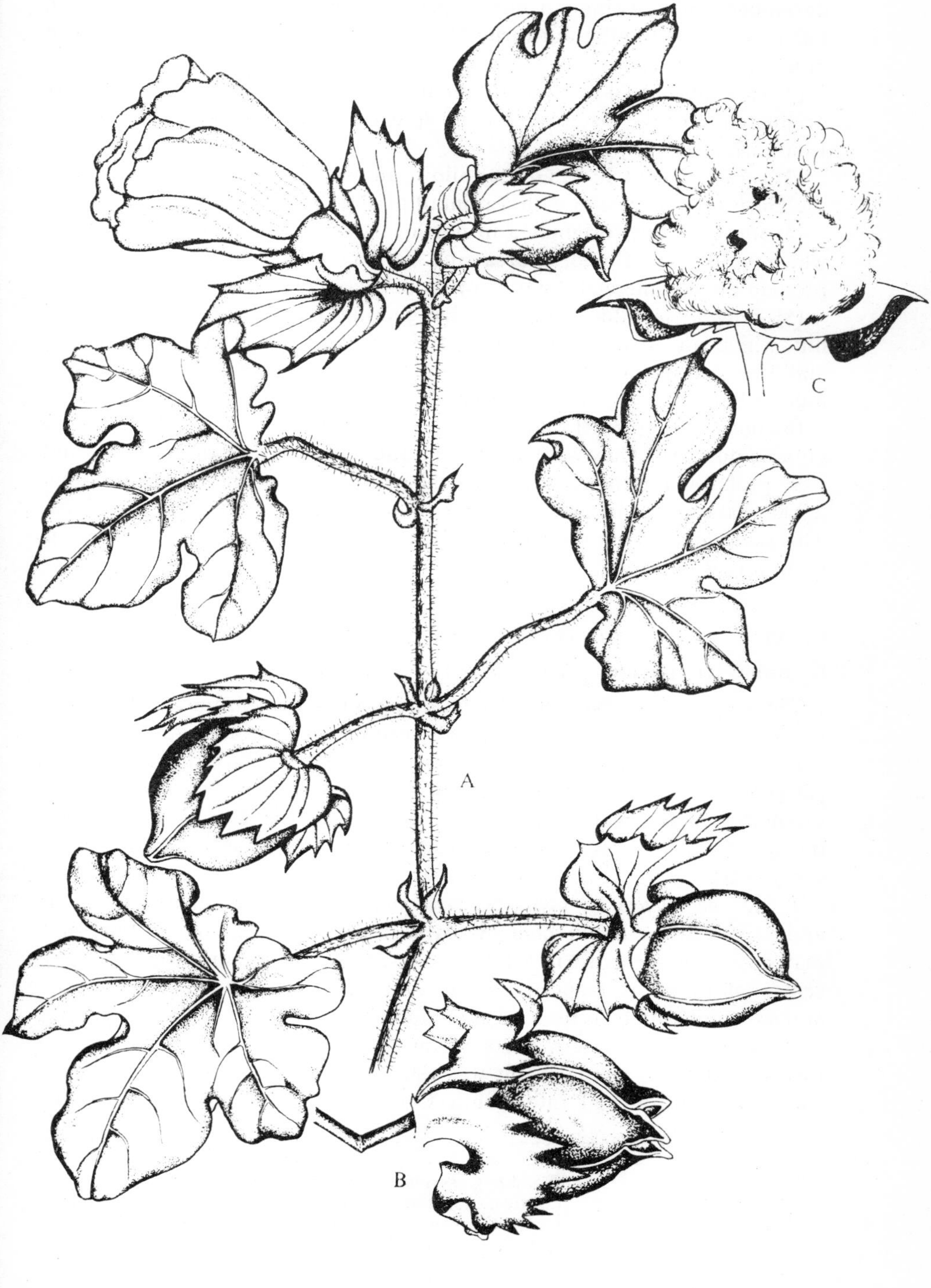
C
A
B

is scanty and coarse and is frequently coloured. Annual forms were developed with medium to fine staple, approaching that of *G. hirsutum*, but with a low ginning out-turn. The annual forms are extensively grown in peninsular India.

(2) *burmanicum* Silow, a predominantly perennial race, which probably originated in north-eastern India and spread to Burma and the East Indies. Lint variable.

(3) *cernuum* (Hutch. & Ghose) Silow, annual long-bolled type grown in the Garo hills of Assam with a rainfall of 150–200 in. per annum.

(4) *sinense* Silow, originated in north-east India and is the annual form which spread to China, Korea and Japan. These early-fruiting forms provide the commercial Asiatic cotton of these countries.

(5) *bengalense* Silow, annual, high-ginning forms, with coarse, short lint, which spread westwards to the Indo-Gangetic plain from East Bengal and Assam. It includes the 'Bengals' and 'Oomras' cottons, which contribute a great bulk of the crop in northern India.

(6) *soudanense* (Watt) Silow, perennial cottons with scanty lint and low ginning out-turn. They were taken from India to Africa at an early date and were probably the cottons grown by the peoples of ancient Meroe in Nubia 2,000 years ago, who were the first to spin and weave cotton in Africa. This race spread from the Sudan to West Africa and is now only grown for domestic use.

C. NEW WORLD LINTED COTTONS

Section VIII: **Hirsuta** (2n = 52) AD GENOME

Three spp. of allopolyploids, containing one set of chromosomes homologous with the Old World linted cottons (A genome), probably *G. herbaceum* and one set homologous with a New World wild species (D genome), probably *G. raimondii*. They are usually believed to have originated in tropical America (see below) and the 2 cultivated species include the best commercial cottons in cultivation and have now spread to every cotton growing country in the world.

1. *G. barbadense* L. (syn. *G. peruvianum* Cav.). Perennial shrubs or annual subshrubs 1–3 m high. Seeds with copious even coat of lint; fuzz may cover the seeds, form a tuft at one end or be absent altogether. The centre of origin is tropical South America. The wild prototype of *G. barbadense* is found in the arid mountainous region of northern Peru and it is in the north-western parts of the continent that variable perennial forms occur. The annual habit was established when seed from the West Indies was introduced into South Carolina in 1786, and gave rise to

---

Fig. 54. **Gossypium aboreum** race **bengalense** : TREE COTTON. **A**, sympodial branch (×½); **B**, unopened boll (×½); **C**, opened boll (×½). After Sastri (1956)—*Wealth of India*, Vol. 4, p. 170.

A
B
C

'Sea Island' cotton, which yields the highest quality lint of any cotton. There are no well-defined geographical varieties of perennial *G. barbadense*, except the two given below:

(1) *brasiliense* (Macf.) Hutch. with large leaves, large flowers, large bolls and with the seeds fused into a kidney-shaped mass. It is grown in eastern tropical South America. It includes the 'Rim de Boi' cotton of north-eastern Brazil.

(2) *darwinii* (Watt) Hutch., a perennial, much-branched shrub, with scanty, irregular, rather fine brown lint which is strongly adherent to the seeds. This variety is endemic in the Galapagos Islands.

2. *G. hirsutum* L. Perennial small trees and shrubs and small annual subshrubs in Central America from Guatemala northwards. Harland (quoted by Purseglove, 1965) states that wild *G. hirsutum* occurs in north-eastern Brazil. Hutchinson recognizes 7 races, of which *palmeri*, *morilli*, *richmondi*, and *yucatanense* occur wild on coastal dunes in Central America, or as relics of primitive cultivation. The other 3 races, which may be given varietal status, are:

(1) *marie-galante* (Watt) Hutch., perennial, late flowering, large shrubs or small trees to 7 m in height, the largest of the cottons, with many large ascending branches from the lower part of the stem. It is highly photoperiodic, flowering only occurring with short days. The capsule is usually tapering with few oil glands. The seeds have a copious coat of lint hairs, sometimes with a coat of fuzz, but this is often limited to a tuft at the tip. It occurs on the coasts of the Spanish Main, from Panama to Trinidad, north to the Greater Antilles and south to northern Brazil. It is found in house-yards and sometimes wild among dry coastal vegetation as on Chacachacare Island off the coast of Trinidad.

(2) *punctatum* (Schum.) Hutch., perennial, bushy shrubs, 1–3 m tall, branching freely from near base. The capsule is small, rounded, with a few oil glands sunk below the smooth surface. It occurs round the coasts of the Gulf of Mexico from Yucatan to Florida and the Bahamas, but is absent in Jamaica and the lesser Antilles. It has wide adaptability, but the bolls are small, the lint is short and has a low ginning out-turn, so is little used as a commercial cotton. Annual races have been developed in West Africa, which have a high resistance to bacterial blight disease, *Xanthomonas malvacearum*. The 'taitense' cotton of the Pacific and north-eastern Australia are sometimes referred to a separate variety (syn. *G. taitense* Parl.)

(3) *latifolium* Hutch., small, annual subshrubs, 1–1·5 m high, with few or no vegetative branches. Capsules are large and rounded with a smooth surface and with few oil glands. The seeds have copious lint and usually a thick coat of fuzz. The centre of origin is probably Mexico. The central American stock is predominantly photoperiodic, but, when taken to the southern United States about 1700, forms capable of fruiting irrespective of day length were selected to give 'Upland' cotton, which now forms the basis of much of the world's commercial cotton.

3. *G. tomentosum* Nutt. ex Seem., a densely tomentose perennial, 1–1·5 m tall, with seeds densely covered with brownish hairs, not readily separated into 2 layers and strongly adherent to the seed. This distinctive species is endemic in the Hawaiian Islands, where it grows on arid, rocky or clay plains not far from the sea.

ORIGIN AND DISTRIBUTION OF *Gossypium* spp.

The origin of the species of *Gossypium*, more particularly the New World cultivated cottons, has been a subject of considerable controversy and unanimity in this matter has not yet been reached.

A. THE WILD LINTLESS DIPLOID SPECIES: As has been shown above, the wild species of *Gossypium* occur in arid regions of the tropics and subtropics of Africa, Asia, Australia and America. They can be divided into well defined sections or groups and five distinct genomes (A–E) have been recognized. Saunders (1961), following Skovsted (1937), considers that the ancestral stock from which the genus arose was itself polyploid (n = 13), perhaps a combination of species with 6 and 7 chromosomes respectively, and that it was of African origin. Evolutionary change in chromosome size occurred and the differentiation of genomes A–E then took place in Central Africa. He believes that the occurrence of species with the D genome in America and the C genome in Australia finds an acceptable explanation in terms of Wegener's theory of continental drift.

Some botanists looking at the general distribution of angiosperms consider that this theory provides more problems than it solves. I believe that a more plausible explanation is by Antarctic land connections in Cretaceous or Tertiary times. Smith (1967) in his study of primitive angiosperm families, which he believes originated in the south-eastern Asiatic–Indo-Malaysian tropics, uses the Antarctic routes to connect South America and perhaps southern Africa for migration from the 'source area'. He adds 'hypotheses involving continental drift, northern origins and southward migrations, and transoceanic land bridges cannot satisfactorily explain the present distribution of primitive angiosperm taxa'. Fryxell (1965) considers that the original dissemination and diversification of the genus occurred in the Mesozoic under mesophytic conditions and that adaptation to more xerophytic environments in which *Gossypium* spp. characteristically occur today began in the early Tertiary.

B. THE OLD WORLD LINTED DIPLOID SPECIES: *G. herbaceum* is more primitive cytogenetically than *G. arboreum*. The mutation for lint is believed to have occurred in *G. herbaceum* race *africanum*, which occurs wild in the dry bush region of southern Africa from Angola and South-west Africa to Mozambique. Hutchinson (1959) considers that race *africanum* was taken at an early date along the ancient trade routes from Sofala to southern Arabia, where cotton was first domesticated and gave rise to the most primitive of cultivated diploid cottons, race *acerifolium*. Hutchinson (1962) states that this race is found as occasional perennial plants in

Ethiopia, southern Arabia and southern Baluchistan. As there was no evidence of spinning and weaving in southern Africa, it seems somewhat surprising that man should have taken race *africanum* to Arabia and there domesticated it. Nicholson (1960) puts forward the hypothesis that A genome cotton was gradually evolved in southern Ethiopia as an ancestral cultivated stock which spread into other regions in association with man and that present day *africanum* and *acerifolium* may be regarded as the separate descendants of a common ancestral cultivated stock. The annual races of *G. herbaceum*, *persicum* and *kuljianum*, developed as cotton was taken further north and had to produce a crop before the onset of winter. Annual races of *G. herbaceum* were not taken to India until about 1780 and these gave rise to the race *wightianum*.

G. arboreum race *indicum* in its more primitive perennial forms in western India is more closely related to *G. herbaceum* race *acerifolium* than other races of *G. arboreum* and must have been developed from it. *G. arboreum* then spread to the Indo-Gangetic alluvial plain giving rise to perennial northern forms, which are distinct from those of peninsular India. To these belong the oldest known specimens of Asiatic cotton, dated about 3000 B.C., found in excavations at Mohenjo-Daro in West Pakistan. Northern perennial *arboreums* spread throughout northern India, eastwards to Burma, Indonesia and the Philippines, giving rise to the race *burmanicum* and westwards across southern Arabia, the northern Sudan, the southern borders of the Sahara to West Africa, giving rise to the race *soudanense*. The latter was probably the cotton grown by the people of Meroë, an ancient Nubian Kingdom (650 B.C.–350 A.D.), who were the first people to spin and weave cotton in Africa. The northern perennial *arboreums* gave rise to annual cottons, including the races *bengalense* and *sinense*. The latter has been grown as a field crop in China since the 11th century.

C. THE NEW WORLD LINTED TETRAPLOID SPECIES: The New World allopolyploid cottons are hybrids between two diploids, one of which was closely related to the Old World linted cottons with the A genome and the other related to New World species with the D genome. Harland (1967, personal communication) thinks that the New World tetraploid cottons are polyphyletic in origin; that *G. barbadense* arose from a cross between an Old World linted cotton and a New World species like *G. raimondii*, and that *G. hirsutum*, which he says occurs wild in north-eastern Brazil, was probably the result of a cross between an Old World linted cotton and a species resembling *G. thurberi*. How the diploid species came together is a matter of conjecture. Harland (1939) suggested that they came together on a land bridge across the Pacific, although little evidence has been adduced for the existence of such a land bridge. Stebbins (1947) has suggested that the diploid Old World parent reached the New World by way of China and Alaska. Hutchinson and Stephens (in Hutchinson, Silow and Stephens, 1947) consider that *G. arboreum* was carried across the Pacific by man and was in pre-Inca Peru where it crossed with wild *G. raimondii* to give *G. barbadense*. In support of their hypothesis they state that 'in the same area an ancient civilization was located which used cotton as its basic textile and spun and wove it with instruments that differed in no essential particular from those used in Asia'. Hutchinson (1959) points out that *G. herbaceum* is nearer than *G. arboreum* to the ancestral line of the New World allopolyploids and now invokes the aid of man to bring this species across the Pacific and, in support of his theory, he quotes the finding of seeds of *G. barbadense* and fragments of primitive fabric and net fragments, dated 2400 B.C., in a pre-maize, pre-ceramic horizon in an ancient Peruvian habitation site at Huaca Prieta in northern Peru. They are said to have been made by fisher folk 'totally unlike the earlier inhabitants of the area', who practised a primitive agriculture. Implicit in the suggestion is the fact that man must have taken *G. tomentosum* in the opposite direction to Hawaii at some later date.

Purseglove (1960; 1963) could not accept the possibility of man taking *G. herbaceum* across the Pacific for the following reasons: (1) the difficulties of such a crossing by man some 4,500 years ago; (2) the voyagers do not seem to have taken any Old World food crops with them, as none had been found in the diggings or were established in the New World in pre-Columbian times; (3) only seeds of *G. barbadense* were found in the excavations and no Old World cottons or remains of them have ever been found in Peru; (4) it is even more difficult to envisage the transport by man of *G. herbaceum* race *africanum*, which occurs wild in southern Africa, than of *G. arboreum*; (5) the distinctive *G. tomentosum*, a wild endemic species in Hawaii, could not have become so differentiated in the time postulated; (6) it is not necessary to invoke the aid of man in the transfer of any other crop between the hemispheres before 1492 (see p. 13).*

Hutchinson (1962) abandons the aid of 'Pacific Regattas' and implies that the allopolyploids might have arisen in eastern South America before the continents drifted apart and that the early Peruvian cotton farmers might have come from the east over the Andes and not from the sea. This involves accepting an ancient and not a recent origin of the allopolyploids. Fryxell (1965) considers that the origin of the New World amphidiploid species took place during the Pleistocene, which coincides with a period of great fluctuations in sea level and hence a mobility of shore lines. He points out that the wild amphidiploid cottons are plants of littoral habitats, that most of them have very hard seeds, that their seeds can float in sea water for at least a year with undiminished viability, and, more recently (Fryxell, 1966, personal communication), that seeds have germinated after three years in sea water. Fryxell concludes that seeds are distributed by sea currents, that *G. tomentosum* reached Hawaii from Central America and the 'taitense' forms of *G. hirsutum* reached Polynesia from South America in the same way. Previously Stephens (1958) had stated that seeds of *G. barba-*

* The Peruvian cultivators had no boats, thus making trans-Pacific transport even more improbable.

dense var. *darwinii* have sufficient buoyancy to have enabled them to reach the Galapagos Islands from the mainland of South America.

Purseglove (1960; 1963) suggested that *G. herbaceum* could have reached South America in Tertiary times *via* Antarctica or that *G. barbadense* could have arisen in Antarctica and then retreated northwards as glaciation advanced. I now consider it more likely that seeds of *G. herbaceum* race *africanum* floated across the Atlantic from Africa to South America, as must have been the case also of the gourd, *Lagenaria siceraria* (q.v.), and it there crossed with a New World diploid to produce *G. barbadense*.

Cotton was grown in South and Central America and the West Indies when the New World was discovered and is recorded by Columbus. These ancient cottons were perennial shrubs and perennial *G. barbadense* spread as far south as northern Argentina. Locally they gave rise to the modern perennial Tanguis crop in Peru, from which annual forms have now been developed. In the West Indies they came into contact with earlier-fruiting, free-seeded types, which spread northwards through Colombia, the Spanish Main and the Greater Antilles. Seed of *G. barbadense* from the West Indies were taken to the coastal areas and islands of South Carolina in 1786 where the annual habit was established. These gave rise to Sea Island cotton, which was taken to some of the islands of the Lesser Antilles in the first decade of the 20th century. The annual habit permitted the establishment of the West Indian industry in the face of pests which rapidly become epidemic on perennial types. Sea Island cotton in the United States was virtually extinguished when the boll weevil, spreading across the Cotton Belt from Mexico, reached the eastern seaboard. Owing to its specialization for extreme length and fineness and the highest quality of all cotton, Sea Island has not spread widely.

Perennial forms of *G. barbadense* spread in post-Columbian times from eastern South America and the Caribbean to West Africa giving rise to the Ishan cotton of Nigeria and thence along trade and slave routes to the Sudan. From this source Jumel in 1820 established the commercial cotton crop in Egypt based on perennials. In 1850 these were crossed with Sea Island cotton to produce the very distinct, fine, annual Egyptian cottons, now grown under irrigation in Egypt and the Sudan.

The Caribbean tree cotton, *G. hirsutum* var. *marie-galante*, was the basis of early cotton cultivation in the West Indies. Apart from its area of natural distribution in Central and South America, this race is only known from Ghana, where it was introduced by Basle missionaries.

The second perennial form of *G. hirsutum*, var. *punctatum*, was introduced into West Africa about the end of the 17th century and spread across the continent south of the Sahara, replacing the Old World cottons for the local spinning and weaving industry, where the climate is too rigorous for Upland cotton. Annual forms were developed.

G. hirsutum var. *latifolium* occurs in southern Mexico and Guatemala with small-fruited forms throughout the area and large-fruited forms near Acala, the latter having given the modern high-yielding cvs now grown in

Arizona and California. Smith and MacNeish (1964) have described an archaeological specimen of a large-fruited cotton from southern Mexico dated 7000–5000 B.C. which is associated with agricultural development.* Cotton of the var. *latifolium* was taken from Mexico to the United States about 1700 and the early-fruiting, annual habit was established. The early introductions had green-fuzzed seeds and were grown in the upper country or interior by settlers for homespun; they became known as 'Uplands' to distinguish them from the black-seeded Sea Island cottons of the coast. Early in the 19th century grey- and white-fuzzed seeds were introduced and also became known as Uplands. With the invention by Whitney of the saw gin in 1793 and the rise of the Lancashire cotton industry, Upland cotton became a major crop and achieved a dominant place in world's cotton production. When supplies from America were cut off to Lancashire during the American Civil War, Upland cotton was introduced into most tropical and subtropical countries of the world. It now forms the basis of all commercial cotton crops of Africa outside the Nile Valley, all those of South America except Peru and northern Brazil, the modern Russian crop, and much of that of northern India and Pakistan.

The Cambodia cottons of southern India can be traced to direct introductions by the Spaniards from Mexico to the Philippines. It is a *latifolium* type which still retains the short-day habit. Its dense hairiness gives it resistance to jassids. It was introduced into India from Cambodia in the second decade of the 20th century.

KEY TO THE CULTIVATED SPECIES

A. Bracteoles entire or dentate; teeth usually less than 3 times as long as broad........................	Section *Herbacea*
B. Bracteoles flaring widely from flower, usually broader than long; upper margin usually with 6–8 teeth; capsule rounded or with prominent shoulders	*G. herbaceum*
BB. Bracteoles closely investing flower, longer than broad, entire or with 3–5 teeth near apex; capsule tapering..................................	*G. arboreum*
AA. Bracteoles deeply lacinate; teeth usually more than 3 times as long as broad........................	Section *Hirsuta*
B. Lvs deeply lacinate for ⅔ length into 3–5 lobes; petals usually bright yellow with basal reddish spot; anthers compactly arranged; pollen deep yellow; capsule coarsely pitted with black oil glands......	*G. barbadense*
BB. Lvs less deeply lacinate for ½ length or less into 3, or rarely 5 lobes; petals usually pale yellow or cream without basal reddish spot; anthers loosely arranged; pollen pale yellow or cream; capsule surface smooth	*G. hirsutum*

* This report now withdrawn and earliest cotton cultivation in Mexico probably about 3500 B.C. (Smith, C. E. (1968). *Econ. Bot.*, **22**, 253–266).

USES

Cotton lint is the most important vegetable fibre in the world today and is woven into fabrics, either alone or combined with other fibres. It was in early use in India, but in many countries it is the latest of the natural fibres to be used. In early times wool was the principal fibre of western and southern Europe, hemp in northern Europe, flax in Egypt, and silk and ramie in China. At the end of the 18th century cotton provided only 4 per cent of the world's raw textiles, compared with 78 per cent from wool. The invention of the saw gin and the development of the factory system, together with the ease of production and adaptability to machine manufacture, caused a rapid expansion in the use of cotton. By 1890 it provided 78·6 per cent of the world's textiles, increasing to 84·2 per cent in the period 1924–1928. Since then the percentage has decreased somewhat due to competition from rayon and other synthetic fibres. Although the bulk is used for textile manufacture, cotton also supplies yarn, cordage, twine and tyre cord.

The seeds yield a semi-drying edible oil which is used in lard substitutes (shortening), as salad and cooking oil, and in margarine manufacture. Low-grade oil is used in the manufacture of soap, lubricants, sulphonated oils and protective coatings. The residual seed cake, decorticated or undecorticated, is an important protein concentrate for livestock. Low-grade cake is used as manure. The whole seed may also be used as cattle feed and manure. Cotton seed hulls are used as roughage for livestock and as bedding, fertilizer and fuel.

The fuzz, which is not removed from the seed during ginning, is cut later to provide linters, which are used in felts, upholstery, mattresses, twine, wicks, carpets, surgical cotton, and in chemical industries such as rayon, plastics, lacquers, paper, photographic films, cellulose explosives and sausage skins. Unspun lint may also be used for these purposes.

CULTIVARS

Commercial cotton cvs can be classified on the length of lint as follows:

1. Short staple under $\frac{13}{16}$ in., usually rather coarse lint; produced mainly by cvs of *G. herbaceum* and *G. arboreum* in India, Pakistan and China; rain-grown.

2. Medium staple, $\frac{7}{8}$–$\frac{31}{32}$ in., mainly from Upland cvs of *G. hirsutum*; about 20 per cent of the world production; rain-grown; main production in the United States and India.

3. Medium long staple, 1–1$\frac{3}{32}$ in., mainly from Upland cvs of *G. hirsutum*; over 60 per cent of the world's cotton production; mainly rain-grown. This comprises about 75 per cent of the crop in the United States, with extensive production elsewhere, particularly in the Soviet Union, Mexico and Brazil.

4. Long staple, 1$\frac{1}{8}$–1$\frac{5}{16}$ in., from the long-stapled Upland cvs and cvs of *G. barbadense;* rain-grown or under irrigation; about 9 per cent of the world's production; with main production in Egypt, California, the Soviet Union, Peru and Uganda.

5. Extra long staple, $1\frac{3}{8}$ in. and over, about 6 per cent of the world's cotton production; from cvs of *G. barbadense* grown under irrigation in Egypt, the Sudan and Peru, and with a small production of Sea Island cotton in the United States and the West Indies.

Over 1,200 cvs had been recorded in the United States up to 1937. Efforts are now being made to limit the number of cvs and 90 per cent of the total cotton acreage is planted to 10 cvs. The American farmers replace the seed regularly from large seed producers, who maintain the purity by selective breeding. Three general groups of Upland cvs are recognized in the United States, but their terms short, medium and long staples do not conform to those used on a world basis given above. (1) Medium staple: about $1\frac{1}{16}$–$1\frac{1}{8}$ in.; medium to small-boll cvs grown in the eastern Cotton Belt, Mississippi Delta and the Gulf coast of Texas, of which the most widely grown are 'Deltapine' and 'Coker 100 Wilt'. (2) Short staple: about 1 in.; large-boll, storm-proof cvs grown in the Plains area of Texas and Oklahoma, of which the most widely grown is 'Lankart'. (3) Long staple: $1\frac{3}{32}$–$1\frac{3}{8}$ in.; large-boll 'Acala' type cvs grown in the irrigated areas of the south-western states; in California the planting of single cv., 'Acala 4-42' is enforced by legislation.

Old America Upland, long-season, medium-staple cvs were introduced into various countries in Africa early in the 20th century and cvs have been bred and selected locally to suit local conditions. These include 'BP52' and 'S47' in Uganda, Uk series in Tanzania, 'U4' and 'A637' in Malawi and 'Allen' in Nigeria.

Extra-long-staple Sea Island cotton has been selected in the West Indies and includes 'St. Vincent Superfine' and 'Montserrat Sea Island' and more recent hybrids. The extra-long-staple, irrigated, Egyptian cottons include 'Karnak', 'Giza 30' and 'Ashmouni'. Cvs of the extra-long-staple, longer-term, irrigated cottons of the Gezira scheme in the Sudan are based on the old Egyptian cv. 'Domains Sakel' in which resistance has been bred to leaf-curl and black-arm.

ECOLOGY

The wild species of *Gossypium* are tropical and subtropical in distribution. Commercial cotton production now extends from 37°N to 32°S in the New World, and from 47°N in the Ukraine to 30°S in the Old World. In its spread to the more temperate regions the annual habit was developed (see below). The crop cannot be grown successfully above 5,000 ft in tropical Africa or above 3,000 ft in India. The northern limit of production in the United States corresponds with the 77°F isotherm and with an average frost-free growing season of about 200 days. The optimum temperature for germination is 94°F, for the growth of seedlings 75–85°F and for later continuous growth 90°F. Low temperature increases the production of vegetative branches and extends the cropping period; high temperature increases the number of fruiting branches and reduces the cropping period.

Cotton is a sun-loving plant and cannot tolerate shade, particularly in the seedling stage. Wild cottons or escapes will not persist under forest conditions. Reduced light intensity, due to prolonged overcast weather, shading from interplanted crops or too dense a stand of cotton, retards flowering and fruiting and increases boll shedding. Perennial cottons are photoperiodic and are medium- to short-day plants. This has been lost in Upland and early annual cottons which are day-neutral.

The crop will not tolerate very heavy rainfall and, where grown as a rain-fed crop, the average rainfall is usually about 40–60 in. In arid areas it is grown with irrigation. The wild species are xerophytic and the ability to withstand drought has persisted in modern cultivated cottons so that they can recover from a dry spell and resume growth and fruiting. However, selection has permitted cultivation in more mesophytic environments. Adequate, but not excessive moisture is required for early vegetative growth; the first flowering period requires relative dryness, otherwise excessive boll shedding ensues; an increase in moisture is required for boll swelling and renewed growth, followed by dry weather for ripening and harvest. Up to 15 per cent more bolls are shed on days when rain falls during flowering. Continued flowering occurs if the bolls do not remain on the plant until the full capacity of the plant has been attained. In irrigated cotton in Egypt water loss through transpiration is 50 tons per acre per day or 3 pints per plant.

Cotton can be grown on a variety of soils from light sandy soils to heavy alluvium and Rendzina-type clays. Soil aeration, moisture and temperature are more important in germination and early growth than for starchy seeds due to the high oil and protein content. At the time of maximum leaf area the nitrogen content of the leaf is 2·5 per cent; thereafter the rate of nitrogen utilization overtakes the supply and internal starvation begins. The number of flowers laid down is dependent upon the nitrogen relationship and further growth of the boll upon carbohydrate supply. Apical growth then ceases, nitrogen uptake from the soil stops and nitrogen is withdrawn from the vegetative parts, particularly the bark. Too much nitrogen increases vegetative growth and delays flowering; too little nitrogen causes excessive boll shedding.

STRUCTURE

Unless otherwise stated the following account is based on American Upland cotton, *Gossypium hirsutum* var. *latifolium*, an annual subshrub, 1–1·5 m tall.

ROOTS: There is a well developed tap-root, usually with 4 rows of lateral roots. The zone of soil penetration is governed by soil structure, height of water-table and the age, size and fruiting of the plant. In deep, alluvial, irrigated soils in California roots reached a depth of 1–2 m when the plants were 20–25 cm high.

STEMS: Branching is dimorphic. Main stem is monopodial with leaves and branches, but no flowers. Two buds occur in each leaf axil, a true axillary bud and an extra-axillary or lateral bud, which may be on either side of the axillary bud. Usually only one bud develops. Lower axillary buds produce vegetative monopodial branches usually carried at an acute angle repeating structure of main stem. Higher on main stem, and also on lateral monopodia, extra-axillary buds grow to give sympodial fruiting branches, which are more horizontal than the monopodia. On these sympodial branches the terminal bud develops into a flower and growth is continued by an axillary bud in the leaf axil, giving the appearance of flowers being borne opposite the leaves. The lower the node on main stem at which first sympodial branch is produced the quicker maturing is the cotton and this varies with the cv. In Upland cottons none to 4 axillary buds grow into monopodia; in Egyptian cottons up to the 7th or 10th node may produce monopodia. Number of vegetative branches is influenced by weather, soil and photoperiod; more are produced in rainy weather on rich soils and on short-day cottons grown during long days. In hot climates growth of the stem usually takes place at night and ceases during the day due to water stress. Length of internodes is controlled by water supply; number of internodes by nitrogen supply. Last formed nodes are shorter than early nodes; later growth of terminal buds ceases due to deflection of carbohydrate and nitrogen from apices to developing bolls.

LEAVES: Variable in size, shape, texture and hairiness. Lvs spirally arranged on monopodia with a phyllotaxy of $\frac{3}{8}$ in Upland cottons and $\frac{1}{3}$ in Old World cottons. Lvs on sympodia have appearance of being alternate in 2 rows. Stipules falcate, 10 × 4 mm, caducous; petiole long; lamina cordate, as broad as they are long, 7·5–15 cm across, usually 3–5 lobed for about half their length, lobes broadly triangular, acuminate, often with multicellular stellate hairs; extrafloral nectaries usually seen on under-surface of 3 main veins. Stomatal counts in Egyptian cotton gave 44–97 per sq mm on upper surface and 116–176 per sq mm on lower surface.

FLOWERS: Floral formula: E 3–4 K(5) C 5 A (100–150) *G* (3–5). First flowers produced 8–10 weeks after planting and are borne singly and terminally on sympodial branches (see above) with 6–8 fls on each fruiting branch of average length. When bracteoles are first visible they are called squares and from this stage to opening of flower takes about 21 days. Flowering follows a spiral course beginning with the innermost bud on the oldest and lowest sympodium and finishing with the outermost bud on the youngest and highest sympodium; the vertical interval, *i.e.* first flower on first sympodium to first flower on second sympodium, is approximately 3 days, and the horizontal interval along each sympodium is approximately 6 days. Bud and flower shedding is a normal phenomenon occurring naturally in some perennials if copious supplies of water are available; this facultative shedding ensures that the fruit is only brought to maturity if the weather is dry. Shedding can also occur in annuals when nutrient supply is insufficient or under other adverse conditions. In all cases shedding is caused by an abscission layer one cell thick near base of peduncle.

Flowering normally continues for two months, but under optimum conditions the bulk of the crop is derived from the first 3–4 weeks of flower production. The time taken from flowering to the opening of the boll is about 45–65 days.

Fls stalked with 3 nectaries near top of peduncle. Epicalyx of 3, occasionally 4 bracteoles, which protect fl. in bud and usually persist until harvest; in *G. hirsutum* bracteoles longer than broad, cordate, lacinate with 7–12 acuminate teeth, more than 3 times as long as broad. Calyx reduced to cup-shaped structure with 5 lobes and with ring-shaped nectary inside and 3 nectaries outside near base. Corolla large, showy, about 5 cm in diameter; petals 5, convoluted in bud, free except where fused to staminal column. Corolla in Upland cvs creamy-white to pale yellow on opening, turning pink towards close of first day, red on second day, withering and shed on third day together with staminal tube and style. Corolla of Sea Island and Egyptian cvs deeper yellow with red or purple spot at base of each petal. Stamens united to form tubular staminal column with 5 vertical ridges to which fine filaments are attached, each ending in small 1-celled reniform anther, pale yellow to deep gold in colour; upper filaments longer near top of staminal tube. In *G. barbadense* anthers closely packed on short filaments which are all about the same length. Pollen grains large, sculptured, spinose, spherical, with numerous germ pores. Ovary superior, of 3–5 united carpels, with several ovules in axile placentation; style inside staminal tube, exserted stigma united, lobed.

FRUITS: Spherical or ovoid leathery capsules, called bolls, 4–6 cm long, splitting on maturity along carpel edges into several valves or locks, exposing linted seeds, which are picked when dry from fully opened bolls to give 'seed cotton'. Bolls of Upland cotton are light green, smooth, with few oil glands, usually 4–5-locular with 5–11 seeds per loculus. Bolls of *G. barbadense* are darker green in colour, pitted with abundant oil glands, usually 3-locular with 5–8 seeds per loculus. Bolls grow to full size in about 25 days after opening of the flowers and seeds develop for a further 25 days before the boll opens, although cv. and environmental differences occur.

SEEDS: Pyriform, about 1 cm long in Upland cotton, dark brown after removal of fuzz. Aborted ovules produce motes, which cause difficulty in spinning. Embryo slow to develop, being visible to the eye on the 15th day; as it develops it uses up the endosperm until only a thin papery perisperm is left surrounding embryo and highly convoluted cotyledons. The weight of 100 seeds of Upland cotton is 10–13 g.

Fig. 55. **Gossypium barbadense :** SEA ISLAND COTTON. **A**, monopodium and sympodium with flower (×½); **B**, flower with bracteole removed (×½); **C**, flower in longitudinal section (×1); **D**, unopened boll (×½).

A
B
C
D
M. WONG

LINT: Epidermis of seeds of cultivated cottons has two types of unicellular hairs, namely, short unconvoluted hairs or fuzz, which is firmly attached to the testa and provides the linters of commerce, and long convoluted hairs or lint, which is usually white in commercial cotton, evenly distributed over the surface and easily removable. Lint is formed from a single cell of the epidermis by rapid outward extension of its external wall. In Egyptian cotton lint grows in length for the first 25 nights, growth being interrupted during the day, with maximum growth on the 15th night. Environmental conditions for 5 days before and after the 15th night determine the lint length; if these are adverse lint length below its inherited nature will be produced. Growth in thickness begins on the 21st night and continues until the bolls open. Spiral bands of cellulose are laid down internally on successive nights. The direction of the spiral may be reversed at any point and this, together with different rates of deposition, produces convolutions when the lint collapses into ribbons through water loss on drying when the bolls open. This causes the locks of seed cotton to expand and fluff out. The periods required for increase in length and thickness vary with the species and cvs. Hairs which have little secondary thickening produce neps in spinning and also react differently to dyes.

ESTABLISHMENT OF THE ANNUAL HABIT: Fruiting in perennial cottons is controlled by: (1) the node at which the first sympodial fruiting branch appears irrespective of the external environment, thus determining the onset of the reproductive phase in the first season; (2) photoperiodicity; being a short-day plant, this determines the reproductive capacity set up in conformity with the season; (3) facultative shedding of the flower buds if copious supplies of water are available, so that fruit is only brought to maturity in dry weather. When most tropical perennial cottons are taken to temperate regions they become highly vegetative and produce few or no fruiting branches when grown during long days. However, in its spread to temperate countries, Upland cotton (*G. hirsutum* var. *latifolium*) developed annual forms under man's selection in which all the periodicity controls were lost. Sympodial fruiting branches were produced low down on the main axis regardless of the day length or the duration of the wet season, the whole of the plant's development being subordinated to the demand for a quick and copious crop. The annual habit was also established in *G. hirsutum* var. *punctatum* in West Africa, Sea Island and Egyptian cvs of *G. barbadense*, and the northern races of *G. herbaceum* and *G. arboreum* in Asia. None of the early annual forms of New and Old World cottons shows successful reversion to the ancestral perennial form and retains the annual form regardless of the conditions under which

---

Fig. 56. **Gossypium hirsutum** var. **latifolium** : UPLAND COTTON. **A**, sympodial branch ($\times\frac{1}{2}$); **B**, flower from side ($\times\frac{2}{3}$); **C**, flower from above ($\times\frac{2}{3}$); **D**, flower in longitudinal section ($\times\frac{2}{3}$); **E**, unopened boll ($\times\frac{2}{3}$); **F**, opened boll ($\times\frac{2}{3}$).

A
B
C
D
E
F

they are grown. Consequently they cannot persist in a wild state, while some of the perennials, which have retained some of the primitive characters, may become established successfully in natural open vegetation.

POLLINATION

The stigma is receptive when the flower opens and pollen is shed at the same time so the majority of flowers are self-pollinated. The flowers are visited by insects and some degree of cross-pollination occurs, normally ranging from 6–25 per cent, but over 50 per cent has been reported, depending on the proximity of the plants and the insect population. The percentage cross-pollination at 100 m is about 0·1 per cent, so this is the minimum distance required to prevent intercrossing. The pollen remains viable for 12 hours and the period from pollination to fertilization is 30 hours. Self pollen is prepotent over foreign pollen in interspecific populations.

In artificial cross-pollination the flowers are emasculated by removing the corolla and staminal tube on the afternoon before the flower opens when the corolla is still twisted and folded. The pistil is then flushed with water to remove any pieces of anthers and the flower is bagged. Pollination is usually done the next morning. Some workers place the anthers in a length of soda straw which is pulled over the stigma and is kept in place by wiring. About 75 per cent of hand-pollinations are successful. For self-pollination the flowers must be bagged or the tip of the corolla may be fastened with fine copper wire, a wire cone, paper clip, rubber band, collodium, nail polish, or small piece of lint wound round and smeared with clay.

GERMINATION

The radicle emerges through the micropyle, followed by the hypocotyl which becomes arched and appears above the ground in 5–15 days. The hypocotyl then straightens lifting out the much folded cotyledons which rapidly expand into broad heart-shaped leaves on short petioles. Seeds of some cvs have a period of dormancy before germination of 2–3 months after harvesting. A good sample should give 85–100 per cent germination, but this is reduced by storing for 2 years or more under humid conditions and the seeds lose viability rapidly if the moisture content exceeds 10 per cent. Seed in which the moisture content has been reduced to 7 per cent can be stored in sealed jars for up to 15 years.

Seeds may be treated with mud or cow dung to make sowing easier. It is usually recommended that fuzzy seeds should be delinted by machine or concentrated sulphuric acid and has the following advantages: (1) reduces the weight and bulk of seed to be transplanted; (2) facilitates uniform sowing; (3) gives quicker and higher germination; (4) seed is easier to treat with fungicidal powders against seed-borne diseases; (5) acid treatment gives partial sterilization of seeds and during washing permits separation of light seeds which float.

Organized seed multiplication and distribution is necessary in order to ensure maintenance of purity. This is done by maintaining a continuous flow of seed from the plant breeder, through multiplication plots and isolated seed production areas, for distribution to the growers. In peasant cultivation ginners are normally required to set aside seed for planting, which should be stored under cool dry well-ventilated conditions, and then to transport it to the seed distribution centres.

CHEMICAL COMPOSITION

Cotton lint contains an average of 94 per cent cellulose, $(C_6H_{10}O_5)_n$, on a dry weight basis. Commercial cottonseed contains approximately 92 per cent dry matter, 16–20 per cent protein, 18–24 per cent oil, 30 per cent carbohydrate, and 22 per cent crude fibre. As obtained from the gin cottonseed includes remnants of unginned lint and the fuzz and yields on an average 16 per cent crude oil, 45·5 per cent cake or meal, 25·5 per cent hulls, 8 per cent linters, with a manufacturing loss of 5 per cent. Decorticated cottonseed cake contains approximately 36 per cent protein and undecorticated cake approximately 21 per cent protein. The cotyledons are rich in resin which gives a blood red colouration with concentrated sulphuric acid and this is used as a standard test for cottonseed cake. The principal pigment in the seed is gossypol, a poisonous phenolic compound, of which the kernels contain 0·4–2·0 per cent, *G. barbadense* having the highest content. Most of the gossypol is rendered harmless on crushing or heating by union with protein, but the meal may still contain minute quantities to which pigs and chickens are sensitive.

PROPAGATION

Commercial cotton is always grown from seed. Perennial cottons, such as the 'Tanguis' cotton of Peru, can be cut back and ratooned. For experimental purposes cotton can be propagated vegetatively by cuttings, budding or grafting.

HUSBANDRY

ESTABLISHMENT: Preparation of the land should be early and only to a depth necessary to control weeds. The seedbed should be clean, fairly fine and firm. Planting should be early, as soon as rainfall occurrence and expectation are adequate for germination and growth of the crop. Peasant-grown cotton commonly suffers from planting being delayed until after that of the food crops. In East Africa cotton is usually the first crop taken after a bush fallow and is grown in rotation with annual food crops. In the Sudan Gezira it is planted after a bare fallow. Cotton is sometimes interplanted with short-term crops such as groundnuts or beans.

In hand planting, cotton is usually sown with a seed rate of 10–12 lb per acre at a depth of about 1 in. with 3–6 seeds per hole in rows or ridges.

Ridges are an advantage as they can be tied to conserve water under dry conditions and aid drainage under wet conditions. The optimum spacing depends on the size and fruitfulness of the plant permitted by local conditions, and on the interactions between cv., soil and fertilizer treatment, climate including light intensity, rainfall and length of growing season, time of planting, and the incidence of pests and diseases. In Egypt and the Sudan with ample and regulated water supply from irrigation, ample sunshine and fertile soil conditions, a close spacing with 2 plants per hill 12 in. apart in rows 26 in. apart is used. In western and central Uganda with low light intensity due to prolonged dull weather, intermittent water supply due to unreliable rainfall, and unpredictable disease and pest attack, rain-fed cotton is planted at a wider spacing of 3×1 ft with 2 plants per hill in order to give well-developed individual plants capable of persistent cropping and with good powers of recovery from drought or pest damage. Thinning is done when the plants are 6–10 in. high and 2 plants per hill are usually left. After-cultivation serves no useful purpose except to control weeds.

In the United States in mechanical planting the rows are usually 40 in. apart with the plants thinned to 2 per hill 8–18 in. apart. The seed rate is about 30 lb per acre of fuzzy seed, 25 lb per acre of mechanically defuzzed seed, or 18–20 lb per acre of chemically defuzzed seed. The seed rate is increased if the crop is thinned mechanically.

MANURING: Much of the cotton in the tropics is grown without the application of manures. Responses are more commonly obtained from nitrogen than from any other nutrient. The usual recommendation is 1–2 cwt of sulphate of ammonia per acre, half applied at planting and half at thinning or just before flowering. On acid soils nitrochalk should be used. Deficiency of nitrogen produces chlorosis, meagre growth and increased boll shedding. Large direct and residual responses have been obtained to phosphates on some soils, when 1–2 cwt of double superphosphate per acre may be applied at planting. Phosphorus deficiency causes dark green leaves, and delayed flowering and fruiting. Responses to potash are not common in the tropics, but are widespread in the United States and 1 cwt of muriate or sulphate of potash may be applied at planting. Potash deficiency produces mottled yellowish leaves with brown spots, the margins of which dry and curl inwards, after which they dry out and are shed prematurely. Placement of fertilizers at planting is better than broadcasting and they are best placed in bands 2 in. wide, $2\frac{1}{2}$ in. from the row and 2 in. below the seed level.

HARVESTING: Picking should be done regularly, so that open cotton is not left in the field which might result in deterioration in the colour and quantity of the lint. Cotton should be picked clean and free from pieces of leaves, bolls, bracteoles and twigs. It is then sorted into clean and stained cotton before marketing. Much of the world's cotton is picked by hand. Harvesting begins about 6 months after planting and 3–4 pickings are usually made. Mechanical harvesting by stripping or spindle-picking began in the United States in the 1940's, and by 1955 23 per cent of the crop was harvested mechanically. In many countries legislation exists for the compulsory uprooting and burning of plants after harvest with the object of minimizing the carry over of pests and diseases from one season to the next.

GINNING AND BALING: In the tropics much of the cotton is grown by smallholders who sell their seed cotton to the ginners. The ginneries may be privately or co-operatively owned. Lint is removed by saw gins for the shorter-stapled cottons and by roller gins for the longer-stapled fine types. The percentage of lint, known as ginning out-turn, is 30–40 per cent. The lint index is the weight of lint from 100 seeds. The lint is baled under pressure and covered with hessian or other material. The bale weight varies from 400 lb in most Commonwealth countries, 500 lb in the United States and 700–800 lb in Egypt.

YIELDS: The average yield of lint per acre varies considerably between countries, ranging from 100–2,000 lb. Irrigated cotton produces the highest yields with up to 3,000 lb per acre in California. Those in Egypt are about 550 lb per acre. Yields of the rain-fed crop in India and East Africa are about 100–200 lb per acre.

LINT QUALITY: Lint is classified according to the following characteristics: (1) length: the normal staple length is the average length of the longer fibres which may be measured by hand sampling, halo measurement, cotton sorters or fibrograph; (2) maturity: lint showing immaturity contains a fair proportion, 25 per cent or more, of thin-walled hairs, which result in neppiness and other manufacturing difficulties; (3) fineness which is associated with long hairs with small cell diameter and properly thickened walls and is an indication of maturity and strength; (4) fibre strength which depends partly on the cross-sectional area of the hair walls and is measured by determining the breaking strain; (5) yarn strength which depends partly on fibre strength and partly on the efficiency with which the individual fibres twist about each other and cling together in spinning, which in turn depends upon hair length, fineness and convolutions; (6) uniformity of lint which is dependent upon growing pure strains or cvs, sorting the seed cotton before ginning, efficient ginning, and classifying the lint into standard grades after ginning; (7) absence of faults such as freedom from foreign matter including bits of leaves, bracteoles and seed-coat fragments, a good colour without staining or spotting from insect or fungal damage and without greyness due to exposure to weather, freedom from neps (small tangled knots of hairs due to immaturity) and motes (aborted ovules or immature seeds).

MAJOR DISEASES

Bacterial blight, caused by *Xanthomonas malvacearum* (E. F. Sm.) Dowson, has now spread to most cotton-growing countries. It produces water-soaked lesions on the cotyledons, leaves and bolls. Those on the leaves later appear as angular discoloured or dead spots bounded by veins (angular leaf spot). Those on the bolls later produce blackened lesions, which cause shedding of young bolls, destruction of one or more locks of older bolls and premature opening. Infection of the petiole and stem produces large blackened lesions and kills the branch terminals (black arm). Losses of up to 50 per cent of the crop may occur. It is carried by seed and plant debris, but not by soil. Mercurial dressing of seeds gives partial control. *G. hirsutum* var. *punctatum* in West Africa is virtually immune and is used in breeding to obtain resistance.

Fusarium wilt, *Fusarium oxysporum* Schlecht. f. *vasinfectum* (Atk.) Synder & Hansen, is now widespread. It is soil-borne and causes death or stunting of the plant with a yellowing and wilting of the leaves and discoloration of woody portion of stem, first appearing as a brownish ring inside the cambium. 'Coker 100 Wilt' is a resistant cv.

Verticillium wilt, caused by *Verticillium alboatrum* Reinke & Berth. in the New World and *V. dahliae* Kleb. in Africa, is soil borne and favoured by cold wet weather and irrigation. It causes stunting, chlorotic mottling of the leaves and shedding of leaves, squares and bolls. Some cvs of *G. barbadense* are fairly resistant.

Leaf curl, a serious virus disease in the Sudan, is transmitted by the white fly, *Bemisia tabaci* (Genn.) (syn. *B. gossypiperda* M. & L.). All parts of the plant are distorted—stems become twisted and spindly, leaves curl and crinkle, veins thicken and lamina has chlorotic spots and streaks. It is controlled by strict plant sanitation and a closed season or by the use of resistant cvs.

MAJOR PESTS

Pearson (1958) lists 150 spp. of cotton pests in Africa. Bollworms, which are the caterpillars of several species of insignificant moths, are among the most serious pests of cotton. They feed in the bolls damaging lint and seed and cause very considerable reduction in yield and quality. The main spp. are: pink bollworm, *Platyedra gossypiella* Saund., widely distributed in Africa, Asia and America; spiny bollworm, *Earias biplaga* Wlk. and *E. insulana* (Boisd.), in Africa and Asia; Sudan or red bollworm, *Diparopsis castanea* Hmps. and *D. watersi* (Roths.), in Africa; cotton bollworm or corn earworm, *Heliothis zea* (Boddie) in America and *H. armigera* (Hubn.) in Africa; South American bollworm, *Sacadodes pyralis* Dyar, in South America. Some measures of control can be obtained by strict phytosanitation, a closed season and an early maturing crop. *Gossypium thurberi* is markedly resistant and an attempt is being made to introduce this into the cultivated cottons.

Boll weevil, *Anthonomus grandis* Boh., which is indigenous in Mexico, was first observed in the United States in 1892 in southern Texas and spread rapidly northwards and eastwards. It is the worst pest in the United States attacking young squares, bolls and terminal buds. As the weevil population increases as the season progresses most of the squares set late will be injured, so early rapid-fruiting cottons have been bred to escape damage.

Leaf, stem and bud sucking bugs which can cause considerable damage to cotton include: jassids, *Empoasca devastans* Dist., in India, *E. lybica* (de Berg), in the Sudan, and *E. facialis* (Jacobi), in the rest of Africa; *Helopeltis* spp., in Africa; *Aphis gossypii* Glov., which is cosmopolitan; *Lygus vosseleri* Popp., in Uganda and the Congo.

Cotton stainers, *Dysdercus* spp., occur in all cotton-growing countries except Egypt and most of the United States. They feed on the seeds and are the vectors of the fungi, *Nematospora* spp., which cause internal boll-rot.

IMPROVEMENT

The cultivated cottons show extensive variability and plasticity compared with the wild perennial species and races. Much of this diversity must have arisen since domestication. This occurred due to the development of the annual habit, the spread through a great diversity of environments and the vast planted populations of the crop. Mutations and genotypes which fail in stationary populations may be preserved in an expanding population and through the agency of man, and there is the opportunity for gene exchange, recombination through crossing, and introgression from other species. There is sufficient natural cross-pollination to maintain heterozygosity which is accompanied by a moderate degree of heterosis. The new variability contributes to the synthesis of new characters which may increase the adaptability to the new conditions under which the crop is grown, which now includes many different types of habitat. Thus characters required by the breeder may be found in the area of spread or at its periphery, rather than at the centre of origin. Examples of this are the resistance to black arm in *G. hirsutum* var. *punctatum* in Nigeria and the hairiness of Cambodia cottons which provides resistance to jassids.

METHODS

(1) INTRODUCTION AND ACCLIMATIZATION: This occurred in the development of the cotton industries of the United States, Africa and elsewhere. Subsequent natural crossing then took place between cvs and races. Introductions are now primarily used for hybridization.

(2) SELECTION: This is used to maintain the purity of existing cvs and for the development of new cvs both from open- and self-pollinated plants. Mass selection is seldom used now except sometimes for the maintenance of cv. type. Primary selections are from single plants of a large commercial crop with large variability and often lead to a large initial advance. Progeny or pedigree selections are made in more uniform material and consist of a careful assessment of a smaller range of material under

controlled breeding. The selections are open- or cross-pollinated in progeny rows. It is usual to mix several similar progeny rows or blocks in each wave of increase, provided none are off-type, in order to produce a wider basis of gene exchange, as stable mixtures of strains provide more flexibility in varietal responses to environmental factors. Selections within selfed lines maintain the highest degree of uniformity. Inbreeding for several years results in a reduction of yield of 10–15 per cent in some cases.

Manning developed a further refinement by a selection index in which it was possible by precise statistical assessment 'to determine the heritability of each of the characters considered to contribute to the yield of a single plant progeny and so to combine the estimates of heritability as to provide a single estimate of net worth on which progenies can be selected' (Hutchinson, 1959). A selection differential was set and only those progenies which pass this limit contribute to the next generation. The population was shifted in the direction desired, maintaining as far as possible variability which is the source of further improvement. As a constant check is maintained on the rate of improvement, it is possible to decide when to abandon the project and put the stock on a maintenance basis and proceed to more variable stocks or hybridization.

(3) HYBRIDIZATION as a result of natural, controlled or artificial pollination. The technique is given under pollination above. The types of hybrids are as follows:

(a) Intervarietal hybridization: cvs and races of a species are cross-fertile, the genotype dissolving in the common gene pool. It is used for the development of new types.

(b) Interspecific hybridization: species within a section of the genus usually give fully fertile F_1's, but extensive genetic breakdown occurs in the F_2's maintaining the species integrity. In spite of the genetic barrier introgression does take place naturally but is a long process. With repeated backcrossing this gene transfer can be achieved more quickly and it permits the integration of a portion of the gene complement of one species without the risk of genetic breakdown and ensures the maintenance of genetic balance in the recurrent parent. It is used for transferring new characters to old types. Where the species barrier is far apart crossing is very difficult or almost impossible to achieve.

(c) Panmixia: by growing a number of cvs together and hand-pollinating between them a highly variable population can be synthesized with free gene exchange to produce a long-term, plant-breeding reserve at little cost. In Uganda a panmixis of 19 *latifolium* cvs from a number of countries was chosen and the more important characters desired from them were high yields, persistent cropping, quality, and resistance to jassids and black arm. The interpollinated bolls were marked and were harvested in bulk. This was repeated annually.

(d) Hybrid vigour has been suggested as a means of increasing yields. Cytoplasmic male sterility has not been found in cotton, but selective gametocides are being developed which prevent pollen development in some cvs but not in others. Thus by interplanting with a cv. not affected and spraying several weeks before blooming, the plants rendered male sterile are cross-pollinated, and the cheap production of hybrid seed should be possible.

AIMS AND RESULTS

(1) MAINTENANCE OF TYPE: This is extremely important where the market demands a complete uniformity and maintenance of set standards, *e.g.* 'Acala' types in the south-western United States, which are maintained by selection within selfed lines, 'Domains Sakel' in the Sudan, 'Sea Island' in the West Indies. In St. Vincent 20 plants of 'Sea Island Superfine V135' were selected each year and grown in progeny rows. These were then bulked and grown for a further 3 years when sufficient seed was available for the whole island. In the 6th year the stock were discarded and crushed for oil and the next wave followed on. Thus 3,000 acres with some 20 million plants were produced from the original 20, thus ensuring extreme uniformity.

(2) YIELD AND QUALITY: High yield of lint of acceptable quality is the ultimate object of the breeder. Yield is determined by the number, size and weight of bolls, the number of loculi per boll and the ginning out-turn, which in turn depends upon the number of seeds per loculus, the size of seed and the lint index. Yields are also affected by diseases, pests, harvesting losses, etc. Quality characters include staple length, fibre and yarn strength, fineness, uniformity and the percentage of mature fibres.

In Uganda using a selection index technique, Manning reported an annual 4 per cent increase in yield over 8 generations in 'BP52', one of the longer staple Upland cvs. There has been an improvement in staple length in the United States, where in 1930 three-quarters of the total acreage was planted with cvs which had a staple length of 1 in. or less; by 1950 these accounted for only one-third of the acreage.

(3) MATURITY: Early and rapid fruiting and uniformity of maturity is usually desirable, particularly in temperate countries, as it permits development during the most favourable weather, reduces losses from diseases and pests, particularly boll weevils in the United States, permits picking before unfavourable weather and killing frost, and facilitates machine harvesting.

Climatological and cropping studies in Uganda show a relationship between sowing date and crop production and that cropping is limited by water strain at the time of the largest leaf area. By breeding a type with a longer vegetative period, it might be possible to obtain the maximum leaf area when water is not limited and to build up food reserves which may be used later when the leaf area–soil moisture relationship becomes critical. An attempt is being made to obtain a cv. which can be planted early and which will not ripen its crop until the end of the rains.

(4) METHOD OF HARVESTING: For hand picking, large bolls which open widely are easier to harvest. For mechanical spindle picking the following characters are desirable: bolls which open widely and permit the seed cotton to fluff out; sufficient storm resistance for the cotton to stay in the loculi; compact plants with bolls well spaced along the stems and set high off the ground; plants which have an early short fruiting period and a natural tendency to shed the leaves or are easy to defoliate by chemical defoliants; plants with smooth leaves and a small or deciduous epicalyx. In mechanical stripping the entire bolls are removed and cvs should be short with short fruiting branches, with early fruiting and maturity, with the bolls borne singly and high off the ground, and with lint firmly attached in the bolls, which may never open fully as in 'Macha'.

(5) RESISTANCE TO DISEASES AND PESTS: Resistance to bacterial blight (*Xanthomonas malvacearum*) is obtained from *G. hirsutum* var. *punctatum* in West Africa and in resistant cvs of var. *latifolium*. The Upland cv. 'Allen' has types which are virtually immune to bacterial wilt. 'Albar 49' and 'Albar 51' are immune cvs developed from 'Allen'. Knight in the Sudan obtained resistance to bacterial wilt in 'Domains Sakel' cv. of *G. barbadense* without serious change in type by recurrent backcrossing after the transference of genes from the *latifolia* and *punctatum* races of *G. hirsutum* and from *G. arboreum*, the latter species conferring complete immunity. It was shown that resistance was due to 6 major identifiable genes and several minor genes.

Resistance to *Fusarium* wilt in the United States was found and cvs such as 'Coker 100 Wilt' and 'Stonewilt' have been bred which are not only wilt resistant, but can compete with other cvs on non-wilt soil. They are also resistant to nematodes. Although cvs are available which are tolerant to *Verticillium* wilt, immunity has not yet been achieved. Resistance to the leaf curl virus in the Sudan has been obtained in 'X1730A', a selection from 'Domains Sakel'.

Jassids are a serious pest in Africa and India. A dense coating of long hairs on leaves and stems provides resistance to the bugs. The Cambodia cottons of southern India are resistant to attack and the extreme hairiness of modern jassid-resistant strains of African Uplands owes much to deliberate hybridization or accidental crossing with Cambodia. One of the first resistant cvs was 'U4' bred by Parnell at Barberton in South Africa, and others were later bred in East Africa and in the Sudan. No resistance has been obtained in the United States to the boll weevil, but early rapid-fruiting cvs have been bred which escape damage.

PRODUCTION

A brief history of the rise of the cotton industry is given under USES. It was only in the 19th century that cotton became the leading fibre for the manufacture of textiles. World production in the 1890's was around 6,000 million lb, which had risen by 1920 to about 12,000 million lb. World production by 1960 had reached nearly 24,000 million lb. The leading producers are the United States with an annual production of over 7,000 million lb, China and the Soviet Union with over 3,000 million lb each, India with over 2,000 million lb, and Egypt and Mexico with over 1,000 million lb each. In British Commonwealth countries in Africa, Uganda produces about 150 million lb, Nigeria about 100 million lb and Tanzania about 75 million lb. Production in the Sudan is about 250 million lb and in the West Indies 1 million lb. The total world acreage is about 80 million acres.

The principal exporting countries are the United States, Egypt, the Soviet Union and Mexico. China uses almost all the cotton she produces, while India imports more than she exports. The principal importing countries are Japan with 1,500 million lb and Western Germany, France, United Kingdom and Italy each with 600–750 million lb annually. The *per capita* consumption of cotton in the United States is 25–30 lb annually, providing about 70 per cent of total fibres used. The annual consumption of cotton in the United States is about 4,000 million lb.

REFERENCES

ARNOLD, M. H. (1963, 1965). The Control of Bacterial Blight in Rain-grown Cotton. *J. Agric. Sci.*, **60**, 415–427, **62**, 29–40.

BALLS, W. L. (1953). *The Yields of a Crop based on an Analysis of Cotton Grown by Irrigation in Egypt.* London: Spon.

BROWN, C. H. (1953). *Egyptian Cotton.* London: Leonard Hill.

BROWN, H. B. and WARE, J. O. (1958). *Cotton.* 3rd Ed. New York: McGraw-Hill.

CARDOZIER, V. R. (1957). *Growing Cotton.* New York: McGraw-Hill.

CHRISTIDIS, B. G. and HARRISON, G. J. (1955). *Cotton Growing Problems.* New York: McGraw-Hill.

EMPIRE COTTON GROWING CORPORATION. *Empire Cotton Growing Review* and *Progress Reports from Experimental Stations.*

FRYXELL, P. A. (1965). Stages in the Evolution of *Gossypium* L. *Adv. Frontiers Pl. Sc.*, **10**, 31–56.

HARLAND, S. C. (1939). *The Genetics of Cotton.* London: Cape.

HUTCHINSON, J. B., SILOW, R. A. and STEPHENS, S. G. (1947). *The Evolution of Gossypium.* London: Oxford Univ. Press.

HUTCHINSON, SIR JOSEPH (1959). *The Application of Genetics to Cotton Improvement.* Cambridge: Univ. Press.

HUTCHINSON, SIR JOSEPH (1962). The History and Relationships of the World's Cottons. *Endeavour*, **21**, 5–15.

LAGIÈRE, R. (1966). *Le Cotonnier.* Paris: Maisonneuve & Larose.

NICHOLSON, G. E. (1960). The Production, History, Uses and Relationships of Cotton (*Gossypium* spp.) in Ethiopia. *Econ. Bot.*, **14**, 3–36.

PEARSON, E. R. (1958). *The Insect Pests of Cotton in Tropical Africa*. London: Emp. Cotton Growing Corp. and Commonw. Inst. of Entomology.

PURSEGLOVE, J. W. (1960). Review of Hutchinson (1959). *Trop. Agriculture, Trin.*, **37**, 245–248.

PURSEGLOVE, J. W. (1963). Some Problems of the Origin and Distribution of Tropical Crops. *Genetica Agraria*, **17**, 105–122.

PURSEGLOVE, J. W. (1965). The Spread of Tropical Crops, in *The Genetics of Colonizing Species*, ed. H. G. Baker and G. L. Stebbins, 375–389. New York: Academic Press.

SAUNDERS, J. H. (1961). *The Wild Species of* Gossypium *and their Evolutionary History*. London: Oxford Univ. Press.

SKOVSTED, A. (1937). Cytological Studies in Cotton. *IV*. Chromosome Conjugation in Interspecific Hybrids. *J. Genet.*, **34**, 97–134.

SMITH, A. C. (1967). The Presence of Primitive Angiosperms in the Amazon Basin and its Significance in indicating Migrational Routes. *Atlas do Simpósio Sõbre a Biota Amazõnica*, Brazil, **4**, 37–59.

SMITH, C. E. and MACNEISH, R. S. (1964). Antiquity of American Polyploid Cotton. *Science*, **143**, 675–676.

STEPHENS, S. G. (1958). Salt Water Tolerance of Seeds of *Gossypium* species as a possible Factor in Seed Dispersal. *Amer. Nat.*, **92**, 83–92.

Hibiscus L. (x = 7, 8–39)

About 200 spp. of herbs, shrubs and trees in the tropical and subtropical regions of the world. Lvs simple, palmately veined. Fls often large and showy, hermaphrodite, axillary, mostly bell-shaped; epicalyx of few to several bracteoles; calyx 5-toothed; petals 5; stamens in central column. Fr. a dry, 5-valved capsule.

USEFUL PRODUCTS

The stems of many species yield fibres of which the most important are *H. cannabinus* (q.v.) and *H. sabdariffa* (q.v.). Other species are used locally for fibres and some of them have been tried experimentally. These include: *H. floccosus* Mast. and *H. macrophyllus* Roxb. in Malaya; *H. lunarifolius* Willd., *H. rostellatus* Guill. & Perr. and *H. squamosus* Hochr. in West Africa; and *H. kitaibelifolius* St. Hil. in Brazil. *H. tiliaceus* L., a small tree of tropical shores throughout the world, yields a fibre which is used for ropes, cordage and mats throughout the Pacific, in Malaysia and in the Sundabans and Andaman Islands.

The fruits of *H. esculentus* (q.v.) and *H. sabdariffa* (q.v.) are edible. The leaves of some species are eaten raw or as pot-herbs, *e.g. H. manihot* L.

The musk-scented seeds of *H. abelmoschus* L. (syn. *Abelmoschus moschatus* (L.) Medic.) are used in perfumery and are exported from Martinique; they are also used medicinally in some places.

Several species are highly decorative. *H. rosa-sinensis* L. which has long been cultivated in China, Japan and the Pacific, is now one of the most widely planted ornamental shrubs throughout the tropics, with various shades of yellow, orange, pink, red and white, and with both single and double flowers. There is also a variegated form, var. *cooperi* Nichols. The East African *H. schizopetalus* Hook f. is also grown as an ornamental and hybrids with *H. rosa-sinensis* occur. *H. mutabilis* L., changeable rose, has long been cultivated in China; in one form the double flowers open white in the morning and have turned red by the evening.

KEY TO ECONOMIC SPECIES

A. Capsule long and tapering .... *H. esculentus*
AA. Capsule ovoid
  B. Calyx fleshy and inflated .... *H. sabdariffa*
  BB. Calyx fibrous and spiny .... *H. cannabinus*

Hibiscus cannabinus L. (2n = 36) KENAF

Also known as Bimli or Bimlipatum jute and Deccan hemp.

USES

Bast fibre obtained from the stem is used in a similar manner to jute, *Corchorus capsularis* (q.v.). The retted fibre strands, 5–10 ft long, are comparable to jute in lustre; they are somewhat coarser and less supple, but are more resistant to rotting. They are used for rope and cordage, for making fishing nets, and can be woven to make coarse sacks, bags and canvas, for which purpose they are often used in admixture with jute. Unretted fibres are made into bags in some countries. The seeds contain about 20 per cent oil which is sometimes extracted and used as a lubricant and for illumination; it is suitable for the manufacture of soap, linoleum and in paints and varnish. The young leaves are sometimes used as a pot-herb.

ORIGIN AND DISTRIBUTION

H. cannabinus is a common wild plant in tropical and subtropical Africa, which is probably its original home. It also appears to be wild or naturalized in Asia. It has now been taken to most tropical and subtropical countries for trial as a fibre crop, mainly in the 1940's.

CULTIVARS

Considerable variation occurs in wild and cultivated plants in habit, colour and thickness of stem, leaf form, flower colour and adaptability to environment. Several botanical varieties have been erected, but as they cross readily they are of doubtful validity. A number of improved cvs have been obtained in India, Cuba and elsewhere.

ECOLOGY

H. cannabinus is grown between 45°N and 30°S. The optimum conditions are temperatures of 60–80°F during the growing season with a rainfall of 20–25 in. distributed evenly over a period of 4–5 months. The plant is photoperiodic, flowering on a shortening day of 12½ hours or less. The crop is grown during long days in order to obtain as much rapid vegetative growth as possible before flowering begins. It thrives best on well-drained, neutral, sandy loams, rich in humus. It will not tolerate waterlogging. It is less exacting in its requirements than jute.

STRUCTURE

The cultivated forms are erect herbaceous annuals, 2·5–4 m tall, with well-developed tap-roots. Stems straight and slender, glabrous or prickly, green, red or purple. Lvs alternate; petioles long, sometimes prickly, red or green; lamina variable in shape and colour, in some cvs all lvs are cordate, in others all deeply divided, in others lower lvs cordate with upper lvs deeply palmately 5–7-lobed; lobes oblong-lanceolate, sinuous, dentate. Fls borne singly in axils of upper lvs, large 7·5–10cm in diameter peduncles very short; epicalyx of 5–10 linear bracteoles less than half length of sepals; calyx persistent, bristly, connate below with prominent gland on mid-rib, tips lanceolate; petals 5, thin, spreading, pale yellow, often with crimson spot at base; staminal column with short filaments and small 1-celled anthers; ovary superior of 5 united carpels, style after emerging from staminal tube tipped with 5 stigmas. Fr. a globose capsule, pointed, about 1 cm long, half length of calyx, bristly with irritant hairs. Seeds numerous, brown, nearly glabrous, wedge-shaped with acute angles, 5 × 3 mm.

POLLINATION

Flowers usually open before daybreak and begin to close about midday. The stigmas do not emerge from the staminal tube until the anthers begin to burst. Self-pollination occurs when the corolla closes. Some cvs are entirely self-pollinated, but up to 4 per cent cross-pollination has been observed in others.

GERMINATION

Under ordinary storage conditions the seeds retain their viability for about 8 months. They may germinate in 72 hours, but take longer at low temperatures.

Fig. 57. **Hibiscus cannabinus : KENAF. A,** flowering shoot (× ⅓); **B,** flower in longitudinal section (× ⅓).

A

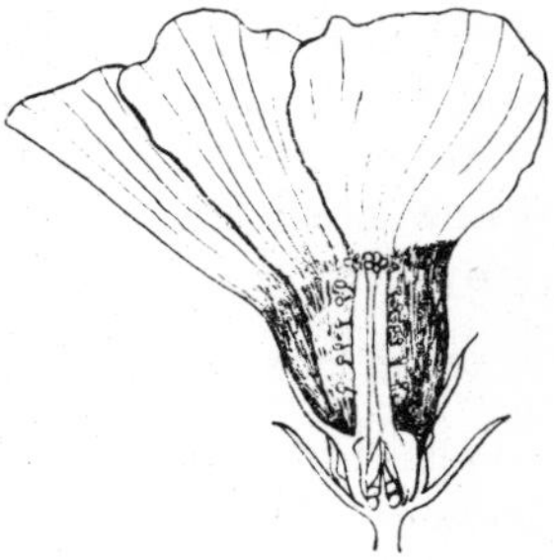

B

HUSBANDRY

For fibre production a seed rate of 25–30 lb per acre in drills 8 in. apart with 2–3 in. between plants in the rows appears to give the highest yields. For seed production a wider spacing of 2 × 1 ft is used. The crop is harvested 3–5 months after sowing. The longer the vegetative phase the greater will be the yield. The best time to harvest is when about 10 flowers are in bloom, at which time the fibre is at its best quality and more easily separable. If harvesting is delayed until seed has set the fibre is coarser and lacks lustre. The stems are cut at ground level and tied into bundles. The leafy tops are cut off and the stems are retted in water in the same way as jute. Retting takes 5–14 days depending on the temperature, etc. The bark is then stripped and is gently beaten to separate the fibres, which are then washed and dried in the sun. The average fibre out-turn is about 4 per cent of the weight of green stems or 16 per cent of the weight of dry stalks. Mechanical decortication with or without retting has been tried in some countries. Yields in India are about 1,000 lb of dry fibre per acre. Elsewhere yields of up to 3,000 lb per acre have been recorded. When grown for seed yields of 700–800 lb per acre of seeds are obtained.

MAJOR DISEASES AND PESTS

The most serious diseases recorded are: a dry rot, *Macrophomina phaseoli* (Maubl.) Ashby; leaf spot, *Cercospora hibisci* Tracy & Earle; leaf blight, *Phyllosticta hibisci* Peck; stem rot, *Diplodia hibiscina* Cke. & Elb.; and anthracnose, *Colletotrichum hibisci* Poll. Kenaf is attacked by root-knot nematodes, *Meloidogyne* spp. Among the insect pests reported are *Argrilus acutus* Thumb. and *Podagrica* spp.

IMPROVEMENT

Improved cvs have been produced in India, Cuba and elsewhere. 'Cuba 2032' is resistant to anthracnose and is an interspecific hybrid with *H. diversifolus* Jacq.

PRODUCTION

India is the largest producer with about 850,000 acres producing about 633 million lb annually. Elsewhere the crop has been introduced, mainly since World War II, and successful production has been recorded from the Soviet Union, China, Thailand, South Africa, Egypt, Mexico and Cuba.

REFERENCES

HAARER, A. E. (1952). *Jute Substitute Fibres.* London: Wheatland Journals.

KIRBY, R. H. (1963). *Vegetable Fibres.* London: Leonard Hill.

Hibiscus esculentus L. (2n = 72–132) LADY'S FINGER, OKRA

Syn. *Abelmoschus esculentus* (L.) Moench.

USES

The tender fruits are used as vegetables either boiled or sliced and fried. They have a high mucilage content and are used in soups and gravies. They are one of the ingredients of callaloo soup, a Trinidadian dish. They are sometimes sliced and dried. The ripe seeds contain about 20 per cent of an edible oil. A mucilaginous preparation from the pod can be used as a plasma replacement or blood-volume expander. Mucilage from the stem and roots is used for clarifying sugarcane juice in gur manufacture in India and for sizing paper in China. An inferior fibre may be obtained from the stem.

ORIGIN AND DISTRIBUTION

H. esculentus originated in tropical Africa and has now been widely spread throughout the tropics.

CULTIVARS

There are a number of named cvs, of which 'Dwarf Prolific' is recommended in Trinidad and 'White Velvet' in Puerto Rico.

ECOLOGY

Okra is essentially tropical and grows well in the lowland tropics. It can be grown on any type of soil, but does best on well-manured loams.

STRUCTURE

Robust erect annual herb, 1–2 m tall. Stems green or tinged red. Lvs alternate; stipules narrow, caducous; petiole hispid, often tinged red, 15–35 cm long; lamina broadly cordate, 10–25 × 10–35 cm, palmately 3–7-lobed, hirsute, serrate, pale green beneath. Fls solitary, axillary; peduncle about 2 cm long; epicalyx of up to 10 narrow hairy bracteoles, up to 1·5 cm long, usually falling before fruit reaches maturity; calyx completely fused as fl. develops, splitting longitudinally as fl. opens, falling with corolla after anthesis, 2–3 cm long; petals 5, yellow with crimson spot on claw, obovate, 5–7 cm long; staminal column united to base of petals, 2–3 cm long, with numerous stamens; ovary superior; stigmas small, 5–9, deep red. Fr. a pyramidal-oblong, beaked capsule, 10–30 × 2–3 cm, longitudinally furrowed, hirsute or glabrous, dehiscing longitudinally when ripe. Seeds dark green to dark brown, rounded, tuberculate, 5 mm in diameter.

CHEMICAL COMPOSITION

The edible portion of the pods contain approximately: water 86·1 per cent; protein 2·2 per cent; fat 0·2 per cent; carbohydrate 9·7 per cent; fibre 1·0 per cent; ash 0·8 per cent.

HUSBANDRY

The usual spacing is 2–3 × 1 ft with 2–3 seeds per hill. Okras respond to organic manures. Harvesting begins 2–3 months after planting and continues for a period of 2 months. The pods are picked while the tips are still tender and break with a snap. Yields of 4,000–5,000 lb of green pods per acre are produced.

MAJOR DISEASES AND PESTS

As *H. cannabinus*. Yellow vein virus occurs in India of which *Bemisia* spp. are the vectors.

IMPROVEMENT

Named cvs have been bred and selected.

PRODUCTION

Okras are grown for local consumption throughout the tropics.

Hibiscus sabdariffa L. (2n = 36, 72) ROSELLE, JAMAICAN SORREL

Two botanical varieties are recognized: var. *sabdariffa*, a bushy branched subshrub with red or green stems and red or pale yellow inflated edible calyces; var. *altissima* Wester, a tall, vigorous, practically unbranched plant, 10–16 ft high, with fibrous, spiny, inedible calyces, grown for fibre.

USES

The red acid succulent calyces of var. *sabdariffa* are boiled with sugar to produce sorrel drink. They are also made into jellies, sauces, chutneys and preserves. The tender leaves and stalks are eaten as salad and as a pot-herb and are used for seasoning curry. The seeds contain oil and are eaten in Africa. Var. *altissima* is grown for fibre in India, Java and the Philippines.

ORIGIN AND DISTRIBUTION

H. sabdariffa is probably a native of West Africa and is now widely cultivated throughout the tropics. It has been in Asia for at least three centuries and was taken early to the New World by the slave trade, reaching Brazil in the 17th century. It was recorded in Jamaica in 1707.

Fig. 58. **Hibiscus esculentus** : OKRA. **A**, flowering and fruiting branch (×¼); **B**, flower from above (×½); **C**, flower in longitudinal section (×½); **D**, fruit in longitudinal section (×½).

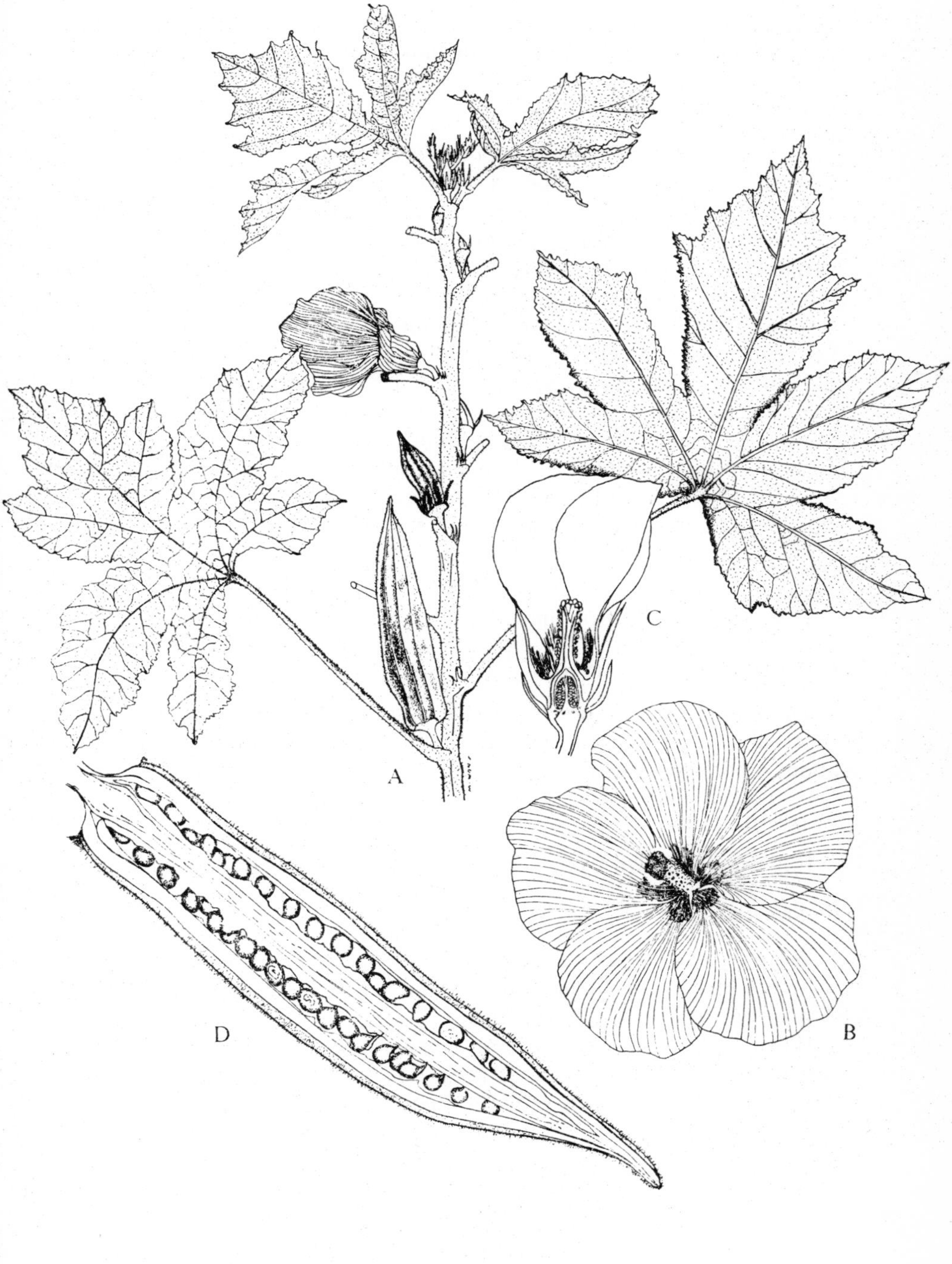

D2

ECOLOGY

H. sabdariffa is confined to the tropics and will tolerate poorer conditions than jute and *H. cannabinus*. It is a short-day plant.

STRUCTURE

This account applies to *H. sabdariffa* var. *sabdariffa*.

An erect, branched, glabrous, annual subshrub, 0·5–3 m high, with a strong tap-root. Stems red or green, lvs alternate; stipules subulate, deciduous, 5–8 mm long; petiole 2–10 cm long, red or green; lamina variable, lower lvs often ovate and undivided, others palmately 3–5-lobed, lobes oblong-lanceolate, 7–15 cm long with middle lobe longest, margin serrate, with gland near base of prominent midrib. Fls solitary, axillary, peduncle short; epicalyx of up to 10 linear fleshy bracteoles, calyx 5-lobed, 1–2 cm long, enlarging and becoming fleshy after anthesis; epicalyx and calyx red, whitish or green; petals 5, pale yellow, usually with maroon spot at base, 3–5 cm long; staminal column 1–2 cm. long with numerous small anthers; stigma 5-lobed. Fr. ovoid capsule, villous, 2–3 cm long, closely invested with enlarged persistent fleshy calyx, dehiscing by 5 valves when ripe. Seeds reniform, dark brown, pilose, 4–6 mm long.

POLLINATION

H. sabdariffa is self-pollinated.

CHEMICAL COMPOSITION

The fruits contain approximately: water 84·5 per cent; protein 1·7 per cent; fat 1·0 per cent; carbohydrate 12 per cent. The calyces contain about 4 per cent of citric acid.

PROPAGATION

It is usually grown from seed, but it can be propagated by cuttings.

HUSBANDRY

When grown for fruit production the spacing is usually 4 × 3 ft. The fruits are picked while tender, plump and fleshy, 15–20 days after flowering. A plant yields about 3 lb of fruit.

When grown for fibre spacing is approximately 8 × 6 in. with a seed rate of about 30 lb per acre. The crop is harvested in the bud stage, 7–8 months after sowing. It is retted in the same manner as *H. cannabinus*.

Fig. 59. **Hibiscus sabdariffa :** ROSELLE. **A,** fruiting shoot (×½); **B,** leaf (×½); **C,** flower from side (×½); **D,** flower from above (×½); **E,** flower in longitudinal section (×½); **F,** fruit in longitudinal section (×½).

D
E
C
F
B
A
M. WONG

MAJOR DISEASES AND PESTS

As *H. cannabinus*. *Renococcus hirsutus* Gr. is said to be a serious pest in India.

IMPROVEMENT

A few improved cvs have been selected.

PRODUCTION

Roselle is grown to a limited extent for its edible fruits in many tropical countries. Var. *altissima* is grown commercially for fibre in India. In 1937 in Java 50,000 acres were estimated to yield 30,000 tons of fibre.

Urena L. (x = 7)

A small tropical genus of about 7 spp., of which *Urena lobata* (q.v.) yields fibre.

Urena lobata L. (2n = 28, 56) ARAMINA FIBRE, CONGO JUTE

USES

Urena lobata is grown as a fibre crop in Brazil, where it is known as aramina fibre and in the Congo under the name of Congo jute. The fibre resembles jute more closely than other jute substitutes; it is as fine as jute and can be spun on the same machinery; it is often used in admixture with jute. The fibre is soft, lustrous and pale yellow and is used for making hessian, ropes, carpets, etc. The plant is widely used in native medicine in Malaya.

ORIGIN AND DISTRIBUTION

Urena lobata is now widely distributed in a wild state throughout the tropics and subtropics of both hemispheres. Its centre of origin is not known with certainty, but it is probably Old World. It can be a troublesome weed, particularly in pastures, and is a declared noxious weed in Fiji.

ECOLOGY

When grown as a crop, it needs a hot and humid climate, and thrives best in full sun on deep, fertile, well-drained soil.

STRUCTURE

A very variable perennial shrub; in its weedy forms low, spreading and branching, but cultivated forms are erect, with few branches, reaching a

---

Fig. 60. **Urena lobata :** ARAMINA. **A,** leafy shoot (×½); **B,** flowering shoot (×½); **C,** inflorescence with fruits (×½); **D,** flower in longitudinal section (×1); **E,** fruit from above (×1); **F,** ripe carpel (×2); **G,** seed (×5),

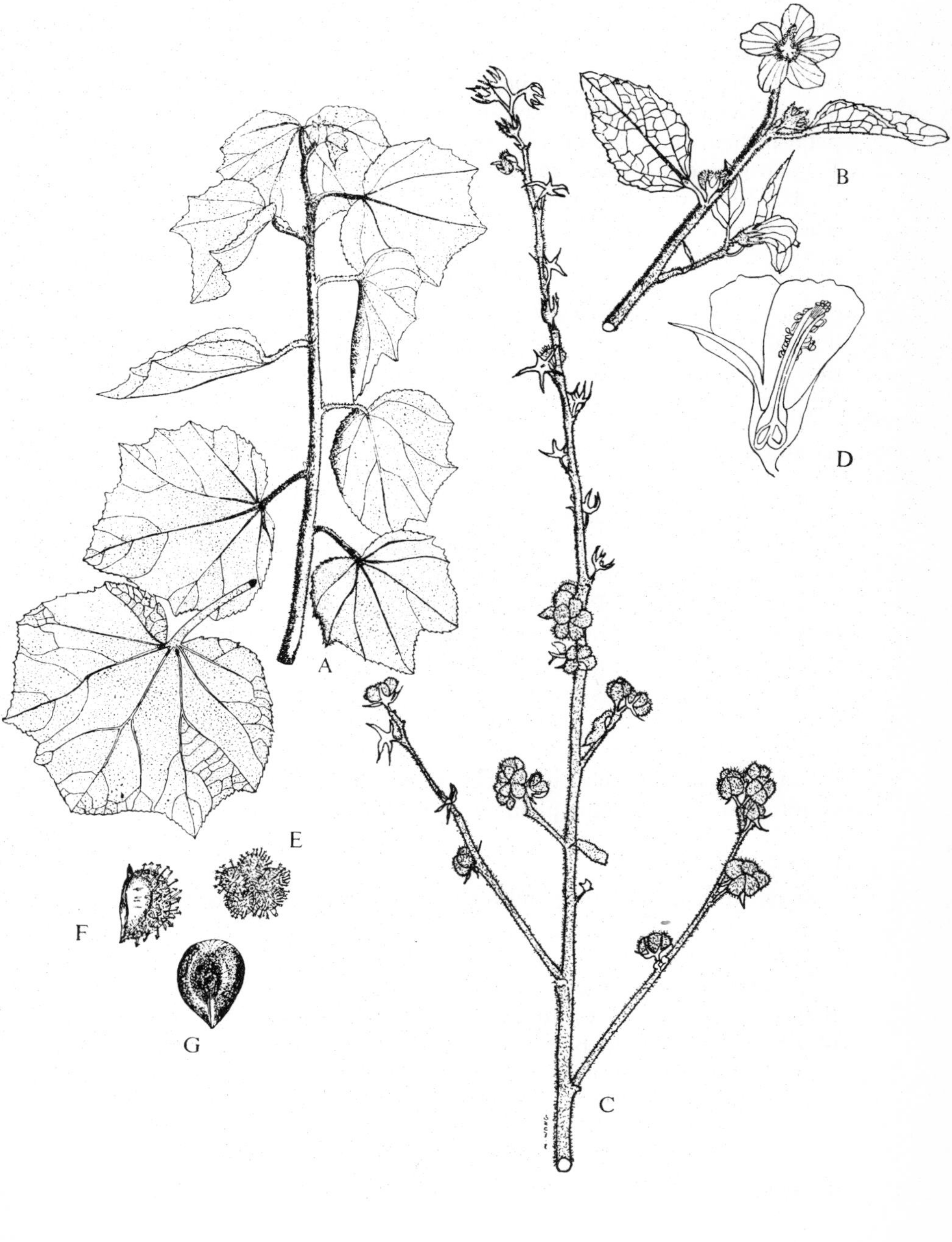
A
B
C
D
E
F
G

height of 4 m or more. All the above ground parts are covered with stellate hairs. Short tap-root with wide-spreading lateral roots. Lvs alternate, very variable in size and ranging from simple, coarsely-toothed simple lvs, lanceolate or broadly ovate, to shallowly or deeply palmately 3–7-lobed, base obtuse or cordate; petiole 1–5 cm long; lamina 3–10 cm long, greyish beneath, with 1–3 slit-like glands at base of main veins. Fls axillary and terminal, solitary or in small clusters; epicalyx of 5 lanceolate pointed bracteoles, united at base, 6 mm long, longer than calyx, persistent in fr.; calyx with 5 short broad lobes; corolla spreading widely of 5 obovate petals, usually pink, about 1·5 cm long, staminal column short; stigmas 10. Fr. consisting of 5 one-seeded carpels, convex, about 5 mm long, bearing thick stellate down as well as short, hooked, anchor-like spines, falling when ripe, indehiscent.

GERMINATION

The seeds are not normally removed from the carpels before sowing and it is advisable that the hooks should be removed by rubbing with sand as they cause the carpels to cling together. Treatment with concentrated sulphuric acid increases the percentage germination.

HUSBANDRY

In the Congo the plants are spaced about 2 in. apart with a seed rate of 60–80 lb per acre. Weeding is necessary in the early stages. Harvesting is done when the plants are flowering, 4–6 months after sowing. The yield of fibre is about 5 per cent of the weight of green stems before retting and yields of 700–900 lb of dry fibre are obtained, with up to 1,350 lb per acre from selected seed. The crop is usually grown as an annual, but in some areas it is grown as a perennial and is ratooned. The Madagascan industry is based on wild plants and cutting is regulated by legislation.

MAJOR PESTS AND DISEASES

In the former French Equatorial Africa, where an attempt was made to establish an industry, the plants were attacked by *Stagonospora urenae*.

IMPROVEMENT

Some work on selection and breeding was carried out in the Congo.

PRODUCTION

Cultivation by peasant farmers in the Congo began in 1929 and some 50,000 acres were planted annually by the early 1950's. The Congo exported 14,000 tons of fibre to Belgium in 1952, but now exports to South Africa. There is a sizeable industry in Brazil, but most of the fibre is used locally, as is the production in Madagascar.

REFERENCE

HAARER, A. E. (1952). *Jute Substitute Fibres.* London: Wheatland Journals.

MORACEAE

About 60 genera and over 1,000 species of trees and shrubs, rarely herbs or vines, occurring mainly in the tropics. Milky juice usually present. Lvs mostly alternate and simple; stipules paired, prominent, often deciduous, leaving a scar. Fls much reduced, often in heads, disks or hollow receptacles, unisexual, actinomorphic, monoecious or dioecious; sepals usually 4; petals absent; stamens usually same in number and opposite to sepals; female fls with superior or inferior ovary, usually 1-celled, 1-ovuled. Frs various, often compound and fleshy. Seeds with or without endosperm.

USEFUL PRODUCTS

Some species of *Artocarpus* (q.v.) and *Ficus* (q.v.) have edible fruits and other uses.

Brief mention is made below of some other plants in the family which are used in various ways.

Antiaris africana Engl. is the bark-cloth tree of West Africa, the inner bark providing a cloth used in Ashanti in Ghana. Its latex has been used as an adulterant of rubber from *Funtumia elastica* Stapf (Apocynaceae).

Antiaris toxicaria (Pers.) Lesch. is the famous upas tree of south-eastern Asia from India to the Philippines. The creamy white latex from the bark of this lofty tree, up to 50 m tall, contains a powerful cardiac glucoside, antiarin, which is used as an arrow poison; to be effective it must enter the blood stream. Cloth is also made from the bark, but care is required in its preparation because of the latex.

Brosimum utile (HBK.) Pittier (syn. *B. galactodendron* Don.) is the cow tree of Venezuela, whose latex can be drunk like milk. *B. alicastrum* Swartz. of Mexico and Jamaica has edible fruits like chestnuts.

Broussonetia papyrifera (L.) Vent., paper mulberry, is a small tree to 12 m tall, which is native in China and Japan. It is planted in the Far East for paper manufacture and for making bark-cloth. It is the source of tapa cloth, which constituted the chief clothing of the Polynesians.

Castilla elastica Cerv. is the source of Panama or Castilloa rubber. It is a rapidly growing, deciduous, forest tree with large velvety leaves about 45 × 15 cm. This was the source of the rubber seen by Columbus and used by the Olmecs and Aztecs. The tree is tapped by making incisions in the bark, but it cannot compete successfully with *Hevea brasiliensis* (q.v.). It was an important source of rubber until 1850. It was extensively planted towards the end of the 19th century in Central America, Trinidad and Tobago, and Java, but most of the plantations were later abandoned. The surviving trees in the New World were the source of the emergency supply of Castilloa rubber used during World War II. *Castilla ulei* Warb. of the Amazon region is the source of caucho rubber. The plants collected by Cross in 1876 in Panama, which were distributed by Sir Joseph Hooker, Director of the Royal Botanic Gardens, Kew, to Ceylon and Singapore were not *C. elastica* as originally thought, but were *C. panamensis* O. F. Cook (syn. *C. markhamiana* Hook. f.).

Chlorophora excelsa (Welw.) Benth., native of tropical Africa, is a majestic dioecious tree up to 50 m tall with a bole of 25 m or more. It is often planted and is one of the most valuable timber trees throughout tropical Africa and is widely used for making furniture. *C. tinctoria* (L.) Gaud., a forest tree of the West Indies, Central and South America, yields the khaki dye fustic, which is obtained from the heart wood.

Morus alba L., white mulberry, is a native of China and is grown to a small extent in many parts of the tropics for its edible fruits which may be eaten raw or cooked. It is grown more extensively, as in India, for its leaves which are fed to silkworms. The bark has been used in China since early times for making paper. The wood is used for the manufacture of sports goods such as rackets and hockey sticks. It is a small tree with thin ovate serrate leaves, often deeply lobed, monoecious or dioecious. The female spikes produce a syncarp, 2·5–5 cm long, containing many drupes enclosed in the fleshy perianth and are white, red or purple in colour. It can be propagated by seeds or cuttings. *M. nigra L.*, black mulberry, is only suited to the higher elevations in the tropics.

Treculia africana Decne., the African breadfruit, is often seen near villages in West Africa. Its fruits, about 45 cm in diameter, have edible seeds which are eaten after boiling or roasting or are made into flour for use in soups.

Artocarpus Forst. (x = 14)

About 50 spp. of monoecious trees in south-eastern Asia and Polynesia. Thick white latex in all parts. Lvs spirally arranged, simple or pinnate; large stipules covering bud. Fls minute and very numerous set on a fleshy rachis. Male inflorescences axillary borne near ends of twigs; fls with small calyx and 1 stamen. Female inflorescences borne singly in leaf axils or on trunk and branches, often large and prickly; fls with a single ovary and long style. Compound fruits are derived from the swollen flower heads. Seeds, when present, large, surrounded by a succulent layer derived from calyx and separated by undeveloped female fls.

USEFUL PRODUCTS

A. altilis (q.v.) and *A. heterophyllus* (q.v.) are cultivated for their edible fruits.

A. champeden (Lour.) Spreng., champedak, occurs wild in Malaya and selected forms of it are cultivated throughout the Malay Archipelago. It is very similar in appearance to *A. heterophyllus*, but can be distinguished from it by the long wiry brown hairs on the leaves and twigs and the fruits are rather smaller. The custardy slimy pulp round the seeds has a very strong stench. Corner (1952) says 'if the Jack has the biggest of fruits, the champedak has the strongest and richest smell of any in creation; the smell is pervasive like that of the durian, but much harsher'. The immature fruits are used in soups, the fresh pulp of the ripe fruits is eaten, and the seeds are eaten after cooking. The fruits of some wild species are also eaten, including the monkey jack, *A. rigida* Bl.

The bark of *A. elastica* Reinw. is used for making bark-cloth by jungle tribes in Malaya and that of *A. tamaran* Becc. is used for making loin-cloths in Borneo.

KEY TO ECONOMIC SPECIES

A. Leaves deeply lobed, 30–60 cm long............. *A. altilis*
AA. Mature leaves entire, 5–20 cm long............. *A. heterophyllus*

Artocarpus altilis (Park.) Fosberg (2n = 56) BREADFRUIT, BREADNUT

Syn. *A. communis* Forst.; *A. incisa* L.f.

USES

The breadfruit is an important staple food in Polynesia. It is now grown throughout the tropics and is used more as a vegetable than as a fruit. The seedless forms are eaten after cooking and may be boiled, baked, roasted, fried or made into soups. In Polynesia it may be combined with other ingredients such as coconut cream and grated coconut meat. They are usually cooked after peeling and slicing. Sometimes the whole fruit is cooked after which the rind and core are removed. Biscuits are also made from it. It can be preserved by cooking and drying for off-season use.

In the seeded forms, known as breadnut (chataigne in the West Indies), the seeds are eaten after boiling or roasting and have a chestnut flavour. The flesh and nuts of the unripe fruits are sometimes cooked and eaten.

A fibre can be obtained from the bark and the latex is used for caulking boats. The leaves may be fed to livestock. It is a very handsome tree and deserves to be more frequently grown as a garden ornamental tree.

ORIGIN AND DISTRIBUTION

The breadfruit is a native of Polynesia where it is of ancient cultivation. Coenan and Barrau (1961) consider that it may be of hybrid origin and that *A. mariannensis* Tréc., which occurs wild in Micronesia, may have contributed to the domesticated species.

The story of its introduction into the New World is well known. Banks, Cook and other travellers in Polynesia brought back descriptions of the tree from which 'bread itself is gathered as a fruit'. West Indian planters hoped that it would provide a staple diet for their slaves and petitioned H.M. George III to mount an expedition for its collection. The King was graciously pleased to comply with their request. Sir Joseph Banks, who had been to Tahiti on Captain Cook's first voyage, 1768–1771, suggested that Lieutenant William Bligh should be sent to Tahiti to collect breadfruit plants and transfer them to the West Indies. Banks had been made President of the Royal Society in 1778, a position he was to occupy for nearly half a century. The British Government chartered a ship. Banks superintended the whole equipment and named her the *Bounty*. Bligh, who had been on Cook's third voyage, 1776–1780, when he was master of the *Resolution*, had also visited Tahiti. It was on this voyage that Captain Cook was killed in Hawaii. The *Bounty* sailed from Spithead on 23rd December, 1787 and arrived at Tahiti on 26th October, 1788, sailing *via* the Cape of Good Hope.

Due to the time taken to propagate the breadfruit (a botanist and a gardener were attached to the expedition), Bligh remained on Tahiti until 4th April, 1789, when he sailed with 1,015 breadfruit and other plants in 774 pots, 39 tubs and 24 boxes. Then occurred the famous mutiny of the *Bounty* led by Fletcher Christian near the Friendly Islands on 28th April, after which Bligh and 18 men performed the astonishing feat of crossing the Pacific in a small open boat landing on Timor on 14th June, 1789, having covered a distance of 3,618 nautical miles. Bligh arrived back in England on 14th March, 1790.

The mutiny created world-wide interest, not only in itself, but also in the breadfruit. Bligh returned to Tahiti in 1792 in the *Providence* and successfully carried breadfruit plants to St. Vincent and Jamaica. The original breadfruit tree planted by Bligh in 1793 still stands in the Botanic Gardens in St. Vincent and in 1966 H.M. Queen Elizabeth II planted a scion from it nearby.

Breadfruit had already reached eastern Malaysia before the Bligh epic. Breadfruit reached Penang about 1802 and Malacca in 1836. It has now been carried throughout the tropics.

CULTIVARS

A number of seedless and seeded cvs are recognized in Polynesia.

ECOLOGY

Breadfruit is a tree of hot humid tropical lowlands and does best in insular climates with an annual rainfall of 60–100 in. and temperatures of 70–90°F. Young plants grow better under shade, but later require full exposure. It can be grown on a variety of soils provided they are of sufficient depth and are not waterlogged.

STRUCTURE

A handsome striking monoecious tree to 20 m in height; evergreen in the ever-wet tropics, deciduous in monsoon countries. Viscid latex in all parts.

ROOTS: Being vegetatively propagated all roots adventitious with dense surface mass of feeding roots.

BRANCHES: Trunk straight; twigs very thick, sparsely hairy, with pronounced leaf and stipule scars and lenticels; buds 10–20 cm long covered with big conical keeled stipules.

LEAVES: Spirally arranged, phyllotaxis $\frac{2}{5}$, large, 30–60 × 20–40 cm, thick, leathery, stiff, dark green and shiny above, pale green and rough below, deeply pinnately cut into 7–9-pointed lobes with narrow sinuses, lf. base entire, cuneate, veins prominent, lf. slightly hairy beneath and on veins above; petiole short and stout, 3–5 cm long.

INFLORESCENCES AND FLOWERS: Inflorescences axillary. Male inflorescences drooping, on stout peduncles, 4–8 cm long, club-shaped, rather spongy, yellow, 15–25 × 3–4 cm; fls minute with single stamen; 2-locular anther exserted. Female inflorescences stiffly upright on stout peduncles, 4–8 cm long, globose or oblong, green, 8–10 × 5–7 cm, fls very numerous embedded in receptacle, calyx tubular, ovary 2-celled, style narrow, stigma 2-lobed protruding above calyx.

FRUITS AND SEEDS: Fr. a syncarp formed from whole inflorescence, oblong to globose, 10–30 cm in diameter, yellowish-green rind reticulately hexagonally marked and may bear short spines in some cvs. The breadfruit is seedless or nearly so with a large central core surrounded by numerous abortive flowers, which when ripe form a moist pulp which is pale yellow or whitish in colour with a distinctive odour; it is mild in flavour and is the edible portion of the fr. In the breadnut the rind is covered with fleshy prickles; there is little edible pulp as this is replaced by the seeds, brownish in colour, rounded or flattened by compression, about 2·5 cm long.

POLLINATION

The male heads have no scent and give off clouds of pollen when tapped so that it seems probable that the flowers are wind-pollinated. Pollen is shed 10–15 days after the emergence of the inflorescence. Female flowers are receptive 3 days after the emergence of the inflorescence from the bracts and open in successive stages. In India hand-pollination results in better fruit set and larger fruits. Singh *et al.* (1963) state that fruit will set without pollination, but the fruits produced are small, and that pollination produces stimulative parthenocarpy rather than fertilization.

GERMINATION

Fresh breadnuts germinate readily, but soon lose their viability and should not be allowed to dry out.

CHEMICAL COMPOSITION

The edible portion, which constitutes about 70 per cent of the fruit, has the approximate composition: water 75·5 per cent; protein 1·3 per cent; fat 0·5 per cent; carbohydrate 20·1 per cent; fibre 1·8 per cent; ash 0·8 per cent.

PROPAGATION

As the breadfruit usually has no seeds it must be propagated vegetatively. This is usually done with root cuttings about 1 in. in diameter and 9 in. long planted horizontally or at an angle in shaded beds. They can also be propagated by root suckers. Breadnuts can be grown from fresh seeds or propagated vegetatively as is done with breadfruit.

HUSBANDRY

The trees are usually planted 30–40 ft apart. The tree grows rapidly and begins to bear at 3–6 years old. The fruit is harvested while it is still firm and before it is fully ripe, 60–90 days after the emergence of the inflorescence. They do not keep for more than a few days. The main cropping season lasts about 4 months. Mature trees will yield up to 700 fruits per year, each weighing 2–10 lb.

MAJOR DISEASES AND PESTS

Breadfruits are rarely attacked by serious diseases or pests, but a soft rot of the fruits caused by *Rhizopus artocarpi* has been reported in India.

IMPROVEMENT

The seedless breadfruit must have been selected by man over the centuries in Polynesia, where the largest number of cvs occur, but little scientific breeding has been done on the crop. The South Pacific Commission has undertaken a regional survey to gain more knowledge of cvs in the area in the hope of selecting early, late or year-round producing cvs in order to extend the cropping season. This survey was initiated in 1958 and cvs from several territories in the South Pacific have been sent to Western Samoa, Tahiti and Fiji for comparative trial (Coenan and Barrau, 1961).

PRODUCTION

Breadfruit is still an important source of food in Pacific islands. It has been introduced into most tropical countries, but has not attained much importance except in the West Indies.

Fig. 61. **Artocarpus altilis: BREADFRUIT. A, shoot with male inflorescence and fruit (×½); B, individual male flowers (much magnified); C, fruit in longitudinal section (×½).**

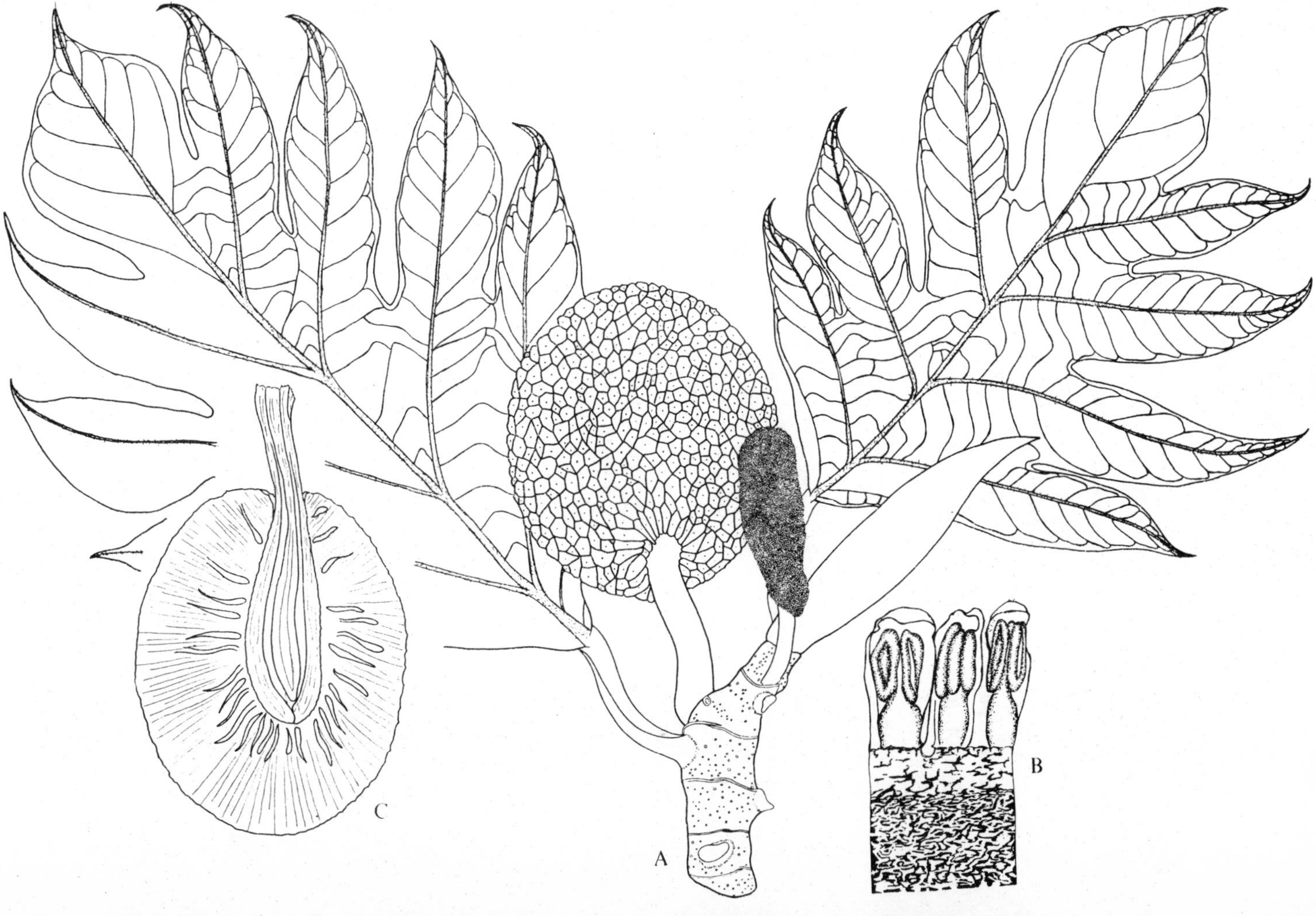
C
A
B

REFERENCES

BARROW, J. (1831). *The Mutiny of the Bounty*. Reprinted London: Blackie, 1961.

COENAN, J. and BARRAU, J. (1961). The Breadfruit Tree in Micronesia. *South Pacific Bull.*, Oct., 37–39.

MOOREHEAD, A. (1966). *The Fatal Impact*. London: Hamish Hamilton.

SINGH, S., KRISHNAMURTHI, S. and KATYAL, S. L. (1963). *Fruit Culture in India*. New Delhi: Ind. Counc. Agric. Resc.

Artocarpus heterophyllus Lam. (2n = 56) JACKFRUIT

Syn. *A. integra* (Thunb.) Merr.; *A. integrifolia* L.f.

USES

Jackfruit is a popular fruit in India and Ceylon. The pulp of the ripe fruit is eaten fresh or preserved in syrup. It is also dried. Burkill (1935) says 'the taste is mawkishly sweet and mousy, agreeable to natives of the East, but not to Europeans'. The juicy flesh round the seeds, which has a pungent odour and may be sweet or acid, is eaten, sometimes in fruit salad. The large seeds are boiled and roasted and have a chestnut flavour; they are also ingredients in many culinary preparations in India. The unripe fruits are used as a vegetable or in soups and are made into pickles. The rind and leaves are fed to livestock. The trees may be used as shade for coffee or areca and as living supports for pepper (*Piper nigrum*, q.v.). The resinous latex is used to mend earthenware and other utensils and as bird-lime. The yellow heart wood is a valuable timber and also yields a yellow dye.

ORIGIN AND DISTRIBUTION

Jackfruit is of very ancient cultivation in India. It appears to have been taken early by the Arabs to the East African coast. It has now spread throughout the tropics.

CULTIVARS

As jackfruits are usually grown from seeds there is considerable variation between trees, particularly in the shape, size and quality of the fruits. Forms with firm and soft flesh are known.

ECOLOGY

It is a tree of the tropical lowlands, but will tolerate higher altitudes and cold better than the breadfruit. It can be grown on a variety of soils, provided they are well drained, but does best on deep alluvial soils.

STRUCTURE

An evergreen monoecious tree to 20 m tall. The delicate tap-root of the seedling is easily damaged so they are difficult to transplant. White resinous latex in all parts. Twigs glabrous. Stipules ovate-triangular, 2–7 cm long. Lvs spirally arranged, phyllotaxis $\frac{2}{5}$, usually glabrous; sapling lvs have 1–2 pairs lobes; adult lvs entire, coriaceous, dark green and shiny above, paler green beneath; petiole 2–4 cm long; lamina 5–20 × 3–12 cm, elliptic to obovate, bluntly tipped, base cuneate, pinnatinerved with 5–8 pairs veins. Male inflorescences axillary on leafy twigs or from trunk and branches, but above female inflorescences, drooping, oblong, ellipsoidal or clavate, 5–10 × 2–3 cm, green; fls numerous, densely crowded, minute, with 2-lobed calyx and single stamen. Female inflorescences borne on short stout twigs on main trunk and branches; peduncle stout with green fleshy ring at apex, larger than male inflorescence; fls small, calyx tubular, ovary oblong and compressed, style terminal and oblique, stigma clavate. Fr. a gigantic syncarp, the biggest of all cultivated fruits, 30–90 × 25–50 cm; barrel- or pear-shaped, glabrous, with short sharp hexagonal fleshy spines, rind yellow, flesh yellow and waxy. Seeds with thick gelatinous yellow covering, large, oblong, about 3 × 2 cm.

POLLINATION

The ripe male inflorescences have a sweet scent of honey and burnt sugar which attracts small flies and beetles and these probably bring about pollination.

GERMINATION

Seeds lose their viability very quickly so they should be planted fresh.

CHEMICAL COMPOSITION

The edible pulp, which comprises about 30 per cent of the fruit, contains approximately: water 73·1 per cent; protein 0·6 per cent; fat 0·6 per cent; carbohydrate 23·4 per cent; fibre 1·8 per cent; ash 0·5 per cent. The seeds, which constitute about 5 per cent of the fruit, contain: water 51·6 per cent; protein 6·6 per cent; fat 0·4 per cent; carbohydrate 38·4 per cent; fibre 1·5 per cent; ash 1·5 per cent.

PROPAGATION

Jackfruit is usually planted from seed which should be sown *in situ* or in containers, as bare-rooted seedlings are difficult to transplant. As the seedlings are very variable it is desirable that selected clones should be propagated vegetatively, which can be done by inarching or grafting on to seedling rootstocks of jackfruit, *A. champeden* or *A. hirsuta* L.

HUSBANDRY

Jackfruits should be planted about 40 ft apart. The 'Singapore Jack' begins fruiting 3 years after planting, but most cvs fruit about the 8th year.

The fruiting season lasts about 4 months. Up to 250 fruits per tree are produced per annum which weigh 20–60 lb each, but fruits weighing 110 lb have been recorded.

MAJOR DISEASES AND PESTS

Few serious diseases and pests have been reported, except for a shoot-borer caterpillar in India.

IMPROVEMENT

Little work seems to have been done on selecting or improving the jackfruit. Some trees produce sweet juicy aromatic fruits; others are nearly dry and sour. The selection and vegetative propagation of clones is possible and efforts should be made to extend the fruiting season.

PRODUCTION

Although the jackfruit is grown sporadically throughout the tropics, it is not much used except in parts of tropical Asia.

Ficus L. (x = 13; 2n = 26) FIGS

A large genus of about 600 or more spp. of trees and shrubs, occasionally climbers, mostly tropical, with the largest number of spp. in the Indo-Malaysian region. Latex is present in all parts.

The tiny flowers are borne inside a cup- or pear-shaped receptacle with a narrow mouth with interlocking scales. The fls are of 3 types: (1) male fls with 1–5 stamens; (2) female fls with an ovary and a long style, each setting 1 seed; (3) gall fls like the female fls, but swollen and with a short style, each containing a fig wasp and not setting seed. In most of the strangling figs the 3 kinds of fls occur inside each fig. In the other spp., which are dioecious, the male and gall flowers are produced inside the same inflorescences on one tree and the female inflorescences on another tree. The fleshy fruit is a fig or syconium formed by the swelling of the entire female inflorescence. In most spp. the figs are borne in the leaf axils or on the twigs just behind the leaves, usually in pairs. Some spp. are cauliflorous. In a few spp. the figs are borne on underground runners.

Cross-pollination is brought about by minute fig wasps, *Blastophaga* spp., which are about 1 mm long. The species of wasp is specific to a given species of *Ficus* or to a few allied species. When the figs with the gall flowers ripen the adult wasps hatch from the ovaries by biting a hole in the ovary wall. The small male wasps, which are wingless and almost quite

---

Fig. 62. **Artocarpus heterophyllus**: JACK FRUIT. **A**, shoot with male inflorescence (×½); **B**, part of male inflorescence in longitudinal section (much magnified); **C**, part of female inflorescence in longitudinal section (much magnified); **D**, fruit (×⅛); **E**, seed (×½).

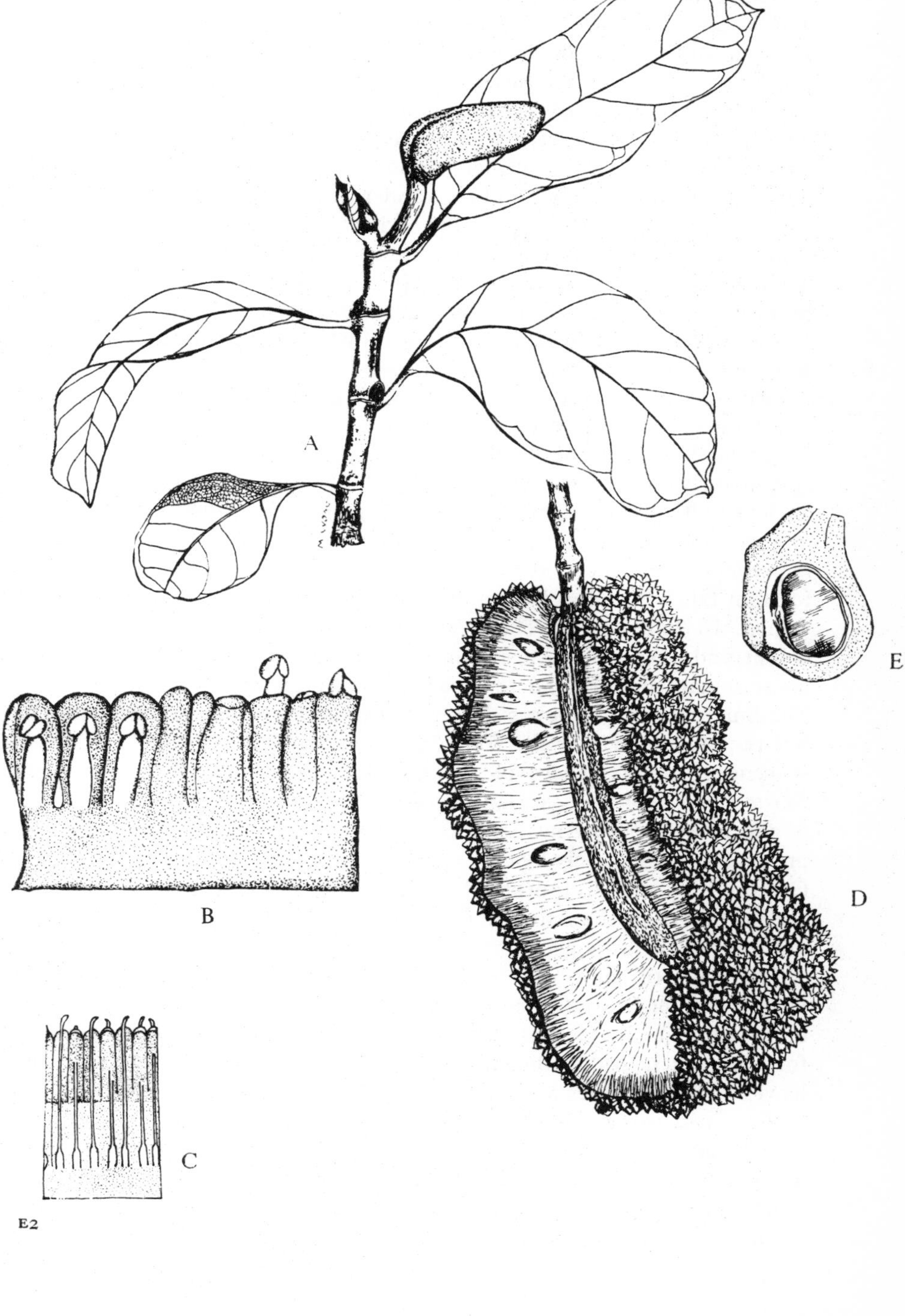

E2

blind, copulate with the somewhat larger females and die without leaving the fig. The impregnated female scrambles out between the scales at the mouth of the fig getting covered with pollen from the male flowers near the mouth. She then flies to another tree to oviposit. If she enters a gall fig she oviposits an egg down each of the short styles into the ovary and continues the process in this and other inflorescences until her stock of eggs is finished and then dies of exhaustion without having eaten any food since she hatched. The eggs then develop into larvae in the ovaries. If the female wasp enters a fig with female flowers, her ovipositor is too short to penetrate the longer styles and by wandering about in a frustrated manner deposits pollen on the styles and brings about cross-pollination. The dehiscence of the stamens and receptivity of the stigmas coincide with the hatching of the wasps. The figs ripen in about three months and are eaten by birds and other animals. The seeds are passed with the faeces and so spread. Species of *Ficus* taken to countries where their particular species of *Blastophaga* is not present do not set seeds.

Ficus carica L., the common cultivated fig, does not do well in the low wet tropics, but can be grown at higher elevations and in areas of low rainfall at the flowering and fruiting seasons. It is a native of Asia Minor and spread early to the Mediterranean region. The fig is a plant of extremely ancient cultivation and was grown in Egypt before 4000 B.C. The fruits may be eaten fresh or stewed and they may be dried or canned. A laxative is prepared from them. It is a large shrub or low-growing deciduous tree to 10 m tall with large leaves, deeply 3–5-lobed, conspicuously palmately veined, 10–20 cm long. The pear-shaped fruits are variable in size and colour and have a high sugar content. The common or Adriatic figs have no staminate flowers and the seedless fruits develop parthenocarpically. The Smyrna figs, which are widely cultivated, have no staminate flowers, but require pollination for fruit development. This is brought about by suspending inflorescences of the wild caprifig in which the fig wasps, *Blastophaga psenes* L., are about to emerge (see above) on the branches of the flowering Smyrna fig, a process known as caprification. Figs are propagated vegetatively by cuttings. The main producing areas are Turkey, Greece, Italy and California. In introducing figs into a new area these should be cvs of the Adriatic fig, unless the caprifig and its fig wasp are also introduced as is necessary for the Smyrna fig.

The fruits of some wild *Ficus* spp. are edible, but are usually rather insipid and tend to be eaten in times of food shortages. The young leaves of some species are used as pot-herbs. Several species are lopped for fodder for livestock in India. Bird-lime is commonly made from the latex. The leaves of some species are smoked along with opium.

Ficus elastica Roxb., the india-rubber fig, occurs wild in India and Malaya and grows into a vast tree. In the wild state it usually starts life as an epiphyte and then strangles the tree on which it grows. It also grows an abundance of aerial roots from the trunk and main limbs which may develop into prop roots. The rubber from the tree has long been used for lining receptacles in north-eastern India. Considerable interest was taken in *F. elastica* as a source of rubber during the 19th century and plantings were made in India, Java and elsewhere. It is tapped by slashing or incising the bark and native methods are very wasteful. It can only be tapped every 3 months. The rubber has a high resin content and the yields are very inferior to *Hevea brasiliensis* (q.v.) with which it cannot compete. Young plants with handsome juvenile leaves are extensively grown as house-plants. *Ficus vogelii* Miq. is used as a source of resinous rubber in West Africa.

Three common strangling figs which begin life as epiphytes and later send out stout prop roots have been widely distributed throughout the tropics. They are:

Ficus benghalensis L., the Indian banyan, a large spreading evergreen tree with massive pillar-roots which extend the tree laterally indefinitely, thus permitting it to develop the biggest crown of any plant in the world. There is a record of a tree near Poona in India with a circumference of 2,000 ft. The fruits are edible.

Ficus benjamina L., a native of India and Malaysia, has graceful drooping branches and is sometimes known as the Ceylon willow.

Ficus religiosa L., a native of India, is the sacred pipal-tree of the Hindus and the sacred bodh-tree of the Buddhists. Buddha became incarnate and attained Nirvana in its shade at Magadha in India. A scion of this tree was planted at Anuradhapura in Ceylon in 288 B.C. and still stands. The tree can be propagated by cuttings and layering when it then stands upon its own trunk. It is frequently planted as a roadside tree in many parts of the tropics. The lac insects feed upon it (see below).

The creeping fig, *Ficus pumila* L., is a native of eastern Asia and often takes the place of ivy in the tropics, rooting flat against walls. The small juvenile leaves which occur where the plant is attached are small, almost sessile, cordate and about 2 cm long. The leaves of the erect fruiting stems, which appear when the plant has reached the top of its support, are much larger, elliptic, 5–10 cm long, and have petioles. The creeping fig is sometimes wrongly called *F. repens*.

Bark-cloth is obtained from the bark of *Ficus thonningii* Bl. and *F. natalensis* Hochst. in East Africa, and *F. nekbudu* Warb. (syn. *F. utilis* Sim) in Mozambique. The bark is removed by making two circular and one longitudinal cuts and is then macerated and beaten by mallets, which causes it to stretch considerably. They can easily be propagated by large stakes. The first two species are frequently planted in banana gardens round Lake Victoria. They are also used as shade trees for coffee and for growing other plants up them.

Several species of *Ficus*, including *F. benghalensis*, *F. religiosa* and *F. cunia* Buch.-Ham., are used as the host plant for the lac-insect, *Laccifer* spp., for the production of lac (shellac), which is an important industry in India and Thailand. Plants of other genera are also used. The minute red larvae feed on the young shoots and secrete a thick resinous fluid which envelops their bodies; the secretions from individual larvae coalesce and form a hard continuous encrustation over the twigs. Just before the adult insects emerge the twigs are harvested and the encrustations are scraped off, dried and processed to yield the lac of commerce. In the 18th and 19th centuries a red dye was produced from it as a substitute for cochineal which is obtained from the insect *Dactylopius coccus* Costa; these are grown on the cactus *Nopalea cochinellifera* Salm-Dyck. With the advent of aniline dyes the importance of the dye declined, but the resin or lac gained recognition as a thermoplastic moulding material and as the basis of varnishes and polishes. Since then lac has found application in diverse industries and has attained an important place in international trade. It was extensively used in the manufacture of gramophone records. Its most extensive use now is in the surface coating industry for protective and decorative purposes. It is also used for stiffening and finishing felt, straw and silk hats.

REFERENCES

CONDIT, I. J. (1947). *The Fig*. Waltham, Mass.: Chronica Botanica.

CONDIT, I. J. and ENDERUD, J. (1956). A Bibliography of the Fig. *Hilgardia*, **25**, 1–663.

CORNER, E. J. H. (1952). *Wayside Trees of Malaya*. 2nd ed. Singapore: Govt. Printing Office.

MYRISTICACEAE

A small family with 16 genera and about 380 spp. of trees, mainly confined to the lowland tropical rain forests of the Old and New World, with most spp. in the Indo-Malaysian region, particularly in New Guinea. Evergreen, aromatic trees with watery pink or red sap in the bark. Lvs alternate, exstipulate, entire, pinnately nerved, often with pellucid dots. Inflorescences axillary; fls small, regular, usually dioecious; calyx usually 3-lobed, cup-shaped; petals absent; stamens 2–40, filaments united into a column, anthers 2-locular; ovary superior, sessile, 1-locular. Fr. fleshy drupe, usually splitting into 2 valves and exposing single seed surrounded by undivided or lacinate coloured aril; seed with copious endosperm, usually ruminate, and small embryo.

USEFUL PRODUCTS

The only economic species is *Myristica fragrans* (q.v.), the nutmeg.

Myristica Gronovius (x = 7)

About 100 spp. extending from India to Polynesia and north-eastern Australia.

USEFUL PRODUCTS

M. malabarica Lam., a tree of the Malabar coast of India, yields Bombay nutmeg and Bombay mace which are used as adulterants of the genuine products from *M. fragrans* (q.v.). They are practically odourless and tasteless. The seeds and mace of *M. argentea* Warb., from Papua and New Guinea, are also used as adulterants.

Myristica fragrans Houtt. (2n = 42) NUTMEG

USES

M. fragrans yields two spices, nutmeg which is the dried seed and mace which is the dried aril. Although the essential oils are the same in both spices, the flavours of them are distinctive. Nutmeg is grated in small quantities to flavour milk dishes, cakes, punches and possets. Mace is favoured for use with savoury dishes, in pickles and ketchups. Nutmeg is used medicinally and is said to have stimulative, carminative, astringent and aphrodisiac properties. It is used in tonics and is official in the British Pharmacopoeia. The husk (pericarp) is made into sweetmeats and jellies in Malaya. The seeds yield a fixed oil, nutmeg butter, which is solid at ordinary temperatures; it is usually obtained from broken seeds and mace which are not good enough for the spice trade, by pressing them between hot plates. It is used in ointments and perfumery. A mixture of volatile essential oils, oil of nutmeg, is obtained by distillation and is used in medicine externally and in perfumery. Both butter and oil contain myristicin, which is narcotic and poisonous, so that nutmegs and mace must be used sparingly, as 4–5 g produce symptoms of poisoning in man.

ORIGIN AND DISTRIBUTION

Nutmeg is a native of the eastern islands of the Moluccas, but it is seldom, if ever, found truly wild, its nearest allies occurring in New Guinea. The spice was not known to the Greeks and the Romans and there is a record of it in Constantinople about 540 A.D. It must have reached India before this time, and probably reached Europe soon afterwards. By the end of the 12th century it was well known in Europe, but was very costly. The spice came from the Moluccas *via* entrepôts in Java and India. As with other spices, the Arabs tried to hide the true source and this was not known until the Portuguese discovered the tree growing in Banda and Amboina in 1512. The Portuguese maintained a monopoly of nutmegs and mace, until ousted by the Dutch in the mid 17th century, and the Dutch retained the monopoly until early in the 19th century. They restricted by force nutmeg cultivation to Banda and Amboina, destroying the trees on the other islands, but fruit pigeons carried the seeds to neighbouring islands. Vast accumulations were made and in order to maintain prices large quantities were burnt in Amsterdam in 1760. The French in 1772 succeeded in introducing the tree into Mauritius and French Guiana. The Dutch monopoly was broken when Christopher Smith, botanist of the British East India Company, who had sailed with Bligh on the *Providence*, was sent to the Moluccas to collect spice plants. By 1802 Smith had sent over 70,000 nutmeg plants to Penang and a few were sent to Kew, Calcutta and Madras. Flourishing plantations were begun in Penang, Malacca and Singapore. Many of these trees were wiped out during the period 1859–1864, probably due to attacks by Scolytid beetles and disease. Sir Stamford Raffles, the founder of Singapore (1819) and the London Zoo, had encouraged production in Bencoolen in Sumatra, then held by the British; in 1820 there were 100,000 trees there. Nutmeg plants were taken to St. Vincent and they were introduced from there to Trinidad in 1824. In 1843 nutmegs were introduced into Grenada, and this crop first influenced the world market about 1865. Nutmegs, together with cloves, were introduced into Zanzibar in 1818 from Mauritius or Rèunion. Economic production is now mainly confined to Indonesia and Grenada.

ECOLOGY

In its original home on Banda, nutmegs are grown on rich volcanic soil and with a non-seasonal climate with 87–142 in. of rain and temperatures of 76–92°F. Cultivation of the crop is largely confined to islands in the hot, humid tropics at altitudes up to 1,500 ft. Nutmegs cannot tolerate waterlogging or excessive drying out of the soil. Shade appears to be beneficial in early growth.

STRUCTURE

A spreading dioecious evergreen tree, 5–13 m high; sometimes attaining 20 m. Root system superficial. Young branchlets very slender. Lvs alternate, exstipulate; petiole about 1 cm long; lamina glabrous, elliptic or oblong-lanceolate, base acute, tip acuminate, dark green, shiny above, paler beneath, 5–15 × 2–7 cm, pinnatinerved with 8–11 pairs lateral veins, aromatic when bruised. Fls typically unisexual-dioecious, but occasional trees occur with male and female fls on the same tree. Hermaphrodite fls have also been recorded. Male and female inflorescences similar, glabrous, axillary in few-flowered umbellate cymes with 1–10 fls in males and 1–3 fls in females; peduncle usually unbranched, 1·0–1·5 cm long; pedicels pale green, 1·0–1·5 cm long. Fls fragrant, pale yellow, waxy, fleshy, glabrous; calyx bell-shaped, nectiferous at base, and with 3 reflexed triangular lobes at top; petals absent. Male fls 5–7 mm long; stamens 8–12 with anthers adnate to a central column by their backs and attached to each other by their sides. Female fls up to 1 cm long; ovary sessile, puberulous, 1-celled, and with very short 2-lipped stigma. Frs fleshy drupes, broadly pyriform, drooping, yellow, smooth, 6–9 cm long, with circumferential longitudinal ridge and persistent remains of stigma; when ripe succulent yellow pericarp splits into 2 valves exposing purplish-brown, shiny seed surrounded by a much lacinate red aril, attached to base of seed. Seeds broadly ovoid, 2–3 cm long; with convoluted dark brown perisperm, lighter coloured endosperm and a very small ovary.

POLLINATION

This is probably effected by small insects.

GERMINATION

The seeds soon lose their viability and should be sown soon after collecting. They are planted in the shell. Seeds which rattle will not germinate as they have dried out. Germination takes 4–6 weeks.

CHEMICAL COMPOSITION

Nutmegs contain approximately: water 9 per cent; carbohydrate 27 per cent; protein 6·5 per cent; fixed oils 33 per cent; essential oils 4·5 per cent. Mace has less fixed oils, 22·5 per cent, and more essential oils, 10 per cent. Nutmeg butter contains about 73 per cent of trimyristin and 13 per cent of essential oils; seeds yield 25–30 per cent of butter. Nutmeg oil, which comprises 7–12 per cent of seed, contains essential oils, of which the principal are 80 per cent of pinene and camphene, and about 4 per cent of myristicin, which is poisonous.

PROPAGATION

Nutmegs are normally propagated by seeds sown in shaded nurseries at a depth of 2–3 in. in rows 12–18 in. apart. The seedlings are transplanted in the field when they are 6 in. high and about 6 months old. Seeds may also be planted in baskets or other containers. In Grenada seeds, which are overlooked during harvesting from fallen fruits, germinate, so that plantations will contain seedlings of different ages, some of which are allowed to mature to replace dead or unthrifty trees.

Due to the almost complete dioecy, seedling progeny will give about 50 per cent of each sex and these cannot be determined until the trees flower at 5–8 years of age. It is customary to cut out the surplus males at this stage, leaving 1 male to 10 females. This results in irregular spacing. Vegetative propagation thus has considerable attraction and it will also permit the selection of high-yielding clones.

Budding has not been very successful. Nichols and Pryde (1958) rooted semi-hardwood and hardwood cuttings, 12–15 in. long from wound shoots and water shoots in Trinidad-type propagating bins (see cocoa). Only about 10 per cent of the cuttings rooted successfully, and this took 6 months. Nichols and Cruickshank (1964) developed a method of approach-grafting seedlings of pencil thickness in perforated polythene bags to similar sized twigs on the female tree. Grafts took about 4 months to unite. The scion is severed above the union and the plants are repotted and hardened off. They also developed a method of marcotting vigorous healthy branches, $\frac{1}{2}$–$\frac{3}{4}$ in. in diameter, but rooting took 6–18 months and the percentage success was less than with the approach grafting. Plants produced by both methods have been issued in Grenada to replace the trees lost during the hurricane 'Janet' in 1955.

HUSBANDRY

Seedlings are usually planted at a spacing of 25 ft. On the removal of the excess males an irregular spacing of about 40 ft apart is achieved. Seedling trees begin bearing in 5–8 years. Marcotted and approach-grafted

---

Fig. 63. **Myristica fragrans:** NUTMEG. **A**, flowering shoot of male tree (×$\frac{1}{2}$); **B**, male flower with part of calyx removed (×3); **C**, female flower with part of calyx removed (×2); **D**, female flower in longitudinal section (×2); **E**, shoot with dehiscing fruit (×$\frac{1}{2}$); **F**, seed surrounded by aril (×$\frac{1}{2}$); **G**, seed in longitudinal section (×$\frac{1}{2}$).

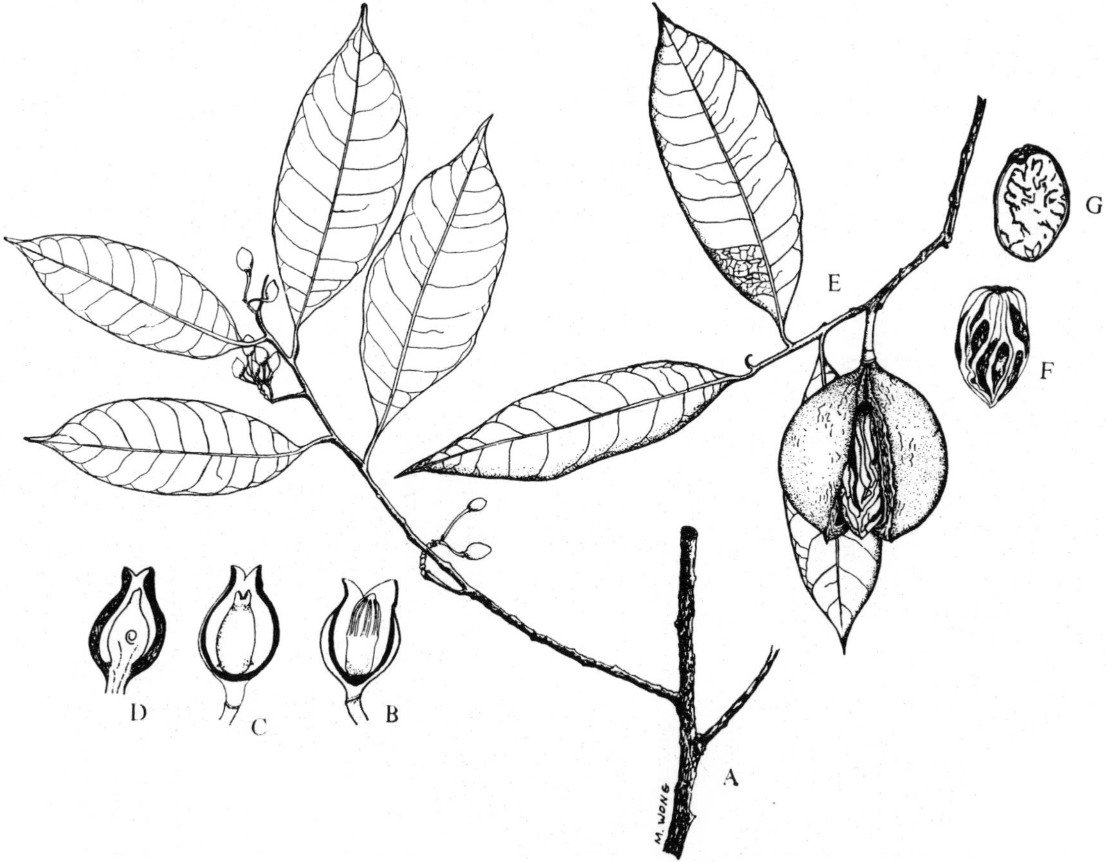
A
B
C
D
E
F
G
M. WONG

trees are now planted 30 ft apart in Grenada, and in some cases the former have borne fruit in 18 months. Trees come into full bearing at 15–20 years and continue for 30–40 years or more. The fruits ripen 6 months after flowering, usually with two peaks of fruiting annually, although some fruits ripen at all times. The fruits may be harvested on the tree after splitting, but more usually after dropping. A tree produces 1,500–2,000 fruits per year. The seed is removed from the pericarp; the mace is taken off and flattened by hand or between boards, care being taken to avoid breakage. In Grenada the mace is dried for 6 months in large wooden bins, saucers of carbon disulphide being used against weevils. The colour changes during curing from bright red to yellow, the brightness of the colour being important. The nutmegs are dried in large wooden trays under cover until they rattle in the shell. They are then shelled by a tap on the end with a wooden mallet; cracking on the side bruises the endosperm. Nuts are graded according to size and vary from 50–120 nuts per lb; round, bold, large nuts fetching the highest price. 100 nutmegs produce 3 oz of mace. In the East nutmegs are sometimes treated with lime before exporting as a protection against insect attack. Yields per acre may vary from 500–1,000 lb of nutmegs and 100–200 lb of mace per annum.

MAJOR DISEASES AND PESTS

Rosellinia spp. attacks nutmegs in Grenada. A scolytid beetle, *Phloeosomus cribratus* Bland. caused much destruction in Penang.

IMPROVEMENT

Little or no work has been done on the improvement of nutmegs by selection and breeding. Now that a satisfactory method of vegetative propagation has been obtained, selection of clonal material should be possible. In addition to high yield, nutmegs should be large, of uniform size, round in shape, light brown in colour, with a low turpene content, and with a thick mace.

PRODUCTION

The earliest production was in the Moluccas, particularly in Banda and Amboina (see ORIGIN AND DISTRIBUTION above). Early in the 19th century plantations were made in Penang, Malacca, Singapore and Sumatra. Exports from Grenada in the West Indies began after 1850. Production in Malaya is now practically non-existent, but Singapore imports considerable quantities from Indonesia and re-exports them. During the period 1951–1955, the average annual exports from Grenada were 84,000 cwt of nutmegs and mace, accounting for about half the total domestic exports; during the same period average annual exports from Indonesia were 75,000 cwt. Exports from Grenada declineu after hurricane 'Janet' in 1955 which destroyed about 90 per cent of the nutmeg trees. The principal areas of production in Indonesia are the Moluccas, particularly Banda, Minahasse in the Celebes, and Sumatra.

REFERENCES

NICHOLS, R. and PRYDE, J. F. P. (1958). The Vegetative Propagation of Nutmeg (*Myristica fragrans*) by Cuttings. *Trop. Agriculture, Trin.*, **35**, 119–129.

NICHOLS, R. and CRUICKSHANK, A. M. (1964). Vegetative Propagation of Nutmeg (*Myristica fragrans*) in Grenada, West Indies. *Trop. Agriculture, Trin.*, **41**, 141–146.

SINCLAIR, J. (1958). A Revision of the Malayan Myristicaceae. *Gdns.' Bull., Singapore*, **16**, 205–466.

MYRTACEAE

A distinctive family with about 75 genera and 3,000 spp. of trees and shrubs, mainly in the tropics and subtropics, and with a large concentration in Australia. With the exception of *Eugenia*, the family appears to be specialized to hot and rather dry climates. Lvs simple, mostly entire and opposite, finely dotted with oil glands, exstipulate. Fls regular, hermaphrodite or polygamous by abortion; calyx tube more or less adnate to ovary with 4–5 lobes often persistent in frs; petals usually 4–5, free, imbricate; stamens numerous inserted on disc, with slender filaments and small bilocular anthers; ovary inferior, 1- to many-locular, style one. Fr. usually a pulpy berry or woody capsule.

USEFUL PRODUCTS

EDIBLE FRUITS: Several species of *Eugenia* (q.v.) have edible fruits. *Feijoa sellowiana* Berg., feijoa, a native of South America, is sometimes cultivated in warm countries for its fruits. *Myrciaria cauliflora* Berg. [syn. *Eugenia cauliflora* (Berg.) DC.], jaboticaba, a native of Brazil, is cultivated for its small, globular, purple, grape-like, cauliflorous fruits, which are eaten fresh and are used in jellies, wines and cordials. *Psidium guajava*, guava (q.v.), is now widely planted throughout the tropics.

ESSENTIAL OILS AND SPICES: *Eugenia caryophyllus*, clove (q.v.), and *Pimenta dioica*, allspice (q.v.), are important spice crops and also yield essential oils. *Eucalyptus* spp. (q.v.) and *Pimenta racemosa*, bay (q.v.), are sources of essential oils.

OTHER USES: *Eucalyptus* spp. (q.v.) provide timber, firewood, kinos and tannins. *Melaleuca leucadendron* L., punk-tree of Malaysia, yields cajeput oil which is distilled from the leaves and is used medicinally; the tree will tolerate saline conditions and can be used for fixing muddy shores; it is extensively used for firewood in Malaya. *Leptospermum laevigatum* F. Muell., Australian tea-tree, is extensively planted for the reclamation of moving sand; the dried leaves are made into tea. *Callistemon* spp., bottle-brush, natives of Australia, are grown as ornamentals, of which *C. citrinus* Stapf is most widely planted in the tropics and warm countries. *Eucalyptus* spp. and *Melaleuca* spp. provide some handsome ornamental trees.

Eucalyptus L'Hér (x = 11; 2n = 22) EUCALYPTS

A large genus with about 600 species and varieties, which provide the most characteristic element of the Australia flora, and with a few species in eastern Malaysia. Small shrubs less than 3 m high to giant trees over 100 m high, the tallest hardwoods in the world; usually evergreen. In smooth-barked spp. the bark is shed regularly leaving a single periderm. Lvs often heterophyllous with juvenile lvs opposite, horizontal, sessile, cordate and glaucous; adult lvs often alternate, vertical, rigid, petiolate, lanceolate and dark green. Fls borne in a dichasium, usually of 1, 3, 7, 11 or 15 fls. Calyx lobes and petals united to form an operculum or cap which dehisces transversely as fls open. Stamens numerous, usually free; ovary 3–6-celled with numerous ovules in each cell. Fr. a woody capsule, dehiscing at the apex by 3–6 valves.

After the operculum has fallen, the stamens unfold and the anthers dehisce. The stigma is receptive a few days later; the flowers are usually insect-pollinated and most species are cross-pollinated. Interspecific hybrids occur.

Eucalyptus spp. have been introduced into most tropical and warm countries, the two most commonly planted spp. being *E. globulus* Labill. and *E. saligna* Sm., which are very quick growing. *E. deglupta* Blume, from New Guinea, has been introduced into the West Indies. Introduced into Brazil during the period 1855–1870, there are now 2 million acres of eucalypts in that country. The genus as a whole will tolerate a wide range of climatic and soil conditions, but they are not commonly grown in the very wet tropics at low altitudes, although the New Guinea and allied spp. may succeed under these conditions.

Eucalyptus oil is distilled from the leaves, but less than 20 spp. have been exploited commercially. Medicinal oil is used in inhalants, embrocations, soaps, gargles, sprays and lozenges; it is an antiseptic and stain remover. It should contain a minimum content of 70 per cent cineole, the active therapeutic agent. The 3 principal spp. now used in Australia for oil extraction are *E. fruticetorum* F. Muell., *E. radiata* Sieb. ex DC. and *E. dives* Schau.; the leaves yield 2–4 per cent of oil. Elsewhere eucalyptus oil is produced in Spain, Portugal, Brazil and the Congo, mainly from *E. globulus* Labill., which yields 0·75 per cent oil; in the Congo this sp. is now being replaced by *E. smithii* R. T. Baker, which is richer in oil. The oil is also used for industrial purposes as a disinfectant and deodorant, and as a source of piperitone from which thymol and menthol are manufactured; it is obtained mainly from *E. dives* and *E. radiata*. The oils from *E. citriodora* Hook., which contains 65–85 per cent of citronellal, and *E. staigeriana* F. Muell. ex F. M. Bail. are used in the perfumery and allied industries; they are grown for this purpose in Java, the Seychelles, the Congo, Brazil and Guatemala.

The eucalypts are an important source of timber; about 60 spp. being used for hardwood in building construction, for railway sleepers, furniture manufacture, etc. They supply large quantities of firewood and building poles and are grown for this purpose in the tropics. The wood is used in the manufacture of newsprint paper, fibreboard and rayon. They are often grown as windbreaks and shelter belts, but many spp. have severe root competition. Some spp. are used for draining swamps. Tannin is extracted from the bark of some eucalypts, particularly *E. wandoo* Blakely and *E. astringens* Maiden; others exude kinos, which are collected, *e.g. E. camaldulensis* Dehnh. Rutin, a vitamin P-like substance used in the treatment of hypertension, is obtained from *E. macrorrhyncha* F. Muell. They are an important source of nectar and pollen for bees. Some spp. are very ornamental and are grown for this purpose.

REFERENCE

PENFOLD, A. R. and WILLIS, J. L. (1961). *The Eucalypts.* London: Leonard Hill.

Eugenia L. (x = 11)

A large genus of about 1,000 spp. of evergreen trees and shrubs in the tropics. Some of the Old World *Eugenia* spp. are now placed in the genus *Syzygium* by some authorities. All parts usually glabrous. Lvs opposite. Fls usually clustered in lf. axils; calyx-cup usually with 4–5 lobes; petals 4–5, often falling as a cap on opening of flowers; stamens numerous, usually with fluffy appearance; ovary inferior, 2–3-celled; fr. a berry crowned by persistent calyx cup, usually with 1–2 large seeds, each with 2 large cotyledons.

USEFUL PRODUCTS

EDIBLE FRUITS: The following spp. have edible, rather insipid fruits and are sometimes planted for this purpose; some of them are used as windbreaks and are planted for ornamental purposes.

E. jambos L. (syn. *Syzygium jambos* (L.) Alston), rose apple, has long been cultivated in the Indo-Malaysian region for its rose-scented frs which may be eaten fresh or made into preserves. It is now spread throughout the tropics. A small tree up to 10 m in height with narrow pointed lvs, 15–20 × 2·5–5 cm; fls about 7 cm in diameter with numerous greenish-white stamens; frs ovoid, about 4 cm in diameter, whitish tinged yellow or pink, with a thin layer of yellow flesh, usually with 1–2 brown seeds, which are often polyembryonic.

E. malaccensis L. (syn. *Syzygium malaccensis* (L.) Merr. & Perry), pomerac or Malay apple, a native of Malaysia, is planted throughout the tropics as an ornamental and windbreak and for its rather insipid frs, which may be eaten fresh or made into preserves. Tree to 20 m, lvs 20–30 × 8–22 cm; fls 5–7 cm in diameter, vivid crimson-pink, with numerous long stamens; frs oblong or pyriform, about 8 cm long, reddish-pink or white striped crimson pink; flesh white; seeds one, large.

E. uniflora L., pitanga or Surinam cherry, is a native of Brazil and is now grown in tropical and subtropical countries, including Florida for its frs which are eaten fresh or made into pies and preserves. It is also used for hedges. Shrubs or small tree to 7 m in height; lvs small, ovate; fls about 1 cm across, cream coloured; frs borne on slender stems, 2–4 cm in diameter, deeply 8-ribbed, red to almost black with flesh of same colour; seeds 1, sometimes 2.

Other species of *Eugenia* which have edible fruits are: *E. aquea* Burm. f. (syn. *Syzygium aqueum* (Burm. f.) Merr. & Perry), watery rose-apple, native of India; *E. cumini* (L.) Druce (syn. *Syzygium cumini* (L.) Skeels), jambolan, a native of Brazil; *E. javanica* Lam. (syn. *Syzygium javanicum* (Lam.) Merr. & Perry), Java or wax apple, a native of Malaysia.

SPICES: *E. caryophyllus* (q.v.) is the clove tree.

OTHER USES: The bark of some *Eugenia* spp. is used in Malaya for tanning, for toughening fishing nets and lines, and for dyes.

Eugenia caryophyllus (Sprengel) Bullock & Harrison CLOVE

Syn. *Caryophyllus aromaticus* L.; *Eugenia aromatica* Kuntze; *E. caryophyllata* Thunb.; *Syzygium aromaticum* (L.) Merr. & Perr.

USES

Cloves, the dried unopened flower buds of *E. caryophyllus*, have been used in the Orient, particularly India and China, for over 2,000 years as a spice, to check tooth decay and to counter halitosis. In the 3rd century B.C. they were used by court officials in China to sweeten their breath when speaking to the Emperor. In Persia and China they were considered to have aphrodisiac properties and 'to create love'. Now in the East, cloves are used as a table spice and in the preparation of curry powder when they are mixed with chillies, cinnamon, turmeric, dill and other spices; they are also used to flavour the betel quid (*pan pati*), in which chopped *Areca* nut (*Areca catechu* L.) mixed with lime, folded in a leaf of betel pepper (*Piper betle* L., q.v.) and fastened with a clove are chewed. In the West cloves are used in mixed spices, in tomato and other sauces, for seasoning sausages, in apple puddings and tarts, to decorate and improve the flavour of hams and in sousing herrings and other fish. Cloves are stuck into oranges to make pomanders. In medicine cloves are stimulative, antispasmodic and carminative. In Java one-third shredded cloves is mixed with two-thirds tobacco in clove cigarettes, which are strong, pungent and aromatic, of which commercial production did not begin until 1916; Indonesia now provides the largest market for Zanzibar cloves. The dried fruits, mother of cloves, are sometimes used as an adulterant and as a spice.

Clove oil, produced by the distillation of cloves, stems (flower stalks) and leaves, is used in the manufacture of perfumes, particularly carnation and wallflower, in soaps, bath salts, etc., as flavouring, in medicine and dentistry, in dentifrices, and as a clearing agent in microscopy. It is also used in the production of vanillin.

ORIGIN AND DISTRIBUTION

Clove trees are indigenous to the small volcanic islands of Ternate, Tidore, Mutir, Makyan and Bachian in the Moluccas. The Chinese obtained the spice by at least the 3rd century B.C. This trade led to planting in Amboina and Ceram. Cloves were imported into Alexandria as early as A.D. 176; by the 4th century A.D. they were well known in the Mediterranean and by the 8th century throughout Europe; they were extremely expensive. Their country of origin remained unknown in Europe for many centuries, as the traders who imported them *via* Java, Malacca and Ceylon, kept this a secret. The Portuguese, early in the 16th century, discovered and occupied the spice islands in the Moluccas and cloves become the monopoly of the royal house of Portugal. The Dutch captured Amboina in 1605 and they retained the monopoly for two centuries; in order to maintain this they ordered the destruction of all trees except on Amboina. However, a secret French expedition to the Moluccas in 1770 resulted in the introduction of seeds and plants from Ceram to Mauritius and Réunion in 1772. Plants were taken to Cayenne about 1789, from where they were introduced into Dominica, Martinique and other West Indian islands. At the beginning of the 19th century, Smith obtained cloves and nutmegs (q.v.) from the Moluccas for planting by the East India Company in Penang. He obtained 15,000 plants in 1800 and by 1802 had sent a total of 55,265 clove plants to Penang. Extensive plantings were made in Penang in the 1820's, but the crop was not so successful in Singapore and Malacca.

A Zanzibar Arab took cloves from Mauritius to Zanzibar in 1818 and these were grown in the royal plantation at Mtoni. As they flourished, Sultan Said bin Sultan forced plantation owners to plant cloves under threat of confiscation of their land, and more than half the area of Zanzibar and Pemba was planted with cloves. As much of the Mauritius and Réunion crops had been destroyed by cyclones, Zanzibar became the leading producer of cloves, a position which it still maintains. In 1872 a hurricane destroyed most of the cloves in Zanzibar, but not in Pemba, and the Sultan of the day ordered the replanting of the crop. Cloves were introduced from Mauritius to the island of Sainte-Marie off the coast of Madagascar in 1827, but extensive production did not begin until 1895 and was extended to the 'Grande Terre' of Madagascar in 1900.

ECOLOGY

Cloves thrive best with insular, maritime climates in the tropics at low altitudes. Continuously humid climates are not so suitable; cloves were more successful in Penang than in Singapore. In the original habitat in the Moluccas, where the trees are semi-wild, annual rainfall is 87–142 in.

and temperatures 76–92°F. Drier weather is desirable for harvesting and drying the crop. In Zanzibar the mean daily maxima and minima temperatures are 84° and 77°F and in Pemba 86° and 76°F; the average annual rainfall in Zanzibar is approximately 60 in. and in Pemba is 80 in., with the main *masika* rains from March–June and the lesser *vuli* rains from October–November. The Madagascar clove-growing areas receive about twice as much rain as Zanzibar and it is more evenly spread throughout the year; the temperatures are rather lower than Zanzibar.

The best soils for cloves in Zanzibar and Pemba are deep, sandy, red, acid loams. Pemba with its better soils, higher rainfall, and more hilly topography gives higher yields than Zanzibar. In Madagascar the poorer soils are used for clove production, the better soils being reserved for coffee and vanilla. Good deep drainage is essential and waterlogging is fatal. Shade in the early stages and windbreaks are desirable.

STRUCTURE

A small evergreen tree to a height of 14 m, conical in shape when young, later becoming cylindrical. The primary tap-root usually remains short, but it produces 2–3 primary sinkers which reach a depth of about 3 m. Horizontal lateral roots grow to a length of about 10 m and produce a surface plate of feeding roots and occasionally send out secondary sinkers to the lower levels. The trunk often forks near base to give 2–3 main erect branches; bark grey, rough; smaller branches sub-erect, very brittle. Lvs opposite, glabrous, aromatic; petiole 2–3 cm long with swollen reddish base; lamina 7–13 × 3–6 cm, entire, elliptic, coriaceous, pellucid-dotted, base cuneate, tip acuminate, densely and obscurely pinnatinerved; young lvs tinged pink, mature lvs dark-green, shiny above, paler beneath; inflorescences terminal, about 5 cm long, of 3–20 fls in paniculate cymes; fls usually borne in groups of 3. In Zanzibar fls produced in two seasons, the *mwaka* crop from July–October and the *vuli* crop from November–January. Bud initials appear 6 months before the buds are ready to harvest. Angled peduncles and short pedicels, about 5 mm long, constitute the clove stems of commerce. Fl. buds 1·3–2·0 cm long. Fls hermaphrodite, with fleshy hypanthium, reddish in colour, surmounted by 4 fleshy triangular sepals; petals 4, imbricate, rounded, tinged red, falling as fl. opens; stamens numerous, small, 5–7 mm long, with slender white filaments and small pale-yellow anthers; ovary inferior, embedded in hypanthium, 2-celled with several ovules; style short, obtuse; stamens and style fall after pollination. Frs, called mother of cloves, fleshy drupes, oblong-obovoid, dark red, about 3 × 1·2 cm, containing 1, rarely 2 seeds. Seeds oblong, about 1·5 cm long, grooved on one surface. Dried spice (fl. buds) about 1·7 cm long in Zanzibar and less than 1·2 cm in Madagascar, dark cinnamon in colour; hypanthium flattened at lower end; sepals prominent; domed, rounded head, paler in colour, of petals and enclosed stamens.

POLLINATION

Flowers are visited by bees and are probably cross-pollinated. No fertile fruits were obtained in Zanzibar when the flowers were bagged. Most of the flowers fall without setting seed.

GERMINATION

Seeds obtained from fleshy fallen fruits germinate quickly and the cotyledons are lifted clear of the soil in 12–14 days. Viability quickly diminishes, but fresh seeds usually give a germination of over 90 per cent.

CHEMICAL COMPOSITION

Commercial cloves contain approximately: clove oil 16–19 per cent; tannin (quercitannic acid) 13 per cent; water 16 per cent; fibrous matter 10 per cent; starch a little. Dried clove stems (inflorescence stalks) contain 5–6·5 per cent clove oil and dried leaves about 3–4 per cent. Fruits, mother of cloves, contain very little oil. The oil is obtained by distillation. Clove oil contains 80–95 per cent of eugenol, with an average of about 85 per cent; oil from the leaves contains 75–88 per cent eugenol. The final flavour depends on minor constituents, particularly esters. Vanillin is made from eugenol.

PROPAGATION

Cloves are propagated by seeds. The early planting of cloves in Zanzibar by the Arabs was done with seedlings raised in nurseries. Later self-sown seedlings were transplanted. The Government then maintained public clove nurseries. Fully ripened, freshly fallen fruits are collected; they are soaked in water and fermented in heaps under wet sacks for 3 days; the seeds are then hulled with the fingers. The seeds should be large and of a fresh olive green colour, with not more than 500 seeds per lb, free from damage and taken from single-seeded fruits. The fruits should be obtained from selected healthy trees of suitable shape, which are regular and heavy bearers. Hulled washed seeds produce better seedlings than unhulled fruits. The seeds are planted in shaded nurseries with the radicle downwards and the upper half of the seed exposed above the soil at a spacing of 8 × 8 in. Watering and shade are reduced when the seedlings are 1 year old to harden them off and they are transplanted to the field at about 15 months old with a ball of earth round the roots.

Vegetative propagation by cuttings, marcotts, layerings and grafts have not been successful in Zanzibar. Due to low meristematic activity, clove

---

Fig. 64. **Eugenia caryophyllus:** CLOVE. **A**, flowering branch (×½); **B**, bud in longitudinal section (×2); **C**, flower in longitudinal section (×2); **D**, fruit (×2); **E**, fruit in longitudinal section (×2); **F**, dried clove (×2).

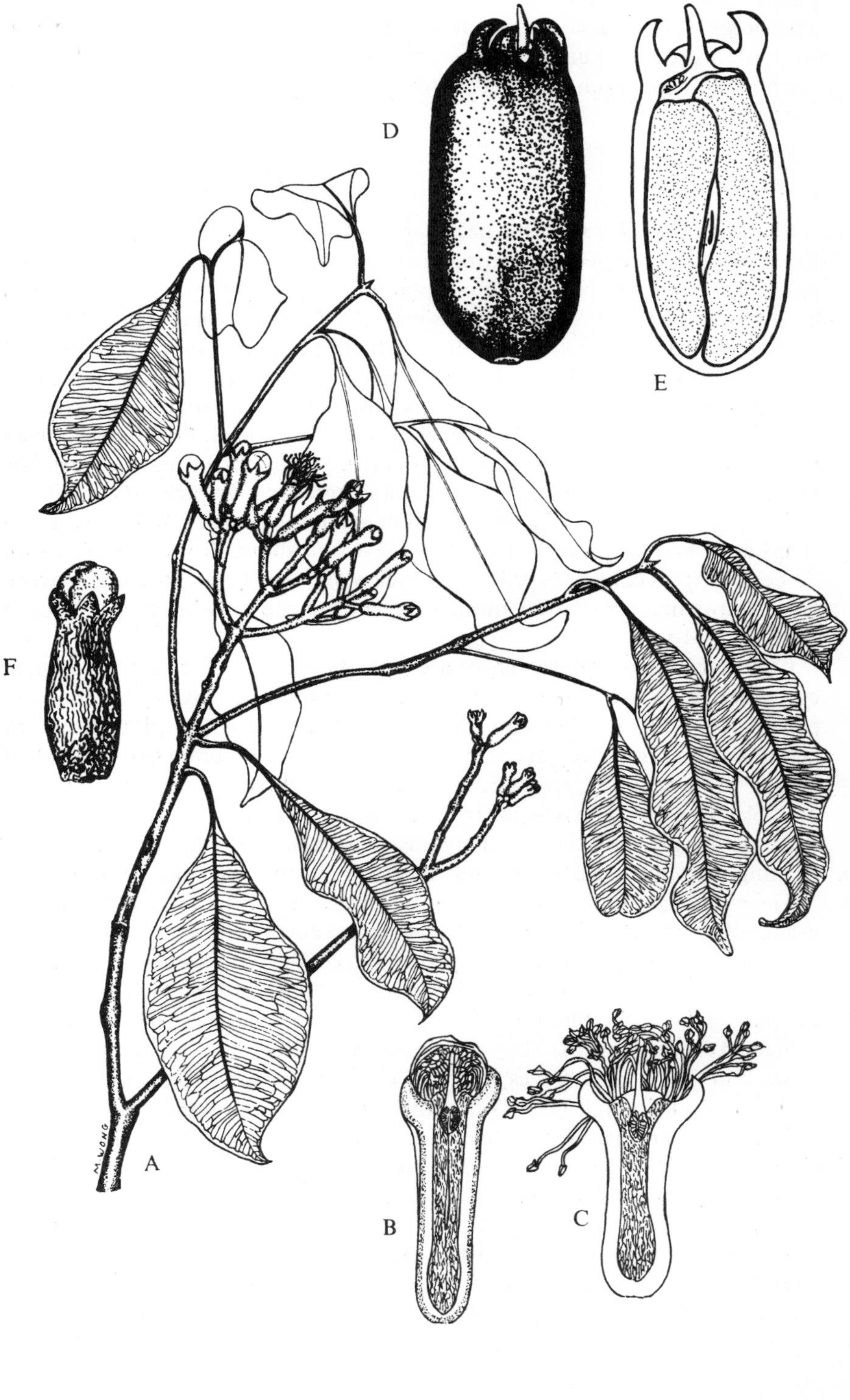
D
E
F
A
B
C

plants are notoriously difficult to propagate vegetatively. Elsewhere occasional success has been reported by layering, approach-grafting and by budding on to guava, *Psidium guajava* (q.v.). It is possible that the method of approach-grafting described for nutmegs (q.v.) might be successful.

HUSBANDRY

Shade and windbreaks are necessary in the early stages of growth, and these may be obtained by cutting lanes through the forest in new land or underplanting when replanting old plantations. They may also be obtained by interplanting with bananas, cassava, or *Gliricidia sepium* (Jacq.) Steud. The usual spacing for cloves is 30 × 30 ft, but they are sometimes planted at 15 × 15 ft or 30 × 15 ft and later thinned. *Imperata cylindrica* Beauv. is the most serious weed of cloves. The young plants are usually ring-weeded. Mulching with dry clove leaves has been found beneficial. Positive responses in bearing trees have been found to nitrogenous fertilizers and organic manures such as coconut meal. Pruning is not done, as picking constitutes a heavy pruning. Trees begin to flower at about 6 years, but subsequent crops are small until full bearing is achieved at about 20 years. Production on healthy trees continues until 70–80 years or even longer. Bearing between years shows much variation and a bumper crop can only be expected about once every 4 years.

Clove clusters or inflorescences are picked by hand before the buds open, but when they are full size and turning pink. The lower branches are picked from the ground; for the rest, branches are lashed together to facilitate climbing away from the trunk and distant branches are pulled nearer by a crook. Care should be taken to break as few branches as possible, this being particularly important in saplings. The clusters are stemmed by pressing the bunch against the palm of the hand with a slight twisting movement when the buds snap off cleanly from the pedicels.

The stemmed cloves are dried in the sun on mats made of palm leaves, usually *Hyphaene* spp., which are placed on cement floors or on the earth. Quick drying produces the best quality spice. The cloves are spread thinly and are stirred frequently. In good weather drying is completed in 4–7 days, but may take a fortnight in cloudy or showery weather. They must be protected from the rain. They may also be dried in the 'sandbath' type of copra kiln. Well-dried cloves will snap cleanly with a sharp click across the thumb nail and weigh about one-third of the green weight; brightness of colour is important. Green cloves left in heaps heat and ferment and the final product will have a whitish, mealy appearance and are known as *khoker* cloves. Abrasion on cement floors without mats also results in an inferior product.

The clove stems are also dried. The green stems are approximately one-fifth the weight of the green cloves taken from them and on drying also give about one-third of their green weight. They are used for the extraction of oil of cloves, yielding about 6 per cent of their dried weight.

The average annual yield of dried cloves per tree is about 7 lb, but in good years yields of 40 lb or more per tree are not uncommon, and over 200 lb has been recorded. Grade I cloves for export for the ordinary spice and Indian trade should be good sound cloves which are bold, reasonably regular, free from mustiness, and which contain not more than 5 per cent of stems, mother of cloves, foreign, superfluous or inferior material and 3 per cent *khoker* cloves, and not more than 16 per cent moisture. A special grade with under 3 per cent of stems and other matter and under 2 per cent of *khoker* cloves, exists to meet the demand of the high-class spice trade. Grade II cloves, with the same specifications as Grade I, but with up to 7 per cent *khoker* cloves, are permitted for the Indonesian market. Grade III cloves with up to 20 per cent *khoker* cloves are confined to exports for the manufacture of clove oil and vanillin.

MAJOR DISEASES AND PESTS

'Sudden death' was known in Zanzibar in the mid 19th century, but did not cause alarm until 1894; by 1953 more than half the mature cloves in Zanzibar island had died of this disease or of die-back (see below). The disease also occurs in Pemba, but not in such epidemic proportions. Physiological disturbance, virus and other causes were suggested, but Nutman and Roberts (1953) showed that it was due to the fungus *Valsa eugeniae* Nutman & Roberts. The disease first attacks the younger rootlets, followed by blocking of the vessels by tyloses in these and larger roots; a rather heavier leaf-fall than usual occurs and the abscissed leaves are lemon-yellow in colour. The final stage is rapid, taking only a few days, in which the leaves wilt and turn brown and the tree dies retaining its dead leaves. The wood is stained a brilliant yellow. Three months after the tree has died perithecia are formed in the region of the collar and later over the trunk and larger branches. The 'slow decline' of young clove trees is a symptom expression of *Valsa* attack when the juvenile resistance of the sapling is beginning to break down.

Die-back, *Cryptosporella eugeniae* Nutman & Roberts, is a virulent wound pathogen, which produces conspicuous red-brown staining of the wood and readily observable fructifications. It can also attack the main stem of young trees and when girdling is complete, causes them to die with symptoms similar to sudden death.

The parasitic alga, *Cephaleuros mycoidea* Karsten, attacks cloves, but caused the worst damage in Malaya.

Termites are a serious pest of young seedlings, both in the nursery and during the first few years in the field. The coccid, *Saissetia eugeniae*, is the commonest insect on cloves in Zanzibar. The red tree ant, *Oecophylla smaragdina* F., which has a vicious sting, is a nuisance during harvesting.

Parasitic Loranthaceae, *Cassytha filiformis* L. (Lauraceae) and strangling figs, *Ficus* spp. (Moraceae), cause damage in Zanzibar and should be removed in the early stages.

According to Maistre (1964) cloves have few important enemies in Madagascar.

IMPROVEMENT

Very little work has been done on the improvement of the clove crop by selection or breeding. As Tidbury (1949) has pointed out 'In Zanzibar no real varietal differences are observed . . . (and) the entire clove population of the Protectorate is derived from a very few trees, perhaps only two or three, after being smuggled in small quantities from the Dutch. It is therefore highly probable that the genotypes of Zanzibar's clove trees are very limited and subject to little variability.' The only mutation observed is one in which overlapping fleshy scales are produced instead of flower buds, and this, of course, cannot be reproduced from seed. Thus it is unlikely that much could be achieved by selection within the Zanzibar population.

If selected clove plants could be obtained from the semi-wild trees in the Moluccas, and a satisfactory method of vegetative propagation could be developed, some positive results might be achieved. Cloves are required with heavy and regular bearing and with resistance to sudden death and other diseases. Grafting on dwarfing stocks would facilitate harvesting.

PRODUCTION

Up to the end of the 18th century the Moluccas were the sole producers of cloves and the Dutch attempted to confine production to the island of Amboina. During the 19th century plantations were made in Zanzibar and Pemba, in Penang, and later in Madagascar. Small quantities were also produced in the Seychelles, Réunion and Sumatra. By the 20th century Zanzibar (including Pemba) and the Malagasy Republic (Madagascar) produced almost all the cloves entering world trade and have continued to do so. Cultivation in Indonesia for domestic consumption has increased.

Average annual exports of cloves from Zanzibar for the period 1897–1929 were 7,777 metric tons and for 1930–1962, 10,188 metric tons. Average annual exports from Madagascar for the same periods were 268 tons and 3,781 tons respectively. During the decade 1951–1960 annual production was 10,048 tons in Zanzibar and 4,249 tons in Madagascar. Average annual exports of clove oil from Madagascar for the years 1930–1962 were 405 metric tons and for the decade 1951–1960, 653 tons, and from Zanzibar for the period 1936–1962, 143 tons and for 1951–1960, 162 tons. The annual consumption of cloves on a world basis is approximately: Indonesia 8,000 metric tons, India 3,000 tons, Malaya 2,000 tons, United States 2,000 tons, Europe and North Africa 3,000 tons, other countries 1,000 tons. Annual imports into the United Kingdom are approximately 300 tons.

REFERENCES

MAISTRE, J. (1964). *Les Plantes à Épices.* Paris: Maisonneuve & Larose.

NUTMAN, F. J. and ROBERTS, F. M. (1953). Investigations into Diseases of the Clove Tree in Zanzibar. *E. Afr. agric. J.*, **18**, 146–154.

TIDBURY, G. E. (1949). *The Clove Tree.* London: Crosby Lockwood.

Pimenta Lindl. (x = 11)

About 5 spp. of aromatic trees in tropical America. *P. dioica* (q.v.) yields the culinary spice pimento or allspice.

P. racemosa (Mill.) J. W. Moore (syn. *P. acris* Kostel.), which is native to the Windward Islands in the West Indies, is the bay tree. Its leaves on distillation yield bay oil, which is used in perfumery and in the preparation of bay rum. Formerly the leaves were distilled in rum and water; now the oil is dissolved in alcohol, with which are mixed various aromatic substances. Bay rum has soothing and antiseptic properties; it is used in toilet preparations and as a hair tonic. A few trees are planted in the Windward Islands, but most of the leaves are obtained from wild trees. Bay rum is manufactured in Trinidad from oil imported from the smaller islands. It is occasionally drunk. The bay is a small tree growing to 10 m and has obovate leathery leaves.

Pimenta dioica (L.) Merrill (2n = 22) ALLSPICE, PIMENTO

Syn. *P. officinalis* Lindl.

USES

The dried unripe fruits provide the culinary spice of commerce, which is considered to combine the flavours of cinnamon, cloves and nutmegs; hence the name allspice. It is used in pickles, ketchups and sausages, and in curing meats. In Jamaica a local drink, pimento dram, is made out of ripe berries and rum. Pimento berry oil is extracted from the dried spice, which yields 3–4·5 per cent of volatile oils. Pimento leaf oil is extracted locally in Jamaica, dried leaves yielding a little over 2 per cent. The oil is used in flavouring essences and perfumes, and as a source of eugenol and vanillin. The spice and oil are stimulant carminatives. Saplings are used as walking sticks and umbrella handles and were formerly exported from Jamaica.

ORIGIN AND DISTRIBUTION

Pimenta dioica is indigenous to the West Indies and Central America, being most abundant in Jamaica; it also occurs wild in Cuba and southern Mexico, and possibly in Haiti and Costa Rica. It has been introduced into other West Indian islands, notably Grenada. Pimento was taken to Ceylon about 1824 and later to Singapore, but it did not do well. Much of the Jamaica crop is produced from semi-wild trees, but it is now being planted.

ECOLOGY

The natural habitat in Jamaica is mainly wet limestone forests, particularly on the Bauxite plateau, which has an annual rainfall of over 60 in. and is well distributed. Elsewhere it is grown with an annual rainfall of 40–100 in. or more. Mean monthly temperatures rarely exceed 80°F. It grows mainly on the Terra Rossa or red limestone soils and on the Rendzina or black marl soils overlying limestone. It requires good drainage. Pimento grows best below 1,000 ft in altitude.

STRUCTURE

A small evergreen tree, usually attaining a height of 7–10 m, but sometimes over 13 m. Trunk slender, much branched at the top; wood very brittle; bark silvery-brown, shed twice a year; young shoots 4-angled. Lvs opposite, coriaceous, aromatic; petiole 1–1·5 cm long; lamina elliptic to elliptic-oblong, glabrous, dark green above and paler and gland-dotted beneath, 6–15 × 3–6 cm; pinnately veined, mid-rib impressed on upper surface and prominent below. Inflorescences stalked many-flowered cymes, 5–15 cm long, in axils of upper lvs. Fls structurally hermaphrodite but functionally dioecious. Barren trees which never produce frs are functionally male, fruiting trees are functionally female. Pedicels about 1 cm long, pale green, pubescent, with small brownish pubescent bracteoles. Fls 8–10 mm in diameter when fully open; calyx tube extending above ovary with 4 spreading rounded lobes about 2 mm long, cream-coloured; petals 4, white, rounded, reflexed, about 4 mm long; stamens numerous, free, about 5 mm long, filaments white, anthers small, cream, basifixed; ovary inferior; style white, shortly pubescent, 5 mm long, with small terminal yellow stigma. In barren male trees stamens more numerous, 80–100, and receptacle broader and flatter and with stamens not so close to style. Bearing female trees have 40–50 stamens and style is somewhat shorter and stigma somewhat larger. Fruit a sub-globose berry, 5–6 mm in diameter, surmounted by persistent calyx; unripe frs green turning dark purple on ripening with sweet pulpy mesocarp. Seeds usually 2 with spirally coiled embryo, very short cotyledons and thick long radicle.

POLLINATION

Chapman (1964) has shown that, although the flowers are structurally hermaphrodite, pimento functions as a dioecious plant and male and female trees occur in approximately equal proportions. The male trees with their more numerous stamens shed copious pollen; insects and possibly

---

Fig. 65. **Pimenta dioica : PIMENTO. A, flowering shoot (×½); B, flower in longitudinal section (×3); C, fruits (×½); D, fruit in longitudinal section (×1); E, fruit in transverse section (×1).**

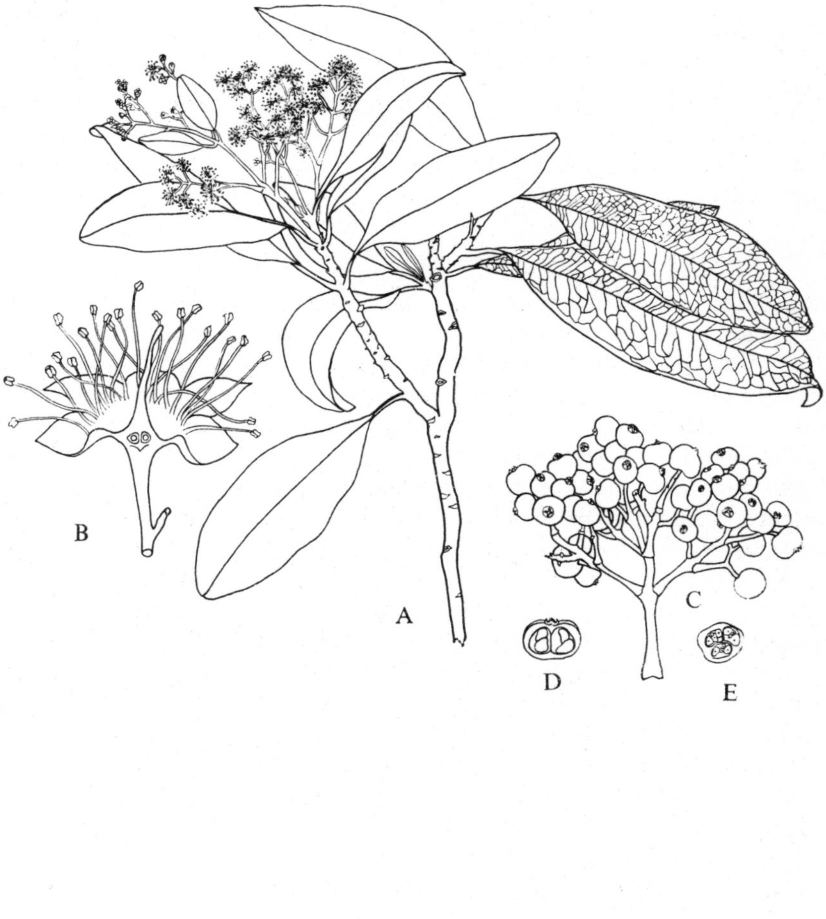
A
B
C
D
E

wind are responsible for the transfer of pollen. The stamens of the female functional trees also shed pollen but it is usually non-viable. Thus some barren trees are essential for pollination.

GERMINATION

The seeds are dispersed by birds and at one time it was thought that the seeds would only germinate after passing through the intestines of birds. However, when seeds, extracted from fresh ripe fruits, are sown, some seeds germinate in 9–10 days and germination continues over an extended period. A short purplish hypocotyl appears above the ground followed by a pair of small opposite seed leaves. The seeds quickly lose their viability.

CHEMICAL COMPOSITION

Dried allspice contains approximately 13 per cent moisture, 3–4·5 per cent volatile oils, 8 per cent or more tannin (quercitannic acid) and up to 25 per cent crude fibre. Berry and leaf oil contains 65–90 per cent phenols, chiefly eugenols.

PROPAGATION

Seedlings are used in commercial planting and the Ministry of Agriculture and Lands in Jamaica maintains nurseries for the distribution of plants. Ripe fruits should be collected from high-yielding trees which fruit regularly and are of good shape. The seeds are extracted by hand or by washing. Freshly washed seeds are dried in the shade and are planted as soon as possible or germination will be greatly reduced. If it is necessary to transport seeds or keep them a few days, it is better to leave them in the fruits. Seeds may be planted in nursery beds (sometimes made of coir waste), boxes or pots, and should be shaded, permitting one-half full sunlight. At the 2- or 4-leaf stage they are transplanted to bamboo pots, baskets or plastic bags. Young plants are planted in the field when they are 10–15 in. high at about 10 months old.

Pimento has been propagated vegetatively by cuttings, budding and approach-grafting, but the percentage of success was low. A dependable method of bud propagation has now been worked out and this will permit the planting of clonal female trees, leaving a few male trees as pollinators.

HUSBANDRY

Much of the pimento in Jamaica is semi-wild plants from self-sown seedlings and the trees are preserved when the land is cleared and put down to pasture. In planting for berry production, the usual spacing is 25–30 ft with 3 seedlings per hole which are reduced to one good plant when they come into bearing. If grown for leaves for the production of leaf oil, they are planted much closer and the trees are trimmed to bushes at the time of harvesting the leaves. It is inadvisable to reap berries and leaves from the same tree. Pimento is often planted in pastures, but it is necessary to protect the young trees from grazing stock and to circle-weed them. Alternatively, the young grove may be intercropped with bananas or other crops, but these should be removed after 5 years. The young trees should be trained to be low and spreading. Under good conditions trees come into bearing in 5–6 years, but neglected trees may take up to 12 years. Trees are in full bearing at 20–25 years and continue to a good age. Nitrogenous manuring may be advantageous, particularly if the crop is grown for leaves.

The berries mature 3–4 months after flowering and are picked when fully developed but still green. The harvesting season in Jamaica is usually July–September. The twigs with the berries are broken or clipped off and the berries are then stripped. Care should be taken to avoid breaking large branches when climbing the trees for harvesting. The berries are dried in the sun on concrete floors; they are raked over 2–3 times per day and should be heaped and covered at night against rain. Drying takes 7–10 days and the berries are then winnowed to remove extraneous material. Tree yields vary between years and a good crop can only be expected every 3 years or so. Dried berries should be brownish-black and rattle when a handful is shaken. The proportion of green to dry berries is approximately 3 : 2. A good picker can harvest about 70 lb of green pimento per day. Young trees 10–15 years old may yield 50 lb of green berries per year, but yields are usually much less and about 2·5–5 lb. Yields of 150 lb for individual mature trees have been recorded.

DISEASES AND PESTS

The most serious disease in Jamaica is the pimento leaf rust—*Puccinia psidii* Wint., which attacks expanding foliage, inflorescences and succulent young stems; it causes defoliation. The incidence is increased by long periods of wet cold weather and is more prevalent at altitudes of over 1,000 ft. Termites can cause damage, as can various wood-boring insects, notably the pimento borer—*Cyrtomerus pilicornis* Fab., which caused wholesale destruction in 1896.

IMPROVEMENT

Little work has been done on the selection and breeding of the crop. With vegetative propagation, the selection of high-yielding clones will obviously be an advantage.

PRODUCTION

Jamaica is the only large producer of pimento. The first pimento reached London in 1601. By about 1755 exports had reached 438,000 lb valued at £21,925. Towards the end of the 19th century annual exports exceeded 10 million lb in some years; the last year when exports exceeded this figure was in 1925 when 13 million lb were exported. Current exports are approximately 4–5 million lb annually, valued at about £1,000,000, and 60,000 lb of leaf oil. The Jamaica Government is the sole exporter and fixes the prices to the grower. West Germany is the principal buyer, followed by the United States and the United Kingdom. The United States is the main market for pimento leaf oil, but Soviet Russia has increased its purchases.

REFERENCES

CHAPMAN, G. P. (1964). Some Aspects of Dioecism in Pimento (Allspice). *Ann. Bot. N.S.* **28**, 451–458.

WARD, J. F. (1961). *Pimento*. Kingston: Govt. Printer.

Psidium L. (x = 11)

About 150 spp. of trees and shrubs native in tropical and subtropical America. Lvs opposite, pinnate-nerved. Fls 1–3 in axillary cymes; calyx 4–5 lobes; petals 4–5 spreading; stamens numerous, free, on broad disc; ovary inferior, 4–5-celled, with numerous ovules. Fr. a berry, usually crowned with persistent calyx lobes.

USEFUL PRODUCTS

Several spp. have edible fruits, of which the most widely cultivated is the common guava, *P. guajava* (q.v.).

The strawberry guava, *P. littorale* Raddi (syn. *P. cattleianum* Sabine), a native of Brazil, is sometimes grown for its small round purplish-red fruits, about 3 cm in diameter; the white flesh has a sweet acid flavour and may be eaten raw or made into jam and jelly. It is a small tree and can be distinguished from *P. guajava* by its terete branchlets and smaller coriaceous glabrous obovate leaves. Its var. *lucidum* Degener has yellow fruits.

Other spp. with edible fruits include: *P. guineense* Sw., indigenous in West Indies and tropical America, and is occasionally cultivated; *P. montanum* Sw., which grows wild in the Jamaica hills; *P. microphyllum* Britton which is grown in Puerto Rico; *P. friedrichsthalianum* (Berg.) Nied., native of Central America, which was introduced into Singapore and Malaya and has very acid fruits.

Psidium guajava L. (2n = 22) GUAVA

USES

The fruits are very variable in size and flavour, ranging from sweet to tart, and all have a characteristic penetrating musky odour, which is more marked in some forms than others, and is largely dissipated by cooking. Ripe juicy sweet types are eaten fresh; the fleshy mesocarp is stewed and is made into pies. The fruits, after the removal of the seeds are made into preserves, jam, jelly, paste, juice and nectar. The greatest commercial use is for jelly and the common sour wild guava makes the best jelly. The paste is made by evaporating the pulp with sugar; it is eaten as a sweetmeat and is known as guava cheese in the West Indies. The fruits are canned, for which purpose they are peeled, halved, the seeds removed and cooked in a light syrup. The juice and nectar are also canned. Guavas contain 2–5 times the vitamin C content of fresh orange juice and dehydrated guava juice powder is a source of vitamin C; it was used to fortify rations of Allied troops during World War II. In some countries the leaves are used medicinally for diarrhoea, for dyeing and for tanning.

ORIGIN AND DISTRIBUTION

P. guajava is indigenous to the American tropics, where it occurs wild and is cultivated. Oviedo in 1526 reported that it was common in many parts of the West Indies and that improved forms were planted by the local peoples. It was taken at an early date by the Spaniards across the Pacific to the Philippines and by the Portuguese from the west to India. It then spread throughout the tropics. It has been naturalized in many countries, being spread by birds; in some places it has become a troublesome weed of pastures, and is a declared noxious weed in Fiji.

CULTIVARS

A number of cvs are recognized, but they can only be maintained by vegetative propagation, as they do not come true from seed. They vary in the colour of the flesh, flavour and vitamin C content.

ECOLOGY

Guavas grow in the tropics from sea level to 5,000 ft and are adaptable over a wide range of climatic and soil conditions. They are susceptible to frost. They can tolerate temporary waterlogging and also high temperatures.

STRUCTURE

Shallow-rooted shrub or small tree, 3–10 m in height, branching close to ground and often producing suckers from roots near base of trunk. Bark smooth, greenish- or reddish-brown, peeling off in thin flakes. Young twigs 4-angled and ridged, pubescent. Lvs opposite, glandular; petiole 3–10 mm long, grooved above; lamina elliptical to oblong, 5–15 × 3–7 cm, glabrous above, finely pubescent beneath, tip somewhat obtuse and mucronate, base rounded or obtuse; lateral veins 10–20 pairs, sunken above, prominent below. Fls axillary, 2·5–3 cm in diameter, solitary or in 2–3-flowered cymes; calyx entire in bud, splitting irregularly into 4–6 lobes, 1–1·5 cm long, reflexed, pubescent, persistent; petals 4–5, white, obovate, concave, reflexed, 1–2 cm long; stamens numerous, inserted in rows on disc, 1–2 cm long, filaments white, anthers pale yellow dehiscing longitudinally; ovary 4–5-locular; style filiform, greenish-yellow, 1·5–2 cm long, exserted above stamens; stigma capitate. Fr. a berry surmounted by calyx lobes, variable, globose, ovoid or pyriform, 4–12 cm long; exocarp pale green to bright yellow; mesocarp fleshy, of varying thickness, white, yellow, pink or red, stone cells usually present. Seeds embedded in pulp, usually numerous, yellowish, bony, reniform, 3–5 mm long; embryo curved.

POLLINATION

Self- and cross-pollination occur. The flowers are visited by bees and other insects. Using the dominant red mesocarp colour as a marker, natural cross-pollination has been shown to be about 35 per cent.

GERMINATION

Seeds retain their viability for about a year and germination takes 2–3 weeks.

CHEMICAL COMPOSITION

The flesh and seeds of guavas contain approximately: water 80 per cent; protein 1 per cent; fat 0·5 per cent; carbohydrate 13 per cent; fibre 5·5 per cent. The total soluble solids in the mesocarp of some improved cvs in Hawaii is about 10 per cent. Guavas are particularly rich in vitamin C; in Florida the ascorbic acid content varied from 23–486 mg per 100 g of fresh fruit; in Hawaii 8 selected vegetatively propagated cvs ranged from 146–492 mg, with an average of 261 mg. Guavas are also a fair source of vitamin A, iron, calcium and phosphorus.

PROPAGATION

In the past most guavas have been grown from seed. The seeds are sown in nurseries or flats at a depth of ½ in. If sown in flats the seedlings are later transferred to containers. In India seedlings are planted in the field when they are 1–1½ ft high at about 1 year old; in Hawaii when 12 in. high at 5–7 months.

Due to the great variability of guavas raised from seed, vegetative propagation of clones is highly desirable. During recent years satisfactory techniques have been worked out for patch budding by the Forkert method (probably the most reliable method), side-veneer grafting, approach-grafting, marcotting and rooting cuttings. They may also be grown from suckers which can be induced by severing the roots 2–3 ft from the trunk. One of the difficulties of budded and grafted guavas is the production of water shoots and suckers from the rootstocks.

Fig. 66. **Psidium guajava : GUAVA. A, shoot with buds (×½); B, flower in longitudinal section (×1); C, fruit (×⅓): D, fruit in longitudinal section (×⅓).**

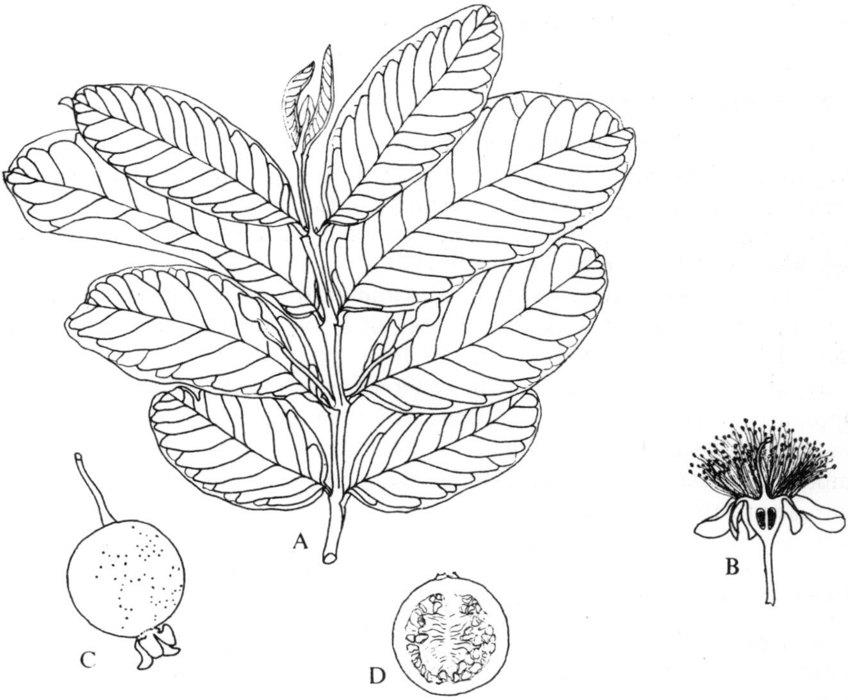
A
B
C
D

HUSBANDRY

The usual spacing is about 20 ft apart. Pruning consists of shaping the tree and the removal of water shoots and suckers. The crop is often irrigated in India and receives organic manures. Guavas respond as well to complete fertilizers as citrus (q.v.), but may require more nitrogen. Zinc, copper, manganese and magnesium deficiencies have been recorded. Trees often bear their first fruits 2 years after transplanting; they are in full bearing at 8 years and cropping may continue for 30 years or more. The fruits mature 5 months after flowering and for distant markets should be picked before they are fully ripe. In India a seedling tree 8–10 years old yields 400–500 fruits weighing 140–180 lb per annum; grafted or layered trees of the same age yield 1,000–2,000 fruits weighing 400–700 lb.

MAJOR DISEASES AND PESTS

A serious wilt of unknown origin is reported from India. *Glomerella cingulata* (Stomem.) Spauld. & Schrenk causes mummification and blackening of immature fruits in Puerto Rico. *Colletotrichum gloeosporioides* Penz causes decay of mature fruits; and the parasitic alga, *Cephaleuros mycoidea* Karsten can be troublesome.

Fruit flies, mealy bugs, scale insects and thrips are reported as causing damage. The most serious pest in Hawaii is the oriental fruit fly, *Dacus dorsalis* Hendel. The Mediterranean fruit fly, *Ceratitis capitata* (Weid.) may also be an important pest in some areas. The guava flies, *Anastrepha* spp., are so bad in Trinidad and parts of the West Indies that it is almost impossible to find a fruit which is not attacked. The larvae of a small moth, *Argyresthia eugeniella* Busch., are reported as causing considerable damage to the fruit.

IMPROVEMENT

Interest has been shown in recent years in the selection and breeding of improved cvs. The standards of selection in Hawaii are: fruit diameter at least 3 in.; diameter of cavity no more than 1·5 in.; fruit weight 7–10 oz; seeds 1–2 per cent; strong pink colour preferred; fruits with a pleasant, palatable, characteristic guava flavour; soluble solids 9–12 per cent; vitamin C 300 mg or more; flesh with few stone cells.

PRODUCTION

Guavas are grown for local consumption in most tropical countries. They are grown commercially in India, where there are about 70,000 acres grown for this purpose; also in Florida, where the first commercial orchard was planted in 1912. Guavas are also grown commercially in Brazil and British Guiana. The Cuban industry is based on wild plants.

REFERENCES

HAMILTON, R. A. and SEAGRAVE-SMITH, H. (1954). Growing Guava for Processing. *Univ. of Hawaii, Ext. Bull.*, **63**.

RUEHLE, G. D. (1948). The Common Guava—A Neglected Fruit with a Promising Future. *Econ. Bot.*, **2**, 306–325.

SINGH, S., KRISHNAMURTHI, S. and KATYAL, S. L. (1963). *Fruit Culture in India*. New Delhi: Ind. Counc. Agric. Resc.

PASSIFLORACEAE

About 12 genera and 500 spp. of herbaceous and woody plants, usually climbing by tendrils, in warm and tropical regions, most abundant in South America. Lvs alternate, stipules usually present. Fls regular, often showy, usually hermaphrodite, axillary; sepals 5; petals 5 or none; corona of 1 or more rows of thread-like filaments and/or scales; stamens usually 5; ovary superior, 1-celled, often on a gynophore. Fr. berry or capsule; seeds with fleshy aril.

USEFUL PRODUCTS

Several *Passiflora* spp. (q.v.) have edible fruit.

Passiflora L. (x = 6, 9, 10)

About 400 spp., mostly native of the New World, with a few in Asia and Australia and 1 sp. in Madagascar. Usually perennial woody vines with unbranched axillary tendrils. Petioles usually gland-bearing. Fls usually large, solitary and axillary; peduncle jointed, usually with 3 bracts; short calyx tube with 5 lobes, often horned at apex; petals 5 or none; corona usually present; stamens 5 with filaments joined to gynophore below; anthers versatile; styles 3; stigma capitate; fr. a many-seeded berry; seeds pitted, netted or transversely grooved. The name passion flower was given by early missionaries in South America in allusion to a fancied representation in the flowers of the implements of the crucifixion.

USEFUL PRODUCTS

P. edulis (q.v.) and *P. quadrangularis* (q.v.) are commonly cultivated throughout the tropics for their edible fruits. Some of the wild species also have edible fruits and a few are occasionally cultivated for these; they include:

P. laurifolia L., variously known as water-lemon, Jamaica honeysuckle, belle apple and pomme de liane, grows wild in thickets and forest borders in the West Indies and north-eastern South America. It is a glabrous woody climber. Lvs oblong, entire, coriaceous, up to 14 × 6 cm. Fls large, handsome, fragrant, 6–10 cm in diameter; sepals and petals dotted with rosy-purple; corona as long as corolla, dark violet banded with white. Fr. ovoid, orange-yellow, rind soft, cavity filled with seeds; aril whitish and pulpy. The pulp is used in the same way as the passion fruit. This species was taken into cultivation in the 17th century and has been widely distributed in the tropics. It grows well in the humid lowlands. It is said to be the best of the passion fruits in Malaya, where it was also tried as a cover crop.

P. ligularis Juss., sweet granadilla, is a native of tropical America. It is a vigorous glabrous climber. Lvs entire, broad-ovate, 10–20 cm long, cordate. Fls 7–10 cm in diameter, sepals and petals greenish-white, corona as long as petals, white with purple bars. Fr. ovoid, hard-shelled, about 7–8 cm in diameter, orange-brown, with edible white aromatic pulp. The fruit is used extensively in the mountainous regions of Mexico and Central America. The plant has become naturalized in Hawaii.

P. mollissima (HBK.) Bailey (syn. *Tacsonia mollissima* HBK.), banana passion fruit, grows wild in the Andes. All vegetative parts of the plant are softly pubescent. Lvs deeply 3-lobed, 10–12 cm long, petiole with 8–10 glands. Fls pink, about 7·5 cm in diameter, corona reduced to a warty rim. Fr. ellipsoidal, downy, yellow, about 7 cm long, with edible pulp. This species is suited to colder conditions. It has been introduced into New Zealand and has become naturalized at elevations of 4,000–5,500 ft in Hawaii.

P. antioquiensis Karst. (syn. *P. van-volxemii* (Lem.) Triana & Planch.) is also known as the banana passion fruit. It is very similar in appearance to *P. mollissima* except that the flowers are bright red. It is a native of Colombia and is grown for its edible fruit at the higher altitudes. It is hardy in the Scilly Islands.

P. foetida L. is a variable weedy species, which grows wild in the West Indies and South America. It has been introduced into many tropical countries of Africa and Asia where it has become naturalized. It is an herbaceous perennial climber with very hairy stems, ovate hairy leaves, usually 3-lobed, about 4–5 cm long, foul smelling when crushed. Fls 4–5 cm in diameter, white with purple and white corona. Fr. ovoid, yellow about 3 cm long. The pulp of the ripe fruit is edible, but little used. The leaves and unripe fruit contain a cyanogenetic glucoside. It has been grown as a cover crop in Malaya and East Africa and is useful in smothering weeds and preventing erosion.

A number of species are grown as garden ornamentals in the tropics and in greenhouses in temperate countries, of which the commonest is *P. caerulea* L., a native of Brazil, and has been hybridized with other species.

KEY TO TWO MAIN CULTIVATED SPECIES

A. Stems not winged; lvs usually trifoliate; frs globose or ovoid, 4–6 cm long . . . . *P. edulis*

AA. Stems winged; lvs entire; frs oblong, 20–30 cm long *P. quadrangularis*

Passiflora edulis Sims (2n = 18) PASSION FRUIT

USES

The pulp and seeds may be eaten directly from the shell or used in fruit salads. The yellow, gelatinous, aromatic pulp round the seeds is used for making jams and jellies and in sherbets. The most important use is for making juice which is used as a beverage.

ORIGIN AND DISTRIBUTION

P. edulis is a native of southern Brazil and was widely distributed throughout the tropics and subtropics during the 19th century. It was introduced into England in 1810. Passion fruit was taken from Australia to Hawaii about 1880, where it became naturalized. It was introduced into Tanzania from Natal in 1896.

CULTIVARS

Two forms are recognized:

P. edulis f. *edulis* is the purple passion fruit, which is the normal form, and has round or egg-shaped fruits, 4–5 cm in diameter, deep purple when ripe. It has a better flavour than the yellow form, especially for eating out-of-hand. It does best at the higher altitudes of about 6,000 ft in the tropics and does not grow well in the wet lowlands.

P. edulis f. *flavicarpa* Degener is the yellow passion fruit. Presumably it originated as a mutation from the purple passion fruit. It has somewhat larger fruits, 5–6 cm in diameter and they are deep canary-yellow when ripe. The pulp is rather more acid. This form is better suited to the tropical lowlands. It was introduced into Hawaii from Australia in 1923.

ECOLOGY

As noted above, the purple passion fruit does best in the highlands in the tropics, while the yellow passion fruit will tolerate lower altitudes. The crop does not fruit well in regions with very heavy rain, as rain at flowering prevents pollination. An average annual rainfall of 30–50 in. is suitable for the crop. It can be grown on a variety of soils, but very heavy, poorly drained soils should be avoided.

STRUCTURE

A vigorous woody perennial climber, up to 15 m long; stems green, grooved, glabrous; tendrils axillary, robust, terete, green, longer than lvs, spirally coiled.

LEAVES: Stipules lanceolate, about 1 cm long; petiole usually glabrous, grooved on upper surface, 2–5 cm long, with 2 circular glands at top; lamina usually deeply palmately 3-lobed, 10–15 × 12–25 cm, but often ovate and undivided on young plants, toothed, teeth with incurved glandular tips, base cordate, lobes ovate-oblong, acuminate.

FLOWERS: Solitary, axillary, fragrant, showy, 7·5–10 cm in diameter; peduncle triangular, 2–5 cm long; bracts 3, near apex of peduncle, leafy, ovate to lanceolate, 1·5–2·5 × 1–2 cm, serrate-glandular. Calyx tubular at base with 5 spreading reflexed lobes, ovate-oblong, 2–3 × 1–2 cm, yellowish-green below, white above, spongy, fleshy, with thorn-like appendage near tip and 0–4 glands on margins. Petals 5, free, alternating with calyx, inserted on throat of calyx, elliptic, 2·5–3 × 0·5–1 cm, white, thin. Corona of 2 outer rows of wavy, threadlike, radiating filaments, 2–3 cm long, purple at base, white above, and several inner rows of short, purple-tipped papillae. Stamens 5; filaments united in a tube round gynophore for about 1 cm and then widely parted for 1 cm; anthers large, versatile, transverse, 2-celled, pale yellow, 1–1·5 × 0·5 cm, hanging downwards below level of ovary. Ovary carried up on a stalk or gynophore, ovoid, pale yellow, 1-locular, with 3 parietal placentas; styles 3, horizontal; clavate with longitudinal furrow, pale green, 1–1·5 cm long.

FRUIT AND SEEDS: Fr. a berry, globose or ovoid, deep purple when ripe, dotted, glabrous, 4–6 cm long; pericarp hard, thin, with greenish mesocarp and white endocarp. Seeds many, attached to peg-like funiculi on ovary wall, surrounded by yellowish aromatic pulpy juicy aril with tart but pleasing flavour; testa blackish, 3-toothed at base, flattened, 5 × 3 mm.

In f. *flavicarpa* stems, lvs and tendrils are tinged reddish or purple, lvs and frs are somewhat larger than in f. *edulis*, the base of corona is a deeper purple, and ripe fruit is bright yellow in colour. In specimens of this form examined in Trinidad both 3-lobed and ovate unlobed lvs occur on the same plant.

POLLINATION

Cross-pollination is done mainly by carpenter bees, *Xylocopa* spp., but the flowers are also visited by honey bees and wasps. Humming birds may assist pollination in the New World. Flowers of the purple passion fruit open at dawn and close at noon; in the yellow passion fruit the flowers open about noon and close in the evening. The pollen is sticky and nectar is secreted at the base of the androgynophore. Akamine and Girolami (1959) working in Hawaii on the yellow passion fruit have shown that most clones are completely self-incompatible and cross-incompatibility may occur in certain clones. They also showed that fruits from flowers cross-pollinated by hand are larger and yield more juice than flowers pollinated naturally and that the number of pollen grains placed on the stigma influences the fruit set percentage, size of fruit, number of matured seeds and juice yield, the positive linear regressions in general being highly significant. Ungerminated pollen grains burst on contact with water, but when the pollen tubes have grown they are not so destroyed. This accounts for failure in pollination during rain. Shortly after the flower opens the erect styles recurve and return to the erect position just before the flower closes. In some flowers, however, the styles remain erect and these have been found to be female sterile but male functional. Thus lack of, or poor fruit set in mutually compatible clones may be caused by insufficient pollination, wetting of pollen, presence of a large percentage of upright-styled flowers, varying flower production and blossom opening time, fruit-fly damage (see below) and combinations of these. The period from pollination to maturity is 60–80 days.

GERMINATION

Seeds may be planted from ripe fruits without removal from the pulp and germinate in 2–4 weeks. Seeds separated from the pulp, washed and dried at room temperatures gave a germination of 85 per cent in Hawaii after storing for 3 months.

HEMICAL COMPOSITION

The pulp and seeds contain approximately: water 72·1 per cent; protein 2·4 per cent; fat 2·8 per cent; carbohydrate 17·3 per cent; fibre 4·2 per cent; ash 1·2 per cent. It is a good source of vitamin C.

PROPAGATION

In commercial production most of the crop is grown from seeds, which are readily available as they are extracted from the pulp during processing. They are usually sown rather thickly in nursery beds and shaded. When two true leaves develop the seedlings are planted in pots, baskets or polythene sleeves. They are planted in the field when about 1 ft tall and 3–4 months after sowing. They may also be propagated by cuttings of 2–3 internodes taken from reasonably matured wood of pencil thickness.

HUSBANDRY

Passion fruits are usually planted 10–20 ft apart and are trained on trellises of posts and wire 6–10 ft apart. Two leaders are usually allowed to grow from each seedling and side-shoots may be removed until the vines reach the wires. Little subsequent pruning is done except to facilitate spraying or to force new growth. Little benefit has been obtained from fertilizers in Kenya. In Hawaii a 10 : 5 : 20 NPK mixture is recommended at the rate of 2 lb per plant. No crop is usually harvested during the first year. Thereafter in Kenya there are two main fruiting seasons

---

Fig. 67. **Passiflora edulis f. flavicarpa :** PASSION FRUIT. **A**, leafy shoot (×⅓); **B**, flower from side (×⅔); **C**, flower from above (×⅔); **D**, flower in longitudinal section (×⅔); **E**, fruit (×⅔); **F**, fruit in longitudinal section (×⅔); **G**, seed (×1½); **H**, seed with pulp (×⅔).

A
B
C
D
E
F
G
H
MWong

per year. The fruits are produced only on new growth. The fruits are allowed to drop on the ground and are picked up every 2–3 days. The fruit can be kept up to a week without deterioration before processing. Average yields in Kenya are 15,000 lb of fruit per acre per annum. The fruits contain 30–40 per cent by weight of juice and 42 lb of fruit are required to produce 1 gallon. In the preparation of juice the pulp is extracted from the fruits and the seed is then separated from the juice by centrifuging. The maximum profitable life of the purple passion fruit in Australia and South Africa is 5–6 years.

MAJOR DISEASES AND PESTS

The most serious disease of passion fruit in Kenya and Australia is woodiness disease, which has been identified in Australia as cucumber virus 1. It is thought that it is transmitted by aphids. The leaves become leathery and malformed; the fruits gradually decrease in size and the rind becomes thick and hard and little pulp is produced. The other serious disease is brown spot, *Alternaria passiflorae*, which causes defoliation and brown spotting of the fruits.

In Hawaii the most serious pests are: the oriental fruit fly, *Dacus dorsalis* Hendel; the melon fly, *D. cucurbitae* Coquillett; and the Mediterranean fruit fly, *Ceratitis capitata* (Wied.). The flies puncture the young fruits and may cause fruits to shrivel and fall if they are very young. Red spider, *Brevipalpus papayensis*, also causes damage.

IMPROVEMENT

For juice production selection is based on flavour, size of fruit, thinness of rind, well-filled cavity and yield of juice. Colour and the content of sugar, acid and vitamins may also be taken into consideration. Resistance to woodiness disease and brown spot is also required. With the yellow passion fruit in Hawaii it was found that oval fruits contain about 10 per cent more juice than round fruits.

PRODUCTION

Passion fruit is grown as a commercial crop in Australia, South Africa, New Zealand and Hawaii. It is also grown commercially in Kenya where production was begun in 1935; it is now produced mainly in the Sotik district at an altitude of 6,000 ft.

REFERENCES

AKAMINE, E. K. and GIROLAMI, G. (1959). Pollination and Fruit Set in the Yellow Passion Fruit. *Hawaii Agr. Exp. Sta. Tech. Bull.*, **39**.

CHAPMAN, T. (1963). Passion Fruit Growing in Kenya. *Econ. Bot.*, **17**, 165–168.

COMMITTEE ON PASSION FRUIT CULTURE (1954). Passion Fruit Culture. *Univ. Hawaii Ext. Circ.*, 354.

Passiflora quadrangularis L. (2n = 18) GIANT GRANADILLA

USES

The flesh and the pulp of the fruit are edible and are eaten alone or in fruit salads. They are used to flavour ice cream and sherbets, to make a cooling drink and in jam. The green unripe fruits may be boiled and eaten as a vegetable. The fleshy tuberous root is usually regarded as poisonous, although they are said to be eaten in Jamaica as a substitute for yams.

ORIGIN AND DISTRIBUTION

P. quadrangularis is a native of tropical South America. It was brought into cultivation in the 18th century and has now been widely distributed throughout the tropics.

ECOLOGY

The giant granadilla grows best in a hot moist climate and does not do well at the higher altitudes where *P. edulis* f. *edulis* gives the best results.

STRUCTURE

A strong glabrous perennial climber with fleshy tuberous roots. Stems sharply quadrangular and 4-winged. Tendrils simple, yellowish green, longer than lvs.

LEAVES: Alternate; stipules ovate to lanceolate, serrate, 2–4 × 1–2 cm; petiole 2–5 cm long, sharply keeled beneath, furrowed above, with 3 pairs of wart-like glands on margins; lamina ovate or elliptic, 10–25 × 8–18 cm, entire, dark-green and shiny above, pale-green beneath, base cordate, tip mucronate, midrib and 10–12 pairs lateral veins prominent.

FLOWERS: Similar in structure to *P. edulis*, axillary, solitary, pendulous, fragrant, showy, 10–12 cm diameter; peduncle to 2 cm long, angled; bracts 3, leafy, at top of peduncle, ovate, serrate, yellowish-green, 2·5–3·5 × 2–3 cm; calyx tubular at base, lobes 5, spreading widely, ovate, thick, spongy, yellowish-green beneath, tinged red above, 3–5 × 1·5–2 cm; petals 5, rather spongy, densely dotted with red, 4–5 × 1·5–2 cm; corona with 2 outer layers of sinuous filaments, 5–7 cm long, white and red at base, white and purple above; internally several rows of papillae, tipped red; stamens 5, connivent with gynophore for about 1 cm; free portion of filaments obliquely erect, about 1 cm long, yellowish-green with purple dots, anthers large, transverse, versatile, pale yellow, hanging downwards; ovary on gynophore, ellipsoidal, 3-furrowed longitudinally, 1·0–1·5 cm long, 1-locular, many ovuled, parietal; styles 3–5, with large terminal globular or reniform stigmas.

FRUITS AND SEEDS: Fr. a fleshy ovate-oblong berry, 20–30 × 10–15 cm, pale yellow or yellowish green when ripe, sometimes tinged pink; pericarp 2·5–4 cm thick; flesh whitish in colour, juicy, rather insipid; cavity filled with numerous seeds surrounded by translucent, whitish, juicy arils. Seeds obovoid, to 1 cm long, flattened, dark brown, 3-toothed at base, margin winged.

POLLINATION

Outside the area of origin poor fruit-set is often reported and is thought to be due to the protandrous nature of the flowers and the absence of the pollinating agents. Hand-pollination is recommended to obtain a reasonable crop.

CHEMICAL COMPOSITION

The flesh (mesocarp and endocarp) contains approximately: water 93·7 per cent; protein 0·7 per cent; fat 0·2 per cent; carbohydrate 4·3 per cent; fibre 0·7 per cent; ash 0·4 per cent. The seeds and aril have about 17·6 per cent carbohydrate.

PROPAGATION

The giant granadilla can be propagated by seeds or cuttings.

HUSBANDRY

The plants are usually planted 8–10 ft apart and trained on rather open overhead trellises which permit the fruit to hang suspended beneath.

MAJOR PESTS AND DISEASES

Rats, bats and birds relish the fruits and it may be necessary to protect the fruit by covering it.

IMPROVEMENTS

Little or no work seems to have been done.

PRODUCTION

The fruit is usually grown for local consumption.

REFERENCE

POPE, W. T. (1935). The Edible Passion Fruit in Hawaii. *Hawaii Agric. Exp. Sta. Bull.*, **74**.

Fig. 68. **Passiflora quadrangularis** : GIANT GRANADILLA. A, vegetative shoot (×⅓); B, flower from above (×½); C, flower in longitudinal section (×½); D, fruit (×⅓); E, seed and aril (×1); F, seed (×2); G, seed in longitudinal section (×2).

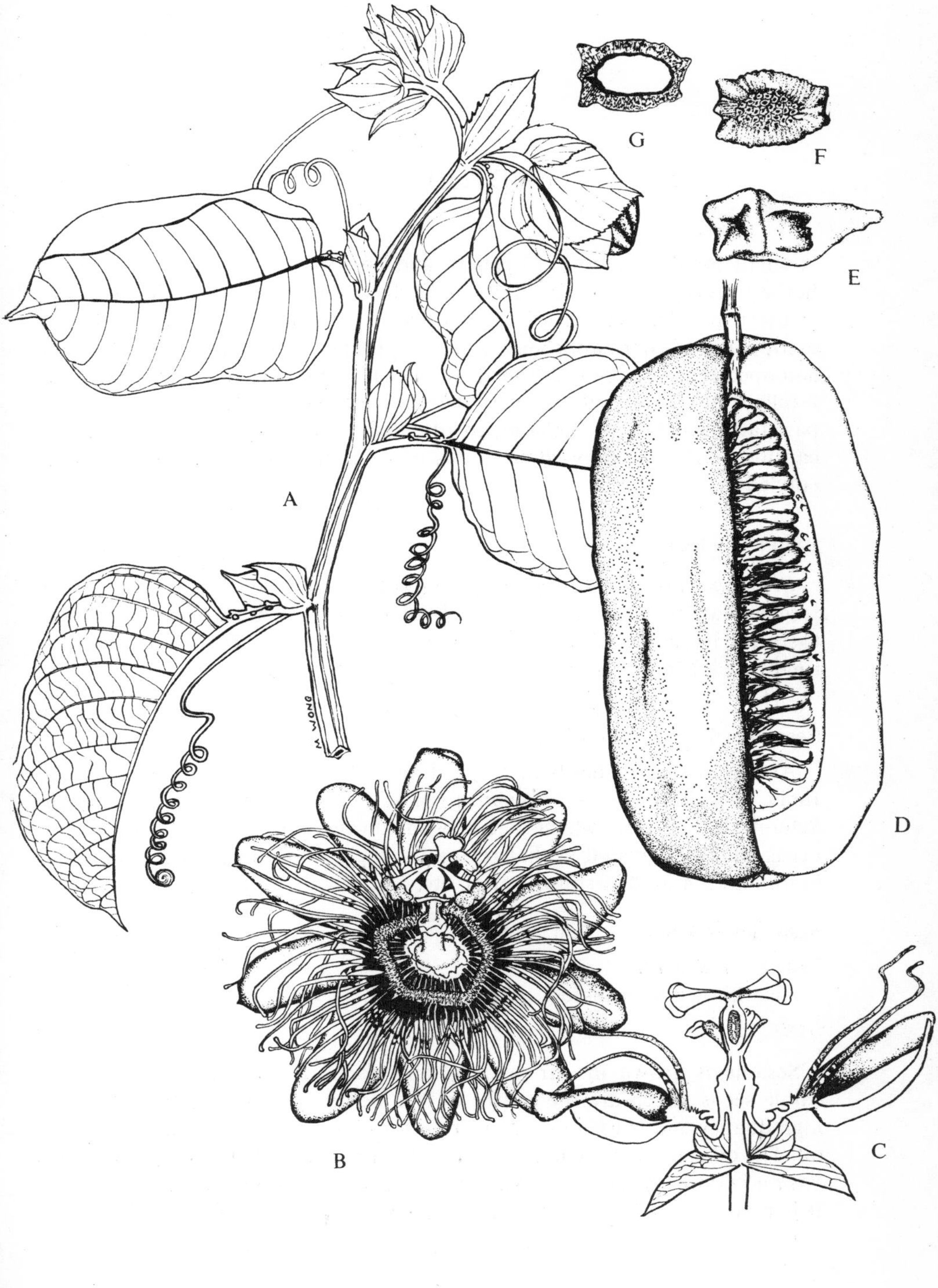

A
B
C
D
E
F
G
M WONG

PEDALIACEAE

A small family of about 15 genera and 60 spp. of annual and perennial herbs occurring mainly in the Old World tropics and subtropics with the greatest number in Africa. Lvs opposite or upper alternate, simple, exstipulate. Fls hermaphrodite, zygomorphic; calyx usually 5-cleft; corolla gamopetalous, 5-lobed, 2-lipped; stamens 4, anthers connivent in pairs, 2-celled; disc hypogynous; ovary usually superior, 1-celled with 2 intrusive parietal placentas, the cells again often divided by spurious septa. Fr. a capsule or nut. Seeds without endosperm; embryo straight with flattened cotyledons.

USEFUL PRODUCTS

Sesamum indicum (q.v.) is an important oil-seed crop. *Ceratotheca sesamoides* Endl. is wild in tropical Africa and is occasionally cultivated there for its edible seeds which contain about 37 per cent oil. It is sometimes used as an adulterant in sesame.

Sesamum L. (x = 8, 13)

About 20 spp. of herbs in tropical and South Africa and with 1 sp. in India. In addition to *S. indicum* (q.v.), *S. alatum* Thonn. and *S. radiatum* Schum. & Thonn., which occur wild throughout tropical Africa, are occasionally cultivated in that region for their edible seeds, which are rich in oil and are used in soups, cakes, etc. and as an adulterant of sesame.

Sesamum indicum L. (2n = 26) SESAME, SIMSIM

Syn. *S. orientale* L.

Also known as beniseed, gingelly and til.

USES

Sesame is grown in Africa, Asia and parts of Latin America for its edible seeds which are the source of sesame oil. The semi-drying oil is of high quality and is used as a substitute for olive oil as a salad and cooking oil. The oil is used in the manufacture of margarine and compound cooking fats and poorer grades in soap and paints, and as a lubricant and illuminant. It is used as a vehicle for medicinal drugs and perfumes and as a synergist for pyrethrum (q.v.) in aerosol sprays. In India it is used as a ghee substitute and for annointing the hair and body, 'a customary bi-weekly practice which in South India is deemed very beneficial to health'. The expressed cake is an excellent, protein-rich, livestock food and people eat it in India and Java, sometimes after fermentation.

The fried seeds are eaten in soups and, mixed with sugar, are a popular sweetmeat in Africa and Asia. The seeds, after removal of the testa, are often scattered on the tops of cakes, bread and pastry. The young leaves are used as a soup vegetable in West Africa. Various parts of the plant are used in native medicines in Africa and Asia. The stems are burnt as fuel.

ORIGIN AND DISTRIBUTION

It seems fairly definite that *S. indicum*, which is of very ancient cultivation, was first taken into cultivation in Africa and that it was taken at a very early date to India where a secondary centre of diversity developed. With the exception of *S. prostratum* Retz. in eastern India, all the wild species of *Sesamum* are in Africa. Zukovskij (1962) states that the crop was unknown in ancient Egypt, but Burkill (1935) says that it was first recorded in Egypt about 1300 B.C. at the time of the expulsion of the Israelites, that it was grown by the Chaldeans and that the Assyrian wars caused an extension of its cultivation. Sesame appears to have reached China in the 1st century A.D. It was taken to the New World by the early slave trade.

CULTIVARS

A large number of cvs and races are known in sesame-growing countries differing in their season of planting, time to maturity, degree of branching, number of flowers per leaf axil which may be 1 or 3, number of loculi in the capsule which may be 4, 6, or 8, and in colour of flower, capsule and seed. Non-shattering cvs are now being developed.

ECOLOGY

Sesame is a crop of the hot dry tropics and is usually grown in areas with an annual rainfall of 20–45 in. Once established it can tolerate short periods of drought, but is very intolerant of waterlogging. It has poor ability to compete with weeds in the early stages. It is not exacting in its soil requirements and does reasonably well on poor soils; sandy loams are preferred. Sesame is very sensitive to day length and both long and short day forms occur.

STRUCTURE

A variable erect annual herb, 1–2 m tall, taking 80–180 days to maturity, covered with glandular hairs and with a somewhat foetid smell.

ROOTS: A pronounced tap-root to 90 cm long and a dense surface mat of feeding roots.

STEMS: Some early cvs are unbranched; late maturing cvs show pronounced branching; stems square, longitudinally furrowed, green, occasionally purple.

LEAVES: Very variable, hairy on both sides, margins ciliate, exstipulate. Lower lvs opposite, ovate, sometimes palmately lobed or palmately compound; petiole about 5 cm long; lamina 8–15 × 6–10 cm, coarsely serrate. Upper lvs alternate or sub-opposite, lanceolate, entire or with 1–few coarse teeth; petiole 1–2 cm long; lamina 5–13 × 1–3 cm.

FLOWERS: Zygomorphic, in axils of upper leaves, borne singly or 2's or 3's. Peduncle short, about 5 mm long with an extra-floral nectary on either side at base, yellow or black, representing aborted fls, each subtended by a linear bract, about 1 cm long. Calyx 6–8 mm long, glandular hairy, 5-partite to near base, lobes narrow, acuminate. Corolla tubular-campanulate, 3–4 cm long, widening upwards, 2-lipped, 5-lobed with middle lower lobe longest, pubescent outside, white, pink or purplish with yellow or purple blotches, spots and stripes on inner surface. Stamens didynamous, inserted near base of corolla, 2 longer stamens 1·5–2 cm long, 2 shorter stamens 1·0–1·5 cm long; filaments slender, basifixed, whitish in colour; anthers lying against inner upper lip of corolla, awl-shaped, pale yellow, with short white or brownish beak. Ovary superior, sessile, many-ovuled, often densely hairy, of 2 united carpels becoming 4–8-locular by intrusive growth of parietal placentas; style slender, cream coloured, 1·5–2 cm long; stigma bifid, hairy, lying between 2 pairs of anthers. Nectiferous disc surrounding ovary. Sepaloid fls with abnormal internal organs have been recorded.

FRUITS AND SEEDS: Capsule up to 3 × 1 cm, glabrous or hirsute, erect, brown or purple, oblong, rectangular in section, deeply grooved, with persistent calyx and conspicuous subulate beak, dehiscing by 2 apical pores, but some cvs indehiscent. Seeds small, 3 × 1·5 mm, ovate, smooth or reticulate, white, yellow, grey, red, brown or black. 100 seeds weigh about 0·3 g.

POLLINATION

Pollen is usually shed before the flowers open. The flowers open early in the morning and at the same time the bifid stigma separates and becomes receptive and is copiously covered with pollen from the stamens. Consequently the great majority of the flowers are self-pollinated. The corolla and stamens are shed 15–20 hours after the flowers open. Bagged flowers set seeds. The flowers are visited by insects so that a negligible amount of cross-pollination can occur.

Fig. 69. **Sesamum indicum : SESAME. A,** flowering shoot (×½); **B,** lower leaf (×½); **C,** flower cut open (×½); **D,** fruit in longitudinal section (×½).

A
B
C
D

GERMINATION

Seedlings appear above the ground in about 3–10 days.

CHEMICAL COMPOSITION

The seeds contain 45–55 per cent oil, 19–25 per cent protein and about 5 per cent water. The oil is of high quality, odourless and not liable to become rancid. The principal unsaturated fatty acids are oleic and linoleic with about 40 per cent of each and about 14 per cent saturated acids. Linolenic acid is not present.

PROPAGATION

The crop is grown from seed.

HUSBANDRY

Sesame is usually grown as a rain-fed crop. When grown in pure stand it is broadcast or drilled in rows about 1 ft apart with a seed rate of about 5 lb per acre. The seedbed should be free of weeds and have a good tilth. The seed is buried by a light scuffling. The seedling stage is precarious and too little or too heavy rain at this time may lead to poor stands. The crop often receives little attention after sowing, but it may be thinned and weeded during early growth.

The time taken to maturity varies from 80–180 days, but most cvs take 100–140 days. The crop is harvested when the bottom capsules turn from green to yellow, but before they open, and when the leaves begin to drop. The stems may be cut near ground level, bound and stooked in the field to ripen the seed. With dehiscent types it is usually better to cut the heads and dry them in bunches hanging downwards on racks for 1–2 weeks. The seeds fall as they ripen onto mats placed below the racks and the heads are also shaken. Threshing can also be done by beating with sticks. Now that undehiscent cvs are available the crop may be harvested mechanically. Previously this was difficult as the capsules ripen progressively upwards, so that seeds were lost from the lower capsules before the upper ones were ripe.

In peasant production yields of 200–600 lb of seed per acre are obtained. Yields of 800–2,000 lb per acre have been reported from Venezuela. Mexican yields vary from 350–1,000 lb per acre.

MAJOR DISEASES AND PESTS

The most serious diseases are leaf spots caused by *Pseudomonas sesami* Malkoff, *Cercospora sesami* Zimm. and *Alternaria* spp., and Fusarium wilt, *Fusarium oxysporum* Schlecht. ex Fr. f. *sesami* Castellani.

The most serious pests in Uganda are a gall midge, *Asphondylia sesami* Felt., which lays its eggs in the ovary of the flower, and the caterpillars of *Antigastra catalaunalis* Dup.

IMPROVEMENT

Apart from a preliminary sorting out of cvs in some countries, little work was done on sesame breeding until after World War II. Since the discovery of a single plant with indehiscent capsules in Venezuela in 1943, attempts have been made with some success to transfer the recessive gene responsible to dehiscent cvs to facilitate combine harvesting. Selection has also been made for single-stemmed plants which ripen more uniformly from base to tip.

Tahir (1964) states that introduced American cvs have given poor yields in the Sudan, and that it has been possible to transfer the non-shattering character, 3 capsules per axil, 8 locules per capsule and short internodes to local cvs. As mixed seed colours reduces market value and white seeds fetch the highest price, it is often necessary to purify local mixtures. As the crop is very largely self-pollinated pure lines can be obtained.

PRODUCTION

World acreage has remained at 11–12 million acres since before World War II and world production is approximately 1·5 million tons. The largest producers are India with 5–6 million acres, China with 2·5 million acres, Burma with 1·5 million acres, the Sudan with 1 million acres and Mexico with 0·5 million acres. Countries with over 100,000 acres are Pakistan, Turkey, Venezuela, Uganda and Nigeria. Most of the crop is consumed in the countries in which it is grown. The biggest exporters are the Sudan and Nigeria who export approximately 80,000 tons and 25,000 tons respectively per annum. The main importing countries are Japan, Italy and Venezuela.

REFERENCES

HAARER, A. E. (1950). Sesame. *World Crops*, **2**, 218–221.

RAM, K. (1931). Studies in Indian Oil Seeds (4) The Types of *Sesamum indicum* DC. *Mem. Dept. Agric. Ind., Bot. Series*, **18**, 127–147.

RHIND, D. and U BA THEIN (1933). The Classification of Burmese Sesamums (*Sesamum orientale* Linn.). *Ind. J. Agric. Sc.*, **3**, 478–495.

TAHIR, W. M. (1964). Sesame Improvement for Mechanized Production in the Central Rainlands of the Sudan. *Emp. J. exp. Agric.*, **32**, 217–224.

MAZZANI, B. (1962). Mejoramiento del Ajonjoli en Venezuela. *Monogr. Fond. nac. Invest. agropec.* No. 3.

PIPERACEAE

About 10 genera and over 1,000 spp. of herbs and shrubs, often climbers, mainly in the tropics throughout the world. The vascular bundles are more or less scattered as in the monocotyledons. Lvs usually alternate, entire, petiolate; stipules adnate to petiole or absent. Fls minute, usually in dense spikes; subtended by bracts; regular; perianth absent, bisexual or unisexual; stamens 2–6, hypogynous; ovary superior, 1-locular, 1-ovuled; stigmas 1–5, short. Frs baccate, dry or fleshy, indehiscent. Seeds small with little endosperm, copious mealy perisperm and minute embryo.

USEFUL PRODUCTS

The large genus *Piper* (q.v.) provides several useful products. A number of herbaceous *Peperomia* spp. are grown as foliage plants.

Piper L. (x = 12, 14, 16)

About 650 spp. in both hemispheres. Herbs, but mostly woody climbers and shrubs. Lvs alternate. Fls mostly unisexual, borne in axils of decurrent bracts, in slender catkin-like spikes; stamens usually 1–4; anthers with 2 distinct cells; stigmas mostly 3.

USEFUL PRODUCTS

SPICES: *P. longum* L., Indian long pepper, grows wild at the foot of the Himalayas and is still grown and used as a spice in India and Ceylon. It was known to the ancient Greeks and Romans, being more highly esteemed by the latter than *P. nigrum*. It no longer enters international trade. The plants are dioecious. The whole spike is used as the spice and consists of minute fruits embedded in a fleshy rachis, 2–3 cm long. It is picked before it is ripe and dried.

P. retrofractum Vahl (syn. *P. officinarum* C.DC.), Javanese long pepper, occurring wild through Malaysia, resembles *P. longum* and is used in a similar manner.

P. cubeba L.f., cubeb, is a native of Indonesia and the dried unripe stalked fruits were used in Europe in the Middle Ages as a spice. Later it was used in medicine.

Other pepper substitutes include *P. clusii* DC. and *P. guineense* Schum. & Thonn. from tropical Africa, *P. longifolium* Ruiz & Pavon from tropical America and *P. saigonense* DC. from Vietnam.

OTHER USES: The leaves of *P. betle* (q.v.) are chewed as a masticatory.

P. methysticum Forst., kava, provides the national beverage of the Polynesians and is used throughout Oceania. It is made from the thick, knotty, greyish-green roots. Formerly the peeled roots were chewed by people with strong teeth and the saliva ejected into bowls constituted the beverage. Now it is usually pulped and fermented in water. It is not an intoxicant, but narcotic, and acts as a sedative, soporific and hypnotic, and is said to bring about pleasant dreams and sensations. It is closely connected with the entire social, political and religious life of the people. The active principle is a viscous resin and it also contains methysticin and a glucoside yangonin. Kava is a bushy shrub, 2–3 m in height, with rounded or cordate leaves.

Piper betle L. (2n = 32) BETEL PEPPER

USES

The leaves are chewed together with betel nut (*Areca catechu* L.) as a masticatory from Zanzibar through India, Malaysia, Indonesia and into Oceania. The habit is of great antiquity and was mentioned by Herodotus in 340 B.C. It is estimated that over 400 million people chew the betel pan. In its simplest form sliced betel nut is wrapped in a betel pepper leaf, smeared with lime and chewed. It is often fastened with a clove and other spices such as cinnamon and cardamom may be added. It is chewed after meals to sweeten the breath and is partaken at all ceremonial gatherings of an auspicious nature and when friends and visitors call. It acts as a gentle stimulant. It blackens the teeth and the spittle is red and so may be considered a substitute for lipstick. It is also extensively used in medicine in the East.

ORIGIN AND DISTRIBUTION

Betel pepper is a native of central and eastern Malaysia and spread in very early time throughout tropical Asia and Malaysia. It was taken at a later date to Madagascar and East Africa.

CULTIVARS

A number of cvs are recognized in India and elsewhere in the East and differ in the shape of the leaves, bleaching quality, softness, pungency and aroma.

ECOLOGY

In its wild state the plant grows in the tropical rain forest. It is cultivated from sea level to 3,000 ft. A good rainfall is required, but it is frequently grown under irrigation. It needs a fertile soil and is usually heavily manured. Heavy clay soils are considered the best for the crop in India. It is grown under shade with protection from the wind.

STRUCTURE

Tall, dioecious, glabrous, woody climber with swollen nodes. Dimorphic branching; orthotropic vegetative branches with adventitious roots which adhere to support; plagiotropic axillary fruiting branches without roots. Lvs coriaceous, smooth; petiole 1·2–2·5 cm long; lamina ovate, size varies with cv. but often about 15 × 8 cm; tip acuminate; base cordate, rounded or oblique; veins 5–9, elevated beneath. Some Indian cvs are sexually sterile and do not normally bear fls; other Indian cvs are male plants only; in Malaysia there is no preponderance of male plants and fruits are plentifully produced. Fls minute, unisexual; male spike cylindrical, blunt; female spike about 5 cm long and 5 mm thick, blunt, brown; ovary embedded in rachis.

POLLINATION

Being dioecious, cross-pollination must occur.

CHEMICAL COMPOSITION

Pungency of leaves is due to volatile oils. Two phenols, betel phenol and chavicol, occur and eugenol is present in Indian leaf. The amount of phenols varies greatly, hence the distinctive taste of the various cvs. Leaves contain 76 per cent moisture and are rich in vitamins B and C.

PROPAGATION

Betel pepper is propagated by cuttings, 12–18 in. long, which are taken from the tips of orthotropic vegetative shoots. They may be planted in nursery beds, but are more usually planted *in situ.* Cuttings usually have 3–5 nodes and the lowest 2 nodes are inserted in the soil.

HUSBANDRY

In Malaya a few plants are grown near the homesteads. In India cultivation of betel is very intensive in small gardens usually less than ¼ acre in area and sometimes only a few yards square. The crop is given a great deal of attention and needs a lot of labour. The cuttings are planted close together, a few inches apart, in pits or long mounds. The vines may be grown up *Areca* palms, or specially planted, quick-growing trees such as *Sesbania grandiflora* Pers. and *Erythrina indica* Lam., or up sticks. In northern India the gardens are usually hedged and covered over with thatch or roofed with jute matting. The crop is usually heavily manured with cattle or sheep manure and is irrigated.

Fig. 70. **Piper betle : BETEL PEPPER. A, flowering shoot of female plant (× ⅓); B, female inflorescence (× 3); C, female inflorescence in longitudinal section (× 6).**

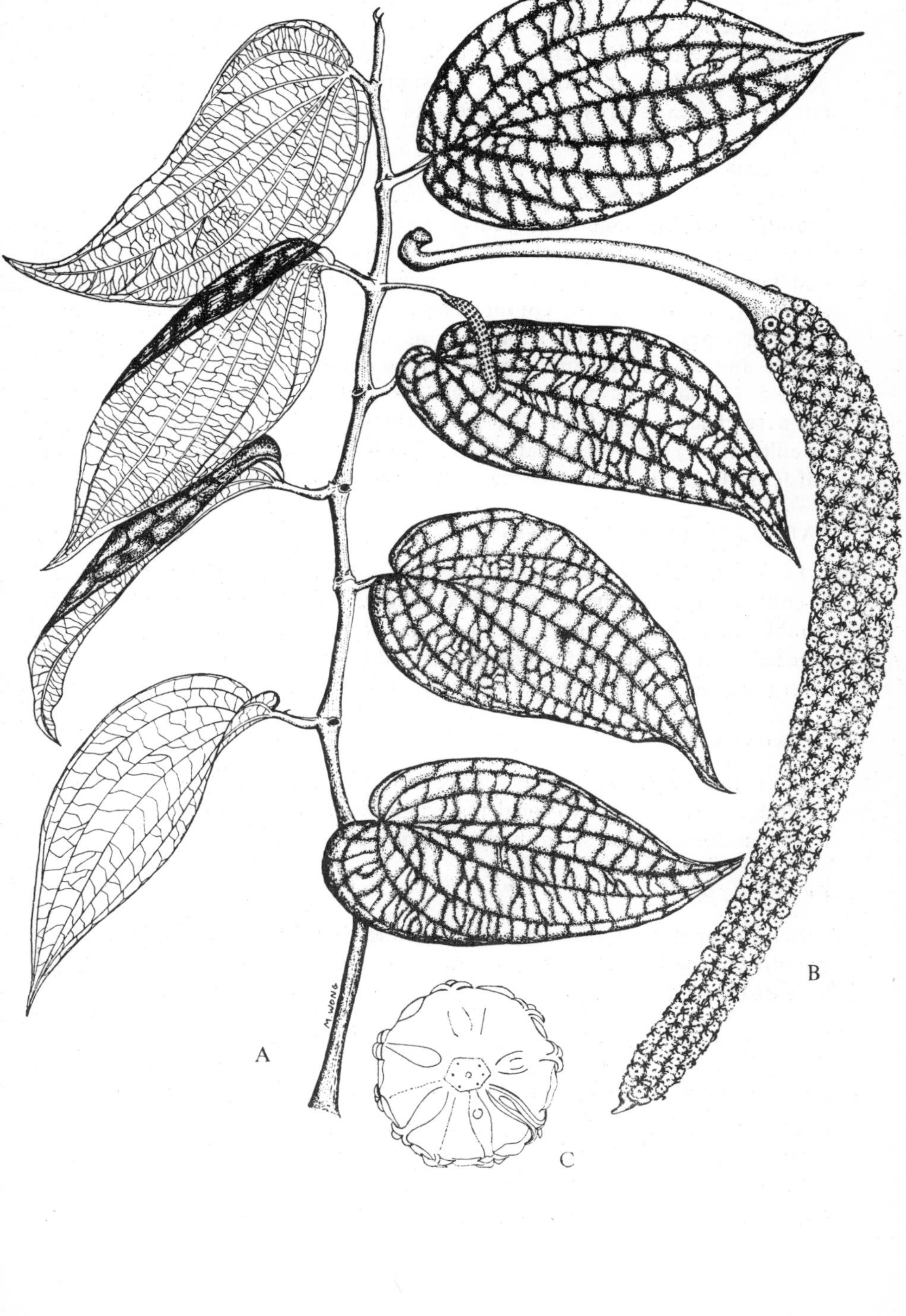
A
B
C
M WONG

Picking may begin 18 months after planting and sometimes not until the 3rd year. Only leaves from the plagiotropic axillary branches are picked. Each vine is usually picked four times per year on a rotational basis. The upper branches produce the best leaves for chewing. The vines are seldom allowed to grow more than 10–12 ft high. In old-established gardens the vines are taken off the supports every year and the bottom 6 ft of 2nd year's growth is coiled and buried in a pit at the bottom of vine. At the 3rd annual lowering the portion buried 2 years previously is severed.

Leaves are harvested by cutting the petiole with a sharpened steel thumbnail and is done in the early morning. The leaves may be bleached by packing them tightly in baskets lined with wet banana leaves or hessian and keeping them in the dark for several days. The leaves become soft and turn a bright yellow colour. The connoisseur's leaf is taken from an upper axillary shoot and should be as large as possible, soft, with thin cuticle and epidermis, and a good yellow colour. Yields vary enormously from 5–150 lacs per acre; in Bengal the yield is about 80 lacs of leaves per acre (1 lac = 100,000). The life of the garden varies with the methods of cultivation; temporary gardens lasting from 3–4 years; semi-permanent gardens 10–12 years; and permanent gardens 30–50 years.

MAJOR DISEASES AND PESTS

Foot rot, *Phytophthora palmivora* (Butl.) Butl., attacks betel pepper and is controlled by applying Bordeaux mixture to the roots. Mildew can be troublesome, but ordinary methods of control are impossible without sacrificing the crop and rendering it unfit for use. The gardens are so carefully tended that pests are not serious.

IMPROVEMENT

Over the centuries growers have selected cvs giving high yields of palatable leaves suited to local taste and environment. Little or no work has been done on the scientific breeding of the crop.

PRODUCTION

Betel pepper does not enter world trade, but is grown in considerable quantity in the East for local consumption. In India betel leaves are taken long distances to markets.

REFERENCES

AIYER, A. K. Y. N. (1947). *Field Crops of India.* Bangalore: Govt. Press.

HOLLIDAY, P. and MOWAT, W. P. (1963). *Foot rot of* Piper nigrum L. (Phytophthora palmivora). Kew: Commonwealth Mycological Institute, Phytopathological Paper No. 5.

RIDLEY, H. N. (1924). *The Flora of the Malay Peninsula.* London: Reeve & Co.

Piper nigrum L. (2n = 52) PEPPER

NOTE

Cayenne, red, green and sweet peppers are *Capsicum* spp. (q.v.) of the family Solanaceae.

USES

Pepper is one of the oldest and most important of all spices. Black pepper is the whole dried fruit; white pepper is the fruit which has been retted in water and the mesocarp removed. Both black and white pepper are ground and used in powdered form; commercial ground pepper is often a blend and can be easily adulterated. Pepper is most widely used as a condiment, the flavour and pungency blending well with most savoury dishes. Its stimulating action on the digestive organs produces an increased flow of saliva and gastric juices. It has extensive culinary uses and is used in pickles, ketchups and sauces, for seasoning dishes, and in sausages, pâtés, etc.

Pepper was mentioned by Theophrastus (372–287 B.C.), who recognized two types, namely black pepper (*Piper nigrum*) and long pepper (*P. longum* (q.v.)), both of which were used by the Greeks and the Romans. It was one of the first articles of commerce between the East and Europe. By the Middle Ages pepper had assumed great importance. It was used to season insipid food and as a preservative in curing meats. Together with other spices, it helped to overcome the odours of bad food and unwashed humanity. It was used in medicine as a carminative and febrifuge, but finds little use in medicine in modern times. Peppercorns were extremely expensive; tribute and rents were often paid in them. This still survives in the term peppercorn rent, but, unlike Medieval times, this now means little or no rent. Arabs brought pepper from the Malabar coast of India, and later from Java, to Arabia, from where it was taken to Alexandria and Venice which waxed rich on the spice trade. A Guild of Pepperers existed in London as early as 1180. Pepper trading between Java and China is recorded in A.D.1200. It was mainly the search for spices and the desire to take part in this lucrative trade which sent Columbus west in 1492, but instead of discovering the spice islands of the east he discovered the West Indies. It sent Vasco da Gama round Africa in 1498 and the spice trade fell into the hands of the Portuguese, who retained the monopoly until the 17th century. Under the Portuguese, Malacca and Goa became the great emporia for pepper. Later the Dutch captured much of the trade, but they never succeeded in maintaining a complete monopoly as they did with cloves (q.v.) and nutmegs (q.v.).

Pepper oil is distilled from the fruits; it has a mild taste and is used in perfumery. The alkaloid piperine is used as a source of synthetic heliotropine.

ORIGIN AND DISTRIBUTION

Piper nigrum is a native of the Western Ghats of India. It occurs wild in the hills of Assam and north Burma, but may have become naturalized in this region from an early introduction. It was probably taken by Hindu colonists to Java between 100 B.C. and A.D. 600. Marco Polo reported pepper in Malaysia in 1280. In the 16th century pepper was grown on the west coast of India with Malabar as the centre, and with smaller production in Java, Sunda, Malacca, Kedah and Siam. It has been grown by Chinese in Borneo for about 300 years. In the 17th and 18th centuries the Dutch established large holdings in Java and Sumatra. Pepper was being grown in Ceylon in the 18th century and later in Cambodia. Early in the 19th century the British made organized planting in Penang, Singapore and parts of Malaya and pepper was usually grown in association with gambier, *Uncaria gambir* (q.v.). During the 20th century pepper has been introduced into most tropical countries and there is now limited production in Brazil and the Malagasy Republic.

CULTIVARS

Wild *Piper nigrum* shows considerable variation in the size of internodes, leaves, inflorescences and fruits and is probably mostly dioecious. Man has selected cvs which are mainly hermaphrodite and these are maintained by cuttings. The greatest number of cvs occur in India, where 'Balancotta' is the most widely grown Malabar pepper, followed by 'Kalluvalli'. 'Balancotta' has large leaves and long spikes and a reputation for high and regular yields; 'Kalluvalli' has small leaves and large fruits. Two cvs are recognized in Sarawak, the high-yielding 'Kuching' which has large leaves and is very susceptible to foot rot, and the smaller leaved 'Sarikei'. Introduced Indian cvs have not yielded well under intensive cultivation in Sarawak and are more suitable for growing up trees. The most commonly cultivated clone in Sumatra is 'Belantung' which is fast growing and has some resistance to foot rot. In Bangka the cvs 'Lampong' and 'Muntok' (syn. 'Bangka') are grown mainly for white pepper.

ECOLOGY

In its wild state pepper grows on the forest-clad slopes of the Western Ghats in a monsoon climate with 70–100 in. of rain per year. It requires a hot wet tropical climate and is usually grown at low altitudes. Pepper areas of Sarawak have an annual rainfall of 100–200 in. with a non-seasonal climate. It cannot stand waterlogging and is usually planted on mounds. Although it can be grown on a variety of soils, the ideal is a well-drained alluvium rich in humus and with a pH above 5·5. Much of the pepper in Sarawak is grown on heavy reddish-brown clays and the gardens are usually made on hillsides.

STRUCTURE

A perennial, glabrous, woody climber to 10 m in height. Under the best cultivation the mature vine has a bushy columnar appearance, 4 m high and 1·5 m in diameter.

ROOTS: The main adventitious roots at base of mature stem number 10–20 and are 3–4 m long and penetrate to a depth of 1–2 m with an extensive mat of surface feeding roots.

BRANCHES: Pepper has dimorphic branching. Orthotropic vegetative climbing stems give the framework of the plant; they become stout, 4–6 cm in diameter, and woody with a thick flake-like bark; internodes 5–12 cm long. At each swollen node there is a leaf, an axillary bud which grows out to give the plagiotropic fruiting branch and short adventitious roots which adhere firmly to the climbing support. Lateral fruiting branches have no roots. Both types of stems branch, but only orthotropic stems will produce further climbing shoots. Xylem in original bundle ring of young climbing stem develops into dichotomously branching plates separated by medullary ray tissue with embedded islands of schlerenchyma.

LEAVES: On both climbing and fruiting branches, alternate, simple; petiole 2–5 cm long, grooved above; 2 lateral stipules sheething petiole turn black and fall early; lamina ovate, coriaceous, 8–20 × 4–12 cm; base oblique, obtuse or rounded, tip acuminate, dark green and shiny above, pale and gland-dotted beneath, veins 5–7.

FLOWERS: Borne opposite lvs on plagiotropic branches in pendent spikes, 3–25 cm long, with up to 150 fls. Fls unisexual (monoecious or dioecious) or hermaphrodite (most cvs the latter), minute, borne in axils of ovate fleshy bracts; perianth absent; stamens small, 2–4, on either side of ovary in hermaphrodite fls, 1 mm long, anthers small with 2 sacs; ovary globose, 1-celled, 1-ovuled, surmounted by 3–5 rather fleshy stigmas covered with papillae, white when receptive later turning brown.

FRUITS: Sessile globose drupe, 4–6 mm in diameter, with pulpy mesocarp, exocarp turning red when ripe, drying black; fruiting spike 5–15 cm long. Seeds 3–4 mm in diameter with minute embryo, little endosperm and copious perisperm. Weight of 100 peppercorns 3–8 g, usually about 4·5 g.

POLLINATION

Martin and Gregory (1962) have shown in Puerto Rico that the flowers are protogynous and that the stigmas are exserted 3–8 days before the dehiscence of the anthers in an hermaphrodite flower. The stigmas may be receptive for up to 10 days with peak receptivity 3–5 days after exsertion. Flowering begins at the base of the spike and continues towards the tip over a period of 7–8 days. Bagged inflorescences produced fruits showing that the hermaphrodite cvs 'Balancotta' and 'Kalluvalli' are self-fertile and that self-pollination can occur without the action of rain or wind. The pollen is in glutinous masses of several to many grains; light rain breaks up the mass and the grains are caught in the papillae of the stigma thus increasing the efficiency of pollen distribution. Pollen is lost to the air, but the poor fruit set in the female clone 'Uthirancotta', 4–50 ft from adequate pollen sources, indicates that wind-pollination is not very efficient.

How far insects assist in pollination is not known. Pollination seems to be mainly confined to individual spikes, so that self-pollination appears to be general in hermaphrodite flowers.

GERMINATION

For commercial production pepper is grown from cuttings, but it may be grown from seed if required. The pulp should be removed and the seeds dried; they are of short viability. Immersion in concentrated sulphuric acid for 2 minutes ensures quicker and more even germination. Fresh seeds should be planted under heavy shade and may give 90 per cent germination in 6 weeks. Seedlings show great variability in the size of the cotyledons and in vigour and a high proportion of abnormal plants is often produced. They may be transplanted into polythene bags 2–4 weeks after germination. Seedling plants take about 7 years to come into bearing and some may have reverted to the unisexual condition.

CHEMICAL COMPOSITION

Pepper contains the alkaloids piperine ($C_{17}H_{19}NO_3$) and piperidine, the latter in much smaller quantity. Freedom from adulteration in ground pepper is ascertained by the percentage of piperine. The pungency of pepper is due to the resin chavicine, which is most abundant in the mesocarp; consequently white pepper is not as pungent as black pepper. The odour is due to a volatile oil containing terpenes. Black pepper contains 8–13 per cent moisture, 22–42 per cent starch, 5–8 per cent piperine, and 1–2·5 per cent volatile oil. White pepper contains more starch, about 50–64 per cent.

PROPAGATION

Pepper is propagated vegetatively by cuttings. In Sarawak, Malaysia and Indonesia these are taken from terminal orthotropic shoots of vines less than 2 years old. Before taking the cutting the terminal bud of the selected shoot is broken off and the leaves and small branches between the 3rd and 7th nodes from the apex are stripped off. After about 10 days, when the terminal bud has regenerated, the shoot is cut below the 7th node and the cuttings, about 2 ft long, may be planted in nurseries or *in situ*. The cuttings are planted obliquely at an angle of about 45°, with 3–4 nodes below the surface, and not deeper than 4–6 in. They are often shaded with fern fronds. If planted direct into the garden the leaves and branches of

---

Fig. 71. **Piper nigrum** : PEPPER. **A**, fruiting branch (×½); **B**, hermaphrodite inflorescence (×3); **C**, hermaphrodite flowers showing stamens (×15); **D**, hermaphrodite flower showing stigmas (×10); **E**, stigmas (×25); **F**, base of inflorescence (×1); **G**, developing fruit (×5); **H**, fruit in transverse section (×5); **I**, dried unripe fruits (×1).

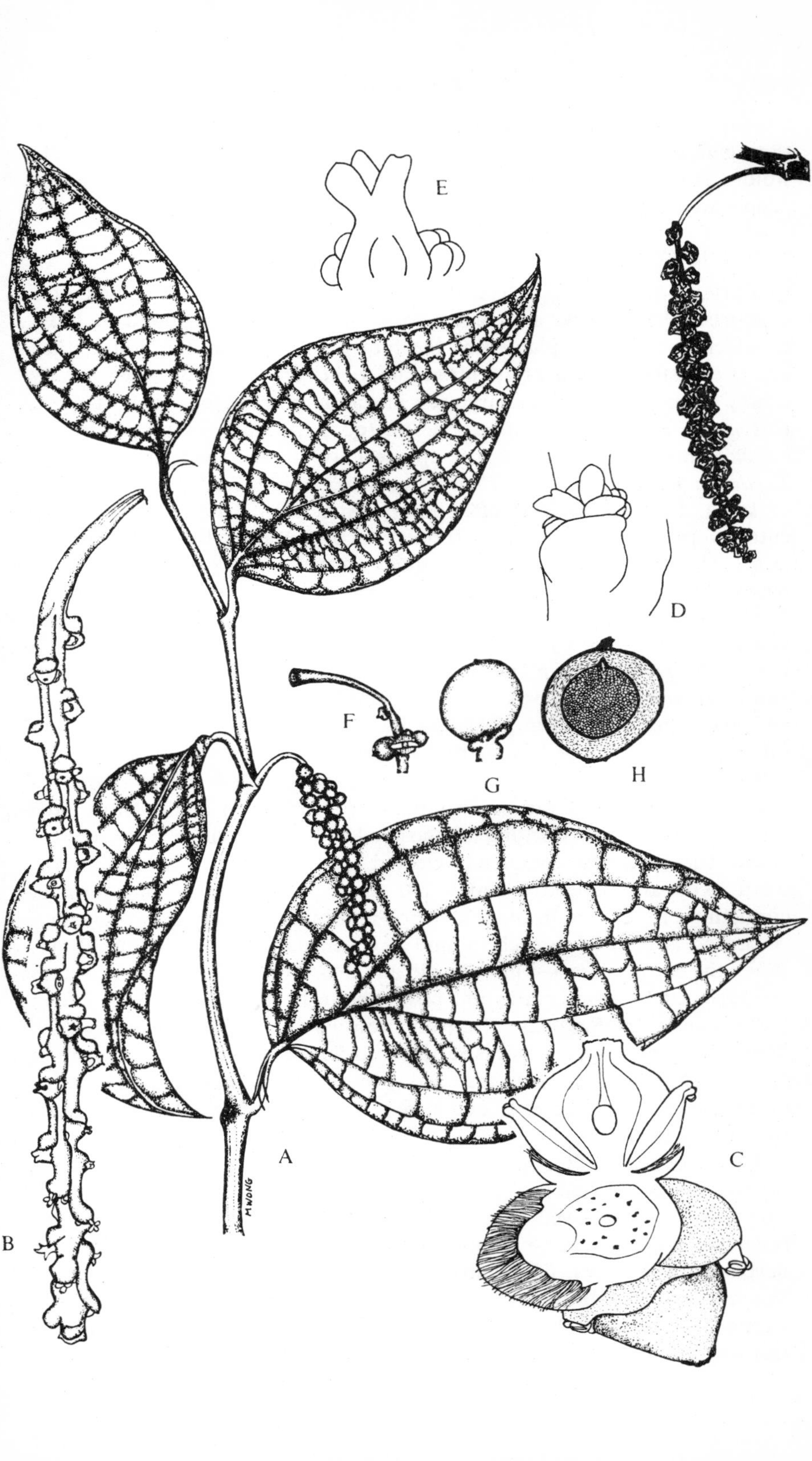
E
D
F
G
H
A
C
B
M WONG

the top 3 nodes are left and the latter are trained up a temporary stake, thus giving one main shoot and 2 lateral orthotropic shoots. In India stolons from the base of the vines are used for planting. Pepper may be propagated by marcots or grafted if required.

HUSBANDRY

CULTIVATION: As pepper is a climbing vine, provision must be made for supports. The simplest and oldest form of cultivation is found in India in which cuttings are planted at the foot of a tree near the house; they are given little subsequent attention and yields are very low. Alternatively it may be grown as a secondary crop in stands of tree crops such as *Areca catechu* L. or *Ceiba pentandra* Gaertn. (q.v.). A more intensive form of cultivation is to grow the pepper up specially planted shade trees such as *Erythrina indica* Lam. or *E. lithosperma* Miq., as is practised in some areas of India. In Sumatra it is often interplanted with coffee. A number of cuttings are planted, which are trained up the support, but no pruning is done. The shade trees should be quick growing and able to withstand heavy pruning. Some manuring and mulching may be done and yields of 2–4 lb of black pepper per vine are obtained. The most intensive method is the one developed by the Chinese in Bangka, Biletong, Sarawak and on the south-east Asia mainland. It consists of growing the pepper, unshaded and clean-weeded, up posts and considerable attention is paid to pruning and manuring. Maximum yields are high with 10–20 lb per vine when in full bearing. Details of this method are given below.

Traditionally 4 acres of primary forest were felled for a garden of ½–1 acre. The topsoil and organic debris (burnt earth) of the entire area was used to make the planting mounds and to replenish them in subsequent years. The garden is completely cleared and mounds 1·5 ft in diameter, 0·5 ft deep at the centre and 7–8 ft apart, are made of burnt earth, consisting of the topsoil, organic debris and humus slowly burnt together. A cutting is planted in each mound as given under PROPAGATION above. When the cuttings are 4–6 months old and 2–3 ft high the temporary stakes are replaced by stout hardwood posts about 12 ft long and sunk 2 ft in the ground. In Sarawak Bornean iron-wood, *Eusideroxylon zwageri* Teijsm. & Binn., is used and is resistant to termite attack. A catch crop of ginger, Capsicum, soya beans, groundnuts or tobacco may be taken during the first 12 months, but is removed well before the vines reach the tops of the posts.

The three main climbing stems of the vines are tied to the posts and are pruned regularly to encourage the development of lateral fruiting branches. The leaves are removed from the nodes of the climbing stems and the latter are pruned back to within 6–9 in. from the ground when they have developed 8–9 nodes. (The cuttings are used for new planting material.) When a further 9–10 nodes have been produced the stems are cut back to within 3–4 nodes of the previous cut. The vines are pruned 7–8 times in this way before they reach the top of the 10 ft support posts. The terminal buds are then removed periodically to prevent further vertical growth. Side branches and leaves near the ground are also removed. During the first 2 years flowering spikes are stripped regularly to prevent premature fruiting and selective leaf plucking is also done to encourage prolific side branching. The first crop is taken in the 3rd year and the production life is 12–15 years. As the gardens are often on slopes and are clean-weeded, serious erosion may occur and it is usually necessary to replenish the mounds from time to time with burnt earth and soil from outside the garden. Contour planting and bench terraces are desirable.

The above method is wasteful of land and the clearing of primary forest for pepper gardens is now rarely permitted in Sarawak. Consequently the grower has to resort to secondary growth or even *lalang* waste, which are often the sites of previous pepper gardens, and yields drop sharply.

Investigations are now being made by the Sarawak Government to place pepper cultivation on a more scientific basis. Planting grass, *Axonopus* sp., between vines has reduced yields, but mulching with cut grass, *Imperata cylindrica*, has increased them. In a small plot, hedge planting with hedges 8 ft apart, 6 ft high and with plants 4 ft apart in the hedge with 7 leading shoots trained up wires has given yields of up to 30,000 lb of green pepper per acre per annum for the first two harvests. This is a 300 per cent increase over pepper grown by the traditional method on poles 10 ft high and 8 ft apart (Ministry of Natural Resources, Sarawak, 1964).

MANURING: During the 19th century pepper in Malaysia was grown in conjunction with gambier, *Uncaria gambir* Roxb., and the extracted waste from the gambier was used as a mulch or manure. When gambier went out of cultivation, increased use was made of burnt earth and wood ashes. Organic manures are extensively used in Sarawak and include guano, prawn and fish refuse, soya-bean cake, and more recently blood-and-bone meal. Up to 5 lb guano per vine is applied four times per year after fruiting begins. Ministry of Agriculture, Sarawak (1964), in a series of fertilizer trials, has shown that there is a marked response to potash on all soil types, that there is a response to nitrogen only at the highest levels of potash, and that there is little or no response to phosphates. Inorganic fertilizer mixtures produced as good yields as organic manures, and when trace elements were added to the inorganic mixture yields were consistently higher. It recommends a tentative mixture of NPK in the proportion of 12 : 5 : 14 with additional magnesium and trace elements (iron, copper, zinc, manganese, boron and molybdenum) as a basic application for all soil types. Potash deficiency causes a black and grey necrosis and nitrogen deficiency a progressive chlorosis.

HARVESTING, PREPARATION AND YIELDS: Vines are not permitted to produce flowering spikes until they are over 2 years old. It takes 5–6 months from the emergence of the flower spike and 4 months after flowering to ripen the fruits, so that the first harvest is 2½–3 years after planting when the vines have reached to the top of their 10 ft supports. Harvesting in Sarawak is from May–July. The entire fruit spike is picked after a few of the fruits have turned red and the rest are yellow or green. The vines are picked over once a week during the harvesting season.

In the production of white pepper the spikes are crushed lightly and then soaked in sacks in slowly running water for 7–10 days. This rots the mesocarp. The spikes are then trampled underfoot to loosen the stalks and skins, after which the pulp and rachis is removed by washing and rubbing with the hands in sieves. The washed peppercorns are dried on mats in the sun, which takes 3–4 days; when fully dry they are difficult to crack between the teeth.

Black pepper is prepared by sun-drying the spikes for 3–4 days. The berries are picked when fully mature but still green. The stalks are removed by rubbing with the hands and winnowing. A better product is obtained if the fruits are scalded in boiling water for a short time before drying. Shed and damaged fruits are also made into black pepper, as are the spikes towards the end of the harvest season when all fruiting spikes, ripe and unripe, are picked in order that the vines will come into bearing and fruit evenly in the following year. The vines are then stripped of all leaves except for the 2–3 terminal leaves of the branches.

Approximately 27 lb of white or 33 lb of black pepper are produced from 100 lb of newly picked green pepper. The proportion of the crop which will be made into white or black pepper will depend on the price differential at the time. With intensive cultivation a vine will yield 2½–4 lb green pepper in its 3rd year, 8–20 lb annually in the 4th–7th years increasing with age, and 4–5 lb annually in the 8th–15th years decreasing with age, after which yields are negligible and the garden is abandoned or replanted.

MAJOR DISEASES

Foot rot, *Phytophthora palmivora* (Butl.) Butl., is the most serious disease of *Piper nigrum* in Malaysia and Indonesia. The first above-ground symptom is a slight droop of the vine, followed by a yellowing of the leaves which drop rapidly. Rapid die-back then occurs and the vine dies. It is caused by destruction of the underground stem and roots in which the cortex is destroyed by a black rot. Initial infection is probably in minor roots. Leaves show brown circular lesions. The disease did much damage in Sumatra in 1925–1935, when Muller recognized the pathogen. It appeared with destructive suddenness in Sarawak in 1952–1953 and by 1956 it was estimated that losses amounted to 7,000 tons, valued during those four years at £1·7 million. By 1960 the incidence had declined to about 5 per cent in Sarawak. Holliday and Mowat (1957 and 1963) discovered that the causative organism of the Sarawak outbreak, first called sudden death, was *Phytophthora palmivora*. The great African snail, *Achatina fulica* Fer., has been shown to be capable of transmitting foot-rot. Resistance has been found in some cvs, notably in 'Belantung'. It is possible to graft high-yielding scions on to root-rot resistant stocks.

Pink disease, *Corticium salmonicolor* Berk. & Br., and white root-rot, *Fomes lignosus* Klotzsch, are common and can cause serious damage in Sarawak, especially the former; while the parasitic alga, *Cephaleuros mycoidea* Karster, caused serious losses before World War II.

MAJOR PESTS

Pepper weevil, *Lophobaris serratipes* Mrshl., which eats holes in the fruit causing them to be shed prematurely can cause considerable damage in Sarawak unless controlled. Two bugs, *Dasynus piperis* China and *Diconocoris hewetti* Dist. also damage the fruits. The root-knot nematode, *Meloidogyne javanica* (Treub.) Chitwood attacks pepper in Sarawak, and the burrowing nematode, *Radopholus similis* (Cobb.) Thorne, causes damage in Bangka.

IMPROVEMENT

Very little work has been done on breeding of *Piper nigrum* although man must, consciously or subconsciously, have been selecting for high-yielding hermaphrodite clones suited to his local environments during the many centuries that he has grown the crop in India and elsewhere. It seems probable that the breeding of high-yielding hermaphrodite clones, with even ripening which is not extended over a long period, would repay the effort involved, with particular attention being paid to foot-rot and other diseases.

PRODUCTION

A brief history of the pepper trade is given under USES above. In the West the nature of the demand for pepper has changed over the centuries. While better means of food preservation have lessened the need for the spice, the canning industry has found new uses. Much of the black pepper is used in the meat processing industry.

Before World War II Indonesia produced 80 per cent of the pepper entering world trade, exporting an average of 50,000 tons per annum during the period 1930–1938. Cambodia was next in importance with 7·3 per cent, followed by India with 5·8 per cent and Sarawak with 4·9 per cent. During this period violent fluctuations in prices occurred. When prices are low gardens are neglected and little new planting is made; high prices produce the opposite effect.

The Japanese occupation of the major producing countries during the war cut off supplies and gardens were abandoned. This, and the troubled post-war political situation in Indonesia, resulted in an acute shortage and extremely high prices being paid for pepper during 1949–1954. Due to this there was greatly increased production in Sarawak, who exported a record 19,800 tons in 1956, which was 40 per cent of the total world exports. Annual world exports of pepper now vary between 50,000–75,000 tons. India is now the biggest exporter, followed by Indonesia and Sarawak. Other countries exporting small quantities are Brazil, Malagasy Republic and Cambodia.

The biggest importer of pepper is the United States, which takes about one-third of the world supply and whose consumption is about 100 g per caput per annum. West Germany and the United Kingdom are the next biggest importers.

REFERENCES

ABRAHAM, P. (1959). Pepper Cultivation in India. *New Delhi, Ministry of Food and Agriculture, Farm Bull.*, **55**.

BLACKLOCK, J. S. (1954). A Short Study of Pepper Culture with Special Reference to Sarawak. *Trop. Agriculture, Trin.*, **31**, 40–56.

DE WAARD, P. W. F. (1964). Pepper Cultivation in Sarawak. *World Crops*, **16**, 3, 24–30.

HOLLIDAY, P. and MOWAT, W. P. (1957). A Root Disease of *Piper nigrum* L. in Sarawak caused by a species of *Phytophthora. Nature, Lond.*, **179**, 543.

HOLLIDAY, P. and MOWAT, W. P. (1963). *Foot Rot of* Piper nigrum L. (Phytophthora palmivora). Kew: Commonwealth Mycological Institute, Phytopathological Paper No. 5.

MAISTRE, J. (1964). *Les plantes à Épices.* Paris: Maisonneuve & Larose.

MARTIN, F. W. and GREGORY, L. E. (1962). Mode of Pollination and Factors affecting Fruit Set of *Piper nigrum* L. in Puerto Rico. *Crop Science*, **5**, 295–299.

MINISTRY OF NATURAL RESOURCES, SARAWAK (1964). *Annual Report of the Research Branch, Department of Agriculture for the Year 1962–1963*, Kuching: Govt. Printing Office.

RUBIACEAE

A large family of some 400 genera and 5,000 spp., mostly trees and shrubs, more rarely herbs, generally distributed, but mostly tropical and particularly abundant in the lower storey of the tropical rain forest. Lvs opposite or whorled, simple, usually entire; stipules always present, most commonly interpetiolar, less commonly intrapetiolar, sometimes leafy or united in sheath, in some cases falling shortly after lvs unfold but leaving linear scar. Fls usually hermaphrodite and regular, solitary to capitate; calyx tubular or cup-shaped, adnate to ovary, 4–5-lobed; corolla tubular, usually with 4–6 lobes, but up to 10 in some spp.; stamens as many as petals and alternating with them, inserted in tube or at its mouth; anthers 2-celled, dehiscing longitudinally; ovary inferior, usually 2-locular, style often slender; some spp. heterostylous; ovules one (Coffeoideae) to many (Cinchonoideae) in each locule. Fr. capsule, berry or drupe, crowned by calyx or with circular calyx scar. Seeds usually with endosperm.

USEFUL PRODUCTS

BEVERAGES: *Coffea* (q.v.) is the most important economic genus, producing the coffee of commerce.

DRUGS: *Cinchona* spp. (q.v.) are the source of quinine. *Cephaelis ipecacuanha* (Stokes) Baill. (syn. *Psychotria ipecacuanha* Stokes), a small subshrub, is the source of ipecacuanha or ipecac obtained from roots, of which the main alkaloid is emetine used in the treatment of amoebic dysentery; it is a native of Brazil and most of the drug is obtained from wild plants from the Matto Grosso. It was introduced into India in 1866 and Malaya in 1876 and these countries have a small commercial production.

TANNINS: Gambier or white cutch is obtained from the leaves and young branches of *Uncaria gambir* (Hunt.) Roxb., a woody climber, indigenous in Malaysia and formerly cultivated in Singapore and Malaya. It is also used in the East as a masticatory in the betel-quid.

ORNAMENTALS: The family provides a number of attractive tropical garden shrubs, *e.g.*, *Gardenia jasminoides* Ellis, native of China; *Ixora* spp., natives of south-eastern Asia and Malaysia; *Mussaenda* spp., natives of

Africa and Asia, in which a sepal in some of the flowers is enlarged, coloured and leaf-like; *M. erythrophylla* Schum. & Thonn. deserves to be more widely grown; *Pentas lanceolata* Schum., a native of tropical Africa; *Portlandia grandiflora* L., native of West Indies; *Rondeletia* spp., native of tropical America; *Warszewiczia coccinea* (Vahl) Kl., the Chaconier, is the national flower of Trinidad and Tobago, a sepal in each group of flowers is enlarged, red, leaflike; in a recently discovered mutant, at least one sepal in each flower is enlarged and it is very handsome.

Cinchona L. (x = 17) QUININE

Evergreen shrubs and trees, indigenous in the Andes from Costa Rica to Bolivia at 1,000–11,000 ft. The nomenclature is confused as many forms and hybrids occur, so it is difficult to give a reasonable estimate of the number of spp. Four spp. are grown for quinine production, but two of these are probably only vars. (see below).

USES

The alkaloid quinine, obtained from the bark, provided the most important antimalarial drug and was used both prophylactically and in the treatment of the disease. It has been largely superseded since World War II by new synthetic drugs. It is also used as a tonic and antiseptic, in some sun-burn lotions and insecticidal preparations; and has a few minor non-pharmaceutical uses.

ORIGIN AND DISTRIBUTION

Quinine was known to the Indians of the Andean region from early times. It is sad that the often-quoted story of the Countess of Cinchon, wife of the Viceroy of Peru and after whom Linnaeus named the genus, being cured of fever in 1636 and later taking the bark to Europe, has now been shown to be a myth. However, the use of *Cinchona* was known to the Jesuits who collected and distributed it under the name of Jesuits' or Peruvian bark. La Condamine and de Jussieu collected botanical specimens in 1739 which were sent to Linnaeus.

Reckless exploitation of the wild trees in South America during the first half of the 19th century led the Dutch and the British to secure seeds and plants and to establish the *Cinchona* plantation industry in the Far East, so that they might be independent of South America, where there was danger of extinction of some spp. A seedling of *C. calisaya* was raised in Paris in 1840 from seeds obtained by Weddell in Bolivia. A plant from this source was taken to Java in 1851 and was planted in the Buitenzorg Botanic Gardens at 850 ft where it died, but not before a cutting from it had been planted at 4,000 ft, where it flourished and was propagated.

A Dutch expedition under Hasskarl obtained plants in South America which were taken to Java in 1854. A British expedition under Sir Clements Markham (see rubber), which included Spruce, Cross and Pritchett, went to the Andes in 1858 and sent seeds and plants to Kew and India in 1860. At this time the localities where the Indians collected the bark were closely guarded secrets. The 125 plants which reached India were planted in the Nilgiri Hills, but died. In 1861 Kew sent seeds to India which gave rise to plantations in Travancore and Sikkim, and to Ceylon, where the Hakgala Botanic Gardens were created as a *Cinchona* nursery. An Australian, Ledger, about this time sent seeds which were sold privately in London and were bought by the Dutch, who sent them to Java. Some 12,000 seedlings were produced from these seeds and gave rise to *C. ledgeriana*, which was found to have the highest quinine content of any material. The several spp. which reached the East were grown, tested and improved. From 1864 India and Java exchanged planting material. The Germans introduced *Cinchona* into Tanganyika during the period 1900–1905. It was introduced into Uganda in 1918.

ECOLOGY

Cinchona spp. occur wild in the Andean montane rain-forests, some extending to the cloud- or mossy-forest near the timberline. They prefer precipitous, well-drained mountain slopes with a well-distributed annual rainfall of over 85 in. Plantings in the East grow best in areas with an average minimum temperature of 57°F and an average maximum temperature of 70°F, a high relative humidity and a high rainfall of 100–150 in. well distributed throughout the year. It grows poorly below 45°F and above 80°F. At low elevations, and where soil moisture is limited, the yield of alkaloids is reduced. It grows best on light, well-drained, virgin forest soils, rich in organic matter, with a pH 4·5–6·5, preferably of volcanic origin; it cannot stand waterlogging. It grows very poorly or not at all on soils that have been exposed to fire, so vegetation should not be burnt off when preparing land for planting.

SYSTEMATICS AND STRUCTURE

Cinchona spp. are evergreen shrubs or trees. Lvs opposite, simple, entire; stipules interpetiolar, deciduous. Inflorescence a terminal panicle. Fls small, fragrant; calyx small, united, with pointed lobes; corolla tubular with 5 spreading lobes with frill of hairs along margins; heterostylous; in microstyled plants 5 exserted anthers alternate with corolla lobes, while bifid stigma reaches half length of corolla tube; in macrostyled plants stamens are half length of corolla tube and stigma is exserted. Fr. a capsule, dehiscing from base upwards with 40–50, small, flat, winged seeds. Floral formula K (5) C (5) A 5 $\overline{G}$ (2).

The following are the economic spp.:

1. *C. calisaya* Wedd. (2n = 34.) Occurs wild at altitudes of 3,500–5,000 ft in south Peru and Bolivia. Large tree with stout, straight trunk. Lvs smooth, thick, oblong-elliptic; stipules oblong, entire, deciduous. Inflorescence a large panicle of numerous pale-pink fls with tubes about 1·2 cm

long. Capsule ovoid, 8–17 mm long. Thick whitish bark yields yellow, Peruvian or Calisaya bark of commerce and collections from wild trees in World War II gave 3·3–5·3 per cent dry wt. quinine and 4·5–6·2 total crystallizable alkaloids; the best sample giving 7·14 per cent. In India it is grown at altitudes of 1,500–3,000 ft.

2. *C. ledgeriana* Moens ex Tremen. (2n = 34.) Probably a high-yielding var. of *C. calisaya*, although some authorities consider it a complex hybrid between *C. calisaya*, *C. succirubra* and *C. lancifolia* Mutis. Obtained from seeds sold by Ledger (see ORIGIN AND DISTRIBUTION above). Weak, straggling, profusely-branched, fast-growing tree, 6–16 m in height. Lvs light-green, thick, elliptical, acuminate, with small grooves in axils of lateral veins on upper surface. Fls pale yellow. Capsule ovoid-lanceolate, 8–15 mm long. Seeds 4–5 × 1 mm; 3,500 per g. In Java flowers at end of rainy season; frs ripen in 8 months in November to January. The most commonly cultivated sp. Thick, brown bark yields Ledger bark of commerce. By selection clones have been produced which give 14–16 per cent quinine. It is somewhat exacting in its requirements. In India it is grown at 3,000–6,000 ft.

3. *C. officinalis* L. (2n = 34.) Indigenous at higher altitudes of 7,000–9,000 ft from Colombia to north Peru, often in montane mossy-forest. Slender tree 6–10 m in height. Lvs small, smooth, shining, ovate-lanceolate, about 10 × 4 cm; petioles reddish. Fls deep pink, 1·4–1·7 cm long, in rather small, dense, terminal panicles. Capsule oblong, 1·5–2·0 cm long. Rough brown bark yields crown or Loxa bark of commerce, which in wartime collection from wild trees gave 0·1–1·6 per cent quinine and 2·4–4·8 total crystallizable alkaloids. It is not grown much commercially.

4. *C. succirubra* Pav. ex Klotzsch (2n = 34.) Probably a form or var. of *C. pubescens* Vahl, a variable sp. which extends from Costa Rica to Bolivia from 3,000–11,000 ft. It was collected by Spruce in Ecuador. Large, erect, rapidly growing tree to 30 m in height; sparsely branched and sheds lower branches; young branches pubescent. Lvs large, thin, light green, elliptic, 40–50 × 30–40 cm; with tufts of hairs in axils of lateral veins on lower surface. Fls rose-pink, 1–1·2 cm long, upper surface of corolla lobes white with pink stripe. Capsules oblong, 2–3 cm long. Seeds 7–10 × 2–3 mm. In Java flowers and sets seeds throughout year. Brown bark with few whitish markings is source of red bark of commerce, but the quinine is rather difficult to extract; it is source of pharmaceutical bark. Hardiest and most easily grown sp. and grown from 2,000–6,000 ft in India. It is used as a root-stock in Java.

In addition to the above, a number of hybrids occur, *e.g.*, *C. ledgeriana* × *C. succirubra* = *C. hybrida*, *C. officinalis* × *C. succirubra* = *C. robusta*, and *C. officinalis* × *C. ledgeriana*.

Fig. 72. **Cinchona calisaya :** QUININE. **A**, inflorescence and leaves (× ½); **B**, microstyled flower (× 1½); **C**, macrostyled flower (× 1½). After *Bot. Mag.*, t. 6052 (1873).

A
B
C

Bark was collected from several other wild spp. in South America during World War II; *C. pitayensis* Wedd. proved high yielding with up to 6 per cent total crystallizable alkaloids.

POLLINATION

Quinine is cross-pollinated by insects, mainly bees, butterflies and flies. Attempts at artificial self-pollination in both types of flowers have been unsuccessful.

GERMINATION

The small winged seeds, approx. 100,000 per oz, require light for germination and so are broadcast on the soil surface. Germination takes 2–3 weeks. They lose viability fairly quickly.

CHEMICAL COMPOSITION

Some 30 alkaloids have been isolated from *Cinchona* spp., of which the most important are quinine ($C_{20}H_{24}N_2O_2$), first extracted in an impure state in 1792 and in pure form in 1820, its isomer quinidine, cinchonine ($C_{19}H_{22}N_2O$) and its isomer cinchonidine. The alkaloids occur in the bark of the roots, trunks and branches, the roots having the greatest content and the branches the least. Stem bark of *C. ledgeriana* in Java contains 4–13·5 per cent quinine, 0·1–0·7 per cent cinchonine, 0·4–1·4 per cent cinchonidine and 0·1–0·7 per cent other alkaloids. *C. succirubra* has a lower proportion of quinine and more of the other alkaloids. Totaquina, prepared from the bark, contains all the alkaloids and should not have less than 70 per cent crystallizable cinchona alkaloids. The alkaloid content is affected by the environment as well as the species and the genotype.

PROPAGATION

Cinchona is usually grown from seeds, but replantings in Java were usually made with high-yielding *C. ledgeriana* patch-budded or veneer-grafted on to *C. succirubra* root-stocks. It can also be grown from apical cuttings.

HUSBANDRY

For climatic and soil conditions see ECOLOGY above.

Graded seed is broadcast on shaded seed-beds and watered with fine mist-spray. The seedlings are transplanted to the main nursery when 1–2 in. high with 2–3 pairs of leaves, 4–5 months after sowing, at spacing of 4–6 in. apart. When vegetatively propagated, budding or grafting is done in the nursery. It is planted in the field when 1–2 ft high at 1–2 years old. Spacing is usually 4 × 4 ft. Saplings are pruned to obtain a single stout stem and bark is obtained from the prunings. Harvesting begins in 4th year and the plants are selectively thinned annually until approximately 25 per cent of original planting is left and the trees are finally uprooted and harvested after 8–12 years, at which stage the bark has the maximum quinine content. After cutting into suitable lengths, roots, stems and branches are beaten with sticks which detaches bark from wood. The bark is then peeled off and dried in the sun, shade or in hot-air kilns. On harvesting bark contains 70 per cent moisture and this is reduced to 10 per cent. In East Africa yields of 9,000–16,000 lb of dried bark per acre over a cycle of 8–10 years have been recorded. In Java the best clones have given about 100 lb quinine sulphate per acre.

DISEASES AND PESTS

Damping-off of seedlings occurs in the nursery by *Rhizoctonia solani* Kuhn and other fungi. Transplants in the field are attacked by root rots caused by *Armillaria* spp. and *Rosellinia arcuata* Petch. *Helopeltis* spp. is the most serious pest.

IMPROVEMENT

Cinchona is cross-pollinated and very heterozygous; thus vegetative propagation is desirable to ensure performance of selected clones. In Java it was shown that *C. ledgeriana* had the highest quinine content of all the species introduced. From 1870–1910 selection of intraspecific Ledger material was carried out. Two trees isolated in 1877 gave 11·0 and 9·4 per cent quinine and these clones were grafted onto *C. succirubra* stocks. Selections were later made for vigour and bark thickness. By World War II yields of quinine sulphate per acre of best clones had increased from 27–36 lb to 90–108 lb. Interspecific hybrids of *C. ledgeriana*, *C. officinalis* and *C. succirubra* were also made and tested.

PRODUCTION

From its discovery and use by the western world until after 1850, the world's supply of *Cinchona* bark was obtained from wild trees in the Andes. Plantations were established in south-east Asia during the second half of the 19th century, mainly in India and Java. South America exported nearly 7 million lb to Europe in 1880, of which 6 million lb came from Colombia; in the same year, 1,170,000 lb were imported from India, 70,000 lb from Java and 21,000 lb from Jamaica. Production in Java increased rapidly; in 1930 Indonesia produced 97 per cent of the world's crop. At the outbreak of World War II Indonesia was producing more quinine than could be marketed profitably and production was controlled by quotas and government regulation of harvesting, marketing, acreages and export of planting material. The 1938 world production of Cinchona bark was 28,786,000 lb, of which Indonesia produced 24,665,000 lb or approximately 85 per cent. India's production of nearly 2 million lb was used locally.

With the capture of Java by the Japanese in 1942, strenuous efforts were made to find alternative sources of supply. Before the outbreak of the war experimental plantings had already been made in Guatemala and Costa Rica and these then received a terrific impetus and plantings were also made in Mexico and Peru, East Africa and the Congo. The United States Government instituted an extensive programme of Cinchona procurement in South America, utilizing available wild stands (Hodge, 1948). In the two years, 1943–1944, some 12½ million lb of dried bark were produced in South America, while in the 3 years 1943–1945 wartime production from Peru alone amounted to over 4 million lb.

Meanwhile extensive search was made for synthetic anti-malarial drugs and this was so successful that they have now largely replaced quinine and are superior to the natural product.

REFERENCES

HODGE, W. H. (1948). Wartime *Cinchona* Procurement in Latin America. *Econ. Bot.*, **2**, 229–257.

MOREAU, R. E. (1945). *An Annotated Bibliography of* Cinchona-*growing from* 1883–1943. Nairobi: Govt. Printer.

POPENOE, W. (1941). *Cinchona* in Guatemala. *Trop. Agriculture, Trin.*, **18**, 70–74.

Coffea L. (x = 11) COFFEE

The taxonomy of the genus is very confused. The number of spp. is given by the various authors as from 25 to over 100. Wellman (1961) lists 64 spp., but a few of these may be considered varieties only. The probable number is about 60 spp. The greatest number of spp. are found in tropical Africa with about 33 spp.; there are about 14 spp. in Madagascar; 3 spp. in Mauritius and Réunion; and about 10 spp. in tropical south-east Asia.

All spp. are woody, ranging from slender, sprawling plants and lianes to shrubs of all sizes and to robust trees. Lvs mostly opposite, rarely in 3's. Fls usually white, in axillary clusters; calyx 4–5-lobed; corolla 4–9-lobed; stamens inserted in or below throat of cylindrical corolla tube. Fr. a drupe, typically 2-seeded with mesocarp, usually fleshy, and with horny endocarp investing seeds.

The most important economic spp. are *C. arabica* (q.v.) which produces about 90 per cent of the world's coffee, *C. canephora* (q.v.) about 9 per cent and *C. liberica* (q.v.) under 1 per cent. The seeds of some wild spp. are used locally, while the following spp. are occasionally grown and used:

C. bengalensis Heyne ex Willd.: occurs wild in Bengal, Burma and Sumatra; occasionally cultivated in India.

C. congensis Froehn.: native of Congo basin; resembles *C. arabica*; has been hybridized with *C. canephora* to give the 'Congusta' coffees of Java. Possibly a form of *C. canephora*.

C. eugenioides S. Moore: native of the Lake Kivu area of the Congo, western Uganda and western Tanzania; resembles a slender form of *C. arabica*; occasionally grown; beans have a low caffeine content.

C. excelsa A. Chev. (syn. *C. dewevrei* De Wild. & Th. Dur. var. *excelsa* A. Chev.): native of West Africa; grows to a large tree; lvs large; frs and seeds small; grown in small amounts in West Africa, the Philippines and Java. It is sometimes included in *C. liberica* (q.v.), but the fruits and seeds are much smaller.

C. racemosa Lour.: native of Mozambique, where it is used locally a small profusely branched shrub with small, red frs.

C. stenophylla G. Don: native of Sierra Leone and Ivory Coast, occasionally cultivated in West Africa; a small tree with blue-black frs when ripe; seeds smaller than *C. arabica*, and of inferior flavour.

C. zanguebariae Lour.: native of Zanzibar, where it was occasionally grown and used, Tanzania and Mozambique; frs and seeds like *C. arabica*.

KEY TO THE THREE MAIN ECONOMIC SPECIES

A. Outer, unshaded lvs small, smooth and glossy, 12–15 × 6 cm. Frs about 1·5 cm long.................... *C. arabica*

AA. Outer, unshaded lvs large, 20 × 10 cm or more

B. Lvs corrugated; frs about 1·2 cm long.......... *C. canephora*

BB. Lvs glossy; frs 2–3 cm long.................... *C. liberica*

Coffea arabica L. (normally 2n = 44) ARABICA COFFEE

USES

The dried beans (seeds) are roasted, ground and brewed to make a stimulating and refreshing beverage. This use was first discovered in Arabia about the middle of the 15th century. The practice of drinking coffee spread throughout the Middle East, reaching Cairo about 1510 and Constantinople about 1550. The first introduction into Europe was in Venice in 1616 and the first record in England is 1650; by 1675 there were about 3,000 coffee houses in England. The source of this coffee was Arabia. Coffee and tea are now the most important beverages in the Western World. Although African in origin, very few indigenous peoples there roast and drink coffee, although consumption is now increasing. It is not very popular in the East. In Ethiopia dried coffee berries have been used as a masticatory since ancient times; ground roasted coffee is also mixed with fat and eaten. The first drink made from coffee was probably an alcoholic one produced by fermenting the sweet pulp of the fruit; even in Arabia today a drink is prepared from the dried coffee pulp. An infusion from the dried leaves is made in Indonesia and Malaya. Coffee pulp and parchment are used as manures and mulches; in India they are occasionally fed to cattle. Coffelite, a type of plastic, can be made from coffee beans.

ORIGIN AND DISTRIBUTION

The home of arabica coffee is the Ethiopian massif where it occurs naturally in forests between 4,500–6,000 ft. Coffee was taken from Ethiopia to Arabia at an unknown date. Wellman (1961) considers the first introduction was about A.D. 575, but Haarer (1962) states that there is no reputable evidence of coffee in Arabia during the 13th century. Certainly it was there in the 15th century when the discovery of brewing coffee was made.

C. arabica was introduced by the Dutch into Java in 1690, although subsequent survivals were from an introduction in 1699. It is usually believed that the planting material came from the Yemen. This coffee was of the variety *typica*, which we should now call *C. arabica* var. *arabica* (see below). A plant from Java was taken to the Amsterdam Botanic Gardens in 1706. It flowered and, being self-fertile, fruited. The Burgomaster of Amsterdam sent a vigorous progeny from this tree in 1713 to Louis XIV in Paris and this was looked after by de Jussieu. Planting material from Amsterdam was sent to Surinam in 1718, from where the French obtained it for Cayenne in 1722 and from there it was taken to Brazil in 1727. Progeny from the Louis XIV plant were sent to Martinique about 1720, but only one plant survived the journey. From Martinique it was taken to Jamaica in 1730, where it gave rise to the cv. 'Blue Mountain'. Subsequently, production from these introductions was spread widely throughout the Caribbean and Central and South America. Thus much of the '*typica*' coffee, which still largely predominates in the New World tropics, originated from one tree and consequently is of very limited genetic variability.

Coffee was introduced into India and Ceylon towards the end of the 17th century, although there are legends of an earlier introduction. The var. *typica*, progeny from the Amsterdam tree, was taken to the Philippines in 1740 and Hawaii in 1825. The French also took it to their African territories. A single tree of this variety from the Edinburgh Botanic Gardens was taken to Nyasaland in 1878 and from there it was introduced into Uganda in 1900 under the name 'Nyasa'.

The variety *bourbon* occurs subspontaneously in Ethiopia. It appears to have been taken by the French to Bourbon (now Réunion) about 1718, but the exact source of this introduction is not known. Progeny from this introduction was later taken to the New World and elsewhere. Catholic missionaries took it to Tanganyika and Kenya at the end of the 19th century and it reached Uganda in 1900. It is also reported that French Fathers imported seed from Aden into Kenya in 1893. This is now known as 'French Mission' cv., and is morphologically distinct from var. *bourbon*.

During its cultivation, mutations, some of which are of economic value, have occurred in both vars, notably in Brazil. So far the extraordinary richness and variability of the wild coffee material in Ethiopia has been little exploited, although some efforts are now being made in this direction.

VARIETIES, MUTANTS AND CULTIVARS

C. arabica is a tetraploid (2n = 44) and is self-fertile. Inbreeding has produced no harmful effects. Mutations, of which over 30 have been recognized, are commoner than in other *Coffea* spp., which are diploid and self-sterile. Krug and Carvalho (1951) working at Campinas in Brazil have made the most complete study to date. They regard all homozygous mutants as new varieties of the species.

Here it is proposed to give the two main botanical varieties and refer briefly to some of the mutants and cultivars.

BOTANICAL VARIETIES:

1. *Coffea arabica* L. var. *arabica*; syn var. *typica* Cramer; 'Nyasa'. This is the type described by Linnaeus in 1753 and thus according to the *International Rules of Botanical Nomenclature* must receive the name var. *arabica*. It is considered to be the primitive form and has the dominant *typica* alleles TT. The origin and subsequent distribution is given above. It grows quickly into a sturdy tree if not topped; it is vigorous; the primary fruiting branches are slender and grow horizontally or nearly so, sometimes drooping later; lvs narrow, bronze-tipped when young. It still provides the bulk of the world's coffee.

2. *C. arabica* L. var. *bourbon* (B. Rodr.) Choussy. This is the var. taken by the French to Réunion (for distribution see above). It was shown to be a double recessive (tt) mutant from var. *arabica*. A somewhat slender tree if not topped; primary fruiting branches borne stiffly at an acute angle, but may be pulled down by fruits later; lvs broader and green-tipped when young. It gives higher yields than var. *arabica* under favourable conditions and was replacing the latter in plantings in Brazil; but new plantings are now mainly cv. 'Mundo Novo' (see below).

SOME MUTANTS:

Angustifolia: elongated, narrow lvs; poor producer.

Caturra: dominant mutant from var. *bourbon*, which it resembles, but plants smaller with shorter internodes; precocious bearer and very productive. Plantings have been made in Colombia.

Cera: recessive from var. *arabica*; seeds with yellow endosperm; produces xenia; used to demonstrate 7–9 per cent natural cross-pollination in *C. arabica*.

Columnaris: tall, cylindrical trees with short fruiting branches; gives heavy yields under shade in Puerto Rico.

Erecta: dominant; fruiting branches erect.

Goiaba: from var. *arabica*; fr. with well-developed persistent calyx.

Laurina: recessive from var. *bourbon*; small conical shrub with small lvs; resistant to drought; gives fair yields of good quality coffee.

Maragogipe: dominant mutant from var. *arabica*, first observed in Brazil in 1870; vigorous, with all parts larger than parent var. and with largest frs and beans in sp.; not very productive. Used in breeding and as a rootstock; sometimes planted commercially in Central America.

Purpurascens: young lvs dark purple; poor yielder.

Semperflorens: recessive from var. *bourbon*; flowers and fruits throughout the year.

Xanthocarpa: recessive from var. *arabica*; ripe frs yellow. 'Amarelo de Botuçatu' cv. of Brazil is a *xanthocarpa* mutant.

All the above are gene mutations; but two others occur due to changes in chromosome number. These are:

Bullata, hexaploid (2n = 66) and octoploid (2n = 88). Lvs thicker and broader than tetraploid, with fewer stomata. Productivity low due to abnormal meiosis.

Monosperma, a diploid (2n = 22) found in both var. *arabica* and var. *bourbon*. Small weak trees, which set only a few frs and these are typically one-seeded. Lvs narrow and thin. This should not be confused with 'peaberry', which are frs with single seeds found on many trees which are normally 2-seeded and is usually due to lack of fertilization of one ovule or subsequent abortion.

Chlorophyll anomalies: variegated plants. Lvs may be of the *albomaculata* type, which are normal except for the variegation and is found frequently in nurseries and plantations. Variegation, however, is often associated with irregular shape and structure of the lvs. Except in one instance, all variegations studied are cytoplasmically inherited.

CULTIVARS

In addition to those mutants which are grown commercially, the following are the best known cultivars:

'Blue Mountain' originated from var. *arabica* in Jamaica and is probably synonymous with it. It is now grown in Kenya, where it is resistant to coffee berry disease, *Colletotrichum coffeanum*, and is planted in areas where control of the disease with fungicides is difficult or uneconomic.

'French Mission' derived from seed imported from Aden by the French Fathers of the Holy Ghost to Bura Mission in Kenya in 1893. This cv. was extensively planted in Kenya and Tanganyika.

'Kent's' originated as a mutation in Mysore, India in 1911; it has the branching habit of var. *bourbon*, but lvs may be bronze-tipped; it has good stamina and gives high yields of good quality coffee; it has some resistance to *Hemileia*, but is very susceptible to coffee berry disease. It was extensively grown in India, but is now being replaced by new *Hemileia*-resistant strains, 'S.288', 'S.333' and 'S.795'. 'Kent's' is also grown in East Africa.

'Mundo Novo' originated in Brazil as a natural cross between var. *arabica* and var. *bourbon*, showing greater resemblance to the latter; it is very vigorous and has a very high-yielding capacity. It is now superseding other cvs in new plantings in Brazil.

'San Ramon' a mutation from var. *arabica*; dwarf conical tree; lvs large; fruit thickly set on branches; drought and wind resistant. It gave rise to the segregate 'Villalobos', which is now being planted in Costa Rica.

NOTE: Recently, new cvs have been obtained from Ethiopia which will become increasingly important; some show marked resistance to almost all strains of *Hemileia*. In Kenya and Tanzania local selections are being tested; 'K7', 'SL6' and 'KP532' are immune to Race II of *Hemileia*.

INTERSPECIFIC HYBRIDS

When tetraploid *C. arabica* is crossed with diploid *Coffea* spp. hybrids are usually triploid (2n = 33) and are largely sterile. A spontaneous hybrid (2n = 44) with *C. liberica* was observed in Java in 1886 and is self-compatible, but it produces many defective spongy seeds. A recent natural cross with *C. dewevrei* is tetraploid, probably with two genomes of the diploid parent; it has been extensively crossed with various genotypes of *C. arabica*.

ECOLOGY

Arabica coffee is an upland species, occurring naturally as an understorey tree in forests between 4,500–6,000 ft in Ethiopia at 6–9°N. The altitude at which commercial coffee is grown depends on the distance from the equator and the proximity to mountain masses, when the upper altitude limit is lower. The ideal conditions are those which permit good crops to be reaped annually without exhaustion and die-back and without the numerous pests and diseases. These conditions are found on the equator at approximately 5,000–6,000 ft, with temperatures of approximately 60–75°F, a rainfall of about 75 in. annually which is well distributed, but with a drier period of 2–3 months for initiation of the flower buds, and with deep, slightly acid, well-drained, fertile loams of lateritic or volcanic origin with reasonable humus content. Periods of mist and low cloud are beneficial; strong winds are detrimental. In the sub-tropics it is grown at sea-level; in Northern Paraná, Brazil, it reaches a latitude of 24°S where it is sometimes killed by frost.

On Mount Kilimanjaro in Tanzania the optimum altitude is 4,500–5,500 ft; in Kenya 5,200–5,800 ft and in Mexico 3,000 ft.

At temperatures above optimum forced rapid growth occurs, with too-early bearing, over-bearing, early exhaustion, die-back and disease attack, particularly *Hemileia* where present. When temperatures are too cold the trees grow slowly, are stunted and uneconomic and may lead to a high production of secondary and tertiary vegetative branches and the physiological condition known as 'hot-and-cold disease'. High and low temperatures may be mitigated to some extent by shade and mulching.

Although coffee is grown with an annual rainfall of under 30 in. to well over 100 in., 60–90 in. give the best conditions. In low rainfall areas it may be irrigated, *e.g.* Yemen; in others, soil moisture plays a completely dominant role, *e.g.* in Kenya east of the Rift Valley, where in areas of limiting rainfall the crop is highly sensitive to weed competition and shade and it is essential to conserve soil moisture by mulching. Much of the coffee in Brazil is grown in full sun, as there is insufficient soil moisture both for the coffee and shade trees. Being evergreen, coffee requires sub-soil water at all times, but the surface feeding roots require a drier period for part of the year to slow up growth, ripen the wood and initiate flower buds. A high water table or heavy clay soil limits root penetration.

Grown under optimum conditions of soil and climate, with a high standard of culture and adequate fertilizers, higher yields are obtained without shade (see cocoa). Where soil conditions are not so favourable, rainfall is excessive, temperatures are too high or too low and possibly where there is a prolonged season of many hours of bright sunlight, it is advisable to use shade to sustain regular yields and prevent over-bearing.

Mes (1957) and Went working at the Earhart Phytotron in California showed that var. *bourbon*, the mutant *caturra* and the cv. 'Mundo Novo' are short-day plants and that more flower buds are initiated at 8- or 10-hour than on 12- or 13-hour photoperiod. They also showed that more flower buds were initiated at day and night temperatures of 23°/17°C than at 26°/20°C and 30°/24°C; at the highest temperature many 'star-flowers' were produced. The buds remained in a dormant condition for two months or more, after which dormancy was broken by rain or submerging the buds in water, flowering then occurred in 8–12 days depending on the temperature. It is predominantly water stress which keeps the flower buds dormant (see below). Dormancy of flower buds can be broken by gibberellic acid sprays. In areas with a marked wet and dry season coffee flowers in flushes; while in areas with even temperature and no dry season flowering flushes are not so obvious. The mutant *semperflorens* which flowers and fruits throughout the year is photoperiodically indeterminate.

In eastern Kenya growth is minimal during the dry season from mid-December to mid-February; the crop is picked in November–December and flower bud differentiation then occurs. Flowering occurs in the 'grass' rains in March and growth ascends to a maximum. During the ensuing rains growth falls rapidly and some flower-bud initiation may take place in the dry season which follows, a smaller flowering then taking place during the short rains which begin in mid-October and with a minor growth period until the rains finish in mid-December. In India, where there is only one dry season and one wet season, there is usually only a single flowering, which occurs with the onset of the rains in March or April.

STRUCTURE

An evergreen, glabrous, glossy-leaved shrub or small tree up to 5 m high when unpruned.

ROOTS: A short, stout tap-root, rarely extending beyond 45 cm; 4–8 axial roots, originating as laterals from the tap-root, go down vertically to 2–3 m or more. Many lateral roots, 1–2 m long, in a horizontal plane, form the surface plate in the first foot of soil. Below these are the lower laterals which ramify evenly and more deeply in the soil. In moist, cool soils the surface mat is better developed; lower laterals predominate in drier, warmer soils. Roots will not penetrate the water-table; in poorly drained clays in Puerto Rico over 90 per cent of roots are in the top 30 cm of soil.

STEMS: Dimorphic branching, due to the different development of 2 buds which occur one above the other in each leaf axil of central orthotropic stem. Under the dominance of the apical bud, only the upper bud in the leaf axil, accessory or extra-axillary bud, grows to produce the lateral or primary branch. The primaries are plagiotropic and arise on the opposite side of each node in succession from the base upwards around the stem. They are borne horizontally in var. *arabica* and at first at about 55° in var. *bourbon*, and are the branches which produce fls and frs, which are seldom ever found on the orthotropic shoots. When the primaries die back or are cut back to the upright, there is no other accessory bud at the node to produce new primaries. Dead primaries are held on the tree for a considerable time unless pruned away.

The lower axillary bud normally remains dormant and does not grow until the main stem has been topped or damaged, when it grows out immediately under the primary branch, often curling round it, to produce an upright orthotropic vegetative shoot (=sucker or water-shoot). If this young shoot is damaged, a second axillary bud may grow. Under certain conditions, such as abnormally high temperatures, orthotropic shoots may grow out spontaneously to give a bushy appearance. The secondary orthotropic stems repeat the vertical vegetative pattern and by topping and/or pruning it is possible to obtain any given number of upright stems. Vegetative orthotropic stems can also be induced by bending over the main axis, when they will grow from each node, and is a method of increasing material for vegetative propagation. The number of orthotropic shoots allowed to develop will depend on the system of pruning adopted. The balance between leaf area and crop is the basic consideration in efficient coffee management.

In axils of lvs on primary or fruiting branches there is a series of 6 buds, borne one above the other. The top bud distal from the petiole is the largest and oldest, while the sixth bud nearest the lf. petiole is the smallest and youngest. Any or all of these buds may develop into inflorescences or into secondary plagiotropic vegetative shoots depending upon the circumstances. Under conditions conducive to flowering, the first 3–4 buds usually develop into inflorescences. Provided they have not passed a certain morphological stage in their development and flower buds have not been initiated, they can all be changed into vegetative shoots by removal of the branch above their point of insertion. Sometimes when not cut back the first and top bud will produce a secondary branch, while the next 2–3 buds will give inflorescences. Damage by *Antestiopsis*, growing coffee at very high, cold altitudes, or boron deficiency will cause prolific development of secondary vegetative shoots giving a fan of branches. Primary branches can grow secondaries and tertiary fruiting branches, but they seldom ever produce orthotropic shoots or suckers. The number of shoots is controlled by pruning.

LEAVES: Opposite, dark green when mature; young lvs bronze-tipped in var. *arabica*; elliptical with prominent acuminate tip; short petiole about 1 cm long; lamina 5–20 × 1·5–7·5 cm, usually about 10–15 × 6 cm; margin sometimes undulate; lateral veins 7–12 pairs. Domatia or small cavities on lower surface of lf. at insertion of lateral veins, give slight protuberance on upper surface. Stipules small, interpetiolar, deltoid with acuminate tip.

INFLORESCENCES: Produced from 1–6 buds in axils of lvs of plagiotropic branches, but usually only from first 3–4 buds. Typically each inflorescence consists of 4 fls in 2 alternate decussate pairs, but all 4 may not develop. At base of inflorescence 2–3 pairs scale lvs or bracteoles, inner ones connate at base of pedicels; buds in their axils may develop to give further inflorescences thus giving flowering on old wood. Fl. buds at first closely packed and covered by a gum-like substance.

FLOWERS: Fragrant white fls in axillary clusters, 2–20 per axil, on primary and secondary branches (see above). Buds initiated and remain dormant when 4–5 mm long until stimulated by rain or wetting. Growth stops before meiosis has taken place in microspore mother cells. At this stage buds suffer from water stress and much of transport of water is through the phloem of the peduncle. Sudden increase in water by rain or wetting increases water content of fl. buds, meiosis takes place and more xylem vessels rapidly develop in peduncles. Corolla then rapidly expands and all fls open simultaneously 8–12 days after wetting, the number of days depending on temperature. Flowering in areas with wet and dry seasons occurs in flushes (see ECOLOGY above). Fl. buds open on sunny days in early morning; after 2 days begin to wither and a few days later all floral parts drop away except ovaries.

Floral formula: K(5) C(5) A5 $\overline{G}$ (2).

Calyx small, cup-shaped; corolla tubular about 1 cm long, typically 5-lobed, lobes spreading about length of tube; stamens usually 5, filaments short, inserted on corolla tube between lobes; anthers wholly exserted, rather shorter than corolla lobes and with filaments attached towards base; ovary inferior with 2 united carpels and 1 ovule per carpel; style long; stigma bifid, exserted.

Under adverse conditions, particularly at high temperatures, abnormal flowers occur, usually called star-flowers, in which petals remain small, fleshy and stiff, and are often greenish in colour. The style is exserted, but there are no functional stamens and they fail to set frs.

Fig. 73. **Coffea arabica** : ARABICA COFFEE. **A**, plagiotropic shoot (×½); **B**, portion of under-surface of leaf showing domatia (×½); **C**, portion of shoot with star flowers (×½); **D**, star flower (×½); **E**, normal flower in longitudinal section (×2); **F**, fruiting node with some of fruits removed (×½); **G**, fruit (×½); **H**, fruit with part of mesocarp removed (×½).

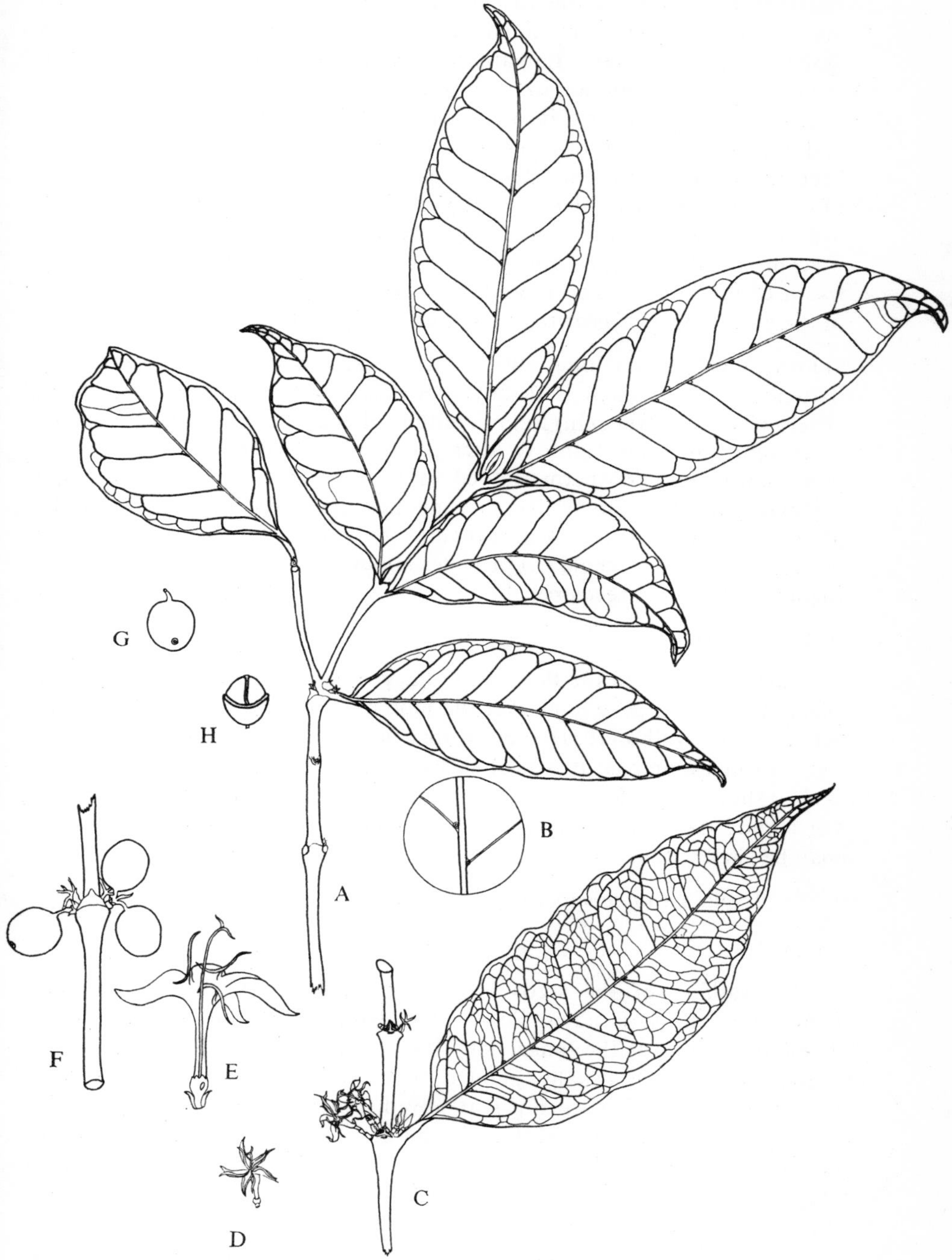

K2

FRUITS: Takes 7–9 months to mature. A drupe, about 1·5 cm long when mature, oval-elliptic, surmounted by calyx scar (calyx persistent in mutant 'goiaba'); pedicel short. Immature frs green, ripening yellow and then crimson (yellow in mutant *xanthocarpa*), dried frs turning black. Consists of smooth, tough outer-skin or exocarp, soft yellowish pulp or mesocarp and greyish-green, fibrous endocarp (= parchment) surrounding seeds. Seeds normally 2; but one only in pea-berry, due to failure of fertilization of one ovule or subsequent abortion and in diploid *monosperma*; frs with 3–4 or even more seeds recorded in mutant *polysperma*.

Approximately 40 per cent of fls set frs and are harvested as mature frs; a certain number never swell, but may persist until harvest; others fall in the early stages of growth, mainly in the first 10 weeks.

SEEDS: 8·5–12·5 mm long, ellipsoidal in shape and pressed together by flattened surface which is deeply grooved; outer surface convex. Thin, silvery testa (= silver skin) follows outline of endosperm, so that fragments often found in ventral groove after preparation. Seeds consist mainly of green corneous endosperm, folded in a peculiar manner, and a small embryo near the base. Polyembryony has been recorded. Dried seeds, after removal of silver skin, provide the coffee beans of commerce. 5–6 lb of cherry (= whole fr.) provides about 1 lb of clean coffee. Approximately 1,000 dried seeds per lb.

POLLINATION

Seeds are set when the flowers are bagged, so self-pollination may take place. Using as a genetic market the mutant *cera* with yellow endosperm, Krug and Carvalho (1951) showed that natural cross-pollination was 7–9 per cent at Campinas; it is probably effected by insects and wind. Pollen is shed shortly after flowers open and the stigmas are immediately receptive. On cloudy days fully developed buds tend to remain closed and pollen may be shed within buds.

NOTE: Diploid *Coffea* spp. are self-sterile.

GERMINATION

The viability of the seeds is comparatively short, depending upon conditions, and it is advisable to plant within 2 months of harvesting. The older the seeds the longer they take to germinate and the viability is reduced. They are usually planted with the parchment attached, but germination is quicker when it is removed. Emergence from soil takes 3–4 weeks at a temperature of 82°F and 6–8 weeks at 68°F. During germination the cotyledons of small embryo enlarge absorbing endosperm; the radicle and hypocotyl then grow out; the hypocotyl emerges from soil in a hooked form and on straightening it pulls out cotyledons still enclosed in parchment and silver-skin. The rounded cotyledons expand tearing the parchment and silver-skin and are at first held closely together, later taking up a horizontal plane.

CHEMICAL COMPOSITION

Raw dried coffee beans have the following approximate percentage composition: water 12; protein 13; fat 12; sugars 9; caffeine 1–1·5; caffetanic acid 9; other water soluble substances 5; cellulose and allied substances 35; ash 4. During roasting water is lost, sugars are partially caramelized and the cellulose partially carbonized. Roasting develops and releases the aroma and flavours, which are water-soluble on infusion. The leaves also contain caffeine.

PROPAGATION

Most commercial arabica coffee is planted from seed. Provided a good strain of seed can be obtained from progeny-tested mother trees, little advantage seems to be obtained from vegetative propagation, but it is of value in increasing clones and for top-working inferior plantations. Buds, grafts and cuttings must be taken from orthotropic shoots, as those from primary fruiting branches do not produce upright growth and cuttings made from them sprawl along the ground. This limits the amount of propagating material available, but this can be increased by grafting the first suckers as scions on to nursery stocks, and as they grow bending them over, pegging them down and pruning away the primaries, so that orthotropic shoots will be produced at each node.

The following methods of vegetative propagation are used:

BUDDING: This is not generally favoured. They are usually budded on nursery stocks not more than 2 years old at a height of 9–12 in. from the ground, using the T or inverted T method of insertion. Buds shoot in 3–4 months and will flower and fruit in about 2½ years.

GRAFTING: Cleft-grafting is most commonly used, although other forms of grafting have also been successful. It is usually done on stocks of about ½ in. diameter and scions, consisting of one or two nodes and internodes, are inserted about 8 in. from ground. Scions are obtained from strongly sprouting suckers which have just become woody. Root-grafting is also successful. Top-working may be used for improving inferior trees and rehabilitating old plantations. Old trees are topped to induce sucker growth and cleft-grafts are then inserted on the suckers.

CUTTINGS: These have the advantage of complete clonal uniformity without stock/scion interaction. Soft wood cuttings are taken from the tips of leaders or suckers of unshaded trees; they are severed in the middle of the internode and are 4–6 in. long with 3–4 nodes. The two bottom leaves are removed and the remainder trimmed. They are dipped in a growth-promoting substance, *e.g.* Hortomone A, and planted 12 in. apart in equal proportions of peat-moss and coarse sand in propagators. They require maximum light without causing the temperature to rise, and 90 per cent humidity. They root in 3–4 months (see cocoa).

HUSBANDRY

NURSERIES: Seedlings are usually raised in nurseries. Seeds should be obtained from ripe berries of selected, high-yielding trees, which should be hand-pulped to avoid damage; light beans are floated off and peaberry and large seeds (elephants) removed. Nurseries should be made in areas with reasonably level ground near water and with deep fertile loam. Beds should be dug to depth of 2 ft; a convenient width is 4 ft which is accessible for weeding and watering; they are usually raised by putting top-soil from the paths on the beds; NPK in mixture of 10 : 15 : 10 or 10 : 10 : 5 may be applied at rate of 1 oz per sq. ft. Overhead shade is usually provided and the beds may also be mulched. Seeds may be planted thickly in beds or seed-boxes at a depth of ½ in. and then transplanted at the cotyledon stage at a spacing of 6–10 in., or the seeds may be planted direct at the spacing required so that no transplanting is done until ready for field. Seedlings may also be grown in bags of polythene or other material. Germination takes 4–8 weeks. Haarer (1962) recommends transplanting to the field at the 6-leaf stage when 6–10 months old. They are sometimes planted as stumps cut back to 9 in. of brown wood, or as plants with 2–6 primary branches when 12–24 months old; these will require wider spacing in beds. It is desirable to select the best seedlings only from the nursery and to transport them to field with earth attached, although they are often planted with bare roots.

PLANTING: In Brazil, when coffee is planted on virgin soils after the removal of forests, direct seeding is done in the field. Holes 10–12 ft apart, and about 18–24 in. square are dug in which 12–20 seeds are sown at 4–6 in. below the general ground level. On germination, the holes are covered with sticks and 4–6 trees are allowed to grow to maturity at each hole. In the State of Sao Paulo, where new coffee plantations are being established on land cleared long ago, transplanting of seedlings is done.

In preparing fields for transplants from the nursery it is desirable to put in soil conservation and drainage work, etc. before planting and to clean cultivate, especially where African couch, *Digitaria scalarum* (Schweinf.) Choiv., and other grasses are a problem, and also where rainfall is marginal. Holes are then lined out; the spacing used depending on type and fertility of soil, vigour of cv. and type of pruning to be adopted; wider spacing is required for multiple-stem and agobiado than the single-stem system. On poor soil, using single-stem pruning or dwarf cv., coffee can be planted 5–6 ft apart, but 8–10 ft is usually recommended using triangular planting. The addition of manure and superphosphate in the planting holes is beneficial. Transplanting should be done early in the rains. The roots should be spread carefully and the seedlings should be planted at the same level as in the nursery. The young plants may be temporarily shaded with palm leaves or other vegetation.

Sun-hedge planting is sometimes recommended in Central America, in which 3 rows are planted 3–4 ft apart, with same distance between plants in the rows, and with hedges 13 ft apart; the 2 outer rows are grown as topped, single-stem plants and central row unpruned. Variations with single or double hedge rows have also been suggested. Temporary shade, *e.g. Crotalaria* or pigeon pea, provides shade between hedges for 2–3 years, after which they are self-shading. They should be mulched and heavily fertilized. The method is not yet proven, but disease and minor-element deficiencies present problems.

SHADE: (See also ECOLOGY above.) The need for shade is still a very controversial question. Most of the world's coffee is planted without shade and there is a growing tendency to eliminate shade and to establish new plantations without it. In a favourable environment and with intensive cultivation, coffee grows satisfactorily without shade, *e.g.* Kona region of Hawaii which produces the world's highest yields. Cocoa behaves in a similar manner. When soil water is limited, shade should not be planted, *e.g.* Brazil and the main coffee areas of Kenya. Unshaded coffee almost invariably gives higher early yields than shaded coffee, but it has a tendency to over-bearing, die-back, biennial bearing, erosion, disease and short productive life. Thus shade may be an advantage when coffee is grown under marginal conditions with a low standard of cultivation. The benefits derived from shade are: (1) extends the productive life of the tree; (2) prevents over-bearing and die-back; (3) gives more even annual cropping; (4) reduces temperature of air and soil at high temperatures and raises it at low temperatures, and provides a favourable microclimate; (5) reduces hail damage and may reduce pests and diseases; (6) reduces evaporation and transpiration; (7) leaf fall will provide mulch, humus and nutrients, particularly if deep-rooting by bringing nutrients from the lower levels in the soil; legumes may fix nitrogen; (8) protects the organic matter of the surface layer from breakdown by exposure to the sun; (9) root systems of the shade trees may assist in drainage and aeration; (10) branches may be lopped for mulch; (11) reduces cost of weeding and helps to keep out grass; (12) shade may provide firewood, timber or other useful products.

Permanent shade trees should be compatible with the crop and should not compete unduly with the coffee for water and nutrients. They should be long-lived and deep-rooting, and with a good root system so that they do not blow over. The branches should be spreading and well above the level of the coffee giving a mottled pattern of moderate shade; they should not be brittle nor break easily. It is preferable that they should be thornless, so that they can be climbed and trimmed. It is an advantage if the shade trees can be planted from large cuttings. They should be free from the pests and diseases which attack the crop. If possible leguminous trees, which are known to fix nitrogen, should be used. In areas with a marked dry season, a deciduous tree which sheds its leaves at this time may be an advantage, as a dense mulch of fallen leaves will help to retain soil moisture, and the transpiration of the shade tree is also checked, although that of the crop will be increased.

Sometimes selected indigenous trees may be left for shade when planting; in other cases they may be planted from seed or cuttings. The species most commonly used are: *Grevillea robusta* A. Cunn. (Proteaceae)—not ideal, as insufficiently spreading and competes for moisture and nutrients; *Albizia* spp. (Mimosoideae), particularly *A. falcata* (L.) Backer (syn. *A. moluccana* Miq.), *A. chinensis* (Osbeck) Merr. (syn. *A. marginata* (Lam.) Benth.), *A. stipulata* (Roxb.) Boivin, and *A. lebbeck* (L.) Benth.; *Erythrina* spp. (Papilionoideae), particularly *E. subumbrans* (Hassk.) Merr. (in India) and *E. glauca* Willd.; *Inga* spp. (Mimosoideae)—the most important coffee shade in Central and South America, particularly *I. edulis* Mart. and *I. vera* Willd.; *Ficus* spp. (Moraceae); *Cordia holstii* Gurke ex Engl. (Boraginaceae); *Entada abyssinica* Steud. ex A. Rich. (Mimosoideae); *Commiphora* spp. (Burseraceae); *Maesopsis eminii* Engl. (Rhamnaceae). The last 5 are commonly used in East Africa.

Small quick-growing trees are sometimes planted until the large shade trees have developed and include: *Leucaena glauca* Benth. (Mimosoideae); *Gliricidia sepium* (Jacq.) Walp. (syn. *G. maculata* HBK.—Papilionoideae); and *Sesbania* spp. (Papilionoideae). Shade may be provided for the first 2 years after planting by growing *Crotalaria* spp. (Papilionoideae) or pigeon peas.

Windbreaks are sometimes planted and the species generally used are *Grevillea robusta*, and *Eucalyptus* spp. (Myrtaceae), but they should not be planted too close to the coffee, as they have a profound root effect. Bananas may also be used.

CARE AND MAINTENANCE: Young coffee can be intercropped with beans, groundnuts or cowpeas during the first 2 years without much disadvantage. All blanks in the coffee should be filled as quickly as possible, as they are difficult to establish as the trees grow taller. Clean weeding increases yields and weeds are particularly deleterious where rainfall is marginal. Grasses are very harmful, notably couch, *Digitaria scalarum* (Schweinf.) Chiov., in East Africa; Bahama grass, *Cynodon dactylon* Pers. (introduced), and *Paspalum fasciculatum* Willd. in Central America. Tillage and clean weeding without adequate soil conservation measures may lead to serious erosion. Weeds, including grasses, may be controlled by herbicides. Cover crops generally have not given satisfactory results with coffee and usually lower yields, although they are better than weed growth.

Mulching is particularly beneficial, giving substantially higher yields, particularly in low rainfall areas. It increases the penetration of water, conserves moisture, reduces soil temperature, adds nutrients, improves soil structure and keeps down weeds. It is best applied before the onset of the rains to a depth of 4 in. of dry grass or other mulch and alternate row mulching sometimes gives better results than a complete cover. Elephant grass, *Pennisetum purpureum* Schum., is the commonest mulch, but other grasses and banana trash are also used. Repeated mulching with grass has produced induced magnesium deficiency in Kenya and is rectified by adding Mg as fertilizer. Mulching provides a fire hazard. In areas of Brazil subject to frost, damage is more noticeable in mulched than in unmulched plantations.

Manuring of coffee varies considerably between countries. In Kenya 45–90 lb N is given per acre per annum and with higher yielding cvs P is also required. In the Kona region of Hawaii with its high yields and intensive cultivation, NPK in the proportion of 10 : 10 : 5 at rate of 1,500 lb per acre per annum is applied to young trees and mature trees are given a 10 : 5 : 20 mixture at rate of 2,000 lb per acre per annum in 4 applications; in the case of very high yields an additional 500–850 lb of ammonium sulphate may be applied. In Central and South America formulations of 10 : 5 : 20, 12 : 5 : 12 and 14 : 14 : 14 are variously recommended at rate of $\frac{1}{2}$–2 lb per tree per annum in 3 applications. Continued application of ammonium sulphate may result in excessive lowering of pH of the soil, so it may be desirable to give nitrogen in some other form. Beginnings have been made in foliar analysis to assess manurial requirements. A heavy crop makes heavy demands on carbohydrates and with a critical need for N and K in the leaves. For mineral deficiency symptoms see below.

Irrigation is seldom used except in the Yemen, but, supplied as required, in Kenya has given increased yields.

The longevity of a coffee plantation depends upon the environmental conditions and management. In Brazil the average longevity is 30–40 years, but trees of 80–100 years old are known and some plantations have lasted for 50–70 years; in others the trees are worn out in 10–15 years or even less.

PRUNING: It is rather surprising to find (Krug, 1959) that nearly 50 per cent of coffee-growing countries do not practise pruning. In Brazil, in view of the prevailing planting system in which 4 or more trees are grown in one planting hole, pruning is limited to the periodic removal of dry wood and excessive suckers. Thus a multiple-stem effect is produced, but from several trees.

A knowledge of the branching system is essential for correct pruning (see above). The aim of pruning is to provide a plentiful supply of healthy, leafy wood on which the following season's crop will be borne, to maintain the correct balance between leaf area and crop, to prevent over-bearing and die-back and to reduce or eliminate biennial bearing. Pruning controls the shape and height of the tree, which affects picking, and controls the crop to be obtained in the following year. It consists of topping to a desired height, cutting out weak and dead branches, removal of surplus suckers or orthotropic branches, removal of weaker, secondary fruiting branches near the trunk to permit the circulation of air, removal of lower branches which trail on the ground, reducing the number of secondary and tertiary fruiting branches, and preventing primary branches from growing too long and spindly and replacing them where necessary with strong secondaries. A light pruning, provided that the tree is vigorous and robust, usually gives the best results. The best time to prune is in the dormant period after harvesting before the new flush of growth begins. It may be necessary to prune after growth has started to control over-bearing.

The main systems of pruning employed are:

Single-stem which may consist of retaining the original seedling stem and keeping it topped at a height of 5–6 ft, removal of all suckers and thinning of secondary fruiting branches by removal of those within 6 in. of upright and usually by removal of alternate secondaries on the remainder of the primary branches. A modification of this is to cap the main stem at knee, waist and chest height, permitting only one of the axillary branches to develop at each capping and finally stop growth at 5–6 ft. This is believed to strengthen the tree and provide sturdier primaries. Capping is done just above the primary branches, and those immediately below the cap are cut back to prevent splitting of the stem.

Multiple-stem pruning consists of topping to provide 2 lateral orthotropic stems and these may again be topped to provide 4 stems. The first topping is sometimes done in the nursery. Sometimes 3 stems only are allowed to grow. Due to weight of fruit, branches bend away from each other and the centre of tree is open. As the lower primaries become exhausted they are cut away and eventually a new sucker may be allowed to grow by cutting back the unproductive upright. This method is easier and cheaper than single-stem pruning and is now extensively practised.

Agobiado is a variation of multiple-stem pruning, in which main stem is bent over and pegged to ground, when orthotropic upright shoots grow out from axillary buds; 3 or 4 of these shoots are then allowed to form the main shoots, and the top of the original shoot may be allowed to grow upwards, or is cut off. The process may be repeated when required.

Candelabro pruning is used in Costa Rica. By 3 prunings of orthotropic stems, 8 main stems are obtained on each tree to produce a candelabra effect. New uprights are produced as required.

Hawaiian in which 4 to 6 verticals are left. When 4 are left, the annual elimination of the oldest one, cut a few inches above the parent stump, allows the complete renewal of the tree every 4 years. When 6 verticals are left, 2 of them are replaced every other year and one each alternate year.

HARVESTING AND PROCESSING: Trees come into bearing 3–4 years after planting and are in full bearing at 6–8 years. Fruits mature 7–9 months after flowering, depending upon climatic conditions. The out-turn of clean dry coffee from ripe cherry is 15–20 per cent. Selective picking of ripe red fruits will produce the highest quality, so that several pickings at intervals of 10–14 days are needed, the crop ripening over a period of several weeks. Over-ripe berries fall from tree. In Brazil all berries are stripped at one time on to ground.

Two methods of processing are used:

Dry method: whole cherries are spread out thinly and dried in sun, with protection from rain when necessary, taking about 15–25 days. Cherries are then hulled. Most of Brazil's coffee is prepared in this way.

Wet method: cherries are pulped as soon as possible after picking and not longer than 24 hours or they begin to ferment. Under- or over-ripe fruits should be kept separate from red-ripe fruits. Pulping removes the exocarp and part of the fleshy mesocarp. Light cherries are floated off and pulped separately. Pulped parchment then passes along grading channels, where it is separated into three grades by specific gravity, and passes to the fermenting tanks. The fermentation by enzymes, yeasts and bacteria removes the mucilage adhering to endocarp and usually takes 12–24 hours. Fermentation may be hastened by adding enzyme preparations or 2 per cent NaOH; the mucilage is sometimes removed mechanically. Fermented parchment is washed and graded and may then be sun-dried or dried in a flow of hot air. Sun-drying takes 8–10 days and the coffee should be protected from rain; it is stirred to give an even and quick drying. Well-prepared beans are bluish-green in colour, and shrink within the parchment shell and silver-skin. The dried parchment is then hulled, polished and graded, these processes being known as curing. Hulling removes the endocarp and testa (silver-skin), final traces of which are removed in polishing, which also puts a shine on surface of bean. Defective beans are sometimes picked out by hand.

Quality is important in arabica coffee and is judged on characteristics of the raw bean, roasted bean and the liquor. Quality factors are:

In raw bean: size of bean—preferably large and of even size and shape in each grade; colour—blue or greyish-blue preferred; brownness should be avoided and is caused by processing over-ripe cherry, over-fermenting, insufficient washing, or inadequate removal of the silver-skin; blotchy beans are caused by uneven drying; absence of defective beans, *e.g.* beans nipped in pulper, insect damage, oxidized beans and 'stinkers' which have been overlooked when washing out tanks, etc.

In roasted bean: general appearance must be even and bright depending mainly on clean, thorough fermentation and good drying; centre cut or groove should stand out clean and white; brown centre cuts indicate under-fermentation.

Liquor: judging of final beverage by expert liquorer. Absence of taint is important. Taints include 'onion flavour', probably due to water used in processing; 'bricky flavour' due to use of insecticides B.H.C. and P.C.P.; 'over-fermented' due to being left in vinegary solution too long during fermenting.

YIELDS: These vary enormously between countries and between plantations. Highest yields are in Hawaii with average yields of over 2,000 lb clean coffee per acre. Elsewhere average yields are as follows: Kenya 896 lb; Costa Rica 850 lb; El Salvador 720 lb; Colombia 450 lb; Brazil 360 lb.

MAJOR DISEASES

Leaf rust, *Hemileia vastatrix* Berk. & Br., appears first as pale green spots, followed by orange-yellow pustules on undersurface of the leaves; defoliation occurring later. It is most serious at low altitudes and when trees are debilitated by over-bearing. It is presumed to be endemic in Ethiopia. It reached Ceylon in 1869 and caused the complete collapse of coffee industry there, which was replaced by tea; it reached Java and Sumatra in 1876, where *C. arabica* was replaced by *C. liberica*, but the latter sp. was susceptible, and was replaced by *C. canephora* introduced in 1900 and which is resistant. Leaf rust has since spread throughout Asia, the Pacific (except Hawaii) and Africa. It was first reported in the Camerouns in 1951 and has now reached the west coast of Africa. Apart from an isolated outbreak in Puerto Rico in 1903, which was promptly eliminated, it has not yet reached the New World, but its presence on the West African coast provides a real danger.* Should it spread to Latin America it would have profound economic and political consequences. Many races of the rust are now known, and resistance to some races has been achieved by breeding in India ('S.288', 'S.333' and 'S.795' which are selections and crosses of cv. 'Kent's'), in Kenya ('K7', 'SL6'), and in Tanzania ('KP532'). Resistance is also found in some cvs recently obtained from Ethiopia. All New World coffees were found to be susceptible to nearly all races of rust. The Coffee Rust Investigation Centre near Lisbon, under Dr B. d'Oliviera, has obtained collections of rust races and coffee cvs from all over the world, and tests for resistance are being made. Fair control is obtained by 2 per cent Burgundy Mixture, 2 : 2 : 40 Bordeaux Mixture or 50 per cent copper formulations and it is best to spray about the time the rains break.

Grey rust, *Hemileia coffeicola* Maubl. & Rog., occurs irregularly over the leaf with spores in grey clumps on the undersurface. It was found first in the Camerouns (1932) and San Thomé. It is most serious at high altitudes.

Coffee berry disease, *Colletotrichum coffeanum* Noack, causes serious losses in East Africa. This strain attacks the berries in the green stage, causing brown sunken spots. Badly affected berries go completely brown to black and the beans inside are destroyed. Brown blight caused by a mild strain produces brown lesions on ripe berries and large brown spots on leaves, which are commonly marginal. Recently it has been shown that Elgon die-back is caused by the same fungus; the first symptom is blackening of the node and petiole, followed by death of the branch above the node. 'Blue Mountain' cv. is reasonably resistant. Coffee berry disease is controlled by fungicides applied at flowering and early fruiting.

American leaf-spot, *Mycena citricolor* (Berk. & Curt.) Sacc., the most serious disease of Central and South America (except Brazil). Fruit bodies are small, golden-yellow agarics and gemmifers. It causes defoliation, branch tips are killed and cherries are also attacked. Peronox affords some measure of control.

Brown-eye spot, *Cercospora coffeicola* Berck. & Cooke causes brown spots on leaves usually with reddish brown margins. It is widespread and causes serious losses in nurseries in Central America; it is also found elsewhere.

*** *Hemileia vastatrix* was first discovered in Bahia and north-eastern Brazil in 1970.**

Die-back is a physiological condition in which the tips of fruiting branches die back and lose leaves due to over-bearing. It is particularly common in young trees at lower elevations with marginal rainfall. Part or all of the crop can be stripped to prevent serious damage and this is advisable on trees bearing their first crop.

MINERAL DEFICIENCIES

Boron: distal ends of older leaves pale olive-green; young leaves malformed, because tips fail to develop; undersurface of main veins becomes corky; production of narrow leathery leaves; death of growing points of fruiting branches with the production of short weak branches giving a witches' broom effect.

Calcium: marginal chlorosis of young leaves, followed by convex cupping of leaves and corky growth on veins.

Iron: clearly defined, contrasting, interveinal chlorosis on young leaves; severe deficiency produces creamy, practically white colour between green veins.

Magnesium: herring-bone pattern of interveinal chlorosis on older leaves. Common. High concentration of K will often cause Mg deficiency, particularly when heavily mulched.

Manganese: pale olive-green terminal leaves with abrupt change to dark green leaves below; mature leaves with yellow mottling on green background.

Nitrogen: general yellowing of whole leaf; particularly noticeable with low leaf N and high sunlight intensity.

Phosphorus: blue-green colour of older leaves, which hang downwards and backwards, followed by patches of lemon-yellow, which may eventually change to red-bronze, autumn-tint colouration; it is rarely seen in the field.

Potash: brown scorching of entire leaf margins, affecting older leaves first and may be followed by shedding of leaves. Heavy drain of K from leaves to fruits.

Zinc: interveinal chlorosis of young leaves, shortened internodes, production of small, deformed, narrow leaves and may be followed by die-back of young twigs. Common.

MAJOR PESTS*

Latin-American countries are remarkably free of damaging pests, although leaf miner, *Leucoptera coffeella* Guer.-Mén., is widely spread and coffee berry borer (see below) occurs in Brazil.

Elsewhere the following pests occur, Africa having the greatest number:

Variegated coffee bug, *Antestiopsis orbitalis* (Westwood), mainly in Africa; suck young berries causing longitudinal brown zebra-striping on parchment; sometimes empty beans; in absence of young fruits the bugs feed on the terminal buds of branches causing fan branching. If more than 1–2 per tree control with pyrethrum or malathion.

---

* See Le Pelley, R. H. (1968). *Pests of Coffee*. London: Longman.

Capsid bug, *Lygus coffeae* China,* in East Africa, feed on flower buds, causing petals to shrivel and styles to elongate and have club-like appearance. If more than 4 per tree control as above.

Berry borer, *Stephanoderes coffeae* Haged.,** may cause serious damage in Africa, Asia and Brazil, by adults boring into seeds where they breed and larvae continue burrowing. Controlled by Dieldrin.

Mealy bugs: Kenya common coffee mealy bug, *Planococcus kenyae* (Le Pelley), and root mealy bug, *P. citri* (Risso). In Kenya the former is largely controlled by the introduced parasite *Anagyrus kivuensis* Comp.; in addition, banding trunks with 1 per cent Dieldrin prevents spread by attendant ants.

Scale insects: green scale, *Coccus* spp.; brown scale, *Saissetia coffeae* (Walk.); white waxy scale, *Ceroplastes brevicauda* Hall; star scale, *Asterolecanium coffeae* Newst. All occur in East Africa and *Coccus* spp. in India. Controlled by Dieldrin banding to prevent spread by ants, white oil sprays, etc.

Stem borers: In Africa yellow-headed borer, *Dirphya nigricornis* (Ol.); white borer, *Anthores leuconotus* Pasc.; black borer, *Apate monacha* Fabr.; in Asia white stem borer, *Xylotrechus quadripes* Chev., (worst pest of coffee in India) and shot-hole borer, *Xyleborus* spp. in Asia. The stem borer, *Bixadus sierricola* (White), is also troublesome in Africa. Controlled by Dieldrin, etc.

Leaf miners: *Leucoptera coffeella* Guer.-Mén. in Central and South America; *L. caffeina* Washb. and *L. meyricki* Ghes. in Africa. Controlled by diazinon.

Nematodes: serious in Indonesia; controlled by grafting onto resistant stocks.

IMPROVEMENT

Most commercial coffee is raised from seed which should be collected from old and proven mother trees. This will at least ensure that trees which are unsuited to the environment will have died out, and also omits those which have shown over-bearing and die-back and ensures the selection of ecotypes suited to the particular locality. Surprisingly little work has been done in most coffee-producing countries on the improvement of *C. arabica* by selection and breeding. Wellman (1961) states 'true breeding programmes in *arabica*, outside of understanding mutant characters, are practically non-existent'. This is true for most countries, but selection and breeding programmes have been carried on at Campinas in Brazil, in India, East Africa and the Congo for some years.

AIMS: (1) Improved yield and vigour. (2) Adaptation to local environment; a cv. which is high-yielding under good conditions and with intensive cultivation may do poorly under marginal conditions or with lower standards of cultivation. (3) Inherent stamina, so as to produce good yields as regularly as possible, without too marked biennial bearing and without

* *Lamprocapsidea coffeae* (China).

** *Hypothenemus hampei* (Ferrari).

exhaustion and die-back. (4) Habit of tree; short internodes for ease of harvesting; suitable branching habit for ease of pruning and harvesting. (5) Time to maturity; precocious bearing is an advantage provided it does not lead to exhaustion. (6) Even ripening so as to reduce the number of pickings. (7) High pulping and hulling out-turn with low proportion of pulp (mesocarp) and parchment (endocarp) and high proportion of clean coffee per sample; empty beans, peaberry and polyspermy are undesirable. (8) High quality; uniform large heavy beans of high specific gravity improves grade; good flavour and aroma, absence of taint; cup quality judged by expert liquorer. (9) Resistance to pests and diseases, particularly *Hemileia vastatrix*.

SELECTION: *C. arabica* is largely self-pollinated. Much of the world's commercial *C. arabica* is of highly uniform genetic composition, *e.g.* var. *arabica* (syn. var *typica*) in the New World, which originated from one tree, while var. *bourbon* developed from a few originals. It is not surprising, therefore, that selection has often produced very disappointing results. In Brazil individual records of 1,107 trees were kept for 19 years in var. *bourbon*. Those that appeared early as undoubtedly outstanding trees were progeny tested and, after 12–15 years, it was difficult to establish any correlation between high-yielding mother trees and that of the progeny. In Tanzania it was found that 20 per cent of the crop was produced by 5 per cent of the trees, 66 per cent of the crop by 25 per cent of the trees, showing that 75 per cent of the trees were uneconomic. In East Africa seedlings from outstanding mother trees showed marked differences in cropping capacity, but seedlings of some trees gave consistently higher yields than others, *e.g.* 'SL28' in Kenya and 'N197' in Tanzania. It is probable that there is more variability in arabica coffee in these countries. Should outstanding progeny-tested trees be discovered these could then be vegetatively propagated as clones. The tendency for marked biennial bearing adds to the difficulty of selection; also it is essential that the size of tree and the space occupied should be taken into consideration. The cv. 'Kent's' was first selected in Mysore, India, in 1911. It quickly became renowned for its vigour and resistance to some races of *Hemileia*. It has been extensively planted in India and East Africa.

MUTATIONS: Over 30 mutations have been recorded in *C. arabica*, some of which have been cultivated or used for hybridizing. (See Mutants above.)

HYBRIDIZATION: Emasculation is done with a pair of scissors with a small nick in each blade and a piece of metal on one handle, so that the blades do not close entirely. By cutting in the middle of the corolla, corolla and stamens are removed leaving the style intact. All other flowers, buds and fruits on the branch are then removed and the branch is bagged. Branches of male parent are also bagged and open male flowers are applied to the females with sterilized forceps; branches of the female parent should then be bagged for about a week.

Inter- and intraspecific hybrids have occurred naturally or have been produced artificially. One of the most important natural hybrids is 'Mundo Novo', which was found and later improved by the Instituto Agronómico at Campinas in Brazil. It has been shown to be a cross between var. *arabica*, cv. 'Sumatra' and var. *bourbon* and selected progeny give the highest yields of *C. arabica* in the New World. It is possible that 'Kent's' may also be of hybrid origin.

Normally interspecific hybrids between *C. arabica* and the diploid spp. are triploid and sterile. The 'Kawisari' coffee in Java (2n = 44) are natural hybrids between *C. arabica* and *C. liberica*; they are self-fertile and show some resistance to *Hemileia*. They are vegetatively propagated as clones on *robusta* stocks. Recently a tetraploid form, probably a natural cross between *C. arabica* and *C. dewevrei* with 2 genomes of the diploid parent, has been found in Brazil and is being extensively crossed with genotypes of *C. arabica*.

Breeding work in India consists of: (1) selection of mother trees and selfing; (2) hybridizing promising parents; (3) inter- and intraspecific hybridization and search for new material; (4) selfing desirable selections. A chance *C. arabica* × *C. liberica* hybrid was found to be practically immune to *Hemileia*. It was backcrossed to *arabica* to give 'S.26', from which selection 'S.288' was produced, which is now in large-scale production in India. 'S.288' was then crossed with 'Kent's' to give 'S.333', a vigorous, rust-resistant and acceptable commercial type.

As explained above the highly uniform genetic composition of much of the world's arabica coffee limits the success of its improvement. Ethiopia's highly important reservoir of germplasm has been little exploited. Recent and future collections in Ethiopia will add greatly to the variability available and they are likely to be extremely important in all future breeding of the crop. The examination and testing of rust races and cvs at the Coffee Rust Investigation Centre in Portugal should help greatly in obtaining resistance to *Hemileia*.

PRODUCTION AND TRADE: All figures are approximate

Prior to 1700 all the coffee for the western world was obtained from Mocha, the habit of drinking coffee having developed in the previous century. Planting was then begun in Java, Ceylon and the West Indies and for 100 years the Netherlands East Indies dominated the trade. Although coffee was introduced into Brazil in 1727, the first exports did not take place until 1800; since 1840 Brazil has dominated coffee production in the world; by 1850 it was producing over 50 per cent of the world's total, by 1900 nearly 80 per cent and the present production is about 45 per cent or 1,600,000 tons, giving over 50 per cent of the total value of exports of that country. Colombia, which made its first important coffee exports (2,326 bags) in 1935, is now the world's second largest producer, with about 450,000 tons and coffee provides about 80 per cent of Colombia's total exports. Coffee is the most important crop in many Latin-American countries; important producers are Mexico 125,000 tons (over 10 per cent of total exports), El Salvador 95,000 tons (90 per cent of total exports), Guatemala 100,000 tons (75 per cent of total exports), Costa Rica 50,000 tons (46 per cent of total exports), and with smaller, but still substantial, quantities in Venezuela, Haiti, Nicaragua, Ecuador and Dominican Republic. Jamaica, which exported 15,000 tons in 1814, now exports only about 900 tons; Trinidad exports about 2,500 tons, but it is mainly robusta.

Although Africa is the home of coffee, prior to World War I, that continent produced only about 1 per cent of the world's coffee or 15,000 tons. By 1944–1945 production had reached 174,000 tons and by 1963 this has increased to over 800,000 tons or some 20 per cent of total world output. About 80 per cent of this coffee is robusta and is in great demand for the manufacture of instant coffee (see *C. canephora* below). The principal African producers are: Ivory Coast 190,000 tons, Angola 180,000 tons, Uganda 135,000 tons, Cameroun 50,000 tons, Malagasy Republic 50,000 tons (all mainly robusta); Congo, including Uranda-Urundi, 180,000 tons, and Tanzania 25,000 tons (arabica and robusta); Kenya 32,000 tons (arabica); and Ethiopia 75,000 tons.

Asia was the first continent to establish commercial plantations of coffee. Much of the arabica was decimated by the *Hemileia* epidemics during the last two decades of the 19th century and many countries changed to robusta or to other crops. In 1937–1938 Asia produced 157,000 tons, after which it was overtaken by Africa. The crop suffered greatly during World War II, particularly in Indonesia as a result of the Japanese occupation, and production for the continent in 1944–1945 was 34,900 tons. This has now increased to 133,000 tons, with Indonesia (85 per cent robusta) and India (50 per cent robusta) as the main producers.

Over-production and wide fluctuations in prices have been features of the world's coffee production during the present century. In an attempt to stabilize the situation over 78 million bags of Brazilian coffee were destroyed during the period 1931–1944. New planting which will result in over-production usually occurs during periods of high world prices and after seasons of serious frost damage in Brazil. Huge coffee stocks are piling up in many producing countries; in 1959 Brazil had 22 million bags stored and, if no frosts occur in Paraná, production in Brazil is expected to increase to over 2 million tons. An attempt is now being made to control production under the International Coffee Agreement.

A record production of 4,450,000 tons was recorded in the 1959–1960 season in the main producing countries. Total exports from all countries is approximately 2,700,000 tons per year, but it should be remembered that considerable quantities may be used locally or may be stockpiled if markets are not available. Although Ceylon now produces some coffee, nevertheless, it is obliged to import some from abroad; India consumes more than half of its production, home consumption increasing from 8,000 tons in 1940 to 28,000 tons in 1958. The countries importing the greatest quantities of coffee are the United States (1,400,000 tons), West Germany (230,000 tons), France (200,000 tons), Italy (100,000 tons), Sweden (80,000 tons), Canada (70,000 tons), Belgium (55,000 tons), Netherlands (80,000 tons), and the United Kingdom (70,000 tons). Consumption in the United States is approximately 16 lb per head per annum, in Western Europe about 7–8 lb per head, and in the United Kingdom 3·5 lb per head per annum; the maximum consumption being in Sweden with over 20 lb per head per annum.

Contrary to popular belief, Krug (1959) has shown that coffee is predominantly a smallholder's crop. He also states that 'in the period 1952 to 1957 coffee was the second most valuable commodity in international trade, surpassed only by petroleum and its products', a position which it still retains. Coffee represents the largest single item imported into the United States, accounting for some 6 per cent by value of the total imports in 1962.

Coffea canephora Pierre ex Froehner (2n = 22) ROBUSTA COFFEE

Syn. *C. robusta* Linden, *C. laurentii* De Wild., *C. maclaudii* A. Chev., *C. arabica* L. var. *stuhlmannii* Warb., *C. bukobensis* Zimm., *C. welwitschii* Pierre ex De Wild., *C. ugandae* Cramer, *C. kouilouensis* Pierre ex De Wild. (= *C. quillou* Wester—'Kouilou' coffee).

The taxonomy of this variable sp. is very confused and for the time being, until further detailed studies have been made, it is probably wisest to lump the above under the general name *C. canephora*.

USES

As *C. arabica* above. It is more neutral in cup quality and has not such a good aroma as *C. arabica*. As it is cheaper to produce and to buy, the percentage of robusta coffee in coffee blends has steadily increased in Europe since World War II, *e.g.* in France it is now over 70 per cent. It is extensively used in the manufacture of 'instant' coffees, and is often the basis of the blend for this purpose, so that its use is rapidly increasing. Baganda and other Uganda tribes grew robusta coffee for chewing long before their discovery by European explorers; the fruits being parboiled and dried; they also use twin beans from one fruit in the blood-brotherhood ceremony.

ORIGIN AND DISTRIBUTION

It occurs wild in African equatorial forests from the west coast to Uganda, chiefly between 10°N and S of the equator and from sea level to 5,000 ft. Africans in Uganda and elsewhere in equatorial Africa had planted it on a small scale before the arrival of the Europeans and they also collected beans from wild trees. The type of *C. canephora*, named by Pierre in 1897, had been collected by Klaine in the Gabon. In 1895 Laurent collected material in Congo basin; in 1898 he identified the sp. correctly as *C. canephora*; in 1900 De Wildman believing it to be a new sp. named it *C. laurentii.* Planting material was taken from the Congo to Belgium and was then distributed by Linden, the director of the Brussels nursery firm L'Horticole Coloniale, under the name *C. robusta.* In the trade it is still known as robusta coffee to distinguish it from arabica. In 1900 Linden sent 150 plants from Brussels to Java, with further importations being made at later dates. In Java it proved vigorous and was also resistant to *Hemileia vastatrix*, which had decimated *C. arabica* and later *C. liberica*, and so was extensively planted. *C. canephora* had earlier been sent to botanic gardens in Europe; Kew sent it to Singapore and Trinidad in 1898 as an undetermined coffee. Since 1900 robusta coffee has been widely distributed throughout the tropics, where it is grown successfully at lower elevations unsuited to *C. arabica* and where *Hemileia* is a serious disease. It is now the most important sp. in tropical Africa and Asia, but it is grown to a very limited extent in the New World, where *C. arabica* still reigns supreme.

VARIETIES AND CULTIVARS

As pointed out above, the taxonomy of *C. canephora* is in a very confused state; this also applies to its vars and cvs. The sp. is self-sterile and as vars and cvs cross readily their classification and identification is difficult. It is proposed, therefore, to refer to the two main forms as distinguished by Thomas (1947) in Uganda.

1. Upright *Robusta* forms. Haarer (1962) considers that this is the type of the sp. and must thus be considered var. *canephora* (syn. var. *robusta*). They have strong upright growth and unpruned seedlings develop into small trees.

2. Spreading *Nganda* forms. These have been named var. *nganda* by Haarer. These have spreading growth, forming dome-shaped shrubs, as typified by the indigenous coffee cultivated by the Baganda; leaves are usually smaller.

Mention must also be made of the 'Congusta' hybrids (syn. 'Conuga') in Java which are hybrids between *C. canephora* and *C. congensis* (2n = 22). It is possible that the latter may be a form of *C. canephora.* The 'Congustas' are suited to higher, moister areas than normal robustas and are usually grafted on to nematode-resistant robusta stocks; they are resistant to *Hemileia.*

ECOLOGY

As might be expected from the wide distribution of the wild plants at varying altitudes and different environments, *C. canephora* is not so specific in its requirements as *C. arabica* and shows a wider range of adaptability. It occurs naturally in fringing forests along rivers, in forest glades and as an understorey tree in dense forest. It is best suited to lower altitudes and is grown from sea level to 5,000 ft, with an optimum altitude in Java of 1,000–2,500 ft, but it grows very well at 4,000 ft near Lake Victoria in Bukoba and Buganda. It is grown with a rainfall from 40–100 in., but 70 in. is probably optimal, and temperatures of 65–90°F. A dry period is desirable for initiation of flower buds and it flowers in the rains in flushes. It is more tolerant of adverse conditions and poor management than *C. arabica*, and most cvs are resistant to *Hemileia*.

STRUCTURE

A robust, glabrous, evergreen shrub or small tree to 10 m in height. Thomas (1947) has recorded specimens over 100 years old on the Sesse Islands in Lake Victoria.

ROOTS: Shallow rooted, with bulk of feeding roots in top 6 in. of soil; short tap-root.

STEMS: Wild trees in light shade spreading and branching near base; in deep shade usually a well-developed trunk. In cultivation erect and spreading cvs distinguished (see vars above). Dimorphic branching (see *C. arabica* above) with upright orthotropic stems and plagiotropic fruiting branches; latter on dying back are shed naturally, often producing umbrella-shaped plants.

LEAVES: Large, 15–30 × 5–15 cm, often corrugated or undulating; oblong-elliptic, shortly acuminate at apex, rounded or broadly cuneate at base; mid-rib flat above, prominent below; domatia not apparent on upper surface; lateral veins 8–13 pairs; stout petiole, 0·8–2 cm long; interpetiolar stipules broadly triangular, pointed, 5 mm long.

INFLORESCENCES: See *C. arabica* above. Arising from buds in axils of lvs on fruiting branches; usually 6 buds in each axil, but normally only 3–4 (occasionally all 6) develop into inflorescences; subtended by leafy bracts and small bracteoles; peduncle short; typically each inflorescence with 6 fls, but sometimes only 2–4 develop.

FLOWERS: White, very fragrant, subsessile; calyx cup-shaped and very short; corolla tube about 1 cm long, lobes 5–7, usually longer than tube, 1·0–1·5 cm long; stamens same number as corolla lobes, anthers exserted, narrow, 1 cm long, fixed towards base to filaments, 0·5 cm long; style 1·5 cm long, stigma bifid, 0·5 cm long, exserted; ovary inferior with 2 united carpels each with 1 ovule.

Fig. 74. **Coffea canephora:** ROBUSTA COFFEE. **A**, shoot (×½); **B**, portion of inflorescence (×½); **C**, flower in longitudinal section (×3); **D**, bracteoles on inflorescence (×½); **E**, fruiting branch with leaves removed (×½); **F**, fruits (×½); **G**, fruit with part of mesocarp removed (×½); **H**, fruit in longitudinal section (×½); **I**, fruit in transverse section (×½).

A
B
C
D
E
F1
F2
G
H
I

FRUIT: A rounded drupe, 0·8–1·5 cm long, usually 1·2 cm, unripe frs green, ripe frs crimson, later drying black and held on tree until harvested; exocarp thin and comparatively little pulp (mesocarp); stout endocarp enclosing seeds. More fruitful than *C. arabica* and 3–6 frs develop from each inflorescence; and as 3–6 inflorescences develop in each axil it is possible to get 20–70 frs per node borne in tight heavy clusters. Peduncle enlarges during fr. development and fr. borne on short pedicel.

SEEDS: 7–9 mm long, average about 8·5 mm; ellipsoidal, pressed together by grooved flattened surface; outer surface convex. Proportion of cherry to dried bean approximately 4·5 : 1. Approximately 1,500 dried beans per lb.

POLLINATION

It is self-sterile and the highest percentage of self-fertilization recorded in the Congo was 0·24 per cent. Sterility is due to the failure in the formation and growth of the pollen tube. Spreading nganda coffee is said to be somewhat more self-fertile. Pollen grains are light and they have been recorded to be blown 100 yd or more, so it is mainly wind-pollinated. It is visited by insects, but these are thought to play an insignificant part in pollination. It is essential to interplant two or more vegetatively propagated clones to ensure good cropping. Little, if any interclonal incompatibility occurs and cross-pollination between clones gives 30–40 per cent fruit set.

GERMINATION

As *C. arabica* above.

CHEMICAL COMPOSITION

See *C. arabica* above. Caffeine content 2·0–2·5 per cent.

PROPAGATION

Most *C. canephora* is grown from seed, which should be obtained from elite populations of high quality clones. In Indonesia it is usually cleft-grafted onto nematode-resistant stocks.

HUSBANDRY

Similar to *C. arabica* (see above), except for the following:

Erect cvs are usually planted at 10 × 10 ft or 12 × 12 ft; spreading nganda types at 15 × 15 ft. It appears to benefit from selective shading, particularly during the early years. In East Africa it is often interplanted with bananas and in Indonesia it is often grown as a catch-crop or inter-cropped with *Hevea* rubber. Single-stem pruning is unsatisfactory, as there is scanty production of secondary branches and the lower primaries usually die and drop off naturally after producing 2–3 crops, thus producing umbrella-shaped growth. Consequently, erect types are grown on multiple-stems, allowing a succession of orthotropic stems to grow up as required. In the shrubby spreading nganda cvs a modified agobiado training is used in Uganda. Seedlings 3–4 ft high are bent over and the shoots which grow from these are in turn bent over to produce more shoots. Stems are then released to grow naturally, producing a fountain effect, no further pruning being done. The branches are pulled out by the weight of frs and more upright branches are then usually produced in centre. Flowering to harvest takes 9–10 months. As the ripe cherries are held on the tree, and not dropped as in *C. arabica*, one harvesting only is made, and the cherries may be allowed to dry on the tree. The cherries are usually sun-dried and are often sold to the curing works in this form. They are then hulled to remove the skin, dried pulp and parchment, and polished to remove silver-skin. A better product, however, is produced by the wet pulping method as described for *C. arabica*. Robusta does not require so much care and attention as arabica and yields are usually greater, as it is more fruitful. Good average yields are 900–1,120 lb of clean coffee per acre. Higher yields have been obtained from selected trees in Uganda and have on occasions exceeded 1 ton per acre per annum. Yields in Indonesia decreased from 600 to under 300 lb per acre in recent years.

DISEASES AND PESTS

These are far less serious in the hardier *C. canephora* than in *C. arabica*. Most cvs and the 'Congusta' hybrids are resistant to *Hemileia vastatrix*, nor is robusta attacked by *Antestiopsis* spp. The root rot, *Armillaria mellea* (Vahl ex Fr.) Kummer, may cause damage; the berry borer, *Hypothenemus hampei* (Ferr.), can cause substantial losses. In India the shot-hole borer, *Xyleborus morstatti* Hag., is a major pest.

IMPROVEMENT

As *C. canephora* is cross-pollinated and its introduction into many countries was on a broader base, often with introductions under different specific names, it exhibits much more variability than *C. arabica*, but with fewer mutations. It is highly self-incompatible, but clonal differences have been noted. Interclonal sterility is rare. Inbreeding in clones showing some degree of self-compatibility reduces yields. Nevertheless, if inbred lines can be produced, crossing of these might produce progeny with marked hybrid vigour and these could then be propagated vegetatively. It is necessary to work out stock/scion compatibility. In Java, 'R.124.01' which is resistant to nematodes, has been found to be one of the best stocks. It would seem that the taxonomy and cytogenetics of *C. canephora* need much more study before substantial progress can be made.

A considerable amount of work on selection and breeding was carried out by the Dutch in Java. Seedlings from selected mother trees were used by Cramer in 1907, but results were not very satisfactory, as the progeny were very heterozygous. Controlled artificial pollinations were first made in 1912 and selected families from these crosses gave 25–50 per cent higher yields than unselected seedling populations. With the development of cleft-grafting in 1916, the best trees were propagated as clones and could be tested. The best clones were then crossed and new mother trees selected from the F_1's and clones from these were tested. Seeds of the best crosses were then increased by bi-clonal seed gardens. As the work progressed mother trees were selected which had a productivity index of at least 3 times the yield of the average of the plantings around them. In addition the following were also taken into consideration: regularity of bearing, bean size, out-turn, habit of tree, vigour and resistance to pests and diseases. After more than 30 years, families have been obtained which gave 50–100 per cent more yield than the original material. It was also found possible to increase bean length by 2·4 mm in 3 generations. Certain selected combinations of clones are now used to produce legitimate seed in Indonesia.

Thomas (1947) in Uganda made selections of old, heavy-yielding trees of 'Nganda' type in native gardens and compared these with selected erect forms. Planted in 1935, the average yield per acre per annum for the period 1937–1946 was 1,390 lb for the erect forms and 1,300 lb for the spreading forms. The erect bushes gave the heaviest yields in the early years and reached peak production 6 years after planting, while the yield of the 'Nganda' selections was still increasing after 10 years. Progenies of one selection gave 1,780 lb per annum over a period of 7 years, and this and others have given over 1 ton per acre in individual years.

Belgian workers in the Congo, using the Java clone 'S.A.34' as their basis for comparison, obtained selections which gave 25–30 per cent increase in 7 years' work. Their two outstanding clones are 'L.147' which is very vigorous and 'L.125' which has beans 11 mm long.

PRODUCTION AND TRADE

Figures for robusta are included under *C. arabica* above. In 1963 the combined export quotas of robusta producers under the International Coffee Agreement amounted to only 617,000 tons. Due to increasing popularity and its use in the manufacture of 'instant' coffees, the demand is likely to exceed this figure. Any further increase would be at the expense of arabica coffee, which is now over-produced. The bulk of robusta coffee is produced in Africa, with the Ivory Coast, Angola, Uganda, Congo and Madagascar exporting the greatest amounts; in Asia Indonesia is an important producer.

Coffea liberica Bull ex Hiern (2n = 22) LIBERICA COFFEE

Syn. *C. abeokutae* Cramer, *C. klainei* Pierre ex De Wild.; *C. excelsa* A. Chev. and *C. dewevrei* De Wild. & Th. Dur. are sometimes included in *C. liberica.*

USES

It is used as a filler with other coffee, as the liquoring quality is not usually of a high standard. It has a bitter flavour which seems to be appreciated in Malaya and the East.

ORIGIN AND DISTRIBUTION

A lowland sp., indigenous near Monrovia in Liberia, but spread by cultivation here and there in West Africa from early times. Specimens were collected by Afzelius in Sierra Leone in 1792; he did not describe it, but suggested it as an article for export. Sir Joseph Hooker, Director of Royal Botanic Gardens, Kew, drew attention to it in 1872. Bull, a Chelsea nurseryman, had planting material of it in 1874. Planting material reached Ceylon and Java in 1873 and Malaya in 1875. In Java seeds from the Buitenzorg Gardens fetched almost their weight in gold and it was used to replace *C. arabica*, which was being killed off by *Hemileia* at the lower altitudes. Alas, *C. liberica* soon became susceptible and was replaced by a natural cross of *C. liberica* × *C. arabica*, which was first discovered in 1885, and later by *C. canephora* introduced in 1900. *C. liberica* was widely distributed throughout the tropics from Kew and elsewhere, *e.g.* Trinidad 1875. It has nowhere become really important and now provides less than 1 per cent of the world's coffee.

ECOLOGY

It is a tree of hot, wet, lowland forests and requires a heavy rainfall and high temperatures and appears to benefit from light shade. It can be grown on a variety of soils, from peat to clays, and can be grown on poorer soils and withstand more neglect than the preceding spp.

STRUCTURE

A robust evergreen shrub or tree, 5–17 m. Dimorphic branches as in *C. arabica.* Lvs large, leathery, dark green, glossy, elliptic-oblong to elliptic-obovate, 15–30 × 5–15 cm; tip shortly acuminate, base cuneate, lateral veins 7–10 pairs with pronounced pits or domatia in axils below appearing as swelling on upper surface; petiole stout, 1–2 cm long, stipules broadly ovate, 3–4 mm long. Inflorescences axillary, usually 1–3 in each axil, each with 1–4 fls. Fls white, star-like, fragrant; calyx small, cup-like; corolla tube 1·5–2 cm long; lobes 6–9, spreading, 1·5–2 cm long; stamens same number as corolla lobes; filaments 6–7 mm, attached to lip of tube; anthers narrow about 1 cm long, dark-brown on undersurface after anthesis; style 2·5 cm long with exserted bifid stigma 0·5 cm long. Frs mature 1 year after flowering, oval, 2–3 × 2–2·5 cm; streaked red when ripe, drying black; skin tough. Seeds about 1·3 cm long; proportion of cherry to clean coffee 10 : 1; 800 seeds per lb.

Unlike *C. arabica* and *C. canephora* which flower in flushes, fls open at irregular intervals. Frs at various stages of ripeness found on tree at one time and are retained when ripe. Fls and frs largest of cultivated coffees.

POLLINATION

Self-sterile. Pollen light and spread by wind and insects.

GERMINATION

As *C. arabica*.

CHEMICAL COMPOSITION

See *C. arabica* above. Caffeine content 1·4–1·6 per cent.

HUSBANDRY

It is usually grown from seeds raised in nurseries. The planting distance is usually 12 × 12 ft. Although light shade is beneficial, it is often grown without any; it is grown under coconuts in the Philippines. Trees are grown on single or multiple stem, but it is often unpruned. In Malaya it is grown on a single stem, topped at a height of 5–6 ft and excessive secondary branches are thinned from time to time. Ripe fruits are usually allowed to dry on the tree before harvesting and they are processed by the dry method. Little reliable information is available on yields, but in Malaya it gives 6–8 cwt clean coffee per acre.

DISEASES AND PESTS

A hardy sp., but it is susceptible to *Hemileia vastatrix*. In Malaya it is attacked by the clear-winged hawk moth, *Cephonodes hylas* L. In Guyana the most serious disease is *Sclerotium coffeicolum* Stah.

IMPROVEMENT

The crop has received little attention. It has been used in hybridization and for rootstocks. The 'Kawisarie' and 'Kalimas' hybrids arose by natural crossing in Java with *C. arabica*. These hybrids are vigorous, rust-resistant and self-fertile but produce many defective spongy seeds. They were cleft-grafted onto nematode resistant robusta stocks as they do not come true from seed. They are tetraploid (2n = 44). Although *C. liberica* is normally diploid (2n = 22) a tetraploid form has been recorded in Java.

PRODUCTION AND TRADE

Liberica coffee is the least important of the cultivated species, providing less than 1 per cent of the world's coffee. It is produced in Liberia, Fernando Po, the Philippines, Malaya, Surinam and Guyana. Small quantities are exported, mainly to Scandinavian countries.

Fig. 75. **Coffea liberica :** LIBERICA COFFEE. **A,** plagiotropic branch with buds (×$\frac{1}{3}$); **B,** leaf, upper-surface (×$\frac{2}{3}$); **C,** flower in longitudinal section (×1); **D,** leaf axil with fruit (×$\frac{1}{3}$); **E,** developing fruit (×$\frac{1}{2}$); **F,** fruit (×$\frac{1}{3}$); **G,** fruit with part of mesocarp removed (×$\frac{1}{3}$); **H,** fruit in longitudinal section (×$\frac{1}{2}$); **I,** fruit in transverse section (×$\frac{1}{2}$); **J,** portion of inflorescence (×$\frac{1}{2}$).

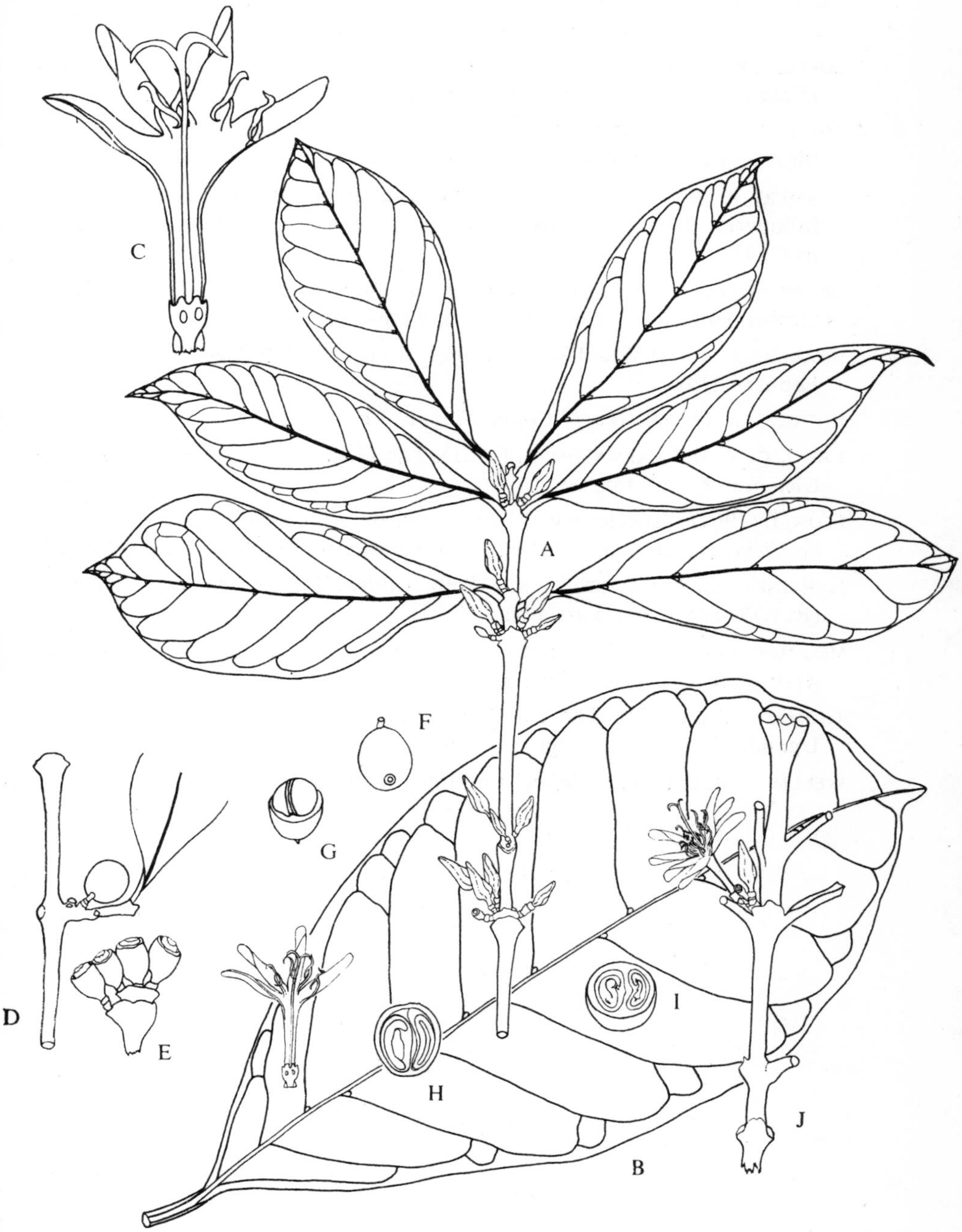
C
A
F
G
D
E
I
H
J
B

REFERENCES

CARVALHO, A. (1958). Recent Advances in our Knowledge of Coffee Trees. 2. Genetics. *Coffee and Tea Industries*, **81**, 30–36.

COFFEE RESEARCH SERVICES, KENYA (1962). *An Atlas of Coffee Pests and Diseases*. Nairobi: Coffee Board of Kenya.

COSTE, R. (1954, 1959). *Cafetos y Cafés en el Mundo*. 3 vols. Paris: Maisonneuve & Larose.

CRAMER, P. J. S. (1957). Review of Literature of Coffee Research in Indonesia. Turrialba, Costa Rica. *Inter-Amer. Inst. Agric. Res., Misc. Bull.*, **15**.

DE MONTOYA, G. P. (1963). *Coffee, Bibliographical List No. 1* (Rev.). Supplement No. 1. Costa Rica: Turrialba.

HAARER, A. E. (1962). *Modern Coffee Production*, 2nd edition. London: Leonard Hill.

KRUG, C. A. (1959). *World Coffee Survey*. Rome. F.A.O.

KRUG, C. A. and CARVALHO, A. (1951). The Genetics of Coffee. *Advances in Genetics*, **4**, 127–158.

LEÓN, J. (1962). Especies y Cultivares (variedades) de Café. *Inst. Inter-Amer. de Ciências Agricolas, Materiales de Enseñanza de Café y Cacao*, No. 23.

MARTINEZ, A. and JAMES, C. N. (1960). *Coffee. Bibliographical List No. 1.* (Rev.) Costa Rica: Turrialba.

MES, M. G. (1957). Studies on the Flowering of *Coffea arabica* L. *Portugaleae Acta Biologica*, **4**, 328–356; **5**, 25–44.

THOMAS, A. S. (1947). The cultivation and selection of Robusta coffee in Uganda. *Emp. J. exp. Agric.*, **15**, 65–81.

WELLMAN, F. L. (1961). *Coffee*. London: Leonard Hill.

RUTACEAE

About 130 genera and 1,500 spp., mainly trees and shrubs, with the largest number in the warmer regions of the world. Usually aromatic with resinous tissues. Lvs simple or compound, alternate or opposite, exstipulate, usually with pellucid oil glands. Fls usually regular and hermaphrodite; sepals 3–5; petals usually 3–5, free; stamens generally as many or twice number of petals, inserted at base round annular disc; ovary superior, usually syncarpous, and often 3–5-locular, styles usually connate. Frs various, usually a capsule or berry.

USEFUL PRODUCTS

EDIBLE FRUITS: The most important are *Citrus* spp. (q.v.).

Aegle marmelos (L.) Corr., Indian bael or bel fruit, is a native of India, cultivated throughout South-east Asia and East Indian Archipelago and introduced into some tropical countries. A thorny, deciduous tree with dimorphic branches, trifoliate lvs, and globose frs, 5–17·5 cm in diameter, with hard, smooth greyish-yellow pericarp and softer flesh which becomes reddish and hard in drying. The flesh is eaten fresh or dried and sherbet is made from it.

Casimiroa edulis La Llave & Lex., white sapote, is a native of highlands of Mexico and Central America and introduced into subtropical areas. It is a spreading tree with palmate lvs and yellowish-green, apple-like frs; soft, yellow, sweet, custard-like pulp is eaten.

Clausena dentata (Willd.) Roem. (syn. *C. willdenowii* Wight & Arn.) of India, has frs the size of cherries tasting like blackcurrants.

C. lansium (Lour.) Skeels, native of south China and now grown in some tropical countries, is a small tree with pinnate lvs and clusters of whitish frs, about 2·5 cm in diameter, with aromatic, slightly acid, edible pulp.

Feronia limonia (L.) Swing., Indian wood apple, native of India and Ceylon and planted in tropical Asia, is a small, deciduous, spiny tree, with pinnate lvs, reddish fls, and globose, whitish frs, 5–6 cm in diameter, with hard woody pericarp and aromatic pulp, which is made into sherbet.

Fortunella japonica (Thunb.) Swing. (syn. *Citrus japonica* Thunb.), round kumquat, and *F. margarita* (Lour.) Swing. (syn. *Citrus margarita*

Lour.), oval kumquat, are widely cultivated in China and Japan. Evergreen, thorny shrubs to 3–4 m in height with simple lvs. They differ from *Citrus* spp. in having 3–5 loculi (usually), each with 2 ovules and very small fruits, about 2·5 cm in diameter, with acid pulp and a sweet edible more or less pulpy skin. The whole fruits may be eaten or may be preserved in syrup or candied. They have a degree of resistance to winter cold and a profound winter dormancy which permits them to pass through warm weather without starting growth. *F. margarita* was introduced into England by Fortune in 1846. *F. polyandra* (Ridl.) Tan. is grown in Malaya. *Fortunella* spp. cross readily and hybrids have been produced with *Citrus aurantifolia* to give the limequats and with *C. reticulata* to give orangequats, as well as trigeneric hybrids such as (*Poncirus trifoliata* × *Citrus sinensis*) × *Fortunella* sp.

OTHER USEFUL SPECIES:

Citropsis spp., which grow wild in tropical Africa, are closely related to *Citrus*; they have pinnate or trifoliate lvs. They can be used as a rootstock for *Citrus*, particularly *C. gilletiana* Swing. & Kell., which is used for this purpose in the Congo and is resistant to foot-rot.

Murraya koenigii (L.) Spreng., curry-leaf tree of India, is grown for its lvs which are used for flavouring curries and chutneys. *M. paniculata* (L.) Jack, orange jasmine, native of South-east Asia, an evergreen shrub with glossy, pinnate lvs and fragrant white fls, is grown throughout the tropics as an ornamental and makes an excellent hedge; it can be propagated by cuttings or seed.

Poncirus trifoliata (L.) Raf., trifoliate orange, is a native of China and has been grown there for thousands of years as an ornamental; it was an early introduction to Japan and has been widely introduced elsewhere. A small thorny tree with trifoliate, deciduous lvs and globose frs, 3–5 cm in diameter; cold-resistant. It is used in Japan as a rootstock for Satsuma orange. Bigeneric hybrids have been obtained between it and most cultivated species of *Citrus* and *Fortunella*. Those with *C. sinensis* are citranges, which have proved vigorous, hardy, disease-resistant rootstocks for *Citrus*. In crosses with *Citrus* spp., the trifoliate lf. is dominant, so by using its pollen it is possible to identify plants from sexual embryos in polyembryonic seeds.

Ruta graveolens L., rue, a native of the Mediterranean region, has been introduced into many tropical countries and is used medicinally.

Triphasia trifolia (Burm. f.) P. Wils., limeberry, probably a native of South-east Asia, has now been spread widely in tropical and subtropical regions. A thorny evergreen shrub with trifoliate lvs, trimerous fls and reddish frs, about 1 cm in diameter. The frs are sometimes made into preserves. It makes an excellent hedge.

Citrus L. (x = 9) CITRUS

Subfamily—Aurantioideae; Tribe—Citreae; Subtribe—Citrinae.

The sub-tribe includes 6 genera which are indigenous in an area extending from north-eastern India and northern Central China to eastern Australia and New Caledonia, which are characterized by having fruits with slender-stalked pulp vesicles filling the space in fruit segments not occupied by seeds and by having at least 4 times as many stamens as petals. All have simple evergreen leaves, except *Poncirus* which has trifoliate, deciduous lvs. The genera are: *Citrus* with 16 spp. (q.v.); *Clymenia polyandra* (Tan.) Swing. (monotypic) in New Ireland; *Eremocitrus glauca* (Lindl.) Swing. (monotypic), Australian desert lime, a xerophytic tree which can stand high concentrations of salt and which grows wild in north-eastern Australia; *Fortunella* with 4 spp., kumquats (see above); *Microcitrus*, Australian wild limes, semixerophytes, with 5 spp. in eastern Australia and 1 sp. in New Guinea; and *Poncirus trifoliata* (L.) Raf. (monotypic), trifoliate orange (see above). The genera may be grafted on one another; intergeneric hybrids are also produced. This may prove important in producing root-stocks and hybrids adapted to special climatic and soil conditions.

The taxonomy of *Citrus* spp. is confused and is complicated by the ease with which they hybridize, by polyembryony, by mutations and the spontaneous production of autotetraploid forms with thick leaves and rinds. Writing on *Citrus* cvs Dodds (1963) states that 'with the exception of the limes, these fruits represent on the whole natural groups of horticultural varieties and to apply the concept of formal taxonomy is more confusing than helpful'. Tanaka has described many new species of *Citrus*, some of which are known to be hybrids. In this account I follow Swingle's classification given in Webber and Batchelor (1948). Two subgenera are recognized:

1. PAPEDA: 6 spp. All have numerous droplets of acrid oil in the pulp-vesicles which render them inedible, and have long petioles very broadly winged; stamens usually free. They occur wild as follows:

C. hystrix DC. and *C. macroptera* Montr. have a wide distribution from South-east Asia to New Caledonia; *C. celebica* Koord. and *C. micrantha* Wester are confined to the Philippines; *C. latipes* (Swing.) Tan. in India and *C. ichangensis* Swing. in China. Hybrids between most of the above and cvs in the next section are known.

2. EUCITRUS: 10 spp.; includes all the commonly cultivated spp. of *Citrus* (see below), all of which have pulp-vesicles filled with acid, subacid or sweet juice, free, or nearly free from oil droplets which is never acrid; wingless petioles or when winged usually narrow; stamens cohering in bundles. In addition to the well-known spp. this subgenus includes *C. indica* Tan. which occurs wild in India and *C. tachibana* (Mak.) Tan., the tachibana orange of China and Japan.

Commonly Cultivated Citrus Species

These are first considered together, but further details of the individual spp. are given below.

USES

The fruits of species with sweet juice are eaten in a fresh state or the segments may be canned. The juice may be extracted and used in squashes and cordials or for flavouring. The waste pulp is used for cattle food. The peel is a source of essential oils, used in flavouring and perfumery, and it also provides pectin. Essential oils may also be obtained from flowers and leaves. Citric acid may be manufactured from *Citrus* spp.

ORIGIN AND DISTRIBUTION

The cultivated spp. are believed to be native of tropical and subtropical regions of South-east Asia, where they have been cultivated since remote times. It seems probable that they originated in the drier monsoon areas rather than the tropical rain forests, as the plants exhibit periods of dormancy and the water-storing hairs (pulp vesicles) would help to nourish the seeds in such an environment; furthermore, *Citrus* do not grow well in the very humid tropics. Man took into cultivations those with edible juice. Natural hybridization then occurred between cvs and spp. producing an array of complex hybrids. Man selected for his use those with the greatest quantity of juice-storing pulp and the most desired flavours. Some of the spp. have never been found in a truly wild state; others have become naturalized in some areas. The cultivated *Citrus* have now spread throughout the tropics and subtropics. Most of the commercial production is now in subtropical regions with a Mediterranean climate, much of the cultivation taking place during the last 100 years.

STRUCTURE

Small evergreen shrubs or trees up to 10 m or more in height with resinous glandular tissue in all parts.

ROOTS: Most *Citrus* spp., when not transplanted, usually have a single tap-root and lateral roots growing horizontally, providing a surface mat of feeding roots with weak development of root hairs. Roots show definite cycles of growth, alternating with that of stems.

STEMS: Most spp. have a single trunk with main branches arising about 1 m above the ground. Wood very hard. Thorns occur in some spp. and are adjacent to buds in lf. axils; in some cvs thorns only develop on rapidly growing shoots. New growth of shoots occurs in definite cycles, usually 3 per year, alternating with periods of dormancy. Buds present on main trunk which may remain dormant for many years or may grow out as water-shoots, particularly when upper part of tree damaged.

LEAVES: Unifoliate, representing single terminal leaflet of pinnate lf., with articulation between leaflet and petiole (except in *C. medica*); wings present on petiole in some spp; stomata confined to lower surface; oil glands present in palisade layer. Lvs usually live for a year or more.

FLOWERS: Singly or in small clusters in axils of leaves on new growth; in subtropics produced in greatest numbers in spring; 2·5–4 cm in diameter, sweet scented; usually hermaphrodite; but male fls with aborted ovaries found in *C. medica* and *C. limon*. Calyx persistent, cup-shaped, with 3–5 projections. Corolla of 4–8 petals, mostly 5, usually white; oil-glands present. Stamens 20–40 in groups; filaments white and those of each group partially united at base, representing a single whorl which has multiplied by splitting; anthers 4-celled, yellow. Nectiferous disc just inside stamens and is surmounted by ovary, which is superior with 8–15 united carpels, each with 2 rows of ovules, usually with 4–6 ovules in axile placentation; short deciduous style with capitate stigma, which secretes sweet fluid when receptive; stylar canals of same number as loculi open on stigma and run down style to ovary and provide path for developing pollen tubes.

FRUITS: A berry known as an hesperidium. Thick leathery peel protects frs from injury and keeps pulp in a sanitary condition and has epidermis with thick cuticle and stomata. Exocarp or flavedo is coloured outer portion of peel consisting of thin-walled parenchyma, rich in chloroplasts and photosynthetic in young fr.; oil glands embedded in tissues in which oil kept under pressure by turgor of surrounding cells; during ripening chlorophyll breaks down, xanthophyll and carotene become dominant and colour changes from green to yellow or orange; colour usually better in subtropics, as often greenish in tropics. Mesocarp or albedo is inner colourless section of peel composed of elongated branched cells with large intercellular spaces; rich in sugars, pectin, vitamin C and glucosides, type of which varies with spp. Endocarp or rag consists of thin transparent membrane surrounding carpels or segments; multicellular hairs grow out from inner carpel walls and fill with juice as frs develop to give pulp vesicles, which fill loculi cavities except for space occupied by seeds; endocarp thus provides edible portion of fr.; juice contains sugars and acids (mainly citric acid); during ripening amount of acids decreases and there is an increase in sugars and aromatic substances, the proportion of acids and sugars varying with spp. In Navel oranges there is partial development of a second whorl of carpels above first which produces umbilicus effect. In some cvs, e.g. 'Washington Navel' orange, parthenocarpic seedless frs are produced for which pollination is not necessary as this cv. has no viable pollen. In most other cvs, which sometimes produce parthenocarpic frs, pollination is necessary for fr. development. Frs take 7–14 months from pollination to maturity. Only a fairly small percentage of fls produce mature frs; fls and young frs may fall from unopened-bud stage to 10–12 weeks after opening of fls.

SEEDS: Number of seeds per fr. and per locule varies greatly with cvs, some being completely seedless (see parthenocarpy above). Endosperm used up as seed develops. With the exception of *C. grandis*, most cvs have polyembryonic seeds, which may consist of sexual or gametic embryo from fertilized egg cell and up to 6 or more asexual or nucellar embryos, which develop from somatic cells of nucellus and therefore have the same genetic constitution as female parent. Gametic embryo often lacks vigour and may be completely suppressed by nucellar embryos. Pollination is usually necessary for the development of nucellar embryos.

POLLINATION

Most *Citrus* cvs are self- and cross-compatible. Some spp. *e.g. C. paradisi*, are somewhat protandrous, which tends to increase self-pollination; in others stamens and stigma mature at same time; the stigma is receptive for 6–8 days. The flowers are entomophilous and are visited by bees which are attracted by white corolla, strong perfume, abundant nectar and sticky pollen; thrips also visit the flowers in great numbers. Fruits usually set on bagged branches of most cvs, showing that they are self-compatible. No viable pollen is produced in some cvs, e.g. 'Washington Navel' orange. In artificial pollination anthers are removed in bud and the flowers are bagged.

GERMINATION

Germination is hypogeal. The radicle forms a thick, fleshy tap-root and secondary roots appear when this is 8–10 cm long. The plumule, which is rudimentary in the resting seed, grows to give the epicotyl and first 2 leaves are formed above surface of soil.

ECOLOGY

Citrus spp. are cultivated from 45°N to 35°S. Most of the commercial crop is grown in subtropical countries between sea level and 2,000 ft; on the equator they do not do well above 6,000 ft. Most cvs will stand light frost for short periods only, susceptibility depending on the state of dormancy of the trees. The degree of hardiness, beginning with the tenderest spp. is as follows: citron, lime, lemon, grapefruit, sweet orange, sour orange, mandarin, kumquat, trifoliate orange. Growth activity is greatly reduced at temperatures below 55°F. They can stand relatively high temperatures of over 100°F. An average annual rainfall of at least 35 in. is required if they are to be grown without irrigation. *Citrus* spp. are unsuited to the very humid tropics and high atmospheric humidity increases the incidence of pests and diseases. Mandarins will tolerate wetter conditions than the other species. *Citrus* spp. are intolerant of high winds and windbreaks should be provided where necessary. No photoperiodic responses have been recorded. They are grown over a wide range of soil types, but a light loamy soil of good fertility is preferred. They will tolerate a pH of 5–8, but are sensitive to certain salts such as boron, sodium carbonate and sodium chloride, and also to waterlogging.

KEY TO CULTIVATED SPECIES OF CITRUS

A. Petioles wingless, without apparent articulation at top; fr. large with very thick peel and scant acid juice . . *C. medica*

AA. Petioles clearly articulate with lamina, usually winged or margined.

B. Petioles narrowly margined. Fl. buds tinged reddish; stamens usually more than 5 times number of petals . . . . *C. limon*

BB. Petioles with broad or narrow wings; fl. buds white; stamens usually 4–5 times number of petals.

C. Fls small, usually 2·5 cm or less across; fr. small, usually 4–6 cm in diameter; juicy, very sour . . . . *C. aurantifolia*

CC. Fls larger; fr. usually much larger, juice sweet, mild or sour.

D. Lvs rather broadly winged.

E. Lvs subcordate, twigs and undersurface of lvs sparsely pubescent; fr. very large, usually 10–20 cm in diameter; pulp vesicles large, separating easily; seed monoembryonic . . . . . *C. grandis*

EE. Lvs not subcordate, glabrous beneath; pulp vesicles coherent; seeds polyembryonic.

F. Fr. large, usually 9–13 cm in diameter; pulp vesicles rather large . . . . *C. paradisi*

FF. Fr. medium sized, 4–6 cm in diameter with hollow core and small pulp vesicles; sour and bitter . . . . *C. aurantium*

DD. Lvs narrowly winged.

E. Lvs usually small and narrow, 4–6 cm long; fr. with loose peel; embryos green . . . . *C. reticulata*

EE. Lvs larger and broader; fr. with adherent peel; embryos white . . . . *C. sinensis*

NOTE

Colour of frs, used in many keys, is unreliable in the tropics, as full colour of ripe frs seldom develops in the tropics and many of them remain green. An attempt has been made to use vegetative characters as far as possible, as it is desirable to identify spp. when they are not fruiting. With mature fruits a key should not be necessary. However, vegetative characters, such as size of lf. and wing on petiole, are very variable, not only between cvs of same sp., but even on same tree, particularly in juvenile stage and on water-shoots. No attempt has been made to include interspecific hybrids.

Citrus aurantifolia (Christm.) Swing. (2n = 18) LIME

Limes are extensively used in the tropics for fresh juice and as flavouring for many foods. Important commercial products are limeade, lime-juice cordial and marmalade; lime oil is prepared from the peel, sometimes by ecuelling; citric acid is made from the fruit. Little fresh fruit is imported by temperate countries, as it does not transport well, and where its place is taken by lemons.

Swingle states that the lime is indigenous to the East Indian Archipelago and that it was taken from there to the mainland, but it is also said to be wild in northern India and may have originated there. It was taken to Europe about the 13th century and by the Spaniards to the New World early in their colonization. It has since spread throughout the tropics, where it is the most commonly cultivated species of the acid *Citrus*. There is extensive commercial production in Mexico, West Indies (particularly Dominica) and Egypt.

Lime is a very distinct species, not closely related to other *Citrus* spp., although it is known to hybridize with them.

A small, much-branched tree up to 5 m in height, usually heavily armed with short, sharp spines. Lvs small, 4–8 × 2–5 cm, ovate-elliptic, margin crenulate; petiole narrowly winged. Inflorescences axillary, 1–7 fld., produced over an extended period. Fls small; petals 4–5, 8–12 × 2·5–4 mm, white; stamens 20–25; ovary with 9–12 loculi. Frs small, oval or globose, 3·5–6 cm in diameter, often with apical papillae; peel thin, adherent, greenish-yellow when ripe; pulp greenish, very acid; seeds small, oval; polyembryonic; cotyledons white.

'Mexican' (syn. 'West Indian') is cv. most commonly grown in the tropics; frs small; very susceptible to wither-tip (*Gloeosporium limetticolum*), but immune to citrus scab (*Elsinoe fawcettii*). Larger fruited 'Tahiti' seedless lime has defective pollen and egg cells and is triploid (2n = 27); almost immune to wither-tip, but susceptible to citrus canker. Mandarin lime and sweet lime are interspecific hybrids probably with *C. reticulata* and *C. medica* respectively.

Limes are more tender than most other *Citrus* spp.; they can survive on poor soil, often under neglected conditions.

Citrus aurantium L. (2n = 18) SOUR OR SEVILLE ORANGE

Sour oranges are too bitter to use as fresh fruits, but are highly regarded for making marmalade; they are used as flavouring and in liqueurs (Curacao). Leaves, flowers and fruits yield volatile bigarade oil used in perfumery. Bergamot fruits, sometimes considered a subsp.—*bergamia* (Risso & Poit.) Wight & Arn., yield bergamot oil. The peel is used medicinally. Sour orange is extensively used as a rootstock for lemon, sweet orange and grapefruit; it is resistant to gummosis, but in combination with sweet orange and grapefruit it is susceptible to tristeza.

This species originated in south-eastern Asia, probably in Cochin China, and then spread eastwards. It was introduced into Europe in the 11th century, five centuries before the introduction of the sweet orange. It was

---

Fig. 76. **Citrus aurantifolia : LIME. A, flowering branch (×⅓); B, flower in longitudinal section (×2); C, stamen (×3); D, fruit in transverse section (×½).**

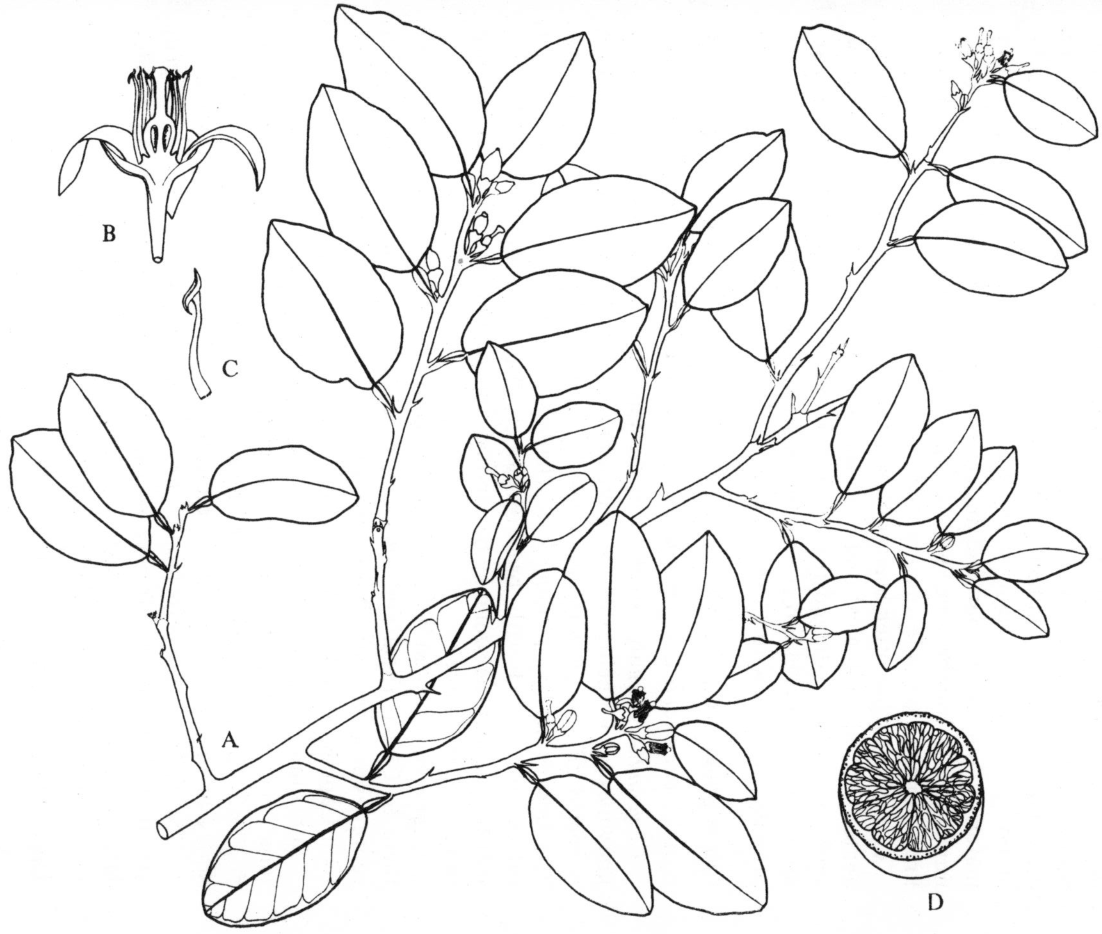
A
B
C
D

one of the first *Citrus* to be taken to the New World and has since spread throughout the tropics and subtropics; it has become naturalized in some areas.

Tree to 10 m high; thorns slender. Lvs medium-sized; petiole 2–3 cm long, rather broadly winged; lamina ovate or elliptic, up to 10 × 7 cm, sometimes crenulate. Fls axillary, large, white, very fragrant, with 5–12 per cent staminate only; stamens 20–25; ovary with 10–12 loculi. Frs subglobose; peel thick, rough, strongly aromatic, often bright orange-red at maturity; pulp very sour and bitter; central core usually hollow; seeds numerous, polyembryonic, with a high proportion of nucellar embryos. Predominant glucoside aurantamarin.

C. aurantium has often been confused with *C. sinensis*, but morphological, anatomical, physiological and chemical differences separate them and the smell and taste are diagnostic. It is hardier than the sweet orange. The var. *myrtifolia* Ker Gawl. arose as a bud mutation and bittersweet oranges are probably hybrids with *C. sinensis*.

Citrus grandis (L.) Osbeck (2n = 18) PUMMELO, SHADDOCK

Syn. *C. decumana* L; *C. maxima* (Burm.) Merr.

C. grandis is highly esteemed in the East as a dessert fruit, where it is known as the Pomelo; the strong membrane (endocarp) round the segments is peeled off and the large pulp vesicles, which fall apart easily, are eaten. The best fruits are produced in Thailand, where the crop is grown on ridges surrounded with brackish water. The fruit may be made into marmalade and the peel may be candied. It is of no commercial importance in temperate countries.

C. grandis is probably a native of Thailand and Malaysia and spread from there to China, India and Persia. It reached Europe in the 12th or 13th centuries as a curiosity. It was introduced into Barbados in the 17th century by a Captain Shaddock, commander of an East India ship.

Spreading spiny tree, 5–15 m in height; young branches pubescent. Lvs large; petiole broadly winged; lamina ovate to elliptic, subcordate, 5–20 × 2–12 cm, undersurface of midrib often pubescent. Fls large, borne singly or in clusters, 3–7 cm in diameter; petals cream coloured; stamens 20–25; ovary with 11–16 loculi. Frs very large, globose or pear-shaped, 10–30 cm in diameter, often yellowish when ripe; peel thick; pulp vesicles large, pale yellow or pink, with sweetish juice. Seeds large, ridged, yellowish, monoembryonic. Predominant glucoside naringin.

Citrus limon (L.) Burm. f. (2n = 18) LEMON

Lemons are not eaten as fresh fruits, but are widely used in the preparation of lemonade, squashes and for culinary and confectionary purposes as

Fig. 77. **Citrus aurantium** : SOUR ORANGE. **A,** leafy shoot (×½); **B,** base of leaf (×½); **C,** flower in longitudinal section (×1); **D,** fruit (×½); **E,** fruit in transverse section (×½).

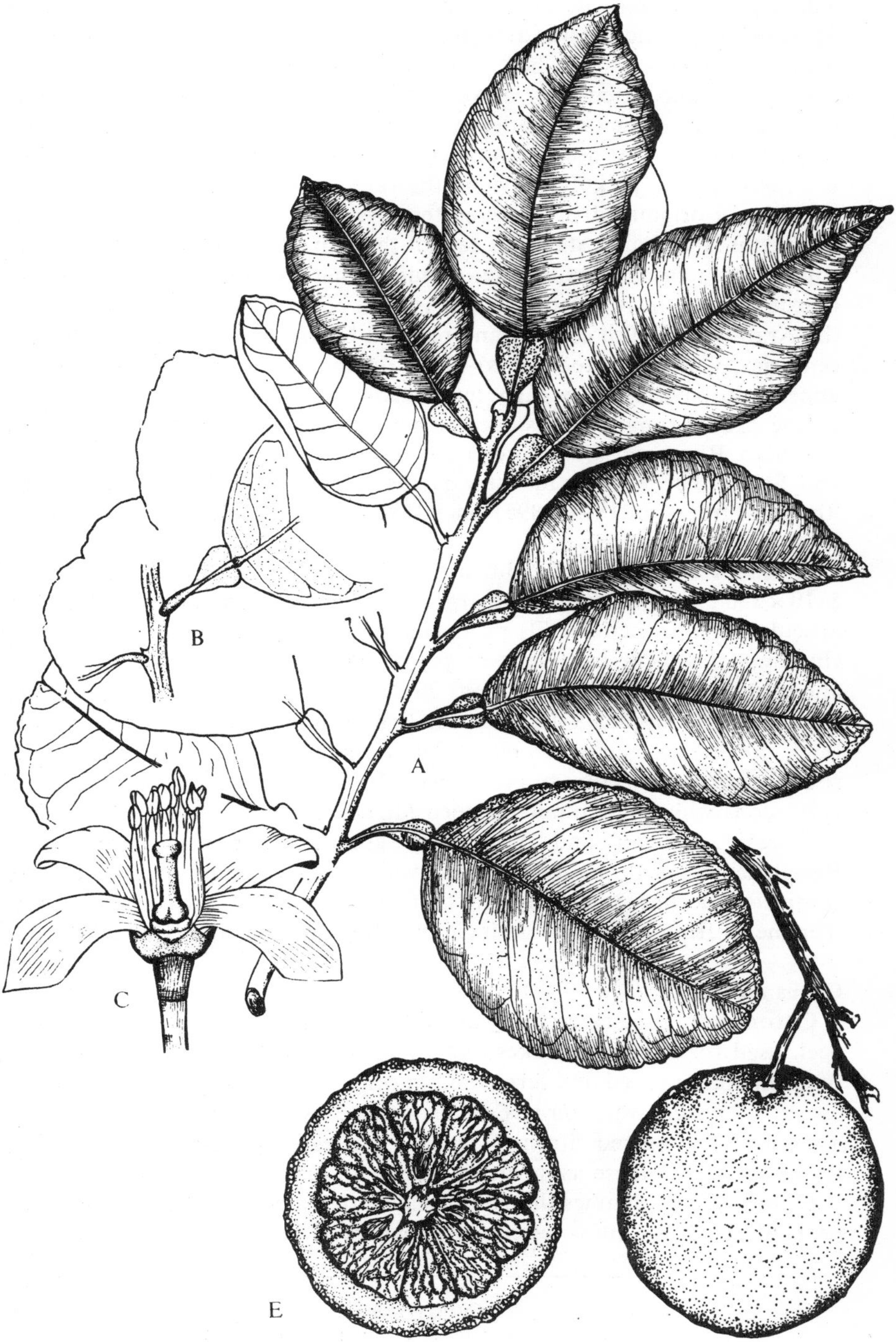
A
B
C
D
E

a flavouring and garnish. They are used in cosmetics and for the production of lemon oil, citric acid and pectin. Candied peel is made from the rind, which is also one of the best sources of Vitamin P. Rough lemons (see below) are used extensively as root-stocks for sweet oranges, grapefruits and mandarins. Lemons are less important than limes in the tropics, but they are extensively used in temperate countries. They may be kept for 6–8 months after picking provided they are cured in dry weather.

C. limon originated in south-eastern Asia, but the exact region is not known; it is variously stated as east of the Himalayas in north Burma and south China or in north-western India, whereas Singh *et al.* (1963) say that it was not introduced into India 'until recent years'. It was known to the Arabs in the 10th century and reached Europe during the 12th or 13th centuries. Columbus took it to Haiti on his second voyage in 1493. It appears to have been an early introduction into East and Central Africa by the Arabs; rough lemons became naturalized along the Mazoe River in Southern Rhodesia. Commercial cultivation was first developed in Italy and Spain. The Florida industry in the United States was destroyed by 'the great freeze' of 1894–1895; since 1890 a considerable industry has been developed in California.

A small tree, 3–6 m in height, with stout, stiff thorns. Lvs ovate, serrate, 5–10 × 3–6 cm; petioles short, margined but not winged and with a distinct articulation with petiole. Fls produced at all seasons, axillary, solitary or clustered, 3·8–5 cm in diameter; petals pink in bud, on opening white above and purplish below; stamens 20–40; ovary with 8–10 loculi. Fr. oval with terminal nipple, 5–10 cm long, light yellow when ripe; peel rather thick, adherent, prominently gland-dotted, slightly rough; pulp pale-yellow, sour; seeds ovoid; polyembryonic with 10–15 per cent nucellar embryos; cotyledons white. Predominant glucoside hesperidin.

'Eureka', 'Lisbon' and 'Villafranca' are the most widely grown cvs. Rough lemon is probably a hybrid with *C. medica* and exhibits almost complete nucellar embryony, so that stocks from it are very uniform. 'Perrine' is from a cross by Swingle of 'Mexican' lime and 'Genoa' lemon.

Citrus medica L. (2n = 18) CITRON

Citrons were used medicinally and as flavouring by the Romans. Candied peel, used for flavouring cakes and confectionery, is made from the rind, the fruits being sliced and fermented in brine before being candied. 'Etrog' citron, *C. medica* var. *ethrog* Engl., is carried by Jews at the Feast of the Tabernacles. Fingered citron, *C. medica* var. *sarcodactylis* (Noot.) Swing., is used by the Chinese as medicine and as an ornamental.

C. medica is sometimes thought to have originated in India, but the more probable area is south-western Asia where it has been cultivated since

Fig. 78. **Citrus grandis** : PUMMELO. A, flowering shoot $(\times \frac{1}{2})$; B, flower in longitudinal section $(\times 1)$; C, fruit $(\times \frac{1}{2})$; D, fruit in transverse section $(\times \frac{1}{2})$.

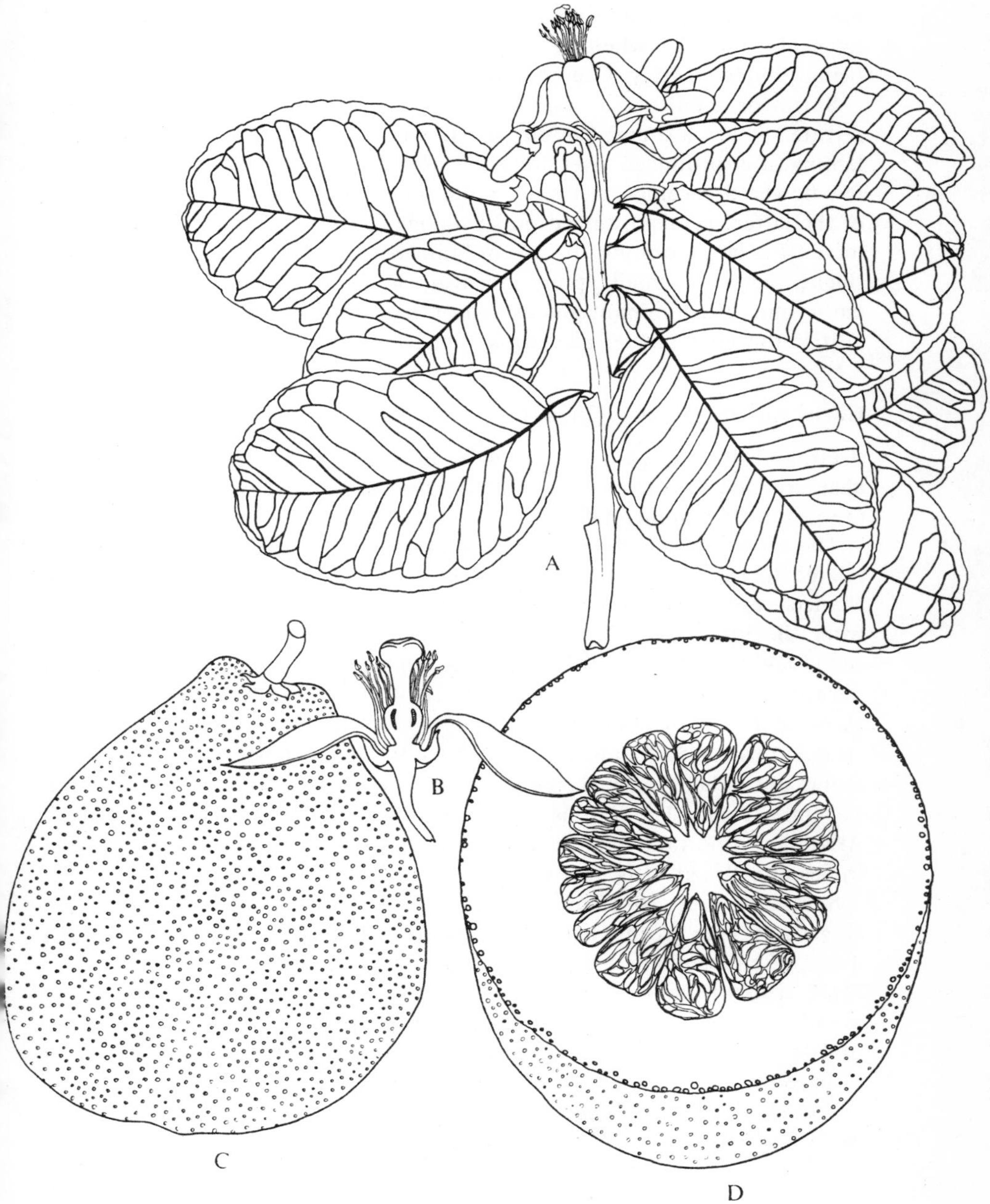
A
B
C
D

ancient times. It was the first of the *Citrus* spp. to be brought to Europe and reached there about 300 B.C. It is mentioned by Theophrastus as the Persian or Median apple; Pliny was the first person to call it *Citrus*, the name later adopted by Linnaeus for the genus. It has been taken to most tropical countries, where it is of little importance, commercial planting being limited to Italy, Greece and Corsica.

Shrub or small tree to about 3 m high with stout spines. Lvs elliptic, serrate, 8–20 × 3–9 cm; petiole short, wingless and not articulated with lamina. Fls produced over long periods, 3–4 cm in diameter, in axillary few-flowered racemes, hermaphrodite and male with a large proportion of the latter; petals 5, tinged pink; stamens 30–40; ovary large with 10–13 loculi. Fr. large, oblong, 10–20 cm long; peel usually bumpy, yellow, very thick; segments small; pulp greenish, sour. Seeds small, white, polyembryonic. Predominant glucoside in mesocarp is hesperidin.

It is the tenderest of all *Citrus* spp. It can be propagated by cuttings.

Citrus paradisi Macf. (2n = 18) GRAPEFRUIT

Grapefruits are used mainly as a breakfast fruit, the juice having a characteristic flavour with a mild bitterness. Canned segments and juice are produced commercially.

The origin of *C. paradisi* is not known with certainty. It is closely related to the pummelo and probably originated in the West Indies as a chance cross between this species and the sweet orange or as a bud mutation from the pummelo. There are records of 'the forbidden fruit or small shaddock' in Barbados in 1750 and in Jamaica in 1789. The first record of the name 'grapefruit' is in Jamaica in 1814. It was introduced into Florida early in the 19th century by Count Odet Philippe, Chief Surgeon of Napoleon's navy. Commercial production did not begin in the United States until 1880; since then its rise to fame as a breakfast fruit has been extremely rapid. The crop is now grown in Florida, California, Texas and Arizona; important exports are made from Israel, South Africa, West Indies and Brazil. It is now grown in all tropical countries.

Spreading tree, 10–15 m in height; twigs glabrous. Lvs smaller than *C. grandis*, pale green when young; petiole rather broadly winged; lamina ovate, often crenulate. Fls axillary, singly or in clusters, 4–5 cm in diameter; petals white, usually 5; stamens 20–25, ovary with 12–14 loculi. Fr. large, usually globose, 8–15 cm in diameter, greenish or pale yellow when ripe; rind thinner and pulp vesicles smaller than *C. grandis*. Seeds white, polyembryonic; cotyledons white. Predominant glucoside naringin.

'Marsh' (syn. 'Marsh Seedless') is the most widely planted cv.; it usually has 0–8 seeds per fruit. 'Thompson' (syn. 'Pink Marsh') with pink flesh and

---

Fig. 79. **Citrus × limon:** ROUGH LEMON. **A**, shoot (× ½); **B**, flower in longitudinal section (× 1); **C**, fruits (× ½); **D**, fruit in transverse section (× ½).

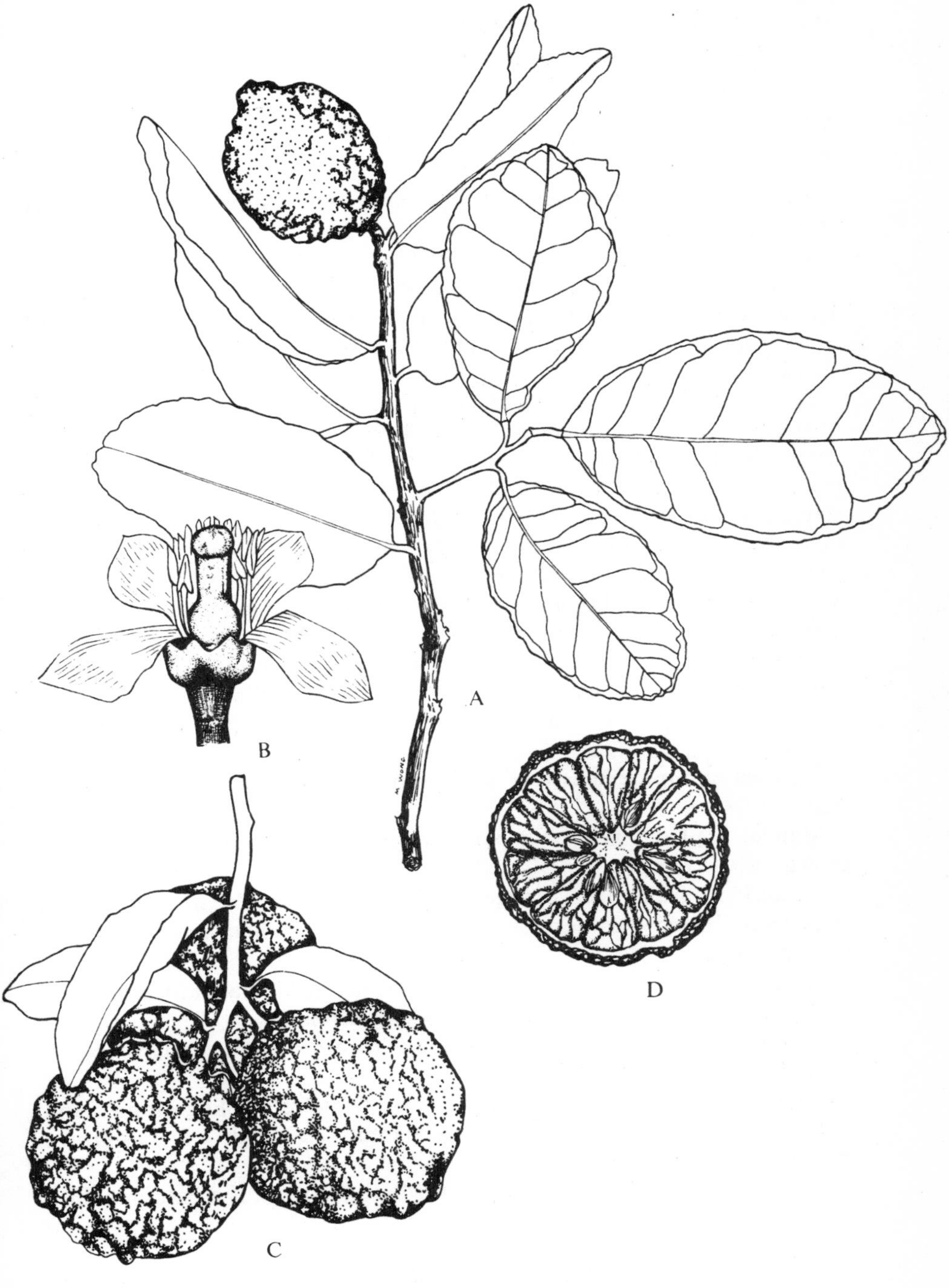
A
B
C
D

3–5 seeds per fruit arose as a bud mutation in Florida. 'Ruby' and 'Webb', two red-fleshed cvs, arose as bud mutations from 'Thompson'.

Interspecific hybrids with *C. reticulata* are tangelos.

Citrus reticulata Blanco (2n = 18) MANDARIN, TANGERINE

Syn. *C. nobilis* Andrews (non Lour.)

C. reticulata is sometimes termed 'a fancy fruit' and is eaten as a dessert fruit; segments are also canned.

It appears to have originated in Cochin China and has long been cultivated in China and Japan. It did not reach Europe until 1805 and the United States until about the middle of the 19th century. It is now grown in most tropical and subtropical countries. It is the hardiest of all the cultivated *Citrus* spp.

A small tree, 2–8 m in height, sometimes spiny, Lvs small and narrow, ovate, elliptic, or lanceolate, 4–8 × 1·5–4 cm, usually crenate, dark shining green above, yellowish-green below; petioles usually narrowly winged or margined. Fls axillary, small, 1·5–2·5 cm in diameter; petals 5, white; stamens about 20; ovary with 10–15 loculi. Frs depressed-globose, 5–8 cm in diameter; peel thin, loose, easily separating from segments, green, yellow or orange-red when ripe; pulp sweet and juicy, orange in colour. Seeds small, polyembryonic; embryos green. Predominant glucoside is tangeretin.

In some countries the terms mandarin and tangerine are used indiscriminately, but it is better to use mandarin for yellow-fruited cvs and tangerine for those with deep orange rind. The cvs can be classified as follows:

Satsuma group: extensively grown in Japan; very hardy and cold-resistant.

Mandarins: fruits yellow or pale orange, *e.g.* cv. 'Emperor' extensively grown in Australia.

Tangerines: fruits deep orange-red, *e.g.* 'Clementine' which originated in Algeria; 'Dancy' is the best known and most highly prized cv. in the United States.

Citrus reticulata var. *austera* Swing. is the sour mandarin and it is probable that the Rangpur lime may belong here.

Interspecific hybrids have been made with other *Citrus* spp., of which the most important are:

Tangelos, *C. reticulata* × *C. paradisi*, with fruits resembling oranges and with a flavour intermediate between the parents, *e.g.* 'Ugli' which is grown in and exported from Jamaica.

Tangors, *C. reticulata* × *C. sinensis*, the 'King' tangor, is a natural hybrid

---

Fig. 80. **Citrus medica :** CITRON. **A**, flowering shoot (×½); **B**, flower in longitudinal section (×½); **C**, fruit (×⅓); **D**, fruit in transverse section (×⅓).

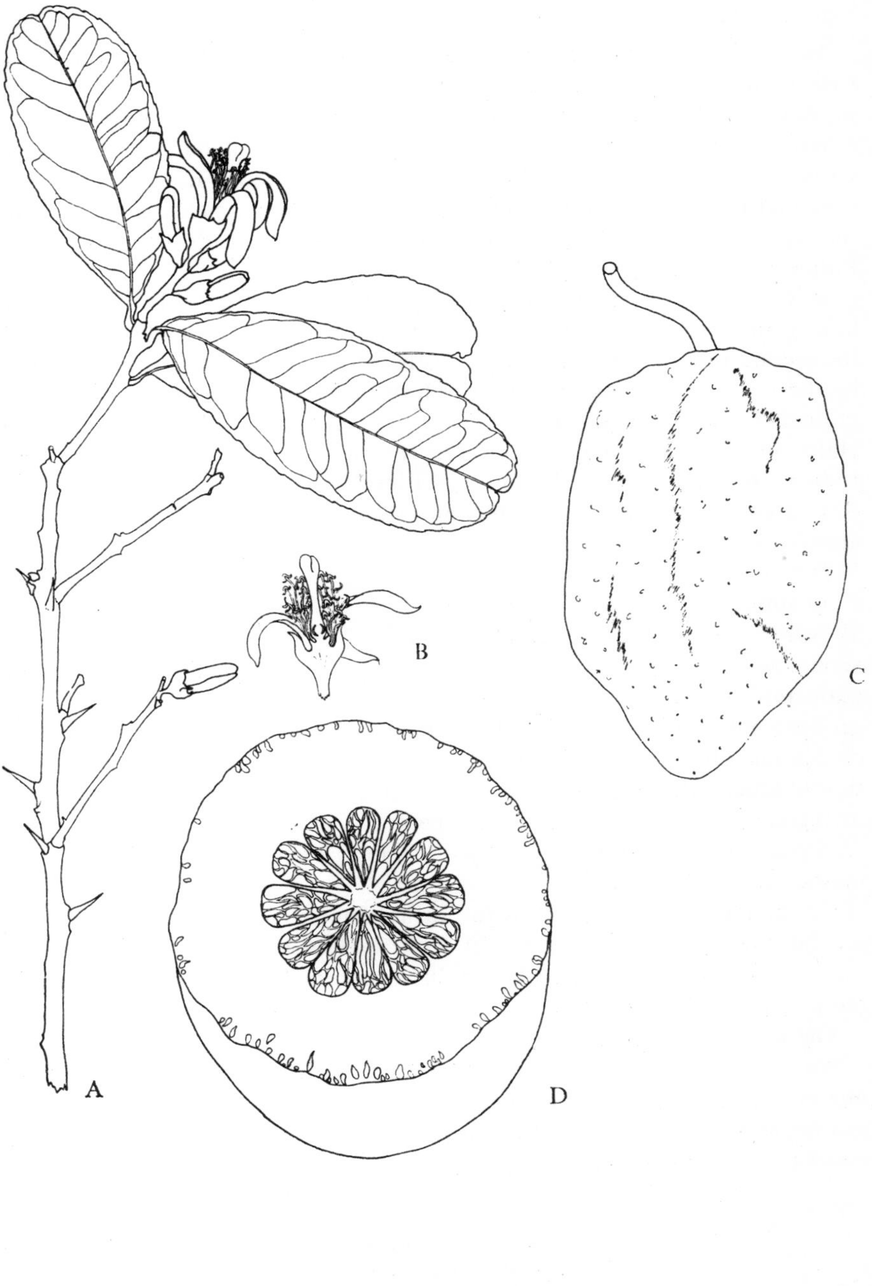
A
B
C
D

from Siam and is the type of *C. nobilis* Lour.; 'Ortanique' is a cv. in the New World.

Citrus sinensis (L.) Osbeck (2n = 18) SWEET ORANGE

Syn. *C. aurantium* L. var. *sinensis* L.

The sweet orange is the most widely grown and most important of all *Citrus* spp. It is eaten as a dessert fruit, but it is not widely canned as the product is not as good as the fresh fruit. Large quantities are now used for making orange juice which is canned; it is also made into squashes. Oranges are used for flavouring and for marmalade. Essential oils are obtained from the plant: neroli oil from the flowers, petitgrain oil from the leaves and orange oil from the peel. Pectin is made from the peel. Pulp after squeezing is used as cattle food.

C. sinensis is native of southern China or Cochin China. It is no longer known in a truly wild state. The sweet orange did not reach Europe until the second half of the 15th century; superior cvs were brought from China by the Portuguese about 1520. Later large orangeries were built in Europe in areas too cold for satisfactory growth. Columbus took orange seeds to Haiti on his second voyage in 1493. From there it spread to the West Indies, to Mexico in 1518 and to Florida in 1565. It is now grown throughout the tropics and subtropics, but does not do well in very wet areas, nor at high altitudes.

Tree 6–12 m high at maturity, twigs angled when young, often with stout spines. Lvs ovate or oval-elliptic, 5–15 × 2–8 cm, glabrous, dark green above, sometimes slightly serrate; petiole 1–2·5 cm long, narrowly winged, articulated. Fls axillary, borne singly or in small racemes, 2–3 cm in diameter, fragrant; petals usually 5, white; stamens 20–25, united into groups; ovary with 10–14 loculi; style slender with globose stigma, soon deciduous. Frs subglobose, 4–12 cm in diameter; peel to 0·5 cm thick, tightly adherent, ripening to an orange colour, but often remaining green in the tropics; pulp juicy, subacid; central axis solid; seeds nil to many, obovoid, polyembryonic; embryos white; predominant glucoside in mesocarp is hesperidin.

Cvs divided into: (1) those with normal fruits; (2) navel oranges with a second row of carpels, opening at fruit apex in an umbilicus; (3) blood oranges with red or red-streaked pulp. Aberrant forms may occur within the groups.

The most important cvs are:

'Valencia', which was introduced into California from a London nurseryman in 1876 and is identical with a cv. grown in Spain. A vigorous grower and prolific. Fruit of medium size with excellent flavour and few seeds (5–6). It is a late cv. ripening in Trinidad in March–July and in

---

Fig. 81. **Citrus paradisi :** GRAPEFRUIT. **A,** vegetative shoot (× ½); **B,** flower in longitudinal section (× 1); **C,** fruit (× ⅓); **D,** fruit in transverse section (× ⅓).

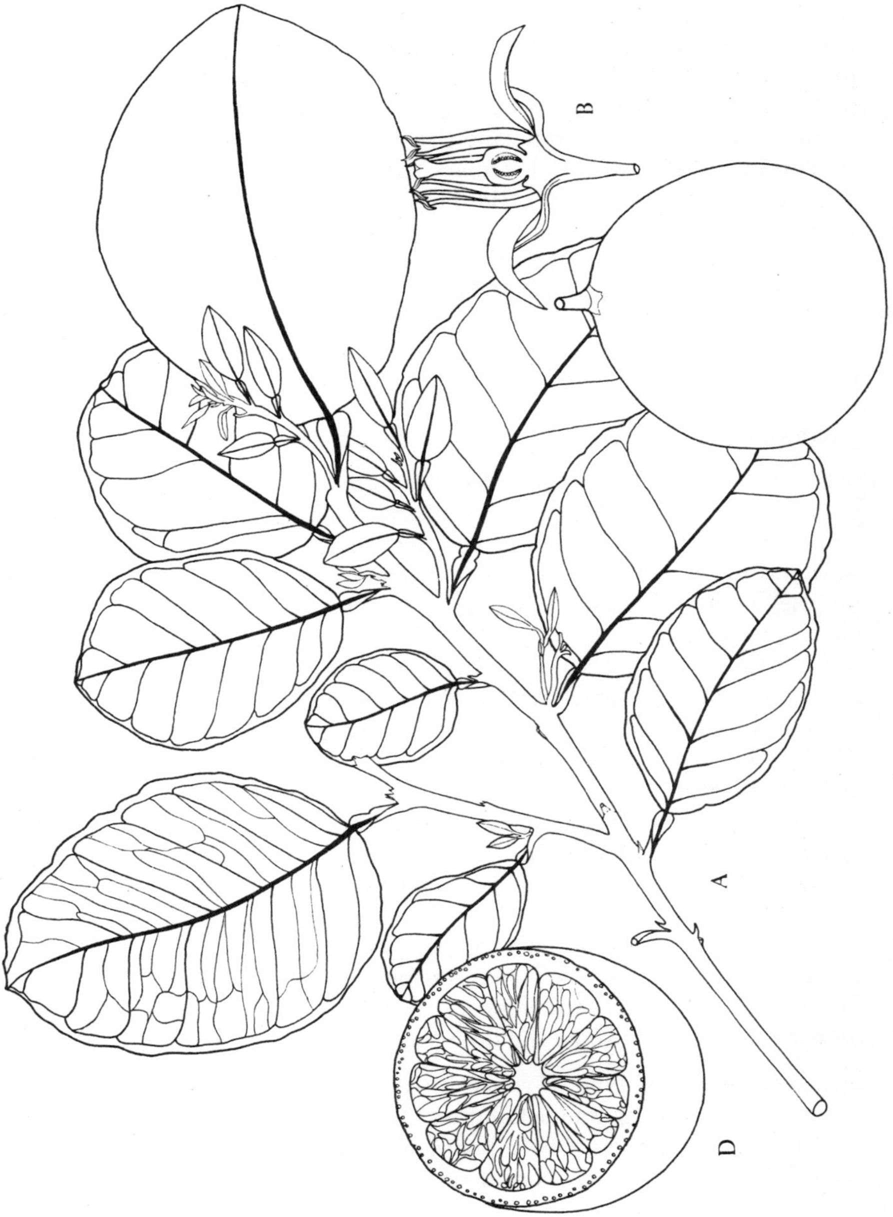
A
B
C
D

California in April–September. It hangs on the tree after the fruit is ripe and is high in total solids. It is the most important commercial cv., particularly in California, South Africa, Australia and the West Indies.

'Washington Navel' probably originated as a bud mutation at Bahia in Brazil early in the 19th century and was taken to Washington in 1870. Fruit larger with navel and rather thick peel which facilitates shipping; juice sweet and of excellent flavour; seedless. It is a mid-season cv. ripening in Trinidad from November–February and in California from November–April. Fruit swelling is erratic and they become very large in wet climates. It is an important cv. in California, South Africa and Australia.

Other important cvs are: 'Hamlin' and 'Parson Brown', early cvs in Florida; 'Pineapple', mid-season cv. in Florida; 'Jaffa' and 'Shamouti' in Palestine. 'Maltese' and 'Ruby' are blood oranges.

Hybrids have been produced with most *Citrus* spp., *Fortunella* spp. and *Poncirus trifoliata*. The most important are the tangors (*C. sinensis* × *C. reticulata*) which bear orange-like fruits often of high quality, e.g. 'King Orange'.

PROPAGATION

SEED: In the past much *Citrus* has been grown from seeds, but it is now customary to propagate them by budding. Due to the presence of zygotic embryos cvs do not reproduce true by seed. Nucellar embryos produce seedlings of the same genetic constitution as buddings from the same tree but they tend to be rather more vigorous, more thorny, and are slower to come into bearing.

Rootstocks, however, are grown from seeds. Seeds should be obtained from good, fully matured fruits growing on vigorous, adult, healthy trees. After removal from the fruits the seeds are washed and partially dried; if allowed to dry out they soon lose their viability; they may be stored in ground charcoal if required. The seeds are broadcast or planted at approximately 1 × 1 in. in well-drained, shaded beds; 1 lb of seed will plant approximately 120 sq. yd of bed. The optimum temperature for germination is 80°–90°F and they will not germinate below 55°F. Seedlings are then transplanted from the seedbeds into nurseries at a spacing of 3–4 × 1 ft or into plastic bags, when about 8 in. high, which is usually 5–6 months after sowing in the tropics. They may be budded after a further 6 months. Small seedlings, which are often of zygotic origin, and those with bent roots or which are off-type should be discarded at transplanting.

ROOTSTOCKS: The type of rootstock used will depend upon the species and cv. which is being budded and on the area in which the crop is grown. The stock and scion must be capable of uniting and producing long-lived productive trees, the first consideration being the yield of good quality

---

Fig. 82. **Citrus reticulata**: TANGERINE. **A**, leafy shoot (× ½); **B**, fruits (× ½); **C**, fruit in transverse section (× ½).

C
B
A

fruits per acre over long periods. The rootstock should be compatible with the scion giving a smooth and even union with the trunk tapering from the base. When rootstock growth is slower than the scion a bulge or overgrowth occurs which is undesirable as it interferes with growth; when rootstock growth is faster than the scion undergrowth or a bulge may occur below the union, which tends to dwarf the tree and produce early fruiting. Other considerations include the effect of the rootstock on the growth of the scion; the period taken to come into bearing; the season of fruiting; the effect on the size, shape, juiciness, grade and yield of fruit; the length of the productive life of the tree; resistance to cold, drought and wind; resistance to diseases, particularly to gummosis, scab, tristeza and other virus diseases.

The following are the most commonly used rootstocks:

SOUR ORANGE: Hardy, with deep tap-root, and is not susceptible to drought; can withstand hurricane injury; one of the best rootstocks for damp heavy soil. Growth is somewhat slow in the early years, but it produces high yields later. Fruits on this stock are of excellent quality, thin skinned and juicy, and are held on the tree to the fullest maturity without deterioration. It is resistant to gummosis, but is susceptible to scab and tristeza. It has 85–95 per cent nucellar embryos. Scion compatibility is good except with limes. In the West Indies, it is the best stock for 'Marsh' grapefruit and 'Valencia' orange. It is extensively used in the Mediterranean region and Florida, and for grapefruit in California. It is a good rootstock for sweet oranges, grapefruits, lemons and tangerines.

ROUGH LEMON: Less hardy than sour orange and more subject to frost damage; it has no tap-root, but develops abundant wide-spreading system of lateral roots; it is best suited to light sandy or loamy soils with good drainage. Produces vigorous early-fruiting trees. Fruits on this stock tend to be large, thick-skinned, more acid and of poorer quality; fruits cannot be held so long in the season without deterioration. It is susceptible to gummosis and scab, but is resistant to tristeza. It has almost 100 per cent nucellar embryos so that it is very uniform. It is extensively used in South Africa and Australia, and for lighter soils and 'Washington Navel' oranges in Florida. It is a good stock for lemons and limes, and is also used for grapefruits and tangerines.

SWEET ORANGE: Less hardy than sour orange, but hardier than rough lemon; it is rather shallow-rooted and suited best to rich well-drained loams. Growth is rather slow in the early years, but the trees are long-lived. Fruits on this stock are of high quality. It is resistant to scab and tristeza, but susceptible to gummosis. It has 70–95 per cent nucellar embryos.

Fig. 83. **Citrus sinensis :** ORANGE CV. 'WASHINGTON NAVEL'. **A**, vegetative shoot (×½); **B**, flower in longitudinal section (×1); **C**, fruit in longitudinal section (×½); **D**, fruit in transverse section (×½).

A
B
C
D

It is compatible with other species and is used in California for oranges and lemons.

OTHER ROOTSTOCKS: Grapefruit attracted some attention as a rootstock for rich, heavy, poorly drained soils. It produced vigorous growth, but yields were poor and it has now largely been discontinued, but a 'wild' type is used for budding limes in Dominica. Rangpur lime, which is easy to transplant, will stand wet conditions, and is resistant to gummosis and tolerant to tristeza; it is used in South America and the Far East. Trifoliate orange (*Poncirus trifoliata*) is the hardiest rootstock for withstanding cold; it will tolerate waterlogging; trees on this stock are dwarf, precocious and prolific; tolerant of tristeza and resistant to gummosis; it is compatible with most *Citrus* spp., and is used for Satsuma oranges in Japan. Citranges (*C. sinensis*×*P. trifoliata*) are now being used in some areas where cold hardiness and resistance to tristeza are necessary. Palestine sweet lime has been used as a rootstock for 'Shamouti' orange in Israel. 'Cleopatra' mandarin is also used as a rootstock in some areas.

It is now possible to identify the various rootstocks by chromatographic methods.

BUDDING: Budwood should be obtained from selected, virus-free, high-yielding trees, with no off-type bud mutations. It should be taken from fairly well-matured wood of the current year's growth, after the shoots have become round (not young angular wood), with buds in the dormant stage. Budwood can be stored if required in moist sawdust for several months. The stock should be about pencil-thickness where it is budded at 6–18 in. above the ground; the bark should slip easily. Scion wood should be the same thickness. Budding is usually done with an inverted T and then bound with budding tape. The bud is inspected 10–21 days after insertion; a green colour indicates a satisfactory take and the bud is unwrapped. If the bud has died the stock may be re-budded. When the bud is about 1 in. long the stock is cut off with a slanting cut just above the bud. Stock suckers should be removed regularly. Buddings are usually staked and are topped at 2–3 ft to induce branching. The lowest branches near the union are removed and 4–6 branches higher up are allowed to develop to give the future framework of the tree. The time from budding to transplanting varies in different countries and is 6–8 months in Trinidad and 1–2 years in California. The branches may be cut back to 6–8 in. before transplanting.

OTHER METHODS OF PROPAGATION: *Citrus* may also be grafted or inarched and some species may be propagated by layers, marcotts or cuttings, but these methods are seldom practised. Trees may be topworked when necessary, but it is seldom economic.

HUSBANDRY

PLANTING: Citrus may be transplanted in the field either with bare roots, or with a ball of earth, or from plastic bags. Using bare roots the transplants are easier to handle and transport, but they require careful handling and should never be allowed to dry out as they are very susceptible to sun and wind injury; they must be planted immediately after digging and require larger planting holes. Planting with bare roots permits inspection for disease, nematodes and damage. Planting with a ball of earth or from plastic bags permits transport over long distances and the transplants may be left out of the ground for a number of days; it is useful in drier areas, but they are heavier and more expensive to transport.

The planting distance will depend on the nature and fertility of the soil, topography, and the type of rootstock and scion. On average soils on the flat in the West Indies with sour orange stock the spacing recommended is 25 ft for grapefruit, 20 ft for orange and lemon, and 15–20 ft for limes and mandarins. Planting may be square, rectangular, triangular or hexagonal and on steep slopes should be on the contour. Drainage and provision for mechanical cultivation, spraying and irrigation when used must be taken into consideration. There must be ample roadways or traces for transport of fruit, etc.

Normal planting holes are 20 in. in depth and diameter and 1 lb of mixed fertilizer may be incorporated at planting. The transplants should never be planted deeper than they were in the nursery. Watering is often done immediately after planting. Windbreaks are necessary for protection from exposure, but shade is not generally used. Young citrus is often intercropped with vegetables or other crops. In high-rainfall areas a permanent cover crop such as kudzu, *Pueraria phaseoloides* (Roxb.) Benth., is desirable, or broad-leaved weeds may be encouraged. These may be kept in check manually by slashing or mechanically; increasing use is now made of herbicides. In drier areas where the cover would compete for water, or when grown under irrigation, the soil is often kept bare.

MANURING: In Florida the following application of fertilizers per box of fruit harvested per annum is recommended: 0·4 lb N, 0·2 lb P_2O_5, 0·4 lb K_2O, and 0·1 lb MgO. In the West Indies nitrogenous fertilizers are the most important and most usually applied at the rate of 1–2 lb (20 per cent N) to young trees and 4–10 lb per mature tree per annum, often in two applications, the first at the beginning of the wet season just before flowering and the second 4–5 months later. Potash affects fruit quality and 1 lb potassium sulphate per tree per annum may be applied. Zinc and magnesium deficiencies (see below) are common and may be controlled by foliar sprays or incorporated with the other fertilizers. Organic manures are beneficial and are strongly recommended where available and economic. Leaf analysis is being used increasingly to indicate fertilizer requirements.

PRUNING: The first shaping of the trees is done in the nursery and will continue for the first 2–3 years in the field. It is essential to remove all suckers, more particularly those from the stock, and to cut out dead wood and Loranthaceae parasites which frequently attack citrus; otherwise little pruning is done in the tropics.

HARVESTING: Fruits should be held on the trees until fully mature, although lemons and limes are usually picked while still green. Colour is seldom an indication of maturity in the tropics as many citrus retain a green colour. Fruits are harvested individually either by clipping off near the calyx with special clippers or by bending and pulling with a slight twist. For the fresh fruit trade it is essential that the fruit should be picked at the correct stage of maturity and be clean and free from blemish and bruises, and of good colour; the latter may be obtained by treating with ethylene after harvest. Some countries have maturity laws and grade regulations. The fruits are placed in boxes and taken to the packing or processing factory.

YIELDS: These vary enormously. Oranges yield 1–10 boxes per tree with 175 fruits per box (Trinidad average 2–3 boxes per annum); grapefruit yield 2–12 boxes per tree with 80 fruits per box (Trinidad average 4–5 boxes); and limes about 1 barrel or 170 lb per tree.

MAJOR DISEASES

Gummosis, foot-rot, mal-di-gomma, *Phytophthora* spp., particularly *P. citrophthora* (Sm. & Sm.) Leon. and *P. parasitica* Dastur, which was first discovered in the Azores in 1834, and is now in all citrus areas. Lesions occur at the union on budded trees and at the crown on seedling trees, on the trunk and branches and on the roots; gum exudes from the lesions. It also causes a brown rot of fruits. The bark is killed to the cambium layer and spreads until the trunk may be girdled and the tree dies. Water-logging and high humidity favour infection. The best control is by using resistant rootstocks such as sour orange and budding high at about 18 in. Diseased areas may be cut out and painted with a fungicide.

Scab, *Elsinoe fawcetti* Bitanc. & Jenk., is widespread in areas with cool wet conditions. It attacks twigs, leaves and fruits producing corky lesions. Sour orange, lemon and some grapefruit cvs are particularly susceptible; sweet orange and lime have marked resistance. It is controlled by pre-blossom and post-blossom copper sprays.

Melanose, *Diaporthe citri* (Fawc.) Wolf (syn. *Phomopsis citri* Fawc.), is widespread, producing brown raised pustules on young twigs, leaves and fruits, more particularly on mature trees of all *Citrus* cvs. Controlled by copper sprays and regular pruning to remove dead diseased wood.

Anthracnose, withertip and blossom-blight of limes, *Gloeosporium limetticolum* Claus., and of oranges, grapefruit and lemons, *Colletotrichum gloeosporioides* Penz. The former fungus causes the tips of twigs of 'Mexican' limes in the West Indies to die, where it is a serious disease and is very difficult to control. The latter fungus attacks branches, leaves and fruits which have become weakened or injured. It is controlled by copper or lime-sulphur sprays and pruning.

Tristeza, a virulent virus infection, which has caused substantial losses in South Africa where it was first discovered, and has since been found in

South America and Australia, with less virulent forms elsewhere. It is transmitted by using diseased budwood and the black citrus aphid, *Aphis citricidus* Kirk., which is the most efficient vector. It causes partial suppression of new growth and yellowing of the leaves, followed by leaf fall and die-back; roots then gradually die and tree dies. The only method of control is the use of resistant rootstocks such as sweet orange or rough lemon; sour orange is susceptible.

Other virus diseases include exocortis, psorosis and xyloporosis, which are controlled by the use of virus-free budwood.

MAJOR PESTS

Scale insects and mealybugs include: Californian red scale, *Aonidiella aurantii* (Mask.); Florida red scale, *Chrysomphalus aonidum* (L.); Mediterranean red scale, *C. dictyospermi* (Morg.); cottony–cushion scale, *Icerya purchasi* Mask.; purple scale, *Lepidosaphes beckii* (Newm.); Glover's or long scale, *L. gloverii* (Pack.); black scale, *Saissetia oleae* (Bern.); rufous scale, *Selenaspidus articulatus* (Morg.); snow scale, *Unaspis citri* (Comst.); citrus mealybug, *Pseudococcus citri* (Risso); and Baker's mealybug, *P. maritimus* (Ehrh.). Most of the above spp. are very widespread and heavy infestations may kill young trees. Biological control by insect predators and parasites and entomogenous fungi and bacteria is possible for some spp., otherwise regular spraying is necessary.

Mites: citrus rust mite, *Phyllocoptruta oleivora* (Ashm.), and citrus red mite, *Paratetranychus citri* (McG.), cause russetting of fruits and leaf and fruit fall. They attack all green parts of plant.

Thrips, *Scirtothrips* spp.; aphids, *Aphis* spp.; and whiteflies, *Dialeurodes* spp. attack citrus in many countries.

Fruitflies: Mediterranean fruitfly, *Ceratitis capitata* (Wied.), is a serious pest in the Mediterranean region, South Africa, Australia and South and Central America; it has twice been eradicated from Florida. Mexican fruitfly, *Anastrepha ludens* (Loew); and South American fruitfly, *A. fraterculus* (Wied.) are important pests in Central and South America respectively.

Moth borer, *Citripestis sagittiferella* Moore, is the most injurious pest of citrus in Malaysia and Indonesia.

Other pests include: parasol or bachac ant, *Atta* spp., in the West Indies and South and Central America; grasshoppers everywhere which attack the leaves; citrus root weevil, *Diaprepes* spp., in the Antilles; and nematodes, more particularly the burrowing nematode, *Radopholus similis* (Cobb) Thorne, and Citrus nematode, *Tylenchulus semipenetrans* Cobb.

MINERAL DEFICIENCIES

The symptoms of the commoner mineral deficiencies are as follows:

Boron: water-soaked spotting of leaves and lumpy fruits.

Copper: causes die-back or exanthema with gum pockets or blisters on young growing branches, leaves and fruits, followed by leaf and fruit drop and distortion of twigs.

Iron: leaves show chlorosis, the veins remaining green.

Magnesium: leaves become yellow all over including veins and in pronounced cases take on a bronzed appearance. It is cured by a dose of Epsom salts.

Zinc: frenching or mottling of leaves with a reduction in size of developing leaves. Controlled by spraying with 5 lb zinc sulphate and $2\frac{1}{2}$ lb of slaked lime in 100 gallons of water at the rate of 1 gallon per tree.

IMPROVEMENT

The following characteristics should be considered in any selection or breeding programme: (1) sustained yield including time to come into bearing and longevity; (2) adaptability to local environment; (3) time of bearing in order to spread fruiting season; (4) resistance to pests and diseases, including viruses; (5) ease of propagation and compatibility with rootstocks; (6) good tree growth, size, shape and thornlessness; (7) fruit quality including shape and size; colour, texture, smoothness and thickness of peel; juiciness, flavour, aroma, texture, vitamin content and soluble solids/acids ratio of pulp; toughness and amount of rag; holding quality on tree; seediness with few or no seeds desirable; keeping quality in transit and storage; for canned juice or concentrate the most important characters are high juice content, low waste content, aroma, suitable soluble solids/acid ratio, ease of extraction, the outward appearance being of little importance.

Any breeding programme must take into consideration the ease with which *Citrus* and allied spp. cross, the great genetic diversity and abundant heterozygosis of cvs, the presence of nucellar embryos, the frequent occurrence of bud mutations, parthenocarpy, the ease with which cvs may be budded onto a variety of rootstocks, and the effect of rootstock on scion.

It will include: the selection of the best mother trees for vegetative propagation; the careful selection of rootstocks for adaptability to local conditions, resistance to disease and for the desired effect upon the scion; the selection of useful bud mutations or sports; hybridization. In the production of controlled hybrids the flowers are emasculated in the bud stage and must be bagged.

Most of the cvs now widely grown are old cvs introduced from other countries, *e.g.* 'Washington Navel' orange, or have arisen as chance seedlings, *e.g.* 'Marsh' grapefruit, or by the selection of useful bud mutations, *e.g.* 'Thompson' grapefruit ('Pink Marsh'), or from natural crosses, *e.g.* 'King' orange. Almost all commercial cvs are diploid ($2n = 18$). Tetraploids are known which have occurred naturally or have been produced artificially, but have poor fruits and low yields. The production of triploids by crossing tetraploids with diploids permits the production of seedless fruits. 'Washington Navel' orange, however, is diploid, parthenocarpic and male sterile. Many interspecific and intergeneric crosses have been made, but few, if any, of them are grown commercially. The 'Morton' citrange was produced from a cross made by Swingle in 1897 of *Poncirus trifoliata* × *Citrus sinensis*.

Although a great deal of work has been done on the selection and breeding of citrus, particularly in the United States, most of the commercial cvs grown today originated in the 19th century. Most of the work has been done in subtropical regions with semi-arid climates where the crop is grown under irrigation. Little work has been done in the wetter tropics and there is a need to select and breed citrus more suited to this environment.

PRODUCTION

The world production of oranges in 1963, about 16 million tons, is double that of before World War II, and may be trebled by 1970. Approximately 15–20 per cent of this production is mandarins, tangerines and other loose-skinned types. The quantity of fresh oranges entering international trade is exceeded only by bananas. The largest producers in descending order are the United States (about 25 per cent of the world total), Brazil, Spain, Italy, Mexico and Israel. The countries exporting the most fresh oranges are Spain, Israel, Mexico and South Africa, the last being the principal off-season supplier to the northern hemisphere. The largest importers of oranges in descending order are West Germany, France and the United Kingdom. Much of the crop is now used for processing of juice and frozen concentrate, two-thirds of the United States crop being used in this way.

Grapefruit production has shown little expansion in recent years and the annual world production is just over 1½ million tons. The United States produces 75–80 per cent of this total and Israel 65–70 per cent of the fresh grapefruit entering world trade. Trinidad, Jamaica and British Honduras are important suppliers of grapefruit and its products to the United Kingdom, which, together with Canada, imports more fresh grapefruit than any other country.

World production of lemons has been increasing fairly steadily and is now (1½ million tons) approaching the level of world grapefruit production. The United States and Italy are the major producers.

Mexico, Jamaica, Dominica and Trinidad are the main exporters of lime juice and other lime products.

Although most *Citrus* spp. originated within the tropics, the only tropical Commonwealth territories exporting citrus are in the West Indies; elsewhere in the tropics citrus is treated mainly as a household crop.

REFERENCES

BOWMAN, F. T. (1956). *Citrus-growing in Australia.* Sydney: Angus and Robertson.

DODDS, K. S. (1963). The Origin of Fruits and Vegetables. *Span*, **6**, 64–67.

HUME, H. H. (1957). *Citrus Fruits*. 2nd edition. New York: Macmillan.

KNORR, L. C., SUIT, R. F. and DUCHARME, E. P. (1957). Handbook of Citrus Diseases in Florida. *Florida Agr. Exp. Sta. Bull.*, **587**.

PRATT, R. M. (1958). *Florida Guide to Citrus Insects, Diseases and Nutritional Disorders in Color*. Florida Agr. Exp. Sta.

WEBBER, H. J. and BATCHELOR, L. D. (1948). *The Citrus Industry*. 2 vols. Los Angeles: Univ. of California Press.

SINGH, S., KRISHNAMURTHI, S. and KATYAL, S. L. (1963). *Fruit Culture in India*. New Delhi: Indian Council of Agric. Resc.

SOLANACEAE

About 75 genera and 2,000 spp. of herbs, shrubs and small trees, generally distributed, but most numerous in the tropics. Lvs alternate, simple, stipules absent. Fls hermaphrodite, mostly actinomorphic; corolla sympetalous, usually 5-lobed, aestivation folded, contorted or valvate; stamens inserted on corolla tube, typically as many as corolla lobes and alternating with them; anthers 2-locular, often connivent, dehiscing lengthwise or by apical pores; ovary 2-locular, sometimes again divided by false septa; style 1, terminal; ovules numerous, axile. Frs a berry or capsule. Seeds with copious endosperm and curved or annular embryo.

USEFUL PRODUCTS

FRUITS AND VEGETABLES: The most important vegetables are: *Lycopersicon esculentum* (q.v.), tomato; *Solanum tuberosum* (q.v.), potato; and *S. melongena* (q.v.), eggplant. The genus *Solanum* also provides a number of minor fruits and vegetables in the tropics. *Capsicum annuum* (q.v.) is the sweet pepper, and together with *C. frutescens* (q.v.) provides chillies, which are used as a spice.

Cyphomandra betacea (Cav.) Sendt., the tree tomato, is a native of Peru and is extensively cultivated in the Andean region. It has been taken to many tropical countries, where it does best at medium to high altitudes. It is a small short-lived tree, 3–6 m high, which starts bearing in its 2nd year. The orange, red or purple, subacid fruits, the size and shape of a hen's egg, are eaten raw, but are usually stewed; they can be made into preserves.

Physalis peruviana L., the cape gooseberry, a native of tropical America, is cultivated in some tropical countries for its globular yellow berries, 1–2 cm in diameter, which are stewed or made into jam. The berry is enclosed in the bladder-like inflated calyx. Species of *Physalis* have long been cultivated and consumed in Mexico.

DRUGS AND NARCOTICS: By far the most important member of the family entering world trade is tobacco, *Nicotiana tabacum* (q.v.).

There are several drug plants of long standing, of which the most important are:

Atropa belladonna L., belladonna, a poisonous plant of southern Europe and Asia Minor, whose leaves and flowers are used to relieve pain. One of its alkaloids is atropine. It is cultivated as a drug plant in Europe, India and the United States.

Datura stramonium L., thorn-apple, a very poisonous plant which is now cosmopolitan throughout the tropics, yields the drug stramonium. The seeds have been found in wheat grown in Kenya, where it is a prohibited plant. It is used for trial by ordeal in Uganda.

Hyoscyamus niger L., henbane, yields hyoscyamine and other alkaloids.

Mandragora officinarum L., mandrake, native of the Mediterranean region, has been used in Europe as a medicine since ancient times and there are many old superstitions connected with it, partly due to the fanciful resemblance of the branched root to the human figure. It should not be confused with the North American *Podophyllum peltatum* L. (Family Berberidaceae), which is also known as mandrake and is used as a drug.

The genus *Datura* has been, and is still extensively used in many parts of the world for its narcotic and hypnotic properties, which produce a sense of illusion and motor disturbances. The species used include *D. stramonium* L. in many areas, *D. arborea* L. and *D. sanguinea* Ruiz & Pav. in the western Amazonian region and *D. innoxia* Mill. by the Aztecs.

The family has many poisonous members.

ORNAMENTALS: A number of showy herbs, shrubs and climbers are grown throughout the tropics belonging to the following genera: *Browallia, Brunfelsia, Cestrum, Datura, Nicotiana, Nierembergia, Petunia, Salpiglossis, Solandra, Solanum* and *Streptosolen.*

Capsicum L. (x = 12) CHILLIES

Also known as bird, capsicum, cayenne, paprika, red and sweet peppers, depending upon the type. They should not be confused with black or white pepper which is the product of *Piper nigrum* (q.v.).

The classification of the genus is very confused. Many species, over 100, and many botanical varieties have been erected from time to time, but most authorities recognize 2 main spp., *C. annuum* (q.v.) and *C. frutescens* (q.v.), and with this I concur. (See SYSTEMATICS below.)

USES

Sweet peppers have the mildest flavour with little of the pungent principle. They are eaten raw in salads and cooked in various ways; they are often stuffed with meat and are also pickled. Paprikas are European cvs with large mild fruits. Spanish paprikas (pimiento) are lacking in pungency; they are preserved and used in cheese preparations and stuffed olives. Hungarian paprika has long pointed fruits and is more pungent. The dried fruits are ground to produce powdered paprika, which is used as a condiment and in cooking; it is a constituent of Hungarian goulash.

Chillies are the dried ripe fruits of pungent forms of *C. annuum*, and sometimes *C. frutescens*. In its powdered form it constitutes red or cayenne pepper. Both chillies and cayenne pepper are used for culinary purposes and for seasonings. African chillies are very pungent; Japanese chillies are rather less pungent. Chillies are widely used throughout the tropics, particularly in India. They are the hot ingredient of curry powder, which is made by grinding roasted dried chillies with turmeric, coriander, cumin, and other spices. Before the introduction of *Capsicum*, *Piper nigrum* was used for this purpose in India. Chillies are extensively used in Central America and are constituents of dishes such as tamales and chile con carne. They are used in pickles. Pepper sauce, such as Tabasco, is made by pickling the pulp in strong vinegar or brine. Extracts of chillies are used in the manufacture of ginger beer and other beverages. Cayenne pepper is incorporated in laying mixtures for poultry.

Capsicum from *C. frutescens* is used in medicine, internally as a powerful stimulant and carminative, and externally as a counter-irritant.

SYSTEMATICS*

The two species can be distinguished as follows:

A. Plants usually annual, fruits borne singly.......... *C. annuum*
AA. Plants perennial, fruits borne in groups.......... *C. frutescens*

Capsicum annuum L. (2n = 24) CHILLIES, RED OR SWEET PEPPERS

There are many cvs, differing from each other in the shape and colour of their fruits, the way in which they are borne, which may be erect or pendent, and in their pungency. Redgrove (1933) and Chittenden (1956) recognize 7 botanical varieties as was worked out by Irish in 1898. These are given below. It is doubtful whether this is justified, as they all intercross and intermediates occur.

var. *abbreviatum* Fingerh. WRINKLED PEPPERS

Fruits generally ovate, wrinkled, 5 cm long or less.

var. *acuminatum* Fingerh. CHILLIES

Fruits linear-oblong, over 9 cm long, usually pointed, pungent. Widely grown in India.

var. *cerasiforme* (Miller) Irish CHERRY PEPPERS

Fruits globose with firm flesh, 1·2–2·5 cm in diameter, red, yellow or purple, pungent.

var. *conoides* (Miller) Irish CONE PEPPERS, TABASCO**

Fruits erect, conical, about 3 cm long, very pungent.

var. *fasciculatum* (Sturt.) Irish CLUSTER PEPPERS

Fruits clustered, erect, slender, about 7·5 cm long, very pungent. As the fruits are not borne singly it is probable that these are forms of *C. frutescens*.

var. *grossum* (L.) Sendt. SWEET PEPPERS, PAPRIKA

Fruits large with basal depression, inflated, red or yellow, flesh thick and mild.

---

* See Heiser, C. B. and Pickersgill, B. (1969). Names for the cultivated *Capsicum* species (Solanaceae). *Taxon*, **18**, 277–83.

** A form of *C. frutescens*.

var. *longum* (DC.) Sendt. LONG PEPPERS

Fruits mostly drooping, tapering at apex, 20–30 cm long, red, yellow or ivory, often mild. Calyx not embracing fruits.

Capsicum frutescens L. (2n = 24) BIRD CHILLIES

Syn. *C. minimum* Roxb.

Shrubby perennial. Inflorescence of several flowers. Fruits small, clustered, erect, conical, pointed, 2–3 cm long, usually red, extremely pungent.

var. *baccatum* (L.) Irish CHERRY CAPSICUM

Fruits globose about 1 cm diameter. Usually grown as an ornamental plant.

Bailey (1948) recognizes one species only, *C. frutescens* (syn. *C. annuum*) and 5 varieties, namely, *cerasiforme*, *conoides*, *fasciculatum*, *grossum* and *longum*.

Heiser and Smith (1953) recognize *C. annuum* and *C. frutescens* as valid spp., as they are difficult to cross and the few F_1 hybrids obtained have been highly sterile. They recognize 2 other spp. of cultivated *Capsicum*:*

C. pubescens Ruiz & Pav., which is very distinct, is grown in Central and South America, and has pubescent leaves, purple corolla lobes, and black wrinkled seeds.

C. pendulum Willd., which is grown in South America, has yellow or tan markings on the corolla and yellow anthers.

Wild species also occur.

ORIGIN AND DISTRIBUTION

Prehistoric *Capsicum* peppers are known from burial sites at Ancon and Huaca Prieta in Peru and were widely spread throughout the New World tropics in pre-Columbian times. It would seem that there was either a diffusion from there to Mexico or an independent origin in the latter centre, where there is a great diversity of cvs. *C. annuum* is not known in a wild state; *C. frutescens* doubtfully so, but it has now become naturalized in many parts of the tropics and is spread by birds. Columbus took back fruits to Spain on his first voyage from the New World. The long viability of the seeds and the ease with which they can be transported assisted in its rapid spread in the tropics and subtropics throughout the world after 1492. Already by 1542 three races were recognized in India. The Spaniards and the Portuguese in their search for pepper were quick to seize this new and much more pungent pepper from the New World and to distribute it widely.

ECOLOGY

Capsicums are grown from sea level to 6,000 ft or more in the tropics. They are killed by frost. They are usually grown as a rain-fed crop, with a

* The following cultivated spp. are now recognized: *C. annuum* L. var. *annuum*, *C. baccatum* L. var. *pendulum* (Willd.) Eshbaugh (syn. *C. pendulum* Willd.), *C. chinense* Jacq., *C. frutescens* L., and *C. pubescens* Rucz & Pav.

rainfall of 25–50 in. Too heavy a rainfall is detrimental, as it leads to a poor fruit set and rotting of the fruit. Waterlogging even for a short time causes leaf shedding. Light loamy soil rich in lime is the best for their cultivation, but they can be grown on a variety of soils provided they are well drained.

STRUCTURE

C. annuum—a very variable herb, or sub-shrub, sometimes woody at base, erect, much branched, 0·5–1·5 m high; grown as an annual. Strong tap-root usually broken or arrested in growth on transplanting and numerous profusely branched laterals develop, extending to 1 m. Main shoot is radial, but later branches are cincinnal, one of the branches at each node remaining undeveloped and subtending bract or bracts are adnate and are carried up a lateral shoot to node above. Lvs very variable in size, simple; petiole 0·5–2·5 cm long; lamina broadly lanceolate to ovate, entire, thin, subglabrous, 1·5–12 × 0·5–7·5 cm, tip acuminate, base cuneate or acute. Fls, usually borne singly, are terminal, but because of form of branching appear to be axillary; pedicels up to 1·5 cm long; calyx campanulate, shortly 5-dentate, 10-ribbed, about 2 mm long, usually enlarging and enclosing base of frs; corolla rotate-campanulate, deeply 5–6 partite, 8–15 mm in diameter, white or greenish; stamens 5–6, inserted near base of corolla, anthers bluish, dehiscing longitudinally; ovary 2-celled, but often multiplying under domestication, style simple, white or purple, stigma capitate. Fr. indehiscent many-seeded berry, pendulous or erect, borne singly at nodes, very variable in size, shape and colour and degree of pungency, linear, conical or globose, 1–30 cm long; unripe fruit green or purplish, ripening to red, orange, yellow, brown, cream, or purplish. Seeds 3–5 mm long, pale yellow.

C. frutescens—perennial subshrub, living 2–3 years. Similar in structure to *C. annuum*, but usually 2 or more erect berries at fruiting nodes, small and narrow, 0·7–2·5 × 0·3–1 cm, red or yellow, extremely pungent. Var. *baccatum* has globose frs.

POLLINATION

Bees and ants visit the flowers. Both self- and cross-pollination occurs, the latter being about 16 per cent. Anthesis takes place some time after the flowers have opened. Flowers remain open for 2–3 days. The percentage fruits set is 40–50 per cent.

GERMINATION

The seeds are very light, about 140 per g, and retain their viability for 2–3 years. They germinate in 6–10 days.

CHEMICAL COMPOSITION

The pungent principle of chillies is capsaicin, $C_{18}H_{27}NO_3$, the decylenic acid derivative of vanillylamine, which is present in the placentas; commercial samples contain about 0·1 per cent. Green chillies contain about 83 per cent moisture, 0·6 per cent fat, 3 per cent protein, 6 per cent carbohydrate and 7 per cent fibre. Capsanthin, $C_{40}H_{58}O_3$, is the most important pigment of paprika. Dried chillies contain about 10 per cent moisture. Chillies are a rich source of vitamin C; *C. annuum* contains 50–280 mg/100 g ascorbic acid and *C. frutescens* 2–50 mg/100 g of ascorbic acid. The former contains 100–1,200 I.U. Vitamin A and the latter 200–20,000 I.U. of Vitamin A.

PROPAGATION

Chillies are grown from seeds, which are sown in nurseries or flats. In the former the seed is broadcast at the rate of 1½ lb per 1,000 sq. ft which produces enough plants for 1 acre. The seedlings are transplanted at 4–5 weeks when they are about 6 in. high. Sometimes they are topped 10 days previously to encourage branching.

HUSBANDRY

The fields receive a thorough cultivation and organic manures are often applied. Spacing is usually 2–3 ft apart or 3 × 1 ft. Flowering begins 1–2 months after planting and it takes a further month to the first picking of green peppers; thereafter ripe fruits are picked at intervals of 1–2 weeks and harvesting continues over a period of about 3 months. The ripe chillies are dried in the sun, taking 3–15 days depending on the weather conditions; 100 lb of fresh fruits produce 25–30 lb of dried chillies. The yield of dry chillies per acre in India is 250–750 lb for rain-fed crops and 1,500–2,500 lb per acre for irrigated crops.

MAJOR DISEASES AND PESTS

The virus diseases, mosaic and leaf-curl, cause serious damage to capsicums in the tropics. It is probable that they are transmitted by the thrip, *Scirtothrips dorsalis* H., which is the most serious pest of the crop in India. Other serious diseases are a fruit-rot caused by *Colletotrichum capsici* (Synd.) Butl. & Bisby and anthracnose caused by *C. nigrum* Ell. & Halst.

Fig. 84. **A. Capsicum annuum : SWEET PEPPER. A1, vegetative shoot (×½); A2, flower (×3); A3, flower in longitudinal section (×5); A4, fruit (×½); A5, fruit in longitudinal section (×½). B. Capsicum frutescens : BIRD CHILLI. B1, flowering shoot (×½); B2, leaf (×½); B3, flower (×3); B4, flower in longitudinal section (×4); B5, fruit (×½); B6, fruit in longitudinal section (×¾); B7, seed (×5).**

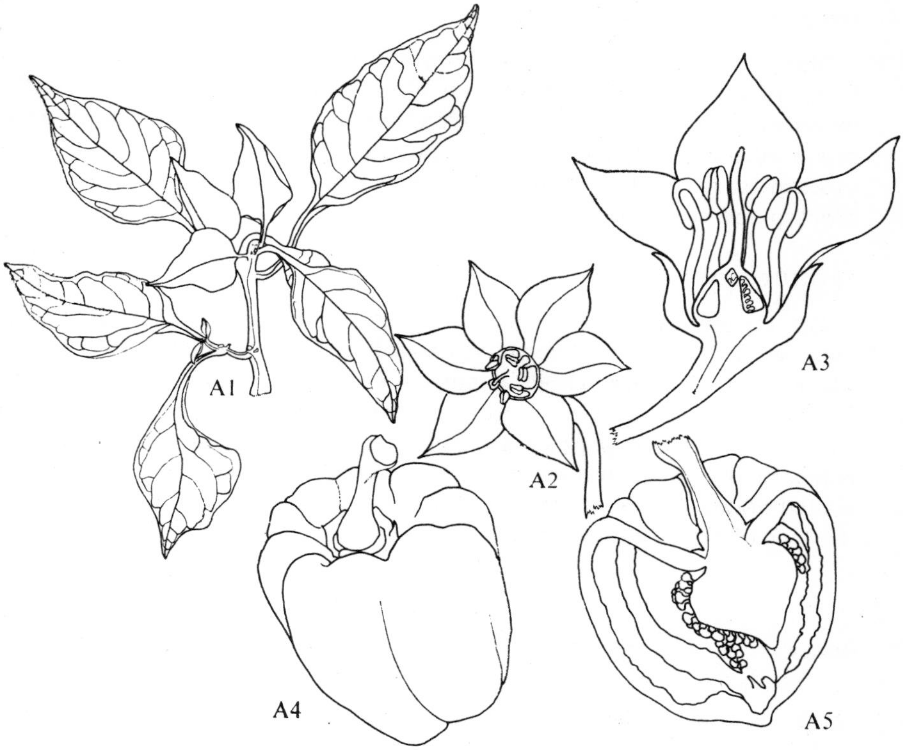
A1
A2
A3
A4
A5

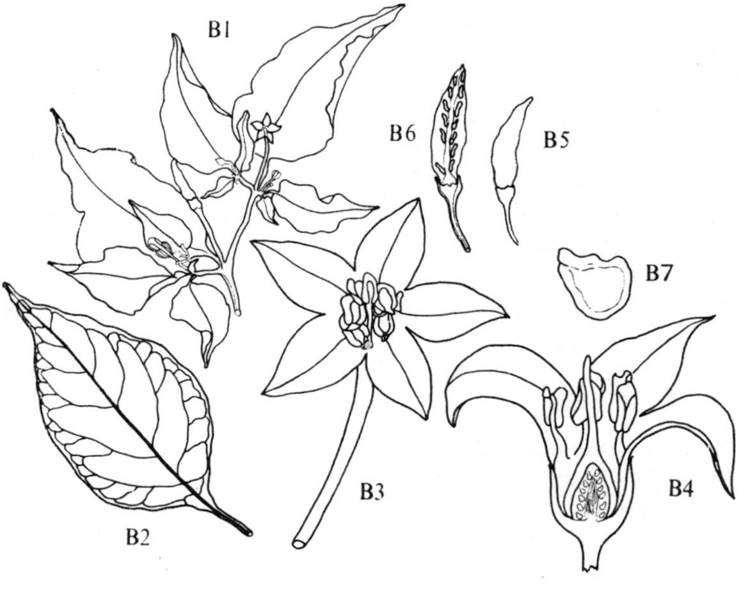
B1
B2
B3
B4
B5
B6
B7

IMPROVEMENT

Many cvs exist and the ease with which they cross can be used for breeding and selecting improved cvs. With the discovery of cytoplasmically inherited male sterility (Peterson, 1958) hybrid F_1 seed can now be cheaply produced. High temperatures accentuate the sterile expression. Red exocarp in the mature fruit and pendent fruits are dominant to yellow exocarp and erect fruits.

PRODUCTION

The biggest exporter of chillies is India, where annual exports have exceeded 200,000 cwt in some years, followed by Thailand with about 100,000 cwt per year. Other countries exporting smaller quantities are Indonesia, Japan, Mexico, Uganda, Kenya, Nigeria and the Sudan. The largest exporters of paprika are Spain and eastern Europe, particularly Hungary. Large quantities of green peppers are grown in the southern United States. Ceylon imports the largest quantities of chillies, followed by the United States and Malaya.

REFERENCES

HEISER, C. B. and SMITH, P. G. (1953). The Cultivated Capsicum Peppers. *Econ. Bot.*, **7**, 214–227.

PETERSON, P. A. (1958). Cytoplasmically inherited male sterility in *Capsicum*. *Am. Naturalist*, **92**, 111–119.

Lycopersicon Mill. (x = 12)

A small genus of 6 spp. of soft herbs, usually perennial. With the exception of *L. esculentum*, which is the cultivated tomato, and *L. cheesmanii*, which is endemic in the Galapagos Islands, they grow wild in a narrow coastal strip extending from Ecuador to northern Chile. The genus is sometimes included in *Solanum* (q.v.), from which it differs in always being unarmed, the leaves are always pinnate or pinnatifid, and the anthers project into a narrow sterile tip and dehisce longitudinally. There are 2 subgenera: *Eulycopersicon* with 2 red-fruited edible spp., *L. esculentum* and *L. pimpinellifolium*, and *Eriopersicon* with 4 green-fruited spp., *L. cheesmanii* Riley, *L. glandulosum* C. H. Muller, *L. hirsutum* Humb. & Bonpl., and *L. peruvianum* Mill.

USEFUL PRODUCTS

L. esculentum (q.v.) is cultivated for its edible fruits.

L. pimpinellifolium (Jusl.) Mill., currant tomato, is a slender-stemmed herb with small red edible 2-celled fruits, about 1 cm in diameter, borne in long trusses; the seeds are glabrous. It grows wild in Peru and Ecuador, where natural hybridization and introgression with *L. esculentum* occurs.

It has now been fairly widely distributed and has been used for crossing with *L. esculentum* to produce some useful cvs which show resistance to certain diseases.

Lycopersicon esculentum Mill. (2n = 24) TOMATO

Syn. *Solanum lycopersicum* L.

USES

The tomato is one of the most important vegetables in most regions of the world, ranking second in importance to potatoes, *Solanum tuberosum*, in many countries. The fruits are eaten raw or cooked. Large quantities of tomatoes are used to produce soup, juice, sauce, ketchup, purée, paste and powder. They are extensively used in the canning industry. Green tomatoes are used for pickles and preserves. Tomatoes grown in the tropics tend to be rather coarse and lacking in flavour compared with those grown in temperate countries. The seeds contain 24 per cent oil and this is extracted from the pulp and residues of the canning industry. This semi-drying oil is used as a salad oil and in the manufacture of margarine and soap. The residual press cake is used for stock feed and fertilizer.

SYSTEMATICS

The following varieties are sometimes recognized:

var. *cerasiforme* (Dun.) Alef., cherry tomato, which occurs wild in Ecuador and Peru and has now been taken throughout the world; it has become naturalized in some tropical countries. Fls 5-partite in long inflorescences; fruits small, about 2 cm in diameter, red or yellow, 2-celled.

var. *pyriforme* Alef., pear tomato, with pear-shaped fruits and 5-partite fls. Indeterminate in growth. Suited to adverse conditions.

var. *commune* Bailey, common tomato. Fls usually 6-partite. Indeterminate in growth.

var. *grandifolium* Bailey, potato-leaved tomato, with large entire leaflets.

var. *validum* Bailey, upright tomato. Plants stout, erect and very compact. Determinate in growth.

Some authorities only recognize var. *cerasiforme* and form *pyriforme* as distinct entities within the species.

ORIGIN AND DISTRIBUTION

Jenkins (1948) and Rick (1956) consider that *Lycopersicon esculentum* var. *cerasiforme*, the putative ancestor of the cultivated tomato, was originally confined to the Peru–Ecuador area and that it spread as a weed throughout much of tropical America, either with or without man's active co-operation. In Mexico, due to its general similarity to the older food plant *Physalis*, it was domesticated. The Spaniards took cultivated forms to Europe soon after the conquest of Mexico (1523) and from there it spread throughout the Old World. Matthiolus describes the tomato in Italy in 1544 and considered it a type of mandrake. It was recorded in England, where it was called the love apple, by Gerard in 1597. Apart from Italy, it does not seem to have been eaten at this time in Europe, but may have been used medicinally. It was taken by the Spaniards across the Pacific to the Philippines and Rumpf found it in eastern Malaysia after 1650. The tomato was introduced into the United States from Europe towards the end of the 18th century. Except for Central America, tomatoes were little cultivated in the tropics until the end of the 19th century.

CULTIVARS

A great many cvs have been developed, particularly in temperate countries, but good cvs suited to the wetter tropics are required. Cvs may be determinate or indeterminate in growth; they vary in the colour, size, shape, flavour and vitamin content of the fruits, the degree of earliness, method of growth which may be erect to sprawling, and in resistance to diseases and root-knot nematode. Cvs have been developed in Hawaii with multiple disease resistance. The cvs at present recommended in Trinidad for wet and dry season production are 'Indian River', 'Floralou' and 'Tehcumseh'.

ECOLOGY

Tomatoes show a wide climatic tolerance and can be grown in the open wherever there is more than 3 months of frost-free weather. They are more successful where there are long sunny periods with light evenly distributed rainfall and night temperatures between 50–68°F. Very wet weather in the tropics with low sunshine incidence and high night temperatures results in excessive vegetative growth at the expense of fruiting and leads to an increase in disease. Yields are higher in the dry season in Trinidad, when irrigation is used, than during the wet season. Tomatoes can be grown at sea-level in the tropics, but usually do better at higher altitudes. A light, free-draining, fertile loam with a pH of 5–7 is the best for the crop, but they can be grown on a variety of soils. Flowering is not affected by the photoperiod. In temperate countries they are extensively grown in heated greenhouses.

STRUCTURE

Variable annual herb, 0·7–2 m tall, erect with thick solid stems, or spreading and later becoming prostrate, coarsely hairy, glandular, with characteristic strong odour. Strong tap-root often damaged at transplanting and a dense system of fibrous and adventitious roots is formed. Branches

---

Fig. 85. **Lycopersicon esculentum : TOMATO. A, leaf (×1); B, flower (×3); C, flower in longitudinal section (×4); D, fruit in transverse section (×⅓).**

B
C
A
D

usually sympodial, although the base of stem may be monopodial. In the former the terminal bud aborts or produces an inflorescence and axis is continued by the development of the axillary bud. Each axis produces leaves at several nodes and then terminates in an inflorescence and this mode of development is then repeated. Small capitate glandular hairs and long pointed trichomes occur on stems, petioles and peduncles. Lvs spirally arranged with a 2/5 phyllotaxy, imparipinnate, 15–30 × 10–25 cm; petiole 3–6 cm long; major pinnae 7–9, opposite and/or alternate, incurled, ovate to oblong, 5–10 cm long, irregularly toothed and sometimes pinnatifid at base (entire in var. *grandifolium*); a variable number of smaller pinnae occur between the larger leaflets. Inflorescences, which arise terminally, are borne opposite and sometimes between leaves, due to method of branching of axis, 4–12 flowered. Fls pendent, hermaphrodite, hypogynous, regular, about 2 cm in diameter; pedicel 1–2 cm long with constricted abscission zone in middle of its length; calyx tube very short with usually 6 narrow pointed lobes, 1 cm long, with glandular and other hairs, sepals persistent and enlarging in fruit; corolla rotate, petals usually 6, about 1 cm long, yellow, stellate, later reflexed, glandular hairy without; stamens usually 6, inserted on short corolla tube, filaments short, anthers 5 mm long, bright yellow, connivant round style, tip prolonged into sterile beak; pistil with several loculi, usually 5–9, due to fasciation under domestication, central fleshy placenta, style pushes up through encircling anthers and may or may not be exserted. Fr. a fleshy berry, hairy when young, glabrous and shiny when ripe, red or yellow, usually globose or depressed at either end, smooth or furrowed, 2–15 cm in diameter. Seeds numerous, reniform, hairy, light brown, 3–5 × 2–4 mm, curved embryo in endosperm; about 350 seeds per g.

In vars *cerasiforme* and *pyriforme* fls pentamerous and ovary bilocular.

POLLINATION

Dehiscence of the anthers is introrsely longitudinal 1–2 days after the opening of the corolla. If the pollen is shed as the style grows up through the anther tube self-fertilization occurs, as is also the case when the style is short and the stigma is not exserted beyond the connivant anthers. In long style forms, the stigma projects beyond the anther cone and cross-pollination can occur. Bumble bees and other solitary bees are the most important pollinating agents. Rick (1958) has shown that in Peru *L. esculentum* is a moderately cross-pollinated species and obtained figures of 25·7 and 14·8 per cent, whereas in Europe and throughout most regions of its cultivation it is almost exclusively self-pollinated. Cross-pollination tends to be higher in tropical than in temperate regions. Growth of the pollen tube is slow and fertilization takes place 50 hours or more after pollination in favourable circumstances. The optimum temperature for the germination of the pollen and the growth of the pollen tube is about 70°F. Parthenocarpic fruits are not uncommon and can be induced by growth hormones.

GERMINATION

Seeds dried and stored in airtight containers will retain their viability for 3–4 years, but viability is usually reduced in the tropics. Germination takes 7–10 days; the primary root emerges and the arched hypocotyl then emerges followed by the liberation from the seed coat of the slender cotyledons.

CHEMICAL COMPOSITION

Ripe tomatoes contain approximately: water 94 per cent; protein 1 per cent; fat 0·1 per cent; carbohydrate 4·3 per cent; fibre 0·6 per cent; vitamin A 250 I.U.; ascorbic acid 25 mg/100 g. The colour of the mature fruits is due to lycopene and carotene, but the former is absent in yellow fruits. Ripe fruits and leaves contain the alkaloid tomatine. The seeds contain 24 per cent of a semi-drying oil.

PROPAGATION

Tomatoes are grown from seed. Ripe fruits are squeezed and the seeds are washed and dried; they may be freed from the pulp by the addition of dilute hydrochloric acid or washing soda. The seeds may be sown in nurseries at 3 × 3 in. and the soil should be sterilized with methyl bromide or formaldehyde. Seedlings are ready for transplanting in 4–6 weeks when they are 6–8 in. high. Seedlings may also be raised in flats or pots.

Tomatoes can be propagated easily from cuttings and can be grafted. Intergeneric grafting is possible and tomatoes have been grafted on to potato, tobacco, various *Solanum* spp., the tree tomato (*Cyphomandra betacea* (q.v.)) and other plants.

HUSBANDRY

In the tropics 3 crops per year can be grown, particularly if irrigation or watering is possible. The land should be thoroughly cultivated before planting and the crop benefits from heavy organic manuring. In the extensive method of production in Trinidad seedlings are planted 3 to a hole, later thinned to a single plant. The usual spacing is 3–4 × 2 ft, but 3 × 1 ft has been found to give higher yields. The crop receives a dressing of 1 cwt of sulphate of ammonia per acre 2 weeks after planting and a further 1 cwt when the first fruits set, together with 1 cwt of muriate of potash. The fields are irrigated in the dry season. Harvesting is spread over a 3–4-week period and yields of 1–6 tons per acre are obtained, being lowest in the wet season. With this method of growing, staking has no effect on yield and pruning reduces yield. Mulching has been shown to be beneficial.

Bharath (1965) recommends intensive planting in Trinidad and growing the crop during the rainy season in open-sided greenhouses, with field production in the dry season. He recommends strawy manure applied at the rate of 80 tons per acre, a closer spacing in double rows alternately

3 and 1½ ft apart, with plants 1–1½ ft apart in the rows; staking and pruning, a high total fertilizer application of 30 cwt per acre, adequate irrigation, and effective control of diseases and pests.

Tomatoes can be grown very successfully using the hydroponic system of cultivation.

For distant markets tomatoes are usually harvested when fully mature, but still green; for local markets they are picked when pink or firm ripe. The fruit should be graded for size and quality; fruits about 3 in. in diameter are usually preferred and they should be firm, evenly coloured and free from damage. Containers for packing tomatoes should be rigid and have adequate ventilation. Tomatoes in the mature green stage can be cold stored at 50–60°F. Average yields in the United States are over 10 tons per acre, but much higher yields can be obtained.

MAJOR DISEASES

Bacterial wilt, *Pseudomonas solanacearum* E. F. Sm., is one of the most serious diseases of tomatoes in the wet tropics, particularly if waterlogging occurs. It produces a rapid wilting without yellowing, followed by death. The pith has a brown waterlogged appearance and disintegrates. Little resistance has yet been found.

Fusarium wilt, *Fusarium bulbigenum* Cooke & Massee var. *lycopersici* (Brushi) Wollenw., causes a wilt accompanied by yellowing of the leaves. The organism enters by the roots; bundles are stained a brownish-yellow.

Sclerotium wilt, *Sclerotium rolfsii* Sacc. causes a quick wilt. The plant is attacked at ground level and white feathery mycelia encircle the stem, followed by the production of sclerotia.

Septoria leaf-spot, *Septoria lycopersici* Speg. produces brown water-soaked spots on the leaves and may cause defoliation.

Grey leaf-spot, *Stemphylium solani* Weber, recently reported in Trinidad and Mauritius, may cause complete defoliation.

Early blight, *Alternaria solani* (Ell. & Mart.) Sorauer attacks leaves, stems, flowers and fruits. It produces brown lesions with concentric rings.

Late blight, *Phytophthora infestans* (Mont.) de Bary, causes serious damage in some countries.

Leaf mould, *Cladosporium fulvum* Cooke, is most serious on greenhouse tomatoes.

Various organisms cause damping-off of seedlings and fruit-rots.

Tomatoes are attacked by a number of virus diseases which include tobacco mosaic, spotted wilt, cucumber mosaic, curly top and tomato yellow top.

MAJOR PESTS

A number of insects attack tomatoes, but they can usually be controlled by spraying. They include: serpentine leaf miner, *Liriomyza stricata* Meig.; flower midge, *Contarinia lycopersici* Felt.; mites, *Hemitarsonemus latus*

Banks; and various caterpillars, particularly the corn ear-worm, *Heliothis armigera* Hubner. Root-knot nematodes, *Meloidogyne* spp., are very common on tomatoes, infection often taking place in the nurseries.

IMPROVEMENT

A great deal of work has been done on the selection, breeding and cytogenetics of tomatoes in temperate countries. With the exception of Hawaii, very little has been done in the tropics, where better flavoured cvs are required, suited to hot wet conditions, and with multiple disease resistance. At the University of the West Indies, Trinidad, an attempt is being made to obtain resistance to bacterial wilt. Growers in Trinidad have selected over the years local hardy cvs which are reasonably well suited to their methods of cultivation.

The characters required in various countries include: compact plants; uniform colour of fruits; fruit size; earliness; multiple disease resistance, including nematodes; resistance to fruit cracking; high vitamin content, particularly vitamin C; cvs suitable for mechanical harvesting.

Resistance to disease is usually controlled by a single dominant gene and has been obtained by interspecific crosses. *L. esculentum* does not cross readily with species in *Eriopersicon* section of the genus, but crosses can be made when the latter are used as the male parents. F_1 hybrids will only mate in backcrosses to *L. esculentum* when they are used as male parents and in backcrosses to the wild parent when they are used as female parents. Many of the interspecific hybrids are self-incompatible. *L. peruvianum* is resistant to a number of diseases and root-knot nematode and has a high vitamin C content. *L. hirsutum* is resistant to *Septoria lycopersici*.

In their spread to the Old World tomatoes became almost entirely self-pollinating. Recessive genes were exposed and fixed and gave rise to many diverse forms. Because of the self-pollination most cvs are pure lines and maintain themselves as such in cultivation. On crossing heterosis may be obtained and F_1 hybrid seed is now being produced. Attempts are being made to breed suitable parents with cytoplasmic male sterility which would reduce the cost of production of F_1 hybrid seed; it may also be possible to sterilize the pollen of the female parent by chemical means. Tetraploids, triploids and haploids have been produced artificially and also occur naturally, but they do not fruit well.

PRODUCTION

The production of tomatoes is increasing in most regions of the world, brought about by increased acreage and increased yields. Mean annual world production per year for the period 1949–1953 was 9,876,000 tons from 1,705,000 acres with an average yield of 5·6 tons per acre. By 1962 acreage had increased to 2,150,000 yielding 16,358,000 tons, an average of 7·6 tons per acre. The United States is by far the largest producer and in 1963 harvested 3,635,928 tons for processing from 246,360 acres and

895,982 tons for the fresh fruit market from 159,840 acres. The value of the crop was $259,596,000. The United States also imports fresh fruits from Mexico. A large proportion of the United Kingdom imports come from the Canary Islands. Bulgaria is the largest European and world exporter of fresh tomatoes. Italy is the largest exporter of processed tomatoes and tomato products. Average annual consumption of fresh tomatoes per head of population is 19·7 lb in Canada, 14 lb in the United Kingdom and 12·2 lb in the United States. If processed tomatoes are taken into consideration, the average annual consumption per head of population in the United States is the equivalent of more than 40 lb of fresh tomatoes.

REFERENCES

BHARATH, S. (1965a). *The Tomato—Trends in Cultivation.* U.W.I.—London M.Sc. Thesis (unpublished).

BHARATH, S. (1965b). Tomato Growing. *J. Agric. Soc. Trin. Tob.*, **65**, 17–33.

CHAPMAN, T. and ACLAND, A. D. (1965). Investigations on Out-of-season Tomato Production in Trinidad. *Trop. Agriculture, Trin.*, **42**, 153–162.

JENKINS, J. A. (1948). The Origin of the Cultivated Tomato. *Econ. Bot.*, **2**, 379–392.

RICK, C. M. (1956). Cytogenetics of the Tomato. *Advances in Genetics*, **3**, 267–382.

RICK, C. M. (1958). The Role of Natural Hybridization in the Derivation of Cultivated Tomatoes of Western South America. *Econ. Bot.*, **12**, 346–367.

Nicotiana L. (x = 9, 10, 12, 16)

Sixty spp., of which 30 spp. are restricted to South America, 6 spp common to both North and South America, 9 spp. in North America, 14 spp. in Australia, and 1 sp. in the South Pacific. South America is the centre of origin of the genus, with subsequent dispersal to North America and to Australia and the South Pacific. Goodspeed (1954) considers that the spread to Australia was via Antarctica during the Upper Oligocene and Lower Miocene, when the climate of Antarctica permitted the development of a temperate to subtropical flora (also see Cotton).

The majority of the species are confined to the temperate zone. They require strong illumination and well-drained soil and tend to occupy newly disturbed soils; they cannot tolerate forest conditions or grasslands and often grow under semi-arid conditions.

Tall, soft-woody shrubs to diminutive annuals, often viscid-pubescent. Lvs alternate, entire. Fls usually in terminal panicles or 1-sided racemes; calyx 5-toothed; corolla tubular, shallowly 5-lobed; stamens 5, free, inserted on corolla tube; ovary bilocular, style terminal. Fr. a capsule with numerous minute seeds.

USEFUL PRODUCTS

N. tabacum (q.v.) is the source of commercial tobacco. *N. rustica* (q.v.) is also grown for tobacco in some areas. In western North America, a number of indigenous spp. have been used for smoking by Indian tribes, notably *N. bigelovei* (Torrey) Watson and *N. attenuata* Torrey ex Watson. Several spp. were in aboriginal use in Australia; they include *N. excelsior* Black, *N. gossei* Domin and *N. ingulba* Black.

Several spp. are grown as ornamentals, the commonest being *N. alata* Link & Otto, a South American sp. with large fragrant white flowers, *N. forgetiana* Hort. ex Hemsley from Brazil with red flowers, and *N. sanderae*, which is an artificial hybrid between *N. alata* and *N. forgetiana*.

KEY TO THE ECONOMIC SPECIES

A. Petiole not winged; corolla greenish-yellow........ *N. rustica*
AA. Lf. sessile or with decurrent petiole which is usually winged; corolla white, pink or red................. *N. tabacum*

Nicotiana rustica L. (2n = 48) NICOTINE TOBACCO

A highly polymorphic sp., long cultivated and unknown in a wild state, with the possible exception of var. *pavonii* (Dunal) Goodspeed, which occurs as a ruderal in the Andes. It is of amphiploid origin and probably originated in Peru and appears to be derived from 2 diploid spp., the progenitors of *N. paniculata* L. and *N. undulata* Ruiz & Pavon, which still occur wild in the area. In pre-Columbian times it was cultivated in Mexico, south-western and eastern United States and eastern Canada, where it was used for smoking, chewing and snuff and was extensively used in rituals. It was the first tobacco to reach England and to be grown and exported by the American colonists. *N. rustica* has now been completely replaced as a commercial source of tobacco for smoking, snuffing and chewing by *N. tabacum*. With its higher nicotine content it is much stronger than *N. tabacum*. Its cultivation is now largely confined to the Soviet Union, where it is grown as a source of nicotinic and citric acids and is also smoked, and northern India, where it is still extensively smoked in cigars and hookahs.

It has been grown as a source of nicotine for insecticidal purposes, but has now been largely superseded by new chemical insecticides. Cvs with high nicotine content have been developed with 4–9·5 per cent nicotine in the cured leaf. These were grown commercially by peasant cultivators in south-western Uganda during the period 1940–1955 and the nicotine was extracted locally and exported. High nicotine content can only be obtained in the tropics at altitudes above 6,000 ft. *N. rustica* is more resistant to cold than *N. tabacum*.

N. rustica is a viscid annual herb, 0·5–1·5 m high, usually single-stemmed. Lvs spirally arranged; petiole unwinged, 5–6 cm long; lamina ovate,

10–30 × 8–25 cm. Fls in many-flowered panicle; calyx pubescent, 0·8–1·5 cm long; corolla greenish-yellow, 1·2–2 cm long, lobes very short; stamens 5 inserted near base of tube and not exserted; capsule ovoid, 0·7–1·5 cm long, with numerous minute seeds.

Nicotiana tabacum L. (2n = 48) TOBACCO

USES

The cured leaf provides a pleasurable and habit-forming narcotic used for smoking, chewing and snuffing. Smoking is now practised throughout the entire world. This use was observed by Columbus in the West Indies in 1492. At this time *N. tabacum* was grown for this purpose in Central America and northern and eastern South America. Tobacco was then taken to Europe, where it was first grown as an ornamental and was used medicinally for the treatment of sores and ulcers, and snuff was taken for headaches. It was not until about the last two decades of the 16th century that the habit of smoking in Europe became at all widespread, the habit of pipe smoking being introduced by sailors returning from the New World. Sir Walter Raleigh did much to popularize pipe smoking at the court of Elizabeth I. By the first decade of the 17th century the habit was widespread and tobacco had been taken to Africa and Asia. In Persia the water-pipe or hookah was used for smoking it. Then followed a reaction wherein kings, potentates and puritans in many countries tried, with little success, to suppress the habit by persecution and taxation. Tobacco is now heavily taxed by many governments and provides a major source of revenue. Cigarettes became popular during World War I and are smoked more than any other form of tobacco. They are now considered to be a contributory cause of lung cancer.

During modern manufacture a variety of flavouring and conditioning materials are used as a sauce in which the leaves are immersed or as a spray. Many of them are closely-guarded secrets of the manufacturers, but they include glycerine, licorice, sugar, molasses and tonka bean (*Dipteryx odorata*, q.v.). In Java shredded cloves (q.v.) are added to the cigarettes. Cigarettes are made mainly from flue-cured tobacco, but a proportion of light air-cured or Turkish tobacco may be added. Three types of air-cured leaf are used in the manufacture of cigars, filler, binder and wrapper, for which special cvs are grown. Pipe tobaccos are usually made from blended flue- and air-cured mixtures. Plugs and twists, which are used for chewing and smoking, are usually made from dark air- and fire-cured tobacco, but some flue-cured may be included. Snuff is prepared by grinding up dark air- and fire-cured leaves.

The alkaloid nicotine is extracted from tobacco waste and was formerly much used as an insecticide. On oxidation it yields nicotinic acid, a constituent of many vitamin preparations. The seeds contain an edible oil, but as only Turkish tobacco is allowed to flower and fruit in commercial production, it is not much utilized.

ORIGIN AND DISTRIBUTION

There is no well-authenticated record of the occurrence of *N. tabacum* in a wild state, but occasional escapes from cultivation are found. Goodspeed (1954) has shown its demonstrable origin involving progenitors of *N. sylvestris* Speg. & Comes (2n = 24) and a member of the section *Tomentosae*, probably *N. otophora* Grisebach (2n = 24). It probably originated in north-western Argentine where the parent species are still in contact. It was in aboriginal cultivation in pre-Columbian times in the West Indies, Mexico, Central America, Colombia, Venezuela, the Guianas and Brazil.

The Spaniards first cultivated *N. tabacum* in Haiti about 1530 with seed obtained from Yucatan. Commercial production was begun in Haiti about 1580 and in Trinidad about 1595. The earliest cultivation in Europe was in France in 1556 with seed from Brazil. It is recorded in Spain in 1559 and in Rome in 1561. Tobacco was taken to England by Sir John Hawkins from Florida in 1565, but this would have been *N. rustica.* Thereafter tobacco spread rapidly throughout Europe. The Spaniards and the Portuguese distributed tobacco widely throughout their spheres of influence and, by the first decade of the 17th century, it was reported as far afield as Africa, India, Japan, the Middle East and the Philippines.

The first commercial cultivation began in Virginia in 1612 and in Maryland about 1631. This early production was *N. rustica*, but was later replaced by *N. tabacum*, with seed obtained from Trinidad. In Europe two types of tobacco were recognized, Virginia and Spanish, the latter coming from the Spanish colonies in the West Indies and South America. Tobacco is now more widely distributed than almost any other crop.

CULTIVARS

Various attempts have been made in the past to distinguish botanical varieties within this highly polymorphic species, but have proved unsatisfactory. It is better to regard the species as an assemblage of cvs exhibiting wide morphological variation and ecological tolerance. The desirable characters of the cured leaf are dependent upon soil and climatic conditions, the cv. grown, and the methods of cultivation and curing employed. This determines the use to which it is put. In most countries, especially the United States where many major official types are recognized, the production of each type is localized in specific geographical regions and the trade obtains tobacco with specific characteristics from the individual regions. Consequently growers in a region tend to cling to the cvs and production methods already in use.

Cvs can be divided into classes according to the method of curing as follows:

FLUE-CURED: 'Orinoco' is the basic type cv. from which most of the other cvs have arisen. They include 'White-Stem Orinoco', 'Virginia

Bright', 'Yellow Special', '400', 'Oxford', 'Dixie Bright', 'Vamorr', 'Vesta', and 'Coker'. New numbered strains of the basic cvs are constantly being developed, *e.g.* 'Coker 140'.

FIRE-CURED: 'Pryor' is the principal type cv. and originated from 'Orinoco'. 'Malawi Dark Western' is grown in Malawi (Edmond, 1965).

AIR-CURED: 'Burley' is the principal type cv., from which 'White Burley', with creamy stalk and midrib was selected in 1865. Disease-resistant cvs of 'Burley' have been developed. 'Southern Maryland', with thin, light-bodied leaf, is one of the oldest tobacco cvs in the United States. A mutation from it produced 'Maryland Mammoth', which is a short-day plant. 'One-sucker', which only produces one sucker in the leaf axils, has dark, coarse, strong leaves.

CIGAR TOBACCOS: These may be divided into three types: (a) Cigar-filler used to form the core of the cigar. Cvs include 'Pennsylvania Broadleaf', 'Spanish', 'Gebhardt', and 'Dutch'. (b) Cigar-binder used to hold the filler in shape. 'Havana' is the principal type variety. (c) Cigar-wrapper, used as the final wrapping of the cigar. The basic cv. is 'Cuban', which has numerous derivatives.

TURKISH TOBACCO: Largely produced in the Middle East and in Rhodesia. The leaves are very small, often only 3–6 in. long and the cured leaf has a very distinctive aroma.

ECOLOGY

N. tabacum originated in the sub-tropics, but is now grown from the equator to as far north as central Sweden and as far south as south Australia. It grows more rapidly in warm climates. It requires a frost-free period of 90–120 days from transplanting to harvesting; at higher latitudes the seedlings are produced in greenhouses. The optimum mean temperature for the growing season is 70–80°F. Strong illumination is required. A minimum of 10 in. of rain is needed during the growing seasons, but 20 in. gives better growth; drier weather is required for ripening and harvesting. Continual rain during the growing season leads to disease and thin, light-weight leaves. A prolonged dry period when nearing maturity causes premature ripening; heavy rain after a dry spell during ripening causes secondary growth and deficiency of gum; in both cases the leaves are difficult to cure. Very dry weather during fire- and air-curing may cause the leaves to dry out too quickly and remain green; too high a humidity may cause barn rot. The crop is very susceptible to severe injury by hail and strong winds.

No crop is more sensitive than tobacco to small variations in soil and they determine the type and use of the leaf produced. They have a profound effect on the flavour of the tobacco and the aroma of the same cv. can be quite different when grown on different soil types. The soil requirements vary according to the class of tobacco grown, but in general the crop requires adequate drainage, moisture retention and aeration. Tobacco will not tolerate waterlogging. Tobacco soils are usually acid with a pH of 5–6·5. A light soil is essential for flue-cured tobacco and light sandy loams are used, ideally with about 70 per cent fine sand and 6–8 per cent of clay. Cigar-wrapper cvs also require a light soil. Fire-cured, dark air-cured and cigar-filler tobaccos are grown on heavier silt or clay loams. Light air-cured and cigar-binder tobaccos are grown on soils intermediate between the two.

Tobacco is not affected by photoperiod, except the 'Mammoth' cvs which have indeterminate growth, continuing to increase in height and produce new leaves during long days, and only flower when exposed to short days.

STRUCTURE

Stout viscid annual or limited perennial herb, 1–3 m high; usually grown as a short-term annual.

ROOTS: Well-developed tap-root, but under commercial cultivation often broken; an extensive fibrous root system then develops from several horizontal laterals. Root development stimulated by topping and suckering.

STEM: Thick, erect, unbranched (except when suckers are allowed to grow); base becoming somewhat woody. Considerable variation in length of internodes. When plant is topped, *i.e.* inflorescence removed in late bud stage, each axillary bud grows out to produce a branch or sucker; these are removed and more are produced and these are also removed.

LEAVES: Simple, spirally arranged, either clockwise or anti-clockwise, with a phyllotaxy of $\frac{3}{8}$, $\frac{5}{13}$, or more rarely $\frac{2}{5}$; angle with stem varies from drooping, horizontal or more erect to an angle of 45°; the number of lvs is usually fairly constant in each cv., 20–30, but may exceed 100 in 'Mammoth' cvs of indeterminate growth. Lvs usually sessile and decurrent; petiole if present usually winged. Lamina typically ovate-lanceolate or elliptic, 5–75 cm long, average about 50 cm, entire, pinnately-nerved; base may be auricled; tip acuminate; multicellular hairs borne in great numbers and may be simple or capitate and glandular, latter secreting gum. Lower lvs somewhat broader and less pointed than upper lvs. Topping and suckering profoundly modifies growth and development of lvs depending on the height and time of topping (see CHEMICAL COMPOSITION and HUSBANDRY below). Age gradient and ripening of lvs is from the base of stem upwards; ripeness indicated by yellowish-green colour in thin lvs and yellow mottling in thick lvs.

FLOWERS: Borne in terminal panicled racemes with up to 150 fls per inflorescence; pedicels 1–2 cm long, bracteate; calyx cylindrical, viscid, 1–2·5 cm long with 5 somewhat unequal pointed teeth; corolla funnel-shaped, 3·5–5·5 cm long, woolly without, expanding into deep cup about

1 cm in diameter with 5 acute lobes, usually pink, rarely white or red; stamens 5, attached to base of corolla tube, of unequal length with 2 longest near mouth of corolla or slightly exserted, 2 slightly shorter and 5th shorter than either pair; anthers small, dehiscing longitudinally; ovary superior, 2-celled, fleshy axile placentation, ovules numerous; style slender; stigma sticky, capitate, at mouth of corolla or slightly exserted. Floral formula K(5) C(5) A5 $\underline{G}$(2).

FRUIT: 2-valved, ovoid capsule, 1·5–2 cm long, almost completely covered by calyx.

SEEDS: Minute, oval to spherical, 0·5 mm long, finely reticulate, light to dark brown, 3–5 layers of endosperm and straight embryo. Produced in enormous numbers with 2,000–8,000 seeds per capsule and up to 1 million per plant; 1,000 seeds weight 0·08 gm = 300,000 per oz.

POLLINATION

The anthers dehisce as the flowers unfold and, as the stigma is then receptive, self-pollination normally takes place. Bees and other insects, as well as humming birds, visit the flowers for nectar and up to 4 per cent cross-pollination can occur. Foreign pollen applied to the stigma after self-pollination may fertilize a proportion of the ovules due to the differential growth rates of the pollen tubes; 2 hours after self-pollination artificially cross-pollination produced 27 per cent of total progeny, and 10 per cent after 8 hours. Cross-incompatibility has been reported. In compatible matings the growth rate of the pollen tube is rapid; in incompatible matings growth is slow and the pollen tube may not extend more than half way down the style after 10 days, which is the maximum life of the flower. The differences in the rates of pollen tube growth are controlled by a series of multiple alleles, S_1, S_2, S_3, etc., and only pollen tubes with an allele different from that in the style will grow at the normal rate.

GERMINATION

The seed is long-lived and will remain viable for 20 years or more when stored under good conditions. Before sowing, the seeds should be cleaned by a blast of air and disinfected for 15 minutes in 1 : 1,000 solution of mercuric chloride or silver nitrate to control angular leaf-spot and wildfire. Seed of good quality after cleaning gives 90 per cent or more germination, for which the optimum temperature is 75–80°F. Germination begins after 5–7 days, when the primary root ruptures the testa, followed by the hypocotyl carrying with it the testa. The cotyledons then emerge and a few days later the first two leaves appear.

Fig. 86. **Nicotiana tabacum** : TOBACCO. **A**, leaf (×$\frac{1}{3}$); **B**, portion of inflorescence (×$\frac{1}{2}$); **C**, flower in longitudinal section (×1); **D**, capsule in longitudinal section (×1).

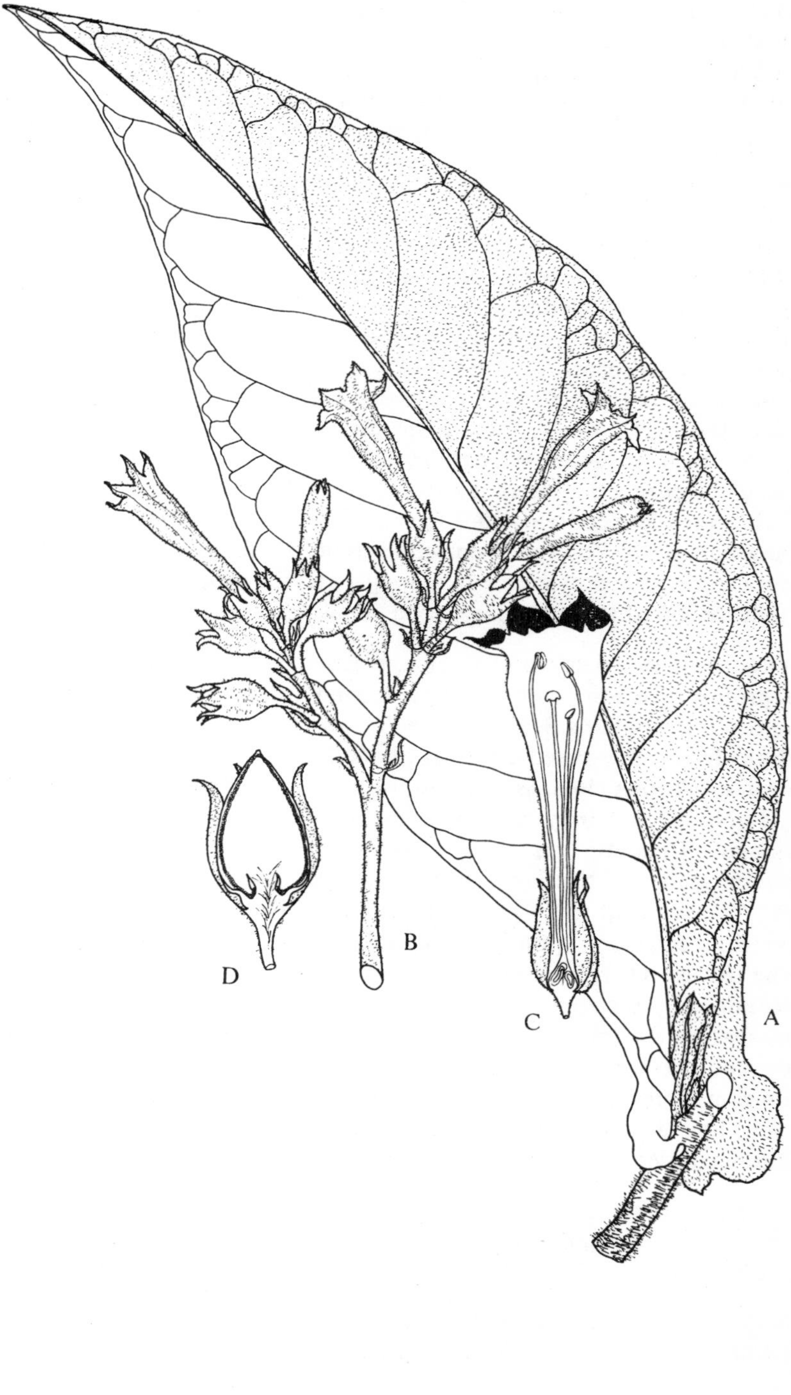
A
B
C
D

CHEMICAL COMPOSITION

WATER: The water content of the leaf decreases during ripening; freshly-harvested leaf contains 85–95 per cent, which is reduced to 80–85 per cent on wilting and to 10–25 per cent on the completion of curing. For export to the United Kingdom the water content should not be less than 10 per cent, as extra duty is charged below this figure.

CARBOHYDRATES: The proportion of carbohydrates to nitrogenous matter increases during ripening, and, because of the high starch content, the tissue, particularly in heavy-bodied leaf, becomes brittle and cracks readily when folded between the fingers (a test for ripeness in conjunction with colour). Carbohydrates comprise 25–50 per cent of the total dry matter of the mature leaf, of which starch and invert sugars are the most important constituents, and are relatively high in flue-cured tobacco and low in cigar types. Respiration continues during the early stages of curing; most of the starch, due to the enzyme diastase, is converted to dextrin and maltose, then to monosaccharides which are oxidized in respiration. During flue-curing the starch is reduced from 20–40 per cent to 5–8 per cent and the cured leaf contains 20–30 per cent sugars, which is further reduced during bulking by continued enzyme activity to about 18 per cent. In air- and fire-curing the protoplasm lives longer and the process continues further so that the cured leaf contains only a small percentage of sugars.

NITROGENOUS COMPOUNDS: In the green leaf most of nitrogen is in the protein fraction, with nicotine second in importance. Green cigar-wrapper leaf contains 3–15 per cent of protein dry weight and is low in carbohydrates, flue-cured contains about 2 per cent and is high in carbohydrates. Protein loss is high during curing. The alkaloid nicotine, $C_{10}H_{14}N_2$, occurs only in the genus *Nicotiana*; it is very poisonous to mammals. Cigarette grades of flue-cured cvs contain 1·5–2·5 per cent; cigarette grades of 'Burley' 3–4 per cent; *N. tabacum* × *N. rustica* 5–7 per cent; *N. rustica* up to 10 per cent. 'Strength' of tobacco and its smoke is primarily dependent upon its nicotine content. Nicotine is most abundant in the leaves, but is present in all parts of the plant except the seeds. It is synthesized by the roots as is shown by grafting tomato on tobacco root-stocks when the former contains nicotine. Low topping causes a marked and rapid accumulation of nicotine in the leaves. Heavy nitrogenous manuring increases the nicotine content. The nicotine content varies with season and location, but cultivar differences are relatively constant. Nicotine is used as a source of nicotinic acid.

ORGANIC ACIDS: Nonvolatile polybasic acids occur in the leaves. During curing citric acid tends to increase; oxalic acid undergoes no change; malic acid shows a decided loss.

POLYPHENOLS: Tannin-like substances occur in important quantities in the leaf and affect the colour and properties of the cured product. On the death of the cells in air- and fire-curing they are oxidized giving red and brown colours to the leaf. In flue-curing the leaf is dried rapidly to a state where this oxidation is prevented, thereby fixing a yellow colouration.

CHLOROPHYLL AND OTHER PIGMENTS: During the early phases of curing the green pigments, chlorophyll a and b, are destroyed by oxidation and yellow pigments, carotene and xanthophyll, become evident. Completion of the yellowing process marks the approximate end of the starvation period. In flue-curing this colour is fixed by the rapid killing and drying of the leaf; in air- and fire-cured leaf the process goes a stage further and the oxidized polyphenols produce a brown colour.

ETHERIAL OILS AND RESINS: These occur in the capitate glandular hairs of the leaf and furnish the aromatic principles of tobacco.

MINERAL CONSTITUENTS: Tobacco has a high content of ash, the quantity in dry leaf ranging from 12–25 per cent. The quantity and composition of the mineral components affect the combustibility and other elements of quality in the leaf.

In passing from the lowest to the topmost leaf of the plant there is a progressive increase in the content of total nitrogen (protein, ammonia and nicotine) and a decrease in total ash, calcium, magnesium and pH of the cured leaf. Leaves cured on the stalk show marked differences in composition as compared with leaves harvested and cured separately, as minerals and organic constituents are translocated from the leaves to the stalks.

The seeds contain 32–42 per cent of a semi-drying oil, containing 60–75 per cent linoleic acid, and 20 per cent protein of high biological quality. The seeds are non-toxic.

PROPAGATION

Commercial tobacco is always grown from seed. The plant can be grafted if required.

HUSBANDRY

SEEDBEDS: The production of good seedlings is the foundation of a good crop. Seedbeds should be sited near clean permanent water. They are usually 20–60 ft long, 3–4 ft wide, raised 3–6 in. by the addition of top soil, and separated by paths 1–2 ft wide. A fine tilth is essential. The soil is sterilized and this may be done by burning brushwood and grass on the surface, or injecting steam, or by the use of chemicals such as formaldehyde, methyl bromide, Vapam or C.B.P. In Rhodesia an NPK mixture in the proportion of 4 : 18 : 4 at the rate of 10 lb per 30 sq. yd of seed bed is recommended to be applied a few days before sowing. The seed is cleaned and sterilized before sowing (see GERMINATION above). The minute seeds should be sown thinly and one level teaspoonful is sufficient for 30 sq. yd of seedbed and this will supply sufficient plants for 1 acre transplanted at 3 × 2 ft. The seed is sown by placing it in a watering can, filling the can with water, stirring it and then watering on through a fine rose. Alternatively it may be sown mixed with sand or ash. The seed is not covered with soil. The seedbeds are shaded with grass on a frame, or with cheese cloth (weave 26 × 22 threads per in.) or saran netting. Beds must be kept damp, but not overwatered. The seedlings are hardened off by gradually opening up the covers. The time taken from sowing to transplanting is 40–60 days.

ESTABLISHMENT: Tobacco should not be grown on the same land for more than 2 years in succession and should be rotated with crops which are not susceptible to eelworm, such as sorghum, millets, maize, or grasses. The land should be thoroughly cultivated before planting, which may be done on ridges or on the flat. The usual spacing is 3–4 × 2–3 ft; wider spacing is required for fire-cured than for flue-cured tobacco, which is usually planted at 3 × 2 ft. Strong rosette-type seedlings, 6–8 in. high, are used for transplanting, which should be done in the late afternoon or early morning if possible, preferably on damp dull days. Transplants may be given temporary shade and may be watered if necessary. Cigar-wrapper tobacco is usually grown under cheesecloth shade.

MANURING: With flue-cured tobacco very fertile soil and too heavy manuring, particularly with nitrogen, will produce heavy leaves which will not give a bright yellow colour on curing. Nitrogen is usually applied at the rate of 10–40 lb N per acre to flue-cured, up to 100 lb per acre for dark air- and fire-cured, and 10–20 lb per acre for Turkish tobacco. Phosphates are applied at the rate of 50–100 lb P_2O_5 per acre and potash at 100–150 lb K_2O per acre. The application of MgO and boron is also recommended. In Central Africa a 6 : 12 : 18 NPK mixture is recommended at the rate of 120–200 lb for flue-cured; in the United States 800–1,200 lb per acre of a 3 : 10 : 6–12 mixture is used.

CULTURAL OPERATIONS: The crop should be kept free of weeds and the soil ridged round the plants. In the past it was customary to remove the lower leaves to a height of 6 in. when the plants were 15–20 in. high, an operation known as priming, but this is now seldom done.

When the flower buds are formed, the inflorescence and the topmost leaves are broken off by hand, an operation which improves the yield and quality. The time and height of topping depends on the class of tobacco grown, the type of soil and its fertility, and the spacing. The kind of weather which is likely to follow topping must also be taken into consideration, and, as it is impossible to judge this with certainty, it constitutes a definite hazard. If it is too dry too many leaves may have been left for normal development; if it is too wet secondary growth may occur and the leaves will become too thick. With fire-cured tobacco 8–10 good leaves are usually left on the plant; 12–20 leaves in the case of flue-cured; air-cured is intermediate. Low topping causes an increase in the size and thickness of the leaves, accumulation of nutrients and nicotine and gives a cured leaf which is heavy, oily, dark in colour, tough and somewhat leathery. If the leaf of flue-cured types is too heavy, topping may be deferred until the flowers open or even until the seeds have begun to set. The time from transplanting to topping is 50–70 days.

Soon after topping suckers are produced due to the growth of axillary buds. These are removed when not more than 3–5 in. long and thereafter suckering is done weekly as more develop. Unless it is done regularly the advantages of topping are lost. Suckering may be prevented by the application of light oils or other chemicals.

HARVESTING: In the United States air- and fire-cured, and most cigar tobacco (except wrapper) are harvested whole by cutting the stem near ground level when the greatest number of the best leaves are at the proper stage of ripeness, which is usually 40–55 days after topping. The leaves are cured on the stems, which are tied to poles and the leaves are wilted before transfer to the curing barn. Flue-cured and cigar-wrapper leaf is harvested individually as the leaves ripen, beginning with the lowest leaves. The lowest leaves ripen 80–100 days after planting and 14–21 days after topping. Two to four leaves are taken at each picking, which is continued at weekly intervals. The leaves are strung back to back in alternate pairs on either side of sticks and are hung in the barn for curing. Peasant-produced air- and fire-cured crops in east and central Africa are often harvested by the priming method.

CURING: No method of curing can induce the development of qualities which are not potentially present, but imperfect curing can destroy these potential qualities. Constant care and attention is required. The process is one of slow starvation; during the yellowing process circulation of the sap is necessary, followed by killing and drying of the leaf. If water is lost too quickly the leaf remains green; if lost too slowly the leaf becomes sponged and barn-rot may occur. Cigarette tobaccos are flue-cured; cigar tobaccos, 'Burley' and 'Maryland' tobaccos are air-cured; dark heavy chewing and plug tobaccos are fire-cured; Turkish tobacco is sun-cured.

(1) FLUE-CURING: This is usually done in brick barns in which heat from wood, oil, gas or other source is supplied in closed flues. Absolute control of temperature and humidity in the barn is essential. A barn 12 × 12 × 16 ft is sufficient for 1–2 acres of crop and takes 1,500–2,000 lb of green leaf at each filling. The usual size of barns is 16 × 16 ft and 20 ft high and it is usual to build them in blocks of 4–6 barns. A barn of this size requires a minimum of 8 acres of crop to fill it with 4,000–6,000 lb of green leaf at the correct stage of ripeness. Plans of the barn and other details are given in Purseglove (1951) and Pirie (1961). 5,000 lb of green leaf produces about 800 lb of cured leaf and the number of curings per crop is 4–8. The time taken to cure a barn is about 5 days, but is dependent on the type of leaf, heavy leaf taking longer than light leaf.

There are three stages in flue-curing:

(a) YELLOWING OF LEAF: After the barn has been filled in one day, all the ventilators are closed and the temperature is maintained at 90–100°F with a relative humidity of 85 per cent. When the leaf starts to yellow at the tips and around the edges the temperature is raised slightly until the yellow colour spreads to the midrib. During this process the temperature is raised gradually to 110–115°F and the ventilators are partly opened during the later stages. The time taken for this stage is 36–48 hours.

(b) FIXING THE COLOUR: Ventilation is gradually increased and the temperature raised in order to kill the leaf, destroy the enzymes, fix the yellow colour, and to dry out the web of the leaf. The temperature is gradually increased from 115–125°F and the process takes 12–20 hours.

(c) DRYING THE LEAF: The drying of the web is completed at 135–140°F with all the ventilators open, after which the bottom ventilators are closed and the midrib is dried out by gradually increasing the temperature to 160°F. This stage takes approximately 50 hours.

Flue-curing tobacco requires considerable skill and experience and it is usual to inspect the barns several times during the night.

(2) FIRE-CURING: This is usually done in log or grass barns. The leaf is hung 4–7 days to yellow and small open fires are then made in pits in the floor to provide smoke and to produce the creosotic and distinctive aroma. The fires are put out at night. During the early stages temperature should not exceed 90°F and when drying should not exceed 125°F. The time taken for curing is 3–4 weeks.

(3) AIR-CURING: This is a natural process and curing under normal atmospheric conditions is done in wooden or grass barns. The leaf should yellow before it dries out, after which the rate of drying is gradually increased by increasing the ventilation. The best temperature for the early wilting is 70–75°F and it should never exceed 110°F even in the final stages. The time taken for curing is 6–8 weeks.

(4) SUN-CURING: This is done by exposing the leaves to the sun and is used for Turkish tobacco, but it is usual to air-cure the leaf for 2–3 days and ferment it in heaps for 24–36 hours, before exposing it to the sun.

BULKING AND GRADING: After curing the leaf is taken from the barn and is bulked on platforms. The bulks should not be more than 6 ft high and should be large enough to allow some fermentation. For bulking the lamina and upper part of the midrib should be pliable and the lower part of the midrib should be only slightly supple. Weights are placed on the bulks. The tobacco should be bulked for at least a month before grading and marketing.

The leaves are then graded according to size, colour and texture. A large number of grades are employed in many countries. The grade will depend partly on the position of the leaf on the plant. The bottom leaves are called lugs, the lower middle leaves cutters, and the upper middle and top leaves leaf. The best flue-cured tobacco should have leaves of medium texture with good elasticity, relatively free from gum, slightly oily, mildly aromatic, and a bright yellow colour without blemish. Fire-cured tobacco should have large, broad leaves with heavy body and texture and good elasticity, and should be dark mahogany in colour without blemish. Light air-cured tobacco should be relatively light in body and texture and light or reddish brown in colour. Cigar-filler should be heavy bodied, with a considerable aroma and a satisfactory burn. Cigar-binder should have finer texture and more elasticity. Cigar-wrapper, the elite of tobaccos, should have thin, silky leaves, with fine veins and texture, good elasticity, free from injury and blemish, and with a satisfactory even colour, aroma and burn. Turkish tobacco has very small leaves, yellow to light reddish brown in colour and with a distinctive aroma.

After grading, the leaves are tied in hands with 12–15 leaves per hand for fire-cured and 20–30 leaves for flue-cured. This is done by wrapping a leaf of the same grade folded into a band round the top of the petioles and then tucking the stem of the leaf inside the hand. Sometimes the leaf is stripped by removing the petiole and midrib before selling. The leaf is then conditioned when it is dried thoroughly and steamed for a few seconds to acquire the correct moisture content which should be about 12 per cent for local sales and as little over 10 per cent as possible for the United Kingdom export market. The tobacco is then baled, hessian usually being used, and is ready for auction or export. The tobacco is left to mature in the bales for 1–2 years before being used for manufacture.

YIELDS: Very variable yields are obtained. They may be as low as 400–500 lb of cured leaf per acre in African peasant cultivation. Average yields per acre are about 1,000 lb or more in Rhodesia, 1,800–1,900 lb in the United States and over 2,000 lb in Europe.

MAJOR DISEASES

Damping-off in seedbeds caused by *Corticium solani* (Prill. & Delacr.) Bourd. & Galz. and *Pythium* spp. can cause losses.

Blue mould, *Peronospora tabacina* Adam, endemic in Australia, is relatively new in the United States, where it is now a serious disease. It is spreading in other regions. Blue mould was introduced into north-western Europe in 1959, and had invaded all the tobacco areas of Europe and the Mediterranean region within 4 years. It appears on the undersurface of the leaves as a blue-grey coating.

Black root rot, *Thielaviopsis basicola* (Berk. & Br.) Ferraris, was one of the most troublesome diseases of tobacco in Canada and the United States until resistant cvs were raised in recent years.

Black shank, *Phytophthora parasitica* Dast. var. *nicotianae* (Breda de Haan) Tucker, is one of the most serious diseases in the United States, Canada, and the East Indies; it has also been reported from Africa. The attack begins at soil level and blackened dead roots and a stem rot are produced.

Granville wilt, *Pseudomonas solanacearum* (E. F. Smith) Dowson, is a very serious disease in many parts of the world, particularly in North America and the Far East. It causes decay of the roots, followed by wilting.

Wildfire, *Pseudomonas tabacum* (Wolf & Foster) Stevens, and angular spot, *P. angulata* (Fromme & Murray) Stapp, are two serious bacterial diseases of tobacco. They produce spotting on the leaves. In wildfire yellow-green spots are produced on the leaves which develop a red-brown central area; in angular spot angular dark brown spots are produced. Disinfecting seeds assists control.

Frog-eye, *Cercospora nicotianae* Ell. & Ev., does little damage in temperate countries, but in the tropics is one of the most serious diseases, particularly in wet weather. Circular spots with white or pale brown centres and narrow dark brown margins are produced on the leaves. It also causes spotting in the barns during the early stages of curing.

Brown spot, *Alternaria longipes* (Ell. & Ev.) Mason, produces brown spots with concentric rings; it can be distinguished from frog-eye by the presence of blackish spots on stem, midrib, pedicels and capsules. It causes serious damage in Africa.

Mildew or white mould, *Erysiphe cichoracearum* DC., can be very serious on mature tobacco in Africa, particularly at higher altitudes. It appears at the end of the growing season on the lower leaves as round powdery patches, which soon coalesce to give a white coating to the upper surface of the leaf.

Mosaic, one of the most infectious of plant viruses with several strains, is very common on tobacco and is cosmopolitan. The common symptom is dark green and yellowish mottling, but it can also cause spotting and scorching. It is transmitted mechanically in handling plants and the use of infected tobacco for smoking, chewing and snuff.

Other viruses include leaf curl, which causes a puckering of the leaves and veins and is transmitted by white flies, *Bemisia* spp.; rosette, which causes cessation of vertical growth and distortion of leaves and is transmitted by *Myzus persicae* (Sulz.); kromnek, which is transmitted by *Frankliniella* spp. (thrips), is serious in South Africa.

Barn rot, *Rhizopus arrhizus* Fischer, causes a rotting of the leaves during curing and is associated with excessive moisture, particularly during air-curing.

The parasitic angiosperms *Orobanche ramosa* L. and *O. minor* Sutton (broom-rape) are parasitic on tobacco in Africa and Asia.

MAJOR PESTS

Any insect which damages the leaves causes a reduction in the value of the crop. The most serious pests are: hornworms, *Protoparce* spp., the most destructive pest in the United States; the tobacco flea-beetle, *Epitrix parvula* Fab.; cutworms, *Agrotis* spp., *Euxoa* spp., *Feltia* spp. and *Lycophotia saucia* Hbn. cause damage in many countries; budworms, *Heliothis* spp., attack young leaves and terminal bud and also damage the capsules; stem borer, *Phthorimaea heliopa* Low, causes damage in Africa. The tobacco beetle, *Lasioderma serricorne* Fab., has a world-wide distribution; both the larva and the mature beetle are destructive pests of cured leaf and manufactured tobacco products. The tobacco moth, *Ephestia elutella* Hbn., also attacks stored tobacco. The tobacco crop is also attacked by mole crickets, grasshoppers, wireworms, aphids and thrips. One of the most serious pests of tobacco in all parts of the world is the root-knot eelworm, *Meloidogyne* spp.

IMPROVEMENT

The genus *Nicotiana* has long been used in fundamental studies of plant genetics, particularly *N. tabacum*, because of (1) the relative size and simple structure of the flowers which are easy to manipulate, (2) large number of seeds produced, (3) longevity of seeds, (4) ease with which self- and cross-pollination can be carried out, (5) pollen viability over several weeks if stored properly, (6) extreme variability of plant characters exhibited, (7) the number of interspecific crosses which can be made.

The great practical difficulty in improving existing cvs is that the quality of the product is the determining factor in successful tobacco production and the more important elements of quality are not measurable or visible in the growing crop. Furthermore, the market in each production area has been established largely on the basis of the cvs already in production, and marked changes in cvs are resisted by the growers and manufacturers of tobacco products. A further difficulty is the profound effect of local soil and climatic conditions on the final product.

METHODS: Many of the cvs now grown are of long standing and improvement includes the introduction and testing of new cvs, selection, and intra-specific and interspecific hybridization. In this work it is necessary to rely on characters which are visible and measurable, while continued back-crossing is relied upon to transfer the assembly of invisible quality. Few mutations of economic importance have appeared, except 'White Burley' in 1864 and the indeterminate growth of 'Mammoth' cvs, which is due to a single recessive.

Interspecific crosses have been made freely within the genus and these have been used to supply disease resistance. The chromosome number within the genus varies from 2n = 18–64. *N. tabacum* L. has been crossed with *N. alata* Link & Otto and *N. langsdorffii* Weinmann (2n = 18); *N. longiflora* Cav. *N. plumbaginifolia* Viv. (2n = 20); *N. glauca* Graham, *N. glutinosa* L., *N. sylvestris* Speg. & Comes, *N. tomentosa* Ruiz & Pav. (2n = 24); *N. suaveolens* Lehm. (2n = 32); *N. bigelovii* (Torrey) Watson, *N. debneyi* Domin, *N. rustica* L. (2n = 48); and others. When the species differ in the number of chromosomes, crosses are more likely to be successful if the female parent has the higher number. In transferring dominant genes for disease resistance in intraspecific crosses the technique used has been to double the chromosome number of the F_1 hybrid by colchicine and to back-cross the amphiploid to the *N. tabacum* cv. By this means sterility in the F_1 can be avoided. Octoploids have been induced, but they show reduced growth, later maturity, smaller and thicker leaves of an abnormal dark-green colour, larger stomata, larger seeds of reduced viability.

TECHNIQUE: In artificial cross-pollination the upper leaves, low flowering branches and open flowers are removed and the inflorescence is covered, usually by a strong paper bag of 16 lb capacity. The inflorescence is dusted with an insecticide against budworm. The corollas of buds which will open next day, as indicated by their pink tips, are slit and the anthers removed. Pollen, which can be dried and stored for several weeks, is applied to the receptive stigma. Any buds are then removed and the inflorescence is re-bagged. The bag is opened from time to time to remove dead flowers.

AIMS AND RESULTS: (1) Yield is not of primary importance if this means any radical alteration of cv. already grown, as each area is expected to produce its own type of tobacco.

(2) Improved field and handling characteristics include: (a) toughness, so that the leaves will stand rough handling; (b) storm resistance to prevent bruising and breakage; (c) resistance to sun scald; (d) uniformity of ripening, particularly for cvs harvested on the stalk; (e) stand-up types with long internodes which reduces damage in the field and during harvesting; (f) fewer and smaller suckers or slow-growing suckers, which is difficult to select for, as the suckers are not produced until the plant is topped and then no seed is produced.

(3) Disease resistance: this has received the major attention in recent years. Resistance to black root-rot, black shank and Granville wilt has been found in some *N. tabacum* cvs. Immunity to black root-rot has been obtained from *N. debneyi* and to black shank from *N. longiflora* and *N. plumbaginifolia*. Some resistance to wildfire has been obtained from *N. longiflora*. Resistance to blue mould has been obtained from *N. debneyi*, and *N. longiflora* and *N. plumbaginifolia* are stated to be immune. *N. glutinosa* has been used to obtain resistance to mosaic and has been incorporated into a number of cvs. Attempts are being made to obtain resistance to nematodes. Intraspecific resistances are frequently polygenic; interspecific resistance to mosaic and wildfire depends on single dominant genes.

(4) Quality: This consists of a complex of characters which are not easy to define or to measure and the trade insists on receiving leaf of prescribed quality and this will vary with the class and type of tobacco. Factors taken into consideration are the size and shape of leaf; the proportion of midrib to web; leaf venation; thickness, density, elasticity, body, gumminess, and grain of leaf; curing properties; burning quality which includes slow rate, evenness and completeness of burn, fire-holding capacity and fine coherent residual ash; strength, aroma, and taste; hygroscopic properties and chemical composition including nicotine and sugar content. Although colour

of the cured leaf has by itself little to do with the actual quality except in appearance, it is the most important single criterion employed in judging quality, since it is closely correlated with various other characters which cannot be seen by simple inspection. Uniformity and depth of colour, lustre and brilliance are important factors in determining the grade within the class and type.

PRODUCTION

Since the beginning of World War I there has been a phenomenal increase in cigarette smoking and a decline in pipe smoking. World production of tobacco in 1930 was approximately 5,000 million lb. Present production is about 7,000 million lb per annum from 7·5 million acres, excluding China, where estimated production is at least 1,000 million lb. The United States is the largest producer with over 2,000 million lb per annum, followed by China and India (800 million lb). Other large producing countries in order of production are Brazil, Japan, Rhodesia, Pakistan and Canada. Turkey, Greece and Bulgaria produce substantial quantities of Turkish or oriental tobacco, a total of about 600 million lb. The main exporters of unmanufactured tobacco are the United States (about 500 million lb), Rhodesia and India. The principal importing countries are the United Kingdom (over 300 million lb), Western Germany, the United States (mainly Turkish tobacco), Soviet Union and the Netherlands (about 120 million lb).

In the United Kingdom consumer expenditure on tobacco is approximately £1,250,000,000 per annum and the consumption in million lb weight of manufactured tobacco is approximately: cigarettes 240, pipe tobacco 33, cigars 2 and snuff 0·8. The customs duty on imported unmanufactured tobacco into the United Kingdom, which was 9s. 6d. in 1931, is now over £2 10s. 0d. per lb. The annual tobacco consumption per adult member of the population is: United States 10·6 lb, Canada 10 lb, Netherlands 8·6 lb, United Kingdom 6·5 lb.

REFERENCES

EDMOND, D. E. (1965). The Development of the Malawi Dark Western Tobacco Variety. *Trop. Agriculture, Trin.*, **42**, 265–271.

GARNER, W. W. (1951). *The Production of Tobacco*. New York: Blackiston.

GOODSPEED, T. H. (1954). *The Genus* Nicotiana. Waltham, Mass.: Chronica Botanica.

HOPKINS, J. C. P. (1956). *Tobacco Diseases with Special Reference to Africa.* Kew: Commw. Mycological Institute.

PIRIE, A. N. (1961). Design and Construction of Flue-cured Tobacco Buildings. *Rhod. Agric. J.*, **58**, 193–203.

POEHLMAN, J. M. (1959). *Breeding Field Crops*. New York: Henry Holt.

PURSEGLOVE, J. W. (1951). *Tobacco in Uganda.* Entebbe: Govt. Printer.

AKEHURST, B. C. (1968). *Tobacco.* London: Longmans, Green.

Solanum L.

One of the largest genera of plants with over 1,000 spp. of erect or climbing herbs, shrubs or rarely small trees in temperate and tropical regions throughout the world. Many are spiny and most are stellate pubescent. Lvs alternate, exstipulate, simple or pinnate. Inflorescences often extra-axillary and opposite lvs. Fls gamopetalous; corolla usually 5-partite; stamens inserted on throat of corolla and same number as lobes, anthers usually connivent round style and often dehiscing by terminal pores; ovary superior, usually 2-celled and many-ovuled; style simple; stigma small and terminal. Fr. a berry with persistent calyx.

USEFUL PRODUCTS

FRUITS: *S. hyporhodium* A. Br. & Bouché, cocona, is a native of the upper Amazon and has red or yellow edible fruits, 3–10 cm in diameter. It is under trial in Puerto Rico and elsewhere and shows promise.

S. muricatum Ait., pepino, which probably originated in Peru, is occasionally cultivated in tropical America for its fruits, 10–15 cm long, which are yellow streaked with purple.

S. quitoense Lam., naranjilla, is a native of the northern Andes and is extensively cultivated in Ecuador and Colombia between 3,000–7,000 ft. It is a sub-shrub, 2–3 m high, with large dentate leaves up to 45 cm long, pale lilac flowers 4 cm in diameter. The globular fruits, about 5 cm in diameter, are bright orange in colour covered with short brittle hairs, which are easily removed by rubbing. The green acid pulp is used for juice and the fruits are made into preserves and pies. It has been introduced into several tropical countries.

VEGETABLES: The species of major importance is the potato, *Solanum tuberosum* (q.v.). The eggplant, *S. melongena* (q.v.), is widely grown as a vegetable.

A number of wild species are grown to a limited extent in the tropics for their leaves which are used as pot-herbs and their immature fruits which are cooked as vegetables and for seasoning other food. These include: *S. aethiopicum* L. in Africa, *S. macrocarpon* L. in Africa and introduced into Malaysia, and *S. nigrum* L., which is widespread as a weed throughout the Old World tropics and is occasionally cultivated, is extensively used as a pot-herb in Africa and Asia, in spite of the fact that it is reputed to be poisonous in Europe. The immature fruits of the following are cooked as vegetables or for seasoning, but their leaves are not used: *S. ferox* L. in south-east Asia, *S. gilo* Raddi in Africa and introduced into South America, *S. incanum* L. in Africa, *S. indicum* L. in Africa and Asia, and *S. torvum* Swartz, which is widely spread throughout the tropics.

OTHER USES: Several species of *Solanum* are used in native medicine throughout the tropics. *S. aculeastrum* Dunal is used as a hedge plant by local peoples in tropical Africa.

ORNAMENTALS: Several showy climbers are grown in tropical gardens, notably *S. jasminoides* Paxt., *S. seaforthianum* Andr. and *S. wendlandii* Hook. f., all of which are from the American tropics.

Solanum melongena L. (2n = 24) BRINJAL, EGG-PLANT, MELONGENE

USES

The cooked fruits provide a useful vegetable in many parts of the tropics. They may be boiled, fried, or stuffed. The unripe fruits are sometimes used in curries.

ORIGIN AND DISTRIBUTION

The egg-plant was first taken into cultivation in India, where wild plants occur. Arabs took it to Spain and Persians to Africa. It has now spread throughout the tropics.

CULTIVARS

A number of named cvs are available.

ECOLOGY

In the tropics egg-plants grow well up to about 3,000 ft. It does best in well-drained, sandy loam.

STRUCTURE

A weakly-perennial, erect, branching herb, 0·5–1·5 m high; grown as an annual; old plants becoming somewhat woody; sometimes somewhat spiny; all parts covered with a grey tomentum. Strong, deeply-penetrating tap-root. Lvs alternate, simple; petioles 2–10 cm long; lamina ovate to ovate-oblong, densely hairy, margin sinuately lobed, apex acute or obtuse, base rounded or cordate, often unequal. Fls solitary or in 2–5 flowered cymes, opposite or sub-opposite lvs, 3–5 cm in diameter; pedicel 1–3 cm long, elongating in fruit; calyx about 2 cm long, spiny, woolly and persistent with 5–7 narrow lobes, enlarging in fruit when it often splits on 1 or 2 sides; corolla with 5–6 broadly-triangular lobes about length of tube, incurved, hairy beneath, glabrous within, purplish-violet in colour; stamens 5–6, about 1 cm long, free, erect, yellow, filaments very short, anthers long and narrow, opening by 2 terminal pores; ovary 2-locular, style simple with capitate lobed green stigma slightly exserted beyond stamens. Frs large pendent berry, ovoid, oblong or obovoid, 5–15 cm long, smooth, shiny, white, yellow, purple or black. Seeds numerous, small, light brown.

POLLINATION

Ochse (1931) in describing the flowers states ‘the female ones solitary, the males in many-flowered inflorescences’. In plants examined by me in which the flowers were borne singly, these were found to be hermaphrodite and produced pollen and isolated plants fruited. Hector (1936) states that the flower drop is very much greater from inflorescences with 2–3 flowers than when borne singly and that the former have a weaker expression of femaleness. Using the pollen of *Petunia violacea* Lindl., seedless fruits have been obtained.

GERMINATION

This takes about 2 weeks and is epigeal.

CHEMICAL COMPOSITION

The fruits contain approximately: water 92 per cent; protein 1 per cent; fats 0·3 per cent; carbohydrates 6 per cent.

PROPAGATION

The egg-plant is usually grown from seed, but axillary shoots can be rooted if required. Where bacterial wilt is prevalent in Puerto Rico, egg-plant is sometimes grafted on *Solanum torvum* Swartz, which is resistant to the disease.

HUSBANDRY

The cultural requirements are practically the same as the tomato (q.v.), but it requires a longer growing period. The seedlings are usually transplanted when 5–7 in. high at a spacing of 3–4 × 2–3 ft.

MAJOR DISEASES AND PESTS

Bacterial wilt, *Pseudomonas solanacearum* E. F. Sm., is a serious disease of egg-plant in the wet tropics, as it is of tomato (q.v.). The fungus, *Phomopsis vexans*, causes damage to leaves, stems and fruits. The two most serious pests in the West Indies are lace-bug, *Corythaica passiflorae* Berg, and flea beetles, *Epitrix parvula* F. and *E. cucumeris* Harris, which produce a shot-hole effect on the leaves.

IMPROVEMENT

Good cvs have been bred in some countries. Heterosis is reported to occur.

PRODUCTION

Egg-plants are grown for local consumption in many parts of the tropics. They are popular in the West Indies.

Fig. 87. **Solanum melongena : EGG-PLANT. A,** shoot (×⅓); **B,** flower (×1½); **C,** flower in longitudinal section (×1); **D,** fruit (×½).

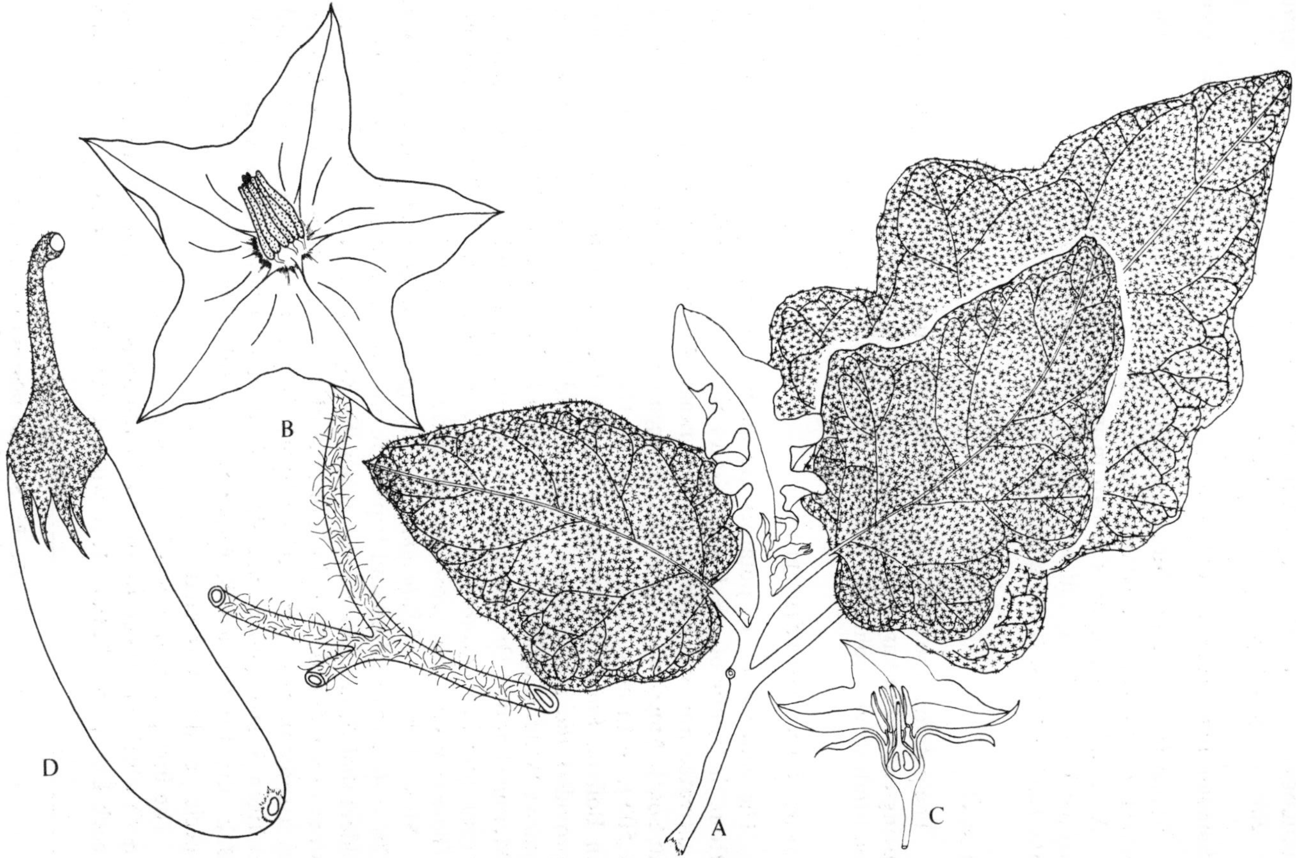
A
B
C
D

REFERENCE

CHILDERS, N. F. *et al.* (1950). Vegetable Gardening in the Tropics. *Puerto Rico Federal Exp. Stat. Circ. 32.*

Solanum tuberosum L. (2n = 48) POTATO

Sometimes called the European or Irish potato to differentiate it from the sweet potato, *Ipomoea batatas* (q.v.).

Although the potato is grown in mountains in the tropics, it is not a tropical crop and, as full accounts are given in books on temperate crops, the full treatment is not given here.

USES

The potato is the most important vegetable in the world today, the tubers being cooked in various ways. It is also used for stock feed. Starch, spirits and industrial alcohol are prepared from it.

ORIGIN AND DISTRIBUTION

The wild species of potatoes are distributed more or less continuously from the south of the United States to southern Chile. The indigenous cultivated potatoes have a narrower range. The main centre of diversity is in South America in the Andes between 10°N and 20°S at altitudes above 6,500 ft, with maximum variability on the altiplano around Lake Titicaca in Bolivia. Potatoes were grown by the Incas and, together with maize, provided the staple food, supplemented by the grain of *Chenopodium quinoa* Willd. (Family: Chenopodiaceae) and the tubers of *Oxalis tuberosa* Molina (Family: Oxalidaceae), *Ullucus tuberosus* Caldas (Family: Basellaceae), and *Tropaeolum tuberosum* Ruiz & Pav. (Family: Tropaeolaceae). These crops are still grown on the sierra.

Most of the cvs of *Solanum tuberosum* grown in the Andes are tetraploids (2n = 48), but diploid and triploid forms are also cultivated. Hawkes considers that *S. tuberosum* arose either directly from an ancestral form of *S. stenotomum* Juz. & Buk. by a process of simple chromosome doubling, or as a spontaneous amphidiploid hybrid between the more ancient diploid cultigen *S. stenotomum*, and the diploid wild species *S. sparsipilum* Juz. & Buk. Brucher favours *S. vernei* Juz. & Buk. (2n = 24) as the ancestral species, as does Cardenas.

The first potato to reach Europe was the sweet potato, *Ipomoea batatas* (q.v.) which was taken to Spain by Columbus. *Solanum tuberosum* did not reach Europe until about 1570 when it was taken to Spain probably from

---

Fig. 88. **Solanum tuberosum** POTATO. **A,** flowering shoot (×½); **B,** leaf (×½); **C,** flower in longitudinal section (×2); **D,** fruit (×½); **E,** fruit in transverse section (×½); **F,** tubers (×½).

A
B
C
D
E
F
M WONG

the northern Andes. It was introduced independently into England in 1586. It is figured in Gerald's herbal of 1597. Sir Walter Raleigh is reputed to have introduced it into Ireland from England. A long time elapsed before its value as human food was recognized and it was probably sheer necessity that drove the Irish peasantry to grow and eat it before 1663; it was in Ireland that it was first grown in quantity and it then spread slowly to other parts of the British Isles. It was not much grown in England before the 19th century. Frederick the Great tried to promote potato cultivation in Prussia in 1744, but it was not really established until the Seven Years' War. Blight, *Phytophthora infestans* (Mont.) de Bary, first appeared in Belgium in 1836 and spread widely in Europe. It caused the great famine in Ireland in 1845 and subsequent years. Resistant cvs were later obtained from South America.

In the United States it was introduced into New England by Irish immigrants in 1719, but it had been introduced into Virginia in 1621 from Bermuda. The Spaniards took the potato to the Philippines at an early date. It reached India in the 17th century and large quantities are now grown there. It had reached Japan by 1766 and about this time was taken by the Dutch to Java. The missionaries probably introduced it into East Africa towards the end of the 19th century and it is now extensively cultivated in the Kenya highlands.

STRUCTURE

Herbaceous branched annual, 0·3–1 m high, with a mass of fine, fibrous, adventitious roots and swollen stem tubers. Lvs pinnate with small interjected leaflets between the main pinnae as in the tomato. Fls white, red or purple. Fr. a small inedible berry.

PROPAGATION

Potatoes are propagated vegetatively by means of tubers (seed potatoes) or parts of tubers which are cut to include at least one bud or eye. They should be cut at right angles to the main axis to remove apical dominance.

CHEMICAL COMPOSITION

The tubers contain approximately: water 80 per cent; protein 2 per cent; carbohydrate (mainly starch) 17 per cent.

PRODUCTION

Potatoes were not extensively planted in Europe until the beginning of the 19th century, but now about 90 per cent of the world's potato crop is grown in Europe. Human consumption in Great Britain is about 15,000 tons per day. Now in volume and value on a world basis, the potato exceeds all other crops, including wheat (Hill, 1952). With the exception of the low tropics, it is grown more universally than any other crop. The largest producers are the Soviet Union and Germany. It grows in cool moist climates in temperate regions. The best temperature for tuber production is 60°F and it is usually stated that no tuberization takes place above 80°F. However, in Trinidad at 10°N the Dutch cvs 'Patrones' and 'Arka' have given up to 10 tons per acre in experimental cultivation when grown in the dry season with irrigation. Long-day cvs should be avoided in the tropics.

REFERENCES

CHAPMAN, T. (1965). Experiments with Irish Potatoes (*Solanum tuberosum*) in Trinidad. *Trop. Agriculture, Trin.*, **42**, 189–198.

DODDS, K. S. (1965). The History and Relationships of Cultivated Potatoes. *Essays on Crop Plant Evolution*, edited by Sir Joseph Hutchinson, 123–144, Cambridge Univ. Press.

GOODING, H. J. (1961). Irish Potatoes (*Solanum tuberosum*) in Trinidad. *J. Agric. Soc. Trin. Tob.*, **61**, 193–211.

HAWKES, J. G. (1967). The History of the Potato. *J. Roy. Hort. Soc.*, **92**, 207–224, 249–262, 288–302.

SALAMAN, R. N. (1949). *The History and Social Influence of the Potato.* Cambridge Univ. Press.

STERCULIACEAE

About 50 genera and 750 spp. of trees and shrubs, rarely herbs, mostly tropical. Stellate hairs usually present. Lvs usually alternate and simple; stipules usually present, mostly deciduous. Fls usually clustered, actinomorphic, hermaphrodite or unisexual; sepals 5, valvate; petals 5 or absent, hypogynous; stamens 5 or more in 2 whorls, joined in column or separate, outer whorl often staminodes, anthers usually bilocular; ovary superior, usually with 2–5 carpels; fruits various.

USEFUL PRODUCTS

BEVERAGES: *Theobroma cacao* (q.v.) is the most important sp., producing cocoa. *Cola* spp. (q.v.) are sometimes used for beverage purposes, but more usually as a masticatory in West Africa.

GUMS: *Sterculia urens* Roxb. yields karaya gum, produced by incisions made in heartwood. It is exported from central India as a substitute for tragacanth and is used in the textile, cosmetic, cigar, paste, and ice-cream industries. *S. setigera* Del. and *S. tragacantha* Lindl., natives of tropical Africa, also yield gum.

Cola Schott & Endl. (x = 10) KOLA

About 60 spp. in tropical Africa, with 51 spp. in West Africa, of which 9 are imperfectly known. Evergreen forest trees, mostly small or moderate in size. In the edible spp. lvs simple, entire, alternate or whorled; perianth white or coloured; apetalous; fls of 2 kinds—male with anthers joined in column and hermaphrodite with double or single ring of anthers at base of ovary, regularly arranged; ovary with several loculi and stigmas, usually 5; ripe frs of star-shaped follicles hanging vertically; seeds up to 5×3 cm with 2 or more cotyledons, small embryo, and no endosperm.

KEY TO CULTIVATED SPECIES

A. Lower lvs alternate, fls not whorled
  B. Lvs curved and twisted, apex long acuminate, lateral veins not prominent; frs usually straight, russet, rough to touch, not tuberculate; seeds with more than 2 cotyledons........................ *C. acuminata*

BB. Lvs flat, apex abruptly acuminate, veins prominent; frs curved with keel and beak, green, smooth to touch, tuberculate; seeds with 2 cotyledons..... *C. nitida*

AA. Lvs in whorls of 3 or 4; fls whorled

B. Lvs usually in whorls of 3, gradually acuminate; lateral veins prominent, close and straight; frs green, tuberculate, beaked; seeds with thin tough testa and more than 2 cotyledons................ *C. anomala*

BB. Lvs usually in whorls of 4, shortly and abruptly acuminate, lateral veins curved; frs not tuberculate; seeds with very thick fleshy testa and 2-6 cotyledons *C. verticillata*

USES

The seeds are chewed as a stimulating narcotic. 'Kola occupies a unique place in the life of West Africa. Not only is it a comestible of everyday use and the sustainer and comforter of the weary traveller and the toiling labourer, but it has also its particular uses in the social life and religious customs of the people.' (Russell, 1955.) It is highly esteemed by the Mohammedans in the drier northern regions of West Africa, who import it from the southern forest regions. In addition to its use as a masticatory, a beverage is made by boiling powdered seeds in water.

SYSTEMATICS, ORIGIN, DISTRIBUTION AND STRUCTURE

Cola acuminata (P. Beauv.) Schott & Endl. ABATA KOLA

Occurs naturally in the forests from Togo and southern Nigeria eastwards and southwards to Gabon, the Congo and Angola. It is extensively planted in western Nigeria where the Yoruba people value it above all other spp. and use it ceremonially and socially.

A slender tree up to 13 m in height, more usually 7–10 m, often branching near base. Lvs sparse and confined to tips of branches, elliptic or oblanceolate; up to 27×11 cm (average $16 \times 5{\cdot}5$ cm), dark green, rather fleshy, often curled, lateral veins not prominent, tip acuminate about 2·5 cm long, often twisted downwards; petiole about 4 cm long. Inflorescence several to many flowered, not whorled. Fls rotate; hermaphrodite fls about 2·5 cm across, male fls smaller; perianth united for nearly half length, white with red splash inside at base. Frs of up to 5 follicles at right angles to peduncle, sessile, straight, brownish in colour, rough to touch, not knobbly, up to 20×6 cm. Seeds up to 14 per carpel, up to $4 \times 2{\cdot}5$ cm; testa thin, white; cotyledons 3–5, usually pink or red, bitter to taste.

Cola anomala K. Schum. BAMENDA KOLA

Occurs in Cameroun, particularly in Bamenda, from 3,000–7,000 ft, where it is cultivated and is the only kola used. Seeds are also exported to the north.

Tree to 30 m, but usually not more than 20 m. Lvs not confined to tips of branches, usually in whorls of 3, elliptic to obovate, about 15 × 6 cm, flat, stiff, leathery, dull green; lateral veins prominent and straight; tip gradually acuminate, straight; petiole 3·5 cm long. Inflorescence few-flowered; fls in whorls, hairy, yellow. Fruiting carpels green, smooth to touch, knobbly, strongly keeled, beaked, up to 12 cm long. Seeds few in each carpel; testa thin, tough; cotyledons 3–6, pink or red.

Cola nitida (Vent.) Schott & Endl. (2n = 40) GBANJA KOLA

Native of the forest zones of Sierra Leone, Ivory Coast and eastwards to Ghana, where it is also planted; it has been extensively planted in western Nigeria since 1912. It was taken by slaves to Jamaica and Brazil about 1630. Plants distributed by Kew to various tropical botanic gardens in the 1880's under the name of '*C. acuminata*' appear to have been this sp.

Tree to 20 m or more in height, but more usually 10–13 m; trunk unbranched for several m. Lvs not confined to tips of branches, rather variable, obovate or elliptic, up to 33 × 13 cm (average 19 × 7 cm), flat with prominent lateral veins, tip abruptly acuminate; petiole about 5 cm long (15 cm has been recorded), with prominent pulvini at base and tip. Inflorescence axillary, not whorled. Fls 2–3 cm long; hermaphrodite fls larger than male fls, about 3–4 cm across, male fls about 2 cm across; perianth divided for at least two-thirds of length, typically white or cream with dark red markings within; forms occur in which inside of perianth is deep maroon or white without red markings (the subsp. *albida* of some authors); androecium with little or no column and almost sessile at base of perianth, reddish in colour; ring of 10 2-lobed anthers. Fruit on short hanging peduncles with up to 7 follicles, ovoid in recurved position, up to 13 × 7 cm; stout dorsal keel, short terminal reflexed beak, outer surface of follicle shining green, smooth to touch, with knobbly tubercules. Fruits mature in 4–5 months. Seeds 4–10 per carpel in 2 rows, ellipsoidal, up to 5 cm long, with thin tough white testa and 2 cotyledons, white, but more commonly pink or red.

Cola verticillata (Thonn.) Stapf ex A. Chev. OWÉ KOLA

Indigenous in forests of Ivory Coast, Ashanti and southern Nigeria to Lower Congo. It was planted on a small scale over a wide area, but yields an inferior product, fruiting 3–4 months before *C. nitida*.

Tree to 25 m, but usually 12–17 m. Lvs usually in whorls of 4, obovate or elliptic, 11–25 × 3·5–9 cm, shortly acuminate; lateral veins curved. Fls whorled with stellate hairs, cream with dark reddish marking within; androecium with short column. Follicles ovoid, borne at right angle to

---

Fig. 89. **Cola nitida** : KOLA. **A**, flowering shoot (×½); **B**, leaf (×½); **C**, portion of inflorescence (×½); **D**, male flower (×1); **E**, hermaphrodite flower (×1); **F**, fruit (×½); **G**, dehiscing fruit (×½).

F
C
A
B
D
E
G

peduncle, nearly straight, 20 × 7 cm, furrow on dorsal surface. Seeds 6–8 per carpel with thick fleshy white testa and 3–4 pink or red cotyledons.

NOTE: The following information is based on *C. nitida*.

ECOLOGY

Cola nitida occurs naturally in the rain forests of West Africa to as far east as Ghana. It is now planted extensively in the forest zone of Ghana and Nigeria in the area 6–7° N with an annual rainfall of 50–70 in. and with well-marked wet and dry seasons. It can be grown on light soils unsuitable for cocoa and on better soils it is often interplanted with cocoa. It grows best on well-drained fertile soils rich in humus. In Ghana wild trees are left when planting cocoa and it is also used as shade for cocoa. Growth is in flushes.

POLLINATION AND INCOMPATIBILITY

Although the anthers of the hermaphrodite flowers contain viable pollen, this is not shed; consequently they are functionally female only. Male and hermaphrodite flowers are often borne on different inflorescences and the male flowers are more numerous. The flowers have a foetid odour which is presumed to attract flies which may be the pollinating agents. When artificially self-pollinated, some trees fail to set fruit, while others are very fruitful, so it would appear that some clones are self-incompatible (see Cocoa). An isolated tree at the University of the West Indies, Trinidad, never sets fruit, while an isolated tree in the Royal Botanic Gardens in Port of Spain sets fruit regularly. Many hermaphrodite flowers wither without setting fruit and there is also a heavy loss of young developing fruits.

GERMINATION AND PROPAGATION

Germination is uneven and takes 7–12 weeks. In native practice it is usually planted at stake. Better germination is obtained if the testa is removed and the rate of germination is increased if the pods are heaped together in the shade and damped regularly for 16 days from harvesting to sowing. The seed should be shallowly planted and not covered with more than ½ in. of soil. The germination rate and percentage is increased by placing in trays and covering with polythene and sacking. On emergence of radicle, seed may then be transferred to baskets or bitumenized cardboard or polythene pots, which may be later used for planting in field, as losses may be numerous if transplanted with naked roots or when the tap-root is damaged.

Cuttings from seedlings and grafted seedlings root readily when taken from shoots that have hardened after a flush in growth; cuttings from mature trees give poor results. Budding gives only 5 per cent take. Whip-, tongue- and saddle-grafting of leafy scion wood from mature trees onto seedling rootstocks in pots is successful if the grafted plant is covered with a polythene bag.

CHEMICAL COMPOSITION

The nut contains about 2 per cent caffeine, a trace of theobromine, a glucoside, kolanin, which is a heart stimulant, and an essential oil. It also contains 9 per cent protein, 2 per cent fat, 74 per cent carbohydrate and 2 per cent fibre.

HUSBANDRY

Kola is usually planted at stake at 20–27 ft apart in food plots prepared from cleared forest land. The maize, yams and cassava act as temporary shade. Growth is rather slow in early years and after 4 years the trees are 6–10 ft in height. The crop receives little or no cultivation after the food crops are abandoned except for slashing weeds. Flowering may occur in the 5th year with a little fruiting in the 7th year. Fair crops are obtained after 12–15 years with full production at 20 years and the tree may continue bearing until 70–100 years old. Nigerian farmers often make cuts in the bark of unproductive trees with cutlasses to induce productivity. The main harvest in Nigeria is from September to January. Ripe fruits before the follicles are split are harvested monthly by cutting with a curved knife blade on the tip of a long pole. The follicles are then split and the seeds removed, which are then fermented in heaps for 5 days, after which the testas are removed and the nuts are washed and cleaned. The nuts are stored in baskets lined with green leaves, which are changed from time to time, and they may be kept for several months without spoiling. They must be examined regularly for kola weevil (see below). There are great differences in productivity between individual trees. Average yields are about 210 saleable nuts per tree or 525 lb per acre.

DISEASES AND PESTS

Attacked by root-rots, *Fomes lignosus* Klotzsch. and *F. noxius* Corner, and thread blight, *Marasmius scandens* Massee, which also attack cocoa. Capsids, *Sahlbergella singularis* Hagl., attack shoots. The most serious pest is the kola weevil, *Balanogastris kolae* (Desbr.), which burrows into the harvested nuts.

IMPROVEMENT

Individual trees show very considerable variation in yield. In 4 years' records of 246 kola trees in Nigeria 46 trees gave no yield at all, 79 per cent of all the trees gave mean annual yields of 0–300 nuts, while the remaining 21 per cent produced 72 per cent of the total yield of the plot. The average was 210 nuts per tree per annum, while the 10 best trees averaged 1,415 nuts and the best tree yielded 2,209 nuts per annum. There was also considerable variation in the size of nuts, ranging from 16·9–40 nuts per lb of fresh seed after removal of the testa, with an average of 27·7 nuts per lb. It is obvious that there is ample room for selection followed by vegetative propagation. The possibility of self-incompatible clones should be borne in mind.

Seeds with white cotyledons are preferred. It has been shown that pink colour is dominant to white, and red is dominant to both white and pink. The colour will thus depend on the genetic constitution of the female and the male pollen parent. A tree which is homozygous for the white colouration may have seeds with pink or red cotyledons as a result of cross-pollination.

PRODUCTION

Kola has not been grown on a large scale outside West Africa, although there are some plantations in Brazil. Nigeria's annual production is estimated at 100,000 tons and that of the Ivory Coast at 30,000 tons. Ghana, Sierra Leone and Liberia also produce kola. Ghana exports about 6,000 tons annually to Nigeria. The main exports are to the northern regions of West Africa and the Sudan, south of the Sahara. The kola trade fixed the old caravan routes of the area, being carried by Housa traders to Timbuktu, Sokoto, Kano and elsewhere. It is now mainly transported by the Nigerian Railway.

REFERENCES

CLAY, D. W. T. (1964). Germination of the Kola Nut (*Cola nitida* (Vent.) Schott and Endl.). *Trop. Agriculture, Trin.*, **41**, 55–60.

CLAY, D. W. T. (1964). Vegetative Propagation of Kola (*Cola nitida* (Vent.) Schott and Endl.). *Trop. Agriculture, Trin.*, **41**, 61–68.

RUSSELL, T. A. (1955). The Kola of Nigeria and the Cameroons. *Trop. Agriculture, Trin.*, **32**, 210–240.

RUSSELL, T. A. (1955). The Kola Nut of West Africa. *World Crops*, **7**, 221–225.

Theobroma L. (x = 10)

Cuatrecasas (1964) lists 22 spp., confined to tropical America, none of them native in the Antilles. Under-storey trees in tropical rain forest and seasonally-flooded sites from sea level to 3,250 ft. With the exception of *T. cacao* and *T. bicolor*, which probably owe their extended range to man, most spp. have a restricted geographical distribution. The genus originated in South America, probably in the Amazon basin, with a second centre of speciation on the Pacific slopes of the Colombian Andes where several new spp. have been more recently discovered.

Trees of variable height, 5–33 m, but usually of short stature, with dimorphic branching. In all spp. except *T. cacao*, which may produce 3–5 branches at the jorquette, the main trunk branches into a regular whorl of 3 main plagiotropic laterals or fan branches; in spp. with epigeal germination a lateral bud develops below the jorquette to produce a new orthotropic shoot or chupon; in spp. with hypogeal germination an adventitious lateral bud below the terminal bud grows vigorously giving the appearance of continuation of the terminal bud. Lvs simple, stipulate, petiolate with pronounced pulvini. Fls small, solitary, paired, or in compressed cymes on trunk and main branches (cauliflorous) or axillary on young branches; sepals 5; petals 5 with ligule; stamens 5 forming short tube with 5 staminodes which may be subulate or petaloid; ovary sessile, with 5 loculi with many ovules; style filiform. Frs large indehiscent drupes; seeds surrounded by pulp derived from ovule integuments, exalbuminous, cotyledons convoluted.

In the closely related genus, *Herrania*, lvs compound-digitate and normally with unbranched habit.

T. cacao (q.v.) is the only sp. of economic importance. *T. bicolor* H. & B., occurring from Mexico to Brazil, is cultivated outside its natural range for the edible pulp round seeds and seeds are used like those of cocoa. The seeds of *T. angustifolium* Moc. & Sesse are sometimes mixed with cocoa in Mexico and Costa Rica; the pulp round the seeds of *T. grandiflorum* (Willd. ex Spreng.) K. Schum. is used for making a drink in parts of Brazil and is also eaten.

Theobroma cacao L. (2n = 20) COCOA

NOTE: The word 'cacao' is often used for the tree and its parts and the word 'cocoa' for the products of manufacture. In this account the word 'cocoa' is used throughout, both for the tree and its products.

USES

The crop is of ancient cultivation in Central America (see below) and the Indians believed it to be of divine origin; hence the generic name *Theobroma* given by Linnaeus, meaning the 'food of the gods'. Well-to-do Indians made a thick beverage by pounding roasted cocoa beans with maize and *Capsicum*. The beans were also used as currency; 100 beans would buy a slave. Tribute to the Aztec emperor from the lowland people was made in cocoa beans. Large quantities were found in Montezuma's palace when he was defeated by Cortez in 1519.

Cocoa beans, obtained from a canoe off Central America, were taken by Columbus to Europe as a curiosity. The Spaniards did not appreciate the Indian method of preparation, but soon learnt to make it palatable by mixing the ground roasted beans with sugar and vanilla; exports to Spain in this form soon began, Cortez recognizing its possible commercial value. Later the unprocessed dry beans were exported and the first chocolate factories were opened in Spain. Cocoa beverage became popular in Italy and France early in the 17th century and soon afterwards in Holland, Germany and England. Chocolate houses appeared alongside the coffee houses and were used as clubs. At this time cocoa was very expensive, over £1 per lb, and consumption was limited to the wealthier classes.

In 1828 van Houten, a Dutch manufacturer, invented a method of expressing much of the fat from the bean, thus making it more palatable and digestible. References to defatted cocoa had been published in France in 1763 and in Italy in 1769. The fermented beans are roasted, when water and acetic acid are driven off, and the characteristic aroma of chocolate is produced. The beans are then cracked to form nibs and the shell (12 per cent) and embryo (1 per cent) are removed by winnowing. The nibs are ground to make the cocoa mass and from this cocoa powder is made by reducing the fat content from 55 to 24 per cent. The expressed fat is used for making chocolate or sold as cocoa butter. In the manufacture of chocolate the cocoa mass is ground with sugar and extra cocoa fat is added; lecithin may also be added to increase fluidity. In the manufacture of milk chocolate, discovered by M. D. Peter in Switzerland in 1876, dried or powdered milk is incorporated. Cocoa butter is also used in cosmetics and pharmaceutical preparations; its melting point is a little below blood heat.

The cocoa shell (testa) is used in stock feed or as manure; it is a source of theobromine, shell fat and vitamin D.

ORIGIN AND DISTRIBUTION

The centre of origin is placed by Cheesman (1944) on the lower eastern equatorial slopes of the Andes, where the greatest range of variation in natural populations exists. Cocoa has been cultivated since ancient times in Central America, from Mexico to the southern Costa Rican border, possibly for over 2,000 years. It is probable that it was never truly wild in this region and was introduced in early times from South America; the theory of divine origin and its use as currency would appear to support this view, as it is unlikely that a wild tree would be held in such high esteem. This Central American cocoa is Criollo cocoa (see below).

Cuatrecasas (1964) assumes that 'in early times a natural population of *Theobroma cacao* was spread throughout the central part of Amazonia —Guiana westwards, and northwards to the south of Mexico (and) that these populations developed into two different forms geographically separated by the Panama isthmus'.

It is possible, however, that the Criollos originated as mutations and the fixing of homozygous recessive characters in populations on the periphery of distribution, and were then maintained through geographic isolation and selection.

After the arrival of the Spaniards cocoa spread rapidly in the New World. Criollo cocoa, presumably from Central America, was taken to Venezuela and Trinidad, introduction into the latter being in 1525. Jamaica, Haiti and the Windward Islands became important producers and there were extensive plantations in Jamaica when the island was captured by the British from Spain in 1655. In 1727 much of the cocoa in Trinidad was destroyed by a 'blast', which may have been a hurricane or an epidemic outbreak of pests or disease. Some 30 years later Trinidad obtained planting material of the darker-beaned cocoa of the Amazonian Forastero type or admixture, possibly from Eastern Venezuela, and this hybridized with the remnants of the original white-beaned Criollo introduction to produce the heterogeneous Trinitario cocoas. The latter were introduced into Venezuela about 1825 and gradually supplanted the Venezuelan Criollo as they were hardier and more productive. Amazonian Forastero cocoas, with cotyledons paler than is normal in this group, were taken over the Andes to Ecuador, and became the Cacao Nacional of that country.

The variability of *Theobroma cacao* decreases as one goes further down the Amazon from the centre of origin on the slopes of the Andes and self-compatible trees are encountered (see below). The Amelonado forms of Amazonian Forastero with dark purple cotyledons were taken into cultivation in Bahia in Brazil. These were introduced by the Spanish and Portuguese to islands in the Gulf of Guinea in the 17th century. One or a few Amelonado pods were taken to Ghana in 1879 from Fernando Po and gave rise to the large West African industry. There is evidence that cocoa was grown at mission stations in Ghana before this date. The West African Amelonado shows little variability, as might be expected from a very limited introduction of self-compatible material, and is very uniform in bean size and flavour.

Cocoa was introduced into South-east Asia by the Spaniards (Philippines in 1670) and the Dutch in the 17th century. The Germans took it to Samoa and New Guinea in the 18th century. Uganda obtained seedlings from the Royal Botanic Gardens, Kew, in 1901.

It has been shown above that the cvs in the different countries varied with the region from which they were introduced and the amount of hybridization. The general names used are based on Venezuelan terminology, namely, Criollo = native, Forastero = foreign, and Trinitario = native of Trinidad.

SYSTEMATICS

Cuatrecasas (1964) recognizes two subspecies of *Theobroma cacao*, of which one has four forms as shown below:

T. cacao L. subsp. *cacao* f. *cacao* = *T. sativa* (Aubl.) Lign. & Le Bey.

The original Central American Criollo and the type of the Linnean sp. Frs oblong, tapering and pointed, surface warty, 5 deep narrow furrows with 5 shallower furrows between, immature pods green or dark red; woody mesocarp thin; seeds rounded in cross section; cotyledons white. Highest quality. Mexico and British Honduras.

T. cacao L. subsp. *cacao* f. *pentagonum* (Bern.) Cuatr. = *T. pentagona* Bern.

'Alligator cocoa'. Frs oblong-oval, about 20×9 cm, 5 ribs, prominent and continuous, surface warty, immature pods reddish yellow, mesocarp not woody; seeds large, round; cotyledons white. High quality. Known only in cultivation, originally in southern Mexico and Central America. Distinctive pod-sculpturing is probably a recessive character, as 'alligator' forms segregate from the selfing of 'non-alligator' cocoas like 'ICS 45'.

T. cacao L. subsp. *cacao* f. *leiocarpum* (Bern.) Ducke = *T. leiocarpa* Bern.
Frs ovoid, shallowly 5-furrowed, almost smooth, shell thin; seeds plump; cotyledons white or pale violet. High quality. Known only in cultivation, mainly on Atlantic coast of Guatemala. A segregate form or mutant. 'Porcelaine Java Criollo' may be this form.

T. cacao L. subsp. *cacao* f. *lacandonense* Cuatr.
A wild half-vine known only from the type locality near Chiapas in Mexico. Frs ovoid-oblong. 10-angled.

T. cacao L. subsp. *sphaerocarpum* (Chev.) Cuatr. = *T. sphaerocarpum* Chev. = *T. leiocarpum* sensu Pittier; 'Amazonian Forastero'; 'Amelonado'.
Frs ovoid, smooth, shallowly 10-furrowed, rounded at both ends; unripe pods green; pericarp very thick, mesocarp woody; seeds ovoid, compressed, cotyledons dark purple. Quality variable, but lower than subsp. *cacao*. Native in South America; found spontaneous in the Hylaea from the Guianas and middle Amazon north and westward to the Andes. Now extensively planted throughout the tropics. The dark purple colour of the cotyledons is dominant to the absence of pigment.

The above forms and subsp. interbreed readily to give fully fertile F_1 hybrids and have given rise to a large number of recognizably distinct local populations.

The most satisfactory classification (Cheesman, 1944) is to divide the cultivated and wild cocoas into 3 main groups, based on the Venezuelan trade names.

1. CRIOLLO: Pods yellow or red when ripe, usually deeply 10 furrowed, often markedly warty, usually conspicuously pointed, pod wall thin in section, so that pod compresses under hand pressure; seeds large, plump and almost round in section; cotyledons white or pale violet, which give a cocoa lacking any astringent characteristics. Beans ferment quickly; yields comparatively low. It produces the highest quality of all cocoas; only small quantities are now available on the world market. It can be subdivided into:

(a) CENTRAL AMERICAN CRIOLLO: Unripe pod wall predominantly green, ripening to yellow. The original cocoa cultivated in Central America and Mexico and subjected to human selection for a long period, probably over 2,000 years, resulting in fixing the recessive white cotyledons, with no astringency and requiring little fermentation. No genuine wild trees known.

(b) VENEZUELAN CRIOLLO: Exhibits greater tree to tree variation in colour, size and shape of pods, consistent with its position nearer the centre of origin. Unripe pod wall usually red. Probably introduced into Venezuela from Central America, followed by slight subsequent admixture.

2. AMAZONIAN FORASTERO: Unripe pods whitish or green, ripening yellow, usually inconspicuously ridged and furrowed, surface often smooth (rough, warty pods in headwaters of Amazon, smoother towards mouth), ends rounded or very bluntly pointed; pod walls relatively thick and often

with a woody layer difficult to cut; seeds flattened; fresh cotyledons deeply pigmented and dark violet in cross-section; usually giving an astringent product. Trees hardier, more vigorous and higher yielding than Criollo types. Occurs naturally throughout the basin of the Amazon and its tributaries; variation decreasing nearer the mouth. Members of the group were taken into cultivation in Brazil and the Amelonado form was subsequently taken to West Africa, where it gives a highly uniform population. The quality is lower than in other forms and it was accident rather than selection that this group now supplies the bulk of the world's cocoa. The Forastero cocoa taken to Ecuador in the 18th century has plumper seeds and paler cotyledons; it became known as Cacao Nacional and in the trade is a 'Fine Forastero'.

3. TRINITARIO: A wide range of hybrids between Criollo and Amazonian Forastero, occurring typically in Trinidad (see above); very heterogeneous and exhibiting a wide range of morphological and physiological characters. Colour of unripe pod may be whitish, green, red or purple, ripening yellow, orange or red, variable in shape and wall thickness; surface ranging from complete smoothness to heavy sculpturing; beans range from plump to flat; pigmentation of cotyledons from white to nearly black. Introduced into Venezuela about 1825 and subsequently into most cocoa-growing countries. Hardier and more productive than Criollo, the best clones combining the vigour of Amazonian Forastero with much of the quality of Criollo; other clones very inferior. Trinitario cocoas are of great importance in breeding, the only definable unit being the clone. In the trade it is regarded as a 'fine' cocoa.

ECOLOGY

In its natural habitat *Theobroma cacao* is a small tree in the lowest storey of the evergreen tropical rain-forest of South America. It usually grows in groups along river banks, where it may often stand for 6 months in water, but the water must be running to supply the necessary oyxgen. Its status as an under-storey tree led to the belief that cocoa must be grown under shade, but, as with many other economic crops, the natural environment where it grows in competition with many other species does not necessarily give the best conditions for growth and high yield when cultivated; heavily-shaded wild cocoa often carries little fruit.

It is a strictly tropical crop. The limits of cultivation are 20°N and S, but the bulk of the crop is grown within 10°N and S. It is grown mainly at low elevations, usually below 1,000 ft, but it is cultivated at up to 4,000 ft in the Venezuelan Andes and at 3,000 ft in Colombia. Areas of tropical evergreen rain-forest and semi-evergreen rain-forest are the most suitable ecological zones for cocoa. The optimum temperature range is 70–90°F, with small seasonal and diurnal range. Rainfall in the cocoa belt varies from 40–100 in., but most cocoa, without irrigation, is grown with a rainfall above 50 in. The rainfall should be well distributed, preferably with 4 in.

or over per month, and with the absence of a marked or intense dry season with less than 2·5 in. per month.

Cocoa can survive in its natural habitat in dense shade which would kill many species, but it can also survive considerable exposure. Photosynthesis is reduced by shading, but is partly compensated for by the larger leaf area under shade. The stomata remain open and transpire freely in full sunlight provided the water supply is adequate. Seedlings grow best under shade with approximately 25 per cent of full sunlight. Self-shading occurs in mature trees, thus modifying the light relations. The amount of light may gradually be increased to 50 per cent as growth occurs in the young trees. Later, provided that the crop has optimal conditions of rainfall, drainage, soil aeration and nutrition, overhead shade may be gradually removed and the crop can be grown in full sunlight when considerable self-shading has been attained. When shade trees are suddenly removed in mature cocoa, or when environmental factors fluctuate excessively or are not optimal, or when diseases and pests attack the tree, the outer layer of the canopy in unshaded cocoa becomes disrupted and leaf-shedding and die-back occur. High winds may cause considerable mechanical damage, particularly to young cocoa trees, and windbreaks are desirable. Flowering and leaf-flushing do not appear to be affected by photoperiod.

In cultivation cocoa requires a well-drained, well-aerated soil with good crumb structure and adequate supplies of water and nutrients. The soil should possess ample root room; it should be deep and easily penetrable by roots to at least 5 ft and with a good top humic layer. The best soils are aggregated clays or loams or sandy loams, often red or reddish-brown in colour, developed over coarse crystalline igneous or metamorphic rocks, or rocks of recent volcanic origin. The best cocoa soils in West Africa are derived from igneous rocks; those in Trinidad are formed from marine sediments containing calcium carbonate and glauconite and rich in potassium and phosphorus; those of the Pacific are mainly of volcanic origin. The optimum pH is around 6·5.

STRUCTURE

Small under-storey tree, 6–8 m, but sometimes reaching 12–14 m in Nacional cocoa in Ecuador and Amelonado in West Africa.

ROOTS: Tap-root of seedling tree grows straight downwards to a depth of 2 m in well-aerated soil, but much less in compact soils or where the permanent water table is high. It may bifurcate on reaching a clay layer. There is a definite collar at the junction of tap-root and trunk and below this most of secondary roots arise to a depth of 15–20 cm and grow out in humic layer to a length of 5–6 m, giving rise to a dense mat of surface-feeding roots. Middle of tap-root usually devoid of secondary roots, but these may arise lower down and either grow upwards to surface mat or downwards into lower layers of soil. There is probably an association with mycorrhiza.

BRANCHES: Tree shows dimorphic branching. Seedlings form a single main stem or chupon, 1–1·5 m high at about 14 months. The terminal bud then breaks up into 3–5 meristems to give the so-called jorquette (the bud being used up in the process), and grow out into lateral plagiotropic fan branches, which may be almost horizontal. Further increase in height is made by an axillary bud just below the jorquette and this produces an orthotropic sucker or chupon which grows up vertically between the fan branches and then repeats the growth pattern by forming another jorquette and a second whorl of fan branches. In this way, several successive chupons arise sympodially, each producing a tier of branches, the lowermost of which tend eventually to dry and fall. Plantation practice varies as to the number of tiers which will be kept, unwanted chupons being removed by pruning.

Trunks or stems below a jorquette produce only chupons; these are morphologically the same as the trunk, are determinate in growth and have a phyllotaxy of $\frac{3}{8}$. Rooted chupon cuttings have the same habit. Fan branches from the jorquette give rise to further fan branches and are indeterminate in growth, bearing lvs alternately in two ranks with a phyllotaxy of $\frac{1}{2}$. Fan branch cuttings may produce chupons when they are pruned or accidentally wounded. The habit of a budded tree depends on the type of bud used—fan or chupon. Both chupon and fan branches bear fls and frs. Mucilage cells occur in the pith and cortex.

LEAVES: Large, simple, dark green when mature; petiole 1–4 cm long, longer on chupons, pubescent, well-marked pulvinus at each end; stipules lanceolate, 5–20 × 1–2 mm, pubescent, early deciduous, except when chupons about to branch when they form a brush at tip; lamina elliptic- or obovate-oblong, entire, usually glabrous, 12–60 × 4–20 cm, being largest in middle of tree where they receive least light, base rounded and obtuse, apex acuminate; main vein prominent; lateral veins pinnate, 9–12 pairs; axil spot correlated with red pod, absence of spot with green pod.

Lvs on fan branches produced in flushes, young lvs being produced in rapid succession by burst of activity of terminal buds, which then return to dormant condition and new growth hardens. Young lvs soft and limp and hang down vertically, pale green or various shades of red depending on clone or cv. Period between flushes depends on temperature and tree's nutrition. If temperature too high or nutrition faulty, flushes appear in rapid succession, which debilitate the tree. Under good conditions lvs persist in healthy functional condition, and new flushes appear only 2–4 times per year. Lvs usually persist through 2 further flushes and are dropped on the third. Stipules of terminal bud leave characteristic scars on twigs when growth resumed so extent of successive flushes is evident on examination.

INFLORESCENCES: Cauliflorous on older leafless wood of main stem and fan branches, never on recent flushes. Much compressed cincinnal cymes with branches greatly reduced, originating from buds in axil of reduced prophylls, which are minute sessile lvs at base of branch arising from axillary bud of an ordinary lf. Branch does not usually grow out, but its shortened and twisted branches broaden into a cushion, which may bear up to 50 fls in one season; peduncles and bracts pubescent. When inflorescence stimulated by witches' broom disease the cushion grows out into leafy shoots.

FLOWERS: Regular, pentamerous, hermaphrodite. Pedicels 1–2 cm long, greenish, whitish or reddish, with sparse hairs and constricted abscission layer at base. Sepals 5, pink or whitish, 7–10 × 1·5–2·5 mm, triangular, rather fleshy, valvate, shortly united at base. Petals 5, smaller than sepals; base obovate 3–4 mm long, expanding into concave, cup-shaped pouch, white with 2 prominent purple guide lines; end of petal spatulate, 2–3 mm long, yellow, bending outwards and backwards and attached to pouch by narrow connective. Androecium with 5 outer staminodes, opposite sepals, erect, pointed, with dark purple centres and whitish ciliate margins and form ring round style; 5 inner fertile stamens with 4 pollen sacks which dehisce longitudinally; filaments bend outwards so that anthers are concealed in pouch of corresponding petal. Ovary superior; carpels 5; ovules numerous, anatropous; deeply penetrating stylar canal makes placentation axile at base of ovary and parietal above; style single, 2–3 mm long, hollow, shorter than surrounding fence of staminodes, divided at tip into 5 stigmas which are often more or less adherent.

FRUITS: Usually considered a drupe, but is commonly called a pod; indehiscent, variable in size, 10–32 cm long, in shape from nearly spherical to cylindrical, pointed or blunt, smooth or warty, with or without 5 or 10 furrows; young pods white, green or red and ripening green, yellow, red or purple (see SYSTEMATICS above). Pericarp (= husk) usually fleshy and thick with mesocarp varying in degree of lignification. Pods attain full size 4–5 months after fertilization and require a further month for ripening when colour may change. 20–90 per cent of fruits fail to develop fully, drying out being most prevalent at 50 days when about 5 cm long.

SEEDS: Usually called beans; 20–60 per pod, Forastero cvs having more than Criollo cvs; usually arranged in 5 rows; variable in size, 2–4 × 1·2–2 cm, in shape ovoid or elliptic, and with white to deep purple cotyledons (see SYSTEMATICS above). Embryo develops late in seed using up endosperm. At maturity seed consists of two large convoluted cotyledons, a small embryo, a thin membrane which is the remains of the endosperm and a leathery testa (= skin or shell). Fresh seeds surrounded by mucilaginous, whitish, sugary, acid pulp which develops from the outer integument of the ovule; this pulp is removed during fermentation and drying. As seeds

Fig. 90. **Theobroma cacao** : COCOA. **A,** jorquette with young fan branches (×¼); **B,** cauliflorous inflorescence (×1); **C,** flower in longitudinal section (×4); **D,** fruit (×½); **E,** seed in longitudinal section (×½); **F,** seed in transverse section (×½).

A
B
C
D
E
F

are of short viability and indehiscent pods remain on tree unless harvested, natural dispersal is by monkeys, squirrels, rats and other animals which break through husk, suck off sweet pulp and disseminate the unpalatable seeds. Seeds constitute 25 per cent by weight of mature fruit; 250–450 dry fermented beans per lb.

POLLINATION

The flowers are ill-adapted for pollination by more regular methods or for self-pollination and they are devoid of scent and nectar; the pollen is too sticky for wind-pollination; the position of the anthers hidden in pouched petals and the ring of staminodes hindering access to the stigma. A certain amount of self-pollination may be effected by crawling insects such as thrips, ants and aphids. When these are excluded, pollination still occurs and some clones are self-incompatible and yet fruit well. Thus some insects which can fly or be carried by wind must be responsible. It is now believed that the main pollinating agents are ceratopogonid midges; in Trinidad *Forcipomyia quasi-ingrami* Macfie and *Lasiohelea nana* Macfie; in Ghana *Forcipomyia ingrami* Carter, *F. ashantii* Ingram & Macfie and *Lasiohelea litoraurea* Ingram & Macfie. These midges appear to be attracted by, and to feed on the purple tissues of the staminodes and guide-lines of the petal pouches and would then present their backs in turn to stigma, style and pollen. Most pollination takes place in the first 2–3 hours after dawn except during or after heavy rain. In Ghana 2–5 per cent of flowers only are pollinated and a fairly large number of these fail to set seed. Later during the season, when flowers are fewer, pollination may rise to 50–75 per cent. Fertilization takes place 7–8 hours after pollination. Unpollinated flowers drop 24 hours after opening. The pollination efficiency is low, but this is compensated for by the large numbers of flowers produced; it is estimated that only one in every 500 flowers matures to a ripe fruit. Fruit setting is further complicated by self- and cross-incompatibility.

In artificial pollination a flower bud, which will open the following day and is recognized by its whitish colour and swollen appearance, is selected and covered with a glass or plastic tube, $5 \times 1{\cdot}5$–2 cm, which is sealed to the bark by plasticine. The plasticine should be applied to embrace the flower pedicel and all crawling insects removed from the bud and its pedicel before the tube is applied. The tube is covered by cheese cloth at the top, kept in place with a rubber band. This ensures circulation of air and exclusion of insects. Small, nylon-gauze, tent-shaped hoods attached with pins may be used instead of the glass tube. The stamens are carefully removed from inside petal bases and the anthers are dabbed directly onto the stigmas, one or two staminodes having been pinched out to give access to the stigmas. Emasculation should not be necessary in cross-pollination if the flowers are not roughly handled. In order to prevent undue shedding and wilting of fruits from these pollinations it is usual to remove all the developing fruits on the tree produced by natural open-pollination. The tubes are removed 24 hours after pollination and in 3–5 days fertilization is ascertained by swelling of the ovary. It may be necessary to spray the developing pods against fungal diseases and to protect them against vermin.

INCOMPATIBILITY

Self-incompatibility in cocoa was first reported in Trinidad by Harland in 1925 and again by Pound in 1932. Self-incompatible Trinitario clones have so far shown themselves to be cross-incompatible, but are cross-compatible with self-compatible trees. The clone 'ICS 1' is self-compatible; 'ICS 60' is self-incompatible.

Cope (1962a) has shown that the site of incompatibility is in the embryo-sac and not in the stigma and the style, thus providing the first example in the Angiosperms of incompatibility of this type. Pollen tubes in incompatible matings grow as fast as those in compatible pollinations and deliver their gametes into the embryo-sacs in a perfectly normal fashion. The embryo-sac is in no way abnormal and incompatibility is due to the failure of the male nuclei to unite with the egg and polar nuclei and is genetically controlled. In incompatible matings the flowers are shed 3–4 days after pollination. In incompatible pollinations the proportion of ovules showing non-fusion averages 25, 50 or 100 per cent. Fusion or non-fusion is controlled by a series of alleles operating at a single locus (S), showing dominance or independence relationships, *e.g.* $S_1 > S_2 = S_3 > S_4 > S_5 > S_6$. They are the same in both male and female parts of the flower so that reciprocal pollinations give the same results. The diploid constitution of both parents decides whether the cross is compatible or incompatible, encounters between the gametes being a random process. Incompatible crosses involve parents with the same dominant allele, *e.g.* $S_{1.2} \times S_{1.3}$, $S_{2.4} \times S_{2.5}$, or a genotype with independent alleles and another where one of these independent alleles is dominant, *e.g.* $S_{2.3} \times S_{2.4}$, $S_{2.3} \times S_{3.5}$. In the case of $S_{1.2}$ selfed or $S_{1.2} \times S_{1.3}$, the proportion of non-fusion ovules is 25 per cent because S_1-bearing gametes meet in one-quarter of all gametic encounters; $S_{2.3}$ selfed ($S_2 = S_3$) gives 50 per cent because one-quarter are between S_2-bearing gametes and one-quarter between S_3-bearing gametes; $S_{2.2}$ selfed or $S_{2.2} \times S_{2.2}$ give 100 per cent non-fusion because of selfing or crossing homozygotes. The S locus appears to have its action both before and after meiosis. Progenies arising from crosses between certain self-compatible clones, *e.g.* 'ICS 1' × 'ICS 45' give all self-incompatible progeny. To account for this it appears that in addition to the S locus two other complementary loci A and B are involved, the role of which is to produce a non-specific precursor to which the S alleles impart their specificities to prevent fusion between gametes carrying the same S allele in certain circumstances. Genotypes homozygous for inactive alleles at one or more of the A, B and S loci will be self-compatible.

Cope (1962b) has shown that self-compatible and self-incompatible genotypes cannot exist in equilibrium in an isolated population, and that, on theoretical grounds, the self-compatible trees will ultimately displace the self-incompatible genotypes, a process which is hastened by greater fruitfulness of the self-compatible trees. Nevertheless, in a wild population there appears to be some strong selective discrimination against self-compatible types. Thus near the centre of origin of the species on the eastern equatorial slopes of the Andes all clones so far collected have proved to be uniformly self-incompatible. The further removed from the centre of origin the greater is the proportion of self-compatible trees, a feature which would help the spread into new habitats. West African Amelonado, which is believed to have originated near the mouth of the Amazon, is uniformly self-compatible.

GERMINATION

Seeds will germinate immediately on reaching maturity and are viable for only a short time, dying quickly when dehydrated or fermented or subjected to extremes of temperature; they are killed in 8 minutes at 4°C and often during air transport. Germination may take place in the pod. Removal of the testa speeds up germination by a few days, but such seeds lose viability rapidly if allowed to dry out. The moisture content of the cotyledons should be kept at about 50 per cent. Under suitable conditions with free access of air and protection against water loss seeds may be stored for 10–13 weeks without losing viability. Seeds may be transported or stored in charcoal powder of 30 per cent moisture content in perforated containers; polythene bags have proved very satisfactory.

Seeds may be rubbed with sand, wood ash or coconut fibre to remove the mucilage before planting, but this is not essential. They are usually planted 1–2 in. deep with the hilum scar downwards or horizontally and germinate in 2–3 weeks. Germination is epigeal. The cotyledons and hypocotyl develop chlorophyll and the first true leaves appear 15–20 days after germination.

CHEMICAL COMPOSITION

	Forastero Cocoa Beans Fresh weight per cent			Cocoa Products Oven dried per cent			
	Cotyledons	Pulp	Testa	Fresh Kernel	Cured and Dried nibs	Cocoa powder	Bar chocolate
Water	35·0	84·5	9·4				
Starch	4·5		46·0	7·7	7·4	22·2	4·1
Sugars	6·0	13·4		1·8	0·7	4·3	54·4*
Fat	31·3		3·8	54·0	57·3	26·5	29·1*
Protein	8·4	0·6	18·0	14·8	6·7	22·2	1·5
Fibre	3·2		13·8				
Theobromine	2·4			2·3	1·7	1·3	1·1
Caffeine	0·8						
Polyphenols	5·2		0·8				
Acids	0·6	0·7					
Inorg. Salts	2·6	0·8	8·2				
Other substances				19·4	26·2	23·5	9·8

From Hardy (1960). * In part added.

The polyphenols consist of catechins, complex tannins, leuco-cyanidins and cyanidin glycosides. Forsyth and Quesnel (1963) state that 'the chocolate aroma precursors are formed immediately after the death of the seed (during fermentation) when the proteins and polyphenol compounds come together, react, and are subjected to the action of hydrolytic enzymes. Later oxidative condensations render the astringent catechins and other polyphenols insoluble. Loss of the bitter purines by exudation also changes the taste characteristics of the final product'.

PROPAGATION

SEED: Propagation by seed is the cheapest method. Seed may be planted at stake, which is the usual custom in West Africa, where 3 seeds per hole are planted and later thinned to one plant. Seedlings may also be grown in nursery beds with natural or artificial shade, which should allow approximately 50 per cent sunlight. Lateral shade and wind protection should also be provided. Seed is usually planted 12 in. apart in the beds, but much closer spacing is used in Ghana. Seedlings may also be raised in baskets made of fibre, cane or bamboo, and have the advantage that there is little root disturbance on transplanting. In Trinidad baskets are made of stems of *Ischnosiphon arouma* (Aubl.) Koern. (Marantaceae). Baskets should be large enough to allow sufficient root room for a 4–6 months' old seedling, the usual time for planting out, and a useful size is 9 in. deep, 7–9 in. wide at the top and 6–7 in. at the base. Polythene bags, 12 × 7 in., with drainage holes in the sides and base have proved very cheap and efficacious, but must be removed before planting. A good potting mixture is 7 parts loam with a pH not higher than 6·5 (a higher pH results in iron chlorosis), 3 parts dried cattle manure, 2 parts sharp sand and 1 oz double superphosphate per basket. As in the nursery, seedlings in baskets or pots must be raised under shade, which is later reduced before transplanting (also see GERMINATION above).

CUTTINGS: Environmental conditions for the successful rooting of cuttings depend upon the following factors: (1) Reduction of the leaf area to cut down transpiration, to prevent mutual over-shading and to economize in space in the bins, but sufficient leaf for the production of enough carbohydrates by photosynthesis to supply the developing roots and continued life of the cutting. Thus with cuttings with 6–8 leaves, 2–3 of the lower leaves are removed and the remainder trimmed by cross-cuts to $\frac{1}{2}$–$\frac{1}{3}$ of the original length. (2) Controlled light and temperature so that the rate of photosynthesis exceeds the rate of respiration. This is given with about 10–15 per cent of incident sunlight obtained by overhead shading and with cloth covers over the glass or polythene in the bins, and a temperature not higher than 87°F, obtained by spraying the cuttings and cloth covers with water. In open bins or beds continuous mist spraying is necessary. Too low a light intensity causes starvation and yellowing of the leaves; too high a light intensity causes destruction of the chlorophyll. (3) 100 per cent humidity of the atmosphere to reduce transpiration and to ensure maximum turgor of the leaf cells; this is obtained by frequent or continuous watering. (4) An optimum air/moisture relationship at the base of the cutting with adequate aeration and free drainage and yet sufficient moisture to maintain cell turgidity. A saturated soil atmosphere, particularly when oxygen is lacking, results in rotting, or the development of rods of callus from the lenticels. Inadequate watering and too high an oxygen concentration leads to callus formation at the base and delay in rooting. The most suitable rooting media are composted sawdust, coconut fibre dust or sand of uniform particle size of 1–2 mm. (5) Absence or control of pathogens by sterilization of rooting medium and application of fungicides. (6) Stimulation of root development by growth hormones; the ends of cuttings being dipped in β-indole-butyric acid and α-naphthalene-acetic acid in equal proportions of 8 mg per ml in 50 per cent alcohol.

NURSERY FOR CUTTINGS: Cuttings are usually obtained from young vigorous healthy trees in gardens specially maintained for the purpose; the trees are planted 5–6 ft apart; they should be free from mineral deficiencies, and should be grown on well-drained, loamy soil of pH 5–6·5, rich in organic matter. Shade trees, *e.g. Gliricidia sepium* or bananas at 10 × 6 ft or *Erythrina poeppigiana*, are planted to give approximately 50 per cent sunlight. No cuttings are taken in 1st year, but 20–30 cuttings per tree in 2nd and 3rd years and 40–50 cuttings per tree in 4th and 5th years can be taken, after which it is usual to replant in 6th–8th year. In taking cuttings 2 buds are left at base of branches to produce new shoots and it is essential to maintain the plants in an actively growing condition. Trees should be given 1–1½ lb of complete fertilizer per tree per annum and foliar sprays of of 1 per cent urea. They should be irrigated in the dry season.

PREPARATION OF CUTTINGS: Cuttings are taken from recently matured flushes of fan branches when the leaves are fully green and wood semi-hard with upper surface of stem brown and lower surface still green. Cuttings are taken at right angles, usually just above a node, but this is not essential, and they should be 5–12 in. long with 3–9 leaves. They should be made early in the morning and placed in water or wrapped in damp cloth or paper until set. The bottom leaf or leaves are removed by a clean cut close to stem and remaining leaves are trimmed (see above), the base dipped in hormone and inserted 2–4 in. deep in the rooting medium with a backward slant in staggered rows with about 30 cuttings per sq. yd. Cuttings may also be rooted directly in baskets with a core of sand, sawdust and other rooting medium, thus obviating the need for transplanting and reducing cost. Single-leaf cuttings may be used where material is limited and these are obtained by cutting up stem cuttings ¼ in. above a node, usually while still green; the hormone dip is used at half strength. The ease of rooting stem and single-leaf cuttings varies with the clone. The average take varies from 50–90 per cent.

PROPAGATORS: Various types are available:

CLOSED-BIN PROPAGATORS (for designs see Hardy, 1960): These consist of a bin or battery of bins of brick, stone, concrete or wood with close fitting lids of glass or polythene, covered with calico to reduce light and which is kept moist. A convenient size of bin is 4–5 ft × 3 ft and 3 ft deep. Holes in the base and a layer of stones facilitate drainage. The stones are covered with pebbles and a 9 in. layer of rooting material is placed on top. Watering is by automatic sprays within the bins or by hand. Overhead shade at 6–7 ft of saran netting, wood slats or split bamboo reduces incident light to 25–30 per cent and this is further reduced by the cloth on the lids.

OPEN-SPRAY BEDS with continuous mist spray during the day and overhead shade as above. This method is cheaper in labour, but it uses more water and is less efficient.

HUMIDIFIED GREENHOUSES with tiers of movable shelves 2 ft apart. A continuous fine spray is produced during the day by a centrifugal humidifier. The greenhouse is shaded to give 25 per cent incident light. A greenhouse 20 × 12 × 8 ft will hold 2,500 plants in 6 in. baskets.

UNDER POLYTHENE SHEETS: Cuttings, usually 2-leafed, are planted in cored baskets (see above) and are covered with polythene sheets weighted at the edges. Overhead shade is also provided. The method has been found to be very efficient in Ghana and it is much cheaper than the previous methods.

IN POLYTHENE BAGS: Cuttings are wrapped round with wet sawdust held in coconut fibre, placed in polythene bags 20 × 11 in. A little water is added and the neck is tied with string. The bags are hung up in a shady place with 7 per cent sunlight. They are ready for planting in the field in 22 days. The method is cheap and economical of labour and transport; it is suitable for the production of small numbers of plants.

HARDENING-OFF AND STORAGE: Rooted cuttings in bins or other propagators are hardened off after 28 days by gradually increasing light and reducing humidity. Except when cuttings have been planted direct into cored baskets (see above), rooted cuttings are potted up in fibre baskets or pots of bamboo, asphalt paper or polythene bags. A good potting mixture is 2 parts of slightly acid soil and 1 part of sawdust compost to which 4 oz sulphate of ammonia, 1 oz superphosphate and 2 oz muriate of potash per bushel of soil has been added 7–10 days prior to use. Hardening-off can be in the propagators or in bins with an expanded metal shelf over water, or in a humidified greenhouse, and takes approximately 10 days. Plants are then moved to the storage area, which may consist of a house with a concrete floor, glass and aluminium roof, and with open sides protected by growing plants. Watering needs careful attention. Lime in the water or an alkaline potting mixture will induce iron chlorosis. Sulphate of ammonia,

2 g per basket, or a foliar spray of 1 per cent urea may be applied every 3–4 weeks. Pests and diseases must be controlled, particularly *Phytophthora* with Bordeaux mixture, 2–2–50. Cuttings should have at least one hardened flush before transplanting into the field, and the usual time taken from cutting to planting is about 6 months. In commercial production the average losses are 20–30 per cent during rooting, 3–10 per cent during hardening and 10–30 per cent in field.

OTHER METHODS OF PROPAGATION: Cocoa can be propagated by budding, the patch-bud method usually being employed. It is highly economical of planting material and budwood can be kept up to 7 days if properly stored, thus permitting transport by air. It is useful for clones which are difficult to propagate by cuttings. Budwood is prepared by cutting petioles and is taken after petiole stumps have fallen. Seedling stocks are usually used, 6–12 months old and 1–3 cm in diameter. It is advisable to bud below the cotyledons scar to prevent chupon growth from the stock.

Saddle- and wedge-grafting of leaf-bearing shoots are possible, as is marcotting, but these methods have little commercial application.

HUSBANDRY

PLANTING: As has been shown above under ECOLOGY, cocoa is an exacting crop in regard to its environment, particularly the soil conditions, so that the amount of land suitable for cocoa is limited. In West Africa the peasant cultivators achieved success by growing cocoa widely and, where it grew well, planting more of it.

Forest areas may be selectively thinned, as is done in West Africa, leaving 2–3 tall trees and 15–20 smaller trees per acre. Nurse shade is established and 3 seeds are planted at stake, approximately 5 ft apart; later thinned to one plant per stand. As the trees grow the stand will be thinned. In some areas such as Trinidad the forest is completely felled, followed by lining, holing and draining. Permanent or temporary shade and ground cover are established before planting the cocoa, which may be seedlings or cuttings. Soil should not fall away from the roots of the transplants as cocoa will not stand planting with bare roots. Cocoa is usually planted at 12 × 12 ft. The optimum spacing will depend upon the environmental conditions and cv. grown. Closer spacing usually gives greater yields in the early years, although yields tend to even out later; it also provides a quicker closed canopy. Close spacing in the early years, followed by selective thinning, will probably give maximum yields over a long period. A spacing of 8 × 8 ft gave higher yields than 12 × 12 ft in experiments in Trinidad in early years. In Ceylon, New Guinea and Samoa spacing is usually 15 × 15 ft. Except when planting at stake, holing is usually advocated and with the addition of organic manure. A Trinidad experiment has shown no advantage of large over small holes. It is important that the soil should not sink on planting or water may collect.

Cocoa is occasionally interplanted with other crops such as bananas, *Hevea* rubber, oil palm or coconuts.

SHADE AND WINDBREAKS: For advantages provided by shade and good characteristics of shade tree spp. see coffee (pp. 471–472).

It is usually considered essential that cocoa should be given adequate shade from planting until it becomes self-shading or permanent overhead shade has become established, and that ground shade should be planted to protect the soil and provide temporary lateral shade. Later, permanent overhead shade is considered necessary, except where the cocoa is growing under very favourable conditions. Shade reduces photosynthesis, transpiration, metabolism and growth and therefore the demand on soil nutrients and so enables a crop to be obtained on soils of lower fertility. On fertile soils with liberal fertilizers, particularly nitrogen, shade may be gradually removed altogether as has been shown in Trinidad, and later in Ghana, provided it is possible to maintain a closed canopy of cocoa (see ECOLOGY above and MANURING below).

Bananas planted at 10 × 10 ft or 12 × 12 ft before planting cocoa are most frequently used as temporary shade, but other plants also used include *Crotalaria* spp., particularly *C. anagyroides* Kunth., *Tephrosia* spp., such as *T. candida* DC., *Leucaena glauca* (L.) Benth. (also used as permanent shade), *Carica papaya* L. (papaya), *Manihot glaziovii* Muell.-Arg. (tree cassava) and *Cajanus cajan* (L.) Millsp. (pigeon pea). Some of the above also provide temporary lateral shade, as do *Xanthosoma* spp. (tannias), *Colocasia* spp. (dasheen and eddoes)—both known as cocoyams in West Africa, and *Manihot esculenta* Crantz (cassava). They also provide good ground shade.

Permanent shade may be provided by selective thinning of the original forest, intercropping with other economic trees or specially planted shade trees, which may be planted before or at the same time as the cocoa. In areas where swollen shoot disease is present, alternate hosts of the virus such as spp. of Sterculiaceae and Bombacaceae should be avoided. Trinidad has relied upon *Erythrina poeppigiana* (Walp.) O. F. Cook, mountain immortelle, on the hillsides and *E. glauca* Willd., swamp immortelle, in the flatter wetter areas. Both have spiny trunks. They are usually planted at 24 × 24 ft and later thinned to 48 × 48 ft. The yield of cocoa under immortelles has been consistently higher than under many other shade trees which have been tried. Immortelles have been shown to have over 4 per cent nitrogen in the root nodules, 2–3 per cent in the leaves and 3–6 per cent in the flowers, flower fall alone contributing 20 lb of nitrogen per acre per annum. Unfortunately *E. poeppigiana* is now attacked by a witches' broom disease, probably caused by the fungus, *Botryodiplodia theobromae* Pat., and *E. glauca* by a bark-destroying fungus, *Calostilbe striispora* (Ell. & Eberh.) Seaver. The falling of dead or diseased trees causes considerable damage to the cocoa. *Gliricidia sepium* (Jacq.) Walp. (madre de cacao) is an alternative; the branches may be cut back before the dry season which results in new flushes and prevents the flowering leafless state.

Elsewhere the following spp. are used as permanent shade: *Erythrina subumbrans* (Haask.) Merr., dadap, in western Samoa and Ceylon, of which thornless forms occur, and which can be planted as large cuttings; *Albizia* spp., *Peltophorum* spp., *Terminalia superba* Engl. & Diels (in Congo), *Inga* spp., *Leucaena glauca* (L.) Benth. and *Musanga cecropioides* R. Br., which is very quick growing.

Cold and drying winds have an adverse effect on cocoa and it is usual to provide windbreaks. These can be provided by leaving strips of forest 50–100 ft wide between fields or by the planting of artificial windbreaks for which the following evergreen spp. may be used: *Dracaena fragrans* Gawl., *Mangifera indica* L. (mango), *Eugenia caryophyllus* (Sprengel) Bullock & Harrison (clove), *E. malaccensis* L. (pomerac), *Myristica fragrans* Houtt. (nutmeg), *Calophyllum antillanum* Britt. (galba), *Manilkara achras* (Mill.) Fosberg (sapodilla) or *Anacardium occidentale* L. (cashew).

CARE AND MAINTENANCE. Hand weeding round the seedlings ensures that the plants can be seen when cutlassing. Ground shade must be controlled and may be removed in the 3rd year. All gaps in the stand of cocoa and permanent shade should be supplied as soon as possible; replants after 5 years are difficult to establish because of shading. It has been shown that there is a highly significant correlation between the diameter of the trees at 3½ years old and subsequent yields. Successful establishment and vigorous early growth is thus of great importance. In the early years overhead shade should be maintained at 50 per cent of full sunlight and should be kept growing more rapidly than the cocoa; temporary overhead shade such as bananas is gradually removed in the 4th year. For further treatment of shade see above.

Weeds have a deleterious effect on the growing cocoa and may be partially controlled by the degree of shade. They may be controlled by cutlassing, but the use of herbicides is superseding this in some areas. When the cocoa is established cultivation is not usually practised; in fact in Trinidad any form of soil cultivation reduced yields. Mulching is sometimes advocated, but in Trinidad a sawdust mulch depressed yields and a bush mulch increased yields, but not sufficiently to be economic.

PRUNING: The main objects of pruning are to allow the development of a framework of branches which will give a tree in the shape of an inverted cone, to remove unwanted growth, and to obtain a closed canopy. Drastic pruning reduces early yields and should be kept to a minimum. In a seedling tree it is usual to leave 3 or 4 branches at the jorquette. Unwanted chupons are removed and growth is controlled at the first or second jorquette. In West Africa upper chupons are usually allowed to grow, thus giving tall trees, but the lowest jorquette branches are usually removed to facilitate movement on the farm. Fan cuttings may be allowed to grow to form 3–4 branches at the base or a chupon may be allowed to come up from ground level and the first fan branches are then removed. Subsequent pruning in seedling and vegetatively propagated trees consists of removing diseased or dead wood, including witches' brooms, cutting out mistletoes (Loranthaceae), removal of basal and unwanted chupons at an early stage, and providing any further shaping that may be necessary. Old and senile trees may be rehabilitated by allowing a chupon to come up from the base, and by putting earth on its base it will develop its own roots.

MANURING: Cocoa grown under shade is seldom manured and manurial trials under these conditions have usually been inconclusive. It is doubtful whether the application of fertilizers would be economic when mature cocoa is grown under adequate shade, although increase in yields of 20–30 per cent has been obtained in Ghana from 40 lb P_2O_5 per acre per annum; other minerals giving no response. In the establishment of cocoa in Trinidad it is recommended that ½ lb per tree of NPK 1 : 1 : 1 mixture should be applied in the 1st year, gradually increasing to 2 lb per tree per annum in the 4th year.

When grown under favourable conditions with no shade spectacular responses to nitrogen have been obtained. Murray in Trinidad has shown that 450 lb per acre of ammonium sulphate has little or no effect under shade, but in one experiment gave 519 lb increase in dry beans per acre in the absence of shade; similar but smaller effects were shown for P and K. Cunningham in Ghana obtained yields of Amelonado planted at 8 × 8 ft of 3091 lb of dry beans per acre with fertilizers and no shade, compared with 1211 lb with shade and the same dressings of fertilizers. He recommends (Cunningham, 1963) annual dressings of 2 cwt urea, 1 cwt triple superphosphate and 4 cwt potassium sulphate per acre and states that this can maintain yields around 2,000 lb dry cocoa per acre in unshaded, vigorous, high-yielding cocoa. The extra expenditure and work associated with clear-felling and growing unshaded cocoa with larger amounts of fertilizer would probably only be justified when yields of 3,000 lb per acre and over are obtained.

HARVESTING: Pods are produced throughout the year, but the main harvest usually begins at the end of the wet season and continues for a period of about 3 months; in West Africa during October–January; in Trinidad during November–February; followed by a minor harvest early in the rains. Ripe pods assume a distinctive colour and the seeds rattle inside; green-podded Amelonado turn yellow and red pods usually turn an orange shade. The pods remain in a suitable condition for harvesting for 2–3 weeks; beans from unripe pods give poor fermentation; over-ripe beans may germinate in the pod. Pods should be cut off with a sharp knife, care being taken not to damage the cushion. The pods may be kept for up to a week before breaking and extracting the beans for fermentation. Pods are usually opened by cutting the husk with a cutlass, but they can also be cracked with a wooden mallet or by striking two pods together. The number of pods required to produce one pound of dry cocoa varies from 7 to 14, depending on the pod size, the thickness of the wall, etc.

FERMENTATION: During this process the mucilage round the seeds is removed, the purple pigment diffuses through the cotyledons, the precursor of the chocolate flavour is produced, and astringency disappears. The time taken to ferment Criollo is less than for Forastero. In West Africa cocoa is fermented in heaps or in baskets, usually covered with banana leaves. The beans are left for 4–7 days depending on the season. They may be left undisturbed or may be turned once or more times. In most other countries fermentation is done in wooden sweat boxes which should not be more than 3 ft deep in order to permit aeration. The other dimensions vary and a good average size is $6 \times 5 \times 3$ ft; they are usually built in batteries. The base should be slatted to permit aeration and free drainage of the sweatings. Fermentation takes 6–7 days. During this process the beans are transferred to a second box after 2–3 days, then to a third box after a further 2–3 days and remain there for a further 2 days. At the C.R.I. in Ghana a method has been devised of fermenting in trays 3×4 ft and 4 in. deep with slatted bottoms and these are stacked to a depth of 10 trays and covered with sacking. With this method Amelonado fermented in 3 days without any mixing and the same trays can be used for drying. Methods have also been devised for fermenting small batches of cocoa for experimental purposes.

During the first 36 hours of fermentation the temperature rises to about 96°F; there is very limited aeration and under these conditions yeasts develop and convert the pulp sugar into alcohol and CO_2. Temperatures then increase to about 120°F. As the pulp cells collapse, air enters the mass and there is a rapid oxidation of the alcohol to acetic acid by acetobacter bacteria. The beans are killed mainly through the penetration of alcohol and acetic acid into the cotyledon tissues. Diffusion from the polyphenol storage cells then occurs followed by enzyme attack on the anthocyanins. Proteins are hydrolysed to amino-acids. The colour of the tissues becomes progressively paler and then pale brown. The brown colour deepens; the cotyledons shrink from the testa and separate. There is a gradual development of aroma and flavour and loss in astringency.

DRYING AND POLISHING: After fermentation the beans are spread on mats, trays, or drying floors and dried in the sun. They should be covered over by mats or by sliding roofs during rain and often for the hottest part of the day. The beans are stirred to ensure uniform drying and in sunny weather this is completed in about 7 days. During the process of drying, enzymic action continues and the moisture content is reduced from 56 to 6 per cent. In some countries artificial driers are used. During the whole process of fermentation and drying the loss in weight is 55–64 per cent.

The beans may be polished mechanically or by feet; the latter is known as dancing the cocoa in Trinidad. The beans are wetted and trampled with bare feet, care being taken not to crush the beans but rather to rub one against the other.

GRADING: Shells, broken beans and extraneous matter may be removed by hand or by machines. The number of dried cured beans per pound is approximately 350–450.

The quality of the cured beans is judged by the following characters: (1) The beans should be plump and of even size and usually of not less than 1 g fermented dry weight, when the amount of shell is about 10 per cent. The proportion of shell increases with small beans and may rise to 12–16 per cent. (2) Shells should be intact, free from mould, of a uniform brown colour and should not be shrivelled. (3) Cotyledons should be separated, friable, with a crisp break, of open texture, chocolate brown in colour and on roasting should develop the characteristic chocolate flavour. (4) They should have a fat content of not less than 55 per cent.

The following characteristics are undesirable: (1) Internal mouldiness or diseased beans. (2) Violet or slaty cotyledons denoting under-fermented beans. (3) Off-flavours denoting improper curing, faulty drying, or picking up of taints during storage and shipping. (4) Broken beans. (5) Flat immature small beans. (6) Germinated beans. (7) Beans attacked by insects such as the tobacco weevil, *Lasioderma serricorne* (F.), the beetle *Araecerus fasciculatus* (Deg.) or the moth *Ephestia cautella* (Wlk.).

Cocoa beans of commerce may be classified into: (1) Fine cocoa such as the Criollo of which negligible quantities are now produced. (2) Flavour cocoas as produced in Central America and the West Indies. The flavour of Trinidad cocoa is probably due to a mixture of a large number of genotypes of Trinitario origin and is somewhat astringent. (3) Ordinary Forastero or Amelonado as produced in West Africa and Brazil, which usually has a bland flavour. Cocoa beans are normally shipped in jute bags.

YIELDS: These vary enormously and depend on local environmental conditions, spacing, age of tree, pests and diseases, and the genotype. In Trinidad average yields of old unselected seedling plantations are about 200 lb dry cocoa per acre and that of selected cuttings 600–1,000 lb. Hybrid seedlings of Trinitario–Amazon parentage may yield over 2,000 lb per acre. In West Africa the average yield of peasant Amelonado is said to be about 200 lb dry cocoa per acre and 600 lb is considered good. Crosses between Upper Amazon, Amelonado and local hybrids have yielded over 2,000 lb per acre. Fertilized unshaded cocoa with the control of weeds, pests and diseases has yielded over 3,000 lb per acre.

MAJOR DISEASES

Black pod, *Phytophthora palmivora* (Butl.) Butl., a facultative parasite and the most widely distributed disease of cocoa. Infection usually at either end of pod causing chocolate-brown necrotic spots which spread rapidly; later turning black with a very sharp line of demarcation between diseased and healthy host tissue; beans partially or wholly destroyed by brown-coloured rot; also affects cherelles and young leaves; it can cause flower-cushion and stem cankers; it also attacks seedlings. Fungus sporulates on surface of pod and infection is mainly by contact, raindrops or wind. Temperatures below 70°F and high humidity are conducive to its spread. Regular and frequent harvesting and removal of infected pods reduce loss. Controlled by copper sprays at intervals of not more than a month when wet, cold conditions prevail.

Witches' broom, *Marasmius perniciosus* Stahel,* affects meristematic tissue in active stage of development on vegetative shoots causing proliferation of branches and hypertrophied shoots with broom effect; infected flower cushions produce vegetative brooms; destroys cherelles and young pods; mature pods become hard and woody and internal tissues destroyed by dry brown rot. The sporophores, small pink mushrooms, are produced during the rainy season on tissues killed by fungus. Confined to the Caribbean and South America; originated in Amazon valley; first reported in Surinam in 1895, in Trinidad in 1928 and in Grenada in 1948. Best controlled by planting resistant clones and by removal of brooms which should be burnt.

Ceratostomella wilt, *Ceratocystis fimbriata* Ell. & Halst., confined to New World and recently spreading in Trinidad. Fungus causes death of branches or of entire tree; dead leaves remain attached to branches for long time and affected wood is discoloured. Usually associated with small beetles, *Xyleborus* spp., which tunnel into wood and also spread the disease.

Monilia disease, *Monilia roreri* Cif. & Par., causes considerable losses in Ecuador and Colombia and has spread to other South American countries. Attacks young pods, producing internal watery rot.

Virus diseases: Swollen shoot was first observed in Ghana in 1936 and Posnette in 1938 showed that it was caused by a virus. Swellings occur on branches and twigs, with greater development of phloem and xylem, followed by die-back. In Ghana swollen shoot killed over 5 million trees per year in period 1939–1945. It also occurs in Nigeria and the Ivory Coast. It has been shown to be a complex of viruses of varying virulence, New Juaben strain being one of the most virulent. All these viruses produce varying leaf pattern symptoms with red bands along veins and chlorotic mosaic effects, causing early senescence and leaf shedding; some strains do not produce swollen shoots; light and dark green mottling may occur on unripe fruit. Transmitted by mealy bugs, of which *Planococcoides* (*Pseudococcus*) *njalensis* (Laing) is the commonest vector, and these are distributed by attendant ants, particularly *Crematogaster striatula* Emery. Alternate hosts include *Cola* spp., *Adansonia digitata* L. (baobab) and *Ceiba pentandra* (L.) Gaertn. (kapok). Controlled by cutting out infected and contact tree; the trunk being severed below ground level to prevent regeneration. Up to 1957, 70 million cocoa trees had been cut out in Ghana. No cvs are immune, but some exhibit tolerance and degree of resistance to infection. Complete eradication is unlikely. Mild virus diseases are recorded in Trinidad and Ceylon.

Cherelle wilt: Although this condition may be caused by fungi (black pod) and insects, unless these are epidemic they are only of secondary importance.

---

* = *Crinipellis perniciosa* (Stahel) Singer.

It is a natural physiological process and is a fruit thinning mechanism, whereby a limited number of pods are brought to maturity depending on the food reserves of the tree. Approximately 20–90 per cent, usually about 80 per cent, of the number of fruits set become dry and shrivel with cherelle wilt and finally turn black; as no abscission layer is formed they usually stay on the tree for some time. Maximum wilt occurs at 50 days when the cherelles are about 6 cm long; during this period most of the nutrients are transported in the xylem of the peduncle and in wilted pods these vessels are blocked by mucilage occlusions, followed by oxidative breakdown within the pericarp. As this period coincides with the division of the zygote nucleus, control must be by the maternal tissues. Wilt may also occur during the next 30 days. During the third stage of a further 50–70 days wilting does not occur and the fruits mature and ripen; food transport in the phloem becomes effective, new vascular bundles are formed in the inner pericarp, together with enlargement of the cells of the vessels, and is initiated by seed auxins.

MINERAL DEFICIENCIES

Nitrogen: Lvs pale or yellowish in colour, reduced in size; older lvs showing tip scorch; younger lvs small with entire lamina, including veins, yellow or almost white; internodes compressed.

Phosphorus: Mature lvs paler towards tip and margin, followed by tip and margin scorch; young lvs reduced in size often with interveinal pallor; stipules frequently persisting after leaf abscission.

Potash: On older lvs pale yellow areas in interveinal areas near lf. margin, quickly becoming necrotic.

Magnesium: On acid soils. Old lvs pale green with necrotic areas between veins near lf. margin, quickly fusing into continual marginal necrosis.

Iron: Induced by pH of 7 or over and free $CaCO_3$. Young lvs show darker green veins against paler green background or green-tinted veins against whitish background; develop tip scorch.

Zinc: Induced by high pH. Young lvs showing prominent, distorted veinlets; lvs often narrow, sometimes sickle-shaped with wavy margin.

MINERAL TOXICITY

Aluminium: On very acid soils. Pale interveinal region near tip of older lvs followed by necrosis and tip scorch.

Manganese: On very acid soils. Youngest lvs with irregular pale areas on darker background; no marginal scorch.

Chlorine: Often near sea. Pale yellow areas along margin, quickly fusing to form continuous scorch.

MAJOR INSECT PESTS*

Capsids are the most important insect pest in Ghana, of which *Distantiella theobroma* (Dist.) and *Sahlbergella singularis* Hagl. cause the most damage. In feeding they produce lesions which become infected with the fungus

* See Entwistle, P. F. (1972). *Pests of Cocoa*. London: Longman.

Calonectria rigidiuscula (Berk. & Br.) Sacc. The effect is most severe on young, tender, green shoots. On mature cocoa damage may be (a) 'blast' in which the fans die, (b) 'stag-headed cocoa' in which a persistent and weak flushing occurs, followed by death of the crown, (c) 'capsid pockets' in which canopies of up to 100 trees are destroyed and chupons grow up which are again attacked. Alternate hosts include *Ceiba pentandra* (L.) Gaertn. and *Cola* spp. Losses from capsids may be 20 per cent or more. Damage is more prevalent when the canopy is incomplete. Controlled by chlorinated hydrocarbons, notably gamma BHC.

Thrips, *Selenothrips rubrocinctus* (Giard), cause damage in the West Indies.

Cocoa beetles, *Steirastoma* spp., may do serious damage to the trunks and branches in some regions.

IMPROVEMENT

AIMS

1. Yield improvement: In many parts of the world cocoa plantings are old and yields are poor on this account, and also because unimproved, unselected seedling material may have been used. Average yields in Trinidad are about 200 lb dry cured beans per acre and consequently cannot pay for better cultural treatment. To do this yields of 800 lb are required. Although some new hybrids have given yields of up to 3,000 lb per acre in experiments, capital is not generally available for replanting or to carry over the required time for new plantings to come into bearing.

2. Resistance to diseases and pests: Particularly to black pod in all regions, to witches' broom, *Monilia* disease and *Ceratocystis fimbriata* in the New World, and to viruses in West Africa.

3. Retention of traditional flavour: Each cocoa-growing area has a characteristic flavour associated with its product which is demanded by the manufacturers for blending purposes. Amelonado has a characteristic flavour, particularly suited for milk chocolate; the 'flavour' cocoas are in demand for higher grade chocolates, *e.g.* Trinidad cocoa.

4. Adaptation to local environment: Some clones will only yield well on good soil and cvs for marginal soils may be required. It may be necessary to breed cvs which will give the optimum yield under shade and those that will give the highest yield without shade and with manuring and improved management.

5. Early and sustained bearing: A tree which is vigorous and precocious and which will maintain high yields for many years under satisfactory conditions is an obvious advantage.

6. Tree shape: Low growing trees for convenience of harvesting and pest control are desirable in West Africa where little pruning is done.

7. Pod size: A tree with large pods with a large proportion of beans would be better than one which produces the same weight of cocoa from a larger number of small pods, and this would reduce harvesting and breaking costs.

8. Bean characteristics: Plump beans of even size, with a low proportion of shell (testa), and which will produce the traditional flavour after curing and roasting are desirable qualities.

METHODS AND RESULTS

SELECTION AND VEGETATIVE PROPAGATION: In the period 1930–1934 Pound made a survey of cocoa in Trinidad and Tobago and selected 100 mother trees, Imperial College selections (ICS), using as his criteria of selection: (a) yield of not less than 1 ton of dry fermented beans per acre per annum over a period of years, based on the yield of the tree and the area it occupied. (b) Wet bean weight of not less than 3·5 g (= about 1·4 g per dry bean). Only the first 90 selections fulfilled both criteria and ICS 90–100 had smaller beans, but were very high-yielding. These selections were made in Trinitario cocoa which is very heterogeneous and the selections included both Criollo types and ranging all the way to Forastero-like types. Due to the selection for large beans many of the selections would have some Criollo ancestry. They included self-compatible and self-incompatible clones.

At the same time Pyke developed in Trinidad the technique of rooting semi-hard wood cuttings (see PROPAGATION above). After the mother trees had been propagated in nurseries, clonal cuttings were taken and planted in a trial in River Estate in 1937. The clones showed large differences in growth, tree size and yielding ability. The best clones such as ICS 1, 39 and 40 were associated with little or no weed growth, while the poor clones did not suppress weeds.

Of the 100 selections, only 15 of them gave good results under River Estate conditions of good drainage and reasonable soils. These clones gave good yields of about 1,000 lb of dry cocoa per acre in the 6th–8th years, reaching a maximum of about 1,400 lb in the 12th year. On the heavy clay soils, more typical of cocoa areas in Trinidad, only 5 of the 15 clones continue to perform well, and of these 4 are near-Criollo types which are very susceptible to witches' broom disease on the vegetative parts, although the pods are only very lightly attacked. Thus of the ICS 100 series there is only one general-purpose clone for use in Trinidad, namely 'ICS 95'; it is self-compatible. These clones either alone or in mixture of a few clones produce the bland chocolate flavour which lacks the astringency of genuine Trinidad cocoa.

In 1937 Pound collected planting material in the Amazon region of Ecuador and Peru and obtained clones which were resistant to witches' broom disease; of these two from the Scavina Estate, SCA 6 and 12, are immune, high-yielding, self-incompatible, but with poor bean size. A clone from the Iquitos area, 'IMC 67', is vigorous and early-yielding, suited to a wide range of conditions, has some resistance to witches' broom, and is self-incompatible. Pound visited this region again in 1942 and made more selections. The Anglo-Colombian Cacao Collecting Expedition of 1952–1953 collected a wide range of material including Forastero types from the tributaries of the Amazon, Criollo crosses from Magdelena and Cauca, and selections from plantations. Clones have also been introduced from Central America, Grenada and from elsewhere.

Since 1945 the Trinidad and Tobago Cocoa Board have distributed over 10 million rooted clonal cuttings, and they usually distribute a mixture in which not more than $\frac{1}{4}$ is of a single clone and at least $\frac{1}{3}$ is self-compatible. These cuttings are expensive to produce and transport; hybrid seed is much cheaper (see below). Cuttings of ICS clones yield earlier and more than ICS hybrid seedlings in the first years of cropping. The most popular vegetatively propagated clones were 'ICS 95' (about 35 per cent of all plantings), 'ICS 1' and 'IMC 67'. Hybrid seedlings are now replacing rooted cuttings.

In Ghana where the Amelonado was from a very limited introduction of self-compatible types, selection has produced little variation in pod or bean characteristics and none has been very high-yielding.

HYBRIDIZATION: Hybrid vigour between parents showing good combining ability can be readily exploited in hybrid cvs. The technique of hand-pollination is given in POLLINATION above. Large numbers of single crosses have been made in Trinidad and the potentialities of the parents have been assessed. Crosses within the ICS series did not produce any high-yielding hybrids and there seems to be little relationship between the performance of a clone as cuttings and that of its progeny in crosses. In an attempt to obtain resistance to witches' broom, 'SCA 6' and 'SCA 12' which are immune to the disease, were crossed with 'ICS 1', 'ICS 6' and 'ICS 60'. The progenies were found to have a high resistance, particularly when 'SCA 6' was one of the parents. Progenies from these crosses have been outstanding in regard to yield, particularly 'ICS 6' × 'SCA 6', which is now being planted commercially. They come into bearing early in 2 years and are capable of yielding at the rate of 1,000 lb fermented dry cocoa per acre in their 4th year and over 3,000 lb per acre in their 7th year. There is an improvement in pod and bean size over the Amazon parents. Mixed samples of beans from the 6 reciprocal crosses have a flavour not entirely Trinidadian in nature, but more acceptable than that of the ICS clones.

By interplanting SCA clones, or any other self-incompatible parent, with the desired cross-compatible male parent in isolated blocks, with 1 of the latter to 4 of the former, all pods on the self-incompatible parent can be reaped for seed purposes. It is estimated that 4 acres of this planting will yield 1 million seeds annually at full bearing, a very much cheaper form of propagation than cuttings. The first distribution of hybrid seedlings began in Trinidad in 1958 and it is likely that hybrid seed will soon supersede all cutting material. If both parents are self-incompatible, but cross-compatible, *e.g.* 'SCA 6' × 'IMC 67', hybrid seed can be obtained from both parents.

In West Africa, crosses between selected upper Amazon clones with the local Amelonado are precocious and vigorous, and have given high yields;

they are of an acceptable flavour. No immunity has been discovered to the virus complex in West Africa, but some upper Amazon clones exhibit tolerance and a degree of resistance to infection.

A long term programme in Trinidad has now begun to produce inbred breeding material, with the aim of breeding single crosses between parents of good general or specific combining ability and the production of parents which are homozygous for disease resistance, etc. Much of the Trinitario cocoa is very heterozygous as a result of hybridization. It is desired to reduce variability, and at the same time maintaining the best features of the parents. In some cases inbreeding has been shown to depress yields, *e.g.* 'ICS 1', but this is not always the case, *e.g.* 'ICS 98'. Polycross seed and synthetic cvs are also being produced (Bartley, 1957).

PRODUCTION AND TRADE

As was shown under USES, exports of cocoa to Spain began shortly after the Spanish conquest of Mexico. Production was then extended to Venezuela and Trinidad. The defatting of cocoa and manufacture of chocolate in the 19th century led to an increasing demand for the crop. For many years cocoa production was confined to the Western Hemisphere. It was introduced into Ghana in 1879. In 1900 world production was approximately 100,000 tons, of which the New World produced 81 per cent and West Africa 16 per cent; the latter now produces over 60 per cent of the world's raw cocoa. World production first exceeded 1 million tons in 1960. Of the world's cocoa, Ghana produces about 30 per cent, Nigeria about 15 per cent, other African countries about 13 per cent, South and Central America 30 per cent, of which Brazil is the biggest producer with 20 per cent, West Indies 5 per cent (Dominican Republic 3·5 per cent, Trinidad 0·7 per cent), and Asia 2 per cent, where the largest producers are New Guinea and Western Samoa.

Over 75 per cent of the world's raw cocoa is consumed in seven countries, of which the United States consumes about 25 per cent, Germany 13 per cent, United Kingdom about 10 per cent and the Netherlands 9 per cent. Europe as a whole takes over 50 per cent and American countries about 40 per cent of the world crop.

Cocoa is produced by comparatively few countries, all of them tropical, but the product is processed and consumed mainly in temperate countries.

REFERENCES

BARTLEY, B. G. (1957). Methods of breeding and seed production in cocoa. *6th Meeting of the Interamerican Cacao Tech., Bahia, 1956.*

COPE, F. W. (1962a). The Mechanism of Pollen Incompatibility in *Theobroma cacao* L. *Heredity*, **17**, 157–182.

—— (1962b). The Effects of Incompatibility and Compatibility on Genotype Proportions of *Theobroma cacao* L. *Heredity*, **17**, 183–195.

CUATRECASAS, J. (1964). Cacao and its Allies: A Taxonomic Revision of the Genus *Theobroma*. *Contrib. U.S. Nat. Herb.*, **35**, 379–614.

CUNNINGHAM, R. K. (1963). What Shade and Fertilizers are Needed for Good Cocoa Production? *Cocoa Growers' Bull.*, **1**, 11–16.

FORSYTH, W. G. C. and QUESNEL, V. C. (1963). The Mechanism of Cocoa Curing. *Advances in Enzymology*, **25**, 457–492.

HARDY, F. (1960). *Cacao Manual.* I.A.I.A.S., Turrialba: Costa Rica.

Reports on Cacao Research, I.C.T.A., Later U.W.I.: Trinidad.

Report and Bulletins of W.A.C.R.I. Tafo: Ghana.

URQUHART, D. H. (1961). *Cocoa.* 2nd edition. London: Longmans, Green.

WILLS, J. B. (1962) ed. *Agriculture and Land Use in Ghana*, London: Oxford Univ. Press.

THEACEAE

(= TERNSTROEMIACEAE)

About 20 genera and 200 spp., trees and shrubs of tropical and warm regions, with the main centres of distribution in America and Asia. Lvs alternate, simple, leathery, pinnately-nerved, mostly evergreen, exstipulate. Fls mostly solitary or few together, usually axillary, hermaphrodite, regular; sepals and petals usually 5–7; stamens numerous, hypogynous; ovary superior, 3–5-locular; fruit capsule or berry with sepals persistent at base.

USEFUL PRODUCTS

The only important economic sp. is tea, *Camellia sinensis* (q.v.). Some spp. belonging to several genera are grown as garden ornamentals, of which *Camellia japonica* L. is the best known.

Camellia L. (x = 15)

About 45 spp. of evergreen shrubs and trees in tropical and subtropical Asia. *C. sinensis* (q.v.) provides the tea of commerce. Cvs. of *C. japonica* L., garden Camellias, have long been cultivated in the Far East as ornamental plants for their beautiful fls and are now extensively grown in temperate countries throughout the world. *C. sasanqua* Thunb. is reported to be cultivated in China and Japan for tea-seed oil, of which the seed kernels contain about 58 per cent.

Camellia sinensis (L.) O. Kuntze (2n = 30) TEA

Syn. *Thea sinensis* L., *T. bohea* L., *T. viridis* L., *Camellia thea* Link, *C. theifera* Griff.

USES

Tea has long been grown in China; the earliest use was probably medicinal, but the leaves have been used as a beverage for 2,000–3,000 years. Green tea is made from the leaves which are steamed and dried without withering and fermenting; for black tea the leaves are withered, rolled, fermented and dried. The Mongols adopted the use from the Chinese and started a caravan trade in tea bricks from China to central Asia and beyond, *via* Siberia. The Japanese early adopted its use. Tea was first brought to Europe in the 16th century, but commercial tea did not reach eastern Europe until after 1650, when coffee drinking was already well established. Its use did not become general until the 18th century and it gradually replaced coffee as the favourite beverage of the British, who spread the tea-drinking habit throughout their sphere of influence. Caffeine is manufactured from tea waste. Soluble instant tea is now made.

ORIGIN AND DISTRIBUTION

The probable centre of origin of tea is near the source of the Irrawaddy River and from there it spread to south-eastern China, Indo-China and Assam. It is variously stated that tea is found wild in Assam and Upper Burma and in south Yunnan and upper Indo-China, but the possibility of these plants being escapes from cultivation must not be overlooked. From the main centres in south-eastern Asia, with traditional peasant cultivation, tea has spread into tropical and subtropical countries and during the 19th century developed into an important plantation industry. It was introduced into Japan at an early date.

Experimental plantings were made in India between 1818–1834 from seed originating in China, but with the discovery of 'wild' tea in Assam and Manipur, commercial plantings were made with these local types from 1836 onwards, the main areas of cultivation being in hilly districts of north-eastern and southern India. Tea was introduced into Java in 1690, but commercial production began with seed from Japan in 1824 and later from China; the tea plantations were unremunerative until Assam types were introduced in 1878. Extensive planting began in Ceylon in the 1870's, replacing coffee which was devastated by *Hemileia vastatrix* (q.v.).

Tea was first successfully introduced into Nyasaland (now Malawi) in 1886 from Kew and the first estate was planted in 1891. Introductions into East Africa at the beginning of the 20th century led to commercial production in the 1920's and 1930's in Kenya, Tanzania and Uganda.

Tea was first planted in Russia in 1846, but the first successful plantations in Georgia were begun in 1895. Tea has been tried in most tropical and subtropical countries, including the New World, and there are small plantings in Argentina, near Santos in Brazil, and in the Andean region of Chile and Peru.

SYSTEMATICS

Various botanical varieties have been erected from time to time but, as they all intercross with ease, these do not seem to be justified. The term 'jat' is used to indicate seed derived from different districts or plantations or to separate types on the basis of foliar characteristics. The cvs may be considered under 2 main groups:

1. China teas, *Camellia sinensis* var. *sinensis*, slow-growing, dwarf trees, with small, erect, comparatively narrow, markedly serrate, dark-green lvs; fls borne singly; low jats; resistant to cold and adverse conditions, but rather low-yielding.

2. Assam teas, *C. sinensis* var. *assamica* (Mast.) Pierre, quick-growing taller trees with large, drooping lvs; fls in clusters of 2–4; high jats; well adapted to tropical conditions. These are sometimes divided into (a) Assam types proper with light green lvs; giving larger yields of better quality tea; less hardy and were restricted to the Brahmaputra valley where climatic conditions are not so severe and (b) Manipuri tea with dark green lvs, drought-resistant, but with poorer yields and quality.

Var. *macrophylla* Makino is a large-leaved triploid ($2n = 45$) from Japan, which gives a bitter decoction. A tetraploid ($2n = 60$) has been found in Russia in a Ceylon type; it is sterile and has large lvs and crown; early maturing; low quality.

Numerous hybrids between China and Assam types are known and are grown around Darjeeling in India. As tea is largely cross-pollinated, and most of the commercial crop is raised from seed, the crop is very heterogeneous. Using chosen vegetative characters commercial tea types form a *cline* extending from plants of Chinese origin to those of undoubted Assam origin.

ECOLOGY

Tea cultivation is confined mainly to the subtropics and the mountainous regions of the tropics. Near the equator elevations where it is grown are usually 4,000–6,000 ft. Tea requires equable temperatures, moderate to high rainfall and high humidity throughout the greater part of the year; it will not tolerate frost. China teas are more tolerant of colder conditions. Best quality tea is produced in a cool climate, *e.g.* Himalayan foothills of north-eastern India. Mean minimum temperatures should not fall below 55°F nor mean maximum temperatures go above 85°F. Rainfall below 45 in. per annum is marginal and should not fall below 2 in. per month for any prolonged period. Hail can cause much damage. The best soils for tea are deep, permeable, well-drained, acid soils and are often tropical red earths. The pH should be between 4·5–6·0. Tea is a calcifuge and is an aluminium accumulator and this element should be available in the soil. *Albizia* spp., *Dissotis* spp. and bracken are sometimes used as indicator plants in selecting soils for tea.

STRUCTURE (see also SYSTEMATICS above)

Under natural conditions tea is a small evergreen tree to 15 m in height (less in China tea), but under cultivation it is usually pruned down to 0·5–1·5 m and trained as a low spreading bush.

ROOTS: A strong tap-root with lateral roots which give rise to a surface mat of feeding roots which lack root hairs when mature, and with an associated endotrophic mycorrhiza. Starch is stored in roots.

SHOOTS AND LEAVES: Seedlings have a main axis with lateral branches from buds in lf. axils. Buds and internodes glabrous or hairy. Lvs alternate, evergreen, obovate-lanceolate, acuminate, leathery, glossy on upper surface, sparsely hairy on lower surface, particularly when immature, serrate, 3–30 cm long, those produced in early stages of growth and after pruning larger than lvs subsequently formed. China jats have smaller and more coarsely serrate lvs than Assam jats. Stomata confined to lower surface. Scleroids of typical form in mesophyll used for checking adulteration. Lvs produced in flushes; during dormant 'banjhi' period young bud 5 mm long formed in axil of terminal lf. of shoot which has attained full size; bud produces 2 scale lvs, first of which usually drops off, followed by small, entire, blunt fish-leaf; normal lvs usually 4 in number then produced which grow to normal flush lvs and internodes elongate; shoot then becomes dormant and new bud develops.

FLOWERS: Axillary, solitary, or in clusters of 2–4, with short pedicels, fragrant, 2·5–4 cm in diameter; calyx persistent with 5–7 sepals; petals 5–7, white or tinged pink, obovate, concave; stamens numerous, 8–12 mm long, with 2-celled yellow anthers; petals and outer stamens united for a short distance at base; ovary superior, hairy, with 4–6 ovules per carpel; styles short, free to almost completely fused, stigmatic lobes 3–5.

FRUITS: Capsule thick-walled, brownish-green, 3-lobed, 3-celled (usually), 1·5–2 cm in diameter, becoming slightly rough at maturity, taking 9–12 months to mature, dehiscing by splitting from apex into 3 valves.

SEEDS: 1–2 per cell, globose or flattened on one surface, 1–1·5 cm in diameter; testa light brown, thin; no endosperm; cotyledons thick, rich in oil; embryo straight. Approximately 230 seeds per lb.

POLLINATION

The flowers are pollinated by insects. It is variously stated that tea is 'virtually self-sterile', that selfing gives 'a much lower percentage of viable seed' and that 'selfing of tea is practised in both Japan and the U.S.S.R. with apparent good success'. It is possible that self-incompatibility may exist in certain pure Assam jats which may be lacking in Chinese jats or Chinese–Assam hybrids. The percentage of fruits with viable seeds produced with natural cross-pollination is about 8 per cent. In cross-pollination by hand the petals and stamens are removed in the bud stage and the flowers are bagged for 3 days after pollination.

GERMINATION

Tea seed retains its viability for a short time only. Germination is epigeal and cotyledons are abscissed after 5–6 months.

Fig. 91. **Camellia sinensis : TEA. A, shoot with flowers and fruits (×½); B, flower in longitudinal section (×3); C, fruits (×1).**

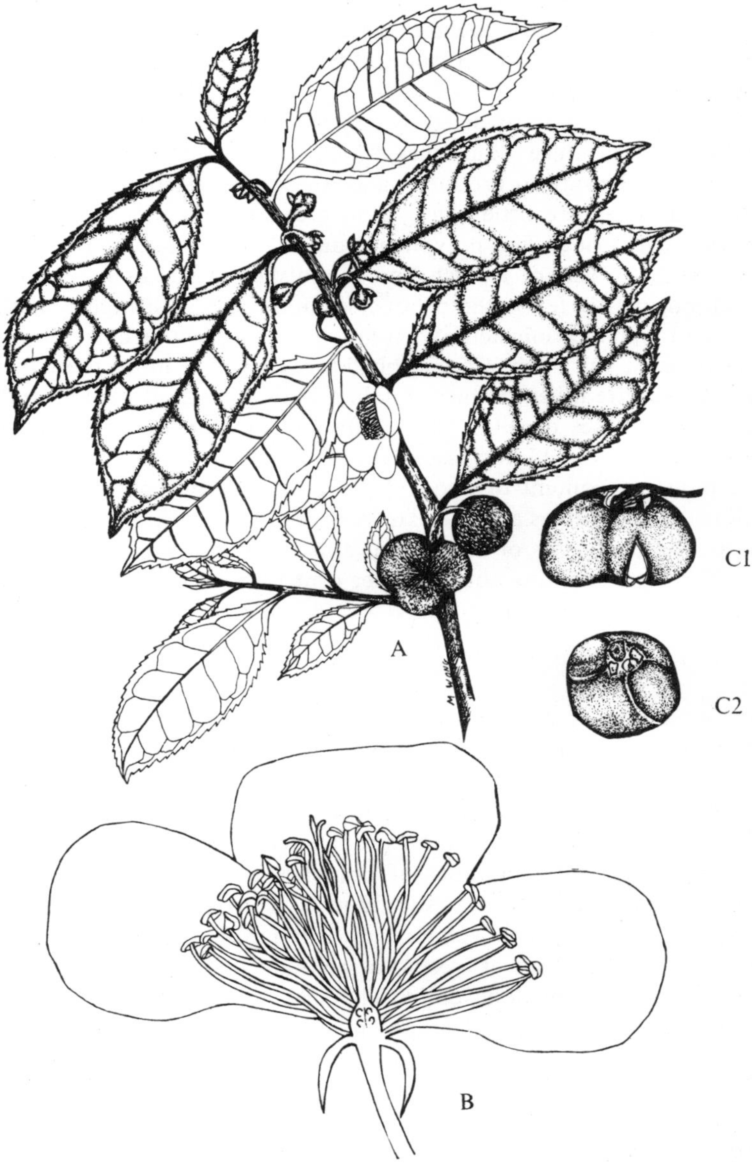
A
B
C1
C2

CHEMICAL COMPOSITION

The most important constituents which give tea its distinctive character as a beverage are polyphenols, caffeine and essential oils. Fresh plucked tea contains 75–80 per cent of water; the soluble and insoluble constituents on an approximate dry weight percentage include: polyphenols 25; protein 20; caffeine 2·5–4·5; crude fibre 27; carbohydrates 4; pectin 6. The polyphenols, derivatives of gallic acid and catechin, but not tannins in the normal sense, are oxidized by enzyme action during maceration and fermentation to produce *o*-quinones, which polymerize to produce coloured astringent condensation products and these are partially extracted in brewing tea. They are little changed in green tea manufacture where no fermentation is done. The polyphenol and caffeine content is highest in the bud and top leaf of a flush and diminishes in successive leaves and stalk. Good quality tea is the product of good leaf with high polyphenol content and high enzymic activity; it also shows a positive correlation with the hairiness of the flush. Coarse plucking, in which more than the bud and 2–3 leaves are taken, reduces quality. Slow growth at high altitudes in the tropics increases quality. The aroma and flavour are also affected by various substances including essential oils. Fresh manufactured tea should have a moisture content of about 3 per cent. The professional tea taster infuses tea for 5 minutes and the extract contains about half the polyphenol products, three-quarters of the caffeine and about half the total extractable solids. The seed contains about 20 per cent of a non-drying oil, but it is not extracted commercially.

PROPAGATION

SEED: Tea is usually propagated by seed from cross-fertilized seed bushes of jats which may have been selected for yield and quality (see IMPROVEMENT). Seed bearers should be planted in special seed gardens with 70–100 trees per acre and take 4–12 years to bear seed. If possible they should be of selected clones which have been propagated vegetatively. The trees are shaped by reducing the number of branches to about 6 when they are 10 ft or so high. Mature bearers should receive fertilizers, *e.g.* 3 lb sulphate of ammonia, 1 lb superphosphate and 1 lb muriate of potash per tree. The plot should be clean of weeds to facilitate the collection of dehisced seed, which should be done daily. The seed is graded and those which pass through a ½ in. mesh and float in water are discarded. Viability is short and the seed should be planted within a few days of gathering. They may be transported in damp powdered charcoal. Fumigation with methyl bromide may kill the seeds unless planted immediately afterwards. Immersion for 30 minutes in water at 52°C destroys borer larvae and enhances germination, which can also be accelerated by removing the testa or by soaking in water and then sun-drying for a few hours.

Much tea has been planted at stake, but it is advisable to grow the seedlings in nurseries. Seeds may be germinated in sand in beds or trays covered with damp hessian; the seeds should not touch each other and they should be protected from sun and wind. On emergence of the radicle, they are planted in baskets, plastic polythene sleeves or other containers, or 6 in. apart in nursery beds. Shade is necessary, but this is gradually reduced. Planting bare-root seedlings 6–12 months old is disastrous, so they should be planted with a ball of earth or in the basket; alternatively, they may be grown on in the nursery for longer periods of up to 3 years and prepared as stumps by loosening the soil with a fork, pulling the plant out like a carrot and cutting the stem 4 in. from the ground mark.

VEGETATIVE PROPAGATION: The usual method is by single internode cuttings taken immediately above a leaf and axillary bud. Green or slightly reddening wood cuttings can be made, the former rooting more easily; old wood and young green shoots should be avoided. The cuttings are planted in shaded propagating beds, the stem being inserted in the soil at a slight angle so that the leaf rests on the surface and the leaves should not overlap. Careful controlled watering is essential. The number of cuttings rooting successfully is usually 80–100 per cent. Rooted cuttings may be planted direct into the field or may be transferred to baskets, polythene sleeves or other containers or may be grown in nursery beds and later cut back and planted as stumps. The time taken from planting the cuttings to planting out at 18 in. high is about 12 months. It is advisable to establish a budwood nursery of selected clones. Each bush will produce about 800 cuttings per year; new growth from young pruned trees is preferred. In Indonesia tea is propagated by budding, usually by the Forkert method (see Rubber).

HUSBANDRY

PLANTING AND CULTIVATION: Virgin forest is usually the most suitable land for tea cultivation, but grassland and areas which have grown other crops are also used. Hut sites should be avoided or dressed with elemental sulphur. The land should be thoroughly cultivated; stumps should be removed and burnt off the area as tea will not grow in patches of wood ash. Soil conservation measures, such as contour terraces or bunds, and adequate drainage should be carried out wherever necessary before the tea is planted. Suppression of weeds is essential and stoloniferous grasses such as *Digitaria scalarum* (Schweinf.) Choiv. in Africa, and *Panicum repens* L. and *Imperata cylindrica* Beauv. in Asia, which are particularly deleterious, should be eradicated. Tea is sometimes interplanted in a nurse crop or with green manure.

The normal spacing is 4–5 ft apart in square or triangular planting, but hedge planting with bushes 2–2½ ft apart in the rows and with 5 ft between rows is gaining in favour as it gives higher initial yields and facilitates mechanical harvesting. Sowing at stake is not recommended and nursery-raised seedlings and stumps are used; the use of clonal cuttings is likely to increase (see PROPAGATION above). The planting hole should be a minimum of 12 in. diameter and 18 in. deep. The young plants may be protected by plucked fronds of bracken or grass, or by planting green manure crops such as *Crotalaria* spp., *Tephrosia* spp., buckwheat (*Fagopyrum esculentum* Moench., family Polygonaceae), or oats (*Avena sativa* L.).

When the plants have become well established deep cultivation which would damage the surface roots should not be done. A light scraping of the soil helps to control weeds, but herbicides will be used increasingly for this purpose. Envelope forking in which a straight fork is inserted into the soil and levered backwards and forwards is practised in some areas.

SHADE AND WINDBREAKS: In most tea-growing countries it has been customary to interplant shade trees, but this is now somewhat controversial. The shade trees were planted 40–50 ft apart and the most commonly used species were *Albizia chinensis* (Osbeck) Merr. and *A. odoratissima* (L.f.) Benth. in Assam, *A. falcata* (L.) Backer in Ceylon, and *Erythrina subumbrans* (Hassk.) Merr. (syn. *E. lithosperma* Miq.) and *Gliricidia sepium* (Jacq.) Walp. in Ceylon and Indonesia. *Grevillea robusta* A. Cunn. was usually planted at higher altitudes in Ceylon and Africa. It has been shown, however, that tea growing under favourable conditions of soil and climate gives considerably higher yields to nitrogenous manuring when grown in full sun (see Cocoa). Shade increases the incidence of blister blight and as a result shade trees have been drastically reduced or removed altogether in southern India and Ceylon. Windbreaks are beneficial.

SHAPING AND PRUNING: These operations are done to turn what is naturally a small tree into a low, wide, spreading bush, to maintain a convenient height for plucking, to induce vigorous vegetative growth and to ensure a continuous supply of flushes. China types, being dwarf and slow growing, require relatively little pruning, whereas Assam teas and Assam–China hybrids have to be kept within bounds.

In frame formation the main leaders of the young shrub are cut back to encourage the development of lateral shoots. Shaping to form a flat-topped table consists of tipping branches and thinning lateral shoots. In Assam it is usual to let the bush grow for 3 years and then cut across the leaders and laterals at 18 in., and cutting back further strong branches showing vigorous apical growth to the level at which branching starts. Thereafter, the bushes are top-pruned annually and a medium prune is given every 12–18 years, cutting back to 18 in. from the ground, to bring down the height of the plucking table. Another method used in Assam and Ceylon is to cut back the entire growth to within 4–6 in. of the ground when the bushes are 3–5 ft in height. A third method used in Ceylon and East Africa is to cut the branches early when the branches at cutting height are pencil thickness, in single-stem plants at 6 in., in double-stem plants at 8 in. and with multiple stems at 10 in. After regeneration further cuts are made at 12–16 in. and the bushes are brought into bearing with a plucking table of not less than 2 ft. After a plucking cycle of 4 years the bushes will be approximately 50 in. high. The most common method used in East Africa is to cut back to 9 in. after 2 years' growth and to 16 in. after a further year.

Maintenance or production pruning is carried out during the remaining life of the bush, until rehabilitation becomes necessary. The pruning cycles are of longer duration at high altitudes and in more temperate climates, the time varying from 2 years in the tropical lowlands to 3–5 years at higher elevations. In Ceylon it was customary to give a medium prune, cutting back to 18–24 in. every 3–5 years, but, with the advent of blister blight, lighter pruning is now practised. In rehabilitation, collar pruning is sometimes done, in which the bole is cut slightly below ground level.

Pruning should be done during a dormant period when this exists, *e.g.* the winter in Assam. In Ceylon and East Africa pruning is usually done immediately following the dry weather. Cuts should be made at an angle of 45° as near as possible to a viable bud on the outside of the stem. All large cuts should be protected with a fungicidal and waterproof paint.

MANURING: A tea crop of 1,000 lb. per acre per annum removes approximately 55 lb of nitrogen, 30 lb of potash and 10 lb of phosphoric acid. Plants 2–3 years old respond well to small doses of nitrogen of the order of 20 lb per acre and this is increased as the bushes get older. Moderate doses of nitrogen, up to 80 lb per acre, give a yield response approximately proportional to the amount applied. In Assam the response is almost linear up to 120 lb of nitrogen and in Malawi up to 80 lb of nitrogen. The upper limit of efficiency is 6–8 lb of manufactured tea for every lb of nitrogen supplied. In East Africa, as a result of factorial experiments, it has been shown that 200 lb of sulphate of ammonia in the year of pruning and 400 lb per annum thereafter will produce up to 6 lb made tea per lb of nitrogen. Chenery (1966) states that up to about 300 lb N per acre responses are linear, and paying responses are obtained up to 500 lb per acre applied in monthly small doses of 40–50 lb N. He also states that very few instances of economic response to phosphatic fertilizers are on record, and that much the same applies to potassic fertilizers. In East Africa it has been shown that in many areas the available potash in the soil may fall below the level required for the maintenance of full tea production.

HARVESTING AND YIELDS: Harvesting consists of plucking the newly grown vegetative shoots composed of the terminal bud and 2–3 leaves immediately below it, together with the intervening stalk. Plucking is usually done by women using the thumb and forefinger. The bud and the youngest leaves have the highest caffeine and polyphenol content and produce the best quality tea. The length of time from planting to first plucking depends on the type of planting material used, about 2 years for stumps and 4 years for seedlings, and on the environment, being earlier at lower elevations. The length of time between flushes and for the plucked shoot to produce a new shoot ready for plucking varies with the plucking system employed and environmental conditions and is usually 70–90 days. However, the plucking table has shoots of varying maturity. Skilful hand-plucking permits only those which are at the right stage of maturity to be harvested. Plucking is usually done every 7–10 days at the lower elevations and every 14 days or so in colder climates. The aim is to remove the maximum amount of good quality leaf compatible with the general well-being of the bush and with sufficient leaves left to maintain normal growth and vigour. The best tea is made from a flush with a bud and 2 leaves. Coarse picking with a bud and 4 leaves yields a lower quality product.

Four pounds of green shoots with a water content of 75–80 per cent produce 1 lb of made tea. Mature trees giving 2 lb of green shoots each per year, yield approximately 1,000 lb of made tea per acre. Yields vary greatly. In Darjeeling yields are about 450 lb per acre, but the tea fetches high prices; the record price paid was 64s. per lb for 6 chests in 1962. At mid and higher altitudes in Ceylon seedling plantations 60–90 years old give average yields of 1,000–1,300 lb per acre on a 3–4-year pruning cycle, but yields of over 2,000 lb per acre are quite common. A yield of over 6,000 lb made tea per acre has been recorded in a plot of clonal tea in Ceylon. In Assam normal yields are 1,100–2,000 lb made tea per acre.

Pluckers harvest 30–80 lb of green shoots per day. The economic life of a tea tree is said to be about 40–50 years, but much of the production in India and Ceylon is from tea 70–100 years old. Mechanical plucking is receiving attention and it is likely that it will be used increasingly as satisfactory methods are developed.

MANUFACTURE: The manufacture of black tea, which forms the bulk of the tea appearing on the world markets, is done in factories on the estates. During the process, which takes about 2 days, enzymic fermentation of the catechin polyphenols occurs and this is arrested when it has gone far enough. Scrupulous hygiene is required at all stages from plucking to packing in order to prevent the picking up of taints and admixture with other matter.

The harvested shoots are first withered by spreading them thinly on jute hessian or nylon net stretched on banks of wire frames (tats) in open lofts or in rooms where ambient or warm air is blown over the leaves; perforated revolving drums are sometimes used. During withering, which takes 18–20 hours on tats and 3 hours in drums, the leaf loses approximately 40 per cent of its water and becomes flaccid and permeable. The leaf is then passed on to rollers and the rolling twists the leaves and breaks them up and liberates the sap in a film over the surface. It ensures that the enzymes and the catechins are thoroughly mixed. Temperature of the leaf increases during rolling and should be between 80–90°F. The leaf mass is broken and sieved, which cools the leaf, aerates the mass and separates the crushed stalks and leaves into particles of reasonably uniform size for fermentation, which has begun during the rolling. The sifted particles are spread out in thin layers to complete the fermentation, which takes about 2 hours. Good air circulation without draughts is required and air conditioning is sometimes used. The leaf becomes a dark copper colour and the typical aroma develops. The fermented tea is treated with a forced draft of hot air, when temperatures rise to 180–200°F, and the tea is dried to a moisture content of 3 per cent. This drying is known as firing.

The dried tea is graded and sorted in a room with a relative humidity of 60–65 per cent and is packed for export in plywood chests lined with aluminium foil. The grading is based on colour and fineness of the particles. The best Orange Pekoe contains a high proportion of buds. The various grades are usually blended in the consuming countries and a blend usually consists of grades from various estates and usually from more than one country. In the production of green tea, which is made in considerable quantity in China and Japan, the enzyme is destroyed by steaming or by rapid drying before fermentation can take place. Oolong tea is a semi-fermented product produced almost exclusively in Taiwan.

MAJOR DISEASES

Blister blight, *Exobasidium vexans* Massee, was first reported in Assam in 1868 where it caused little damage. It spread to Darjeeling in 1908 causing a severe attack. It was reported in Japan in 1912 and Indo-China in 1930 and appeared in south India and Ceylon in 1946, in Sumatra in 1949 and Java in 1951. It is now endemic throughout tea-growing areas in Asia, where it has reached epidemic proportions, but has not yet been reported in Africa. Only young leaves are infectible. It first appears as a translucent spot which later forms a blister, concave on the infected surface (usually the upper) and convex on the reverse. The convex surface becomes grey and then white, and myriads of basidiospores are liberated, the full cycle taking about 1 month. Buds can be attacked and whole shoots may die. The blisters become infected with other leaf-rotting fungi such as *Colletotrichum camelliae* Massee, *Pestalotia theae* Saw. and *Calonectria theae* Loos, all of which can attack tea leaves on their own. Cool, moist, still air favours infection, as do higher elevations. Light pruning only in dry weather and reduction or removal of shade assist control, which may be obtained by a mist spray of copper fungicides or nickel chloride. Resistance has been found in some cvs.

Root diseases, *Armillaria mellea* (Vahl ex Fr.) Kummer in Africa, *Poria hypolateritia* Berk., *Ustulina deusta* (Hoffm. ex Fr.) Lind and *Rosellinia arcuata* Petch in Ceylon, south India and Indonesia, cause damage and are controlled by uprooting and destroying diseased stumps of tea and other trees. *Corticium invisum* Petch and *C. theae* Bernard are economically important in India.

Red rust caused by the alga *Cephaleuros parasiticus* Karst. is widespread in India and Ceylon.

A virus disease causes phloem necrosis in Asia.

MAJOR PESTS

Tea mosquito bugs, *Helopeltis* spp., are serious pests of tea causing necrotic spots followed by holes in the lvs and some spp. cause stem cankers.

Tea tortrix, *Homona coffearia* (Nietner), was serious in Ceylon, but was controlled by the Ichneumon parasite, *Macrocentrus homonae* Nixon.

Red spider, *Oligonychus coffeae* (Nietner), is widespread in tea and a serious pest in north-east India.

Stem borers, *Xyleborus fornicatus fornicatior* Egg. and *Xylosandrus compactus* Eichh., are serious pests in Ceylon.

Eelworms, *Meloidogyne* spp. and *Pratylenchus* spp., can cause damage.

MINERAL DEFICIENCIES

Tea yellows in Malawi and Tanzania is caused by sulphur deficiency and results in mottling of the leaves followed by chlorosis, the edges then become necrotic and leaf fall may occur. It is controlled by applications of ammonium sulphate.

Magnesium deficiency, coupled with manganese excess, is widely spread in the Kericho District of Kenya and gives a chlorosis of the older leaves with the main veins showing dark green.

Phosphate and potash deficiency have been reported in Ceylon; in the former the leaves are very dark green and become rough and thick with age; in the latter there is a marginal scorch; zinc deficiency has also been recorded and produces very short internodes and dwarfed leaves with interveinal chlorosis and crenulated margins.

IMPROVEMENT

Much work remains to be done on the selection and breeding of tea. Most tea has been planted from seed and, as the flowers are cross-pollinated, it is very heterogeneous. Much of the yield of an individual field is contributed by a relatively small number of plants. In Ceylon the number of pluckable shoots on 1,515 bushes, randomly selected, varied from 20 to 1,000 and less than 5 per cent of the bushes gave counts above the mean. There is a considerable variability on which to base selection, including hybrids of China and Assam types. It should be possible to combine the aroma of China tea with the hardiness, vigour and large leaves of Assam tea. Clones can be propagated vegetatively (see above). Biclonal seed may be obtained, if required, by planting two clones in the seed garden (see Rubber).

The final aim is a high yield of tea of acceptable quality from clones which may be successfully propagated vegetatively. Yielding capacity is based on the yield per unit area of bush surface which is dependent upon the number of plucking points and the size of the shoots. The following characters may be used in selection: vigour, with a bush which comes into plucking quickly and gives continuously high yields; adaptability to local environment including drought resistance for dry areas and frost resistance where required; resistance to pests and diseases, particularly blister blight; hairiness of terminal bud denoting high polyphenol content; bushes with spreading habit and tight plucking tables and with ample leaves below the plucking tables; few dormant buds (banjhi) and without tendency to flower; evenness of flush; large, heavy shoots with long internodes and without markedly erect leaves, as these are more difficult to pick; flexible leaves which are easier to roll and leaves that ferment easily, and which are a good colour in the finished product giving an infusion of the correct colour, aroma and astringency; clones which root easily from cuttings. Selection will involve visual selection in the field, in which pluckers with an intimate knowledge of the bushes may assist, testing in the nursery for rooting ability of the cuttings, testing in field trials, and tasting.

The following standards were set in Java in the selection of mother trees: average production at least 300 per cent above the average for the entire plantation; minimum average of 50 g processed leaves and terminals per pluck; consistent yields from one plucking test to the next; commercially attractive types; aggregate number of plants not more than 1 per cent of those in the planting. In selections from about 1 million trees in west Java only 0·2 per cent produced 3 times the average of the plantings from which they came.

PRODUCTION AND TRADE

Black tea, which provides the bulk of the world's supply and requires machine manufacture, is usually grown in large plantations, but small holdings with the harvested tea being taken to central factories are increasing, particularly in East Africa. Green tea, which is produced mainly in China and Japan, is predominantly peasant-produced and much of it is consumed locally.

The world's production of tea is over 2,000 million lb. India is the largest producer, followed by Ceylon, both countries exporting nearly 500 million lb annually. Tea supplies about 65 per cent of Ceylon's total domestic merchandise exports. Other countries exporting substantial quantities of tea in order of importance are China, Indonesia, Kenya, Malawi, Taiwan, Mozambique, Japan, Soviet Russia, Argentina, Uganda and Tanzania.

By far the largest consumer of tea is the United Kingdom which takes more than half of the world's imports and where the *per capita* consumption is nearly 10 lb per year. The tea-drinking habit has become well established wherever British people have made their homes and consumption is high in other Commonwealth countries. The second largest importer is the United States, but they consume only 0·7 lb per head. It is of interest to compare this with coffee consumption which is about 3 lb per head in the United Kingdom and 16 lb per head in the United States. Other countries importing large quantities of tea, given in decreasing order, are: Australia, Iraq, Egypt, Canada, South Africa, Morocco and the Netherlands.

For some years production of tea was controlled by an International Tea Agreement, but production is now unrestricted.

REFERENCES

CHENERY, E. M. (1964a). The objectives of the Tea Research Institute of East Africa. *Tea*, **4**, *4*, 14–18.

CHENERY, E. M. (1964b). Tea and Tea Research in Ceylon and India. *Tea*, **5**, *2*, 15–23.

CHENERY, E. M. (1966). Factors Limiting Crop Production: 4 Tea. *Span*, **9**, 45–48.

EDEN, T. (1965). *Tea*. 2nd edition. London: Longmans, Green.

HARLER, C. R. (1963). *Tea Manufacture*. London: Oxford Univ. Press.

HARLER, C. R. (1964). *The Culture and Marketing of Tea*. 3rd Edition. London: Oxford Univ. Press.

JOURNALS: *Tea*: Journal of the Tea Boards of East Africa; *Tea Quarterly:* Journal of the Tea Research Institute of Ceylon, Talawakelle; '*Two and a Bud*': Newsletter of The Indian Tea Association.

ANNUAL REPORTS OF: *The Tea Research Institute of Ceylon, Talawakelle; The Indian Tea Association, Tochlai; The Tea Research Station, Malawi; The Tea Research Institute of East Africa, Kericho; United Planters Association of South India.*

TILIACEAE

About 40 genera and 400 spp. of trees and shrubs, rarely herbs, in warm and tropical regions, with a few in temperate zones. Bark fibrous with mucilaginous properties; lvs usually alternate, simple; fls regular, usually hermaphrodite; sepals usually 5, valvate; petals present or absent, free; stamens numerous, free or in bundles of 5–10, anthers 2-locular; ovary superior, 2–10-locular with 1 to many ovules per loculus, placentation axile.

USEFUL PRODUCTS

Several spp. yield fibres, of which jute, *Corchorus* spp. (q.v.), are the most important. *Triumfetta* spp. have been tried commercially on a small scale and are widely used in tropical Africa locally for fibres, particularly *T. cordifolia* A. Rich., *T. rhomboidea* Jacq. and *T. tomentosa* Boj. *Cephalonema polyandrum* K. Schum. has been used for commercial fibre production in the Congo and is known in the trade as punga. *Clappertonia ficifolia* (Willd.) Decne. (syn. *Honckenya ficifolia* Willd.), a native shrub of tropical Africa, has been shown to yield a satisfactory fibre, but has not been exploited commercially.

Corchorus L. (x = 7)

About 40 spp. of herbs and sub-shrubs, widely dispersed in the tropics. *C. capsularis* (q.v.) and *C. olitorius* (q.v.) are the source of commercial jute. The latter is also used as a pot-herb in tropical Africa, as are *C. aestuans* L., *C. tridens* L. and *C. trilocularis* L.

KEY TO THE CULTIVATED SPECIES

A. Capsule globose; seeds brown.................. *C. capsularis*
AA. Capsule long, cylindrical; seeds dark greyish-blue.. *C. olitorius*

Corchorus capsularis L. (2n = 14) WHITE JUTE

Herbaceous annual with straight slender stems, 3–4 m in height under cultivation, branching near top. The best fibres occur in outer portion of stem, interspersed with thin-walled tissue of phloem. Lvs alternate, light green; stipules 2, linear, up to 1 cm long, deciduous; petiole 1–2 cm long;

lamina lanceolate, 5–12 × 2–5 cm, serrate, 2 lower teeth prolonged into fine pointed auricles, tip acuminate. Fls solitary or in few-flowered cymes, opposite lvs; sepals usually 5, free, narrow, 4–5 mm long; petals usually 5, yellow, 4–5 mm long; short corona separates petals from insertion of stamens; stamens 10 to many, free, filaments short, anthers small, bilobed; ovary superior, 5-locular with numerous ovules, style short with flattened stigma. Fr. globose capsule, 1·2–2 cm in diameter, wrinkled, 10-ridged, flattened on top, dehiscing loculicidally into 5 valves, without transverse partitions between seeds. Seeds small, 2–3 mm long, oval, pointed, concave on one surface, copper brown in colour, 300 per g.

Corchorus olitorius L. (2n = 14) TOSSA JUTE, JEW'S MALLOW

Similar to *C. capsularis*, but taller when grown as a fibre plant, lvs larger in Trinidad material up to 20 × 7 cm, fls larger with sepals and petals 7–8 mm long and deeper yellow in colour. Fr. a long cylindrical, 10-ridged, beaked capsule, 5–10 × 0·5–0·8 cm, dehiscing by 5 valves and with transverse septa between seeds. Seeds pyramidal, smaller than *C. capsularis*, 1–2 mm long, dark greyish-blue in colour, 500 per g.

The form grown as a vegetable is a much smaller plant, not growing much over 30 cm in height, and is fairly well branched.

USES

The soft bast fibres are second in importance as a textile fibre to cotton; they are weaker than hemp or flax. The use of jute was insignificant until Dundee mill-owners discovered the technique of spinning it about 1838. About 75 per cent of the world's jute is used for coarse woven fabrics, hessian, sacking or burlap, used in the manufacture of sacks and bags. Jute is also used for twines, carpet yarns, cloth backing for linoleum and carpets, tailors' padding, screens to protect plants, deck-chair cloth, awnings and tents. *C. capsularis* is more commonly grown for fibre than *C. olitorius* and comprises about 75 per cent of the acreage under jute in Bengal. *C. olitorius* is extensively grown as a vegetable, eaten as spinach, in the Middle East, Egypt, the Sudan and tropical Africa; it is very mucilaginous.

ORIGIN AND DISTRIBUTION

C. capsularis occurs wild in southern China and it is probable that it was brought from there to India and Bangladesh, where the main centre of production is in the Ganges–Brahmaputra delta. The crop has been tried in many tropical countries but, with the exception of Brazil, has seldom

Fig. 92. **Corchorus capsularis : WHITE JUTE. A, portion of plant (× ½); B, flower in longitudinal section (× 5); C, fruit in longitudinal section (× ½).**

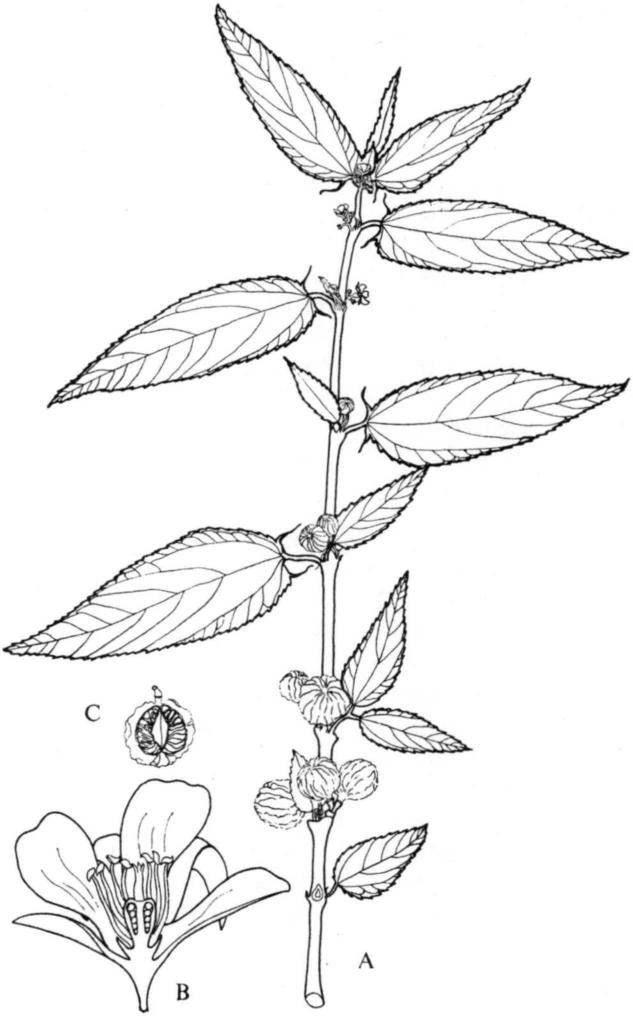
A
B
C

been successful. *C. olitorius* occurs wild in Asia and Africa and has become naturalized as an escape from cultivation in most tropical regions.

CULTIVARS

It has not been found possible to hybridize the two commercial species. A number of improved cvs, both early and late maturing, of *C. capsularis* and *C. olitorius*, have been bred in India.

ECOLOGY

Much of the jute crop is grown in deltaic areas at low altitudes with high temperatures of 75–95°F, a high humidity of about 90 per cent, a rainfall of over 40 in. per annum, so distributed that the young plants have enough moisture, but the bulk of the fall taking place when the crop is more mature, and with fertile alluvial soils. A sufficient supply of clean retting water is also required. Young plants of both species are sensitive to waterlogging. Later *C. capsularis* will tolerate flooding during the monsoon and the water may be 6 ft deep. *C. olitorius* will not thrive under these conditions and is grown on higher land.

HUSBANDRY

A fine seed-bed is required. On land on which silt is deposited annually manuring is not usually practised; elsewhere cowdung, wood ash and 20–40 lb N per acre have proved beneficial. In the lowlands in Bengal *C. capsularis* is usually planted in March or April and is grown in rotation with rice. On the higher land *C. olitorius* is planted in May. The crop is usually broadcast with a seed rate of about 10 lb per acre for *C. capsularis* and 6 lb per acre for *C. olitorius*. Provided there is sufficient soil moisture, seeds germinate in 3–4 days. Jute responds to early weeding and thinning and the plants are thinned to 4–6 in. apart when 6–9 in. high. A thick stand is necessary to obtain tall straight stems.

Harvesting usually begins 100–130 days after planting when about 50 per cent of the plants are in pod. The plants are cut close to the ground and, when the land is flooded, the cutters have to dive under the water to perform this operation. The cut stems are tied into bundles about 8 in. in diameter after the leaves have been allowed to fall and the stems are then retted by submerging them in water, slow running water being the most effective for the purpose. The soft tissues of the cortex are broken down by micro-organisms and the fibre bundles separate from the central woody portion of the stem. With a water temperature of 85°F retting is completed in about 10 days, but takes longer at lower temperatures.

Fig. 93. **Corchorus olitorius : TOSSA JUTE. A,** portion of plant (×½); **B,** flower in longitudinal section (×5); **C,** fruit in longitudinal section (×½).

A
B
C

In stripping the fibre the butt ends of the stems are beaten with a mallet, the fibres at the base of 6–10 stems are wrapped round the fingers and the stems are jerked backwards and forwards in the water. The stripped fibres are then washed and dried in the sun for 2–3 days. The green stem contains 4·5–7·5 per cent of fibre with an average of about 6 per cent. The average yield of fibre per acre is about 1,500 lb, but up to 2,500 lb per acre has been obtained with improved cvs. The quality of the fibre is judged by its strength, fineness, colour, uniformity of colour, lustre, length, and proportion of roots. A part of the crop is left for seed production, which is harvested 4–6 weeks after the fibre crop, and yields about 250–300 lb of seed per acre.

MAJOR DISEASES AND PESTS

The stem rot, *Macrophomina phaseoli* (Maubl.) Ashby, is the most serious and widespread disease of jute, affecting plants at all stages of growth. Soft rot, *Corticium rolfsii* (Sacc.) Curzi, is a soil-borne disease affecting the collar region. The most serious pests are a yellow mite, *Hemitarsonemus latus* Banks, and the caterpillars of *Laphygma exigua* Hubn. and *Diascrisia obliqua* Wlk.

IMPROVEMENT

Improved cvs have been selected from both of the cultivated species in India.

PRODUCTION

Introduction into world trade was from Bengal by the East India Company at the end of the 18th century, but extensive production did not begin until about the middle of the 19th century. About 90 per cent of the world's jute is now produced by peasant farmers in the Ganges–Brahmaputra delta in India and Bangladesh, with annual production varying from 1–2 million tons during the present century from 3–4 million acres. Much of the crop is now manufactured locally in these countries; raw jute and jute manufactures provide about 50 per cent of Pakistan's exports and one-fifth of the value of India's total exports. The United Kingdom is the third largest producer of jute manufactures. The United States is the largest importer of manufactured jute. Since World War II China has started growing jute along the Yangtse River and elsewhere. Under China's Twelve-Year Plan production is expected to be over 500,000 tons per annum by 1967. Jute is also grown in Taiwan. Attempts to grow jute in Brazil were begun in 1932 but it is only in recent years that the industry has become firmly established and about 70,000 acres are now grown in the Amazon delta producing about 40,000 tons of fibre annually, the bulk of which is retained in the country and is used by the local sack factories. Experiments on mechanized production of jute in British Guiana (now Guyana) were carried out during the period 1952–1958, but were then abandoned.

REFERENCES

BANERJEE, B. (1955). Jute—Especially as Produced in West Bengal. *Econ. Bot.* **9**, 151–174.

KIRBY, R. H. (1963). *Vegetable Fibres*. London: Leonard Hill.

Annual Reports of the Jute Agricultural Research Institute, India.

URTICACEAE

About 40 genera and 500 spp. in many parts of the world, chiefly tropical, of herbs and shrubs, and a few soft-wooded trees. Stinging hairs often present; stems often fibrous; lvs simple, alternate or opposite; fls very small, inconspicuous, unisexual; calyx 4–5-lobed, often enlarged in fruit; stamens 4–5, opposite calyx lobes, filament inflexed in bud; ovary 1-celled and 1-ovuled; fr. usually a dry achene.

USEFUL PRODUCTS

Fibre is obtained from cultivated *Boehmeria nivea* (q.v.) and some wild species. *Pellionia* spp. and *Pilea* spp. are grown as ornamental foliage plants. Some noxious weeds occur.

Boehmeria Jacq. (x = 7, 13)

About 50 spp. of shrubs and herbs, mostly of warm regions, allied to *Urtica*, but without stinging hairs. Several species yield fibres, of which *B. nivea* (q.v.) is the most important.

Boehmeria nivea (L.) Gaud. (2n = 14) RAMIE, RHEA, CHINA GRASS

Two varieties are recognized:

(1) var. *nivea*, a native of China and Japan; leaves green above and with thick white felt of hairs below.

(2) var. *tenacissima* Miq. (syn. *B. utilis* Bl.), a native of Malaya; leaves smaller, green on both sides; better suited to tropical conditions.

USES

The stem yields a bast fibre which is one of the longest, strongest, most lustrous and most durable of plant fibres, highly resistant to water, but somewhat lacking in elasticity and flexibility. It has a much higher tensile strength than cotton and this increases on wetting. Ramie is used in a similar manner to flax and hemp. It is made into twine, thread, nets, sail cloth, gas mantles, shoe laces and parachute harness. The fibre is spun and the cloth, known as grass cloth or Chinese linen, is used for clothing, tablecloths, mats, etc. Since the beginning of the 19th century ramie has received a lot of publicity and a great deal of time and money has been spent trying to establish industries in some countries, but it is still not grown on a large scale outside China and Japan. It is difficult to decorticate and to separate the fibres from the gummy pectin which coats them.

ORIGIN AND DISTRIBUTION

B. nivea is indigenous in eastern Asia, from Japan, down the eastern part of China, to Malaysia. It has been cultivated in China since ancient times and is also grown in Japan and the Philippines. It has been grown experimentally in most tropical countries and in the southern United States, but has never been very successful. Commercial production was attempted in British Honduras.

ECOLOGY

The crop needs a warm, moist climate; excessive rainfall or drought affects the quality of the fibre. It is an exhausting crop and needs rich, loamy soil, heavy manuring, and freedom from waterlogging. It will tolerate partial shade. Ramie grows well on the Everglades peat soils in Florida.

STRUCTURE

An erect, monoecious perennial, 1–3 m high, with rhizomes and tuberous storage roots. Lvs alternate, long-petioled, broadly ovate, serrate, abruptly acuminate, 7·5–15 × 5–10 cm; with felty-white hairs on the under-surface in var. *nivea*, which are absent in var. *tenacissima*. Fls small in axillary panicles, unisexual, sepals 5, petals absent. Male fls in the lower part of the inflorescence have 5 stamens and a rudimentary ovary. Female fls in the upper part of the inflorescence with 1-celled, 1-seeded ovary and a slender style, hairy on one side. Fr. a small achene, invested by dry calyx, brownish-yellow, about 1 mm in length.

POLLINATION

The male flowers open first and the flowers are wind-pollinated.

CHEMICAL COMPOSITION

Leaves and tops contain 20–24 per cent protein (dry weight) and may be fed to livestock.

PROPAGATION

Ramie can be grown from seed, but is usually propagated vegetatively by rhizome cuttings, 6–9 in. long, planted at a depth of 2–3 in. It can also be propagated by suckers and stem cuttings.

HUSBANDRY

Rhizome cuttings are planted at a spacing of 4 × 1 ft in fertile, well-cultivated, well-manured soil. Straight unbranched stems, 4–6 ft high, are required. Stems are first cut at about 10 months, but this first cutting is not often used for fibre. Thereafter, the stems are harvested 2–3 times per year. They are cut when the inflorescence is forming and the stems are beginning to turn brown. The fibre is not separated from the stems by retting, but must be removed by hand as is done in China or by decorticating machines as is practised in Florida. The heavy coating of gum may be removed by treatment with soap solution, lime or chemicals. The crop yields about 20 tons of stems per acre per annum, giving 1 ton of dried ribbons, which in turn yields about ½ ton of degummed fibre. Yields in China are about 800 lb of fibre per acre per annum. There is an increase in yields in the 2nd and 3rd years. The crop is replanted after 7–10 years, but the crop may persist for 20 years in parts of China.

MAJOR DISEASES AND PESTS

Rosellinia necatrix Prill. is reported as causing damage in Japan and *Rhizoctonia solani* Kuhn. in Florida. The caterpillar, *Cocytodes caerulea* Guen., attacks plants in Japan.

PRODUCTION

The main producer is China, who is thought to produce about 100,000 tons per year. China exports fibre to Japan and Europe. Other countries producing ramie include Japan, Taiwan, the Philippines and Brazil. Elsewhere it has seldom been remunerative because of the high cost of production, due to the necessity for frequent manuring, the difficulty of decorticating and degumming the fibre, and the large amount of hand labour required.

REFERENCES

KIRBY, R. H. (1963). *Vegetable Fibres*. London: Leonard Hill.

SEARLE, C. C. *et al.* (1953). Agronomic Studies of Ramie in the Florida Everglades. *Univ. Fla Agric. Exp. Stat. Bull.*, 525.

WILLIMOT, S. G. (1954). Ramie Fibre, its Cultivation and Development. *World Crops*, **6**, 405–408.

Fig. 94. **Boehmeria nivea : RAMIE. A, leafy shoot (×½); B, leaf and inflorescence (×½); C, male flower (×10); D, female flower (×10); E, fruit (×10).**

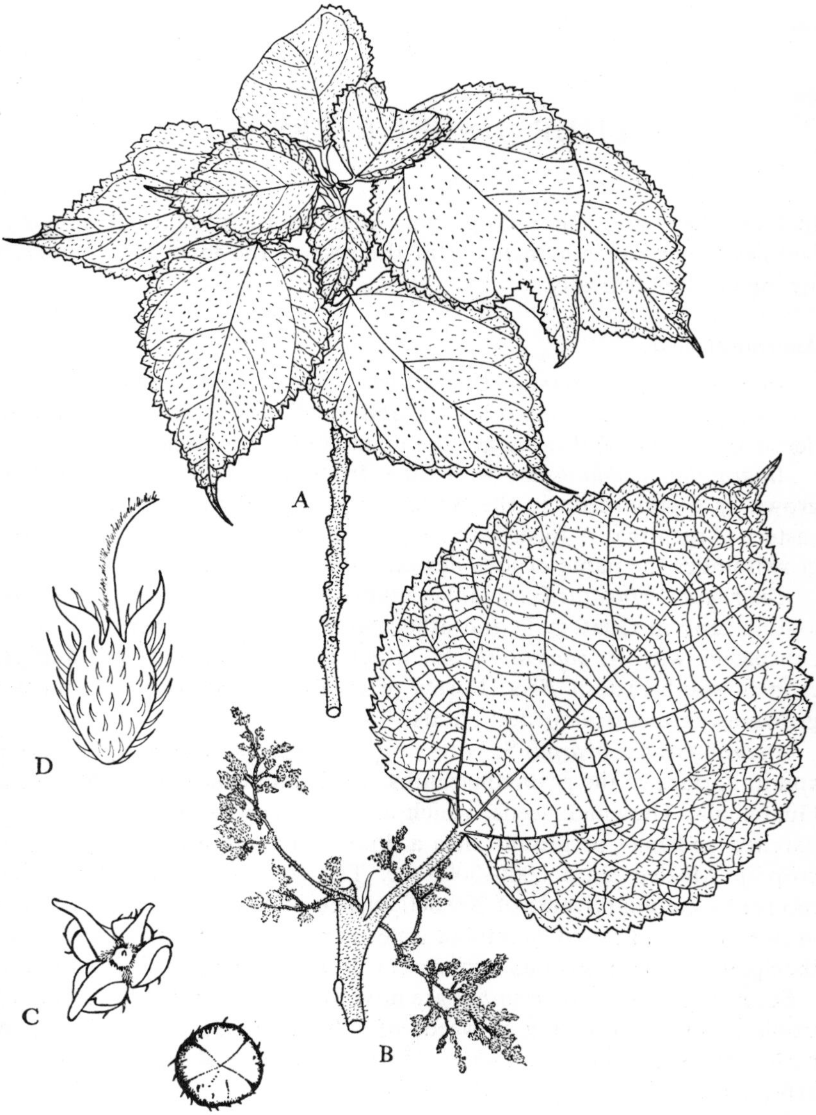
A
B
C
D
E

OTHER USEFUL PRODUCTS

Included here are minor crops and other useful products belonging to families which are not included in the rest of the text. The families are arranged alphabetically.

Amaranthaceae

Amaranthus spp., AMARANTHS, grown for grain and spinach, are herbaceous annuals, with simple leaves, and small chaffy flowers arranged in dense spikes. The following are the main cultivated species:

Amaranthus caudatus L., with long drooping tail-like inflorescences, is grown as a grain crop in the Andean region of Peru, Bolivia and north-eastern Argentina. A form with red flower spikes is a garden ornamental grown under the name of love-lies-bleeding.

Amaranthus cruentus L., with lax inflorescences, is occasionally grown as a grain crop in Guatemala and other parts of Central America. It may be conspecific with *A. paniculatus* L., which is used as a pot-herb and grain crop in south-eastern Asia. Forms of *A. paniculatus* with red spikes are grown as garden ornamental plants.

Amaranthus leucocarpus S. Wats., with stiff spikes, is the most widespread and important of the grain amaranths, and is still cultivated in Mexico and Guatemala. The small seeds, which are produced in great quantities, are parched and may be ground into a flour. It has been an important grain crop in Mexico since 5000–3000 B.C. The Aztec emperor, Montezuma, received an annual tribute of 200,000 bushels of amaranth seed. The decline in production in post-Columbian times is partly due to its suppression by the Spaniards because of its important role in Mexican religious ceremonies.

Several species of *Amaranthus* are now widely distributed throughout the tropics, usually occurring as weeds of cultivation. Some are used as pot-herbs; these include: *A. dubius* Mart. ex Thell.; *A. gangeticus* L.; *A. hybridus* L.

The family includes a number of widely grown garden ornamental species such as: *Alternanthera amoena* Voss., a dwarf perennial, native of Brazil, with small yellow, red, green or variegated leaves, widely used for edging in East Africa and elsewhere in the tropics; *Amaranthus tricolor* L., with a terminal head of bright coloured leaves, which may be red, purple, yellow or variegated; *Celosia cristata* L., with red or yellow inflorescences, often showing fasciation; *Gomphrena globosa* L. with small globular flower heads which may be white, red or mauve.

Annonaceae

Annona spp., which yield edible fruits, are native to tropical America. They were taken to most parts of the tropics after the discovery of the New World. They are small trees to 7 m tall; lvs alternate, simple, entire; fls yellowish, hermaphrodite; sepals 3; petals 6 or 3, stamens numerous, pistils many, 1-ovuled; fr. a fleshy syncarp formed by the fusion of the pistils and receptacle, with hard seeds embedded in soft, custard-like, whitish, edible pulp. They are seldom grown in commercial orchards, but are often planted in gardens or near houses. They are usually grown from seed, but they can be budded or grafted onto seedling rootstocks. They can be grown on a wide variety of soils, but will not tolerate waterlogging. They come into bearing in 3–5 years. The flowers are protogynous and the stigmas have usually lost their receptivity before the pollen is shed. They are pollinated by insects, usually beetles. Fruit setting is often poor and is usually greatly increased by hand-pollination. The ripe fruits are soft and perishable and ferment quickly. Consequently, they are difficult to transport and are not exported, except in the form of nectar. The flesh is usually eaten fresh and is often used in ice-creams and sherbets. The following are the principal cultivated species:

Annona cherimolia Mill., CHERIMOYA, is a small deciduous tree, which occurs naturally in Andean valleys of Ecuador and Peru. Archaeological remains have been found in prehistoric Peru. It is said to be the most delicious of the *Annona* fruits. The cherimoya is not suited to the low hot tropics and can only be grown at the higher altitudes. It can be distinguished from the other cultivated spp. by the brown velvety tomentum on the under-surface of the leaves. The roundish, heart-shaped, green fruits, 8–15 cm in diameter, with fingerprint depressions, contain about 18 per cent sugars.

Annona muricata L., SOURSOP, is a small evergreen tree widely cultivated from Central America to coastal valleys in Peru and is now spread throughout the lowland tropics. The dark-green ovoid fruits, 15–25 cm long, are covered with recurved fleshy spines. The fruits are often distorted and kidney-shaped due to some of the ovules not being fertilized. The white, woolly, rather acid, aromatic, juicy flesh, containing many black seeds, is best strained and used for drinks and ice-cream.

Annona squamosa L., SWEETSOP OR SUGAR APPLE, is known in the East as CUSTARD APPLE. It is a native of the West Indies and South America and is now widely grown throughout the tropics at low and medium altitudes. The yellowish-green, heart-shaped fruits, 7–10 cm in diameter, are covered with rounded, fleshy tubercules, which represent the loosely-cohering carpels, and separate readily when ripe. The fruit surface has a white or bluish bloom. The white, custard-like, sweet, granular pulp surrounds the small, brown, glossy seeds. It is used mainly as a dessert fruit and contains

16–18 per cent sugars. The ATEMOYA is a cross between *A. squamosa* and *A. cherimolia*.

Other species sometimes cultivated are: *A. reticulata* L., BULLOCK'S HEART, known in the West Indies as CUSTARD APPLE, with inferior, rounded, yellowish-red fruits, 7–12 cm in diameter, with nearly smooth skin; *A. diversifolia* Saff., ILAMA; *A. montana* Macfad., MOUNTAIN SOURSOP.

Cananga odorata (Lam.) Hook. f. & Thoms., YLANG-YLANG, grows wild throughout Malaysia and has now been introduced into many tropical countries. An evergreen tree, 3–30 m high, with drooping branches; lvs elliptic, 7–20 × 4–10 cm; fls in hanging clusters from older wood and on branches with lvs, very fragrant; petals about 7 cm long, strap-shaped, wavy, yellow; frs in bunches of 4–12 from each individual fl., clustered on hardened receptacle, olive-like, blackish, several-seeded. Two essential oils are obtained from the fully-opened flowers: ylang-ylang oil, the first portion of the distillate, extensively used in high-class perfumery; and cananga oil, the remainder of the distillate, used in cheap perfumery and for scenting soaps. The oils may also be extracted by solvents. Cultivated plants begin to flower at 1½–2 years old; they give 10–11 lb of flowers per tree per annum at 4 years old and 20–24 lb per annum at 10 years old, The total yield of oils is 1·5–2·5 per cent on the weight of fresh flowers, with ylang-ylang and cananga oils in about equal proportions. The composition of the oils seems to be influenced by environment, as well as physiological races. Cultivated plants are topped at 10 ft to facilitate harvesting. The flowers are picked before or at dawn as the scent is dissipated by heat. Production of ylang-ylang was formerly confined to the Philippines, but was produced later in Java. Now Réunion, where the tree was introduced in 1770, has the virtual monopoly.

Apocynaceae

Carissa grandiflora A. DC., NATAL PLUM, is a native of South Africa. It is a large shrub, heavily armed with sharp bifid thorns; lvs opposite, ovate, thick, glossy; fls white, fragrant, tubular, to 5 cm in diameter; frs ovoid to ellipsoid, to 5 cm long, reddish, with reddish pulp and white milky latex. The ripe fruits are used for making jelly and as a substitute for cranberry sauce. It makes an attractive ornamental shrub. *C. edulis* Vahl, from tropical Africa, and *C. carandas* L., from India, also have edible fruits. All three species may be used for hedges.

Dyera costulata (Miq.) Hook. f., JELUTONG, is a vast emergent tree of the Malaysian forests. These trees, which have a very large flow of latex,

---

Fig. 95. **A. Annona muricata :** SOURSOP. **A1**, leafy shoot (×¼); **A2**, leaf (×½); **A3**, flower (×½); **A4**, flower in longitudinal section (×½); **A5**, fruit (×¼). **B. Annona squamosa :** SWEETSOP. **B1**, flowering shoot (×½); **B2**, fruit (×½). **C. Cananga odorata :** YLANG-YLANG. **C1**, flowering shoot (×⅙); **C2**, leaf (×½); **C3**, flower (×½); **C4**, flower in longitudinal section (×½); **C5**, fruit (×½); **C6**, carpel in longitudinal section (×½).

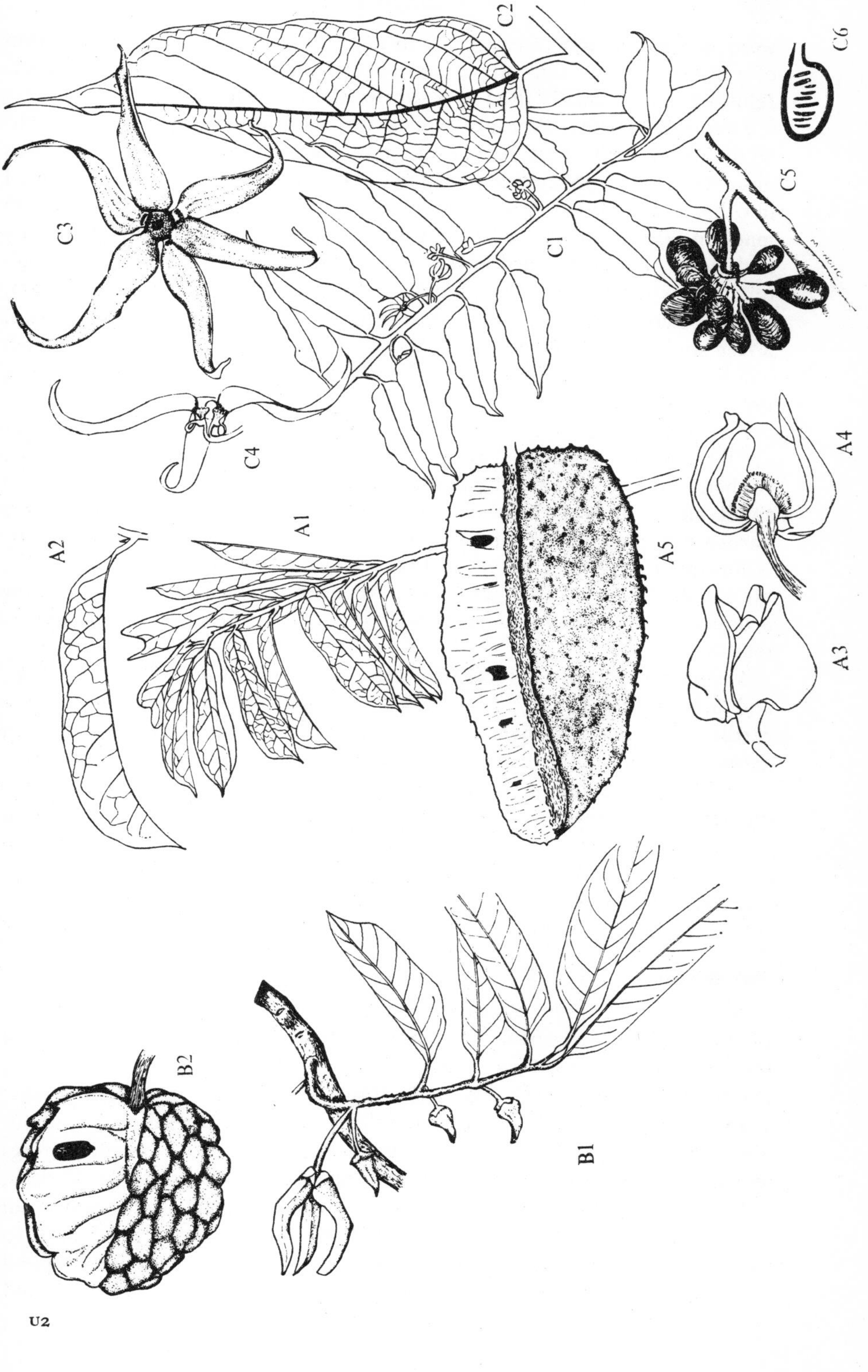

U2

were used as a source of rubber up to 1915. The latex has a high proportion of gums and resins and only 20 per cent of the total dry weight is rubber. Jelutong is now chiefly used as a substitute for chicle (q.v.).

Funtumia elastica (Preuss) Stapf, LAGOS SILK RUBBER, is a forest tree to 30 m, in tropical West Africa, extending eastwards to Uganda. It was discovered as a rubber tree in Ghana about 1883 and was later exploited in other West African territories. Plantations on a considerable scale were made in Ghana, Nigeria and the Camerouns. The latex coagulates readily and yields about one-third of its weight of pure rubber, which is of a high quality. As the highest yield is only 58 lb dry rubber per acre per annum, it cannot compete with *Hevea brasiliensis* (q.v.). *F. africana* (Benth.) Stapf, a tree similar to *F. elastica* and often growing with it, yields a latex which will not coagulate and, if mixed with that of *F. elastica*, results in a valueless product.

Landolphia heudelotii A. DC., *L. owariensis* P. Beauv. and *L. kirkii* Dyer are large woody lianes of tropical Africa and are the source of Landolphia rubber, which is obtained by pulling down the vines and cutting them into small pieces. It was the procurement of this rubber which caused such hardship and harsh treatment in the Congo in the reign of Leopold II.

Rauvolfia serpentina Benth. in India, and other Asian and African spp., are used in local medicines. They are now used commercially for the production of the drug reserpine which is used in the treatment of hypertension.

Strophanthus gratus (Hook.) Franch., *S. hispidus* DC., *S. kombe* Oliv., and *S. sarmentosus* DC., which are lianes in tropical Africa, contain the cardiac glucoside strophanthin used as a heart stimulant. They are used in arrow poisons and have been found to be a source of the drug cortisone.

The family contains a number of ornamental shrubs and climbers, including: *Allamanda cathartica* L., *Beaumontia grandiflora* Wall., *Kopsia fruticosa* A. DC., *Catharanthus roseus* (L.) G. Don, (PERIWINKLE), *Nerium oleander* L. (OLEANDER), *Odontadenia grandiflora* (G. F. W. Meyer) Miq., *Plumeria rubra* L. (FRANGIPANI), *Tabernaemontana coronaria* (Jacq.) Willd. (CREPE JASMINE), and *Thevetia peruviana* (Pers.) K. Schum. (LUCKY NUT).

Basellaceae

Basella alba L. (syn. *B. rubra* L.), INDIAN SPINACH, is probably a native of Asia and is now grown throughout the tropics for its tender stems and leaves which are used as spinach. Green and red-tinted cvs occur. A perennial, twining, glabrous herb; lvs alternate, ovate, rather fleshy, 7–15 cm long; fls small, whitish in axillary spikes; frs ovoid, fleshy, about 8 mm in diameter. This spinach is propagated by seeds or stem cuttings and is grown with a support. The first leaves are picked after 80–90 days.

Ullucus tuberosus Caldas, ULLUCU, is endemic in the Andes and is an important food crop at high altitudes from Bolivia and central Peru to Ecuador and Colombia. The plant is very resistant to frost and gives good yields of edible tubers. A twining herb with alternate, cordate lvs and axillary inflorescences. The tubers may be either small and round or elongated and curved; they are usually pale magenta or yellow in colour. They are prepared and eaten like potatoes; they may be dried and stored.

Bignoniaceae

Crescentia cujete L., CALABASH, is a native of tropical America and the West Indies, and is now widespread throughout the tropics. It is a small tree, 5–8 m high, with short trunk and long spreading branches; lvs lanceolate, up to 20 × 5 cm, borne in tufts; fls tubular, whitish, yellowish or greenish, 4–7 cm long, borne on old wood; frs large, globular or oval, green, up to 25 cm in diameter, with hard dry walls when ripe; seeds 8 × 4 mm. The flowers are probably bat-pollinated. In this connection, I have observed that the bifid stigma closes on touching. The thin durable woody shells of the fruits, which can be polished and carved, are used as containers and other domestic utensils, as ornaments, and for percussion musical instruments. Calabashes have been found in archaeological sites in Peru. Orchids are often cultivated on blocks of calabash wood, which has soft and spongy bark, and the wood is soft but tough. The tree is easily grown from seeds or cuttings. Calabashes should not be confused with gourds, which are obtained from the cucurbit, *Lagenaria siceraria* (q.v.).

Parmentiera edulis DC., CUACHILOTE, is a small tree to 10 m high, with trifoliate leaves, which is a native of Central America. It is cultivated there and in Mexico for its ridged, reddish-yellow fruits, 10–15 × 3–5 cm, the sweet flesh of which is eaten raw or variously cooked. *P. cereifera* Seem., CANDLE TREE, a native of Central America, is occasionally cultivated in the West Indies and elsewhere in the tropics. The cauliferous white flowers, borne on the trunk and branches, are followed by long thin cylindrical yellow fruits about 25 cm long.

The family has a number of beautiful trees and climbers which are widely grown as ornamentals throughout the tropics. These include: *Doxantha unguis-cati* (L.) Miers, CAT'S CLAW CREEPER; *Jacaranda mimosifolia* G. Don, JACARANDA; *Kigelia africana* (Lam.) Benth., SAUSAGE TREE; *Pandorea pandorana* (Andr.) van Steenis; *Podranea ricasoliana* (Tanf.) Sprague; *Pyrostegia venusta* (Ker-Gawl.) Miers, GOLDEN SHOWER; *Saritaea magnifica* (Sprague ex van Steenis) Dugand; *Spathodea campanulata* Beauv., TULIP TREE; *Tabebuia rosea* (Bertol.) DC., PINK POUI; *T. serratifolia* (Vahl) Nicholson, YELLOW POUI; *Tecoma stans* (L.) HBK., YELLOW BELLS; *Tecomaria capensis* (Thunb.) Spach, CAPE HONEYSUCKLE.

Bixaceae

Bixa orellana L., ANNATTO, is a native of tropical America and the West Indies, and now widely introduced throughout the tropics, becoming naturalized in some areas. A shrub or small tree to about 5 m high; lvs cordate, stalked, acuminate, sometimes with reddish veins; lamina 8–24 × 5–15 cm, fls about 6 cm in diameter in terminal erect panicles; sepals 5, deciduous; petals 5, white or pale pink; stamens numerous;

fr. a heart-shaped capsule, about 5 cm long, covered with soft spines, 2-valved, usually red or occasionally green, drying brown; seeds numerous, about 5 mm in diameter, with thin pulpy vermilion skin. The dye annatto is obtained from the pulp round the seeds by macerating them in water. The pigment, which settles to the bottom, is dried into cakes. The yield is 5–6 per cent by weight of the seeds. It has been replaced as a dye for fabrics by the aniline dye Congo red, but is still used for colouring food-stuffs such as butter, margarine, cheese and chocolate. It has long been used by Indians in tropical America as warpaint and also for decorating their bodies. It is often grown as a hedge plant.

Celastraceae

Catha edulis Forsk., KHAT, occurs wild in East Africa at altitudes of 5,000–8,000 ft, and is cultivated in Ethiopia and Arabia. It is a small tree, usually about 6 m high; lvs opposite, elliptic, serrate, 5–10 × 1–4 cm; fls small, white, in axillary cymes about 5 cm long; capsules oblong, pendulous, about 8 mm long, 3-valved. The buds and leaves contain an alkaloid and are chewed in a fresh or dried condition as a stimulative in north-eastern Africa and Arabia. Too much is said to have an intoxicating effect, finally causing coma and death.

Chenopodiaceae

Beta vulgaris L. subsp. *vulgaris*, BEETROOT, is derived from the wild *B. vulgaris* subsp. *maritima* (L.) Thell., which grows wild on sea-shores in Britain and through Europe and Asia to the East Indies. Included in the subsp. *vulgaris* are CHARD, SPINACH BEET, SUGAR BEET and MANGEL or MANGEL-WURZEL. They are usually biennial glabrous herbs with conspicuously swollen roots at junction with stem; lvs often ovate and cordate, dark green or reddish, frequently forming rosette habit. Inflorescence in small 3–4-flowered cymes arranged on a spike, each subtended by a small narrow leaf; perianth of 5 segments, becoming thicker towards base as fruits ripen; stamens 5; ovary sunk in disc; stigmas 3. Fruits mostly an aggregate formed by the cohesion of 2 or more fruits held together by swollen perianth bases and forming an irregular dry body. All forms appear to be interfertile and are wind-pollinated.

In the beetroot the roots are usually a deep red colour and may be globular or cylindrical. They are eaten after boiling, either as a vegetable, but more usually cold in salads; they are also pickled and canned. They have been cultivated in Europe since the beginning of the Christian era. Cvs recommended for the tropics are 'Crimson Globe' and 'Detroit Dark

---

Fig. 96. **Crescentia cujete : CALABASH. A, young tree; B, leafy shoot (×⅓); C, flower (×½); D, corolla cut open (×½); E, stigma closed (×½); F, stigma open (×½); G, seed (×2); H, fruit (×½); I, fruit in longitudinal section (×½).**

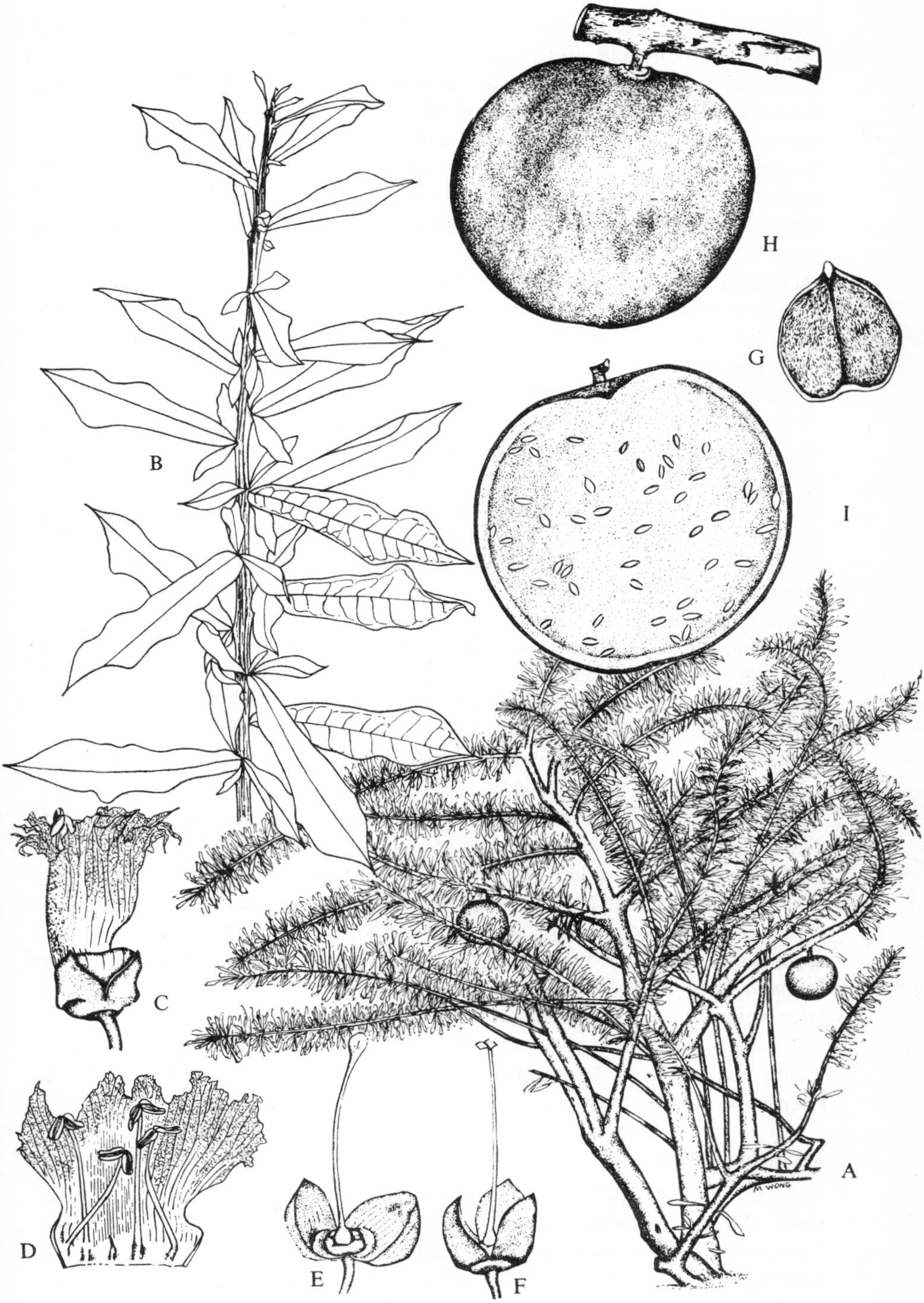

A
B
C
D
E
F
G
H
I
M WONG

Red'. They are usually sown in rows 9–12 in. apart and the plants later thinned to about 6 in. in the rows.

In chard and spinach beet, which are used for spinach, the leaves have been developed at the expense of the roots and usually have thickened midribs. Swiss chard cvs suitable for the tropics are 'Fordhook Giant' and 'Lucullus'. Mangels were developed from chard and have been an important cattle food in Europe since the 16th century. Sugar beet is a white-rooted biennial, which has been developed from the mangel. By selection the sugar content has been raised from 5 to over 20 per cent. Cultivation was encouraged in Europe during the Napoleonic wars. They are now extensively cultivated in Europe and to a lesser extent in the United States and New Zealand. About one-third of the world's production of sugar is from sugar beet.

Chenopodium ambrosioides L., WORMSEED OR MEXICAN TEA, is a native of tropical America and has now become naturalized in other parts of the tropics. It is an erect branched aromatic herb to 3 m or more in height, glandular-pubescent; lvs oblong, 5–8 cm long, lower lvs toothed, upper lvs usually entire; fls minute in axillary clusters on slender spikes. The glandular hairs contain a volatile oil, which is extracted by steam distillation. It is used as a vermifuge in the treatment of hookworm and other intestinal parasites.

Chenopodium quinoa Willd., QUINOA, is a native of Peru and was used in large quantities by the ancient Incas. It has been recovered from several other archaeological sites in western South America. It is still widely grown in Ecuador, Peru and Bolivia and furnishes a staple food, replacing maize at the higher altitudes in the Andes up to 13,000 ft. It is an annual herb, 1–2 m in height, which matures in 5–6 months. The seeds, about 2 mm in diameter, may be white, red or black. They contain about: water 11 per cent; protein 12 per cent; fat 6 per cent; carbohydrate 68 per cent. The whole seeds are used in soups; they are ground into flour which is made into bread or cakes; an alcoholic beverage is also made from them. The ash from the stems is often combined with the leaves of coca (q.v.) for chewing. A related species, *C. nuttalliae* Safford, was grown in pre-Columbian times in Mexico. It is of interest to note that seeds of *C. album* L. have been recovered from Iron Age settlements in Europe, but it was later abandoned as a cultivated crop.

Spinacia oleracea L., SPINACH, is a native of south-western Asia, and is widely cultivated in temperate countries, where it is used for greens. It cannot be grown successfully in the lowland tropics.

Erythroxylaceae

Erythroxylum spp., COCA, are native shrubs in tropical and subtropical South America. The two most important species economically are *E. coca* Lam., which is grown at the higher altitudes in the Andes in Peru, Argentina, Bolivia, Colombia and Brazil, and *E. novogranatense* (Morris) Hieron., which occurs in the same general area, but at lower altitudes from sea level to 5,000 ft. The latter species grows well in the low, hot, wet tropics. They are leafy shrubs, 1–2 m in height, with reddish-brown bark. Lvs elliptic, light green, 4–7 × 3–4 cm; fls small, white, in axillary clusters; frs small ovoid drupes, reddish-orange, 2–8 mm long. *E. novogranatense* closely resembles *E. coca*, but has smaller leaves which are blunt at the tips.

The use of coca leaves as a stimulating masticatory has been widespread among people of the Amazon basin for many centuries. By the time of the rise of the Inca empire its use had become well established in the central Andes. Later Jesuits encouraged its cultivation. A wad of dried leaves is chewed and formed into a small quid held against the cheek, to which is added lime or alkaline ashes of plants such as *Chenopodium quinoa* (q.v.). The narcotic acts directly on the central nervous system causing psychic exaltation and permits people to resist physical and mental fatigue for periods without food or drink. The narcotic principle is the alkaloid cocaine, $C_{17}H_{21}ON_4$, of which the leaves contain about 1 per cent. Cocaine was first isolated in 1860 and has been used in medicine since 1884 as a local anaesthetic and as a tonic for the digestive and nervous systems. It is a habit-forming drug and its abuse has led to legislation for its control in many countries.

Coca can be propagated by cuttings, but is usually grown from seed planted in nurseries. Seedlings are transplanted to the field at a spacing of about 6 ft, when they are 8–10 in. high, at about a year old. The first crop of leaves, which should be stiff, ripe and easily detached, is gathered 1–3 years after planting. The leaves are picked 3–4 times per year and are dried quickly. Yields of 1,500–2,000 lb of dried leaves per acre per annum are reported. Plantings are usually renewed after about 20 years. *E. novogranatense* was widely distributed to the tropics by the Royal Botanic Gardens, Kew, about 1870. The main source of supply of cocaine was Peru, Ceylon and Java.

Guttiferae

Garcinia mangostana L., MANGOSTEEN, is acclaimed by some to be the most delicious of all tropical fruits. It is a native of Malaysia and has seldom been successfully established outside this area. It requires a hot and humid climate. A slow-growing, glabrous, evergreen tree to 15 m in height, with yellow latex in all its parts; lvs dark green, leathery, opposite, short-stalked, elliptic-oblong with acuminate tips, 15–25 × 6–10 cm; fls unisexual-dioecious, but only female trees with infertile staminodes have been found in Malaya and Java. Fls borne terminally on branchlets, 5–6 cm in diameter; sepals 4 in 2 pairs, inner pair reddish; petals 4, yellowish, edged red, falling early; ovary 4–8-celled; stigma sessile with as many lobes as cells of ovary; fr. subglobose berry, usually parthenocarpic, 4–7 cm in diameter, with persistent calyx and stigma lobes; pericarp purple, tough, thick, with yellowish resin; 4–8 translucent white fleshy segments with very delicate flavour; seeds 0–3, formed from nucellar tissue, about 2 cm long.

Mangosteens are difficult to establish as the seeds have a low and short viability and must be planted within a few days after removal from the fruits. Vegetative propagation is seldom successful. Seedlings are transplanted at about 2 years old with a large ball of earth, as they have a long tap-root with few laterals. Trees seldom begin to fruit before they are 10–15 years old. The average yield of 200–500 fruits per tree is common, but in good years 1,000–2,000 fruits have been obtained. The fruits are harvested at maturity and care is necessary to prevent damage. They are best eaten fresh.

Several other species of *Garcinia* have edible fruit, but they are not of as good quality as the mangosteen. They include *G. dulcis* (Roxb.) Kurz, *G. livingstonei* T. Anders., and *G. xanthochymus* Hook. f.

Garcinia hanburyi Hook. f. yields the yellow dye gamboge obtained from a gum resin which exudes from incisions made in the bark.

Mammea americana L., MAMMEY APPLE, is an evergreen tree, to 15 m high, which is native to tropical America and the West Indies. Lvs thick, glossy, obovate, obtuse, about 12 × 7 cm; fls white, fragrant, 5 cm in diameter; frs ovoid, 10–15 cm in diameter, with brown thick skin. The orange-coloured flesh is eaten cooked and is made into preserves.

Labiatae

Coleus amboinicus Lour. (syn. *C. aromaticus* Benth.), INDIAN BORAGE, is a small, hairy, rather succulent herb, about 50 cm high, with ovate, serrate leaves and racemes of small purple flowers. It is a native of Indonesia. It is grown in south-eastern Asia and West Indies for its aromatic leaves, which are used in stuffing and for flavouring meats. They are used as a substitute for sage (*Salvia officinalis*, (q.v.)), and borage (*Borago officinalis* L.).

Coleus parviflorus Benth. (syn. *C. tuberosus* Benth.) is cultivated in south-eastern Asia for its small, dark brown, aromatic tubers, which are used as a substitute for potatoes (*Solanum tuberosum*, q.v.). It is a small herbaceous annual with succulent stems and aromatic leaves. It is propagated by suckers obtained from sprouted tubers. It produces a crop in about 6 months and yields 3,000–6,000 lb of tubers per acre.

The genus *Coleus* has several ornamental species, of which *C. blumei* Benth., a native of Indonesia, is grown for its highly coloured leaves; it has become naturalized in Trinidad and elsewhere.

Hyptis spicigera Lam. is cultivated in parts of tropical Africa in a similar way to *Sesamum indicum* (q.v.). It is a tall erect herb with small white flowers. The small black seeds yield over 20 per cent of a drying oil.

Mentha spp., MINT, have been introduced into the tropics for use as pot-herbs. Some of them yield essential oils. They usually grow best at the

Fig. 97. **Erythroxylum novogranatense** COCA. **A**, flowering shoot (×½); **B**, flower (×6); **C**, flower in longitudinal section (×8); **D**, fruit (×4); **E**, petal (×6).

A
B
C
D
E

higher altitudes. The species are very variable and hybridize freely. The most important species are:

M. arvensis L. var. *piperascens* Malinv., JAPANESE MINT, which is cultivated in Japan and Brazil, is the chief commercial source of menthol.

M. × *piperita* L. (*M. aquatica* L. × *M. spicata* L.), PEPPERMINT, a glabrous, perennial, strongly scented herb, occurring spontaneously or cultivated in temperate regions, and introduced in parts of the tropics. Peppermint oil is obtained by steam distillation. It is used in confectionery and liqueurs, and in pharmaceutical and dental preparations.

M. spicata L. (syn. *M. viridis* L.), SPEARMINT, a glabrous stoloniferous perennial, is a native of temperate Europe. It has now been widely introduced throughout the world. The fresh and dried leaves are used for mint sauce and jelly, and to flavour foods. Spearmint oil is used in chewing-gum, toothpaste, and in confectionary and pharmaceutical preparations.

Ocimum basilicum L. (syn. *O. americanum* L.), BASIL, is a native of the Old World tropics, and has now been widely distributed. It is a stout, bushy, aromatic herb, with white flowers in loose racemes. The plant yields basil oil, which is used in the perfume industry and for scenting soaps. The leaves are used in the tropics for flavouring sauces, soups and other foods. *O. canum* Sims is used in a similar way. *O. sanctum* L. is a sacred plant of the Hindus and has been carried by them to many parts of the world, including Trinidad.

Plectranthus esculentus N. E. Br. (syn. *Coleus dazo* A. Chev., *C. esculentus* (N. E. Br.) G. Tayl., HAUSA POTATO, is cultivated in tropical Africa for its edible tubers in a similar way to *Coleus parviflorus* (see above), to which it is closely allied. It is a perennial with herbaceous erect stems, pilose with whitish hairs, and yellow flowers about 1·5 cm long.

Pogostemon cablin (Blanco) Benth., PATCHOULI, is a small shrub, about 1 m high, which is a native of the Philippines. It used to be grown in Singapore, but the main production is now in Sumatra, the Seychelles, Madagascar and Brazil. The plant rarely flowers in cultivation and is propagated by stem cuttings. The young shoots are harvested every 6 months or so and may be continued for 2–3 years. The essential oil is extracted from the dry shoots. It is one of the best fixatives for heavy perfumes and is used in soaps, hair tonics and tobacco. It gives the characteristic odour to cashmere shawls and carpets.

Salvia officinalis L., SAGE, is the well-known culinary herb of Mediterranean origin. It does not grow well in the lowland tropics. The genus contains a number of ornamental species, including *S. splendens* Sell. ex Roem. & Schult., the scarlet salvia, which is a native of Brazil.

Thymus vulgaris L., THYME, a low shrub with minute leaves, is a native of the Mediterranean region and has now been widely spread through the tropics. The fresh or dried shoots are used as seasoning in stuffing and other foodstuffs. The plant yields an oil which is used in perfumery. Thymol, a derivative of the oil, is antiseptic and is used in tooth-pastes and pharmacy. *Lippia micromeria* Schau. (family Verbenaceae), SPANISH

THYME, a native shrub of South America, is used as a substitute for thyme in parts of the tropics.

Lecythidaceae

Bertholletia excelsa Humb. & Bonpl., BRAZIL NUT, is a large tree, to a height of 40 m, growing wild in the Amazon forests of South America. Lvs alternate, short-stalked, oblong, leathery, 25–50 × 10–15 cm; fls pale yellow, about 5 cm in diameter, in upright panicles; sepals 2; petals 6 unequal; stamens numerous, united on a thick flap on lower side of flower; style short; fruit globose, woody, brown, 12–15 cm in diameter, with thick outer shell; seeds 12–24, angular, woody, large. The trees begin fruiting at 10–20 years old and the fruits take over a year to ripen. The Indians in the Amazon basin use the kernels as food and also for oil, of which they contain 60–70 per cent. Nearly all the world's supply of Brazil nuts, of over 50,000 tons per year, is obtained from wild trees in South America; they are exported mainly to Europe and the United States.

Lecythis zabucajo Aubl., SAPUCAIA NUT, PARADISE NUT, MONKEY POT, is a large forest tree from the Guianas and Brazil. Lvs elliptic, about 15 cm long, acuminate, shed towards end of dry season; fls white in terminal racemes; frs urn-like, thick, woody, pendent, about 20 × 25 cm, with thick woody lid, which becomes detached when fruits mature leaving nuts dangling inside by slender fleshy funiculi. When funiculi rot the seeds fall to the ground. The funiculi are often eaten by bats, who discard the nuts. The nuts, about 5 × 2·5 cm, have brown shells, which are softer and less sharply angled than Brazil nuts. The kernels have an excellent flavour and are thought to be superior to Brazil nuts. The empty fruits are baited with sugar and other substances to trap monkeys, which are said not to be able to withdraw their heads or hands when inserted! I tried it once for this purpose, but without success. A little commercial planting has been done in Trinidad. Other species of *Lecythis* yielding edible nuts are *L. elliptica* Kunth, *L. ollaria* L. and *L. usitatis* Miers.

Malpighiaceae

Banisteriopsis spp., CAAPI, are lianes of the forest of north-western Amazon basin, which are used as narcotics by the Indian population. A beverage is made from the macerated stems which is an excitant and induces unnatural courage. The species used are *B. caapi* (Spruce) Morton, *B. inebrians* Morton, *B. quitensis* (Niedenzu) Morton, and the allied *Tetrapterys* sp.

Malpighia glabra L. (syn. *M. punicifolia* L.), BARBADOS CHERRY, is a native of the West Indies and from northern South America to southern Texas. It has been introduced into other parts of the tropics and subtropics. A dense, spreading, glabrous shrub, 2–5 m in height; lvs opposite, shiny, dark green above, ovate to elliptic, 2·5–7·5 cm long, shortly stalked; fls 1–2 cm in diameter, in axillary cymes, hermaphrodite; calyx with 6–10 large sessile glands; petals 5, fringed, slender-clawed, pink or red; stamens 10; fr. a bright red, juicy drupe, depressed-ovoid, obscurely 3-lobed, 1–3 cm in diameter, borne 1–3 in leaf axils on short pedicels; seeds 3, triangular, ridged. The acid fruits are made into preserves and jellies. They are one of the richest sources of ascorbic acid, containing 1,000–4,000 mg per 100 g of edible matter. The juice is used commercially to enrich other fruit juices low in vitamin C. The unripe fruits have the highest content of ascorbic acid. Sizable plantings have been made in Puerto Rico for juice production. Barbados cherry may be propagated vegetatively by cuttings, budding grafting. The usual spacing is about 10 ft. It makes a good hedge, as does *M. coccigera* L.

Oxalidaceae

Averrhoa bilimbi L., BILIMBI, is native to the monsoon regions of Malaysia and is often associated in a wild state with teak, *Tectona grandis* L. f. (family Verbenaceae). It is now widely introduced in the tropics. Bilimbi is a small tree to 15 m in height; lvs imparipinnate; leaflets 10–20 pairs, oblong, acute, 5–10 cm long; fls dark red, to 2 cm long, from trunk and older branches; frs resembling small cucumbers, cylindrical, 5–7·5 cm long, faintly 5-angled, green; seeds small, embedded in pulp, about 8 mm long. The very acid fruits are used for making pickles, curries and preserves.

Averrhoa carambola L., CARAMBOLA, occurs wild in Indonesia and is now widely spread throughout the tropics. It is an attractive small tree, 5–12 m tall; leaves imparipinnate; leaflets 3–5 pairs, ovate, 2–9 × 1–4 cm, terminal leaflet largest; fls borne on leafy twigs, axillary, or just behind lvs, in clusters, rose-coloured, about 8 mm long; frs a fleshy berry, 8–12 × 3–6 cm, acutely 5-angled and star-shaped in cross-section, ripening translucent yellow, with crisp, juicy, aromatic flesh; seeds ovoid, compressed, light brown, to 1 cm long. Forms with sweet and acid fruits occur. The fruits are used in fruit salads and for tarts, preserves and drinks. They may also be used for cleaning brassware. The carambola can be grown from seed or propagated vegetatively.

Averrhoa has now been placed in a new family Averrhoaceae by Hutchinson (1959).

Oxalis tuberosa Mol., OCA, is cultivated in the Andes from Colombia to Bolivia at high altitudes, where it is second in importance as a root crop to the potato, *Solanum tuberosum* (q.v.). It is a small, erect, branching, rather succulent herb, with trifoliate leaves and orange-yellow flowers. The edible tubers have a smooth skin bearing scales which cover long, deep eyes. They contain calcium oxalate crystals and have to be cured in the sun before they are eaten. They should not be confused with those of *Tropaeolum*

Fig. 98. **Averrhoa carambola : CARAMBOLA. A, leafy shoot ($\times\frac{1}{2}$); B, inflorescence ($\times\frac{1}{2}$); C, flower ($\times 3$); D, flower in longitudinal section ($\times 4$); E, fruits ($\times\frac{1}{2}$); F, fruit in transverse section ($\times\frac{1}{2}$); G, seed ($\times\frac{1}{2}$).**

B
C
E
A
D
F
G

tuberosum Ruiz & Pav., ANU, (family Tropaeolaceae), which they closely resemble and which are grown in the same area. These are borne on a glabrous, twining herb, with 5-lobed leaves and reddish flowers, similar to, but smaller than the garden nasturtium.

Papaveraceae

Papaver somniferum L., OPIUM POPPY, is a cultigen derived from *P. setigerum* DC. in Asia Minor. Opium was known to the ancient Greeks. Dioscorides gives a full account of its preparation. The early trade passed into the hands of the Arabs and Persians who carried it to the East. It had reached India and China by the 8th century A.D. It was early exploited by the Portuguese, the Dutch and the English in turn, but attempts were made later to regulate the production and trade. The main areas of cultivation are now in India, China, Asia Minor and the Balkans. The plant will not thrive in the wet tropical lowlands.

It is an erect, annual, glaucous herb, with latex in all parts, 30–100 cm tall; lvs undulate, ovate-oblong, often shallowly pinnately lobed, coarsely toothed, lower lvs shortly stalked, upper lvs sessile, clasping stem; fls hermaphrodite, actinomorphic, with whorled parts, up to 18 cm in diameter; sepals 2, falling before fl. opens; petals 2+2; stamens numerous with bluish anthers; ovary superior, 1-celled, with very numerous ovules; stigma disc with deep marginal lobes; fr. a capsule, when dehiscent opening by valves or pores; seeds small with minute embryo in oily endosperm. In the subsp. *somniferum*, which is grown for opium, petals are white and unspotted; capsule very large, ovoid and indehiscent; seeds white. In the subsp. *hortense* Husserot, which is grown for seed, petals are pale lilac, spotted at base; capsule globular and dehiscent; seeds dark grey to black.

The seeds of subsp. *hortense* yield an edible drying oil. They are often scattered on cakes and bread. Opium is obtained by making incisions in the capsules of subsp. *somniferum* shortly after the petals fall; the latex hardens on exposure to air and is scraped off and moulded into balls or cakes. Crude opium contains about 20 alkaloids, of which the most important are morphine, codeine and heroin. Opium is used medicinally as a sedative and to relieve pain. It is extensively used in India and China as a narcotic and its misuse has deleterious effects, physically, mentally and morally; it is habit-forming. The immediate effects are pleasurable, inducing alluring dreams and visions. Continued use leads to loss of will power and finally in delirium and death. In India the opium is usually eaten; in China it is smoked.

Polygonaceae

Fagopyrum esculentum Moench (syn: *F. sagittatum* Gilib.), BUCKWHEAT, is a native of Central Asia. It was introduced into Europe during the Middle Ages, where it is now mainly cultivated, especially in the Soviet Union. It is normally a plant of cool, moist, temperate regions, but is grown in India and the tropics at the higher altitudes. It is an erect, annual herb, to 1 m in height, with alternate, sagittate, acuminate lvs, 3–7 cm long; fls in axillary and terminal cymes, pinkish-white, self-sterile; fr. a 3-cornered achene, about 6 mm long. It is used mainly in the form of flour for making bread, pancakes and porridge, and also as stock and poultry feed. Rutin, a glucoside used in the treatment of capillary fragility and hypertension, is obtained from the leaves and flowers.

Rheum rhaponticum L., RHUBARB, is a native of south-eastern Russia and only reached Britain in the 16th century. It is a perennial herb with a large rhizome and large radicle leaves. The succulent petioles are used for pies and sauces. It can only be grown at the higher altitudes in the tropics. *R. officinale* Baill. and *R. palmatum* L., natives of China and Tibet, are the sources of the drug rhubarb, which is obtained from the rhizomes.

Proteaceae

Macadamia ternifolia F. Muell.,* MACADAMIA OF QUEENSLAND NUT, is a small tree to 15 m tall, native of eastern Australia. Lvs, borne in whorls of 3 or 4, are glabrous, leathery and lanceolate, to 30 cm or more in length; fls creamy-white in terminal and axillary racemes; fr. with fleshy husk and single spherical seed, to 2 cm diameter, with hard brown shell. The seed or 'nut' contains over 70 per cent fat and is highly esteemed, whether eaten raw or roasted. Commercial cultivation is undertaken in Hawaii. It has been introduced into many tropical countries, but does not do very well in the tropical lowlands. *M. tetraphylla* L. Johnson is also cultivated.

Punicaceae

Punica granatum L., POMEGRANATE, is a native of Iran. It was grown in the hanging gardens of Babylon and was known in ancient Egypt. It was spread early round the Mediterranean and eastwards to India and China. It has now been taken to most parts of the tropics and subtropics. The best quality fruits are produced in areas with cool winters and hot dry summers and it does not fruit well in very humid climates. It is usually grown as a bush, 2–4 m in height, and is deciduous in the cooler part of its range. Lvs opposite, short-petioled, glabrous, shining, dark green, oblong, 4–8 cm long; fls orange-red, 4–6 cm in diameter; calyx campanulate with 5–7 lobes; petals 5–7, wrinkled; stamens numerous; ovary inferior, 3–7-celled; fr. a berry, leathery-skinned, spherical, 5–12 cm in diameter, brownish-yellow to red, surmounted by persistent calyx; seeds numerous surrounded by pink juicy pulp. This acid pulp is the edible portion. Pomegranates are used as a salad or table fruit and in beverages. The roots, rind and seeds are used medicinally. They may be easily propagated by cuttings of 1-year-old wood, about 12 in. long, usually taken from suckers at base of main stem. The bushes begin to fruit in the 4th year and the fruits ripen about 6 months after flowering. One of the most popular cvs is 'Spanish Ruby'. A type with double flowers, which does not set fruit, is grown as an ornamental; others with white or yellow flowers are known.

---

* Syn: *M. integrifolia.*

Rhamnaceae

Zizyphus mauritiana Lam. (syn. *Z. jujuba* (L.) Lam. non Mill.), INDIAN JUJUBE, is widespread in the drier parts of tropical Africa and in Asia. It is cultivated throughout India where it has a long history. It is a small thorny evergreen tree, 3–12 m tall; young branches and under-surface of leaves densely pubescent; single or paired thorns at base of leaf; lvs elliptic, 2–6 × 1·5–5 cm, rounded or emarginate at apex, 3-veined from base; fls small, greenish, in clusters in leaf axils; frs ovoid, orange to brown, 2–3 cm long, with edible acid pulp and hard central stone. The fruit is eaten fresh or dried and used as dessert; it is also candied. It makes a refreshing drink. Cakes, resembling gingerbread, are made from the dried and fermented pulp in the Sudan. The fruits, which are a rich source of vitamin C, vary greatly in quality. The tree may be propagated by budding. *Z. mauritiana* should not be confused with *Z. jujuba* Mill., CHINESE JUJUBE, a tree of temperate climates, which has been cultivated in China for at least 4,000 years.

Sapindaceae

Blighia sapida Koenig, AKEE, is an evergreen, polygamous tree, 7–25 m high, occurring wild in the forests of West Africa, where it is often planted. It is also planted in Jamaica, where it was introduced late in the 18th century, and has now become naturalized. Lvs pinnate; leaflets 3–5 pairs, upper ones largest, obovate-oblong, 4–18 × 2–9 cm; racemes axillary, pubescent, 4–15 cm long; fls small greenish-white, fragrant, about 5 mm long; capsule yellow and red when ripe, obovoid, pendent, about 6 × 3 cm, with rounded valves, which split at maturity to expose 3 shining, black, oblong seeds, surrounded by a fleshy, cream-coloured aril. The aril is eaten in West Africa and the West Indies, mainly in Jamaica. It may be eaten raw, but usually after cooking, when it resembles scrambled eggs. Great care is required in its preparation as the pink raphe attaching the aril to the seed is highly poisonous. The peptide, hypoglycin A, also occurs in unripe arils, and only those from naturally-opened fruits should be eaten; those from unripe, damaged or fallen fruits should *not* be eaten. Akee poisoning, which causes vomiting, has resulted in casualties in Jamaica. Seedlings begin to fruit in about 5 years. The genus is named after Capt. Bligh of H.M.S. *Bounty*.

Litchi chinensis Sonn. (syn. *Nephelium litchi* Camb.), LITCHI, a dense polygamous evergreen tree, about 10 m high, is a native of southern China. It has been widely introduced throughout the tropics, but only does well at the higher altitudes. Lvs paripinnate, glossy; leaflets 2–4 pairs, elliptic-

---

Fig. 99. **Punica granatum : POMEGRANATE. A, flowering branch (×½); B, flowers (×½); C, flower in longitudinal section (×½); D, fruit (×½); E, fruit in longitudinal section (×½).**

M. WONG
A
B
C
D
E

oblong, coriaceous, 8–15 × 3–4 cm long; fls small, pale greenish-yellow, in large terminal panicles; frs globose, about 3 cm diameter, pendent in loose clusters, covered with tubercles, usually red when ripe; seed dark brown covered with white, fleshy, juicy, translucent aril. The agreeable, sweet-acid aril is eaten fresh and is also canned in syrup. In China the aril is also dried to produce 'litchi nuts'. It is usually propagated vegetatively, mainly by air-layering, and such trees come into bearing in 4–6 years. In the lowland tropics the trees often grow well but fail to fruit, for which a cool dry season is required.

Nephelium lappaceum L., RAMBUTAN, an evergreen, bushy, dioecious tree, to 20 m high, is native to and cultivated throughout the lowlands of Malaysia. For reasons unknown, it seldom seems to be successful or grown outside its native area, although it is one of the best known fruits of the East. Lvs pinnate; leaflets 2–4 pairs, elliptic, 5–20 × 3–10 cm; fls in axillary panicles, which superficially appear terminal, small, greenish-white, about 4 mm in diameter; frs pendent in clusters, red or occasionally yellow, 4–6 × 2–4 cm, covered with soft spines, 1–1·5 cm long; seed covered with whitish aril. In good cvs the edible aril is plump, sweet and juicy and is highly esteemed. Seedling trees fruit in 5–6 years, but wherever possible it should be propagated vegetatively by marcotts or budding. *N. mutabile* Bl., PULASAN, is also grown in Indonesia, Malaya and Thailand for its edible arils.

Paullinia cupana HBK., GUARANA, is a large woody climber of the Amazon basin. The grated seeds, which have three times as much caffeine as coffee, are extensively used as a beverage in Brazil and the plant is sometimes cultivated as a small bush, particularly in the Matto Grosso. The bark of *P. yoco* Schultes & Killip, YOCO, is used for a beverage among the Indians of southern Colombia and adjacent Peru and Ecuador.

Schleichera oleosa (Lour.) Merr., LAC TREE, is used as a host of the lac insect (q.v.) in India. The seeds yield an edible fat, which is also used for illumination and hair-oil; it is probably the original macassar oil.

Sapotaceae

Butyrospermum paradoxum (Gaertn. f.) Hepper subsp. *parkii* (G. Don.) Hepper (syn. *B. parkii* (G. Don.) Kotschy; *Vitellaria paradoxa* Gaertn. f.), SHEA BUTTER TREE, is abundant in the savannas areas of West Africa, where it is often protected and inherited. A small deciduous tree, 7–13 m high, with lvs clustered at end of branchlets; lvs oblong, 10–25 × 5–8 cm; fls white, clustered at ends of shoots, about 1 cm long, calyx and corolla 8–10-merous; staminodes petaloid; stamens 8–10; fr. ellipsoidal, 4–5 cm long, with fleshy pulp and usually 1-seeded; seeds about 5–2·5 cm, shining, dark-brown with white scar down one side The seeds are removed after

---

Fig. 100. **Blighia sapida : AKEE. A**, flowering shoot (×½); **B**, flower in longitudinal section (×4); **C**, fruit (×½); **D**, dehiscing fruit (×½); **E**, seed and aril (×½).

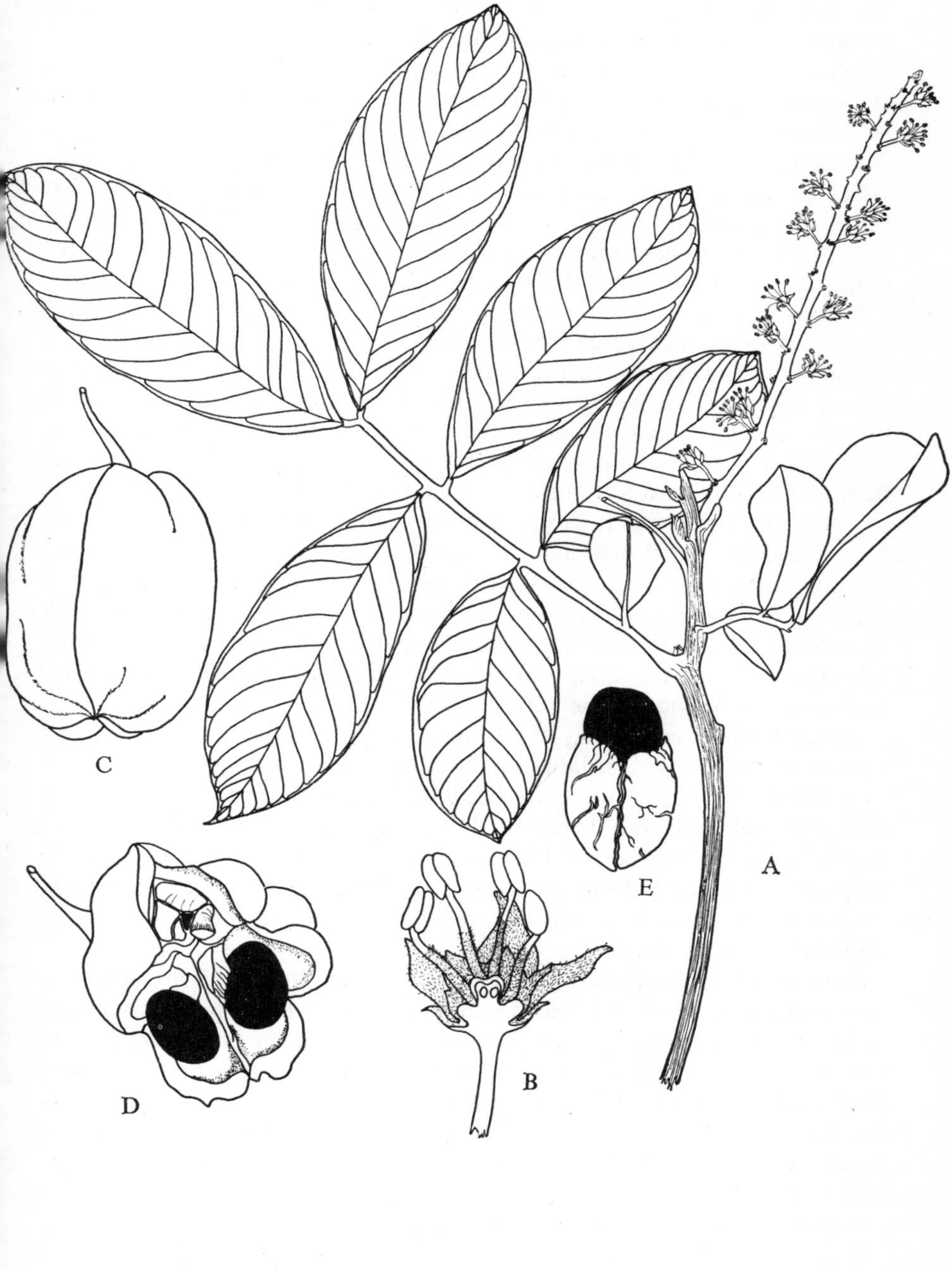
C
A
E
D
B

decomposition of the pulp or after drying and contain 45–60 per cent fat and 9 per cent protein. The fat, extracted locally, is known as shea butter and is used as a cooking fat, illuminant, medicinal ointment, hairdressing and for soap. Shea oil is obtained from nuts exported to Europe and is used for soap and candlemaking, in cosmetics, and as a constituent of fillings for chocolate creams. Shea nut production in West Africa has been estimated at 0·5 million tons per year. The principal exporters of nuts are Nigeria, Ghana, Senegal, Mali, the Ivory Coast, Upper Volta and Dahomey. Holland and Belgium used to be the main importers, but, in recent years, most of the produce goes to the United Kingdom, Japan and Denmark. Seedlings produce a long tap-root, which makes transplanting difficult, and they are best planted *in situ*. Trees start bearing fruits at 12–15 years and take 30 years to mature. The var. *nilotica* (Kotschy) Pierre ex Engl. occurs in northern Uganda and the fruits are an important food item of Nilotic tribes. Uganda has also exported nuts.

Calocarpum sapota (Jacq.) Merr. (syn. *C. mammosum* (L.) Pierre), MAMMEY SAPOTE, is widely cultivated from Mexico to northern South America and in the West Indies. It forms a large tree to 25 m tall, with prominently-veined, obovate lvs, to 30 × 10 cm, closely crowded at the ends of the branches; small whitish fls produced in great numbers along branchlets; frs ovoid, 7·5–15 cm long, scurfy, russet-brown, with reddish, sweet, spicy flesh and usually with a single large brown seed. The fruits are eaten fresh or made into preserves. It should not be confused with: the GREEN SAPOTE, *C. viride* Pitt; the WHITE SAPOTE, *Casimiroa edulis* Llave & Lex. (q.v.) (family Rutaceae); and the BLACK SAPOTE, *Diospyros ebenaster* Retz (family Ebenaceae); all these species have edible fruits and are native of Mexico and Central America. The last two species and the hog plum, *Spondias mombin* L. (q.v.) (family Anacardiaceae), were of very early cultivation in Mexico in the period 5,000–3,000 B.C.

Chrysophyllum cainito L., STAR APPLE, native of the West Indies and Central America, is a striking ornamental tree to 12 m in height, with very attractive oblong-lanceolate lvs, 8–15 × 5 cm, shiny dark green above and silky golden-brown beneath; fls small, purplish-white; frs green or purple, smooth, globose, to 10 cm in diameter, with white, sweet, edible pulp, in which are embedded several small hard brown glossy seeds. The pulp is usually eaten fresh after the removal of the skin which contains an unpleasant-tasting latex.

Lucuma bifera Mol., EGG FRUIT, an evergreen tree, 8–10 m high, is a native of Peru, where it is cultivated for the dry, mealy pulp of its dark yellow, globose fruits, 7–10 cm in diameter. It was an important part of the diet of ancient Peruvians and is frequently found in archaeological remains. *Lucuma nervosa* A. DC., a native of north-eastern South America, is cultivated in Brazil for its egg-shaped, orange-yellow fruits, of which the mealy pulp is eaten. *L. salicifolia* HBK. is used in a similar way in Mexico and Central America. The last 2 species are now included in *Pouteria campechiana* (HBK.) Baenhi.

Madhuca longifolia (Koenig) Macb. (syn. *M. indica* J. F. Gmel.), MAHUA, occurs wild and is planted throughout India. It is an evergreen tree; lvs elliptic, 8–20 × 3–4 cm, clustered near ends of branches; fls in dense fascicles also near ends of branches; corolla tubular, yellowish, 1·5 cm long, caduous; fr. an ovoid berry up to 5 cm long, turning yellow when ripe, with 1–4 brown ovoid shining seeds, 2·5–3·5 cm long. The seeds contain about 50 per cent of a soft yellow oil, moura fat or mahua butter, which is used locally in India for edible and cooking purposes, but mainly in the manufacture of soap. The press cake is unfit for animal feed, but is used as fertilizer. Indian oil mills crush 15,000–30,000 tons of seed per year. Prior to World War I an average of 30,000 tons of mahua seed per annum was exported to Europe, but now the bulk of the crop is used in India. The succulent fallen corollas are eaten raw or cooked and in the preparation of distilled liqueurs.

Manilkara achras (Mill.) Fosberg (syn. *Achras zapota* L.; *Manilkara zapotilla* (Jacq.) Gilly), SAPODILLA OR CHIKU, an evergreen forest tree to 20 m high, of Mexico and Central America, is now widely grown throughout the tropics. It was cultivated in the West Indies in pre-Columbian times, and Oviedo, who was there from 1513–1525, considered it the best of all fruits. It was taken early by the Spaniards to the Philippines and from there spread westwards to Malaysia. Lvs elliptic to obovate, 5–15 × 2·5–6 cm; fls solitary in leaf axils, about 1 cm in diameter, produced over a long season; sepals 6, tomentulose; petals 6; staminodes 6, petaloid; stamens 6; fruits globose to ovoid, 5–10 cm in diameter, greyish to rusty brown, with yellowish-brown flesh; seeds 0–12, hard, black, easily separated from pulp; white latex in all parts. The mawkish, ripe fruits are much appreciated by many people as a dessert fruit; unripe fruits are astringent. Sapodillas begin to fruit 3–4 years after planting; the fruit takes 4 months to mature from flowering; mature 30-year-old trees may be expected to yield 2,500–3,000 fruits per annum. Although trees are usually grown from seed, they are easily propagated by marcotts, inarching, grafting or budding. Chicle, used in the manufacture of chewing gum, is made from the latex which is obtained by tapping the trunk every 2–3 years. The latex contains 20–40 per cent gum, which is made by boiling the hardened latex. South-eastern Mexico, Guatemala and British Honduras are the principal producers of chicle, most of which is obtained from wild trees, which grow gregariously in some places. Chicle, which becomes plastic at mouth temperature, was chewed by the Aztecs. In addition to chewing gum, so widely used in the United States, chicle is used in dentistry. The wood is very durable and was used in the construction of Maya temples.

Manilkara bidentata (A. DC.) Chev. (syn. *Mimusops balata* Pierre), BALATA, a native of Trinidad and South America, yields balata, a non-elastic rubber, which is obtained from the latex, for which the wild trees are tapped about three times per year. It is used in the manufacture of machine belting, as a substitute for chicle, and for the small models sold to tourists. The tree may reach a height of 40 m. The very sweet, sticky pulp of the fruits is much liked by children.

Palaquium gutta (Hook.) Burck, GUTTA-PERCHA, an evergreen tree to 30 m high, occurs wild throughout the forests of Malaysia. It is the principal source of gutta-percha. Attempts have been made from time to time to establish plantations. Lvs 8–20 × 2·5–7·5 cm, obovate, dark green above, coppery-brown hairs beneath; fls pale green, about 2 cm in diameter; fruits oblong, pointed, 2–3 cm long. Gutta-percha is obtained from the latex, which may be obtained by tapping living trees, but usually after felling the trees. The coagulated latex is then boiled. Gutta-percha is hard at ordinary temperatures, but softens at higher temperatures. It is an exceedingly poor conductor of electricity and is used for insulation, particularly of submarine cables. It is also used for splints, supports, golf balls and in dentistry.

Umbelliferae

Anethum graveolens L. (syn. *Peucedanum graveolens* (L.) Hiern), DILL, is a native of Eurasia and was grown in Greece, Rome and ancient Palestine. It is now grown in Europe, the United States and India and sparingly in the tropics. The Indian plant is sometimes placed in a separate species, *A. sowa* Kurz, which has longer fruits than the European species. The green plant is used fresh as a flavouring for soups, sauces, and other culinary purposes. The seeds are used as a substitute for caraway seeds, as a flavouring in curry powder and medicinally as a source of dill-water, especially useful for flatulence in babies. In the United States dill is used chiefly for flavouring pickles. A glabrous annual or biennial herb, 50–100 cm tall; lvs finely dissected, 3–4-pinnately divided into long, narrowly linear segments; umbels to 15 cm in diameter of small yellow fls; fr. elliptic, flattened and ribbed dorsally with distinct narrow wings, 4 mm long. The seeds yield about 3 per cent of an essential oil, which is rich in carvone.

Apium graveolens L. var. *dulce* (Mill.) DC. (syn. *A. dulce* Mill.), CELERY, is derived from *A. graveolens* L., which occurs wild as a marsh plant throughout temperate Europe and Asia. It is widely grown in temperate countries for the young swollen petioles which are eaten after blanching, either raw or as flavouring. The tough outer petioles are the basis of celery soup. Celery was formerly grown in the Old World for its foliage which was used for flavouring and as a garnish. In the lowland tropics the plants do not attain a large size and the leaves and petioles are usually only suitable for flavouring. The cv. recommended in Trinidad is 'Giant Pascal', but it does better at higher altitudes. Celery requires a rich sandy loam and lots of water. The dry ripe fruits are used for flavouring and also medicinally. They yield 2–3 per cent of a volatile oil with a persistent characteristic odour of the plant. It is used for flavouring salt and as a

---

Fig. 101. **Manilkara achras** : SAPODILLA. **A**, leafy shoot (×½); **B**, inflorescence (×½); **C**, flower in longitudinal section (×3); **D**, fruit in longitudinal section (×½).

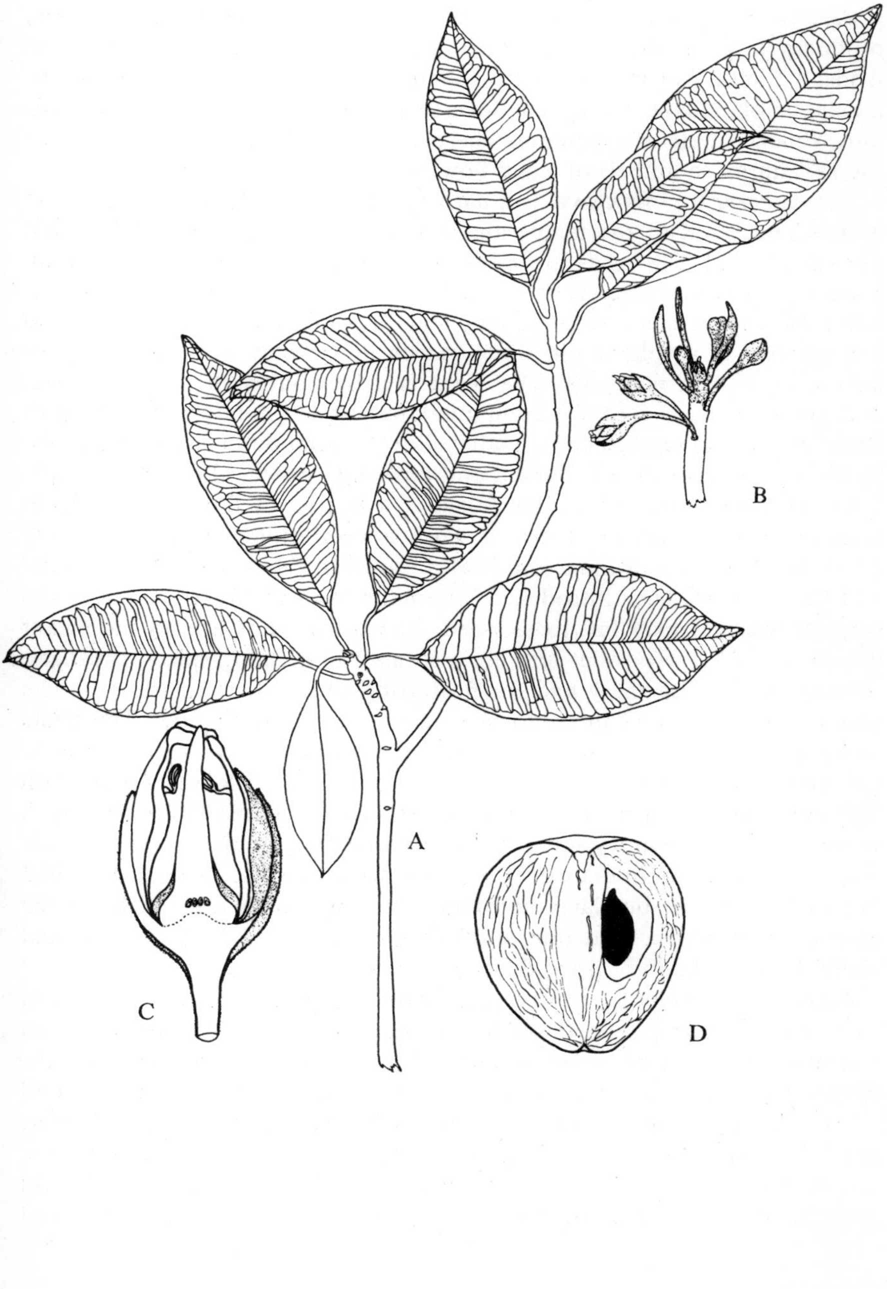
A
B
C
D

fixative and an ingredient of novel perfumes. The leaf stalks take about 9 months from seed to harvesting. Celery is a strong-smelling, glabrous herb to 1 m tall; root leaves pinnately compound; flowering stem grooved and jointed; fls white, very small, in small compound umbels. The var. *rapaceum* (Mill.) DC., CELERIAC, has thickened, turnip-like, edible roots and the leaf stalks are not developed.

Arracacia xanthorrhiza Bancr. (syn. *A. esculenta* DC.), ARRACACHA, produces large fleshy roots, which are an important starchy food in Andean South America. It is cultivated in the temperate regions of the highlands of Bolivia, Peru, Ecuador and Colombia, extending northwards into Venezuela. It is a perennial herb to 1 m high, with deeply divided leaves, and loose umbels of small flowers.

Carum carvi L., CARAWAY, is a native of Europe and western Asia and is now cultivated in the temperate regions, extending to the Sudan and northern India. The seeds are used as a spice and for flavouring bread, biscuits, cakes and cheese. It is used in the manufacture of the liqueur Kümmel and in sausage seasoning and pickling spice. The seeds yield 3–8 per cent of a volatile oil, which is rich in carvone. The oil is used for flavouring and in medicine. The plant is a biennial herb, 30–80 cm tall, with thick tuberous root; lvs compound, 4–5 times pinnately divided, with linear cylindrical segments; fls small, yellow, in compound terminal umbels; 'seeds' (mericarps) slightly curved and tapering, brown, about $5 \times 1{\cdot}5$ mm.

Coriandrum sativum L., CORIANDER, is a native of the Mediterranean region, where it has been grown since ancient times. It is extensively cultivated as a crop in India and is also grown in Europe, the Middle East and Brazil. It has been introduced into many tropical countries, but in the tropical lowland may fail to set seed. The dried fruits are an important ingredient of curry powder and are also used in pickling spices, sausages, seasonings, confectionery and for flavouring spirits, particularly gin. The young plants are used in chutneys, sauces, curries and soups. The fruits contain a volatile oil, which is used for flavouring and in medicine. Coriander is an annual herb, 30–70 cm high; lower leaves broad with crenately-lobed margins, upper leaves finely cut with linear lobes; fls small, white or pink in compound terminal umbels; frs globular, yellow-brown, ribbed, about 3 mm in diameter, 2-seeded, with an unpleasant smell of bed bugs when unripe, but later becoming pleasantly aromatic. In India the crop is sown with a seed rate of 10–20 lb, matures in 3–3½ months, and yields 1,000–2,000 lb of dried fruits per acre.

Cuminum cyminum L., CUMIN, is a native of the Levant, and is grown in south-eastern Europe, north Africa, India and China. The 'seeds' are an essential ingredient of curry powder and are used in mixed spices and for flavouring soups, sausages, pickles, cheese, bread and cakes. The essential oil is used in perfumery and for flavouring beverages. It is a small, slender, much-branched, annual herb, about 25 cm in height; lvs 2–3-partite, linear, bluish-green; fls white or pink in small compound umbels; frs greyish, about 8 mm long, narrow, tapering at both ends, compressed laterally, with ridges with papillose hairs. The plants come into bearing 60–90 days from sowing.

Daucus carota L. subsp. *sativus* (Hoffm.) Thell., CARROT, is derived from the wild carrot, subsp. *carota*, which occurs naturally in Europe, Asia and Africa. It is of ancient cultivation in the Mediterranean region and has now spread throughout the world. It was a favourite vegetable in England in the 16th century and was taken to Virginia in 1609. The roots are used as vegetables and for putting in soups, stews, curries and other dishes; the grated root is used in salads; tender roots are pickled. Carrots are canned and may also be dehydrated. The roots are a rich source of carotene and carotene concentrates are prepared from them. They may be pressed for juice, which is sometimes blended with orange juice. The roots and tops may be fed to livestock. The seeds contain an essential oil which is used for flavouring and in perfumery. The roots contain about: water 88·3 per cent; protein 0·8 per cent; fat 0·4 per cent; carbohydrate 8·9 per cent; fibre 0·8 per cent; ash 0·8 per cent.

The cultivated carrot has a swollen tap-root, which may be short and stumpy or a long tapering cone. Those with a smooth, tender, bright red or orange flesh, with a minimum of core are preferred. It is an erect biennial; 30–100 cm in height, with solid stem; lvs 3-pinnate, segments pinnatifid, lobes lanceolate; petiole usually long; umbels 3–7 cm in diameter, becoming concave in fruit; bracts 7–13; fls usually white; fr. oblong-ovoid, 3–4 mm long, primary ridges ciliate; secondary ridges with hooked spines.

Carrots grow best in a well-drained, light loam, and, in wet weather, should be sown on raised beds and ridges. The seed is sown in drills, about 1 ft apart, at a seed rate of about 4 lb per acre. The seedlings may be thinned to 1 in. apart. The roots should be harvested while still young. Carrots can be stored for several months without loss in quality. In the lowland tropics very early small cvs used for forcing in cooler countries, such as 'Early Horn' and 'Early Gem', give the best results. At higher altitudes 'Nantes' and 'Chatenay' are preferred. 'Danvers Half Long' is recommended in Trinidad.

Foeniculum vulgare Mill. (syn. *F. officinale* Gaertn.), FENNEL, is a native of the Mediterranean region, where it has been cultivated since early times. It has become naturalized in many temperate countries and can be grown in the tropics. It is widely cultivated in India, from where 1,000–2,000 tons of fennel seed are exported annually. All parts of the plant are aromatic. The leaves are used as a pot-herb, more particularly in fish sauces and garnishing. The petioles are used in salads and those of var. *dulce* (Mill.) Thell. are blanched and used as a vegetable. The fruits are used as a flavouring in soups, meat dishes, sauces, etc. and are official in many pharmacopoeias. The fruits yield a volatile oil, of which the main constituent is anethole. The oil is used as a flavouring agent. Fennel is a stout erect glaucous perennial, with very much divided lvs with hair-like segments; fls 1–2 mm in diameter, yellow, in compound terminal umbels; frs ovoid, 5–6 mm long, greenish-brown, with prominent ridges.

Pastinaca sativa L., PARSNIP, occurs wild in Europe and is now cultivated in many temperate countries for its tap-root, which is eaten as a vegetable. It was used by the Greeks and Romans. It does not succeed in tropical lowlands, but is occasionally grown at the higher altitudes. It is a biennial herb, 30–150 cm tall, with hollow stem; lvs simply pinnate, segments, ovate, lobed and serrate; fls minute, yellow; frs 5–8 mm long, ovoid, compressed, with winged margin.

Petroselinum crispum (Mill.) Nym. ex A. W. Hill (syn. *Carum petroselinum* Benth.), PARSLEY, is one of the most familiar and widely cultivated of the garden herbs. It is a native of southern Europe and is now naturalized in most temperate countries. It was used for flavouring by the Greeks and Romans. It can be grown successfully throughout the tropics, but tends to die out rather quickly near the equator. Parsley is one of the most useful of all herbs. The leaves are used as a garnish and for flavouring sauces, soups, omelets and stuffing. They are a rich source of vitamin C. The plant yields a volatile oil, of which the active principle is apiol. It is erect glabrous biennial or short-lived perennial, 30–70 cm tall, with a stout tap-root; rosette lvs in a dense tuft, dark green, shiny, 3-pinnate; leaflets 1–2 cm long, cuneate, lobed, often much crisped; umbels 2–5 cm in diameter, flat-topped; fls yellowish, 2 mm in diameter; frs ovoid, laterally compressed, 2–3 mm long; carpels with 5 slender ridges. The seed, which is usually slow to germinate, can be sown in boxes and seedlings transplanted when they have 6 leaves at a spacing of 3–6 in. and are thinned as required. As the flowering shoots develop they should be cut off at the base. The cv. 'Extra Double Curled Paramount' is recommended in Trinidad.

Pimpinella anisum L., ANISE, was well known to the ancient Egyptians, Hebrews, Greeks and Romans and was highly valued in the Middle Ages as a medicine. It is a native of the Mediterranean region and is now cultivated in Europe, Asia Minor, India and Mexico. It does not grow satisfactorily in the tropical lowlands. The seeds are used for flavouring curries, sweets, confectionery and liqueurs such as anisette. They yield an essential oil on distillation which is used in medicine, perfumery, soaps and beverages. Anise is a pubescent annual herb, about 60 cm high; basal lvs simple with coarse irregular teeth; stem lvs compound-pinnate; fls yellowish, small, in compound umbels; frs ovate, ribbed.

Fig. 102. **Daucus carota : CARROT. A, plant showing root and flowers (×½); B, leaf bases (×½); C, leaf (×½); D, flower (×10).**

A
B
C
D
M. WONG

GENERAL REFERENCES

References on 'the origin and spread of tropical crops' are given at the end of Chapter 2. Selected references on the individual crops are given at the end of each crop. The following additional books have been referred to in writing this work.

AINSWORTH, G. C. and BISBY, G. R. (1961). *Dictionary of the Fungi.* Kew: Commonw. Mycological Inst.

ANDERSON, E. (1952). *Plants, Man and Life.* Boston: Little, Brown.

BAILEY, L. H. (1949). *Manual of Cultivated Plants.* Revised ed. New York: Macmillan.

BAILEY, L. H. and BAILEY, E. Z. (1941). *Hortus Second.* New York: Macmillan.

BAKER, H. G. (1965). *Plants and Civilization.* Belmont: Wadsworth Publ. Co.

BROWN, W. H. (1951–1958). *Useful Plants of the Philippines.* 3 vols. Rep. Phil. Dept. Agric. Comm., Tech. Bull. 10.

BURKILL, I. H. (1935). *Dictionary of the Economic Products of the Malay Peninsula.* 2 vols. London: Crown Agents.

CHANDLER, W. H. (1958). *Evergreen Orchards.* 2nd. ed. Philadelphia: Lea and Febiger.

CHILDERS, N. F. *et al.* (1950). *Vegetable Gardening in the Tropics.* Puerto Rico: Fed. Exp. Stat. Circular 32.

CHITTENDEN, F. J. (1956). *The Royal Horticultural Society Dictionary of Gardening.* 4 vols. and supplement revised by P. M. Synge. Oxford: Clarendon Press.

COBLEY, L. S. (1956). *An Introduction to the Botany of Tropical Crops.* London: Longmans, Green.

COMMONWEALTH ECONOMIC COMMITTEE: The following reviews in their Commodity Series published annually: *Fruit; Grain Crops; Industrial Fibres; Plantation Crops; Vegetable Oils and Oilseeds.*

COMMONWEALTH INSTITUTE OF ENTOMOLOGY (1952). *The Injurious Insects of the British Commonwealth.* London: Commonw. Inst. Ent.

CORNER, E. J. H. (1952). *Wayside Trees of Malaya.* 2 vols. 2nd ed. Singapore: Govt. Printer.

DAHLGREN, B. E. (1947). *Tropical and Subtropical Fruits*. Chicago: Nat. History Museum.

DALZIEL, J. M. (1937). *Useful Plants of West Africa*. London: Crown Agents.

DARLINGTON, C. D. and WYLIE, A. P. (1955). *Chromosome Atlas of Flowering Plants*. 2nd. ed. London: Allen and Unwin.

ECKEY, E. W. (1954). *Vegetable Fats and Oils*. New York: Reinhold Publ. Corp.

FENNAH, R. G. (1947). *The Insect Pests of Food-crops in the Lesser Antilles*. Agric. Depts. Grenada and Antigua.

GARNER, R. J. (1958). *The Grafter's Handbook*. 2nd. ed. London: Faber and Faber.

GREY, W. (1965). *Caribbean Cookery*. London: Collins.

GUENTHER, E. (1948–1952). *The Essential Oils*. 6 vols. New York: Van Nostrand.

HAYWOOD, H. E. (1938). *The Structure of Economic Plants*. New York: Macmillan.

HECTOR, J. M. (1936). *Introduction to the Botany of Field Crops*. Vol. 2. Johannesburg: Central News Agency.

HERKLOTS, G. A. C. (1947). *Vegetable Cultivation in Hong Kong*. Hong Kong: South China Morning Post.

HILL, A. F. (1952). *Economic Botany*. 2nd. ed. New York: McGraw-Hill.

HOWES, F. N. (1948). *Nuts*. London: Faber and Faber.

HOWES, F. N. (1949). *Vegetable Gums and Resins*. Waltham: Chronica Botanica.

HUTCHINSON, J. (1959). *The Families of Flowering Plants*. Vol. 1, *Dicotyledons*. 2nd ed. London: Oxford Univ. Press.

HUTCHINSON, J. (1964, 1966). *The Genera of Flowering Plants*. Vols. 1 and 2. London: Oxford Univ. Press.

HUTCHINSON, J. and MELVILLE, R. (1948). *The Story of Plants and their Uses to Man*. London: Gawthorn.

JACOB, A. and VON UEXKÜLL, H. (1963). *Fertilizer Use: Nutrition and Manuring of Tropical Crops*. 3rd ed. Hannover: Verlagsgesellschaft für Ackerbau mbH.

KENNARD, W. C. and WINTERS, H. F. (1960). *Some Fruits and Nuts for the Tropics*. Washington: U.S.D.A. Misc. Publ. 801.

KIRBY, R. H. (1963). *Vegetable Fibres*. London: Leonard Hill.

KLAGES, K. H. W. (1942). *Ecological Crop Geography*. New York: Macmillan.

MACMILLAN, H. F. (1949). *Tropical Planting and Gardening*. 5th. ed. London: Macmillan.

MAISTRE, J. (1964). Les Plantes à Épices. Paris: Maisonneuve et Larose.

MANJUNATH, B. L. (1948) ed. *The Wealth of India: Raw Materials. Vol. 1.* Delhi: Counc. Sci. and Indust. Resc.

MASSAL, R. and BARRAU, J. (1956). *Food Plants of the South Sea Islands.* New Caledonia: South Pacific Commission.

MATHESON, J. K. and BOVILL, E. W. (1950). *East African Agriculture.* London: Oxford Univ. Press.

MERRILL, E. D. (1945). *Plant Life in the Pacific World.* New York: Macmillan.

MILSUM, J. N. and GRIST, D. H. (1941). *Vegetable Gardening in Malaya.* Kuala Lumpur: Agric. Dept. S.S. & F.M.S.

NICHOLS, H. A. and HOLLAND, J. H. (1940). *A Textbook of Tropical Agriculture.* London: Macmillan.

OCHSE, J. J. and VAN DEN BRINK, R. C. B. (1931). *Fruits and Fruticulture in the Dutch East Indies.* Batavia: Kolff and Co.

OCHSE, J. J. and VAN DEN BRINK, R. C. B. (1931). *Vegetables of the Dutch East Indies.* Buitenzorg: Archipel. Drukkerij.

OCHSE, J. J., SOULE, M. J., DIJKMAN, M. J. and WEHLBURG, C. (1961). *Tropical and Subtropical Agriculture.* 2 vols. New York: Macmillan.

PARRY, J. W. (1945). *The Spice Handbook.* New York: Chemical Publishing Co.

PARRY, J. W. (1953). *The Story of Spices.* New York: Chemical Publishing Co.

PLATT, B. S. (1962). *Tables of Representative Values of Foods Commonly Used in Tropical Countries.* London: H.M.S.O.

POEHLMAN, J. M. (1959). *Breeding Field Crops.* New York: Henry Holt.

POPENOE, W. (1920). *Manual of Tropical and Subtropical Fruits.* New York: Macmillan.

REDGROVE, H. S. (1933). *Spices and Condiments.* London: Pitman.

RIDLEY, H. N. (1912). *Spices.* London: Macmillan.

SAMPSON, H. C. (1936). *Cultivated Crop Plants of the British Empire and the Anglo-Egyptian Sudan.* Royal Botanic Gardens, Kew: Bull. Misc. Information, Additional Series XII.

SASTRI, B. N. (1950–1962) ed. *The Wealth of India: Raw Materials.* Vols. 2–6. Delhi: Counc. Sci. and Indust. Resc.

SAUNDERS, L. H. (1940). *Vegetable Gardening in the Tropics.* London: Oxford Univ. Press.

SCHERY, R. W. (1952). *Plants and Man.* Englewood Cliffs: Prentice Hall.

SINGH, S., KRISHNAMURTHI, S. and KATYAL, S. L. (1963). *Fruit Culture in India.* Delhi: Indian Counc. Agric. Resc.

THOMAS, A. (1965). *Gardening in Hot Countries.* London: Faber and Faber.

THOMPSON, H. C. (1949). *Vegetable Crops*. 4th ed. New York: McGraw-Hill.

TINDALL, H. D. (1965). *Fruits and Vegetables in West Africa*. Rome: FAO.

TOTHILL, J. D. (1940). ed. *Agriculture in Uganda*. London: Oxford Univ. Press.

TOTHILL, J. D. (1948). ed. *Agriculture in the Sudan*. London: Oxford Univ. Press.

UPHOF, J. C. T. (1959). *Dictionary of Economic Plants*. Weinheim: Engelmann.

WATT, G. (1889–1896). *A Dictionary of the Economic Products of India*. 6 vols. Calcutta: Govt. of India.

WATT, G. (1908). *The Commercial Products of India*. London: John Murray.

WEBSTER, C. C. and WILSON, P. N. (1966). *Agriculture in the Tropics*. London: Longmans, Green.

WILLIAMS, R. O. (1949). *Useful and Ornamental Plants in Zanzibar and Pemba*. Zanzibar: Govt. Printer.

WILLIAMS, R. O. and WILLIAMS, R. O. JNR. (1951). *The Useful and Ornamental Plants in Trinidad and Tobago*. 4th ed. Port-of-Spain: Guardian Commercial Printery.

WILLIS, J. G. (1966). *Dictionary of the Flowering Plants and Ferns*. 7th ed. revised H. K. Airy-Shaw. Cambridge University Press.

WILLS, J. G. (1962). ed. *Agriculture and Land Use in Ghana*. London: Oxford Univ. Press.

WINTON, A. L. and WINTON, K. B. (1932–1939). *The Structure and Composition of Foods*. 4 vols. London: John Wiley.

WOOD, R. C. (1957). *A Note-Book of Tropical Agriculture*. 6th ed. Trinidad: Imperial College of Tropical Agriculture.

WOOT-TSUEN, W. L. and FLORES, M. (1961). *Food Composition Table for Use in Latin America*. Bethseda: INCAP-ICNND.

WOOT-TSUEN, W. L., PECOT, R. K. and WATT, B. K. (1962). *Composition of Foods Used in Far Eastern Countries*. U.S.D.A. Agric. Handbook, 34.

WRIGLEY, G. (1961). *Tropical Agriculture*. London: Batsford.

YEGNA NARAYAN AIYER, A. K. (1947). *Field Crops of India*. Bangalore: Govt. Press.

APPENDIX

Dicotyledons referred to in text

Synonyms given in italics

SCIENTIFIC NAME		COMMON NAME	USE	ORIGIN	PAGE
Abelmoschus *Medik.*	MALVACEAE				
esculentus (L.) Moench. = Hibiscus esculentus *L.*					368
moschatus (L.) Medik. = Hibiscus abelmoschas *L.*					365
Abrus *L.*	LEGUMINOSAE (P)				
precatorius *L.*		JUMBIE BEAD	beads	Tropics	217
Abutilon *Mill.*	MALVACEAE				
avicennae Gaertn. = A. theophrasti *Medic.*					333
theophrasti *Medic.*		CHINA JUTE	fibre	Asia	333
Acacia *Mill.*	LEGUMINOSAE (M)				208
arabica (Lam.) Willd. = A. nilotica (*L.*) [*Willd. ex*] *Del.*					209
baileyana *F. Muell.*		WATTLE	ornamental	Australia	215
catechu (*L.f.*) *Willd.*		CUTCH	tannin	India	209
dealbata *Link*		SILVER WATTLE	tannin	Australia	210
decurrens (*Wendl.*) *Willd.*		GREEN WATTLE	tannin	Australia	210
var. *mollis* Lindl. = A. mearnsii *De Wild.*					210
farnesiana (*L.*) *Willd.*		CASSIE FLOWER	perfume	T. America	209
mearnsii *De Wild.*		BLACK WATTLE	tannin	Australia	210
nilotica (*L.*) [*Willd. ex*] *Del.*		BABUL	tannin, gum	Africa	209
pycnantha *Benth.*		GOLDEN WATTLE	tannin	Australia	210
senegal (*L.*) *Willd.*		GUM ARABIC	gum	Africa	209
seyal *Del.*		SHITTIM	gum	Africa	209
spp.			browse	Africa	209
Acalypha *L.*	EUPHORBIACEAE				
hispida *Burm. f.*		RED HOT CAT TAIL	ornamental	E. Indies	139
wilkesiana *Muell.-Arg.*		COPPER LEAF	ornamental	Pacific	139
Achras *L.*	SAPOTACEAE				647
zapota L. = Manilkara achras (*Mill.*) *Fosberg*					
Adansonia *L.*	BOMBACACEAE				
digitata *L.*		BAOBAB	ed. fruit	Africa	33
Aegle *Correa*	RUTACEAE				
marmelos (*L.*) *Correa*		INDIAN BAEL	ed. fruit	India	493
Aeschynomene *L.*	LEGUMINOSAE (P)				
americana *L.*		JOINT VETCH	cover crop	T. America	220
elaphroxylon (*Guill. & Perr.*) *Taub.*		AMBATCH	rafts, floats	Africa	217

SCIENTIFIC NAME	COMMON NAME	USE	ORIGIN	PAGE
Ageratum *L.* COMPOSITAE		ornamentals	T. America	53
Albizia *Durazz.* LEGUMINOSAE (M)				
chinensis (*Osbeck*) *Merr.*		shade	Asia	472
falcata (*L.*) *Backer*		shade	Asia	472
lebbeck (*L.*) *Benth.*		shade	Asia	472
marginata (Lam.) Benth. = A. chinensis (*Osbeck*) *Merr.*				472
moluccana Miq. = A. falcata (*L.*) *Backer*				472
odoratissima (*L.f.*) *Benth.*		shade	Asia	606
spp.		shade		208
stipulata (Roxb.) Boivin = A. chinensis (*Osbeck*) *Merr.*				472
Aleurites *Forst.* EUPHORBIACEAE				140
fordii *Hemsl.*	TUNG	oil	China	140
moluccana (*L.*) *Willd.*	CANDLENUT	oil	Malaya	140
montana (*Lour.*) *Wils.*	TUNG	oil	China	140
trisperma *Blanco*		oil	Philippines	140
Allamanda *L.* APOCYNACEAE				
carthartica *L.*	ALLAMANDA	ornamental	S. America	628
Alternanthera *Forsk.* AMARANTHACEAE				
amoena *Voss.*	RABBIT MEAT	ornamental	Brazil	624
Alysicarpus *Neck.* ex Desv. LEGUMINOSAE (P)				
vaginalis (*L.*) *DC.*	ALYCE CLOVER	cover crop	Asia	220
AMARANTHACEAE				624
Amaranthus *L.* AMARANTHACEAE				
caudatus *L.*	GRAIN AMARANTH	grain	S. America	624
cruentus *L.*	GRAIN AMARANTH	grain	C. America	624
dubius *Mart. ex Thell.*		spinach	Tropics	624
gangeticus *L.*		spinach	Asia	624
hybridus *L.*		spinach	Tropics	624
leucocarpus *S. Wats.*	GRAIN AMARANTH	grain	Mexico	624
paniculatus *L.*		spinach	Asia	624
spp.	AMARANTHS			624
tricolor *L.*		ornamental	Asia	624
Amherstia *Wall.* LEGUMINOSAE (C)				
nobilis *Wall.*	PRIDE OF BURMA	ornamental	Burma	202
ANACARDIACEAE				18
Anacardium *L.* ANACARDIACEAE				19
occidentale *L.*	CASHEW	ed. nut	T. America	19
Anethum *L.* UMBELLIFERAE				
graveolens *L.*	DILL	cul. herb.	Eurasia	648
Annona *L.* ANNONACEAE				625
cherimolia *Mill.*	CHERIMOYA	ed. fruit	S. America	625
diversifolia *Saff.*	ILAMA	ed. fruit	T. America	626
montana *Macfad.*	MOUNTAIN SOURSOP	ed. fruit	C. America	626
muricata *L.*	SOURSOP	ed. fruit	T. America	625
reticulata *L.*	BULLOCK'S HEART	ed. fruit	T. America	626
squamosa *L.*	SUGAR APPLE	ed. fruit	T. America	625
ANNONACEAE				625
Anthemis *L.* COMPOSITAE				
nobilis *L.*	CHAMOMILE	drug	Europe	53

SCIENTIFIC NAME		COMMON NAME	USE	ORIGIN	PAGE
Antiaris *Lesch.*	MORACEAE				
africana *Engl.*		WEST AFRICAN BARK CLOTH	bark cloth	W. Africa	377
toxicaria (*Pers.*) *Lesch.*		UPAS TREE	arrow poison	Asia	377
Antidesma *L.*	EUPHORBIACEAE				
bunius (*L.*) *Spreng.*		BIGNAY	ed. fruit	Asia	139
Apium *L.*	UMBELLIFERAE				
dulce Mill. = A. graveolens *L.* var. dulce (*Mill.*) *DC.*					648
graveolens *L.*		WILD CELERY		Eurasia	648
var. dulce (*Mill.*) *DC.*		CELERY	vegetable	Eurasia	648
var. rapaceum (*Mill.*) *DC.*		CELERIAC	vegetable	Eurasia	648
APOCYNACEAE					626
Arachis *L.*	LEGUMINOSAE (P)				224
hypogaea *L.*		GROUNDNUT	oil seed	S. America	225
Argyreia *Lour.*	CONVOLVULACEAE				
speciosa *Sweet*		ELEPHANT CLIMBER	ornamental	S.E. Asia	78
Armoracia *Gilib.*	CRUCIFERAE				
rusticana *Gaertn.*		HORSE RADISH	condiment	S.E. Europe	90
Arracacia *Bancr.*	UMBELLIFERAE				
esculenta DC. = A. xanthorrhiza *Bancr.*					650
xanthorrhiza *Bancr.*		ARRACACHA	ed. tuber	S. America	650
Artemisia *L.*	COMPOSITAE				
absinthium *L.*		WORMWOOD	flavouring	Europe	53
cina *Berg.*			drug	Russia	53
dracunculus *L.*		TARRAGON	ed. tuber	W. Asia	53
maritima *L.*			drug	Pakistan	53
Artocarpus *Forst.*	MORACEAE				378
altilis (*Park.*) *Fosberg*		BREADFRUIT BREADNUT	ed. fruit	Polynesia	379
champeden (*Lour.*) *Spreng.*		CHAMPEDAK	ed. fruit	Malaya	379
communis Forst. = A. altilis (*Park.*) *Fosberg.*					379
elastica *Reinw.*			bark cloth	Malaya	379
heterophyllus *Lam.*		JACKFRUIT	ed. fruit	India	384
hirsuta *Lam.*				India	385
incisa L.f. = A. altilis (*Park.*) *Fosberg*					379
integrifolia L.f. = A. heterophyllus *Lam.*					384
integra (Thurb.) Merr. = A. heterophyllus *Lam.*					384
mariannensis *Tréc*		WILD BREADFRUIT	ed. fruit	Micronesia	379
rigida *Bl.*		MONKEY JACK	ed. fruit	Malaysia	379
tamaran *Becc.*			bark cloth	Borneo	379
Aster *L.*	COMPOSITAE		ornamental	world-wide	53
Astragalus *L.*	LEGUMINOSAE (P)				
gummifer *Labill.*		GUM TRAGACANTH	gum	W. Asia	217
Atropa *L.*	SOLANACEAE				
belladonna *L.*		BELLADONNA	drug	Eurasia	524
Averrhoa *L.*	OXALIDACEAE				
bilimbi *L.*		BILIMBI	ed. fruit	Malaya	638
carambola *L.*		CARAMBOLA	ed. fruit	Indonesia	638
Baccaurea *Lour.*	EUPHORBIACEAE				
motleyana *Muell.-Arg.*		RAMBAI	ed. fruit	S.E. Asia	139
sapida *Muell.-Arg.*			ed. fruit	S.E. Asia	139

SCIENTIFIC NAME	COMMON NAME	USE	ORIGIN	PAGE
Banisteriopsis *C.B. Rob. & Small* MALPIGHIACEAE				
caapi (*Spruce*) *Morton*	CAAPI	narcotic	S. America	637
inebrians *Morton*	CAAPI	narcotic	S. America	637
quitensis (*Niedenzu*) *Morton*	CAAPI	narcotic	S. America	637
Baphia Afzel. ex *Lodd.* LEGUMINOSAE (P)				
nitida Afzel. ex *Lodd.*	CAMWOOD	dye	W. Africa	217
Basella *L.* BASELLACEAE				628
alba L.	INDIAN SPINACH	spinach	T. Asia	628
rubra L. = B. alba *L.*				
BASELLACEAE				628
Bauhinia *L.* LEGUMINOSAE (C)		ornamental	Tropics	202
Beaumontia *Wall.* APOCYNACEAE				
grandiflora *Wall.*	HERALD'S TRUMPET	ornamental	T. Asia	628
Benincasa *Savi* CUCURBITACEAE				
cerifera Savi = B. hispida (*Thunb.*) *Cogn.*				
hispida (*Thunb.*) *Cogn.*	WAX GOURD	vegetable	Java	101
Bertholletia *Humb. & Bonpl.* LECYTHIDACEAE				101
excelsa *Humb. & Bonpl.*	BRAZIL NUT	ed. nut; oil	S. America	637
Beta *L.* CHENOPODIACEAE				
vulgaris *L.*				630
subsp. maritima (*L.*) *Thell.*	WILD BEET		Eurasia	630
subsp. vulgaris	BEETROOT	ed. root	Europe	630
BIGNONIACEAE				629
Bixa *L.* BIXACEAE				
orellana *L.*	ANNATTO	dye	T. America	629
BIXACEAE				629
Blighia *Koenig* SAPINDACEAE				
sapida *Koenig*	AKEE	ed. aril	W. Africa	642
Boehmeria *Jacq.* URTICACEAE				620
nivea (*L.*) *Gaud.*	RAMIE	fibre	S.E. Asia	620
var. nivea	RAMIE			620
var. tenacissima *Miq.*	RAMIE			620
utilis Bl. = B. nivea (*L.*) *Gaud.* var tenacissima *Miq.*				620
BOMBACACEAE				33
Bombax *L.* BOMBACACEAE				
buonopozense *P. Beauv.*	SILK COTTON	kapok	W. Africa	33
malabaricum *DC.*	SILK COTTON	kapok	India	33
pentandrum L. = Ceiba pentandra (*L.*) *Gaertn.*				33
Borago *L.* BORAGINACEAE				
officinalis *L.*	BORAGE	cul. herb	Europe	634
Bouea *Meissn.* ANACARDIACEAE		ed. fruits	Asia	18
Brachystegia *Benth.* LEGUMINOSAE (C)		woodland	T. Africa	202
Brassica *L.* CRUCIFERAE				90
alba (*L.*) *Rabenh.*	WHITE MUSTARD	condiment	Mediterra-nean	91
campestris *L.*	FIELD MUSTARD	oil seed	India	91
var. sarson *Prain*	INDIAN COLZA	oil seed	India	91
var. toria *Duthie & Fuller*	INDIAN RAPE	oil seed	India	91

SCIENTIFIC NAME		COMMON NAME	USE	ORIGIN	PAGE
Brassica—continued					
caulorapa Pasq. = B. oleracea *L.* var. gongylodes *L.*					95
chinensis *L.*		CHINESE CABBAGE	vegetable	Asia	91
var. pekinensis (*Rupr.*) *Sun.*		PE-TSAI	salad	China	91
hirta Moench. = B. alba (*L.*) *Rabenh.*					91
juncea (*L.*) *Czern. & Coss.*		INDIAN MUSTARD	oil seed	Africa	91
napobrassica (*L.*) *Mill.*		RUTABAGA; SWEDE	vegetable	Europe	92
napus *L.*		RAPE	oil seed	Mediterranean	92
var. *napobrassica* (L.) Rchb. = B. napobrassica (*L.*) *Rchb.*					92
nigra (*L.*) *Koch.*		BLACK MUSTARD	condiment	Eurasia	92
oleracea *L.*		WILD CABBAGE		Europe	94
var. acephala *DC.*		BORECOLE	vegetable	Europe	94
var. botrytis *L.*		BROCCOLI	vegetable	Europe	94
var. capitata *L.*		CABBAGE	vegetable	Europe	95
var. gemmifera *Zenk.*		BRUSSELS SPROUT	vegetable	Europe	95
var. gongylodes *L.*		KOHLRABI	vegetable	Europe	95
var. italica *Plenck*		SPROUTING BROCCOLI	vegetable	Europe	95
rapa *L.*		TURNIP	vegetable	Europe	95
Brosimum *Sw.*	MORACEAE				
alicastrum *Sw.*			ed. fruit	Mexico	377
galactodendron Don. = B. utile (*HBK*) *Pittier*					377
utile (*HBK*) *Pittier*		COW TREE	ed. latex	Venezuela	377
Broussonetia L'Hérit ex *Vent.*	MORACEAE				
papyrifera (*L.*) *Vent.*		PAPER MULBERRY	bark cloth	China	377
Browallia *L.*	SOLANACEAE		ornamental	T. America	524
Brunfelsia *L.*	SOLANACEAE		ornamental	T. America	524
Buchanania *Spreng.*	ANACARDIACEAE				
lanzan *Spreng.*			ed. nut	India	18
Butyrospermum *Kotschy*	SAPOTACEAE				
paradoxum (*Gaertn. f.*) *Hepper*					
subsp. nilotica (*Kotschy*) *Pierre ex. Engl.*		SHEA BUTTER	ed. fat	Uganda	646
subsp. parkii (*G. Don*) *Hepper*		SHEA BUTTER	ed. fat	W. Africa	644
parkii (G. Don.) Kotschy = B. paradoxum (*Gaertn. f.*) *Hepper*					644
Caesalpinia *L.*	LEGUMINOSAE (C)				
coriaria (*Jacq.*) *Willd.*		DIVI-DIVI	tannin	T. America	202
echinata *Lam.*		BRAZILWOOD	dye	S. America	202
sappan *L.*		SAPPANWOOD	dye	S.E. Asia	202
spinosa (*Mol.*) *Ktze.*		TARA	tannin	T. America	202
pulcherrima (*L.*) *Sw.*		PRIDE OF BARBADOS	ornamental	T. America	202
CAESALPINIACEAE					199
CAESALPINIOIDEAE					201
Cajanus *DC.*	LEGUMINOSAE (P)				236
cajan (*L.*) *Millsp.*		PIGEON PEA	pulse	Africa	236
var. bicolor *DC.*		ARHAR CVS PIGEON PEA			237
var. flavus *DC.*		TUR CVS PIGEON PEA			237
indicus Spreng. = C. cajan (*L.*) *Millsp.*					236
Calendula *L.*	COMPOSITAE		ornamental	Eurasia	53

SCIENTIFIC NAME		COMMON NAME	USE	ORIGIN	PAGE
Callistemon *R. Br.*	MYRTACEAE				
citrinus (*Curt.*) *Stapf*		BOTTLE BRUSH	ornamental	Australia	398
spp.		BOTTLE BRUSH	ornamental	Australia	398
Calonyction *Choisy*	CONVOLVULACEAE				79
aculeatum (*L.*) House = Ipomoea alba *L.*					
Calophyllum *L.*	GUTTIFERAE				
antillanum *Britt.*		GALBA	windbreak	W. Indies	588
Calopogonium *Desv.*	LEGUMINOSAE (P)				
mucunoides *Desv.*			cover crop	T. America	218
Camellia *L.*	THEACEAE				599
japonica *L.*		GARDEN CAMELLIA	shrub	Far East	599
sasanqua *Thunb.*			oil	China	599
sinensis (*L.*) *O. Kuntze*		TEA	beverage	Burma	599
var. assamica (*Mast.*) *Pierre*		ASSAM TEA			601
var. macrophylla *Makino*		TETRAPLOID TEA			601
var. sinensis		CHINA TEA			601
spp.			ornamental	Asia	599
thea Link = C. sinensis (*L.*) *O. Kuntze*					599
theifera Griff. = C. sinensis (*L.*) *O. Kuntze*					599
Cananga *Hook. f. & Thoms.*	ANNONACEAE				
odorata (*Lam.*) *Hook. f. & Thoms.*		YLANG-YLANG	perfume	Malaysia	626
Canavalia *DC.*	LEGUMINOSAE (P)				242
ensiformis (*L.*) *DC.*		JACK BEAN	pulse	C. America	242
gladiata (*Jacq.*) *DC.*		SWORD BEAN	pulse	Old World	245
maritima (*Aubl.*) *Thou.*			strand plant	Tropics	242
plagiosperma *Piper*			pulse	S. America	245
rosea (Sw.) DC. = C. maritima (*Aubl.*) *Thou.*					242
virosa (Roxb.) *Wight & Arn.*				T. Asia	245
CANNABIDACEAE					40
Cannabis *L.*	CANNABIDACEAE				
sativa *L.*		HEMP	fibre, narcotic	C. Asia	40
Capsicum *L.*	SOLANACEAE				524
annuum *L.*		CHILLIES	condiments, spice	T. America	525
var. abbreviatum *Fingerh.*		WRINKLED PEPPER			525
var. acuminata *Fingerh.*		CHILLI			525
var. cerasiforme (*Miller*) *Irish*		CHERRY PEPPER			525
var. conoides (*Miller*) *Irish*		CONE PEPPER			525
var. fasciculatum (*Sturt.*) *Irish*		CLUSTER PEPPER			525
var. grossum *Sendt.*		SWEET PEPPER			525
var. longum *Sendt.*		LONG PEPPER			526
baccatum L. = C. frutescens *L.*					526
frutescens L.		BIRD PEPPER	condiment	T. America	526
var. baccatum (*L.*) *Irish*		CHERRY CAPSICUM			526
minimum Roxb. = C. frutescens *L.*					526
pendulum *Willd.*		CHILLI	spice	S. America	526
pubescens *Ruiz & Pav.*		CHILLI	spice	C. & S. America	526
Carica *L.*	CARICACEAE				45
candamarcensis *Hook. f.*		MOUNTAIN PAPAYA	ed. fruit	S. America	45
papaya *L.*		PAPAYA	ed. fruit	C. America	45
peltata *Hook. & Arn.*		WILD PAPAYA		C. America	45

SCIENTIFIC NAME		COMMON NAME	USE	ORIGIN	PAGE
CARICACEAE					45
Carissa *L.*	APOCYNACEAE				
carandas *L.*			ed. fruit	T. India	626
edulis *Vahl*			ed. fruit	T. Africa	626
grandiflora *A. DC.*		NATAL PLUM	ed. fruit	S. Africa	626
Carthamus *L.*	COMPOSITAE				54
lunatus *L.*				Afro-Asia	54
oxyacantha *Bieb.*				Asia	54
tinctorius *L.*		SAFFLOWER	oil seed	Afro-Asia	54
Carum *L.*	UMBELLIFERAE				
petroselinum Benth. = Petroselinum crispum (*Mill.*) *Nym. ex A. W. Hill*					652
carvi *L.*		CARAWAY	cul. herb	Eurasia	650
Caryophyllus *L.*	MYRTACEAE				
aromaticus L. = Eugenia caryophyllus (*Sprengel*) *Bullock and Harrison*					401
Casimiroa *Llave & Lex.*	RUTACEAE				
edulis *Llave & Lex.*		WHITE SAPOTE	ed. fruit	C. America	493
Cassia *L.*	LEGUMINOSAE (C)				
acutifolia Del. = C. senna *L.*					202
alata *L.*		RINGWORM BUSH	drug	Tropics	204
angustifolia *Vahl*		INDIAN SENNA	drug	India	203
auriculata *L.*		AVARAM	tannin	India	204
didymobotrya *Fresen.*			purgative	E. Africa	203
fistula *L.*		INDIAN LABURNUM	purgative	India	203
occidentalis *L.*		NEGRO COFFEE	coffee subst.	T. America	203
senna *L.*		ALEXANDRIAN SENNA	drug	N.E. Africa	203
siamea *Lam.*			firewood	T. Asia	203
spp.			ornamental		203
Cassytha *L.*	LAURACEAE				
filiformis *L.*			parasite	Tropics	187
Castilla *Cerv.*	MORACEAE				377
elastica *Cerv.*		CASTILLOA RUBBER	rubber	C. America	377
markhamiana Hook. f. = C. panamensis *O.F. Cook*					378
panamensis *O.F. Cook*			rubber	C. America	378
ulei *Warb.*		CAUCHO RUBBER	rubber	S. America	378
Catha *Forsk.*	CELASTRACEAE				
edulis *Forsk.*		KHAT	narcotic	E. Africa	630
Catharanthus *G. Don*	APOCYNACEAE				
roseus (*L.*) *G. Don*		PERIWINKLE	ornamental	T. America	628
Ceiba *Mill.*	BOMBACACEAE				34
acuminata (*S. Wats.*) *Rose*			fibre	C. America	34
aesculifolia (*HBK.*)					34
Britton & E.G. Baker			fibre	C. America	34
pentandra (*L.*) *Gaertn.*		KAPOK	fibre	T. America	34
var. caribaea (*DC.*) *Bakh.*		KAPOK	fibre	T. America	35
var. guineensis (*Thonn.*) *H.G. Baker*		KAPOK	fibre	W. Africa	35
var. *indica* (DC.) Bakh. = C. pentandra (*L.*) *Gaertn.* var. pentandra					35
var. pentandra		KAPOK	fibre	T. Africa	35

SCIENTIFIC NAME		COMMON NAME	USE	ORIGIN	PAGE
CELASTRACEAE					630
Celosia *L.*	AMARANTHACEAE				
cristata *L.*		COCKSCOMB	ornamental	Tropics	624
Centaurea *L.*	COMPOSITAE		ornamental	Old World	53
Centrosema (*DC.*) Benth.	LEGUMINOSAE (P)				
plumieri (*Turp.*) *Benth.*			cover crop	T. America	218
pubescens *Benth.*			cover crop	T. America	218
Cephaelis *Sw.*	RUBIACEAE				
ipecacuanha (*Stokes*) *Baill.*		IPECACUANHA	drug	Brazil	451
Cephalonema *K. Schum.*	TILIACEAE				
polyandrum *K. Schum.*		PUNGA	fibre	Congo	613
Ceratotheca *Endl.*	PEDALIACEAE				
sesamoides *Endl.*			oil seed	T. Africa	430
Cestrum *L.*	SOLANACEAE		ornamental	T. America	524
CHENOPODIACEAE					630
Chenopodium *L.*	CHENOPODIACEAE				
album *L.*		FAT HEN	grain	Europe	632
ambrosioides *L.*		WORMSEED	vermifuge	T. America	632
nuttalliae *Safford*			grain crop	C. America	632
quinoa *Willd.*		QUINOA	grain crop	Peru	632
Chlorophora *Gaud.*	MORACEAE				
excelsa (*Welw.*) *Benth.*		MUVULE	timber	T. Africa	378
tinctoria (*L.*) *Gaud.*		FUSTIC	dye	W. Indies	378
Chrysanthemum *L.*	COMPOSITAE				
cinerariaefolium (*Trev.*) *Bocc.*		PYRETHRUM	insecticide	Dalmatia	58
coccineum *Willd.*		PERSIAN INSECT FLOWER	insecticide	Persia	58
indicum *L.*		CHRYSANTHEMUMS	ornamental	China	58
marschallii *Aschers*		CAUCASIAN INSECT FLOWER	insecticide	Caucasia	58
morifolium *Ramat.*		CHRYSANTHEMUMS	ornamental	China	58
roseum Adam. = C. coccineum *Willd.*			ornamental		58
sinense Sabine = C. morifolium *Ramat.*					
Chrysophyllum *L.*	SAPOTACEAE				
cainito *L.*		STAR APPLE	ed. fruit	C. America	646
Cicer *L.*	LEGUMINOSAE (P)				
arietinum *L.*		CHICK PEA	pulse	T. Asia	246
Cichorium *L.*	COMPOSITAE				
endivia *L.*		ENDIVE	salad veg.	Mediterranean	52
intybus *L.*		CHICORY	vegetable	Europe	53
Cinchona *L.*	RUBIACEAE				452
calisaya *Wedd.*		QUININE	drug	S. America	453
lancifolia *Mutis*		QUININE	drug	S. America	454
ledgeriana *Moens ex Trimen.*		QUININE	drug	S. America	454
officinalis *L.*		QUININE	drug	S. America	454
pitayensis *Wedd.*		QUININE	drug	S. America	456
pubescens *Vahl*		QUININE	drug	C. & S. America	454
succirubra *Pav. ex Klotzsch*		QUININE	drug	S. America	454
Cinnamomum *Blume*	LAURACEAE				187
burmanni (*Nees*) *Blume*		PADANG CASSIA	spice	Indonesia	187
camphora (*L.*) *Nees & Eberm.*		CAMPHOR	oil	E. Asia	188

SCIENTIFIC NAME	COMMON NAME	USE	ORIGIN	PAGE
Cinnamomum—continued				
cassia (*Nees*) *Nees ex Blume*	CASSIA	spice	Burma	187
massoia *Schewe*	MASSOIA BARK	spice	New Guinea	188
oliveri *Bailey*	OLIVER'S BARK	spice	Australia	188
tamala (*Buck.-Ham.*) *Nees & Eberm.*	INDIAN CASSIA	spice	India	187
zeylanicum *Blume*	CINNAMON	spice	Ceylon	188
Citropsis (Engl.) *Swing. & Kellerman* RUTACEAE			T. Africa	494
gilletiana *Swing. & Kell.*		rootstock	Congo	494
Citrullus *Schrad.* ex Eckl. & Zeyh. CUCURBITACEAE				102
colocynthis (*L.*) *Schrad.*	COLOCYNTH	drug	India	102
lanatus (*Thunb.*) *Mansf.*	WATER MELON	ed. fruit	Africa	102
var. fistulosus (*Stocks*) *Duthie & Fuller*		vegetable	India	103
vulgaris Schrad. = C. lanatus (*Thunb.*) *Mansf.*				102
Citrus *L.* RUTACEAE				495
aurantifolia (*Christm.*) *Swing.*	LIME	ed. fruit	E. Indies	499
aurantium *L.*	SOUR ORANGE	ed. fruit	S.E. Asia	500
subsp. bergamia (*Risso & Poit.*) *Wight & Arn.*				500
var. myrtifolia *Ker-Gawl.*				502
var. *sinensis* L. = C. sinensis (*L.*) *Osbeck*				510
celebica *Koord.*			Philippines	495
decumanus L. = C. grandis (*L.*) *Osbeck*				502
grandis (*L.*) *Osbeck*	PUMMELO	ed. fruit	Malaysia	502
hystrix *DC.*			S.E. Asia	495
ichangensis *Swing.*			China	495
indica *Tan.*			India	495
japonica Thunb. = Fortunella japonica (*Thunb.*) *Swing*				493
latipes (*Swing.*) *Tan.*			India	495
limon (*L.*) *Burm. f.*	LEMON	ed. fruit	S.E. Asia	502
macroptera *Montr.*			S.E. Asia	495
margarita Lour. = Fortunella margarita (*Lour.*) *Swing.*				493
maxima (Burm.) *Merr.* = C. grandis (*L.*) *Osbeck*				502
medica *L.*	CITRON	ed. fruit	S.W. Asia	504
var. ethrog *Engl.*	ETROG CITRON			504
var. sarcodactylis (*Noot.*) *Swing.*	FINGERED CITRON			504
micrantha *Wester*			Philippines	495
nobilis Andrews *non* Lour. = C. reticulata *Blanco*				508
paradisi *Macf.*	GRAPEFRUIT	ed. fruit	W. Indies	506
reticulata *Blanco*	MANDARIN	ed. fruit	China	508
var. austera *Swing.*	SOUR MANDARIN			508
reticulata × C. sinensis	TANGOR	ed. fruit		508
reticulata × C. paradisi	TANGELO	ed. fruit		508
sinensis (*L.*) *Osbeck*	SWEET ORANGE	ed. fruit	S. China	510
spp.	CITRUS			495
tachibana (*Mak.*) *Tan.*	TACHIBANA ORANGE	ed. fruit	China	495
Clappertonia *Meisn.* TILIACEAE				
ficifolia (*Willd.*) *Decne.*		fibre	T. Africa	613
Clausena *Burm. f.* RUTACEAE				
dentata (*Willd.*) *Roem.*		ed. fruit	India	493
lansium (*Lour.*) *Skeels*		ed. fruit	S. China	493
willdenowii Wight & Arn. = C. dentata (*Willd.*) *Roem.*				493

SCIENTIFIC NAME	COMMON NAME	USE	ORIGIN	PAGE
Clianthus *Banks & Soland.* LEGUMINOSAE (P)				
dampieri *Cunn.*	GLORY PEA	ornamental	Australia	223
Clitoria *L.* LEGUMINOSAE (P)				
laurifolia *Poir.*	BUTTERFLY PEA	cover crop	T. America	220
ternatea *L.*	BUTTERFLY PEA	cover crop	T. America	220
Clymenia (*Tan.*) *Swing.* RUTACEAE				
polyandra (*Tan.*) *Swing.*			New Ireland	495
Cochlearia *L.* CRUCIFERAE				
armoracia L. = Armoracia rusticana *Gaertn.*				90
Codiaeum *Juss.* EUPHORBIACEAE				
variegatum (*L.*) *Blume*	GARDEN CROTON	ornamental	Moluccas	139
Coffea *L.* RUBIACEAE				458
abeokutae Cramer = C. liberica *Bull ex Hiern*				488
arabica *L.*	ARABICA COFFEE	beverage	Ethiopia	459
var. arabica	ARABICA COFFEE			461
var. bourbon (*B. Rodr.*) *Choisy*	ARABICA COFFEE			461
var. *stuhlmannii* Warb. = C. canephora *Pierre ex Froehner*				482
var. *typica* Cramer = C. arabica *L.* var. arabica				461
bengalensis *Heyne & Willd.*	COFFEE	beverage	India	458
bukobensis Zimm. = C. canephora *Pierre ex Froehner*				482
canephora *Pierre ex Froehner*	ROBUSTA COFFEE	beverage	T. Africa	482
var. canephora	ROBUSTA COFFEE			483
var. nganda *Haarer*	ROBUSTA COFFEE			483
var. *robusta* = C. canephora *Pierre & Froehner* var. canephora				483
congensis *Froehner*	COFFEE	beverage	Congo	458
dewevrei *De Wild. & Th. Dur.*	COFFEE	beverage	W. Africa	459
var. excelsa *A. Chev.*				
eugenioides *S. Moore*	COFFEE	beverage	E. Congo	458
excelsa *A. Chev.*	COFFEE	beverage	W. Africa	459
klainei Pierre ex De Wild. = C. liberica *Bull ex Hiern*				488
kouilouensis Pierre ex De Wild. = C. canephora *Pierre ex Froehner*				482
laurentii De Wild. = C. canephora *Pierre ex Froehner*				482
liberica *Bull ex Hiern*	LIBERICA COFFEE	beverage	Liberia	488
maclaudii A. Chev. = C. canephora *Pierre ex Froehner*				482
quillou Wester = C. canephora *Pierre ex Froehner*				482
racemosa *Lour*	COFFEE	beverage	Mozambique	459
robusta Linden = C. canephora *Pierre ex Froehner*				482
stenophylla *G. Don*	COFFEE	beverage	W. Africa	459
ugandae Cramer = C. canephora *Pierre ex Froehner*				482
welwitschii Pierre ex De Wild. = C. canephora *Pierre ex Froehner*				482
zanguebariae *Lour.*	COFFEE	beverage	Zanzibar	459

SCIENTIFIC NAME	COMMON NAME	USE	ORIGIN	PAGE
Cola *Schott & Endl.* STERCULIACEAE				564
acuminata (*P. Beauv.*) *Schott & Endl.*	ABATA KOLA	masticatory	W. Africa	565
anomala *K. Schum.*	BAMENDA KOLA	masticatory	Cameroun	565
nitida (*Vent.*) *Schott & Endl.*	GBANJA KOLA	masticatory	W. Africa	566
verticillata (*Thonn.*) *Stapf ex. A. Chev.*	OWE KOLA	masticatory	W. Africa	566
Coleus *Lour.* LABIATAE				634
amboinicus *Lour.*	INDIAN BORAGE	herb	Indonesia	634
aromaticus Benth. = C. amboinicus *Lour.*				634
blumei *Benth.*	COLEUS	ornamental	Indonesia	634
dazo A. Chev. = Plectranthus esculentus *N.E. Br.*				636
esculentus (N.E. Br.) G. Tayl. = Plectranthus esculentus *N.E. Br.*				636
parviflorus *Benth.*		ed. tuber	India	634
tuberosus Benth. = C. parviflorus *Benth.*				634
Colocynthis *Mill.* CUCURBITACEAE				102
citrullus (L.) O. Ktze. = Citrullus lanatus (*Thunb.*) *Mansf.*				
Commiphora *Jacq.* BURSERACEAE		shade trees	Africa & Asia	472
COMPOSITAE				52
CONVOLVULACEAE				78
Copaifera *L.* LEGUMINOSAE (C)				
officinalis *L.*	COPAIBA	balsam	Venezuela	202
spp.	COPAL	copals	T. Africa	202
Corchorus *L.* TILIACEAE				613
aestuans *L.*		pot-herb	T. Africa	613
capsularis *L.*	WHITE JUTE	fibre	S. China	613
olitorius *L.*	TUSSA JUTE	fibre	Africa & Asia	614
spp.		fibre		613
tridens *L.*		pot-herb	T. Africa	613
trilocularis *L.*		pot-herb	T. Africa	613
Cordia *L.* BORAGINACEAE				
holstii *Gurke ex Engl.*		shade tree	E. Africa	473
Coreopsis *L.* COMPOSITAE		ornamental	Africa, America	53
Coriandrum *L.* UMBELLIFERAE				
sativum *L.*	CORIANDER	cul. herb.	Mediter-ranean	650
Cosmos *Cav.* COMPOSITAE		ornamental	T. America	53
Crescentia *L.* BIGNONIACEAE				
cujete *L.*	CALABASH	utensils	T. America	629
Crotalaria *L.* LEGUMINOSAE (P)				
anagyroides *Kunth.*		green manure	T. America	250
burkeana *Benth.*		poisonous	S. Africa	250
dura *Wood & Evans*		poisonous	S. Africa	250
intermedia *Kotschy*		green manure	T. Africa	250
juncea *L.*	SUNN HEMP	fibre	India	250
maypurensis *HBK.*		poisonous	Guyana	250
mucronata *Desv.*		green manure	Tropics	250

SCIENTIFIC NAME		COMMON NAME	USE	ORIGIN	PAGE
Crotalaria—continued					
retusa *L.*			poisonous	Tropics	250
spp.			cover crop	Tropics	250
striata DC.					250
= C. mucronata *Desv.*					250
usaramoensis Bak. f.					
= C. zanzibarica *Benth.*					250
zanzibarica *Benth.*			green manure	E. Africa	250
Croton *L.*	EUPHORBIACEAE				
tiglium *L.*		PURGING CROTON	croton oil	S.E. Asia	139
CRUCIFERAE					89
Cryptostegia *R. Br.*	ASCLEPIADACEAE		rubber	T. Africa	148
Cucumis *L.*	CUCURBITACEAE				107
anguria *L.*		WEST INDIAN GHERKIN	vegetable	T. Africa	108
hardwickii *Royle*		WILD CUCUMBER		India	114
melo *L.*		MELON	ed. fruit	Africa	110
sativus *L.*		CUCUMBER	salad veg.	India	114
Cucurbita *L.*	CUCURBITACEAE				116
andreana *Naud.*		WILD PUMPKIN		Argentina	119
ficifolia *Bouché*		MALABAR GOURD	ed. fruit	T. America	118
maxima *Duch. ex Lam.*		PUMPKIN	vegetable	S. America	119
var. maxima		WINTER SQUASH	vegetable		119
var. turbaniformis *Alef.*		TURBAN SQUASH	vegetable		119
mixta *Pang.*		PUMPKIN	vegetable	Mexico	120
moschata (*Duch. ex Lam.*) *Duch. ex Poir.*		PUMPKIN	vegetable	C. America	119
pepo *L.*		MARROW	vegetable	Mexico	122
var. medullosa *Alef.*		VEGETABLE MARROW	vegetable		122
var. melopepo *Alef.*		BUSH SQUASH	vegetable		122
var. ovifera *Alef.*		ORNAMENTAL GOURDS	ornamental	T. America	118
var. pepo		FIELD PUMPKIN	vegetable		122
texana *Gray*		WILD MARROW		Texas	122
CUCURBITACEAE					100
Cuminum *L.*	UMBELLIFERAE				
cyminum *L.*		CUMIN	cul. herb	Eurasia	650
Cuscuta *L.*	CONVOLVULACEAE	DODDERS	parasites	World-wide	78
Cyamopsis *DC.*	LEGUMINOSAE (P)				
tetragonoloba (*L.*) *Taub.*		CLUSTER BEAN	ed. pods	India	255
psoraloides DC.					255
= C. tetragonoloba (*L.*) *Taub.*					
Cyclanthera *Schrad.*	CUCURBITACEAE				
pedata *Schrad.*					
var. edulis *Schrad.*		WILD CUCUMBER	ed. fruit	C. America	100
Cynara *L.*	COMPOSITAE				
scolymus *L.*		GLOBE ARTICHOKE	vegetable	Europe	53
Cyphomandra Mart. ex *Sendt.*	SOLANACEAE				
betacea (*Cav.*) *Sendt.*		TREE TOMATO	ed. fruits	Peru	523
Dahlia *Cav.*	COMPOSITAE	DAHLIAS	ornamental	C. America	53
Daniellia *Benn.*	LEGUMINOSAE (C)	COPAL TREES	copals	T. Africa	202
Datura *L.*	SOLANACEAE				
arborea *L.*			hypnotic	S. America	524
innoxia *Mill.*			hypnotic	Mexico	524
sanguinea *Ruiz & Pav.*			hypnotic	S. America	524
spp.		THORN APPLES	narcotics	Tropics	524
stramonium *L.*		THORN APPLE	hypnotic	Tropics	524

SCIENTIFIC NAME		COMMON NAME	USE	ORIGIN	PAGE
Daucus *L.*	UMBELLIFERAE				
carota *L.* subsp. carota		WILD CARROT		Eurasia	651
subsp. sativus (*Hoffm.*) *Thell.*		CARROT	ed. root	Eurasia	651
Delonix *Raf.*	LEGUMINOSAE (C)				
regia (*Boj. ex Hook.*) *Raf.*		FLAMBOYANT	ornamental	Madagascar	202
Derris *Lour*	LEGUMINOSAE (P)				256
elliptica *Benth.*		DERRIS	insecticide	Indo-Malaysia	256
ferruginea *Benth.*		DERRIS	insecticide	India	256
malaccensis *Prain*		DERRIS	insecticide	Malaysia	256
Desmodium *Desv.*	LEGUMINOSAE (P)	TICK CLOVERS	cover crops	Tropics	220
Diospyros *L.*	EBENACEAE				
ebenaster *Retz.*		BLACK SAPOTE	ed. fruit	C. America	646
Dipteryx *Schreb.*	LEGUMINOSAE (P)				258
odorata (*Aubl.*) *Willd.*		TONKA BEAN	flavouring	S. America	258
oppositifolia (*Aubl.*) *Willd.*		TONKA BEAN	flavouring	S. America	258
punctata *Blake*			flavouring	S. America	258
rosea *Spruce*			flavouring	S. America	258
Dissotis *Benth.*	MELASTOMACEAE		indicators	Africa and Asia	601
Dolichos *L.*	LEGUMINOSAE (P)				263
biflorus auct. *non* Linn. = D. uniflorus *Lam.*					263
catjung Burm. = Vigna unguiculata (*L.*) *Walp.*					322
erosus L. = Pachyrrhizus erosus (*L.*) *Urban*					281
lablab L. = Lablab niger *Medik.*					273
sesquipedalis L. = Vigna sesquipedalis (*L.*) *Fruw.*					322
sinensis L. = Vigna sinensis (*L.*) *Savi ex Hassk.*					322
unguiculatus L. = Vigna unguiculata (*L.*) *Walp.*					322
uniflorus *Lam.*		HORSEGRAM	pulse crop	Old World	263
Doxantha *Miers*	BIGNONIACEAE				
unguis-cati (*L.*) *Miers*		CAT'S CLAW CREEPER	ornamental	T. American	629
Dracontomelon *Blume*	ANACARDIACEAE				
mangiferum *Blume*		ARGUS PHEASANT TREE	ed. fruit	Indo-Malay.	18
Durio Adans.	BOMBACACEAE				
zibethinus Murr.		DURIAN	ed. fruit	Malaysia	33
Dyera *Hook. f.*	APOCYNACEAE				
costulata (*Miq.*) *Hook. f.*		JELUTONG	latex	Malaysia	626
Entada *Adans.*	LEGUMINOSAE (M)				
abyssinica *Stend. ex A. Rich.*			shade tree	E. Africa	472
Eremocitrus *Swing.*	RUTACEAE				
glauca (*Lindl.*) *Swing*		AUSTRALIAN DESERT LIME	ed. fruit	Australia	495
Eriodendron *DC.*	BOMBACACEAE				
anfractuosum DC. = Ceiba pentandra (*L.*) *Gaertn.*					34

SCIENTIFIC NAME	COMMON NAME	USE	ORIGIN	PAGE
Ervum *L.* LEGUMINOSAE (P)				
lens. L.				279
= Lens esculenta *Moench*				
Erythrina *L.* LEGUMINOSAE (P)				
abyssinica *Lam.*		shade tree	E. Africa	223
glauca *Willd.*	SWAMP IMMORTELLE	shade tree	T. America	223
indica *Lam.*		shade tree	T. Asia	223
lithosperma Miq.				223
= E. subumbrans (*Hassk.*) *Merr.*				
micropteryx Poepp.				223
= E. poeppigiana (*Walp.*) *O.F. Cook*				
poeppigiana (*Walp.*) *O.F. Cook*	MOUNTAIN IMMORTELLE	shade tree	T. America	223
spp.		shade trees		223
subumbrans (*Hassk.*) *Merr.*		shade tree	T. Asia	223
ERYTHROXYLACEAE				632
Erythroxylum P.Br. ERYTHROXYLACEAE				
coca *Lam.*	COCA	narcotic	S. America	632
novogranatense (*Morris*) *Hieron.*	COCA	narcotic	S. America	632
Eucalyptus *L'Hérit.* MYRTACEAE				399
astringens *Maiden*		tannin	Australia	400
camaldulensis *Dehnh.*		kino	Australia	400
citriodora *Hook.*		essential oil	Australia	399
deglupta *Blume*		timber	New Guinea	399
dives *Schau.*		essential oil	Australia	399
fruticetorum *F. Muell.*		essential oil	Australia	399
globulus *Labill.*	BLUE GUM	essential oil	Australia	399
macrorrhyncha *F. Muell.*		rutin	Australia	400
radiata *Sieb. ex DC.*		essential oil	Australia	399
saligna *Sm.*		timber	Australia	399
smithii *R. T. Baker*		essential oil	Australia	399
spp.	EUCALYPTUS	windbreaks		399
staigeriana *F. Muell. ex. F. M. Bail.*		essential oil	Australia	399
wandoo *Blakely*		tannin	Australia	400
Eugenia *L.* MYRTACEAE				400
aquea *Burm. f.*	WATERY ROSE APPLE	ed. fruit	India	401
aromatica Kuntze				401
= E. caryophyllus (*Sprengel*) *Bullock & Harrison*				
caryophyllata Thunb.				401
= E. caryophyllus (*Sprengel*) *Bullock & Harrison*				
caryophyllus (*Sprengel*) *Bullock & Harrison*	CLOVE	spice	Moluccas	401
cauliflora (Berg.) DC.				398
= Myrciaria cauliflora *Berg.*				
cuminii (*L.*) *Druce*	JAMBOLAN	ed. fruit	Brazil	401
jambos *L.*	ROSE APPLE	ed. fruit	Indo-Malaya	400
javanica *Lam.*	JAVA APPLE	ed. fruit	Malaysia	401
malaccensis *L.*	POMERAC	ed. fruit	Malaysia	400
uniflora *L.*	PITANGA CHERRY	ed. fruit	Brazil	401
Euphorbia *L.* EUPHORBIACEAE				
antisyphilitica *Zucc.*	CANDELILLA	wax	C. America	139
milii *Ch. des Moulins*	CHRIST'S THORN	ornamental	Madagascar	139
pulcherrima *Willd. ex Klotzsch*	POINSETTIA	ornamental	Mexico	139
tirucalli *L.*	MILK BUSH	hedge plant	Africa	140
EUPHORBIACEAE				139
Eusideroxylon LAURACEAE *Teijsm & Binn.*				
zwageri *Teijsm & Binn.*	BORNEAN IRON-WOOD	poles	Borneo	446

SCIENTIFIC NAME		COMMON NAME	USE	ORIGIN	PAGE
Exogonium *Choisy*	CONVOLVULACEAE				
purga (*Hayne*) *Lindl.*		JALAP	purgative	Mexico	78
Faba *Adans.*	LEGUMINOSAE (P)				
vulgaris Moench = Vicia faba *L.*					319
FABACEAE					199
FABOIDEAE					199
Fagopyrum *Mill.*	POLYGONACEAE				
esculentum *Moench*		BUCKWHEAT	ed. grain	Europe	640
Feijoa *Berg.*	MYRTACEAE				
sellowiana *Berg.*		FEIJOA	ed. fruit	S. America	398
Feronia *Correa*	RUTACEAE				
limonia (*L.*) *Swing*		INDIAN WOOD APPLE	ed. fruit	India	493
Ficus *L.*	MORACEAE				386
benghalensis *L.*		INDIAN BANYAN	ornamental	India	389
benjamina *L.*		CEYLON WILLOW	ornamental	India	389
carica *L.*		FIG	ed. fruit	Asia	388
cunia *Buch.-Ham.*			host for lac insect	India	389
elastica *Roxb.*		INDIA-RUBBER FIG	rubber	India	388
natalensis *Hochst.*		BARK CLOTH	bark cloth	E. Africa	389
nekbudu *Warb.*		BARK CLOTH	bark cloth	Mozambique	389
pumila *L.*		CREEPING FIG	ornamental	E. Asia	389
religiosa *L.*		PIPAL TREE	ornamental	India	389
repens Rottl. = F. pumila *L.*					389
spp.			shade tree	T. Africa	389
thonningii *Bl.*		BARK CLOTH	bark cloth	E. Africa	389
utilis Sim. = F. nekbudu *Warb.*					389
vogelii (*Miq.*) *Miq.*			rubber	W. Africa	389
Flemingia *Roxb. ex Ait.*	LEGUMINOSAE (P)				
congesta Roxb. = Moghania macrophylla (*Willd.*) *O. Ktze.*					221
Foeniculum *Mill.*	UMBELLIFERAE				
officinale Gaertn. = F. vulgare *Mill.*					651
vulgare *Mill.*		FENNEL	herb	Mediterranean	651
Fortunella *Swing.*	RUTACEAE				
japonica (*Thunb.*) *Swing.*		ROUND KUMQUAT	ed. fruit	China	493
margarita (*Lour.*) *Swing.*		OVAL KUMQUAT	ed. fruit	China	493
polyandra (*Ridl.*) *Tan.*		KUMQUAT	ed. fruit	Malaya	494
spp.		KUMQUATS	ed. fruit	E. Asia	494
Funtumia *Stapf*	APOCYNACEAE				
africana (*Benth.*) *Stapf*				T. Africa	628
elastica (*Preuss.*) *Stapf*		LAGOS SILK RUBBER	rubber	T. Africa	628
latifolia (Stapf) Schltr. = F. africana (*Benth.*) *Stapf*					628
Gaillardia *Foug.*	COMPOSITAE		ornamental	N. America	53
Garcinia *L.*	GUITTIFERAE				
dulcis (*Roxb.*) *Kurz*			ed. fruit	T. Asia	634
livingstonei *T. Anders*			ed. fruit	E. Africa	634
mangostana *L.*		MANGOSTEEN	ed. fruit	Malaysia	633
xanthochymus *Hook. f.*			ed. fruit	T. Asia	634
Gardenia *Ellis*	RUBIACEAE				
jasminoides *Ellis*		GARDENIA	ornamental	China	451
Gerbera *L. ex Cass*	COMPOSITAE		ornamental	Africa	53

SCIENTIFIC NAME	COMMON NAME	USE	ORIGIN	PAGE
Gliricidia *HBK.* LEGUMINOSAE (P)				222
maculata Steud.				
= G. sepium (*Jacq.*) *Walp.*				
sepium (*Jacq.*) *Walp.*	NICARAGUAN COCOA SHADE	shade tree	T. America	222
Gluta *L.* ANACARDIACEAE				
renghas *L.*	RENGAS	vesicant	Malaya	19
spp.	RENGAS	vesicants	Malaya	19
Glycine *L.* LEGUMINOSAE (P)				264
glacilis *Skvortzor*		oil seed	E. Asia	266
hispida (Moench) Maxim.				265
= G. max (*L.*) *Merr.*				
javanica *L.*		fodder	T. Africa & Asia	265
max (*L.*) *Merr.*	SOYA BEAN	pulse	E. Asia	265
soya Sieb. & Zucc.				265
= G. max (*L.*) *Merr.*				
tomentosa *Benth.*	WILD SOYA BEAN		S. China	266
ussuriensis *Regal & Maack*	WILD SOYA BEAN		E. Asia	266
Glycyrrhiza *L.* LEGUMINOSAE (P)				
glabra *L.*	LIQUORICE	drug	Asia	217
Gomphrena *L.* AMARANTHACEAE				
globosa *L.*	BACHELORS' BUTTONS	ornamental	India	625
Gossypium *L.* MALVACEAE				333
anomalum *Waw. & Pey.*	WILD COTTON		Africa	335
aboreum *L.*	ASIATIC COTTON	fibre	Asia	336
race bengalense *Silow*	COTTON	fibre	Asia	338
race burmanicum *Silow*	COTTON	fibre	Asia	388
race cernuum (*Hutch. & Ghose*) *Silow*	COTTON	fibre	Assam	338
race indicum (*Tod.*) *Silow*	COTTON	fibre	W. India	336
race sinense *Silow*	COTTON	fibre	E. Asia	338
race soudanense (*Watt*) *Silow*	COTTON	fibre	Sudan	338
aridum (*Rose & Standl.*) *Skovsted*	WILD COTTON		Mexico	335
areysianum (*Deflers*) *Hutch.*	WILD COTTON		Aden	336
armourianum *Kearn.*	WILD COTTON		California	335
australe *F. Muell.*	WILD COTTON		Australia	335
barbadense *L.*	SEA ISLAND COTTON	fibre	W. Indies	338
var. brasiliense (*Macf.*) *Hutch.*	COTTON	fibre	S. America	340
var. darwinii (*Watt*) *Hutch.*	WILD COTTON		Galapagos	340
barbosanum *Phill. & Clem.*	WILD COTTON		Cape Verde	335
gossypioides (*Ulb.*) *Standl.*	WILD COTTON		Mexico	335
harknessii *Brandeg.*	WILD COTTON		California	335
herbaceum *L.*	ASIATIC COTTON	fibre	Africa & Asia	336
race acerifolium (*Guill. & Perr.*) *Chev.*	COTTON		Ethiopia, Arabia	336
race africanum (*Watt*) *Hutch. & Ghose*	COTTON		S. Africa	336
race kuljianum *Hutch.*	COTTON	fibre	Russia	336
race persicum *Hutch.*	COTTON	fibre	Persia	336
race wightianum (*Tod.*) *Hutch.*	COTTON	fibre	W. Indies	336
hirsutum *L.*	COTTON	fibre	Mexico	340
var. latifolium (*Murr.*) *Hutch.*	UPLAND COTTON	fibre	Mexico	340
var. marie-galante (*Watt*) *Hutch.*	TREE COTTON	fibre	W. Indies	340
race morilli (*Cook & Hubb.*) *Hutch.*	WILD COTTON		C. America	340
race palmeri (*Watt*) *Hutch.*	WILD COTTON		C. America	340
var. punctatum (*Schum.*) *Hutch.*	COTTON	fibre	W. Africa	340
race richmondi *Hutch.*	WILD COTTON		C. America	340
race yucatanense *Hutch.*	WILD COTTON		C. America	340
incanum (*Schwartz*) *Hillcoat*	WILD COTTON		Aden	336

SCIENTIFIC NAME	COMMON NAME	USE	ORIGIN	PAGE
Gossypium—continued				
klotzschianum *Anders.*	WILD COTTON			
var. davidsonii (*Kell.*) *Hutch.*	WILD COTTON		California	335
var. klotzschianum	WILD COTTON		Galapagos	335
lobatum *Gentry*	WILD COTTON		Mexico	335
longicalyx *Hutch. & Lee*	WILD COTTON		E. Africa	336
peruvianum Cav.				
= G. barbadense *L.*				338
raimondii *Ulb.*	WILD COTTON		N. Peru	335
robinsonii *F. Muell.*	WILD COTTON		Australia	335
somalense (*Gurke*) *Hutch.*	WILD COTTON		Somalia	335
stocksii *Mast.*	WILD COTTON		Arabia	335
sturtii *F. Muell.*	WILD COTTON		Australia	335
thurberi *Tod.*	WILD COTTON		Mexico	335
tomentosum *Nutt. ex Seem.*	WILD COTTON		Hawaii	341
trilobum (*DC.*) *Kearn.*	WILD COTTON		Mexico	335
triphyllum *Hochr.*	WILD COTTON		S.W. Africa	335
Grevillea *R. Br.* PROTEACEAE				
robusta *A. Cunn.*	SILKY OAK	shade tree	E. Australia	472
Guizotia *Cass.* COMPOSITAE				65
abyssinica (*L.f.*) *Cass.*	NIGER SEED	oil seed	T. Africa	65
GUTTIFERAE				633
Haematoxylon *L.* LEGUMINOSAE (C)				
campechianum *L.*	LOGWOOD	dye	Jamaica	202
Helianthus *L.* COMPOSITAE		ornamental	N. America	68
annuus *L.*	SUNFLOWER	oil seed	N. America	69
var. annuus	WILD SUNFLOWER		N. America	69
var. jaegeri *Heiser*	SUNFLOWER			69
var. lenticularis (*Dougl.*) *Ckll.*	SUNFLOWER		N. America	69
var. macrocarpus (*DC.*) *Ckll.*	SUNFLOWER	oil seed	N. America	69
decapetalus *L.*	PER. SUNFLOWER	ornamental	N. America	68
rigidus (*Cass.*) *Desf.*	PER. SUNFLOWER	ornamental	N. America	68
tuberosus *L.*	JERUSALEM ARTICHOKE	ed. tuber	N. America	68
Herrania *Goudot* STERCULIACEAE			T. America	571
Hevea *Aubl.* EUPHORBIACEAE				144
benthamiana *Muell.-Arg.*		rubber	S. America	145
brasiliensis (*Willd. ex Adr. de Juss.*) *Muell.-Arg.*	PARA RUBBER	rubber	Amazon Basin	146
camporum *Ducke*			S. America	145
discolor (Spruce ex Benth.) Muell.-Arg. = Hevea spruceana (*Benth.*) *Muell.-Arg.*				146
guianensis *Aubl.*				
var. lutea (*Spruce ex Benth.*) *Ducke & R.E. Schultes*			S. America	145
microphylla *Ule*			S. America	145
nitida *Mart. ex Muell.-Arg.*			S. America	145
var. toxicodendroides (*R.E. Schultes & Venton*) *R.E. Schultes*				
pauciflora (*Spruce ex Benth.*) *Muell.-Arg.*			S. America	145
var. coriacea *Ducke*				
rigidifolia (*Spruce ex Benth.*) *Muell.-Arg.*			S. America	146
spruceana (*Benth.*) *Muell.-Arg.*			S. America	146
Hibiscus *L.* MALVACEAE				364
abelmoschus *L.*	MUSK MALLOW	perfume	India	365
cannabinus *L.*	KENAF	fibre	T. Africa	365
esculentus *L.*	OKRA	vegetable	T. Africa	368
floccosus *Mast.*		fibre	Malaya	364
kitaibelifolius *St. Hil.*		fibre	Brazil	364

SCIENTIFIC NAME		COMMON NAME	USE	ORIGIN	PAGE
Hibiscus—continued					
lunariifolius *Willd.*			fibre	W. Africa	364
macrophyllus *Roxb.*			fibre	Malaya	364
manihot *L.*			pot herb	E. Asia	364
mutabilis *L.*		CHANGEABLE ROSE	ornamental	China	365
rosa-sinensis *L.*		HIBISCUS	ornamental	China	365
var. cooperi *Nichols*		VARIEGATED HIBISCUS	ornamental	E. Asia	365
rostellatus *Guill. & Perr.*			fibre	W. Africa	364
sabdariffa *L.*		ROSELLE	soft drink	W. Africa	370
var. altissima *Wester*		ROSELLE	fibre		370
var. sabdariffa		ROSELLE	drink	W. Africa	370
schizopetalus (*Mast.*) *Hook.f.*			ornamental	E. Africa	365
squamosus *Hochr.*			fibre	W. Africa	364
tiliaceus *L.*			fibre	Tropics	364
Honckenya *Willd.*	TILIACEAE				613
ficifolia Willd. = Clappertonia ficifolia (*Willd.*) *Decne*					
Humulus *L.*	CANNABIDACEAE				
lupulus *L.*		EUROPEAN HOP	hops	Europe	40
Hymenaea *L.*	LEGUMINOSAE (C)				
courbaril *L.*		LOCUST TREE	copals	T. America	202
Hyoscyamus *L.*	SOLANACEAE				
niger *L.*		HENBANE	drug	Europe	524
Hyptis *Jacq.*	LABIATAE				
spicigera *Lam.*			oil seed	T. Africa	634
Indiofera *L.*	LEGUMINOSAE (P)				217
arrecta *Hochst. ex A. Rich.*		INDIGO	dye	T. Africa	217
hendecaphylla Jacq. = I. spicata *Forsk.*					222
spicata *Forsk.*			forage	T. Africa	222
sumatrana *Gaertn.*		INDIGO	dye	Asia	217
spp.			dyes	Tropics	217
tinctoria *L.*		INDIGO	dye	India	217
Inga *Mill.*	LEGUMINOSAE (M)				
edulis *Mart.*			shade tree	T. America	472
laurina (*Sw.*) *Willd.*		SACKYSAC	shade tree	C. America	208
spp.			shade tree	Tropics	472
vera *Willd.*			shade tree	T. America	472
Ipomoea *L.*	CONVOLVULACEAE				78
alba *L.*		MOON FLOWER	ornamental	T. America	79
aquatica *Forsk.*			spinach	Tropics	78
batatas (*L.*) *Lam.*		SWEET POTATO	ed. tuber	T. America	79
bona-nox L. = I. alba *L.*					79
cairica (*L.*) *Sweet*			ornamental	Tropics	79
eriocarpa			spinach	India	79
fastigiata (Roxb.) Sweet = I. tiliacea (*Willd.*) *Choisy*					80
horsfalliae *Hook.*			ornamental	T. America	79
leari (*Hook.*) *Paxt.*		MORNING GLORY	ornamental	T. America	79
nil (*L.*) *Roth*		MORNING GLORY	ornamental	Tropics	79
pes-caprae (*L.*) *Sweet*			strand plant	Tropics	79
pes-tigridis *L.*			fodder	India	79
purga Hayne = Exogonium purga (*Hayne*) *Lindl.*					78
purpurea (*L.*) *Roth*		COMMON MORNING GLORY	ornamental	T. America	79
quamoclit *L.*			ornamental	Tropics	79
reptans Poir. = I. acquatica *Forsk.*					78
stolonifera (*Cyrill.*) *J.F. Gmel.*			strand plant	Tropics	79
tiliacea (*Willd.*) *Choisy*				T. America	80
tricolor *Cav.*		MORNING GLORY	ornamental	T. America	79
trifida *G. Don*				Mexico	80

SCIENTIFIC NAME		COMMON NAME	USE	ORIGIN	PAGE
Isoberlinia *Craib & Stapf.*	LEGUMINOSAE (C)		woodland	T. Africa	202
Ixora L.	RUBIACEAE	IXORAS	ornamental	S.E. Asia	451
Jacaranda *Juss.*	BIGNONIACEAE				
mimosifolia *G. Don*		JACARANDA	ornamental	S. America	629
Jatropha *L.*	EUPHORBIACEAE		ornamental	Tropics	140
Julbernardia *Pellegr.*	LEGUMINOSAE (C)		woodland trees	T. Africa	202
Kerstingiella *Harms.*	LEGUMINOSAE (P)				
geocarpa *Harms.*		KERSTING'S GROUNDNUT	pulse crop	W. Africa	329
Kigelia *DC.*	BIGNONIACEAE				
africana (*Lam.*) *Benth.*		SAUSAGE TREE	ornamental	T. Africa	629
Kopsia *Blume*	APOCYNACEAE				
fruticosa *A.DC.*		PINK KOPSIA	ornamental	Burma	628
LABIATAE					634
Lablab *Adans.*	LEGUMINOSAE (P)				273
niger *Medik.*		HYACINTH BEAN	pulse crop	T. Asia	273
vulgaris Savi = L. niger *Medik*					273
Lactuca *L.*	COMPOSITAE				74
indica *L.*			vegetable	China	74
saligna *L.*		WILD LETTUCE		Europe	74
sativa *L.*		LETTUCE	salad veg.	Eurasia	74
var. *angustina* Irish = L. sativa *L.* var. asparagina *Bailey*					76
var. asparagina *Bailey*		ASPARAGUS LETTUCE			76
var. capitata *L.*		CABBAGE LETTUCE			74
var. crispa *L.*		LEAF LETTUCE			76
var. longifolia *Lam.*		COS LETTUCE			76
serriola *L.*		WILD LETTUCE		Eurasia	74
virosa *L.*			drug	Europe	74
Lagenaria *Seringe*	CUCURBITACEAE				124
leucantha (Duch.) Rusby = L. siceraria (*Molina*) *Standl.*					124
siceraria (*Molina*) *Standl.*		BOTTLE GOURD	containers	Africa	124
vulgaris Ser. = L. siceraria (*Molina*) *Standl.*					124
Landolphia *P. Beauv.*	APOCYNACEAE				628
heudelotti *A.DC.*		LANDOLPHIA RUBBER	rubber	T. Africa	628
kirkii *Dyer*		LANDOLPHIA RUBBER	rubber	T. Africa	628
owariensis *P. Beauv.*		LANDOLPHIA RUBBER	rubber	T. Africa	628
Lathyrus *L.*	LEGUMINOSAE (P)				276
latifolius *L.*		EVERLASTING PEA	ornamental	Europe	276
odoratus *L.*		SWEET PEA	ornamental	Europe	276
sativus *L.*		GRASS PEA	pulse crop	Eurasia	276
LAURACEAE					187
Laurus *L.*	LAURACEAE				
nobilis *L.*		SWEET BAY	flavouring	Asia Minor	187
LECYTHIDACEAE				,	637
Lecythis *L.*	LECYTHIDACEAE				
elliptica *Kunth*		SAPUCAIA NUT	ed. nut	T. America	637
ollaria *L.*		SAPUCAIA NUT	ed. nut	T. America	637
usitatis *Miers*		SAPUCAIA NUT	ed. nut	T. America	637
zabucajo *Aubl.*		SAPUCAIA NUT	ed. nut	S. America	637

SCIENTIFIC NAME		COMMON NAME	USE	ORIGIN	PAGE
LEGUMINOSAE					199
Lens *Mill.*	LEGUMINOSAE (P)				279
culinaris Medic. = L. esculenta *Moench*					279
esculenta *Moench*		LENTIL	pulse crop	Eurasia	279
subsp. macrospermae (*Baumg.*) *Barul.*					279
subsp. microspermae *Barul.*					279
Lepidium *L.*	CRUCIFERAE				
sativum *L.*		GARDEN CRESS	salad veg.	Levant	96
Leptospermum *Forst.*	MYRTACEAE				
laevigatum *F. Muell.*		AUSTRALIAN TEA TREE	beverage	Australia	398
Lespedeza *Michx.*	LEGUMINOSAE (P)		cover crop	T. Asia	222
Leucaena *Benth.*	LEGUMINOSAE (M)				
glauca (*L.*) *Benth.*			shade tree	T. America	587
Lippia *L.*	VERBENACEAE				
micromeria *Schau.*		SPANISH THYME	herb	S. America	636
Litchi *Sonn.*	SAPINDACEAE				
chinensis *Sonn.*		LITCHI	ed. aril	China	642
Lochnera *Reichb.*	APOCYNACEAE				
rosea (L.) Reichb. = Catharanthus roseus (*L.*) *G. Don*					628
Lonchocarpus *Kunth.*	LEGUMINOSAE (P)				
nicou (*Aubl.*) *DC.*			insecticide	Guyana	258
urucu *Killip & Smith*			insecticide	Brazil	258
utilis *Killip & Smith*			insecticide	Peru	258
Loranthus *L.*	LORANTHACEAE	BIRD VINES	parasites	Tropics	517
LOTOIDEAE					199
Lotus *L.*	LEGUMINOSAE (P)				
tetragonolobus *L.*		ASPARAGUS PEA	ed. pods	Europe	315
Lucuma *Molina*	SAPOTACEAE				
bifera *Molina*		EGG FRUIT	ed. fruit	Peru	646
nervosa *A.DC.*		EGG FRUIT	ed. fruit	S. America	646
salicifolia *HBK.*		EGG FRUIT	ed. fruit	Mexico	646
Luffa *Mill.*	CUCURBITACEAE				128
acutangula (*L.*) *Roxb.*		ANGLED LOOFAH	ed. fruit	India	128
aegyptiaca Mill. = L. cylindrica (*L.*) *M.J. Roem.*					129
cylindrica (*L.*) *M.J. Roem.*		SMOOTH LOOFAH	sponges	T. Asia	129
Lupinus *L.*	LEGUMINOSAE (P)		cover crops		222
termis *Forsk.*		EGYPTIAN LUPIN	cover crop	Sudan	222
Lycopersicon *Mill.*	SOLANACEAE				530
cheesmanii *Riley*		WILD TOMATO		Galapogos	530
esculentum *Mill.*		TOMATO	ed. fruit	T. America	531
var. cerasiforme (*Dun.*) *Alef.*		CHERRY TOMATO	ed. fruit	Ecuador	531
var. commune *Bailey*		COMMON TOMATO	ed. fruit		531
var. grandifolium *Bailey*		POTATO TOMATO	ed. fruit		531
var. pyriforme *Alef.*		PEAR TOMATO	ed. fruit		531
var. validum *Bailey*		UPRIGHT TOMATO	ed. fruit		531
glandulosum *C.H. Muller*				S. America	530
hirsutum *Humb. & Bonpl.*				S. America	530
peruvianum *Mill.*				S. America	530
pimpinellifolium (*Jusl.*) *Mill.*		CURRANT TOMATO	ed. fruit	S. America	530

SCIENTIFIC NAME		COMMON NAME	USE	ORIGIN	PAGE
Macadamia *F. Muell.*	PROTEACEAE				
ternifolia *F. Muell*		MACADAMIA NUT	ed. nut	Australia	641
tetraphylla *L. Johnson*		MACADAMIA NUT	ed. nut	Australia	641
Madhuca *J. F. Gmel.*	SAPOTACEAE				
indica J. F. Gmel. = M. longifolia (*Koenig*) *Macb.*					647
longifolia (*Koenig*) *Macb.*		MAHUA	oil seed	India	647
Maesopsis *Engl.*	RHAMNACEAE				
eminii *Engl.*			shade tree	E. Africa	472
Malachra *L.*	MALVACEAE		fibres	Tropics	333
Malpighia *L.*	MALPIGHIACEAE				
coccigera *L.*			hedge	T. America	637
glabra *L.*		BARBADOS CHERRY	ed. fruit	T. America	637
punicifolia L. = M. glabra *L.*					637
MALPIGHIACEAE					637
MALVACEAE					333
Mammea *L.*	GUTTIFERAE				
americana *L.*		MAMMEY APPLE	ed. fruit	T. America	634
Mandragora *L.*	SOLANACEAE				
officinarum *L.*		MANDRAKE	drug	Mediter-ranean	524
Mangifera *L.*	ANACARDIACEAE				24
caesia *Jack*			ed. fruit	T. Asia	24
fœtida *Lour.*			ed. fruit	T. Asia	24
indica *L.*		MANGO	ed. fruit	Indo-Burma	24
lagenifera *Griff.*			ed. fruit	T. Asia	24
odorata *Griff.*			ed. fruit	T. Asia	24
zeylanica *Hook.f.*			ed. fruit	T. Asia	24
Manihot *Mill.*	EUPHORBIACEAE				171
aipi Pohl = M. esculenta *Crantz*					172
catingae *Ule*				T. America	179
dichotoma *Ule*		JEQUIE-MANICOBA RUBBER	rubber	Brazil	172
dulcis Pax = M. esculenta *Crantz*					172
esculenta *Crantz*		CASSAVA	ed. tuber	T. America	172
glaziovii *Muell.-Arg.*		CEARA-RUBBER	rubber	S. America	171
melanobasis *Muell.-Arg.*				T. America	179
palmata Muell.-Arg. = M. esuulenta *Crantz*					172
piauhyense *Ule*			rubber	Brazil	172
saxicola *Lanj.*				T. America	173
utilissima Pohl = M. esculenta *Crantz*					172
Manilkara Adans.	SAPOTACEAE				
achras (*Mill.*) *Fosberg*		SAPODILLA	ed. fruit	C. America	647
bidentata (*A.DC.*) *A. Chev.*		BALATA	balata rubber	S. America	647
zapotilla (Jacq.) Gilly = M. achras (*Mill.*) *Fosberg*					647
Matricaria *L.*	COMPOSITAE				
chamomilla *L.*		CHAMOMILE	drug	Europe	53
Medicago *L.*	LEGUMINOSAE (P)				
sativa *L.*		ALFALFA	fodder	Mediter-ranean	222
Melaleuca *L.*	MYRTACEAE				
leucadendron *L.*		PUNK TREE	cajeput oil	Malaysia	398

SCIENTIFIC NAME		COMMON NAME	USE	ORIGIN	PAGE
Melanochyla *Hook.f.*	ANACARDIACEAE	RENGAS	vesicants	Malaya	19
Melanorrhoea *Wall.*	ANACARDIACEAE	RENGAS	vesicants	Malaya	19
usitata *Wall.*		BURMESE LACQUER	lacquer	Burma	19
Mentha *L.*	LABIATAE	MINTS			
arvensis *L.*					634
var. piperascens *Malinv.*		JAPANESE MINT	menthol	Japan	636
×piperita *L.* (M. spicata *L.*× M. aquatica *L.*)		PEPPERMINT	essential oil	Temp. regions	636
spicata *L.*		SPEARMINT	cul. herb.	Europe	636
viridis L. = M. spicata *L.*					636
Micrandra *Benth.*	EUPHORBIACEAE				
minor *Benth.*			rubber	S. America	146
spp.			ed. seeds	S. America	146
Microcitrus *Swing.*	RUTACEAE	AUSTRALIAN WILD LIMES	ed. fruits	Australia	495
Mikania *Willd.*	COMPOSITAE				
scandens (L.) *Willd.*			cover crop	Tropics	53
Millettia *Wight & Arn.*	LEGUMINOSAE (P)		ornamentals	Old World	223
MIMOSACEAE					199
MIMOSOIDEAE					206
Mimusops *L.*	SAPOTACEAE				647
balata Pierre = Manilkara bidentata (*A.DC.*) A. *Chev.*					
Moghania *St. Hil.*	LEGUMINOSAE (P)				
macrophylla (*Willd.*) *O. Ktze.*		WILD HOPS	cover crop	T. Asia	220
Momordica L.	CUCURBITACEAE				132
balsamina *L.*		BALSAM APPLE	vegetable	Tropics	132
charantia *L.*		BITTER GOURD	vegetable	Tropics	132
cochinchinensis *Spreng.*			vegetable	T. Asia	132
dioica *Roxb. ex Willd.*			vegetable	T. Asia	132
Mora *Schomb. ex Benth.*	LEGUMINOSAE (C)				
excelsa *Benth.*		MORA	timber	S. America	202
MORACEAE					377
Moringa *Adans.*	MORINGACEAE				
oleifera *Lam.*		HORSE-RADISH TREE	condiment	India	90
MORINGACEAE					90
Morus *L.*	MORACEAE				
alba *L.*		WHITE MULBERRY	ed. fruit	China	378
nigra *L.*		BLACK MULBERRY	ed. fruit	Asia	378
Mucuna Adans.	LEGUMINOSAE (P)				
bennettii *F. Muell.*		NEW GUINEA CREEPER	ornamental	New Guinea	223
pruriens DC. var *utilis* Wall. = Stizolobium aterrimum *Piper & Tracy*					220
rostrata *Benth. Koen. ex. L.*			ornamental	T. America	223
Murraya	RUTACEAE				
koenigii (*L.*) *Spreng.*		CURRY-LEAF TREE	flavouring	India	494
paniculata (*L.*) *Jack*		ORANGE JASMINE	ornamental	S.E. Asia	494
Musanga *R. Br.*	MORACEAE				
cecropioides *R. Br.*			shade tree	E. Africa	588
Mussaenda *L.*	RUBIACEAE				
erythrophylla *Schum. & Thonn.*		RED MUSSAENDA	ornamental	T. Africa	452
spp.			ornamental	Africa, Asia	451

SCIENTIFIC NAME	COMMON NAME	USE	ORIGIN	PAGE
Myrciaria *Berg.* MYRTACEAE				
caulifera (*DC.*) *Berg.*	JABOTICABA	ed. fruit	Brazil	398
Myristica *Gronovius* MYRISTICACEAE				391
argentea *Warb.*	PAPUA NUTMEG	adulterant	New Guinea	391
fragrans *Houtt.*	NUTMEG	spice	Moluccas	391
malabarica *Lam.*	BOMBAY NUTMEG	adulterant	India	391
MYRISTICACEAE				391
MYRTACEAE				398
Myroxylon *L.f.* LEGUMINOSAE (P)				
balsamum (*L.*) *Harms.*	BALSAM OF TOLU	drug	S. America	217
pereirae (*Royle*) *Klotzsch*	BALSAM OF PERU	drug	S. America	217
Nasturtium *R. Br.* CRUCIFERAE				
microphyllum (*Boenn.*) *Rchb.*	WATERCRESS	salad veg.	Europe	96
officinale *R. Br.*	WATERCRESS	salad veg.	Eurasia	96
Nephelium *L.* SAPINDACEAE				
lappaceum *L.*	RAMBUTAN	ed. aril	Malaysia	644
litchi Camb. = Litchi chinensis *Sonn.*				642
mutabile *Bl.*	PULASAN	ed. aril	Malaysia	644
Nerium *L.* APOCYNACEAE				
oleander *L.*	OLEANDER	ornamental	Mediterranean	628
Nicotiana *L.* SOLANACEAE				538
alata *Link & Otto*	TOBACCO FLOWER	ornamental	S. America	539
attenuata *Torrey ex Watson*	WILD TOBACCO	smoking	N. America	539
bigelovei (*Torrey*) *Watson*	WILD TOBACCO	smoking	N. America	539
debneyi *Domin*			Australia	553
excelsior *Black*		smoking	Australia	539
forgetiana *Hort. ex Hemsley*		ornamental	Brazil	539
glauca *Graham*			S. America	553
glutinosa *L.*			S. America	553
gossei *Domin*		smoking	Australia	539
ingulba *Black*		smoking	Australia	539
langsdorffii *Weinmann*			S. America	533
longiflora *Cav.*			Chile	553
otophora *Grisebach*			Argentina	541
paniculata *L.*			Peru	539
plumbaginifolia *Viv.*			S. America	553
rustica *L.*	NICOTINE TOBACCO	narcotic	S. America	539
var. pavonii (*Dunal*) *Goodspeed*			Andes	539
sanderae *Hort.*		ornamental		539
suaveolens *Lehm.*			Australia	553
sylvestris *Speg. & Comes*			Argentina	541
tabacum *L.*	TOBACCO	narcotic	S. America	540
tomentosa *Ruiz & Pavon*			Peru	553
undulata *Ruiz & Pavon*			Peru	539
Nierembergia *Ruiz & Pav.* SOLANACEAE		ornamental	T. America	524
Nopalea *Salm-Dyck* CACTACEAE				
cochinellifera *Salm-Dyck*	COCHINEAL CACTUS	cochineal	T. America	390
Ocimum *L.* LABIATAE				
americanum L. = O. basilicum *L.*				636
basilicum *L.*	SWEET BASIL	cul. herb	Tropics	636
canum *Sims*	HOARY BASIL	cul. herb	Tropics	636
sanctum *L.*	HOLY BASIL	holy herb	India	636
Ocotea *Aubl.* LAURACEAE				
rodioei (*Schomb.*) *Mez*	GREENHEART	timber	S. America	187
Odontadenia *Benth.* APOCYNACEAE				
grandiflora (*G.F.W. Meyer*) *Miq.*		ornamental	T. America	628

SCIENTIFIC NAME		COMMON NAME	USE	ORIGIN	PAGE
Operculina *Silva Manso*	CONVOLVULACEAE				
alata (*Ham.*) *Urban*		WOODEN ROSE	ornamental	T. America	78
Orobanche *L.*	SCROPHULARIACEAE				
minor *Sutton*		BROOMRAPE	parasite	Africa	552
ramosa *L.*		BROOMRAPE	parasite	Africa	552
OXALIDACEAE					638
Oxalis *L.*	OXALIDACEAE				
tuberosa *Mol.*		OCA	ed. tuber	S. America	638
Pachyrrhizus *Rich. ex DC.*	LEGUMINOSAE (P)				281
ahipa (*Wedd.*) *Perodi*		YAM BEAN	ed. tuber	Bolivia	281
angulatus Rich. ex DC. = P. erosus (*L.*) *Urban*					281
bulbosus (L.) Kurz = P. erosus (*L.*) *Urban*					281
erosus (*L.*) *Urban*		YAM BEAN	ed. tuber	Mexico	281
tuberosus (*Lam.*) *Spreng.*		YAM BEAN	ed. tuber	S. America	282
Palaquium *Blanco*	SAPOTACEAE				
gutta (*Hook.*) *Burck*		GUTTA-PERCHA	gutta percha	Malaysia	648
Pandorea *Spach*	BIGNONIACEAE				
pandorana (*Andr.*) *van Steenis*			ornamental	Indonesia	629
Papaver *L.*	PAPAVERACEAE				
setigerum *DC.*		WILD POPPY		Asia Minor	640
somniferum *L.*					640
subsp. hortense *Husserot*		POPPY	ed. oil	S.W. Asia	640
subsp. somniferum		OPIUM POPPY	drug	S. Europe	640
PAPAVERACEAE					640
PAPILIONACEAE					199
PAPILIONATAE					215
PAPILIONOIDEAE					215
Parkia *R. Br.*	LEGUMINOSAE (M)				
filicoidea *Welw. ex Oliv.*		AFRICAN LOCUST BEAN	ed. fruit	W. Africa	208
speciosa *Hassk.*			flavouring	Malaysia	208
Parmentiera *DC.*	BIGNONIACEAE				
cereifera *Seem*		CANDLE TREE	ornamental	C. America	629
edulis *DC.*		CUACHILOTE	ed. fruit	C. America	629
Parthenium *L.*	COMPOSITAE				
argentatum *A. Gray*		GUAYULE	rubber	Mexico	53
Passiflora *L.*	PASSIFLORACEAE				420
antioquiensis *Karst.*		BANANA PASSION FRUIT	ed. fruit	Colombia	421
caerulea *L.*		PASSION FLOWER	ornamental	Brazil	421
edulis *Sims*		PASSION FRUIT	ed. fruit	Brazil	422
f. edulis		PURPLE PASSION FRUIT	ed. fruit	S. America	422
f. flavicarpa *Degener*		YELLOW PASSION FRUIT	ed. fruit	S. America	422
foetida *L.*		WILD PASSION FRUIT		T. America	421
laurifolia *L.*		WATER LEMON	ed. fruit	S. America	420
ligularis *Juss.*		SWEET GRANADILLA	ed. fruit	T. America	421
mollissima (*HBK.*) *Bailey*		BANANA PASSION FRUIT	ed. fruit	Andes	421
quadrangularis *L.*		GIANT GRANADILLA	ed. fruit	S. America	427
van-volxemii (Lem.) Triana & Planch. = P. antioquiensis *Karst.*					421
PASSIFLORACEAE					420
Pastinaca *L.*	UMBELLIFERAE				
sativa *L.*		PARSNIP	ed. root	Europe	652

SCIENTIFIC NAME	COMMON NAME	USE	ORIGIN	PAGE
Paullinia *L.* SAPINDACEAE				
cupana *HBK.*	GUARANA	beverage	S. America	644
yoco *Schultes & Killip*	YOCO	beverage	Colombia	644
Pavonia *Cav.* MALVACEAE		fibres	Tropics	333
PEDALIACEAE				430
Pellionia *Gaud.* URTICACEAE		ornamental	Tropics	620
Peltogyne *Vogel* LEGUMINOSAE (C)				
paniculata *Benth.*	PURPLEHEART	timber	S. America	202
Peltophorum *Walp.* LEGUMINOSAE (C)				
pterocarpum (*DC.*) *K. Heyne*	YELLOW FLAME	ornamental	Malaysia	202
spp.		shade	Tropics	588
Pentas *Benth.* RUBIACEAE				
lanceolata *Schum.*		ornamental	T. Africa	452
Peperomia *Ruiz & Pav.* PIPERACEAE		ornamental	Tropics	436
Persea *Mill.* LAURACEAE				192
americana *Mill.*	AVOCADO	ed. fruit	C. America	192
var. drymifolia *Mez*	AVOCADO		Mexico	193
drymifolia Cham. & Schlecht. = P. americana *Mill.*				193
gratissima Gaertn. = P. americana *Mill.*				192
indica *Spreng.*			Canary Islands	192
Petroselinum *Hill* UMBELLIFERAE				
crispum (*Mill.*) *Nym. ex. A. W. Hill*	PARSLEY	cul. herb	S. Europe	652
Petunia *Juss.* SOLANACEAE	PETUNIAS	ornamental	S. America	524
violacea *Lindl.*	PETUNIA	ornamental	S. America	558
Peucedanum *L.* UMBELLIFERAE				648
graveolens (L.) Hiern = Anethum graveolens *L.*				
Phaseolus *L.* LEGUMINOSAE (P)				284
aconitifolius *Jacq.*	MAT BEAN	pulse	Asia	286
acutifolius *Gray*				287
var. latifolius *Freem.*	TEPARY BEAN	pulse	Mexico	287
angularis (*Willd.*) *Wight*	ADZUKI BEAN	pulse	Japan	289
arborigeus *Burkhart*			T. America	305
aureus *Roxb.*	GREEN GRAM	pulse	India	290
calcaratus *Roxb.*	RICE BEAN	pulse	Asia	294
coccineus *L.*	SCARLET RUNNER BEAN	pulse	C. America	295
cylindricus L. = Vigna unguiculata (*L.*) *Walp.*				322
inamoenus L. = P. lunatus *L.*				296
lathyroides *L.*		cover crop	T. America	285
limensis Macf. = P. lunatus *L.*				296
lunatus *L.*	LIMA BEAN	pulse crop	T. America	296
lunatus macrocarpus L.f. = P. lunatus *L.*				296
metcalfei *Woot & Standl.*		cover crop	T. America	285
multiflorus Willd. = P. coccineus *L.*				295
mungo *L.*	BLACK GRAM	pulse crop	India	301
radiatus *L.*			India	291
sublobatus *Roxb.*			India	301
trilobus Ait. = P. aconitifolius *Jacq.*				286
trinervius *Heyne*			India	301
vulgaris *L.*	COMMON BEAN	pulse crop	T. America	304

SCIENTIFIC NAME		COMMON NAME	USE	ORIGIN	PAGE
Phyllanthus *L.*	EUPHORBIACEAE				
acidus (*L.*) *Skeels*		OTAHEITE GOOSEBERRY	ed. fruit	Madagascar	139
emblica *L.*		EMBLIC	ed. fruit	T. Asia	139
Physalis *L.*	SOLANACEAE				
peruviana *L.*		CAPE GOOSEBERRY	ed. fruit	T. America	523
Pilea *Lindl.*	URTICACEAE		ornamental	Tropics	620
Pimenta *Lindl.*	MYRTACEAE				409
acris Kostel. = P. racemosa (*Mill.*) *J.W. Moore*					409
dioica (*L.*) *Merr.*		PIMENTO	spice	Jamaica	409
officinalis Lindl. = P. dioica (*L.*) *Merr.*					
racemosa (*Mill.*) *J.W. Moore*		BAY	essential oil	W. Indies	409
Pimpinella *L.*	UMBELLIFERAE				
anisum *L.*		ANISE	cul. herb	Mediter-ranean	652
Piper *L.*	PIPERACEAE				436
betle *L.*		BETEL PEPPER	masticatory	Malaysia	437
clusii *DC.*			pepper subst.	T. Africa	437
cubeba *L.f.*		CUBEB	spice	Indonesia	436
guineense *Schum. & Thonn.*			pepper subst.	T. Africa	437
longifolium *Ruiz & Pavon*			pepper subst.	T. Amer.	437
longum *L.*		INDIAN LONG PEPPER	spice	India	436
methysticum *Forst.*		KAVA	beverage	Polynesia	437
nigrum *L.*		PEPPER	spice	India	441
officianarum C. DC. = P. retrofractum *Vahl*					436
retrofractum *Vahl*		JAVANESE LONG PEPPER	spice	Malaysia	436
saigonense *DC.*			pepper subst.	Vietnam	437
PIPERACEAE					436
Pistacia *L.*	ANACARDIACEAE				
cabulica *Stocks.*		BOMBAY MASTIC	mastic	India	19
lentiscus *L.*		MASTIC	mastic	Mediter-ranean	19
vera *L.*		PISTACHIO	ed. nut	W. Asia	18
Pisum *L.*	LEGUMINOSAE (P)				310
arvense L. = P. sativum *L.*					311
elatius *Steven*		WILD PEA	weed	Eurasia	311
hortense Aschers & Graebn. = P. sativum *L.*					311
sativum *L.*		PEA	pulse crop	Europe	311
subsp. *arvense* (L.) Poir = P. sativum *L.*					311
subsp. *hortense* Poir = P. sativum *L.*					311
var. *humile* Poir = P. sativum *L.*					311
var. *macrocarpon* Ser. = P. sativum *L.*					311
Pithecellobium *Mart.*	LEGUMINOSAE (M)				
dulce (*Roxb.*) *Benth.*			browse	T. America	208
saman (Jacq.) Benth. = Samanea saman (*Jacq.*) *Merr.*					208

SCIENTIFIC NAME		COMMON NAME	USE	ORIGIN	PAGE
Plectranthus L'Hérit.	LABIATAE				
esculentus *N.E. Br.*		HAUSA POTATO	ed. tubers	T. Africa	636
Plumeria *L.*	APOCYNACEAE				
rubra *L.*		FRANGIPANI	ornamental	C. America	628
Podranea *Spach*	BIGNONIACEAE				
ricasoliana (*Tanf.*) *Sprague*			ornamental	S. Africa	629
Pogostemon *Desf.*	LABIATAE				
cablin (*Blanco*) *Benth.*		PATCHOULI	essential oil	Philippines	636
POLYGONACEAE					640
Poncirus *Raf.*	RUTACEAE				
trifoliata (*L.*) *Raf.*		TRIFOLIATE ORANGE	ornamental	China	494
trifoliata × Citrus sinensis		CITRANGE	ed. fruit		494
Porana *Burm.*	CONVOLVULACEAE				
paniculata *Roxb.*		BRIDAL WREATH	ornamental	India	78
Portlandia *P. Br.*	RUBIACEAE				
grandiflora *P. Br.*			ornamental	W. Indies	452
Pouteria *Aubl.*	SAPOTACEAE				646
campechiana (HBK.) Baenhi = Lucuma nervosa *A.DC.* and L. salicifolia *HBK.*					
Prosopsis *L.*	LEGUMINOSAE (M)		browse	Tropics	208
PROTEACEAE					641
Psidium *L.*	MYRTACEAE				414
cattleianum Sabine = P. littorale *Raddi*					414
friedrichsthalianum (*Berg.*) *Nied.*			ed. fruit	C. America	414
guajava *L.*		GUAVA	ed. fruit	T. America	414
guineese *Sw.*			ed. fruit	T. America	414
littorale *Raddi*		STRAWBERRY GUAVA	ed. fruit	Brazil	414
var. lucidum *Degener*		STRAWBERRY GUAVA (YELLOW)			414
microphyllum *Britton*			ed. fruit	Puerto Rico	414
montanum *Sw.*		WILD GUAVA	ed. fruit	Jamaica	414
Psophocarpus *Neck.*	LEGUMINOSAE (P)				315
longipedunculatus Haask. = P. palustris *Desv.*					315
palmettorum Guill. & Perr. = P. palustris *Desv.*					315
palustris *Desv.*			ed. pods	T. Africa	315
tetragonolobus (*L.*) *DC.*		GOA BEAN	vegetable	T. Asia	315
Psychotria *L.*	RUBIACEAE				
ipecacuanha Stokes = Cephaelis ipecacuanha (*Stokes*) *Baill.*					451
Pterocarpus *Jacq.*	LEGUMINOSAE (P)				
erinaceus *Poir.*		BARWOOD	dye	W. Africa	217
santalinus *L.f.*		RED SANDERSWOOD	dye	E. Indies	217
soyauxii *Taub.*		BARWOOD	dye	W. Africa	217
Pueraria *DC.*	LEGUMINOSAE (P)				
phaseoloides (*Roxb.*) *Benth.*		TROPICAL KUDZU	cover crop	Malaysia	218
thunbergiana (*Sieb. & Zucc.*) *Benth.*		KUDZU	cover crop	China	220
Punica *L.*	PUNICACEAE				
granatum *L.*		POMEGRANATE	ed. fruit	Iran	641
PUNICACEAE					641
Pyrethrum *Hall.*	COMPOSITAE				58
cinerariaefolium Trev. = Chrysanthemum cinerariaefolium (*Trev.*) *Bocc.*					

SCIENTIFIC NAME	COMMON NAME	USE	ORIGIN	PAGE
Pyrostegia *Presl* BIGNONIACEAE				
venusta (*Ker-Gawl.*) *Miers*	GOLDEN SHOWER	ornamental	Brazil	629
Quamoclit *Moench* CONVOLVULACEAE				
pennata (Desv.) Boj. = Ipomoea quamoclit *L.*				79
Raphanus *L.* CRUCIFERAE				
caudatus *L.*	RAT-TAILED RADISH	vegetable	India	97
sativus *L.*	RADISH	salad	W. Asia	96
var. longipinnatus *Bailey*	JAPANESE RADISH	vegetable	E. Asia	97
RHAMNACEAE				642
Rheum *L.* POLYGONACEAE				
officinale *Baill.*	CHINESE RHUBARB	drug	China	641
palmatum *L.*	CHINESE RHUBARB	drug	China	641
rhaponticum *L.*	RHUBARB	ed. petioles	Russia	641
Rhus *L.* ANACARDIACEAE		tannin	U.S.A.	18
toxicodendron *L.*	POISON IVY	vesicant	U.S.A.	19
verniciflua *Stokes*	CHINESE LACQUER	lacquer	China	18
Ricinus *L.* EUPHORBIACEAE				
communis *L.*	CASTOR	oil	Africa	180
Rivea *Choisy* CONVOLVULACEAE				
corymbosa (*L.*) *Hall f.*		narcotic	Mexico	78
Rondeletia *L.* RUBIACEAE		ornamental	T. America	452
RUBIACEAE				451
Ruta *L.* RUTACEAE				
graveolens *L.*	RUE	medicinal	Mediterranean	495
RUTACEAE				493
Salpiglossis *Ruiz & Pav.* SOLANACEAE		ornamental	Chile	524
Salvia *L.* LABIATAE				
officinalis *L.*	SAGE	cul. herb	Mediterranean	636
splendens *Ker-Gawl.*	SCARLET SALVIA	ornamental	Brazil	636
Samanea (*Benth.*) *Merr.* LEGUMINOSAE (M)				
saman (*Jacq.*) *Merr.*	SAMAN	shade	S. America	208
SAPINDACEAE				642
SAPOTACEAE				644
Saritaea *Dugand* BIGNONIACEAE				
magnifica (*Strague ex van Steenis*) *Dugand*		ornamental	S. America	629
Sassafras *Trew* LAURACEAE				
albidum (*Nutt.*) *Nees*	SASSAFRAS	drug	N. America	187
Schinopsis *Engl.* ANACARDIACEAE				
balansae *Engelm.*	QUEBRACHO	tannin	S. America	18
lorentzii (*Griseb.*) *Engl.*	QUEBRACHO	tannin	S. America	18
Schinus *L.* ANACARDIACEAE				
molle *L.*	PEPPER TREE	mastic	S. America	19
Schleichera *Willd.* SAPINDACEAE				
oleosa (*Lour.*) *Merr.*	LAC TREE	edible oil	India	644
Sechium *P. Br.* CUCURBITACEAE				
edule (*Jacq.*) *Swartz*	CHOYOTE	vegetable	Mexico	134
Senecio *L.* COMPOSITAE		ornamental	World-wide	53

SCIENTIFIC NAME	COMMON NAME	USE	ORIGIN	PAGE
Sesamum *L.* PEDALIACEAE				430
alatum *Thonn.*		oil seed	T. Africa	430
indicum *L.*	SESAME	oil seed	T. Africa	430
orientale L. = S. indicum *L.*				430
prostratum *Retz.*			India	431
radiatum *Schum. & Thonn.*		oil seed	T. Africa	430
Sesbania *Scop.* LEGUMINOSAE (P)				222
aculeata *Pers.*		gaur gum	Pakistan	255
grandiflora (*L.*) *Poir.*		ed. flower	India	222
spp.		cover crop	Tropics	222
Sicana *Naud.* CUCURBITACEAE				
odorifera (*Vell.*) *Naud.*		vegetable	S. America	100
Sida *L.* MALVACEAE		fibre	Tropics	333
Sinapis *L.* CRUCIFERAE				91
alba L.= Brassica alba (*L.*) *Rabenh.*				
Siphonia *Rich.* EUPHORBIACEAE				144
brasiliensis HBK. = Hevea brasiliensis (*Willd. ex Adr. de Juss.*) *Muell. Arg.*				
Soja *Moench* LEGUMINOSAE (P)				265
max (L.) Piper = Glycine max (*L.*) *Merr.*				
SOLANACEAE				523
Solandra *Sw.* SOLANACEAE	CHALICE VINES	ornamental	T. America	524
Solanum *L.* SOLANACEAE				556
aculeastrum *Dunal*		hedge	T. Africa	556
aethiopicum *L.*		vegetable	T. Africa	556
ferox *L.*		vegetable	S. E. Asia	556
gilo *Raddi*		vegetable	T. Africa	556
hyporhodium *A.Br. & Bouche*	COCONA	ed. fruit	S. America	556
incanum *L.*		vegetable	T. Africa	556
indicum *L.*		vegetable	Africa & India	566
jasminoides *Paxt.*		ornamental	T. America	557
lycopersicum L. = Lycopersicon esculentum *Mill.*				531
macrocarpon *L.*		vegetable	T. Africa	556
melongena *L.*	EGG PLANT	vegetable	India	557
muricatum *Ait.*	PAPINO	ed. fruit	Peru	556
nigrum *L.*		pot-herb	Tropics	556
quitoense *Lam.*	NARANJILLA	ed. fruit	Ecuador	556
seaforthianum Andr.		ornamental	T. America	557
sparsipilum (Britt.) *Juz. & Buk.*			S. America	560
stenotomum *Juz. & Buk.*			S. America	560
torvum *Sw.*		vegetable	Tropics	556
tuberosum *L.*	POTATO	ed. tuber	Andes	560
vernei *Juz. & Buk.*			S. America	560
wendlandii *Hook. f.*		ornamental	T. America	557
Spathodea *P. Beauv.* BIGNONIACEAE				
campanulata *P. Beauv.*	TULIP TREE	ornamental	T. Africa	629
Spinacia *L.* CHENOPODIACEAE				
oleracea *L.*	SPINACH	spinach	S.W. Asia	632
Spondias *L.* ANACARDIACEAE				
cytherea *Sonn.*	GOLDEN APPLE	ed. fruit	Polynesia	18
dulcis Forst. = S. cytherea *Sonn.*				18
lutea L. = S. mombin *L.*				18
mombin *L.*	HOG PLUM	ed. fruit	T. America	18
purpurea *L.*	RED MOMBIN	ed. fruit	T. America	18

SCIENTIFIC NAME	COMMON NAME	USE	ORIGIN	PAGE
Sterculia *L.* STERCULIACEAE				
setigera *Del.*		gum	T. Africa	564
tragacantha *Lindl.*		gum	T. Africa	564
urens *Roxb.*	KATAYA GUM	gum	India	564
STERCULIACEAE				564
Stizolobium *P. Br.* LEGUMINOSAE (P)				
aterrimum *Piper & Tracy*	BENGAL BEAN	cover crop	T. Asia	220
deeringianum *Bort.*	FLORIDA VELVET BEAN	cover crop	Asia	220
Streptosolen *Miers* SOLANACEAE		ornamental	S. America	524
Striga *Lour.* SCROPHULARIACEAE				
gesnerioides (*Willd.*) *Vatke*		parasite	Africa	328
Strongylodon *Vog.* LEGUMINOSAE (P)				
macrobotrys *A. Gray*	JADE VINE	ornamental	Philippines	223
Strophanthus *A.DC.* APOCYNACEAE				
gratus (*Hook.*) *Franch.*	STROPHANTHUS	drug	T. Africa	628
hispidus *DC.*	STROPHANTHUS	drug	T. Africa	628
kombe *Oliv.*	STROPHANTHUS	drug	T. Africa	628
sarmentosus *DC.*	STROPHANTHUS	drug	T. Africa	628
Stylosanthes *Sw.* LEGUMINOSAE (P)				
erecta *P. Beauv.*		pasture legume	W. Africa	222
guianensis (*Aubl.*) *Sw.*	BRAZILIAN LUCERNE	pasture legume	S. America	222
guineensis Schum. & Thonn. = S. erecta *P. Beauv.*				222
Syzygium *Gaertn.* MYRTACEAE				
aquem (Burm. f.) Merr. & Perry = Eugenia aquea *Burm. f.*				401
aromaticum (L.) Merr. & Perry = Eugenia caryophyllus (*Sprengel*) *Bullock & Harrison*				401
cumini (L.) Skeels = Eugenia cumini (*L.*) *Druce*				401
jambos (L) Alston = Eugenia jambos *L.*				400
javanicum (Lam.) Merr. & Perry = Eugenia javanica *Lam.*				401
malaccensis (L.) Merr. & Perry = Eugenia malaccensis *L.*				400
Tabernaemontana *L.* APOCYNACEAE				
coronaria (*Jacq.*) *Willd.*	CREPE JASMIN	ornamental	E. Indies	628
Tabebuia *Gomes ex DC.* BIGNONIACEAE				
rosea (*Bertol.*) *DC.*	PINK POU	ornamental	C. & S. America	629
serratifolia (*Vahl*) *Nicholson*	YELLOW POU	ornamental	C. & S. America	629
Tacsonia *Juss.* PASSIFLORACEAE				
mollissima HBK. = Passiflora mollissima (*HBK.*) *Bailey*				421
Tagetes *L.* COMPOSITAE	MARIGOLDS	ornamental	New World	53
Tamarindus *L.* LEGUMINOSAE (C)				
indica *L.*	TAMARIND	ed. fruit	T. Africa	204
Taraxacum *Weber* COMPOSITAE				
kok-saghyz *Rodin*	RUSSIAN DANDELION	rubber	Russia	53
Tecoma *Juss.* BIGNONIACEAE				
stans (*L.*) *HBK.*	YELLOW BELLS	ornamental	T. America	629

SCIENTIFIC NAME		COMMON NAME	USE	ORIGIN	PAGE
Tecomaria *Spach*	BIGNONIACEAE				
capensis (*Thunb.*) *Spach.*		CAPE HONEYSUCKLE	ornamental	S. Africa	629
Tectona *L.f.*	VERBENACEAE				
grandis *L.f.*		TEAK	timber	Burma	638
Tephrosia *Pers.*	LEGUMINOSAE (P)				
candida (*Roxb.*) *DC.*			cover crop	India	222
spp.			cover crop	Tropics	222
vogelii *Hook. f.*			cover crop	T. Africa	222
Terminalia *L.*	COMBRETACEAE				
superba *Engl. & Diels*			shade tree	Congo	588
TERNSTROEMIACEAE					599
Tetragonolobus *Scop.*	LEGUMINOSAE (P)				
purpureus Moench. = Lotus tertagonolobus *L.*					315
Tetrapterys *Cav.*	MALPIGHIACEAE	CAAPI	narcotic	S. America	637
Thea *L.*	THEACEAE				
bohea L. = Camellia sinensis (*L.*) *O. Kuntze*					599
sinensis L. = Camellia sinensis (*L.*) *O. Kuntze*					599
viridis L. = Camellia sinensis (*L.*) *O. Kuntze*					599
THEACEAE					599
Theobroma *L.*	STERCULIACEAE				570
angustifolium *Moc. & Sesse*			ed. pulp	Mexico	571
bicolor *H. & B.*			ed. pulp	Mexico	571
cacao *L.*		COCOA	beverage	S. America	571
subsp. cacao					573
f. cacao		CENTRAL AMERICAN CRIOLLO COCOA	beverage	C. America	573
f. lacondonese *Cuatr.*				Mexico	574
f. leiocarpum (*Bern.*) *Ducke*			beverage	Guatemala	574
f. pentagonum (*Bern.*) *Cuatr.*		ALLIGATOR COCOA	beverage	C. America	574
subsp. sphaerocarpum (*Chev.*) *Cuatr.*		AMAZONIAN FORESTERO COCOA	beverage	S. America	574
grandiflorum (*Willd. ex Spreng.*) *K. Schum.*			ed. pulp	Brazil	571
leiocarpa Bern. = T. cacao *L.* subsp. cacao f. leiocarpum (*Bern.*) *Ducke*					574
leiocarpum sensu Pittier = T. cacao *L.* subsp. sphaerocarpum (*Chev.*) *Cuatr.*					574
pentagona Bern. = T. cacao *L.* subsp. cacao f. pentagonum (*Bern.*) *Cuatr.*					573
sativa (Aubl.) Lign. & Le Bey = T. cacao *L.* subsp. cacao f. cacao					573
sphaerocarpum Chev. = T. cacao *L.* subsp. sphaerocarpum (*Chev.*) *Cuatr.*					574
Thespesia *Soland. ex Corr*	MALVACEAE		fibre	Asia & Africa	333
Thevetia (*L.*) *Juss.*	APOCYNACEAE				
peruviana (*Pers.*) *K. Schum.*		LUCKY NUT	ornamental	C. America	628
Thymus *L.*	LABIATAE				
vulgaris *L.*		THYME	cul. herb	Mediterranean	636

SCIENTIFIC NAME		COMMON NAME	USE	ORIGIN	PAGE
TILIACEAE					613
Tithonia *Desf.*	COMPOSITAE		ornamental	C. America	53
Tragopogon *L.*	COMPOSITAE				
porrifolius *L.*		SALSIFY	ed. tuber	Eurasia	53
Treculia *Decne.*	MORACEAE				
africana *Decne.*		AFRICAN BREADFRUIT	ed. fruit	W. Africa	378
Trichosanthes *L.*	CUCURBITACEAE				
anguina L. = T. cucumerina *L.*					136
cucumerina *L.*		SNAKE GOURD	vegetable	T. Asia	136
Trifolium *L.*	LEGUMINOSAE (P)	CLOVERS	pasture legume	Old World	222
Trigonella *L.*	LEGUMINOSAE (P)				
foenum-graecum *L.*		FENUGREEK	flavouring	Eurasia	217
Triphasia *Lour.*	RUTACEAE				
trifolia (*Burm. f.*) *P. Wils.*		LIMEBERRY	ed. fruit	S.E. Asia	494
Triumfetta *L.*	TILIACEAE				
cordifolia *A. Rich.*			fibre	T. Africa	613
rhomboidea *Jacq.*			fibre	T. Africa	613
tomentosa *Boj.*			fibre	T. Africa	613
Tropaeolum *L.*	TROPAEOLACEAE				
tuberosum *Ruiz & Pav.*		ANU	ed. fruit	S. America	639
Ullucus *Caldas*	BASELLACEAE				
tuberosus *Caldas*		ULLUCU	ed. tuber	Andes	628
UMBELLIFERAE					648
Uncaria *Schreb.*	RUBIACEAE				
gambir (*Hunt.*) *Roxb.*		GAMBIER	tannin	Malaysia	451
Urena *L.*	MALVACEAE				
lobata *L.*		ARAMINA	fibre	Tropics	374
Urtica *L.*	URTICACEAE	NETTLES	fibre	Old World	620
URTICACEAE					
Vaupesia *R. E. Schultes*	EUPHORBIACEAE				
cataractarum *R.E. Schultes*			ed. seeds	S. America	146
Vicia *L.*	LEGUMINOSAE (P)				318
faba *L.*		BROAD BEAN	pulse crop	Eurasia	319
subsp. *eu-faba* = V. faba *L.*					319
var. *equina* Pers. = V. faba *L.*		HORSE BEAN			319
var. *major* Harz. = V. faba *L.*		BROAD BEAN			319
var. *minor* Beck. = V. faba L.		PIGEON BEAN			319
subsp. *paufijuga* Mur. = V. faba *L.*					319
pliniana *Trabut*		WILD BROAD BEAN		Algeria	319
sativa *L.*		COMMON VETCH	cover crop	Europe	318
Vigna *Savi*	LEGUMINOSAE (P)				321
aureus (Roxb.) Hepper = Phaseolus aureus *Roxb.*					290
baoulensis A. Chev. = V. unguiculata (*L.*) *Walp.*					322
catjung (Burm.) Walp. = V. unguiculata (*L.*) *Walp.*					322
cylindrica (L.) Skeels = V. unguiculata (*L.*) *Walp.*					322
hosei (*Craib*) *Backer*			cover crop	E. Indies	321
mungo (L.) Hepper = Phaseolus mungo *L.*					301
sesquipedalis (*L.*) *Fruw.*		ASPARAGUS PEA	vegetable	T. Africa	322

SCIENTIFIC NAME		COMMON NAME	USE	ORIGIN	PAGE
Vigna—continued					
sinensis (*L.*) *Savi ex Hassk.*		COWPEA	pulse crop	T. Africa	322
var. *cylindricus*					322
= V. unguiculata (*L.*) *Walp.*					
var. *sesquipedalis* (L.) Koern.					322
= V. sesquipedalis (*L.*) *Fruw.*					
unguiculata (*L.*) *Walp.*		COWPEA	pulse crop	T. Africa	321
vexillata (*L.*) *Benth.*			ed. tuber	T. Africa	321
Vitellaria *Gaertn. f.*	SAPOTACEAE				
paradoxa Gaertn. f.					
= Butyrospermum paradoxum (*Gaertn. f.*) *Hepper*					644
Voandzeia *Thouars*	LEGUMINOSAE (P)				
subterranea (L.) *Thouars*		BAMBARA GROUNDNUT	pulse crop	W. Africa	329
Warszewiczia *Kl.*	RUBIACEAE				
coccinea (*Vahl*) *Kl.*		CHACONIER	ornamental	Trinidad	452
Wissadula *Medic.*	MALVACEAE		fibre	Tropics	333
Zinnia *L.*	COMPOSITAE	ZINNIA	ornamental	T. America	53
Zizyphus *Mill.*	RHAMNACEAE				
jujuba *Mill.*		CHINESE JUJUBE	ed. fruit	China	642
mauritiana *Lam.*		INDIAN JUJUBE	ed. fruit	Africa, Asia	642
Zornia *J.F. Gmel.*	LEGUMINOSAE (P)		cover crop	Tropics	222

INDEX

The scientific (Latin) names of dicotyledonous genera and species referred to in the text are given alphabetically in the Appendix, which thus provides an index for these names. Crops dealt with in detail are given below in heavy type, under which are the headings for the information on these crops. Page numbers of the illustrations of the crops are given in heavy type. Cultivar names are not listed in this index, but will be found in the text under this heading of the crops concerned. The common names of the plants referred to in the text, and the common and scientific names of the major diseases and pests are included in this index.